Medicinal Plants: Biodiversity, Sustainable Utilization and Conservation

Shaik Mahammad Khasim
Chunlin Long
Kanchit Thammasiri • Henrik Lutken
Editors

Medicinal Plants: Biodiversity, Sustainable Utilization and Conservation

Springer

Editors
Shaik Mahammad Khasim
Department of Botany and Microbiology
Acharya Nagarjuna University
Guntur, Andhra Pradesh, India

Chunlin Long
College of Life and Environmental Sciences
Minzu University of China
Beijing, China

Kanchit Thammasiri
Department of Plant Science
Mahidol University, Thailand
Bangkok, Thailand

Henrik Lutken
Department of Plant and Environmental Sciences
University of Copenhagen
Taastrup, Denmark

ISBN 978-981-15-1635-1 ISBN 978-981-15-1636-8 (eBook)
https://doi.org/10.1007/978-981-15-1636-8

This Springer imprint is published by the registered company Springer Nature Singapore Pte Ltd.
The registered company address is: 152 Beach Road, #21-01/04 Gateway East, Singapore 189721, Singapore

Preface

Plant wealth has been a source of medicine since the inception of human civilization. In spite of tremendous development in the field of synthetic drugs and antibiotics during the twenty-first century, plants play a vital role in modern as well as traditional medicine across the globe. According to WHO, 3.5 billion people have been depending on traditional medicine because of its safety features and effective curing of diseases. In recent past, there was a paradigm shift towards the herbal medicine for the reason that the pronounced adverse effects of many synthetic drugs and chemicals. Over 35,000 plants have been used in various human cultures around the world, while about 20,000 plants are marked for medicines and cosmetics. India is at the forefront in using and exporting herbal drugs because it enjoys the unique position with diverse flora spread over the entire Indian subcontinent. Ayurveda, Unani, and Homeopathic systems of medicine rely heavily on medicinal plants only. There is a need to have sustainable scientific cultivation and extensive phytochemical research. Similarly, there is a need to protect the wild genes of medicinal plants as these are at the verge of extinction due to unsustainable exploitation. No doubt, there is a lot of scope for organized sectors such as phytochemical, pharmaceutical, and herbal drug industries in India. Therefore, proper documentation of this group of plants is need of the hour.

The edited volume on *Medicinal Plants: Biodiversity, Sustainable Utilization and Conservation* is an outcome of Proceedings of International Symposium on 'Biodiversity of Medicinal Plants & Orchids: Emerging Trends and Challenges' held during 9–11 February, 2018 at Acharya Nagarjuna University, Guntur, India, convened by Dr. S.M. Khasim and sponsored by UGC, DST, and CSIR. It aims to report on the state of the art of scientific investigations that have been going on during the last half century on medicinal plants. All papers contained in the book are peer reviewed. Further, the manuscripts were reviewed by editors and editorial board of International Symposium, and those papers judged suitable for publication following the authors' consideration of reviewer suggestions appeared in this edited book.

In view of the importance of medicinal plants globally for their large-scale cultivation and emerging value for the human health, we felt the necessity of this first comprehensive compilation by International experts. To inculcate the basic knowledge and recent trends among the researchers and teachers, the broad spectrum of medicinal plants and their sustainable utilization have been dealt in the book.

The book contains four parts, (1) Biodiversity and Conservation, (2) Ethnobotany and Ethnomedicine, (3) Bioactive compounds from plants and microbes, and (4) Biotechnology. All four parts contain 49 papers authored by eminent scientists/professors of India as well as abroad.

In Part I, biodiversity and application of sea weeds as a resource of medicine for humanity have been discussed; there are two sea weed biodiversity hotspots in India viz., Gulf of Kutch (Gujarat) and Gulf of Mannar (Tamil Nadu) representing species richness with high endemism. Another chapter in this part deals about biodiversity of mangroves in Eastern Ghats of India, particularly Gautami–Godavari estuary of Andhra Pradesh. A comprehensive data on biodiversity of medicinal plants in Eastern Ghats of Andhra Pradesh has been documented in another chapter. Some new species from Eastern Ghats of Andhra Pradesh were well documented in this chapter. In Part II, structural design and establishment of database application system for Miao medicinal plants in Guizhou Province (China) have been explained. The Miao ethnomedicine is regarded as one of the most famous traditional medicines in China. Promoting a complementary in situ and ex situ conservation strategies for medicinal plants of the Qiandongnan Miao and Dong autonomous prefecture is highly recommended. In another chapter, Shui communities of Guizhou, Southwest China, their traditional knowledge of herbal medicinal plants was discussed. Dr. Tapan Mukherjee (former Scientist CSIR-NISCAIR, New Delhi, India) stressed the importance of documentation and protection of traditional knowledge in India and abroad. In Part III, established immunoassays for the determination of Ginsenosides were discussed, and they must be useful for quality control of various ginseng medicines. In another chapter in this part, elicitation of flavonoids in *Kalanchoe pinnata* by *Agrobacterium rhizogenes*-mediated transformation has been documented; various elicitation strategies were discussed with respect to the enhancement of bioactive compounds in *K. pinnata* leaves in this paper. GC-MS profile of secondary metabolites from *Cassia occidentalis* and *Coldenia procumbens* were dealt by some workers. Other topics on veterinary medicinal plants of Andaman and Nicobar Islands and herbal medicine market in China were elaborated in this part. In Part IV, molecular and cytogenetical approaches for genetic diversity analysis of wild and cultivated *Curcuma* from North-East India have been discussed. The high proportion of polymorphism observed in the gene pool of *Curcuma* would be significant in identification of novel genotypes for breeding programs and, their subsequent introduction and usage for therapeutic purposes. In another paper, standard protocols for micropropagation of banana cultivars from India and testing of their genetic fidelity using DNA markers were discussed. DNA barcodes of *Phyllanthus amarus* and physicochemical quality of potable water in the agency areas of India were well documented by some workers.

This book serves as a reference book for the researchers, teachers, and students of Biotechnology, Botany, Microbiology, Biochemistry, and Pharmaceutical sciences. It will be of equal interest to pharmaceutical industry, medical scientists particularly Ayurveda and Allopathy, agricultural scientists, and policy makers.

We would like to express our sense of gratitude to all the contributors from India and abroad for accepting our invitation to contribute chapters and for not only sharing their knowledge, but also for admirably integrating their expertise in composing the chapters of various aspects. We greatly acknowledge Dr. Hiroyuki Tanaka (Fukuoka, Japan), Dr. Tomofumi Miyamoto (Fukuoka, Japan), Prof. V.S. Raju (Warangal, India), Prof. T. Pullaiah (Anantapur, India), Prof. M. Venkaiah (Visakhapatnam, India), Dr. N. Sanjappa (Former Director General, Botanical Survey of India, Howrah, India), Prof. P.B. Kavikishore (Hyderabad, India), and Dr. Tapan Mukherjee (NISCAIR, New Delhi), for sparing their valuable time for bringing the final shape of this edited book.

We also appreciate the support received from colleagues and research scholars particularly Mr. Md. Rahamtulla, Department of Botany and Microbiology, Acharya Nagarjuna University in the Word processing of manuscript.

We profusely thank Ms. Akanksha Tyagi and Ms. RaagaiPriya ChandraSekaran at Springer Nature, and their team for their unstinted support and execution of this project very effectively.

Guntur, Andhra Pradesh, India	Shaik Mahammad Khasim
Beijing, China	Chunlin Long
Bangkok, Thailand	Kanchit Thammasiri
Copenhagen, Denmark	Henrik Lutken

The book has now been revised with an Index at the end of the book.

Contents

Editors and Contributors

About the Editors

Shaik Mahammad Khasim (Acharya Nagarjuna University, India) graduated from Andhra University, Visakhapatnam and holds a Post-Graduate degree from Sardar Patel University, Vallabh Vidyanagar, Gujarat. He obtained his Ph.D. in 1987 from Nagarjuna University and specialized in Orchid taxonomy, anatomy and ecology. He made significant contribution in medicinal plants and Orchid biology, especially molecular characterization, Ethnomedicine and phytochemistry of Angiosperms. He published 60 research papers in National and International journals. He is recipient of Usha Vij Memorial Award 2018 of The Orchid Society of India (TOSI) for his outstanding contribution in Orchid Art and Science, and Associate Fellow of A.P. Academy of Science, India. He is Founder Member of TOSI, Life Member of Indian Science Congress Association (ISCA), Calcutta (India) and Member of the International Society for Horticultural Science (Belgium).

Chunlin Long from Minzu University of China, Beijing, graduated from Hunan Normal University, China and holds Post-Graduate degree from Kunming Institute of Botany, Chinese Academy of Sciences, China. He obtained his Ph.D. in 2002 from Gifu University, Japan. He made significant contribution in fields of agrobiodiversity, ethnobotany and ethnoecology, phytochemistry and natural products, and plant taxonomy and phylogeny. He published 356 papers in national and international journals He is a prolific writer and authored 22 books and 17 patents of innovations in his kit. He established international collaborations with 12 countries. He has been awarded 17 prizes including First-Grade prizes by Yunnan Provincial Government, Chinese Academy of Sciences, and Society of Ethnobotanists. He is also the Vice-President of Chinese Society of Ethnobotany and Vice-Chair of Ethnoecology, Ecological Society of China.

Kanchit Thammasiri (Mahidol University, Bangkok, Thailand) graduated in Agriculture from Kasetsart University, Bangkok and holds Post-Graduation in Agriculture from the same university. He made significant contribution in the fields of Cryopreservation, breeding and tissue culture of orchids and plant genetic resources & conservation. His current research involves Oorchid breeding and micropropagation of selected genera, such as cattleya alliance, dendrobiums, phalaenopsis and vandaceous-type orchids and cryopreservation of Thai orchid species. Presently he is holding the position of Associate Professor, Department of Plant Science, Faculty of Science, Mahidol University, Bangkok, Thailand.

Henrik Lütken (University of Copenhagen, Denmark) Post-Graduated from University of Copenhagen, Denmark. He obtained his Ph.D. in 2007 from the same university. He has during his Ph.D. and PostDoc conducted studies abroad in Germany (Max-Planck Institute, Golm), South Africa (University of Stellenbosch) and Norway (Bioforsk, Aas). He made significant contribution in the field of Genetic engineering, Molecular biology of Ornamental plants, Tissue culture and biotransformation. He published 45 research papers in International journals His outstanding contribution on Agrobacterium rhizogenes got patented in 2016 with United States Patent Trademark Office. He has been the Coordinator of the Crop Science Section Lab-Unit since 2012 and also the Coordinator of DFFResearch Project 1 and also Knowledge Voucher funded by Government of Denmark. He actively participated in Workshop in 'Protected Skills of Competitive Project Proposal', Helsinki, Finland and 'Work Environment Course for Leaders', Denmark.

Contributors

K. Ammani Department of Botany and Microbiology, Acharya Nagarjuna University, Guntur, Andhra Pradesh, India

A. Amrutha Valli Department of Botany and Microbiology, Acharya Nagarjuna University, Guntur, Andhra Pradesh, India

O. Aniel Kumar Department of Botany, Andhra University, Visakhapatnam, Andhra Pradesh, India

D. Aparna All India Co-ordinated Research Project on Medicinal and Aromatic Plants and Betelvine, Horticultural Research Station, Dr. Y.S.R. Horticultural University, Venkataramannagudem, Andhra Pradesh, India

P. Asha Bhanu Department of Chemistry, Acharya Nagarjuna University, Guntur, Andhra Pradesh, India

Thirupathi Azmeera Centre for Lipid Research, CSIR—Indian Institute of Chemical Technology, Hyderabad, Telangana, India

Gregorio Barba-Espín Section for Crop Sciences, Department of Plant and Environmental Sciences, Faculty of Science, University of Copenhagen, Taastrup, Denmark

S. K. M. Basha Vikrama Simhapuri University PG Centre, Kavali, Andhra Pradesh, India

P. Bhagyasree Department of Pharmaceutics, University College of Pharmaceutical Sciences, Acharya Nagarjuna University, Guntur, Andhra Pradesh, India

C. V. S. Bhaskar Department of Botany, V.R. College, Nellore, Andhra Pradesh, India

V. P. Bhatt Department of Botany, Govt. PG College, Karanprayag, Uttarakhand, India

S. Chandra Kala Department of Botany and Microbiology, Acharya Nagarjuna University, Guntur, Andhra Pradesh, India

Lars Porskjær Christensen Section of Biotechnology, Department of Chemical Engineering, Biotechnology and Environmental Technology, University of Southern Denmark, Odense, Denmark

Soumoni De Department of Museology, Rabindra Bharati University, Kolkata, India

Amit Kumar Dixit Regional Ayurveda Research Institute for Metabolic Disorders (RARIMD), CCRAS, Ministry of AYUSH, Bangalore, India

S. H. Doddamani Regional Ayurveda Research Institute for Metabolic Disorders, (CCRAS, Ministry of AYUSH), Bengaluru, India

Rime El-Houri Section of Biotechnology, Department of Chemical Engineering, Biotechnology and Environmental Technology, University of Southern Denmark, Odense, Denmark

Qiong Fang College of Life and Environmental Sciences, Minzu University of China, Beijing, China

Key Laboratory of Ethnomedicine, Minzu University of China, Ministry of Education, Beijing, China

Amalia Fkiara Section for Crop Sciences, Department of Plant and Environmental Sciences, Faculty of Science, University of Copenhagen, Taastrup, Denmark

Saifuldeen Ahmed Hasan Department of Botany and Microbiology, Acharya Nagarjuna University, Guntur, Andhra Pradesh, India

Malik Aabid Hussain Plant Tissue Culture Laboratory, Department of Botany, Sant Gadge Baba Amravati University, Amravati, Maharashtra, India

Shaik Jaffar Department of Biochemistry, Acharya Nagarjuna University, Guntur, Andhra Pradesh, India

G. Jawahar Department of Genetics, Osmania University, Hyderabad, India

Yuanyuan Ji College of Life and Environmental Sciences, Minzu University of China, Beijing, China

Key Laboratory of Ethnomedicine, Minzu University of China, Ministry of Education, Beijing, China

S. Kameswaran Department of Botany, Vikrama Simhapuri University Post Graduation Centre, Kavali, Andhra Pradesh, India

A. Kantha Sri Department of Botany, M.G.P.G. College, Guntur, Andhra Pradesh, India

S. Karuppusamy Department of Botany, The Madura College, Madurai, Tamil Nadu, India

P. B. Kavi Kishor Department of Genetics, Osmania University, Hyderabad, India

Bhattiprolu Kesava Rao Department of Chemistry, Acharya Nagarjuna University, Guntur, Andhra Pradesh, India

Ahmed Abdul Haleem Khan Department of Botany, Telangana University, Nizamabad, Telangana, India

S. M. Khasim Department of Botany and Microbiology, Acharya Nagarjuna University, Guntur, Andhra Pradesh, India

P. Komaraiah Department of Oncologic Sciences, USA Mitchell Cancer Institute, University of South Alabama, Mobile, AL, USA

G. Krishna Department of Microbiology, Kakatiya University, Warangal, Telangana, India

M. Krishna Rao S.R.V.B.S.J.B. Maharanee College, Peddapuram, Andhra Pradesh, India

N. Krishna Department of Botany and Microbiology, Acharya Nagarjuna University, Guntur, Andhra Pradesh, India

A. Lalithamba Department of Botany, DKW College, Nellore, Andhra Pradesh, India

Judith Mary Lamo Department of Biotechnology, St. Anthony's College, Shillong, India

P. Latha Department of Crop Physiology, RARS, Tirupathi, India

K. Laxmi Sowmya Department of Botany, Andhra University, Visakhapatnam, Andhra Pradesh, India

Qiyi Lei College of Life and Environmental Sciences, Minzu University of China, Beijing, China

College of Life and Environmental Sciences, Kaili University, Kaili, Guizhou, China

Key Laboratory of Ethnomedicine, Minzu University of China, Ministry of Education, Beijing, China

Sizhao Liu College of Life and Environmental Sciences, Minzu University of China, Beijing, China

Key Laboratory of Ethnomedicine, Minzu University of China, Ministry of Education, Beijing, China

College of Life and Environmental Sciences, Kaili University, Kaili, Guizhou, China

Wenqiang Liu College of Information Engineering, Minzu University of China, Beijing, China

Chunlin Long College of Life and Environmental Sciences, Minzu University of China, Beijing, China

Key Laboratory of Ethnomedicine, Minzu University of China, Ministry of Education, Beijing, China

Kunming Institute of Botany, Chinese Academy of Sciences, Kunming, China

Henrik Lütken Section for Crop Sciences, Department of Plant and Environmental Sciences, Faculty of Science, University of Copenhagen, Taastrup, Denmark

D. Madhavi Department of Genetics, Osmania University, Hyderabad, India

K. Mallikarjuna Department of Botany and Microbiology, Acharya Nagarjuna University, Guntur, Andhra Pradesh, India

M. Mohan Krishna Reddy Analytical Chemistry and Mass Spectrometry Division, CSIR—Indian Institute of Chemical Technology, Hyderabad, Telangana, India

Tapan K. Mukherjee Indian Journal of Traditional Knowledge, CSIR-NISCAIR, New Delhi, India

Renate Müller Section for Crop Sciences, Department of Plant and Environmental Sciences, Faculty of Science, University of Copenhagen, Taastrup, Denmark

L. Mutyala Naidu Department of Botany, Adikavi Nannaya University, Rajamahendravaram, Andhra Pradesh, India

Mukta Nagpurkar Post Graduate Research Centre, Department of Botany, Modern College of Arts, Science and Commerce, Pune, Maharashtra, India

G. M. Narasimha Rao Department of Botany, Andhra University, Visakhapatnam, India

Varsha Nitin Nathar Plant Tissue Culture Laboratory, Department of Botany, Sant Gadge Baba Amravati University, Amravati, Maharashtra, India

S. B. Padal Department of Botany, Andhra University, Visakhapatnam, Andhra Pradesh, India

T. Padmavathi Department of Botany, UCST, Adikavi Nannaya University, Rajamahendravaram, Andhra Pradesh, India

Krishan Pal Department of Botany, Sri Venkateshwara University, Gajraula, Uttar Pradesh, India

Neeta M. Patil Department of Botany, Modern College of Arts, Science and Commerce, Pune, Maharashtra, India

P. Pavani Department of Botany and Microbiology, Acharya Nagarjuna University, Guntur, Andhra Pradesh, India

C. Periyasamy Department of Botany, Pasumpon Muthuramalinga Thevar College, Melaneelithanallur, Manonmanium Sundaranar University, Tirunelveli, Tamil Nadu, India

K. Perumal Murugappa Chettiar Research Centre (MCRC), Chennai, Tamil Nadu, India

J. Prakasa Rao Department of Botany, Andhra University, Visakhapatnam, Andhra Pradesh, India

R. Prameela Department of Botany, M. R. College for Women, Vizianangaram, Andhra Pradesh, India

M. S. K. Prasad Department of Biochemistry, Kakatiya University, Warangal, Andhra Pradesh, India

B. Prathibha Devi Department of Botany, Osmania University, Hyderabad, India

T. Pullaiah Department of Botany, Sri Krishnadevaraya University, Anantapuramu, Andhra Pradesh, India

Santosh Kumar Rai Department of Botany, Sikkim University, Gangtok, Sikkim, India

B. Raja Rajeswari Department of Botany and Microbiology, Acharya Nagarjuna University, Guntur, Andhra Pradesh, India

G. Rajasheker Department of Genetics, Osmania University, Hyderabad, India

P. Rama Devi All India Co-ordinated Research Project on Medicinal and Aromatic Plants and Betelvine, Horticultural Research Station, Dr. Y.S.R. Horticultural University, Venkataramannagudem, Andhra Pradesh, India

V. S. Rama Krishna Ganduri Department of Biotechnology, Acharya Nagarjuna University, Guntur, Andhra Pradesh, India

Department of Biotechnology, K L University, Guntur, Andhra Pradesh, India

S. Rama Rao Plant Biotechnology Laboratory, Department of Biotechnology and Bioinformatics, North-Eastern Hill University, Shillong, India

V. Rama Rao Regional Ayurveda Research Institute for Metabolic Disorders (RARIMD), CCRAS, Ministry of AYUSH, Bangalore, India

N. Rama Swamy Department of Botany, Kakatiya University, Warangal, Telangana, India

J. Ramudu Department of Botany, VSR Government Degree and Post-graduate College, Movva, Andhra Pradesh, India

M. Ravindra Babu Horticultural Research Station, Dr. Y S R Horticultural University, Venkataramannagudem, Andhra Pradesh, India

Dinesh Singh Rawat Department of Botany & Microbiology, HNB Garhwal University, Srinagar Garhwal, Uttarakhand, India

M. Reddi Sekhar Department of Genetics and Plant breeding, SV Agriculture College, Tirupathi, Andhra Pradesh, India

P. Rohini Department of Pharmaceutics, University College of Pharmaceutical Sciences, Acharya Nagarjuna University, Guntur, Andhra Pradesh, India

Syed Sadarunnisa Department of Plantation, Spices, Medicinal and Aromatic Crops, College of Horticulture (Dr. Y.S.R. Horticultural University), Anantharajupeta, Andhra Pradesh, India

N. Sadhana Reddy Analytical Chemistry and Mass Spectrometry Division, CSIR—Indian Institute of Chemical Technology, Hyderabad, Telangana, India

T. Samatha Department of Botany, Kakatiya University, Warangal, Telangana, India

S. Sandhya Rani Department of Botany, Sri Krishnadevaraya University, Anantapuramu, Andhra Pradesh, India

D. Sandhya Deepika Department of Botany, Andhra University, Visakhapatnam, Andhra Pradesh, India

M. Santosh Kumari Department of Botany and Microbiology, Acharya Nagarjuna University, Guntur, Andhra Pradesh, India

R. Satish Babu Department of Biotechnology, National Institute of Technology, Warangal, Telangana, India

K. Satyavathi Department of Botany, Andhra University, Visakhapatnam, Andhra Pradesh, India

Shiva Shanker Kaki Centre for Lipid Research, CSIR—Indian Institute of Chemical Technology, Hyderabad, Telangana, India

M. N. Shubhashree Regional Ayurveda Research Institute for Metabolic Disorders (RARIMD), CCRAS, Ministry of AYUSH, Bangalore, India

M. A. Singara Charya Department of Microbiology, Kakatiya University, Warangal, Telangana, India

P. Sita Kumari Department of Chemistry, K.T.R. Women's College, Gudivada, India

P. Siva Kumar Reddy Research and Development Centre, Bharathiyar University, Coimbatore, Tamil Nadu, India

K. Sri Rama Murthy R&D Center for Conservation Biology and Plant Biotechnology, Shivashakti Biotechnologies Limited, Hyderabad, Telangana, India

P. V. Subba Rao Aquaculture Foundation of India, Madurai, Tamil Nadu, India

P. Sudhakar Department of Biotechnology, Acharya Nagarjuna University, Guntur, Andhra Pradesh, India

K. Sujatha Department of Chemistry, K.T.R. Women's College, Gudivada, India

Jai Sunder Division of Animal Science, ICAR-Central Island Research Institute, Port Blair, A&N Islands, India

S. Suneetha Department of Plantation, Spices, Medicinal and Aromatic crops, College of Horticulture (Dr. Y.S.R Horticultural University), Anantharajupeta, Andhra Pradesh, India

Sateesh Suthari Department of Plant Sciences, University of Hyderabad, Hyderabad, Telangana, India

B. Swapna Department of Botany, Vikrama Simhapuri University Post-Graduate Centre, Kavali, Andhra Pradesh, India

G. Swapna Department of Chemistry, University College of Sciences, Acharya Nagarjuna University, Guntur, Andhra Pradesh, India

Mahendra Tamang Department of Botany, Sri Venkateshwara University, Gajraula, Uttar Pradesh, India

Sujatha Tamilvanan Division of Animal Science, ICAR-Central Island Research Institute, Port Blair, A&N Islands, India

Hiroyuki Tanaka Department of Pharmacognosy, Graduate School of Pharmaceutical Sciences, Kyushu University, Fukuoka, Japan

B. Tanuja Priya Department of Plantation, Spices, Medicinal and Aromatic Crops, College of Horticulture (Dr. Y.S.R. Horticultural University), Anantharajupeta, Andhra Pradesh, India

M. Tarakeswara Naidu Department of Botany, Andhra University, Visakhapatnam, Andhra Pradesh, India

Department of Plant Sciences, University of Hyderabad, Hyderabad, Telangana, India

K. Thammasiri Department of Plant Science, Faculty of Science, Mahidol University, Bangkok, Thailand

M. Uma Devi Department of Botany, Osmania University, Hyderabad, India

Gundabolu Usha Rani Department of Chemistry, Acharya Nagarjuna University, Guntur, Andhra Pradesh, India

M. Ushakiranmayi Department of Botany and Microbiology, Acharya Nagarjuna University, Guntur, Andhra Pradesh, India

M. Venkaiah Department of Botany, Andhra University, Visakhapatnam, Andhra Pradesh, India

Department of Plant Sciences, University of Hyderabad, Hyderabad, Telangana, India

R. Venkatesh Department of Botany, Dr. L.H.R. Govt. Degree College, Mylavaram, Andhra Pradesh, India

M. Vijayalakshmi Department of Botany and Microbiology, Acharya Nagarjuna University, Guntur, Andhra Pradesh, India

Daniel Regie Wahlang Plant Biotechnology Laboratory, Department of Biotechnology and Bioinformatics, North-Eastern Hill University, Shillong, India

Beixi Zhang College of Life and Environmental Sciences, Minzu University of China, Beijing, China

Key Laboratory of Ethnomedicine, Minzu University of China, Ministry of Education, Beijing, China

College of Life and Environmental Sciences, Kaili University, Kaili, Guizhou, China

Jiangju Zhou College of Life and Environmental Sciences, Kaili University, Kaili, Guizhou, China

Part I

Biodiversity and Conservation

Biodiversity of Medicinal Plants in the Eastern Ghats of Northern Andhra Pradesh, India

1

M. Venkaiah, J. Prakasa Rao, and R. Prameela

Abstract

During our field inventories we have collected 172 medicinal plants to treat 131 diseases in the Eastern Ghats of northern Andhra Pradesh. All taxa are given in alphabetical order along with details of plant parts and diseases cured.

Keywords

Conservation · Diseases · Human · Plant diversity · Tribes

1.1 Introduction

India, by virtue of vast variety of physiography and climatic conditions, is uniquely bestowed with enormous biological diversity. In India forests with their flora and fauna play an important role in the life of man. Indian forests are enriched with medicinal diversity. Medicinal plants are receiving an enormous amount of attention today. The World Health Organization (WHO) has listed 20,000 medicinal plants globally; India's contribution is 15–20%. According to the WHO estimation, about 80% of the population in developing countries depends directly on plants for its medicines (Gupta and Chadha 1995). In India, of the 17,000 species of higher plants, 7500 are known for medicinal uses (Shiva 1996). Currently, approximately 25% of drugs are derived from plants, and many others are synthetic analogues built on prototype compounds isolated from plant species in modern pharmacopoeia (Prajapati et al. 2003). In India, the knowledge of traditional herbal medicine is synonymous with its rich cultural heritage and was found in Vedic literature,

M. Venkaiah (✉) · J. P. Rao
Department of Botany, Andhra University, Visakhapatnam, India

R. Prameela
Department of Botany, M R College for Women, Vizianagaram, Andhra Pradesh, India

S. M. Khasim et al. (eds.), *Medicinal Plants: Biodiversity, Sustainable Utilization and Conservation*, https://doi.org/10.1007/978-981-15-1636-8_1

particularly the Rigveda, Charak Samhita and Susruta Samhita (Vedprakash 1991). Medicinal plants are now under great pressure due to their excessive collection or exploitation. The degree of threat to natural populations of medicinal plants has increased because more than 90% of medicinal plant raw material for herbal industries in India and also for export is drawn from natural habitat. In India, about 2000 drugs used are of plant origin. In the last few decades, overexploitation of forest resources has led to species loss.

The Eastern Ghats are located between 11°03′ to 22°03″ N and 77°02′ to 87°02″ E covering 224,290 sq. km. The Eastern Ghats of northern Andhra Pradesh include Srikakulam, Vizianagaram, Visakhapatnam, East Godavari and West Godavari districts and are with dense and rich forests and enriched with more fertile soil, higher rain fall and better conditions for plant growth (Fig. 1.1). The vegetation in broad sense is of tropical deciduous type. Local tribes like Jatapu, Bagat, Savara, Konda reddy, Gadaba, Konda Dora, Samanta, Khond, Muka Dora, Valmiki, Konda Kammara, etc. are depending upon the forest produces especially on medicinal plants for their health.

During the period of 2012–2016 explorations were undertaken covered pre-monsoon, monsoon and post-monsoon periods to record medicinal plant diversity. All plant taxa were identified up to species level with the help of floras and literature (Gamble and Fischer 1915; Pullaiah 1997; Pullaiah and Chennaiah 1997; Pullaiah and Alimoulali 1997; Rao et al. 1999). All the plant species were arranged according to alphabetical order with their family, habit, medicinal importance and plant parts used (Table 1.1). Some of the plant photographs were provided for easy identification (Figs. 1.2 and 1.3).

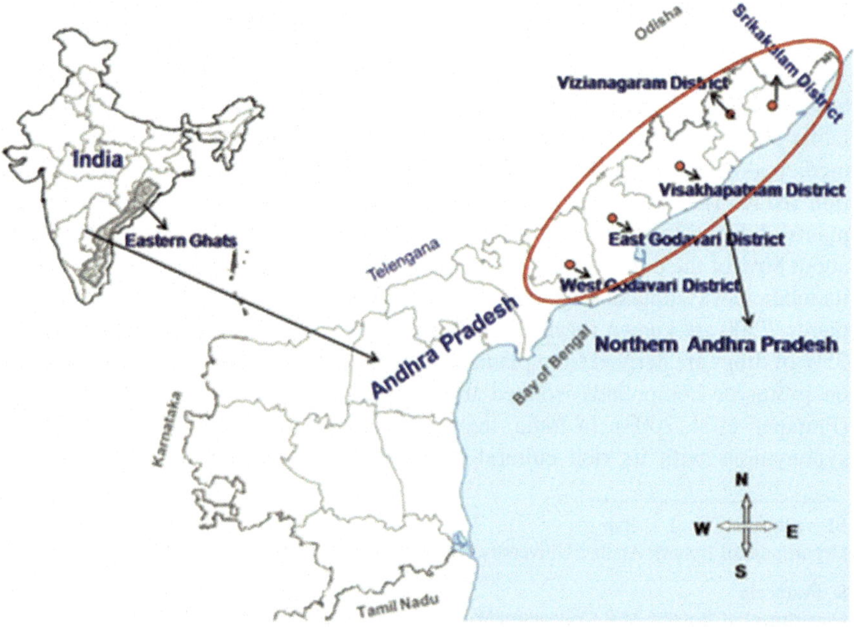

Fig. 1.1 Map of the study area

Table 1.1 List of medicinal plants and their details in the study area

S. No.	Name of the plant	Family	Habit	Part used	Diseases
1	*Abrus precatorius* L.	Fabaceae	Climber	Seed, leaf, root	Abortion, cough, dysentery, epilepsy, rheumatoid arthritis
2	*Acacia chundra* (Roxb. ex Rottl.) Willd.	Mimosaceae	Tree	Leaf	Blisters, boils
3	*Acacia sinuata* (Lour.) Merr.	Mimosaceae	Climber	Seed, tender leaf, fruit	Cataract, dandruff, ophthalmic
4	*Acalypha indica* L.	Euphorbiaceae	Herb	Leaf	Jaundice, scorpion sting
5	*Achyranthes aspera* L.	Amaranthaceae	Herb	Seed, leaf	Antidote to animal bite, antiemetic, dysentery, boils, chicken pox, cough, jaundice
6	*Acorus calamus* L.	Araceae	Herb	Rhizome	Cold, constipation, rheumatoid arthritis
7	*Aegle marmelos* L. Correa	Rutaceae	Tree	Stem bark, leaf, fruit	Cholera, cold, diabetes, diarrhoea
8	*Aerva lanata* (L.) Juss	Amaranthaceae	Herb	Root, whole plant	Headache, kidney stones, leucorrhoea
9	*Aganosma caryophyllata* (Roxb. ex. Sims).	Apocynaceae	Climber	Seed, latex	Cuts, wounds, sprains
10	*Ailanthus excelsa* Roxb.	Simarubaceae	Tree	Bark	Asthma
11	*Alangium salvifolium* (L.f.) Wang.	Alangiaceae	Tree	Leaf, stem bark	Rheumatoid arthritis, paralysis
12	*Alstonia venenata* R.Br.	Apocynaceae	Shrub	Stem bark	Anthelmintic
13	*Alternanthera sessilis* (L.) DC.	Amaranthaceae	Herb	Whole plant	Bone fracture, emetics, eye complaints
14	*Amorphophallus paeoniifolius* (Dennst.)	Araceae	Herb	Corm	Bone fracture, helminthiasis
15	*Andrographis paniculata* (Burm.f.)	Acanthaceae	Herb	Whole plant	Asthma, deworming, diabetes, leucorrhoea, malaria

(continued)

Table 1.1 (continued)

S. No.	Name of the plant	Family	Habit	Part used	Diseases
16	*Anogeissus acuminata* Wall.	Combretaceae	Tree	Stem bark	Dysentery
17	*Anogeissus latifolia* Roxb.	Combretaceae	Tree	Stem bark, seed, gum	Cough, snake bite, urinary problems
18	*Argemone mexicana* L.	Papaveraceae	Herb	Root,	Fever, malaria, spermatorrhoea
19	*Argyreia nervosa* (Burm. f.)	Convolvulaceae	Climber	Leaf, root	Boils, rheumatoid arthritis
20	*Arisaema tortuosum* (Wall.)	Araceae	Herb	Tuber	Headache, rheumatoid arthritis
21	*Aristolochia indica* L.	Aristolochiaceae	Climber	Root	Diarrhoea, snake bite
22	*Asparagus racemosus* Willd.	Liliaceae	Climber	Tuberous root	Urinary problems, diabetes, fertility, tumours
23	*Azadirachta indica* A. Juss.	Meliaceae	Tree	All parts	Allergy, amoebic dysentery, chicken pox, cold, cough, constipation, contraception, contraction of uterus, diabetes, eczema, kidney stones, leprosy, mental disorders, rat bite, rheumatism, ringworm, spermatorrhoea, stomach pain, jaundice
24	*Azima tetracantha* lam.	Salvadoraceae	Shrub	Root, leaf	Asthma, rheumatism
25	*Baliospermum montanum Muell. Arg.*	Euphorbiaceae	Herb	Seed, root	Rheumatism, stomach ache
26	*Barleria strigosa* Willd.	Acanthaceae	Shrub	Root	Tuberculosis
27	*Barringtonia acutangula* L.	Barringtoniaceae	Tree	Leaf, stem bark	Headache, peripheral neuritis, rheumatism
28	*Bauhinia purpurea* L.	Caesalpiniaceae	Tree	Bark, root	Dysentery, rheumatoid arthritis

(continued)

Table 1.1 (continued)

S. No.	Name of the plant	Family	Habit	Part used	Diseases
29	*Bauhinia racemosa* Lam.	Caesalpiniaceae	Tree	Stem bark, root bark, tender leaf	Asthma, Diarrhoea, ophthalmic diseases
30	*Bauhinia vahlii* Wight & Arm.	Caesalpiniaceae	Climber	Root, bark, leaf	Dysentery, syphilis, wounds
31	*Bombax ceiba* L.	Bombacaceae	Tree	Root, seed	Leucorrhoea, menstrual disorders, skin diseases
32	*Bridelia retusa* (L.) Spreng	Euphorbiaceae	Tree	Stem bark	Chest pain, rheumatoid arthritis
33	*Buchanania lanzan* Spreng	Anacardiaceae	Tree	Stem bark	Boils, diarrhoea
34	*Butea monosperma* Lam.	Fabaceae	Tree	Stem bark	Antifertility, postnatal care
35	*Butea superba* Roxb.	Fabaceae	Climber	Stem bark, root	Haematuria, rheumatoid arthritis
36	*Caesalpinia bonduc* L.	Caesalpiniaceae	Shrub	Seed, leaf	Abortion, epilepsy
37	*Canavalia gladiata* (Jacq.) DC.	Fabaceae	Climber	Root	Diarrhoea
38	*Canthium parviflorum* Lam.,	Rubiaceae	Shrub	Stem bark, leaf	Rheumatoid arthritis, constipation
39	*Capparis zeylanica* L.	Capparaceae	Climber	Root bark	Earache, paralysis, tuberculosis
40	*Cardiospermum halicacabum* L.	Sapindaceae	Climber	Leaf, root	Burns, leucorrhoea
41	*Carissa spinarum* L.	Apocynaceae	Shrub	Root	Leucorrhoea, snake bite, stomach pain
42	*Caryota urens* L.	Arecaeae	Tree	Inflorescence, nut	Aphrodisiac, dandruff
43	*Cassia fistula* L.	Caesalpiniaceae	Tree	Stem bark, fruit	Chest pain, conjuctivitis, dysentery, dystopia
44	*Celastrus paniculatus* Wild.	Celastraceae	Climber	Root bark, seed	Leucorrhoea, rheumatoid arthritis

(continued)

Table 1.1 (continued)

S. No.	Name of the plant	Family	Habit	Part used	Diseases
45	*Centella asiatica* (L.) Urban.	Apiaceae	Herb	Whole plant	Anaemia, HIV, jaundice, memory
46	*Chamaecrista absus* (L.) H.S.Irwin & Barneby	Caesalpiniaceae	Herb	Flower, seed	Asthma, cough
47	*Chlorophytum arundinaceum* Baker	Liliaceae	Herb	Tuber	Hydrocele, wounds, ulcers
48	*Chloroxylon swietenia* DC.	Flindersiaceae	Tree	Stem bark, leaf	Mosquito repellant, scorpion sting, rheumatoid arthritis
49	*Cissampelos pareira* L.	Menispermaceae	Climber	Leaf	Chest pain, intestinal worms
50	*Cissus quadrangularis* (L.)	Vitaceae	Climber	Stem, leaf	Fever, headache
51	*Clerodendrum serratum* L	Verbenaceae	Shrub	Root	Analgesic, fever, menstrual disorders
52	*Cocculus hirsutus* (L.) Diels	Menispermaceae	Climber	Root	Rheumatoid arthritis
53	*Cochlospermum religiosum* (L.)	Cochlospermaceae	Tree	Stem bark, leaf	Bone fracture, dandruff
54	*Coldenia procumbens* L.	Boraginaceae	Herb	Whole plant	Cuts, eczema, epilepsy, psoriasis
55	*Costus speciosus* Koenig	Costaceae	Herb	Rhizome	Abortion, chicken pox
56	*Crataeva magna* (Lour.) DC.	Capparaceae	Tree	Root bark, stem bark	Gastric troubles, urinary troubles
57	*Crotalaria verrucosa* L.	Fabaceae	Herb	Leaf	Impetigo, scabies
58	*Cryptolepis buchanani* Roem.	Asclepidiaceae	Climber	Root, leaf	Diarrhoea, galactagogue, irregular menstruation
59	*Curculigo orchioides Gaertn*	Hypoxidaceae	Herb	Tubers root	Cuts, irregular menstruation, piles
60	*Curcuma pseudomontana Graham.*	Zingiberaceae	Herb	Rhizome	Diabetes, galactagogue, jaundice

(continued)

Table 1.1 (continued)

S. No.	Name of the plant	Family	Habit	Part used	Diseases
61	*Cyanthillium cinereum* (L.) H. Rob.	Asteraceae	Herb	Seed, root	Leucoderma, malaria
62	*Dalbergia latifolia* Roxb.	Fabaceae	Tree	Stem bark, leaf	Fever, leprosy, leucoderma
63	*Dalbergia paniculata* Roxb.	Fabaceae	Tree	Leaf	Swellings
64	*Dendrophthoe falcata* (L.f.) Ett.	Loranthaceae	Parasite	Whole plant	Asthma, leucorrhoea, bone fracture
65	*Desmodium pulchellum* L.	Fabaceae	Shrub	Root	Epilepsy
66	*Dillenia pentagyna* Roxb:	Dilleniaceae	Tree	Stem bark	Rheumatoid arthritis
67	*Dioscorea bulbifera* L.	Dioscoredaceae	Climber	Tuber	Sterility
68	*Dioscorea oppositifolia* L.	Dioscoredaceae	Climber	Tuber	Fractures
69	*Dioscorea pentaphylla* L.	Dioscoredaceae	Climber	Tuber	Rheumatoid arthritis
70	*Diospyros chloroxylon* Roxb.	Ebenaceae	Tree	Leaf, stem bark, tender leaf	Diarrhoea, fractures, menstrual disorders
71	*Diospyros melanoxylon* Roxb.	Ebenaceae	Tree	Stem bark, tender leaf, unripe fruit	Cold, cough, diarrhoea, fractures
72	*Diplocyclos palmatus* (L.)	Cucurbitaceae	Climber	Leaf, fruit	Fertility, ticks
73	*Dryptes roxburghii* (Wall.) Hurusawa	Euphorbiaceae	Tree	Leaf, seeds	Cold, fever, rheumatoid arthritis
74	*Eclipta prostrata* (L.) L. Mant	Asteraceae	Herb	Whole plant	Boils, burns, constipation, diarrhoea, dysentery, eczema, hair fall, high blood pressure, jaundice, pimples, urinary tract infections
75	*Elephantopus scaber* L.	Asteraceae	Herb	Root	Anthelmintic, diarrhoea, eczema, stomach pain

(continued)

Table 1.1 (continued)

S. No.	Name of the plant	Family	Habit	Part used	Diseases
76	*Elytraria acaulis* (L.f.)	Acanthaceae	Herb	Root, leaf	Diabetes, boils, menstrual disorders
77	*Entada pursaetha* DC	Mimosaceae	Climber	Seed	Rheumatoid arthritis
78	*Erythrina suberosa* Roxb.	Fabaceae	Tree	Root	Dysentery
79	*Erythrina variegata* L.	Fabaceae	Tree	Leaf	Cold, cough, intestinal worms
80	*Evolvulus alsinoides* L.	Convolvulaceae	Herb	Leaf	Jaundice
81	*Evolvulus nummularius* (L.)	Convolvulaceae	Herb	Whole plant	Burns, fever
82	*Ficus bengalensis* L.	Moraceae	Tree	Root, stem bark, latex	Boils, leucorrhoea, rheumatism
83	*Ficus racemosa* L.	Moraceae	Tree	Stem bark, fruit, latex	Diarrhoea, leucorrhoea, rheumatism
84	*Ficus religiosa* L.	Moraceae	Tree	Stem bark	Diarrhoea, gonorrhoea, paralysis
85	*Flacourtia indica* (Burm. f.) Merr.	Flacourtiaceae	Shrub	Root, leaf	Bronchial allergy, jaundice
86	*Garuga pinnata* Roxb.	Burseraceae	Tree	Stem bark	Stomach pain
87	*Gloriosa superba* L.	Liliaceae	Climber	Leaf, tuber	Asthma, rheumatism
88	*Glycosmis pentaphylla* (Retz) DC	Rutaceae	Shrub	Fruit, root	Conjuctivitis, peripheral neuritis
89	*Gmelina arborea* Roxb	Verbenaceae	Tree	Stem bark, leaf	Chest pain, cough, gonorrhoea, ulcer
90	*Gmelina asiatica* L.	Verbenaceae	Shrub	Fruits	Dandruff, leprosy, toothache
91	*Grewia hirsuta* Vahl	Tiliaceae	Shrub	Root	Leucorrhoea
92	*Grewia tiliaefolia* Vahl	Tiliaceae	Tree	Leaf, root bark	Lice control, fractures
93	*Gymnema sylvestre* (Retz.) R. Br.	Asclepidiaceae	Climber	Root, leaf	Cobra bite, diabetes, dysentery, galactagogue, tumours

(continued)

Table 1.1 (continued)

S. No.	Name of the plant	Family	Habit	Part used	Diseases
94	*Haldinia cordifolia* (Roxb.) Ridsd.	Rubiaceae	Tree	Stem bark, leaf	Leucorrhoea, scorpion sting
95	*Helicteres isora* L.	Sterculiaceae	Shrub	Fruit, root	Dysentery, snakebite
96	*Hemidesmus indicus* (L.) R. Br.	Asclepidiaceae	Climber	Root	Diarrhoea, fever, galactagogue, menstrual disorders
97	*Holarrhena pubescens* Wall.	Apocynaceae	Tree	Bark, root	Asthma, dysentery, stomach pain
98	*Holoptelea integrifolia* Roxb.	Ulmaceae	Tree	Leaf, stem bark	Leprosy, skin diseases, rheumatoid arthritis
99	*Hybanthus enneaspermus* L.	Violaceae	Herb	Whole plant	Aphrodisiac, rheumatoid arthritis
100	*Ichnocarpus frutescens* (L.) R.Br.	Apocynaceae	Climber	Root	Epilepsy, haemorrhage
101	*Ixora pavetta* Andrews.	Rubiaceae	Tree	Root, stem bark, leaf	Jaundice, muscle pain, skin disease
102	*Jatropha curcas* L.	Euphorbiaceae	Shrub	Latex, stem bark, root	Burns, cough, dental problems, rheumatoid arthritis
103	*Jatropha gossypifolia* L.	Euphorbiaceae	Shrub	Root	Bone fracture, leucorrhoea, paralysis
104	*Lagerstroemia parviflora* Roxb.	Lythraceae	Tree	Leaf, root bark, stem bark,	Dysentery, fever, rheumatism, stomach pain
105	*Lannea coromandelica* Houtt.	Anacardiaceae	Tree	Stem bark, gum	Gastric troubles, mosquito repellant
106	*Lasia spinosa* (L.) Thwait	Araceae	Herb	Rhizome	Rheumatoid arthritis
107	*Leonotis nepetifolia* (L.)	Lamiaceae	Herb	Inflorescence, flower	Breast pain, burns, wounds
108	*Leucas aspera* (Willd.) Link.	Lamiaceae	Herb	Leaf	Rheumatoid arthritis
109	*Madhuca indica* J. Gmel.	Sapotaceae	Tree	Flower	Asthma

(continued)

Table 1.1 (continued)

S. No.	Name of the plant	Family	Habit	Part used	Diseases
110	*Mallotus philippensis* (Lam.) Muell.	Euphorbiaceae	Tree	Fruit, leaf, bark, fruit, seed	Anthelmintic, bruises, eruptions, wounds
111	*Mangifera indica* L.	Anacardiaceae	Tree	Gum, resin, bud, stem leaf, flower	Boils, jaundice, leucorrhoea
112	*Manilkara hexandra* (Roxb.)	Sapotaceae	Tree	Stem bark, root	Body pain, headache, piles, throat infection
113	*Memecylon umbellatum* Burm. f.	Melastomaceae	Shrub	Root bark	Leucorrhoea
114	*Mucuna pruriens* (L.)	Fabaceae	Climber	Root, seeds	Dysmenorrhoea, epilepsy, aphrodisiac
115	*Naravelia zeylanica* (L.) DC	Ranunculaceae	Climber	Leaf, stem	Cold, toothache
116	*Naringi crenulata* (Roxb.)	Rutaceae	Tree	Stem bark, fruit	Dysentery, intestinal worms
117	*Oroxylum indicum* (L.) Benth.	Bignoniaceae	Tree	Root bark, stem bark	Antifertility, dysentery, leucorrhoea
118	*Orthosiphon rubicundus* Don	Lamiaceae	Herb	Root tuber	Diarrhoea, gonorrhoea, piles
119	*Pavetta indica* L.	Rubiaceae	Shrub	Leaf, root bark, stem bark	Blisters, epilepsy, jaundice, urinary problems
120	*Pergularia daemia* (Forssk.) Chiov.	Asclepidiaceae	Climber	Leaf, root	Bone fracture, muscle pains, stomach pain
121	*Phyllanthus amarus* Schum.	Euphorbiaceae	Herb	Whole plant	Jaundice, scalp infection, scorpion sting, toothache
122	*Phyllanthus emblica* L.	Euphorbiaceae	Tree	Stem, leaf, fruit,	Scurvy, skin disease, stomach ulcer
123	*Piper longum* L.	Piperaceae	Climber	Stem, root	Cold, cough, emetics, toothache
124	*Plumbago zeylanica* L.	Plumbaginaceae	Shrub	Root	Abortion, fits, ring worms
125	*Pterocarpus marsupium* Roxb.	Fabaceae	Tree	Stem bark	Conception, dysentery, piles

(continued)

Table 1.1 (continued)

S. No.	Name of the plant	Family	Habit	Part used	Diseases
126	*Pterospermum xylocarpum* (Gaertn.)	Sterculiaceae	Tree	Bark, leaf, stem bark	Asthma, dandruff, lice control, whooping cough, asthma
127	*Pueraria tuberosa* Roxb.	Fabaceae	Climber	Tuber	Peptic ulcer, rheumatoid arthritis
128	*Rauvolfia serpentina* L.	Apocynaceae	Shrub	Root	Fever, heart pain, snakebite, stomach pain
129	*Rauvolfia tetraphylla* L.	Apocynaceae	Shrub	Root bark	Blood pressure
130	*Rubia cordifolia* L.	Rubiaceae	Herb	Root	Eczema, giddiness, inflammation, stomach pain, ulcers
131	*Saccharum spontaneum* L	Poaceae	Herb	Root	Eruptions on the skin
132	*Saraca asoca* (Roxb.) de Wilde	Caesalpiniaceae	Tree	Flower, bark	Syphilis, irregular menstruation
133	*Schleichera oleosa* (Lour.) Oken	Sapindaceae	Tree	Root bark, stem bark	Blood purification, bone fracture, rheumatoid arthritis, snake bite
134	*Semecarpus anacardium* L.f	Anacardiaceae	Tree	Stem bark, pericarp of fruit, seed,	Abdominal swellings, cracks, rheumatoid arthritis, ulcers, wounds
135	*Senna alata* (L.) Roxb.	Caesalpiniaceae	Shrub	Flower, leaf	Asthma, ringworm
136	*Senna auriculata* (L.) Roxb.	Caesalpiniaceae	Shrub	Leaf	Burns, ophthalmic diseases
137	*Senna occidentalis* (L.) Link	Caesalpiniaceae	Herb	Root, leaf, seed	Anthelmintic, cough, asthma, skin diseases
138	*Shorea robusta* Gaertn. f.	Dipterocarpaceae	Tree	Gum resin, seed	Amoebic dysentery, chest pain, chicken pox, dysentery, gonorrhoea
139	*Sida acuta* Burm.f.	Malvaceae	Herb	Root, leaf	Boils, cuts, wounds, scorpion sting

(continued)

Table 1.1 (continued)

S. No.	Name of the plant	Family	Habit	Part used	Diseases
140	*Sida cordata* (Burm.f.) Borrss	Malvaceae	Herb	Whole plant	Dysentery, paralysis
141	*Sida cordifolia* L.	Malvaceae	Herb	Leaf, seed	Dysentery, dysmenorrhoea, sexual potency
142	*Smilax zeylanica* L.	Smilacaceae	Climber	Tuber	Paralysis, sperm production, ulcers, wounds
143	*Solanum nigrum* L.	Solanaceae	Herb	Whole plant	Gonorrhoea, swellings
144	*Solanum surattense* Burm.f.	Solanaceae	Herb	Root bark, seed	Jaundice, toothache
145	*Soymida febrifuga* (Roxb.) A.Juss	Meliaceae	Tree	Root, stem bark	Dysmenorrhoea, dyspepsia, rheumatism
146	*Sterblus aspera* Lour	Moraceae	Tree	Root	Rheumatoid arthritis
147	*Sterculia urens* Roxb.	Sterculiaceae	Tree	Root bark, stem bark	Fertility, rheumatoid arthritis
148	*Strychnos nux-vomica* L.	Loganiaceae	Tree	Stem bark	Asthma, dysentery
149	*Strychnos potatorum* L.f.	Loganiaceae	Tree	Seed	Blood pressure, cough
150	*Syzygium cumini* L. Skeels	Myrtaceae	Tree	Stem bark, seed, leaf	Burns, cough, diabetes, dysentery
151	*Tephrosia purpurea* (L.) Pers	Fabaceae	Herb	Root	Fever, paralysis, stomach pain
152	*Terminalia arjuna* Roxb.	Combretaceae	Tree	Leaf, stem bark	Leucoderma, diabetes, heart problems
153	*Terminalia bellica* (Gaertn) Roxb.	Combretaceae	Tree	Fruit	Leucorrhoea
154	*Terminalia chebula* Retz.	Combretaceae	Tree	Fruit, stem bark	Cough, constipation
155	*Tinospora cordifolia* (Wild) Miers	Menispermaceae	Climber	Stem, root	HIV, general weakness, rheumatoid arthritis
156	*Toddalia asiatica* L.	Rutaceae	Shrub	Root	Dysentery, fox bite, mad dog bite
157	*Trema orientalis* L.	Ulmaceae	Tree	Root bark, leaf	Epilepsy, mouth disease

(continued)

Table 1.1 (continued)

S. No.	Name of the plant	Family	Habit	Part used	Diseases
158	*Tribulus terrestris L.*	Zygophyllaceae	Herb	Whole plant	Jaundice, sterility, urinary problem
159	*Trichosanthes tricuspidata* Lour	Cucurbitaceae	Climber	Tuber, seed	Earache, stomach pain
160	*Tylophora indica* Burn.f.Merr	Asclepidiaceae	Climber	Tender leaf, root	Asthma, dysentery
161	*Urena lobata L.*	Malvaceae	Herb	Root, leaf	Anthelmintic, dandruff
162	*Vanda tessellata* Roxb.	Orchidaceae	Epiphyte	Root, leaf	Fractures, skin disease
163	*Vetiveria zizanioides* (L.) Nash	Poaceae	Herb	Root	Allergy, dysuria
164	*Viscum articulatum* Burm.f.	Loranthaceae	Parasite	Whole plant	Bone fractures
165	*Vitex nugundo* L.	Verbenaceae	Shrub	Leaf	Body swellings, head ache
166	*Woodfordia fruticosa* L.	Lythraceae	Shrub	Flower, bark	Diarrhoea, jaundice, leucorrhoea
167	*Wrightia arborea* (Dennst) Mabb	Apocynaceae	Tree	Root bark, latex	Menstrual disorders, Snake bite
168	*Wrightia tinctoria* (Roxb.) R.Br.	Apocynaceae	Tree	Latex, bark	Asthma, obesity
169	*Xanthium strumarium* L.	Asteraceae	Herb	Root	Boils, Cancer
170	*Xylia xylocarpa* (Roxb.) Taub.	Mimosaceae	Tree	Root bark	Gonorrhoea
171	*Zingiber roseum* (Roxb.) Rose.	Zingiberaceae	Herb	Rhizome	Leucoderma, rheumatoid arthritis
172	*Ziziphus rugosa Lam*	Rhamnaceae	Shrub	Leaf	Diabetes

Fig. 1.2 Some important medicinal plants in the study area

Fig. 1.3 Some medicinal plants in the study area

1.2 Biodiversity of Medicinal Plants in the Eastern Ghats of Andhra Pradesh

A total of 172 species of 145 genera belonging to 68 families' medicinal plants were identified to treat 131 diseases in the study area. Among the 68 families Fabaceae is the dominant family with highest number of species with14, followed by Caesalpiniaceae with10, two families Apocynaceae and Euphorbiaceae with 9 species, 3 families Asclepidiaceae, Combretaceae and Rubiaceae with 5 species, 8 families with 4 species, another 8 families with 3 species, 14 families with 2 species and remaining 31 families with single species were reported. Among the 145 genera *Bauhinia, Dioscorea, Ficus, Senna, Sida* and *Terminalia* are the dominant genera representing 3 species each, followed by 15 genera with 2 species each and remaining 124 genera with only single species each. Habit wise trees 65, shrubs 26, climbers 34, herbs 44, parasites 2 and epiphyte 1 were reported (Fig. 1.4). All plants belong to angiosperms which contributes 153 dicots and 19 monocots. *Azadirachta indica* is used for highest number of diseases—19, followed by *Eclipta prostrata* used for 11 diseases, *Achyranthes aspera* for 7 diseases, 6 plants *Abrus precatorius, Shorea robusta, Pterospermum xylocarpum, Andrographis paniculata, Gymnema sylvestre* (Fig. 1.2) and *Rubia cordifolia* for 5 diseases, 22 plants for 4 diseases, 52 plants for 3 diseases, 61 plants for 2 diseases and 28 species for single disease (Fig. 1.5; Table 1.1).

The present study dealt with flora of five districts of northern Andhra Pradesh, such as Srikakulam, Vizianagaram, Visakhapatnam, East Godavari and West Godavari districts. A total of 172 medicinal plants of 145 genera belonging to 68 families are indicating the rich plant diversity in this region and it reveals that local tribes are well utilizing these plants in their daily life to treat various diseases. The present study also supports that Fabaceae, Caesalpiniaceae, Apocynaceae, Euphorbiaceae, Combretaceae, Rubiaceae, Asteraceae, Malvaceae, Moraceae, Rutaceae and Verbenaceae are the dominant families in almost all types of forests (Sandhyarani et al. 2007; Sukumaran and Jeeva 2008; Rasigam and Parathasarthy 2009; Sahu et al. 2010). Out of the 145 genera, *Ficus, Bauhinia, Terminalia* and *Diospyros* are the dominant genera with their species number and it supports that these are predominant genera in other forest areas of India including Eastern Ghats of northern Andhra Pradesh (Sandhyarani et al. 2007; Rasigam and Parathasarthy 2009). *Azadirachta indica, Eclipta prostrata* and *Achyranthes aspera* are used for highest number of diseases, 19, 11, and 7, respectively, and it indicates that these have high potency in treating various diseases.

1.3 Conclusions

Information about 172 medicinal plants used for 131 various diseases by indigenous communities in the region are given in this study. Among the 172 medicinal plants some of the rare plants need to be protected through in situ and ex situ conservation strategies. The government and non-governmental organizations should involve

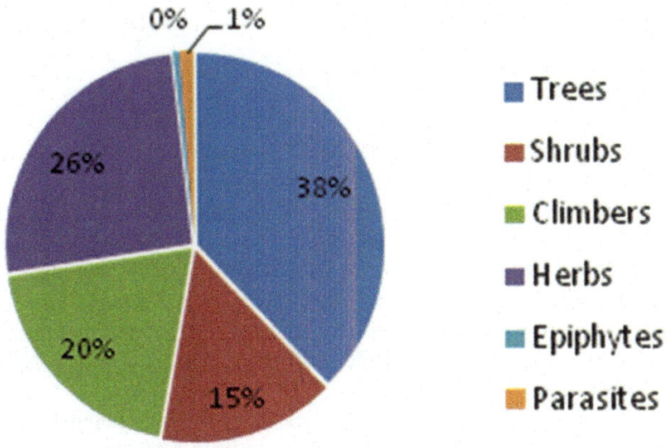

Fig. 1.4 Habit wise % of the medicinal plants in the study area

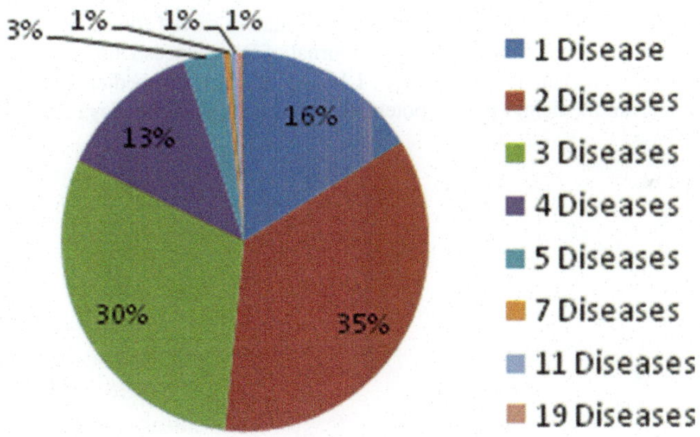

Fig. 1.5 Percentage of medicinal plants used for diseases

creating awareness in the local tribes for the proper conservation and sustainable utilization of the medicinal plants. There is need for investigation on disease-based studies on plants to get more valuable information from the local tribes in the various tribal belts in India and also other parts of the country.

Acknowledgements Authors are grateful to local tribes and forest officials for their great help in the field studies. Authors are also thankful to Dr. S.B. Padal, Professor in Botany, Andhra University, and Prof. B. Sujata, Head, Department of Botany, Andhra University, for their encouragement during the studies.

References

Gamble JS, Fischer CEC (1915–1935) Flora of the Presidency of Madras. London (Reprint ed. 1957, BSI, Calcutta)

Gupta R, Chadha KL (1995) Medicinal and aromatic plants in India. In: Advances in horticulture, medicinal and aromatic plants. Malhotra Publishing House, New Delhi

Prajapati ND, Purohit SS, Sharma AK, Kumar T (2003) A handbook of medicinal plants. Agrobios, Jodhpur

Pullaiah T (1997) Flora of Andhra Pradesh, vol III. Scientific Publishers, Jodhpur

Pullaiah T, Alimoulali D (1997) Flora of Andhra Pradesh, vol II. Scientific Publishers, Jodhpur

Pullaiah T, Chennaiah E (1997) Flora of Andhra Pradesh, vol I. Scientific Publishers, Jodhpur

Rao RS, Sudhakar S, Venkanna P (1999) Flora of east Godavari District. INTACH, A.P. State Chapter, Hyderbad

Rasigam L, Parathasarthy N (2009) Tree species diversity and population structure across major forest formation and disturbance categories in little Andaman Island, India. Trop Ecol 50(1):89–102

Sahu SC, Dhal NK, Bhadra AK (2010) Arboreal taxa diversity of tropical forests of Gandhamardan Hill range, eastern Ghats, India: an approach to sustainable biodiversity conservation. Taiwania 55(3):208–215

Sandhyarani S, Sri Rama Murthy K, Pullaiah T (2007) Tree flora in eastern Ghats of southern peninsular, India. Res J Bot 2(4):176–185

Shiva MP (1996) Inventory of forestry resources for sustainable management and biodiversity conservation. IndusPublishing Company, New Delhi

Sukumaran S, Jeeva S (2008) A floristic study on miniature sacred forests at Agastheeshwaram, southern peninsular India. European Asian Journal of Biosciences 2:66–72

Vedprakash (1991) In: Samant SS, Dhar U, LMS P (eds) Indian medicinal plant: current status in Himalayan medicinal plants: potential and prospects. Gramodaya Prakashan, Nainital, pp 45–63

Biodiversity, Conservation and Medicinal Uses of Seaweeds: The Glimpses

2

P. V. Subba Rao and Chellaiah Periyasamy

Abstract

Seaweeds are marine benthic marcophytes and have multifarious uses. Biodiversity mainly refers to number of species present in a particular habitat. Seaweed flora of India is highly diversified consisting of 1153 species and forms 12.81% of world seaweed species. Conservation aspects, viz., in situ and ex situ, are discussed. Traditional uses of seaweeds as medicine are reviewed. Consumption of seaweeds helps to combat common health alignments as they possess antioxidant, antimicrobial, antibiotic and anti-inflammatory compounds. Moreover certain diseases, viz., cancer, inflammation, arthritis, diabetes, hypertension and cardiovascular, are prevented by eating seaweeds since they contain biologically active compounds like phlorotannins, carotenoids, alginic acid and fucoidan in addition to proteins, carbohydrates and fibre.

Keywords

Antibacterial · Antifungal · Antiviral · Biodiversity · Conservation · Medicinal · Seaweeds · Vermifuge

P. V. S. Rao (✉)
Aquaculture Foundation of India, Madurai, Tamil Nadu, India

Chellaiah Periyasamy
Department of Botany, Pasumpon Muthuramalinga Thevar College, Melaneelithanallur, Manonmanium Sundaranar University, Tirunelveli, Tamil Nadu, India

S. M. Khasim et al. (eds.), *Medicinal Plants: Biodiversity, Sustainable Utilization and Conservation*, https://doi.org/10.1007/978-981-15-1636-8_2

2.1 Introduction

Seaweeds refer to any large marine benthic algae that are multicellular, macrothallic, and thus differentiated from most algae that are of microscopic size (Smith 1944). They are non-flowering primitive plants and inhabit shallow or nearshore waters of ocean, sea, estuaries and also the brackish water environments, wherever dead corals, rocks, stones, pebbles and any other suitable substrata are available for their attachment. They are crucial primary producers in oceanic aquatic food web. They are rich in both minerals and essential trace elements, which are the raw materials for the pharmaceutical and cosmetics industries (Chapman and Chapman 1980; Subba Rao and Mantri 2006). Further they are very low in fat although they contain high percentages of mono- and polyunsaturated fatty acids, and are very rich in carbohydrates (mainly dietary fibres), proteins with all the essential amino acids, and vitamins, including vitamins A, C, E, and those of the complex B, which are usually absent in land vegetables (Mac Artain et al. 2008). Additionally, the value of edible seaweeds in human nutrition is also based on their richness in several minerals like sodium (Na), magnesium (Mg), phosphorous(P), potassium(K), iodine(I), iron (Fe) and zinc (Zn) (Bocanegra et al. 2009). Mineral contents from seaweeds in relation to health benefits and risks for consumers are evaluated by Circuncisao et al. (2018) and observed that the valuable mineral contents of seaweeds grant them great potential for application in food industry as ingredients for developing functional foods and they can be used as NaCl replacers in common foods.

Moreover they form a very important and economic renewable resource in marine environment and have been a part of human civilization from time immemorial. The reports on uses of seaweeds have been cited as early as 2500 years ago in Chinese literature (c.f. Tseng 2004). They have been used for food, feed and fodder from time immemorial and contain carbohydrates, proteins, vitamins and inorganic minerals. Besides, they also contain colloidal polysaccharides, viz., agar, algin and carrageenan of industrial significance (Chapman and Chapman 1980; Levring et al. 1969; Mc Hugh 2003; Tseng 1981). Algae have been part of the human diet for thousands of years based on archaeological evidence from 14,000 YBP in Chile (Dillehay et al. 2008) and on early written accounts, e.g. in China, 300 A.D., and in Ireland, 600 A.D (Aaronson 1986; Craigie 2011; Gantar and Svircev 2008; Newton 1951; Tseng 1981; Turner 2003). Seaweeds, otherwise called "Sea vegetables" or "Marine vegetables", are traditionally consumed in Asia, where for centuries they have been regarded as food for kings and gods (Lee 2008). They were also known for their natural healing (foods) and even gourmet cuisine markets of the western world. Even though their use was realized as sources of gelling or thickening agents in western countries long back, people only recently started enjoying the taste and nutritional value of these vegetables (Escrig and Goni Combrodon 1999). Over 90% of the species of marine plants are algae (Dring 1982) and roughly 50% of the global photosynthesis on the plant group is algal derived (John 1994). Thus, every second molecule of oxygen we inhale is produced by an alga, and every second molecule of carbon dioxide we exhale is reused by an alga (Melkonian 1995). Recent research on bioactive properties of molecules extracted from seaweeds pointed to new opportunities particularly in the field of medicine using them as a CO_2 sink (Muraoka 2004) or even as biofuel.

2.2 Biodiversity

The most common measure of biodiversity is number of species in a given area. Biodiversity also refers to the variety and variability of life, genetic, species, community, ecosystem, biome and biosphere and this diversity unevenly spreads on the globe with latitudinal gradient in species richness with a marked decrease from tropical to polar region (Dobzhansky 1950; Hutchinson 1959; Wright et al. 2007). However this is not the case with seaweeds as the marine flora does not show the highest species diversity in contrast to marine fauna in tropical regions, and coasts that are rich in algae and situated in cold and warm temperate regions or in tropical regions all have about 600–800 species (Lüning 1990). Bolton (1994) has worked on seaweed diversity global patterns and has concluded without any evidence of a peak for seaweed species numbers in tropical latitudes supporting the view of Luning. He further suggested two probable reasons, viz., (1) lack of study in many areas and (2) tropical regions are relatively inhospitable.

Biological diversity or biodiversity provides information base for scientific research and guides immediate and long-term management policy and decision-making. Further it provides insight for conservation and identification of economically important products from wild species and also to understand the ecosystem processes needed for mankind (Subba Rao 2007, 2012).

Seaweed flora of India is highly diversified and comprises mostly of tropical species but boreal, temperate and subtropical elements have also been reported. Biodiversity studies in India during the early stages have emanated mostly from the works of Iyengar and Børgesen and subsequently Srinivasan, Thivy, Umamaheswara Rao, Untawale and Krishnamurthy have contributed much and the works of all these persons have been compiled in the form of checklist by Krishnamurthy and Joshi (1970) in which 525 species are accounted. Following this, Sahoo (2010) and Oza and Zaidi (2001) reported 770 and 830 species of marine algae in their documents. India (08.04–37.06 N and 68.07–97.25 E), a tropical South Asian country with 7516 km coastline, is one of the 12 mega biodiversity countries of the world and accounts for 7.8% of the recorded species of the world with only 2.5% of land area. In all, 1153 species have been encountered (Subba Rao and Mantri 2006) and forms only 12.81% of the 9900 species reported in the world (Khan et al. 2009). The coasts of Tamil Nadu and Gujarat represent very high seaweed diversity (Subba Rao and Mantri 2006; Subba Rao 2007) and the reason probably might be the existence of rich radioactive deposits probably lying underneath the seas, and the radiations of which might come out due to recurrent tectonic activities as well as volcanic eruptions over a long geological periods emitting radio isotopes of uranium, thorium and plutonium and the radiations emitted by these radioactive materials could cause mutations, thus contributing to rich diversity. This concept almost agrees with the one advocated by Sharma and Thokchom (2008) for rich floral diversity along the Western Ghats (Kerala) with radioactive thorium deposits and along Domiosiat area of Meghalaya with uranium deposits.

Most of the biodiversity studies have been carried out on the temperate shores of Europe, the USA and Australia (Bolton 1994.). Thailand represented 350 species (Dawson 1954; Egerod 1974; Velasquez and Lewmanomont 1975) whereas Malaysia recorded 250 species (Phang 1998). A list of marine algae from Thailand has been compiled by Lewmanomont et al. (1995). A report on the biodiversity studies in Southeast Asia consisting of six countries, viz., Japan, Thailand, Malaysia, Indonesia, Philippines and Vietnam, under a field research project operated by the Japan Society for the Promotion of Science (JSPS) on Coastal Marine Science during 2001–2010 has been compiled by Kawaguchi and Hayashizaki (2011) in which 50 species in Thailand, 32 species in Malaysia and 218 species in Indonesia are reported. Until last century, over 900 species of seaweeds have been reported in the Philippines (Kawaguchi and Hayashizaki 2011). Dawson (1954) recoded 200 seaweed species from Vietnam while Pham (2006) and Nguyen and Huynh (1993) reported 500 and 300 seaweeds species from South Vietnam and North Vietnam, respectively. Subsequently 31 species from the Philippines by Baleta and Nalleb (2016) and 60 species from Thailand by Anchana et al. (2011) have been recorded.

Biodiversity hotspots are biogeographic regions that essentially contain high levels of species diversity with threatened extinction and destruction. Besides, these hotspots represent species richness and highly endemic areas which are invariably prone to abiotic pressures like human and animal interference (Mittermeier et al. 1999; Myers et al. 2000).

2.3 Conservation

Biodiversity conservation is mandatory as many species tend to lose much of their genetic potential by reducing in size, number and variety which may be attributed to habitat fragmentation and destruction, invasion of introduced species, and overexploitation of living resources (Miller et al. 1992). This conservation provides for sustainable development coupled with rational and judicial utilization of resources, which is essential to cope with the increasing needs of the world such as food, fodder, health and many unforeseen benefits. And ultimately the conservation of biological diversity leads to unimaginable advantages for the ecosystem as well as for mankind at present and in the days to come (Subba Rao 2012).

Conservation could be done in two ways, viz., in situ (natural habitat) and ex situ (away from natural habitat), by adopting different strategies. In situ conservation aims at conserving wild elite germplasm in its natural habitat. On the other hand, ex situ conservation aims at conserving genetic diversity of rare, endangered and vulnerable species away from their natural habitats, i.e. in the laboratory by tissue culture techniques to preserve the germplasm and subsequently introducing the same in their natural habitats which is a must for rare and endangered species. Recent development in vitro propagation protocols, viz., somaclonal variations and somatic embryogenesis, could enhance long-term seedling viability by using freeze-drying technique, probably at relatively lesser temperatures below 30–35 °C. Cryogenic preservation as well as hybridization strategies between inter-specific and inter-generic population through conventional and molecular means would also achieve conservation objectives. Moreover adaptation of eco-corrective measures would yield fruitful results in conservation to a large extent (Sharma and Thokchom 2008).

There has been very little effort or progress regarding conservation of seaweeds and the initiatives remain patchy which is compounded by lack of trained seaweed taxonomists. Seaweed Invasion can be considered a serious threat to biodiversity if exotic species become established in indigenous communities. Declaring the particular area as Bioreserve like Marine National Parks in India, first in Gulf of Kutch (Arabian sea), second in Gulf of Mannar (Bay of Bengal) and the third in Andaman & Nicobar Islands (Andaman sea) or by cultivating elite germplasm of the economic seaweeds or otherwise in a conducive environment in the sea provides in situ conservation of seaweeds. Alternately to culture the seaweeds in the laboratory either by tissue culture or by spore culture to induce a callus which can be conserved over a prolonged period and the plantlets obtained from the callus as and when required may be used to large-scale propagation and production of biomass in natural habitat in the sea leading to ex situ conservation (Subba Rao 2007, 2012).

2.4 Medicinal Uses

The earliest records on the direct utilization of seaweeds as traditional medicine go back to the days of Emperor Shen Nung who, in 2700 B.C., documented medicinal uses of seaweeds in a Chinese herbal (Moi 1987). The Chinese *Materia Medica,* published in the eighth century A.D., listed many algae used in medicine—e.g. in the treatment of goitre, for wound-healing and for reducing hypertension (Chapman and Chapman 1980).

In Mediterranean Europe, the Greek physician, Stephanopoli in 1775 used red seaweed *Alsidium helminthochorton* that was found on the rocky shores of Corsica as an efficient vermifuge (Chapman and Chapman 1980). The Hawaiians have also, from days immemorial, used the seaweed *Hypnea nidifica* for curing stomach ailments (Reed 1906). In Indonesia, *Hypnea musciformis* was used as a vermifuge from the very ancient times (Zaneveld 1959).

In New Zealand the seaweed *Durvillea* was used as vermifuge and also for the treatment of scabies. In Latin America South American Indians used *Sargassum bacciferum* to cure goitre and kidney disorders (Schwmmer and Schwimmer 1955). In many of the Caribbean Islands, especially in Cuba, *S. vulgare* was widely used as vermifuge (Chapman and Chapman 1980). In the Philippines *Ulva pertusa* for wound healing (Nuqui 1987) and in Malaysia *Acetabularia major* for treatment of gall stones and *Chondria armata* as a vermifuge (Moi 1987) were used. In China and Hong Kong the following seaweeds were used: species of *Sargassum* to cure goitre, fever, coughs and various tumours, *Digenea simplex* as vermifuge, *Lithothamnium pacificum* as an expectorant, to reduce fever and to inhibit tumours and *Caloglossa leprieurii* as an anthelminthic agent (Tseng 1983; Win Shin-Sun 1987).

In the Mediterranean, Western Europe and North America, seaweeds used were *Hypnea musciformis* and *Palmaria palmata* as vermifuges; *Dictyopteris polypodioides* for curing lung diseases; and *Laminaria digitata* (introduced in 1682 by Dr. C.F. Sloan) for use as a cervical dilator to facilitate baby delivery (Chapman and Chapman 1980; Hale and Pion 1972). Other medicinal uses of seaweeds include utilization of *Porphyra* sp. in the Philippines, under the name "gamet", as an aphrodisiac, *Laminaria japonica* as a cure for menstrual troubles and *Laminaria saccharina* to cure syphilis in China (Chapman and Chapman 1980; Nuqui 1987).

Consumption of unprocessed dried seaweeds may yield many healing benefits. No land plant even remotely approaches seaweeds as sources of metabolically required essential minerals (Bergner 1997). Minerals constitute 20 to 50% dry weight of seaweeds (Kazutosi 2002). The single most important mineral provided by seaweeds is iodine and the other abundant minerals include potassium, sodium, calcium, magnesium, zinc, copper, chloride, sulphur, phosphorous, vanadium, cobalt, manganese, selenium, bromine, arsenic, iron and fluorine (Chapman and Chapman 1980; Levring et al. 1969). Potassium is essential for even minimal nerve and muscle functioning and as a cross membrane transporter ion for neurotransmitters and hormones. Selenium like its partner in thyroid hormone helps in metabolism (Ryan Drum 2005).

Algin has great therapeutic value as a heavy metal detoxifying agent. When edible brown seaweeds are added to the diet as a component, algin or sodium alginate present in the seaweeds binds heavy metals present in the food steam and carry them out with stool, since algin is generally not digestible (Schechter et al. 1997). Regular eating of even small amounts of brown seaweeds will aid in an ongoing metal detoxification practice, reducing the quenching of enzymes by heavy metals. If a patient possesses relatively high levels of toxic metals in hair, blood or urine, addition of 3–5 g of brown seaweed to the daily diet will help to remove the same metals from the body but not from the hair (Ryan Drum 2005). Regular algin rich seaweed (3–5 g) consumption may probably prevent or suppress age-related dementia and perhaps Alzheimer's and also helps to slow the bioaccumulation of neurotoxic metals. Fucoidan from brown seaweed (3–5 g) in the pre-surgical patient diet for a week or two seems to reduce the intensity of blood loss and vascular bed collapse shock during and after surgery and the mechanism for this positive effect is unclear (Ryan Drum 2005).

The reality on the following parameters remains to be elucidated on the consumption of seaweeds: (1) understanding of nutritional composition across algal species, geographical regions and seasons, all of which can substantially affect their dietary value, (2) quantifying which fractions of algal foods are bioavailable to humans, and which factors influence how food constituents are released, ranging from food preparation through genetic differentiation in the gut microbiome and (3) to understand how algal nutritional and functional constituents interact in human metabolism (Wells et al. 2017).

Many species of seaweeds are found to be biologically active against many of the common disease pathogens. Biologically active members of green algae (Chlorophyta) include species of *Codium, Halimeda, Ulva, Cladophoropsis, Caulerpa* and *Enteromorpha,* and those of brown algae (Phaeophyta) with antimicrobial activity include species of *Dictyopteris, Zonaria, Ecklonia, Durvillea, Dictyota, Sargassum* and *Turbinaria.* Red algae (Rhodophyta) like the species of *Chondria, Digenea, Laurencia, Caloglossa, Grateloupia, Hypnea* and *Murayella* showed microbial activity (Baker 1987; Chapman and Chapman 1980; Sivapalan 1987; Tseng 1983).

Algae have been the source of about 35% of the newly discovered chemicals between 1977 and 1987, followed by sponges (29%) and cnidarians (22%) (Ireland et al. 1993). The discovery of new products from seaweeds has decreased since 1995 and attention has now shifted to marine micro-organisms (Kelecom 2002).

Recent studies on species of *Sargassum* that were traditionally used in Chinese herbal medicine for the treatment of cancer by Yamamoto et al. (1974) and Yamamoto et al. (1977, 1982) showed that the extract from *Sargassum fulvellum* was active against leukaemia and sarcoma tumour cells implanted on mice. Several species of *Sargassum* have also been found to contain extracts which are very active against bacteria, including *Staphylococcus aureus, Escherichia coli* and *Salmonella* spp. The active antibacterial constituent of *Sargassum kjellmanianum* has been found to be cyclopentenone (Fenical 1983). Modern physicians make frequent uses of algin as an adsorbent in wound dressing and as a haemostatic in brain and thoracic surgery (Schwmmer and Schwimmer 1955).

The discovery of metabolites with biological activities from macroalgae has increased significantly during the last three decades and very few products with real potential have been identified or developed. Substances like sulphated polysaccharides as antiviral substances, halogenated furanones from *Delisea pulchra* as antifouling compounds, and kahalalide F from a species of *Bryopsis* as a possible treatment of lung cancer, tumours and AIDS currently receive most attention from pharmaceutical companies for use in drug development (Smit 2004).

In recent years, pharmaceutical firms have started looking towards marine organisms, including seaweeds, in their search for new drugs from natural products. These products are also increasingly being used in medical and biochemical research. Prior to the 1950s, the medicinal properties of seaweeds were restricted to traditional and folk medicines (Lincoln et al. 1991). Sulphated polysaccharides from red algae, viz., galactan sulphate from *Aghardhiella tenera* (Witvrouw et al. 1994) and xylomannan sulphate from *Nothogenia fastigiata* (Damonte et al. 1994; Kolender et al. 1995), were tested against human immune deficiency virus (HIV), herpes simplex virus (HSV) 1 and 2 types, and respiratory syncytial virus (RSV) and these polysaccharides were found to be active during the first stage of the RNA virus replication when the virus adsorbs onto the surface of the cell (De Clercq 2000).

Carlucci et al. (1997, 1999a, b) observed that λ-carrageenan and partially cyclized μ/ʹι-carrageenan from *Gigartina skottsbergii* have potent antiviral effects against different strains of HSV types 1 and 2 during the virus adsorption stage. A carrageenan-based vaginal microbicide called Carraguard has been shown to block HIV and other sexually transmitted diseases in vitro. Carraguard entered phase III clinical trials involving 6000 non-pregnant, HIV-negative women in South Africa and Botswana in 2003 (Spieler 2002). A sulphated polysaccharide from *Schizymenia pacifica* inhibits HIV reverse transcriptase in vitro (Nakashima et al. 1987a, b), a later stage in HIV replication.

Fucoidan has potent antiviral properties towards viruses such as RSV (Malhotra et al. 2003), HIV (Sugawara et al. 1989), HSV types 1 and 2 and human cytomegalovirus (Feldman et al. 1999; Majczak et al. 2003; Ponce et al. 2003). It also enhances new blood vessel formation by modulating the expression of surface proteins (Matou et al. 2002), and lipogenic activity has been demonstrated for lectins from *Codium fragile* in isolated rat and hamster adipocytes (Ng et al. 1989).

Red seaweeds contain phycobiliproteins having antioxidant properties that help in the prevention or treatment of neurodegenerative diseases caused by oxidative stress (Alzheimer's and Parkinson's) as well as in the treatment of gastric ulcers and cancers. Besides, polyunsaturated fatty acids, viz., omega 3 fatty acids and omega 6 fatty acids, play a critical role in the prevention of cardio vascular diseases, osteoarthritis and also diabetes (Valko 2007).

Seaweeds are excellent source of known vitamins such as A, B especially B12, C, D, E and K, as well as essential amino acids. Nori (*Porphyra*) is very rich in vitamins A and C (Ryan Drum 2005). Seaweed polyphenols known as phlorotannins possess antioxidant activity and polyphenols extracted from brown and red seaweeds have shown antioxidant activity (Nakamura et al. 1996). Carotenoids are powerful antioxidants. Recent studies have shown the correlation between a diet rich in carotenoids and a diminishing risk of cardiovascular disease, cancer (β carotene, lycopene) as well as ophthalmological diseases (lutein, zeaxanthin). Antioxidant property of red seaweed carotenoids in preventing much pathology linked to oxidative stress is explained (Valko 2007; Yan et al. 1998).

Chemicals responsible for antibiotic activities are widespread in macroalgae. Macromolecule recognition processes are common in cells and their specificity is their most important characteristic. Many research programmes exploit recognition events and these have become focus areas of research in biology, chemistry, medicine and pharmacology. Biological reactions that involve recognition events include processes such as cell agglutination and coagulation, the stimulation of cell migration and fertilization.

2.5 Recommendations

- Taxonomists are absolutely necessary to correctly identify the seaweed.
- Autecology of required species is to be studied.
- Periodic assessment of seaweed resources.
- Assessment of production rates of dominant species.
- Long-term monitoring of temporal changes in seaweed-dominated communities.
- To bring awareness by academics and NGOs among the coastal communities to manage practices for invasive species.
- To consider seaweed farming as a tool for coastal resource management which will ultimately protect biodiversity.
- Periodic assessment of biodiversity in time and space.

- Monitoring of biological diversity to avoid undesirable effects on environment.

Acknowledgement The authors would like to appreciate the cordial and healthy discussions on various aspects of algae with late Prof V. Krishnamurthy (during his life time), a Legendary Algologist and Walking Encyclopedia of Algae, which encouraged us to write this review article, and we are ever grateful to him.

References

Aaronson S (1986) A role for algae as human food in antiquity. Food Foodways 1(3):311–315

Anchana P, Supattra P, Anuchit D, Bongkot W, Sutinee S (2011) Diversity and distribution of seaweed at Khanom-Mu Ko Thale Tai National Park, Nakhon Si Thammarat Province, Thailand. Songklanakarin J Sci Technol 33(6):633–640

Baker JT (1987) The search for pharmaceuticals from marine algae. In: Furtado JI, Wereko-Brobby CY (eds) Tropical marine algal resources of the Asia-Pacific region: a status report. Commonwealth Science Council Technical Publication Series. Nr. 181, CSC (87) EPP- 4: 9–23

Baleta FN, Nalleb JP (2016) Species composition, abundance and diversity of seaweeds along the intertidal zone of Nangaramoan, San Vicente, Sta. Ana, Cagayan, Philippines. AACL Bioflux 9(2):250–259

Bergner R (1997) What is psychopathology? And so what? Clinical pshychotherapy. New Idea Pshychol 22:127–141

Bocanegra A, Bastida S, Benedí J, Ródenas S, Sánchez-Muniz FJ (2009) Characteristics and nutritional and cardiovascular-health properties of seaweeds. J Med Food 12(2):236–258

Bolton JJ (1994) Global seaweed diversity: patterns and Anomolies. Bot Mar 37(3):241–245

Carlucci MJ, Pujol CA, Clancia M, Noseda MD, Matulewicz MC, Damonte EB, Cerezo AS (1997) Antiherpetic anticoagulant properties of carrageenans from the red seaweed *Gigartina skottsbergii* and their cyclized derivatives: correlation between structure and biological activity. Int J Biol Macromol 20:97–105

Carlucci MJ, Ciancia M, Matulewicz MC, Cerezo AS, Damonte EB (1999a) Antiherpetic activity and mode of action of natural carrageenans of diverse structural types. Antivir Res 43:93–102

Carlucci MJ, Scolaro LA, Damonte EB (1999b) Inhibitory action of natural carrageenans on *Herpes simplex* virus infection of mouse astrocytes. Chemotheraphy 45:429–436

Chapman VJ, Chapman DJ (1980) Seaweed and their uses. Chapman and Hall Edition, New York

Circuncisao A, Catarino MD, Cardoso SM, Silva AMS (2018) Minerals from macro algae origin: health benefits and risks for consumers. Mar Drugs 16:400

Craigie JS (2011) Seaweed extracts stimuli in plant science and agriculture. J Appl Phycol 23:371

Damonte EB, Neyts J, Pujol CA, Snoeck R, Andrei G, Ikeda S, Witvrouw M, Reymen D, Haines H, Matulewicz MC, Cerezo A, Coto CE, De Clercq E (1994) Antiviral activity of a sulfated polysaccharide from the red seaweed *Nothogenia fastigiata*. Biochem Pharmacol 47:2187–2192

Dawson EY (1954) Note on tropical Pacific marine algae. Bullet South Calif Acad Sci 53:P7

De Clercq E (2000) Current lead natural products for the Chemotheraphy of human immunodeficiency virus (HIV) infection. Med Res Rev 20:323–349

Dillehay TD, Ramírez C, Pino M, Collins MB, Rossen J, Pino-Navarro JD, Dobzhansky (2008) Monte Verde: seaweed, food, medicine, and the peopling of South America. Science 320(5877):784–786

Dobzhansky T (1950) Genetics of natural populations XIX: origin of heterosis through natural selection in populations of Drosophila pseudoobscura. Genetics 35:288–302

Dring MJ (1982) The biology of marine plants. Edward Arnold, London

Ryan Drum (2005) Island herbs and B.C. Kelp, Warldron Island, WA 98297–0025

Egerod L (1974) Report on the marine algae collected on the Fifth Thai-Danish Expedition of 1966, Chlorophyceae. Bot Mar 17:130–157

Escrig JA, Goni Combrodon I (1999) Nutritional evaluation and physiological effects of edible seaweeds. Arch Latinoam Nutr 49(2):114–120

Feldman SC, Reynaldi S, Stortz CA, Cerezo AS, Damont EB (1999) Antiviral properties of fucoidan fractions from *Leathesia difformis*. Phytomed 6:335–340

Fenical W (1983) Investigation of benthic marine algae as a source of new pharmaceuticals and agricultural chemicals. In: Tseng CK (ed) Proc. Joint China-U.S. Phycol. Symp. Science Press, Beijing, pp 497–521

Gantar M, Svircev Z (2008) Microalgae and cyanobacteria: food for thought. J Phycol 44:260–268

Hale RW, Pion RJ (1972) Laminaria: An underutilized clinical adjunct. Clinical Obstet Gynaeol 15:829–850

Hutchinson GE (1959) Homage to Santa Rosalía, or why are there so many kinds of animals? Am Nat 93:145–159

Ireland CM, Copp BR, Foster MP, McDonald LA, Radisky DC, Swersey JC (1993) Biomedical potential of marine natural products. In: Attaway DH, Zaborsky OR (eds) Marine biotechnology, Pharmaceutical and bioactive natural products, vol 1. Plenum Press, NY, pp 1–43

John DM (1994) Biodiversity and conservation: an algal perspective. The Phycologist 38:3–15

Kawaguchi S, Hayashizaki (2011) Biodiversity studies on seaweeds and sea grasses in thecoastal waters of Southeast Asia (Project-3: seaweed/ seagrass group). In: Nishida S, Fortes MD, Miyazaki N (eds) Coastal marine science in Southeast Asia - synthesis report of the Core University Program of the Japan Society for the Promotion of Science: coastal marine science (2001–2010), pp 49–57

Kazutosi N (2002) Seaweeds Kaiso: Bountiful harvest from the seas. In: Sustenance for health and wellbeing. Japan Seaweed Association, Kochi

Kelecom A (2002) Secondary metabolites from marine microorganisms. An Acad Bras Cienc 74:151–170

Khan W, Usha PR, Sowmyalakshmi S, Jithesh MN, Prasanth RD, Hodes M, Critchley AT, Craigie JS, Norrie J, Balakrishnan P (2009) Seaweed extracts as biostimulants of plant growth and development. J Plant Growth Regul 28:386–399

Kolender AA, Matulewicz MC, Cerezo AS (1995) Structural analysis of antiviral sulfated α-D - (1, 3)-linked mannans. Carbohydr Res 273:179–185

Krishnamurthy V, Joshi HV (1970) A check-list of Indian marine algae. CSMCRI Publication, Bhavnagar

Lee RE (2008) Phycology. Cambridge University Press, Cambridge

Levring T, Hoppe HA, Schmid OJ (1969) Maine algae - a survey of research and utilization. Cram de Grutjter 1:421

Lewmanomont K, Wongrat L, Supanwanid C (1995) Algae in Thailand. Integrated Promotion Technology Co., Lts, Thailand

Lincoln RA, Strupinski K, Walker JM (1991) Bioactive compounds from algae. Life Chem Rep 8:97–183

Lüning K (1990) Seaweeds. Their environment, biogeography and ecophysiology. Wiley Interscience, New York

Mac Artain P, Gill CIR, Brooks M, CampbellR RIR (2008) Nutritional value of edible seaweeds. Nutr Rev 65:535–543

Majczak GAH, Richartz RRTB, Duarte MER, Noseda MD (2003) Antiherpetic activity of heterofucans isolated from *Sargassum stenophyllum* (Fucales, Phaeophyta). In: Chapman ARO, Anderson RJ, Vreeland VJ, Davison IR (eds) Proceedings of the 17[th]International seaweed symposium. Oxford University Press, Oxford, pp 169–174

Malhotra R, Ward M, Bright H, Priest R, Foster MR, Hurle M, Blair E, Bird M (2003) Isolation and characterisation of potential respiratory syncytial virus receptor(s) on epithelial cells. Microbes Infect 5:123–133

Matou S, Helley D, Chabut D, Bros A, Fischer A-M (2002) Effect of fucoidan on fibroblast growth factor-2-induced angiogenesis *in vitro*. Thromb Res 106:213–221

Mc Hugh DJ (2003) A guide to seaweed Industry. FAO Fisheries Technical paper. No. 441, Rome
Melkonian M (1995) Introduction: XI - XIII. In: Wiessner W, Schnepf E, Starr RC (eds) Algae, environment and human affairs. Biopress Ltd., Bristol
Miller KR, Raid WV, Barber CV (1992) The global biodiversity strategy and its significance for sustainable agriculture. In: Swaminathan MS, Jena S (eds) Biodiversity-implications for global food security. Macmillan Publications, Madras, p 326
Mittermeier RA, Myers N, Mittermeier CG (1999) HotSpots: earth's biologically richest and most endangered terrestrial ecoregions. CEMEX, Mexico City, p 430
Moi PS (1987) Marine algal resources in peninsular Malaysia. In: Futado JI, Wereko-Brobby CY (eds) Tropical marine algal resources of the Asia-Pacific Region: a status report. Commonwealth Science Council Technical Publication Series, Nr., 181, CSC (87) EPP-4: 69–76
Muraoka D (2004) Seaweed resources as a source of carbon fixation. Bullet Fisher Res Agency Suppl 1:59–63
Myers N, Mittermeier RA, Mittermeier CG, Fonseca GAB, Kent J (2000) Biodiversity hotspots for conservation priorities. Nature 403:853–858
Nakamura, Nakamura T, Nagayama K, Uchida K, Tanaka R (1996) Antioxidant activity of phlorotannins isolated from the Brown alga *Eisenia bicyclis*. Fish Sci 62(6):923–926
Nakashima H, Kido Y, Kobayashi N, Motoki Y, Neushul M, Yamamoto N (1987a) Antiretroviral activity in a marine red alga reverse transcriptase inhibition by an aqueous extract of *Schizymenia pacifica*. J Cancer Res Clin Onc 113:413–416
Nakashima H, Kido Y, Kobayashi N, Motoki Y, Neushul M, Yamamoto N (1987b) Purification and characterization of an avian myeloblastosis and human immunodeficiency virus reverse transcriptase inhibitor, sulfated polysaccharides extracted from sea algae. Antimicrob Agents Chemother 31:1524–1528
Newton L (1951) Seaweed utilization. Sampson Low, London
Ng TB, Li WW, Yeung HW (1989) Effects of lectins with various carbohydrate binding specificities on lipid metabolism in isolated rat and hamster adipocytes. Int J Biochem 21:149–155
Nguyen HD, Huynh QN (1993) Rong bien Vietnam (marine algae of North Vietnam). Nha xuat ban KH-KT, Hanoi, p 364
Nuqui CR (1987) The major groups of Philippine seaweeds. In: Furtado JI, Wereko-Brobby CY (eds) Tropical marine algal resources of the Asia-Pacific region: a status report. CSC Techn. Publ. Ser. Nr. 181, CSC (87) EPP-4: 91–100
Oza RM, Zaidi SH (2001) A revised check list of Indian marine algae. CSMCRI, Bhavnagar
Pham HT (2006) Study on the variation of seagrass population in coastal waters of Khanh Hoa Province. Vietnam Coast Mar Sci 30:167–173
Phang SM (1998) The seaweed resources of Malaysia. In: Crithery AT, Ohno M (eds) Seaweed resources of the world. JICA, Tokyo, pp 79–91
Ponce NMA, Pujol CA, Damonte EB, Flores ML, Stortz CA (2003) Fucoidans from the brown seaweed *Adenocystis utricularis*: extraction methods, antiviral activity and structural studies. Carbohydr Res 338:153–165
Reed MS (1906) The economic seaweeds of Hawaii and their food value. Ann. Rept. Hawaii Agric. Exp Dermatol 1906:61–88
Sahoo D (2010) Common seaweeds of India. I. K. International Pvt Ltd, New Delhi
Schechter SR, Norman SC, Oelsner G (1997) Fighting radiation and chemical pollutants with foods, herbs and vitamins: documented natural remedies that boost your immunity and detoxify, 2nd ed.
Schwmmer M, Schwimmer D (1955) The role of algae and plankton in medicine. Grune and Stratton, Inc., New York
Sharma GJ, Thokchom DS (2008) Biodiversity hotspots: last refuge for mankind? In: Proc. Natl. Semi. on conservetion of bioresources. Manipur University, Imphal, pp 55–60
Sivapalan A (1987) Marine algae of Sri Lanka. In: Furtado JI, Wereko-Brobby CY (eds) Tropical marine algal resources of the Asian- Pacific region: a status report. CSC Techn. Publ. Ser., Nr. 181, CSC (87) EPP-4, 105–107

Smit AJ (2004) Medicinal and pharmaceutical uses of seaweed natural products: a review. J Appl Phycol 16:245–262
Smith GM (1944) Marine algae of the monterey peninsula. Stanford Univ, California
Spieler R (2002) Seaweed compound's anti-HIV efficacy will be tested in Southern Africa. Lancet 359:1675
Subba Rao PV (2007). Seaweed diversity culture and cultivation. In: National Symposium on conservation and valuation of Marine Biodiversity, pp 387–393
Subba Rao PV (2012) Seaweed biodiversity and conservation. Proc. Symp. On Biodiversity Status & Conservation Strategies with reference to NE India. Manipur Univ, Manipur, pp 1–7
Subba Rao PV, Mantri VA (2006) Indian seaweed resources and sustainable utilization: scenario at the dawn of a new century. Curr Sci 91(2):164–174
Sugawara I, Itoh W, Kimura S, Mori S, Shimada K (1989) Further characterization of sulfated homopolysaccharides as anti-HIVagents. Experientia 45:996–998
Tseng CK (1981) Commercial cultivation. In: Lobban CS, Wynne MJ (eds) The biology of seaweeds. Univ. of Calif. Press, Berkeley, pp 71–80
Tseng CK (1983) Common seaweeds in China. Science Press, Beijing, p 316
Tseng CK (2004) Past, present and future of phycology in China. Hydrobiol 512:11–20
Turner N (2003) Women & plants gender relations in biodiversity management and conservation Patricia L. Howard CHAPTER 7 'passing on the news': Women's work, traditional knowledge and plant resource management in indigenous societies of North-Western North America
Valko M (2007) Free radicals and antioxidants in normal physiological functions and human disease. Int J Biochem Cell Biol 39:44–84
Velasquez GT, Lewmanomont K (1975) A checklist on the study of benthic marine algae of Thailand. Kasetsart Univ Fish Res Bull 8:1–25
Wells KR, Hamede Douglas K, Kerlin Andrew H, Paul S, Hohenlohe Menna A, Jones Hamish E, McCallum I (2017) Infection of the fittest: devil facial tumour disease has greatest effect on individuals with highest reproductive output. Ecol Lett 20:770–778
Win Shin-Sun RSS (1987) Marine algal resources of Hong Kong. In: Furtado JI, Wereko-Brobby CY (eds) Tropical marine algal resources of the Asian Pacific region: a status Report.CSC Techn. Publ. Ser., Nr. 181, CSC (87) EPP - 4:31–40
Witvrouw M, Este JA, Mateu MQ, Reymen D, Andrei G, Snoeck R, Ikeda S, Pauwels R, Bianchini NV, Desmyter J, de Clercq E (1994) Activity of a sulfated polysaccharide extracted from the red seaweed *Aghardhiella tenera* against human immunodeficiency virus and other enveloped viruses. Antiviral Chem Chemother 5:297–303
Wright JT, McKenzie LA, Gribben PE (2007) A decline in the abundance and condition of a native bivalve associated with Caulerpa taxifolia invasion. Mar Freshw Res 58:263–272
Yamamoto I, Nagumo T, Yagi K, Tomingaga N, Aoiki M (1974) Antitumor effect of seaweeds. I. Antitumor effect of extracts from *Sargassum* and *Laminaria*. Jpn J Exp Med 44:543–546
Yamamoto I, Nagumo T, Fujihara M, Takahashi M, Ando Y, Okada M, Kawai K (1977) Antitumor effect of seaweeds. II. Fractionation and partial characterisation of the polysaccharide with antitumor activity from *Sargassum fulvellum*. Jpn J Exp Med 47:133–140
Yamamoto I, Takahashi M, Mamura E, Maruyama H (1982) Antitumor activity of crude extracts from edible marine algae against L-1210 leukemia. Bot Mar 25:455–457
Yan XT, Nagata T, Fan X (1998) Antioxidative activities in some common seaweeds. Plant Foods Hum Nutr 52:253–262
Zaneveld JS (1959) The utilization of marine algae in tropical South and East Asia. Econ Bot 13:89–131

Tree Flora of Andhra Pradesh, India

3

K. Sri Rama Murthy, S. Sandhya Rani, S. Karuppusamy, A. Lalithamba, and T. Pullaiah

Abstract

The state of Andhra Pradesh is endowed with rich biodiversity. Extensive field and literature survey of trees in Andhra Pradesh yielded 601 tree taxa under 295 genera belonging to 75 families out of the estimated 2900 species of flowering plants in Andhra Pradesh. Euphorbiaceae (49) followed by Rubiaceae (44), Mimosaceae (32), Moraceae (29), Rutaceae (26), Tiliaceae (20), Verbenaceae and Bignoniaceae (19), Meliaceae (18), Fabaceae, Annonaceae, Sterculiaceae (15), Cordiaceae, Ebenaceae, Combretaceae, Lauraceae (13) are the largest families based on the species number. The dominant genera include *Ficus* with 24 species, *Grewia* 19, *Acacia* 15, *Diospyros* 12, *Terminalia* 11, *Cassia* 9 and *Cordia* 8 species. Endemic tree species distributed in different regions of Andhra Pradesh are *Actinodaphne madraspatana, Alphonsea madraspatana, Albizia thompsonii, Boswellia ovalifoliolata, Bridelia cinerascens, Cordia domestica, Croton scabiosus, Dimorphocalyx kurnoolensis, Eriolaena lushingtonii, Hildegardia populifolia, Lasiococca comberi, Premna hamiltonii, Pterocarpus santalinus, Shorea tumbuggaia, Syzygium alternifolium, Terminalia pallida, Wendlandia gamblei, Toona ciliata* var. *brevipetiolata,* etc. A brief account on topography, geology, climate, vegetation pattern, tree flora analysis and distribution pattern of trees in various regions of Andhra Pradesh are provided.

K. S. R. Murthy
R&D Center for Conservation Biology and Plant Biotechnology, Shivashakti Biotechnologies Limited, Hyderabad, Telangana, India

S. S. Rani · T. Pullaiah (✉)
Department of Botany, Sri Krishnadevaraya University, Anantapuramu, Andhra Pradesh, India

S. Karuppusamy
Department of Botany, The Madura College, Madurai, Tamil Nadu, India

A. Lalithamba
Department of Botany, DKW College, Nellore, Andhra Pradesh, India

S. M. Khasim et al. (eds.), *Medicinal Plants: Biodiversity, Sustainable Utilization and Conservation*, https://doi.org/10.1007/978-981-15-1636-8_3

Keywords
Andhra Pradesh · Trees · Vegetation · Endemic trees

3.1 Introduction

The close bonding of humans with trees may be traced back to their arboreal ancestry. Reposing under a tree is like being cuddled by a mother-comfortable and secured. Trees are major components of forests. By sequestering carbon, releasing oxygen and losing a large amount of water vapour, trees profoundly influence the environment. Trees bind the soil, recycle minerals and regulate the course of rainwater. Unparalleled in the variety of useful products they give and the wide range of organisms they feed and shelter, trees are essential for maintaining the health of several ecosystems. Above all, trees are friendly, beautiful and invoke awe and admiration. They are the largest and longest-lived immobile organisms and often an amazing diversity of forms. Trees symbolize benevolence, fertility and mobility.

3.2 Location and Characteristics of the Study Area

Geometrically the state of Andhra Pradesh lies between the latitudes 12°37′ N and 19°55' N and longitude 76°45' E and 84°46' E (Fig. 3.1). Located in southern India, Andhra Pradesh is bounded on the south by Tamil Nadu, west by Karnataka, northwest by Telangana, north by Orissa and east by the Bay of Bengal. The total area of the state is 1,60,205 km^2. It has 974 km sea coastline along the Bay of Bengal.

The 13 districts of the state are generally grouped into two geographically distinct regions called (1) Circars or Coastal Andhra with nine districts, i.e. Srikakulam, Vizianagaram, Visakhapatnam, East Godavari, West Godavari, Krishna, Guntur, Prakasam and Nellore and (2) Rayalaseema with four districts, i.e. Kurnool, Kadapa (formerly called Cuddapah), Anantapuramu (formerly called Anantapur) and Chittoor. The new capital city of Andhra Pradesh is proposed at Amaravathi in Guntur District, north of Guntur City.

According to Andhra Pradesh Forest Department information brochure, the forest area in the state is 22,862 km^2, forming 14.27% of the state's territory. This is far below the 33% stipulated in National Forest Policy. The depletion of forests all over the world at an estimated rate of 2,45,000 km^2 per year is one of the most alarming aspects of present day biosphere tendencies. Due to such an extensive and unabated destruction of forests, mankind is losing many valuable plants even before we come to know that they exist. Causes of threats to the flora have generally been grouped into two categories, viz., (1) natural and (2) man-made. The natural causes include floods, earthquakes, landslides, natural competition between species, biology of the species such as lack of pollination and natural regeneration, diseases, etc. In man-made threats could be included deliberate destruction of habitat (such as by mining, dam constructions and conversion of forests), excessive grazing, over-exploitation, etc.

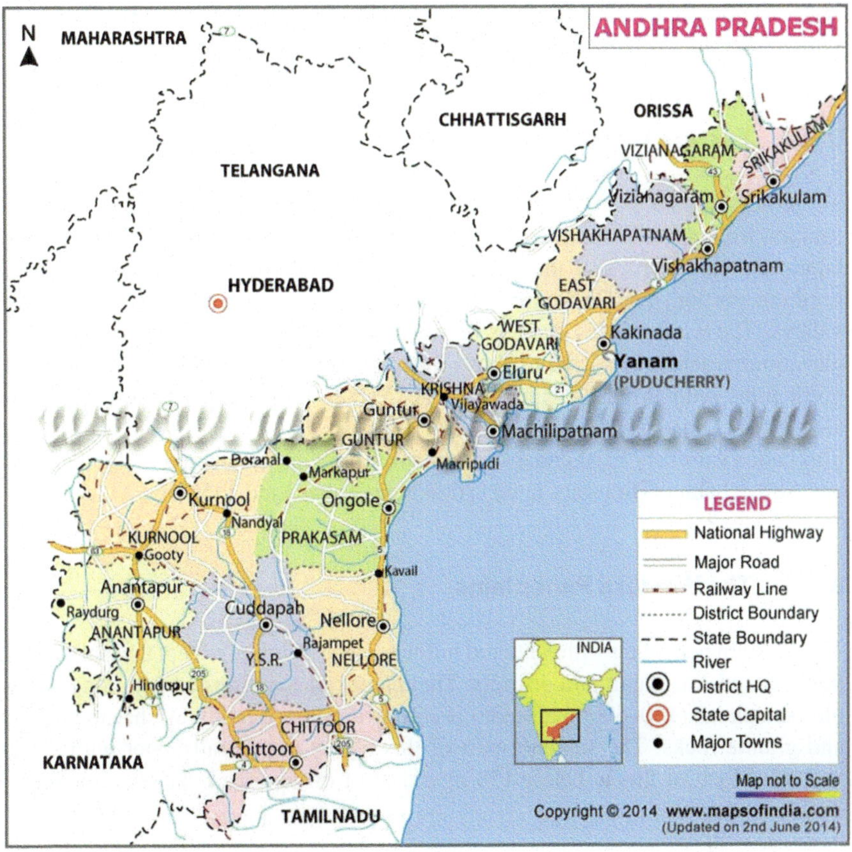

Fig. 3.1 Map of India and map of Andhra Pradesh

3.3 Geographical Division

Geographically, the whole state can be divided into the Coastal plains, the Eastern Ghats and the Western peneplains.

3.3.1 The Coastal Plains

This region stretches along the coast from northern end to southern, i.e. from Srikakulam district to Nellore district. The northern portion of the coastal plain is narrow with an average width ranging between 30 and 40 km. The middle part is wider than northern portion, with an average between 70 and 75 km and in some places 100 km even. It includes the shallow fresh water lake of Kolleru which is a natural depression between Godavari and the Krishna deltas. The width of the coastal plain in the southern part is between 50 and 60 km. The Simhachalam cliff

(244 m), Dolphin's nose (375 m), Kondapalli (573 m), Kolleru lake (259 km^2), Pulicat lake and Sriharikota island dot this coastal-plain region.

3.3.2 Eastern Ghats

These hills, unlike that of Western Ghats of India, do not form a continuous range running from north to south between the coastal plain and the plateau. The Eastern Ghats are the outcrops of the lower Vindhyan and the Cuddapah systems. In the northern portion, the Ghats are highly dissected and intervened by a number of valleys. The width of this place is between 60 and 70 km, and in many parts, the elevations reach above 1200 m above mean sea level. The Papikonda range in the north joins with Simhachalam hill range. Yarada hills extend towards the coast along Visakhapatnam. Veering further south-west they form the Kadapa range of hills called variously Palakondas, Velikondas, Erramalais, Nallamalais, Lankamalais and Seshachalam. They vary in elevation between 600 and 1350 m.

3.3.3 The Western Peneplains

The Western peneplain with scattered hillocks cover Kurnool except the Nallamalais portion, and Anantapuramu districts. The interior plateau formed with long belt of old peneplains (150–600 m altitude) is chiefly developed on the Archaean gneisses and granite rocks. The southern part of this region is generally poor with scanty rainfall. Much of this is below 150 m and most of the area is largely covered by deciduous forests.

3.3.3.1 Vegetation

Biodiversity is fully expressed in the flora, fauna and vegetation. Tree flora is rich and varied and distributed under the following types of vegetation.

1. Tropical evergreen forests

   This type of vegetation is seen in very few valleys in small patch of forests in Laksmipuram of Visakhapatanam in Northern Eastern Ghats. The common trees are *Cinnamomum zeylanicum, Elaeocarpus serratus, Ixora notoniana, Meliosma microcarpa, Symplocos laurina, Toona ciliata,* etc.
2. ***Tropical semievergreen forests*** (moist deciduous forests and mixed with evergreen elements)

   These are prevalent in moist valleys of Sapparla, Dharakonda, Galikonda, Tanjavanam, Minumuluru, some areas near Anantagiri, Nulakamaddi, Maredumilli of Northern Eastern Ghats, trees such as *Magnolia champaca, Mangifera indica, Artocarpus lakoocha, Dillenia pentagyna, Firmiana colorata, Bridelia tomentosa, Xylia xylocarpa, Mesua nagassarium, Polyalthia cerasoides, Macaranga peltata, Pittosporum napaulense, Phoebe lanceolata, Murraya koenigii,* etc. are the dominant elements of semievergreen forests.

3. ***Tropical moist deciduous forests***: subdivided for convenience into
   (a) **Northern subtropical deciduous forests (Sal Forests):** This type of forest is found in parts of the districts of Srikakulam and north-eastern border area of Vizianagaram district. In the Sal forests *Shorea robusta* predominates and is associated with *Syzygium cumini, Xylia xylocarpa, Haldina cordifolia* (Fig. 3.3e), *Terminalia coriacea, Pterocarpus marsupium, Anogeissus latifolia, Albizia procera, Madhuca longifolia, Phyllanthus emblica, Lagerstroemia parviflora, Schleichera oleosa, Cleistanthus collinus, Buchanania lanzan* (Fig. 3.2c), *Dillenia pentagyna, Diospyros melanoxylon, Mallotus philippensis, Careya arborea, Litsea glutinosa, Syzygium operculatum,* etc.
   (b) **Southern Indian tropical moist deciduous forests** (Non-Sal forests): These are found in parts of the Rampa agency, Maredumilli areas of East Godavari district, parts of West Godavari district, small patches of Nallamalais, Talakona and some parts of Seshachalam hill ranges of Middle Eastern Ghats and parts of Southern Eastern Ghats. Dominant tree elements are *Terminalia alata, Xylia xylocarpa, Anogeissus latifolia, Dillenia pentagyna, Pterocarpus marsupium, Mangifera indica, Dalbergia latifolia, Terminalia chebula, Kavalama urens (Sterculia urens)* (Fig. 3.4c), *Mitragyna parviflora* (Fig. 3.4d), *Albizia odoratissima* (Fig. 3.2a), *Bridelia airyshawii, Schrebera swieteniodes, Careya arborea, Grewia tiliifolia, Polyalthia cerasoides, Kydia calycina, Semecarpus anacardium,* etc.
   (c) **Southern tropical moist deciduous Riverian forests**: These are spread in very limited areas along banks and the dried river beds, represent a distinct eco-type comprising tree species of *Terminalia arjuna, Pongamia pinnata, Tamarindus indica, Anogeissus acuminata, Barringtonia acutangula* and also mixed with characteristic species on the sandy and rocky bouldered river beds such as *Homonoia riparia, Tamarix ericoides* and *Syzygium heyneanum.*
4. Tropical dry deciduous forests:

   These forests are found in almost all regions of Andhra Pradesh. These are divided into teak-bearing forest and non-teak-bearing forests.
   (a) **Teak-bearing forests:** These are distributed in Northern and middle Eastern Ghats of Andhra Pradesh and dominated by valuable timber tree species, i.e. *Tectona grandis*. Associating elements are *Anogeissus latifolia, Pterocarpus marsupium, Terminalia chebula, Terminalia bellirica, Garuga pinnata, Bridelia airyshawii, Cassia fistula,* etc.
   (b) **Non-teak-bearing forests**: These are distributed along the Seshachalam hill ranges of middle Eastern Ghats. Interestingly these forests having important and very valuable endemic tree such as *Pterocarpus santalinus* and also *Shorea tumbuggaia, Syzygium alternifolium, Boswellia ovalifoliolata*, prominent in some areas and mixed up with other species like *Terminalia pallida, Shorea roxburghii, Phyllanthus emblica, Anogeissus latifolia, Terminalia paniculata, T. alata, Chloroxylon swietenia, Dolichandrone arcuata, Wrightia tinctoria, Vitex altissima,* etc.

Fig. 3.2 (**a**) *Albizia odoratissima* (L.f.) Benth., (**b**) *Azadirachta indica* A. Juss., (**c**) *Buchanania lanzan* Spreng., (**d**) *Butea monosperma* (Lam.) Taub., (**e**) *Cochlospermum religiosum* (L.) Alston, (**f**) *Croton scabiosus* Beddome

5. Mixed dry deciduous forests
   (a) **Northern mixed dry deciduous forests:** These are distributed in restricted areas in Northern Eastern Ghats. *Shorea robusta* may or may not be present but *Boswellia serrata* is normally common and associated with *Terminalia alata, Bombax ceiba, Hymenodictyon excelsum, Pterocarpus marsupium, Kavalama urens (Sterculia urens), Alangium salvifolium, Mallotus philippensis, Cassia fistula,* etc.
   (b) **Southern mixed dry deciduous forests:** These are found in drier areas of all regions of Eastern Ghats. Dominant tree elements are *Anogeissus latifolia, Chloroxylon swietenia, Diospyros melanoxylon, Gardenia gummifera, Albizia odoratissima, Hardwickia binata, Pterospermum xylocarpum, Helicteres isora, Catunaregam spinosa, Flacourtia ramontchi,* etc.

6. Dry Savannah forests

   These, formed mostly as a result of biotic interference, are scattered throughout the area, covered with stunted tree species of *Phyllanthus emblica, Terminalia chebula, Pterocarpus marsupium,* etc.
7. Scrub forests

   These are considered to be a result of intensive biotic interference. These are seen in all along the Andhra Pradesh in smaller areas. These are dominated by thorny species of *Acacias, Capparis sepiaria, Ziziphus mauritiana, Z. oenoplia, Z. xylopyrus, Euphorbia antiquorum, E. tirucalli, Flacourtia sepiaria,* and *Catunaregam spinosa,* and also associated with non-thorny drought-resistant species like *Dolichandrone falcata, Wrightia tinctoria, Dodonaea viscosa, Cassia fistula,* etc.
8. Tropical dry evergreen forests

   This type of forest is found in South Kadapa, Sriharikota and Mamandur valley in Seshachalam hill ranges. The forest is characterized by tree species like *Manilkara hexandra, Memecylon umbellatum, Syzygium cumini, Albizia amara, A. lebbeck, Strychnos nux-vomica* (Fig. 3.6c), *Sapindus emarginatus* (Fig. 3.5d), *Drypetes sepiaria, Pterospermum canescens, Drypetes ferrea, Garcinia spicata, Cordia dichotoma, Flacourtia indica,* etc.

3.4 Tree Flora Analysis

An extensive field study was conducted during 1993–2018 in different hills of Andhra Pradesh Ghats. Analysis of the tree diversity available within the Andhra Pradesh and literature (Hooker 1872-1897; Gamble and Fischer 1915-1935; Pullaiah and Sandhya Rani 1999; Pullaiah and Sri Rama Murthy 2001, 2018; Pullaiah and Muralidhara Rao 2002; Pullaiah et al. 2007, 2011, 2018a, 2018b; Pullaiah 2018; Pullaiah and Karuppusamy 2018) revealed altogether 601 taxa including 561 species and 5 varieties under 295 genera belonging to 80 families. The details of the number of species, genera and families under each plant group are shown in Table 3.1.

It is evident from the data that the ratio of species belonging to monocotyledons and dicotyledons is 1:530. Dominant families with 13 or more tree species are Euphorbiaceae (49), Rubiaceae (44), Mimosaceae (32), Moraceae (29), Rutaceae (26), Caesalpiniaceae (24), Tiliaceae (20), Bignoniaceae and Verbenaceae (19), Meliaceae (18), Fabaceae, Annonaceae, Sterculiaceae 15 each, and Lauraceae, Combretaceae, Myrtaceae, Ebenaceae, Cordiaceae 13 each. It is interesting that Euphorbiaceae occupies the first position; this indicates the wide range of growth adaptability and distribution of the various members of the family. 23 out of 75 families are represented by single genus with one species each.

Some of the arborescent genera having more than 3 arborescent species occurring in Andhra Pradesh are *Ficus* with 24 species, *Grewia* 19, *Acacia* 15, *Diospyros* 12, *Terminalia* 11,*Cassia* 9, *Cordia* 8, *Mallotus, Syzygium, Wendlandia, Vitex* 8 each, *Litsea, Citrus* 7 each, *Antidesma, Bridelia, Euphorbia, Glochidion, Bauhinia* 6 each, *Maytenus, Ehretia, Erythrina, Ziziphus, Zanthoxylum, Aglaia, Memecylon,*

Table 3.1 Statistical analysis of trees of Andhra Pradesh

	Species		Genera		Families	
Group	Total	Percentage	Total	Percentage	Total	Percentage
Dicots	**590**	**98.18**	**287**	**97.24**	**72**	**96.00**
Polypetalae	329	55.82	164	57.45	46	63.89
Gamopetalae	141	25.13	73	25.18	14	19.44
Monochlamydae	112	19.06	50	17.38	12	16.67
Monocots	**11**	**1.85**	**8**	**02.76**	**3**	**04.00**
Total	601	100	295	100	75	100

Antidesma 5 each and *Cinnamomum, Litsea, Chionanthus, Premna, Annona, Polyalthia, Dolichandrone, Phyllanthus, Caesalpinia, Dalbergia, Morinda, Gardenia*, 4 species each.

3.5 Phytogeographical Distributional Pattern of Tree Taxa

Andhra Pradesh is the storehouse of 601 tree species (details given in Tables 3.2 and 3.3), of which 253 tree species are commonly seen throughout Andhra Pradesh like species of *Kavalama urens (Sterculia urens), Azadirachta indica, Boswellia serrata, Pterocarpus marsupium, Dalbergia* spp., *Diospyros melanoxylon, Haldina cordifolia, Gyrocarpus americanus, Tectona grandis,* etc.

3.6 Endemic Tree Taxa to Andhra Pradesh

Of the estimated 17,000 species of angiosperms in India, about 1932 taxa are endemic to Peninsular India (Ahmedullah and Nayar 1987). As many as 77 taxa are endemic to Andhra Pradesh, of which 16 are tree taxa. The taxonomically and phytogeographically interesting endemic tree species occurring in Andhra Pradesh are given below.

Actinodaphne madraspatana Bedd (Lauraceae). Small- to medium-sized tree, leaves simple, glaucous beneath, flowers yellow, dioecious, fruit ellipsoid; distributed in Seshachalam hill ranges (Kadapa-Chittoor districts) at 600–800 m altitudes.

Alphonsea madraspatana Bedd (Lauraceae). Middle-sized tree, flowers yellow; found throughout Andhra Pradesh but rare in distribution at 700–1000 m altitude.

Boswellia ovalifoliolata Balakr. & Henry (Burseraceae). Deciduous tree, leaves imparipinnate, flowers pinkish-white in lax panicles, drupe 3-gonous, seeds winged; distribution is confined to Seshachalam hills (in Kadapa, Chittoor districts) at 600–900 m.

Cordia domestica Roth (Cordiaceae). Small tree, leaves simple, flowers yellowish-white, drupe ellipsoid, yellowish or pinkish-yellow; found in Guvvalacheruvu of Kadapa district at about 1000 m altitude.

Table 3.2 Trees in Andhra Pradesh

S. No.	Botanical name	Family	Tree height	Status	Remarks
1	*Acacia auriculiformis* A. Cunn. ex Benth.	Mimosaceae	Medium	Social forestry	Planted as avenue tree and also grown in forests as social forestry plantation.
2	*Acacia campbellii* Arnott	Mimosaceae	Small	Common	Open forests
3	*Acacia catechu* (L.f.) Willd.	Mimosaceae	Medium	Common	Foot hills of scrub and deciduous forests
4	*Acacia decurrens* (Wendl.) Willd.	Mimosaceae	Small	Introduced	Naturalized
5	*Acacia eburnea* (L.f.) Willd.	Mimosaceae	Small	Common	Scrub forest
6	*Acacia farnesiana* Willd.	Mimosaceae	Small	Occasional	Open dry deciduous forests
7	*Acacia ferruginea* DC.	Mimosaceae	Small	Occasional	Dry deciduous forests
8	*Acacia horrida* (L.f.) Willd.	Mimosaceae	Small	Common	Scrub and foot hills of dry deciduous forests
9	*Acacia leucophloea* (Roxb.) Willd.	Mimosaceae	Medium	Common	Out skirts of dry deciduous forests and scrubs
10	*Acacia mangium* Willd.	Mimosaceae	Medium	Introduced	Planted as avenue tree and also grown as social forestry plantation.
11	*Acacia nilotica* (L.) Willd. ssp. *indica* Benth. (*A. arabica* Willd.)	Mimosaceae	Medium	Common	Scrub forest and waste lands. Planted as plantation crop in afforestation programmes.
12	*Acacia planifrons* Wight & Arn.	Mimosaceae	Small	Occasional	Scrub and dry deciduous forests
13	*Acacia polyacantha* Willd.	Mimosaceae	Medium	Occasional	Dry deciduous forests
14	*Acacia tomentosa* Willd.	Mimosaceae	Small	Rare	Dry deciduous forests
15	*Acacia torta* (Roxb.) Craib	Mimosaceae	Small	Occasional	Open dry deciduous forests
16	*Acer laurinum* Hassk.	Aceraceae	Large	Rare	Moist deciduous forests

(continued)

Table 3.2 (continued)

S. No.	Botanical name	Family	Tree height	Status	Remarks
17	*Acronychia pedunculata* (L.) Miq.	Rutaceae	Small	Rare	Dry deciduous forests
18	*Actinodaphne madraspatana* Bedd. ex Hook.f.	Lauraceae	Medium	Common	Deciduous forests
19	*Adansonia digitata* L.	Bombacaceae	Large	Introduced	Planted on road sides
20	*Adenanthera pavonina* L.	Mimosaceae	Medium	Cultivated	Planted in gardens and as avenue tree
21	*Aegiceras corniculatum* (L.) Blanco.	Myrsinaceae	Small	Occasional	Mangrove forest
22	*Aegle marmelos* (L.) Correa	Rutaceae	Medium	Planted	Deciduous forests
23	*Aglaia almeidai* Sandhyarani, Sriramamurthy & Pullaiah	Meliaceae	Large	Rare	Semievergreen forests
24	*Aglaia elaeagnoidea* (Juss.) Benth.	Meliaceae	Medium	Rare	Moist deciduous forests
25	*Aglaia lawii* (Wight) C. J. Saldanha	Meliaceae	Medium	Rare	Semievergreen
26	*Ailanthus excelsa* Roxb.	Simaroubaceae	Large	Occasional	Deciduous forests
27	*Ailanthus triphysa* (Dennst.) Alston	Simaroubaceae	Large	Rare	Deciduous forests
28	*Alangium chinense* (Lour.) Harms	Alangiaceae	Small	Rare	Moist deciduous forests
29	*Alangium salvifolium* (L.f.) Wangerin ssp. *salvifolium*	Alangiaceae	Small	common	Deciduous forests outskirts near streams and ponds
30	*Albizia amara* (Roxb.) Boivin	Mimosaceae	Small	Common	Dry deciduous forests
31	*Albizia chinensis* (Osbeck) Merr.	Mimosaceae	Medium	Occasional	Dry deciduous forests along river banks
32	*Albizia lebbeck* (L.) Willd.	Mimosaceae	Medium	Common	Planted along road sides and rarely found in dry deciduous forests

(continued)

Table 3.2 (continued)

S. No.	Botanical name	Family	Tree height	Status	Remarks
33	*Albizia odoratissima* (L.f.) Benth. var. *odoratissima* Fig. 3.2a	Mimosaceae	Medium	Common	Dry deciduous forests
34	*Albizia procera* (Roxb.) Benth.	Mimosaceae	Large	Occasional	Deciduous forests
35	*Albizia saman* (Jacq.) F. Muell. (*Samanea saman* (Jacq.) Merr.)	Mimosaceae	Medium	Avenue tree	Planted for shade along avenues
36	*Albizia thompsonii* Brandis	Mimosaceae	Medium	Occasional	Deciduous forests
37	*Alchornea mollis* Muell.-Arg.	Euphorbiaceae	Small	Occasional	Deciduous forests
38	*Aleurites moluccana* (L.) Willd.	Euphorbiaceae	Small	Cultivated	Planted in gardens
39	*Allophylus cobbe* (L.) Raeusch	Sapindaceae	Small	Common	Deciduous forests
40	*Alphonsea madraspatana* Bedd.	Annonaceae	Medium	Rare	Semievergreen
41	*Alphonsea sclerocarpa* Thw.	Annonaceae	Medium	Occasional	Deciduous forests, along streams in deep valleys
42	*Alstonia scholaris* (L.) R.Br	Apocynaceae	Large	Rare	Deciduous forests, preferably moist situations
43	*Alstonia venenata* R. Br	Apocynaceae	Small	Occasional	Deciduous forests
44	*Amoora canarana* (Turez.) Hiern	Meliaceae	Medium	Rare	Semievergreen forest
45	*Anacardium occidentale* L.	Anacardiaceae	Medium	Cultivated	Introduced but naturalized
46	*Annona cherimola* Miller	Annonaceae	Small	Cultivated	Planted
47	*Annona muricata* L.	Annonaceae	Small	Cultivated	Planted
48	*Annona reticulata* L.	Annonaceae	Small	Cultivated	Planted
49	*Annona squamosa* L.	Annonaceae	Small	Wild and Cultivated	Rocky hills in dry deciduous forests and plains

(continued)

Table 3.2 (continued)

S. No.	Botanical name	Family	Tree height	Status	Remarks
50	*Anogeissus acuminata* (Roxb. ex DC.)	Combretaceae	Large	Rare	Deciduous forests usually on river banks
51	*Anogeissus latifolia* (Roxb. ex DC.) Wall ex Guill. & Perr.	Combretaceae	Large	Common	Deciduous forests
52	*Antidesma acidum* Retz	Stilagiaceae	Small	Common	Deciduous forests
53	*Antidesma bunius* Spreng.	Stilagiaceae	Small	Rare	Deciduous forests
54	*Antidesma ghaesembilla* Gaertn.	Stilagiaceae	Small	Common	Deciduous forests
55	*Antidesma menasu* Miq.	Stilagiaceae	Medium	Occasional	Deciduous forests
56	*Antidesma zeylanicum* Lam.	Stilagiaceae	Small	Occasional	Deciduous forests
57	*Aphanamixis polystachya* (Wall.) Parker	Meliaceae	Very large	Rare	Semievergreen forest
58	*Areca catechu* L.	Arecaceae	Large	Cultivated	Planted
59	*Aridisia depressa* Clarke	Myrsinaceae	Small	Rare	Moist deciduous forests
60	*Aridisia solanacea* Roxb.	Myrsinaceae	Small	Common	Moist deciduous forests
61	*Artocarpus heterophyllus* Lam.	Moraceae	Large	Cultivated	Hills of high elevations.
62	*Artocarpus lakucha* Buch-Ham.	Moraceae	Large	Rare	Semievergreen forests
63	*Atalantia monophylla* (L.) Correa	Rutaceae	Small	Common	Deciduous forests
64	*Atalantia racemosa* Wight & Arn.	Rutaceae	Small	Rare	Deciduous forests
65	*Averrhoa carambola* L.	Averrhoaceae	Small	Cultivated	Planted in hills as handsome ornamental tree
66	*Avicennia officinalis* L.	Verbenaceae	Small	Rare	Mangrove forests
67	*Azadirachta indica* A. Juss. Fig. 3.2b	Meliaceae	Medium	Common	Plains, Villages outskirts, rarely in forests
68	*Balanites aegyptiaca* (L.) Del.	Balanitaceae	Small	Common	Plains

(continued)

Table 3.2 (continued)

S. No.	Botanical name	Family	Tree height	Status	Remarks
69	*Bambusa arundinacea* (Retz.) Roxb.	Poaceae	Large	Common	Deciduous forests
70	*Bambusa tulda* Roxb.	Poaceae	Large	Common	Deciduous forests
71	*Barringtonia acutangula* (L.) Gaertn. subsp. *acutangula*	Barringtoniaceae	Medium	Common	In plains along streams or swamps
72	*Bauhinia malabarica* Roxb.	Caesalpiniaceae	Medium to large	Occasional	Semievergreen and deciduous
73	*Bauhinia purpurea* L	Caesalpiniaceae	Medium	Common	Planted and also grows in deciduous forests
74	*Bauhinia racemosa* Lam	Caesalpiniaceae	Small	Common	Along the hill slopes and forest paths of deciduous forests
75	*Bauhinia semla* Wunderin (*B. retusa* Ham. ex Roxb.)	Caesalpiniaceae	Small	Rare	Slopes and hills of dry deciduous forests
76	*Bauhinia tomentosa* L.	Caesalpiniaceae	Small	Occasional	Often planted for ornamental
77	*Bauhinia variegata* L. var. *variegata*	Caesalpiniaceae	Medium	Common	Often planted for ornamental
78	*Beilschmiedia roxburghiana* Nees	Lauraceae	Small	Rare	Deciduous forests
79	*Beilschmiedia sikkimensis* King ex Hook.f.	Lauraceae	Small	Occasional	Deciduous forests
80	*Benkera malabarica* (Lam.) Tirveng. (*Randia malabarica* Lam.)	Rubiaceae	Small	Common	Deciduous forests
81	*Berrya cordifolia* (Willd.) Burret	Tiliaceae	Medium	Introducted	Deciduous forests
82	*Bischofia javanica* Blume	Stilagiaceae	Medium	Occasional	Deciduous forests
83	*Bixa orellana* L.	Bixaceae	Small	Cultivated	Exotic, occasionally wild as an escape
84	*Boehmeria platyphylla* Don	Urticaceae	Small	Rare	Semievergreen and moist deciduous

(continued)

Table 3.2 (continued)

S. No.	Botanical name	Family	Tree height	Status	Remarks
85	*Bombax ceiba* L.	Bombacaceae	Large	Cultivated	Along the steams in mixed deciduous forests
86	*Borassus flabellifer* L.	Arecaceae	Large	Common	Wild and cultivated in plains from the coast
87	*Boswellia ovalifoliolata* Balakr. & Henry	Burseraceae	Medium	Rare	Deciduous forests
88	*Boswellia serrata* Roxb. ex Colebr.	Burseraceae	Medium	Common	Deciduous forests
89	*Breynia vitis-idaea* (Burm. f) C. Fischer	Euphorbiaceae	Small	Common	Deciduous forests
90	*Bridelia cinerascens* Gehrm.	Euphorbiaceae	Small	Common	Deciduous forests
91	*Bridelia crenulata* Roxb.	Euphorbiaceae	Medium	Rare	Moist evergreen forests
92	*Bridelia glauca* Blume	Euphorbiaceae	Medium	Occasional	Semievergreen along streams
93	*Bridelia montana* (Roxb.) Willd.	Euphorbiaceae	Small	Common	Open dry deciduous forests
94	*Bridelia retusa* (L.) Spreng.	Euphorbiaceae	Medium	Common	Deciduous forests
95	*Bridelia tomentosa* (Bail) Blume	Euphorbiaceae	Small	Occasional	Deciduous forests
96	*Brownea coccinea* Jacq.	Caesalpiniaceae	Small	Planted	Planted in gardens
97	*Bruguiera cylindrica* (L.) Blume	Rhizophoraceae	Medium	Common	Mangrove forests
98	*Bruguiera gymnorrhiza* (L.) Savigny	Rhizophoraceae	Medium	Common	Mangrove forest
99	*Buchanania axillaris* (Desr.) Ramam.	Anacardiaceae	Small	Occasional	Dry deciduous forests on hills
100	*Buchanania lanzan* Spreng. Fig. 3.2c	Anacardiaceae	Medium	Common	Dry deciduous forests

(continued)

Table 3.2 (continued)

S. No.	Botanical name	Family	Tree height	Status	Remarks
101	*Butea monosperma* (Lam.) Taub (*B. frondosa* Koen. ex Roxb.) Fig. 3.2d	Fabaceae	Small to medium	Common	Deciduous forests
102	*Caesalpinia coriaria* (Jacq.) Willd.	Caesalpiniaceae	Medium	Occasional	Deciduous forests
103	*Caesalpinia pulcherrima* (L.) Sw.	Caesalpiniaceae	Small	Cultivated	In gardens also as an escape
104	*Calliandra inermis* (L.) Druce.	Mimosaceae	Small	Planted	Planted in gardens
105	*Callicarpa arborea* Roxb.	Verbenaceae	Small	Common	Deciduous forests
106	*Callicarpa tomentosa* (L.) Murray	Verbenaceae	Small	Rare	Deciduous forests
107	*Callistemon citrinus* (Curtis) Skeels	Myrtaceae	Small	Cultivated	Planted as ornamental tree
108	*Calophyllum inophyllum* L.	Clusiaceae	Medium	Cultivated	Planted as an avenue tree
109	*Cananga odorata* (Lam.) Hook. F.	Annonaceae	Medium	Planted	Ornamental
110	*Canthium dicoccum* (Lam.) Teijsm & Binn var. *dicoccum*	Rubiaceae	Small	Common	Deciduous forests
111	*Canthium dicoccum* (Wight) Sant. & Merch. var. *umbellatum* (Wight) Sant. & Merch.	Rubiaceae	Small	Common	Deciduous forests
112	*Canthium parviflorum* Lam.	Rubiaceae	Small	Common	Deciduous forests
113	*Capparis grandis* L.f.	Capparaceae	Small	Occasional	Foot hills of deciduous forests
114	*Capparis olacifolia* Hook. f. & Thomson	Capparaceae	Small	Rare	Along grassy slopes
115	*Carallia brachiata* (Lour.) Merr.	Rhizophoraceae	Medium	Rare	Mangrove forest
116	*Careya arborea* Roxb.	Barringtoniaceae	Medium	Common	Deciduous forests

(continued)

Table 3.2 (continued)

S. No.	Botanical name	Family	Tree height	Status	Remarks
117	*Carica papaya* L.	Caricaceae	Small	Cultivated	Planted in backyards and also grown as fruit crop
118	*Caryota urens* L.	Arecaceae	Large	Wild, Cultivated	Shady valleys of deciduous forests
119	*Cascabela thevetia* (L.) Lippold	Apocynaceae	Small	Cultivated, common	Planted in gardens for its showy flowers
120	*Casearia graveolens* Dalz.	Flacourtiaceae	Small	Common	Deciduous forests, open hills, valleys and ravines
121	*Casearia ovata* (Lam.) Willd.	Flacourtiaceae	Small	Occasional	Deciduous forests
122	*Casearia tomentosa*Roxb.	Flacourtiaceae	Small	Occasional	Deciduous forests
123	*Casearia zeylinica* (Gaertn.) Thw.	Flacourtiaceae	Small	Common	Dry evergreen forests
124	*Cassia fistula* L.	Caesalpiniaceae	Small	Common	Plains and deciduous forests
125	*Cassia grandis* L.f.	Caesalpiniaceae	Small	Planted	Planted as avenue tree
126	*Cassia javanica* L.	Caesalpiniaceae	Medium	Planted	Planted as avenue tree
127	*Cassia nodosa* Buch.-Ham ex Roxb.	Caesalpiniaceae	Medium	Planted	Planted as avenue tree
128	*Cassia renigera* Wall. ex Benth.	Caesalpiniaceae	Medium	Planted	Planted as avenue tree
129	*Cassia roxburghii* DC.	Caesalpiniaceae	Small	Occasioanl	Planted on road sides, gardens
130	*Cassia spectabilis* DC.	Caesalpiniaceae	Medium	Planted	Planted as avenue tree
131	*Cassia suffruticosa* Koen. ex Roth	Caesalpiniaceae	Small	Planted	Planted as avenue tree
132	*Cassia surattensis* Burm. f.	Caesalpiniaceae	Small	Planted	As an avenue tree
133	*Cassine glauca* (Rottb.) Kuntze	Celastraceae	Small	Common	Deciduous forests
134	*Casuarina equisetifolia* Forst. & Forst. f.	Casurinaceae	Medium	Planted	Planted in gardens

(continued)

Table 3.2 (continued)

S. No.	Botanical name	Family	Tree height	Status	Remarks
135	*Catunaregam spinosa* (Thunb.) Tirveng. (*Randia dumetorum* (Retz.) Poir. var. *floribunda* (DC.) Gamble)	Rubiaceae	Small	Common	Dry deciduous forests
136	*Catunaregam tomentosa* (Bl. ex DC.) Tirveng.	Rubiaceae	Small	Common	Deciduous forests
137	*Ceiba pentandra* (L.) Gaertn.	Bombacaceae	Medium	Introduced	Planted
138	*Celtis philippensis* Blanco ex Wight	Ulmaceae	Medium	Common	Deciduous forests
139	*Celtis tetrandra* Roxb.	Ulmaceae	Medium	Occasional	Semievergreen and moist deciduous forests
140	*Celtis timorensis* Spanoghe	Ulmaceae	Medium	Occsasional	Semievergreen and deciduous
141	*Cerbera odollana* Gaertn.	Apocynaceae	Small	Occasional	Planted as an ornamental
142	*Ceriops decandra* (Griff.) Ding	Rhizophoraceae	Small	Common	Mangrove forest
143	*Ceriops tagal* (Perr.) Robins	Rhizophoraceae	Small	Common	Mangrove forest
144	*Ceriscoides turgida* (Roxb.) Tirveng. (*Gardenia turgida* Roxb.)	Rubiaceae	Small	Occasional	Deciduous forests
145	*Chionanthus intermedius* (Wight) F. Muell. (*Linociera intermedia* Wight)	Oleaceae	Small	Rare	Deciduous forests
146	*Chionanthus mala-elengi* (Dennst.) P. S. Green (*Linociera malabarica* Wall ex. G. Don)	Oleaceae	Small	Rare	Deciduous forests
14	*Chionanthus ramiflorus* Roxb.	Oleaceae	Small	Rare	Deciduous forests
148	*Chionanthus zeylanicus* L. (*Linociera zeylanica* (L.) Gamble)	Oleaceae	Small	Common	Deciduous forests

(continued)

Table 3.2 (continued)

S. No.	Botanical name	Family	Tree height	Status	Remarks
149	*Chloroxylon swietenia* DC.	Flindersiaceae	Medium	Common	Deciduous forests
150	*Chrysophyllum roxburghii* Don	Sapotaceae	Medium	Rare	Semievergreen forests
151	*Chukrasia tabularis* Adr. Juss.	Meliaceae	Large	Rare	Mixed moist deciduous forests
152	*Cinnamomum camphora* (L.) J. S. Presl	Lauraceae	Small	Cultivated	Planted in agency areas
153	*Cinnamomum caudatum* Nees	Lauraceae	Small	Rare	Moist deciduous forests
154	*Cipadessa baccifera* (Roth) Miq.	Meliaceae	Small	Common	Scrubs and open forests
155	*Citharexylum subserratum* Sw.	Verbenaceae	Small	Planted	Planted in gardens
156	*Citrus aurantium* L.	Rutaceae	Small	Cultivated	Horticulture crop
157	*Citrus aurantifolia* (Christon. & Panz.) Swingle	Rutaceae	Small	Planted	Horticulture crop
158	*Citrus grandis* (L.) Osbeck	Rutaceae	Small	Cultivated	Horticulture crop
159	*Citrus limon* (L.) Burm. f.	Rutaceae	Small	Cultivated	Horticulture crop
160	*Citrus medica* L.	Rutaceae	Small	Cultivated	Horticulture crop
161	*Citrus reticulata* Blanco	Rutaceae	Samll	Cultivated	Horticulture crop
162	*Citrus sinensis* (L.) Osbeck	Rutaceae	Medium	Cultivated	Horticulture crop
163	*Clausena heptaphylla* (Roxb.) Wight & Arn.	Rutaceae	Small	Occasional	Deciduous forests
164	*Cleistanthus collinus* (Roxb.) Hook f.	Euphorbiaceae	Small	Common	Deciduous forests
165	*Cleistanthus patulus* (Roxb.) Muell-Arg.	Euphorbiaceae	Small	Occasional	Deciduous forests chiefly in ravines
166	*Cochlospermum religiosum* (L.) Alston Fig. 3.2e	Cochlospermaceae	Large	Common	Open hilly dry deciduous forests
167	*Cocos nucifera* L.	Arecaceae	Large	Cultivated	Planted near the houses and fields

(continued)

Table 3.2 (continued)

S. No.	Botanical name	Family	Tree height	Status	Remarks
168	*Commiphora berryi* (Arn.) Engler	Burseraceae	Small	Rare	Dry hill slopes of deciduous and scrub forests
169	*Commiphora caudata* (Wight & Arn.) Engler	Burseraceae	Medium	Common	Deciduous forests
170	*Commiphora madagascariensis* Jacq.	Burseraceae	Small	Cultivated	Planted in house premises
171	*Cordia dichotoma* Forst.	Cordiaceae	Small	Common	Fallow lands and agriculture fields
172	*Cordia domestica* Roth	Cordiaceae	Small	Rare	Scrub jungles
173	*Cordia evolutior* (Clarke) Gamble	Cordiaceae	Medium	Rare	Dry forests
174	*Cordia gharaf* (Forssk.) Ehrenb.	Cordiaceae	Small	Rare	Dry forests
175	*Cordia macleodii* Hook. f. & Thoms.	Cordiaceae	Small	Rare	Deciduous forests
176	*Cordia monoica* Roxb.	Cordiaceae	Small	Occasional	Deciduous forests and foothills
177	*Cordia sebestena* L.	Cordiaceae	Small	Cultivated	Planted in gardens
178	*Cordia wallichii* G. Don	Cordiaceae	Medium	Occasional	Deciduous forests
179	*Couroupita guianensis* Aubl.	Barringtoniaceae	Medium	Cultivated	Hills, planted in gardens
180	*Crataeva adansonii* DC.	Capparaceae	Small	Occasional	Deciduous forests
181	*Crataeva magna* (Lour.) DC.	Capparaceae	Small	Occasional	Often planted, frequent along river banks
182	*Croton roxburghii* Balakr.	Euphorbiaceae	Small	Rare	Dry forests
183	*Croton scabiosus* Beddome Fig. 3.2f	Euphorbiaceae	Small	Common	Open dry deciduous forests
184	*Cudrania cochinchinensis* (Lour.) Kodo & Masam	Moraceae	Medium	Rare	Semievergreen forests
185	*Dalbergia lanceolaria* L.f.	Fabaceae	Medium	Common	Deciduous forests
186	*Dalbergia latifolia* Roxb.	Fabaceae	Medium to large	Common	Deciduous forests
187	*Dalbergia paniculata* Roxb	Fabaceae	Medium	Common	Deciduous forests

(continued)

Table 3.2 (continued)

S. No.	Botanical name	Family	Tree height	Status	Remarks
188	*Dalbergia sissoo* Roxb.	Fabaceae	Medium	Cultivated	Plains
189	*Debregeasia longifolia* (Burm. f.) Wedd.	Urticaceae	Small	Rare	Semievergreen and moist deciduous forests
190	*Deccania pubescens* (Roth) Tirveng.var. *candolleana* Wight & Arn.(*Randia candolleana* Wight & Arn.) Fig. 3.3a	Rubiaceae	Small	Common	Deciduous forests
191	*Delonix elata* Gamble	Caesalpiniaceae	Medium	Planted	As an avenue tree
192	*Delonix regia* Raf.	Caesalpiniaceae	Medium	Cultivated	Planted in gardens and as avenue tree
193	*Dendrocalamus strictus* (Roxb.) Nees	Poaceae	Large	Common	Deciduous forests
194	*Dichrostachys cinerea* (L.) Wight & Arn.	Mimosaceae	Small	Common	Scrub jungles and open forests
195	*Dillenia aurea* Smith	Dilleniaceae	Small	Rare	Deciduous forests
196	*Dillenia bracteata* Wight Fig. 3.3b	Dilleniaceae	Medium	Rare	Mixed deciduous forests near streams
197	*Dillenia indica* L.	Dilleniaceae	Large	Occasional	Moist deciduous forests near streams
198	*Dillenia pentagyna* Roxb.	Dilleniaceae	Large	Occasional	Deciduous forests
199	*Dimorphocalyx glabellus* Thw.	Euphorbiaceae	Small	Occasional	Deciduous forests and scrubs
200	*Dimorphocalyx kurnoolensis* V. Raju & Pullaiah	Euphorbiaceae	Small	Rare	Foot hills along the streams
201	*Diospyros assamilis* Bedd.	Ebenaceae	Large	Rare	Semievergreen forests
202	*Diospyros candolleana*	Ebenaceae	Small	Rare	Semiever forests
203	*Diospyros chloroxylon* Roxb.	Ebenaceae	Small	Common	Deciduous forests and scrubs
204	*Diospyros cordifolia* Roxb.	Ebenaceae	Small	Rare	Deciduous forests

(continued)

Table 3.2 (continued)

S. No.	Botanical name	Family	Tree height	Status	Remarks
205	*Diospyros ebenum* J. Koenig ex Retz.	Ebenaceae	Small	Rare	Dry evergreen forests
206	*Diospyros exsculpta* Buch.-Ham.	Ebenaceae	Medium	Occasional	Deciduous forests
207	*Diospyros malabarica* (Desr.) Kostel	Ebenaceae	Medium	Rare	Moist deciduous forests
208	*Diospyros melanoxylon* Roxb. Fig. 3.3c	Ebenaceae	Small	Common	Deciduous forests
209	*Diospyros montana* Roxb.	Ebenaceae	Small	Common	Deciduous forests
210	*Diospyros ovalifolia* Wight	Ebenaceae	Medium	Occasional	Deciduous forests
211	*Diospyros sylvatica* Roxb.	Ebenaceae	Medium	Occasional	Deciduous forests
212	*Dodonaea viscosa* (L.) Jacq.	Sapindaceae	Small	Common	Deciduous forests
213	*Dolichandrone arcuata* (Wight) Clarke	Bignoniaceae	Medium	Common	Deciduous forests
214	*Dolichandrone atrovirens* (Roth) Sprague	Bignoniaceae	Small	Common	Deciduous forests
215	*Dolichandrone falcata* (Wall. ex DC.) Seem.	Bignoniaceae	Small	Common	Deciduous forests in dry localities
216	*Dolichandrone falcata* Seem var. *lawii* (Seem) Haines	Bignoniaceae	Small	Rare	Dry decidiuous forests
217	*Dolichandrone spathacea* (L.f.) K.Schum	Bignoniaceae	Small	Rare	Deciduous forests along the margins of streams
219	*Drypetes assamica* (Hook. f.) Pax. & Hoffm.	Euphorbiaceae	Small	Rare	Deciduous forests along the river
219	*Drypetes roxburghii* (Wall.) Hurusawa	Euphorbiaceae	Medium	Occasional	Deciduous forests
220	*Drypetes sepiaria* (Wight & Arn.) Pax & Hoffm	Euphorbiaceae	Small	Occasional	Along streams of moist deciduous forests
221	*Ehretia acuminata* R. Br.	Cordiaceae	Small	Rare	Deciduous forests

(continued)

Table 3.2 (continued)

S. No.	Botanical name	Family	Tree height	Status	Remarks
222	*Ehretia aspera* Willd.	Cordiaceae	Small	Occasional	Deciduous forests
223	*Ehretia canarensis* (Clarke) Gamble	Cordiaceae	Small	Rare	Dry forests
224	*Ehretia laevis* Roxb.	Cordiaceae	Small	Common	Mixed dry deciduous forests
225	*Ehretia pubescens* Benth.	Cordiaceae	Small	Occasional	Deciduous forests
226	*Elaeagnus caudata* Schlecht ex Momiyama	Elaegnaceae	Small	Rare	Semievergreen forest
227	*Elaeocarpus lanceaefolius* Roxb.	Elaeocarpaceae	Large	Rare	Semievergreen forest
228	*Elaeocarpus lucidus* Roxb.	Elaeocarpaceae	Large	Rare	Moist deciduous forest
229	*Elaeocarpus tectorius* (Laur.) Poir	Elaeocarpaceae	Large	Occasional	Moist deciduous forests
230	*Embelia tsjeriam-cottam* DC.	Myrsinaceae	Small	Occasional	Moist deciduous forests
231	*Embelia villosa* Wall.	Myrsinaceae	Small	Occasional	Moist deciduous forests
232	*Epiprinus mallotiformis* (Muell.-Arg.) Croizat	Euphorbiaceae	Small	Rare	Valleys of moist deciduous forests
233	*Eravatamia divaricata* (L.) Burkill	Apocynaceae	Small	Common	Grown as an ornamental
234	*Erioglossum rubiginosum* Blume	Sapindaceae	Small	Rare	Deciduous forests
235	*Eriolaena hookeriana* Wight & Arn.	Sterculiaceae	Small	Rare	Deciduous forests
236	*Eriolaena lushingtonil* Dunn	Sterculiaceae	Small	Rare	Deciduous forests
237	*Eriolaena quinquelocularis* (Wight & Arn.) Cleghorn	Sterculiaceae	Small	Occasional	Deciduous forests
238	*Erythrina fusca* Lour	Fabaceae	Small	Rare	Deciduous forests
239	*Erythrina stricta* Roxb.	Fabaceae	Medium	Occasional	Deciduous forests

(continued)

Table 3.2 (continued)

S. No.	Botanical name	Family	Tree height	Status	Remarks
240	*Erythrina suberosa* Roxb.	Fabaceae	Medium	Occasional	Deciduous forests
241	*Erythrina subumbrans* (Hassk.) Merr	Fabaceae	Large	Introduced	Deciduous forests
242	*Erythrina variegata* L. var. *orientalis* Merr.	Fabaceae	Medium		Plains and hills
243	*Erythroxylum monogynum* Roxb.	Erythroxylaceae	Small	Common	Dry deciduous and scrub forests
244	*Eucalyptus camaldulensis* Dehn.	Myrtaceae	Large	Introduced	Planted as social forestry plant
245	*Eucalyptus globulus* Labill.	Myrtaceae	Large	Introduced	Planted as social forestry plant
246	*Eugenia roxburghii* DC.	Myrtaceae	Small	Occasional	On sandy gound
247	*Euodia lunuakenda* (Gaertn.) Merr.	Rutaceae	Medium	Rare	Deciduous forests
248	*Euonymus glaber* Roxb.	Celastraceae	Medium	Occasional	Deciduous forests near streams
249	*Euphorbia antiquorum* L.	Euphorbiaceae	Small	Common	Foot hills of dry deciduous and scrub forests
250	*Euphorbia barnhartii* Croizat	Euphorbiaceae	Small	Occasional	Dry deciduous and scrub forests
251	*Euphorbia ligularia* Roxb.	Euphorbiaceae	Small	Occasional	Village outskirts
252	*Euphorbia nivulia* Buch-Ham	Euphorbiaceae	Small	Common	Open forests
253	*Euphorbia tirucalli* L.	Euphorbiaceae	Small	Common	Village outskirts and roadsides
254	*Euphorbia tortillis* Rottl. ex Ainslie	Euphorbiaceae	Small	Occasional	Foot hills of scrub and deciduous forests
255	*Excoecaria agallocha* L.	Euphorbiaceae	Small	Common	Mangrove forests
256	*Ficus amplissima* Smith	Moraceae	Large	Common	Deciduous forests
257	*Ficus arnottiana* (Miq.) Miq	Moraceae	Small	Common	Deciduous forests
258	*Ficus asperrima* Roxb.	Moraceae	Medium	Rare	Deciduous forests
259	*Ficus auriculata* Lour.	Moraceae	Medium	Rare	Semievergreen forests along streams

(continued)

Table 3.2 (continued)

S. No.	Botanical name	Family	Tree height	Status	Remarks
260	*Ficus benghalensis* L.	Moraceae	Large	Common	Plains, road sides, village surroundings
261	*Ficus benjamina* L.	Moraceae	Large	Cultivated	Deciduous forests
262	*Ficus carica* L.	Moraceae	Small	Cultivated	Grown as commercial crop in plains
263	*Ficus dalhousiae* Miq.	Moraceae	Medium	Rare	Ravines of moist deciduous forests
264	*Ficus elastica* L.	Moraceae	Small	Cultivated	Grown as ornamental tree
265	*Ficus exasperata* Vahl	Moraceae	Small	Occasional	Mixed dry deciduous forests
266	*Ficus hispida* L.f.	Moraceae	Small	Common	Deciduous forests and village outskirts
267	*Ficus microcarpa* L.f.	Moraceae	Medium	Common	Deciduous forests, often planted as avenue tree
268	*Ficus mollis* Vahl	Moraceae	Large	Common	Deciduous forests
269	*Ficus nervosa* Heyne ex Roth	Moraceae	Large	Occasional	Moist deciduous forests
270	*Ficus oligodon* Miq.	Moraceae	Medium	Rare	Semievergreen forests
271	*Ficus palmata* Forssk.	Moraceae	Small	Rare	Deciduous forests
272	*Ficus racemosa* L. Fig. 3.3d	Moraceae	Medium	Common	Along river banks of deciduous forests
273	*Ficus religiosa* L.	Moraceae	Medium	Common	Plains, along road sides and near temples
274	*Ficus rumphii* Blume	Moraceae	Medium	Occasional	Moist deciduous forests
275	*Ficus semicordata* Buch-Ham ex Smith	Moraceae	Medium	Common	Moist deciduous forests
276	*Ficus talboti* King	Moraceae	Large	Occasional	Deciduous forests
277	*Ficus tinctoria* Forster f.	Moraceae	Medium	Common	Open dry deciduous forests
278	*Ficus tsjakela* Rheede ex Burm.f	Moraceae	Large	Common	Deciduous forests
279	*Ficus virens* Ait.	Moraceae	Large	Common	Deciduous forests
280	*Firmiana colorata* (Roxb.) R.Br.	Sterculiaceae	Medium	Ocassional	Deciduous forests

(continued)

Table 3.2 (continued)

S. No.	Botanical name	Family	Tree height	Status	Remarks
281	*Flacourtia indica* (Burm. f.) Merr.	Flacourtiaceae	Small	Occasional	Deciduous forests
282	*Flacourtia jangomas* (Lour.) Raeusch.	Flacourtiaceae	Small	Rare	Deciduous forests
283	*Flacourtia ramontchi* L.	Flacourtiaceae	Medium	Occasional	Deciduous forests
284	*Garcinia spicata* (Wight & Arn.) Hook. f.	Clusiaceae	Medium	Occasional	Along the coast
285	*Garcinia xanthochymus* Hook. f. ex T. And.	Clusiaceae	Large	Rare	Moist deciduous forests
286	*Gardenia gummifera* L.f	Rubiaceae	Small	Common	Deciduous forests
287	*Gardenia jasminoides* Ellis	Rubiaceae	Small	Cultivated	Planted as garden plant
288	*Gardenia latifolia* Ait.	Rubiaceae	Medium	Common	Deciduous forests
289	*Gardenia resinifera* Roth (*G. lucida* Roxb.)	Rubiaceae	Small	Occasional	Deciduous forests
290	*Garuga pinnata* Roxb.	Burseraceae	Medium	Common	Deciduous forests
291	*Givotia moluccana* (L.) Sreem.	Euphorbiaceae	Medium	Occasional	Deciduous forests
292	*Gliricidia sepium* (Jacq.) Kunth	Fabaceae	Small	Avenue tree	Plains and hills
293	*Glochidion candolleanum* (Wight & Arn.) Chakrab. & M. Gangop.	Euphorbiaceae	Medium	Rare	Mixed forests
294	*Glochidion ellipticum* Wight	Euphorbiaceae	Small	Occasional	Moist deciduous forests
295	*Glochidion tirupathiense* Rasingam et al.	Euphorbiaceae	Small	Rare	Along stream banks in moist deciduous forests
296	*Glochidion tomentosum* Dalz.	Euphorbiaceae	Small	Occasional	Deciduous forests
297	*Glochidion velutinum* Wight	Euphorbiaceae	Small	Occasional	Along the streams of moist deciduous forests
298	*Glochidion zeylanicum* (Gaertn.) Juss	Euphorbiaceae	Small	Common	Deciduous forests

(continued)

Table 3.2 (continued)

S. No.	Botanical name	Family	Tree height	Status	Remarks
299	*Glycosmis mauritiana* (Lam.) Tanaka	Rutaceae	Small	Common	Deciduous forests
300	*Glycosmis pentaphylla* (Retz.) DC.	Rutaceae	Small	Common	Deciduous forests
301	*Gmelina arborea* Roxb.	Verbenaceae	Medium	Common	Deciduous forests and scrub
302	*Gmelina asiatica* L.	Verbenaceae	Medium	Common	Deciduous forests and scrub
303	*Grevillea robusta* A. Cunn.	Proteaceae	Large	Cultivated	Hills
304	*Grewia abutilifolia* Vent. ex Juss.	Tiliaceae	Small	Common	Deciduous forests
305	*Grewia asiatica* L.	Tiliaceae	Small	Rare	Deciduous forests
306	*Grewia bracteata* Heyne	Tiliaceae	Small	Occasional	Deciduous forests
307	*Grewia damine* Gaertn.	Tiliaceae	Small	Common	Deciduous forests
308	*Grewia elastica* Royle	Tiliaceae	Small		Deciduous forests
309	*Grewia eriocarpa* Juss	Tiliaceae	Small	Rare	Deciduous forests
310	*Grewia flavescens* Juss.	Tiliaceae	Small	Common	Deciduous forests
311	*Grewia glabra* Blume	Tiliaceae	Small	Occasional	Deciduous forests
312	*Grewia heterotricha* Mast.	Tiliaceae	Small	Rare	Deciduous forests
313	*Grewia laevigata* Vahl	Tiliaceae	Small	Occasional	Deciduous forests
314	*Grewia obtusa* Wall. ex Dunn	Tiliaceae	Small	Occasional	Deciduous forests
315	*Grewia orbiculata* Rottler	Tiliaceae	Small	Common	Deciduous forests
316	*Grewia orientalis* L.	Tiliaceae	Small	Common	Deciduous forests
317	*Grewia polygama* Roxb.	Tiliaceae	Small		Deciduous forests
318	*Grewia rothii* DC.	Tiliaceae	Small	Common	Deciduous forests
319	*Grewia serrulata* DC.	Tiliaceae	Small	Common	Deciduous forests
320	*Grewia tenax* (Forssk.) Fiori	Tiliaceae	Small	Occasional	Deciduous forests

(continued)

Table 3.2 (continued)

S. No.	Botanical name	Family	Tree height	Status	Remarks
321	*Grewia tiliifolia* Vahl	Tiliaceae	Small	Common	Deciduous forests
322	*Grewia villosa* Willd.	Tiliaceae	Small	Occasional	Deciduous forests
323	*Guazuma ulmifolia* Lam.	Sterculiaceae	Medium	Introduced	Planted
324	*Guettarda speciosa* L.	Rubiaceae	Medium	Cultivated	Planted
325	*Gyrocarpus americanus* Jacq. (*G. jacquinii* Roxb.)	Hernandiaceae	Medium	Common	Deciduous forests
326	*Haldina cordifolia* (Roxb.) Ridsd. Fig. 3.3e	Rubiaceae	Small	Common	Deciduous forests
327	*Hardwickia binata* Roxb.	Caesalpiniaceae	Large	occasional	Scrub and deciduous forests
328	*Helicteres isora* L.	Sterculiaceae	Small	Common	Deciduous forests
329	*Heritiera littoralis* Dryand ex Ait.	Sterculiaceae	Medium	Ocassional	Mangrove forest
330	*Heterophragma quadriloculare* (Roxb.) Schum. (*H. roxburghii* DC.)	Bignoniaceae	Medium	Occasional	Deciduousforests
331	*Hibiscus platanifolius* (Willd.) Sweet	Malvaceae	Small	Occasional	Deciduous forests
332	*Hibiscus tiliaceus* L.	Malvaceae	Small	Occasional	Mangrove forest
333	*Hildegardia populifolia* (Roxb.) Schott. & Endl. Fig. 3.3f	Sterculiaceae	Small	Rare	Deciduous forests
334	*Holarrhena pubescens* (Buch.-Ham.) Wall. ex G. Don	Apocynaceae	Small	Common	Deciduous forests
335	*Holoptelia integrifolia* (Roxb.) Planchon Fig. 3.4a	Ulmaceae	Large	Common	Deciduous
336	*Homalium ceylanicum* (Gard.) Benth.	Flacourtiaceae	Medium	Rare	Deciduous forests
337	*Homalium nepalense* Benth.	Flacourtiaceae	Small	Common	Moist deciduous forests

(continued)

Table 3.2 (continued)

S. No.	Botanical name	Family	Tree height	Status	Remarks
338	*Lasiococca comberi* Haines	Euphorbiaceae	Small	Occasional	Along rocky ravines
339	*Hymenodictyon obovatum* Wall.	Rubiaceae	Small	Rare	Deciduous forests
340	*Hymenodictyon orixense* (Roxb.) Mabb.	Rubiaceae	Large	Common	Deciduous forests
341	*Isonandra villosa* Wight Fig. 3.4b	Sapotaceae	Medium	Rare	Deciduous forests
342	*Ixora brachiata* Roxb. ex DC.	Rubiaceae	Small	Occasional	Deciduous forests
343	*Ixora finlaysoniana* Wall. ex G. Don	Rubiaceae	Small	Cultivated	Planted in gardens as ornamental plant
344	*Ixora pavetta* Andrews	Rubiaceae	Small	Common	Deciduous forests
345	*Ixora undulata* Roxb.	Rubiaceae	Small		Deciduous forests
346	*Jacaranda mimosifolia* DC.	Bignoniaceae	Medium	Occasional	Hills, planted in gardens
347	*Jatropha curcas* L.	Euphorbiaceae	Small	Common	Semi wild; also grown as hedge plant.
348	*Kavalama urens* (Roxb.) Raf. (*Sterculia urens* Roxb.) Fig. 3.4c	Sterculiaceae	Medium	Common	Deciduous forests
349	*Kigelia africana* (Lam.) Benth.	Bignoniaceae	Medium	Avenue	Grown as avenue plant
350	*Kleinhovia hospita* L.	Sterculiaceae	Small	Planted	Planted as an ornamental tree
351	*Knema attenuata* (Hook. & Thw.) Warb.	Myristicaceae	Medium	Rare	Moist deciduous forests
352	*Kydia calycina* Roxb.	Malvaceae	Small	Occasional	Deciduous forests
353	*Lagerstroemia lanceolata* Wall.	Lythraceae	Medium	Common	Deciduous and dry evergreen forests
354	*Lagerstroemia parviflora* Roxb.	Lythraceae	Medium	Common	Deciduous forests
355	*Lagerstroemia speciosa* Pers.	Lythraceae	Medium	Cultivated	Planted as an ornamental tree
356	*Lannea coromandelica* (Houtt.) Merr.	Anacardiaceae	Medium	Common	Dry deciduous forests

(continued)

Table 3.2 (continued)

S. No.	Botanical name	Family	Tree height	Status	Remarks
357	*Laportea crenulata* Gaud.	Urticaceae	Small	Occasional	Semievergreen and deciduous forests
358	*Laurocerasus jenkinsii* (Hook. f. & Thomson) Browicz	Rosaceae	Medium	Rare	Moist deciduous forests
359	*Lawsonia inermis* L.	Lythraceae	Small	Cultivated	Plains, grows as hedge plant
360	*Leea indica* (Burm. f) Merr.	Leeaceae	Small	Occasional	Deciduous forests
361	*Lepisanthes rubiginosa* (Roxb.) Leenh	Sapindaceae	Small	Common	Deciduous forests
362	*Lepisanthes tetraphylla* (Vahl) Radlk.	Sapindaceae	Small	Common	Deciduous forests
363	*Leucaena leucocephala* (Lam.) De Wit (*L. latisiliqua* (L.) Gillis, *L. glauca* Willd.)	Mimosaceae	Small	Cultivated	Plains, recommended for plantation in variety of soils
364	*Ligustrum gamblei* Ramam.	Oleaceae	Small	Rare	Deciduous forests
365	*Ligustrum lucidum* Ait.	Oleaceae	Small	Cultivated	
366	*Limonia acidissima* L.(*Feronia elephantum* Correa)	Rutaceae	Medium	Common	Deciduous forests
367	*Litsea deccanensis* Gamble	Lauraceae	Medium	Rare	Deciduous forests
368	*Litsea glutinosa* (Lour) Robinson	Lauraceae	Small	Occasional	Moist deciduous forests
369	*Litsea laeta* (Nees) Hook. f.	Lauraceae	Small	Rare	Moist deciduous forests
370	*Litsea monopetala* (Roxb.) Pers.	Lauraceae	Medium	Occasional	Deciduous forests
371	*Maba buxifolia* (Rotb.) A. L. Juss. (*Diospyros ferrea* (Willd.) Bakh)	Ebenaceae	Small	Common	Deciduous forests
372	*Maba neilgherrensis* Wight	Ebenaceae	Medium	Rare	Deciduous forests

(continued)

Table 3.2 (continued)

S. No.	Botanical name	Family	Tree height	Status	Remarks
373	*Macaranga indica* Wt.	Euphorbiaceae	Medium	Rare	Ravines of semievergreen forests
374	*Macaranga peltata* (Roxb.) Muell.-Arg.	Euphorbiaceae	Medium	Occasional	Along moist valleys and hills of moist deciduous forests
375	*Madhuca indica* J. F. Gmel.	Sapotaceae	Medium	Common	Mixed deciduous forests
376	*Madhuca longifolia* (J. Koenig) Macbr.	Sapotaceae	Medium	Ocassional	Open deciduous forests
377	*Maerua apetala* (Roth) Jacobs Fig. 3.4e	Capparaceae	Small	Occasional	Foot hills and dry deciduous forests
378	*Magnolia champaca* (L.) Baill. ex Pierre (*Michelia champaca* L.)	Magnoliaceae	Medium	Occasional	On hills and in temples for the sake of fragran flowers
379	*Mallotus philippensis* (Lam) Muell.-Arg.	Euphorbiaceae	Medium	Common	Deciduous forests
380	*Mallotus philippensis* (Lam.) Muell.-Arg. var. *tomentosus* Gamble	Euphorbiaceae	Medium	Occasional	Deciduous forests
381	*Mallotus resinosus* (Blanco) Merr.	Euphorbiaceae	Small	Rare	Deciduous forests
382	*Mallotus rhamnifolius* Muell-Arg.	Euphorbiaceae	Small	Occasional	Moist deciduous forests near the streams
383	*Mammea surgia* (Buch.-Ham. ex Roxb.) Kosterm	Clusiaceae	Medium	Cultivated	Planted as an avenue tree and in house hold garden
384	*Mangifera indica* L.	Anacardiaceae	Medium	Cultivated, also wild	In forest hills; cultivated for its fruit
385	*Manihot glaziovii* Muell.	Euphorbiaceae	Small	Cultivated	Introduced, planted in gardens
386	*Manilkara hexandra* (Roxb.) Dubard	Sapotaceae	Small	Common	Deciduous forests
387	*Manilkara roxburghiana* (Wight) Dubard	Sapotaceae	Large	Rare	Dry evergreen forests

(continued)

Table 3.2 (continued)

S. No.	Botanical name	Family	Tree height	Status	Remarks
388	*Manilkara zapota* (L.) P. Royen	Sapotaceae	Medium	Cultivated	In plains and lower slopes for its fruits
389	*Margaritaria indica* (Dalz.) Airy Shaw	Euphorbiaceae	Medium	Occasional	Semievergreen forests along the streams
390	*Maytenus bailadillana* (Narayan. & Mooney) Raju & Babu	Celastraceae	Small	Rare	Deciduous forests
391	*Maytenus emarginata* (Willd.) Ding Hou	Celastraceae	Small	Common	Scrub forests
392	*Maytenus heyneana* (Roth) Raju & Babu	Celastraceae	Small	Rare	Deciduous forests
393	*Maytenus hookeri* Loes	Celastraceae	Small	Common	Deciduous forests
394	*Maytenus rufa* (Wall.) Hara.	Celastraceae	Small	Common	Deciduous forests
395	*Melia azedarach* L.	Meliaceae	Medium	Planted	Planted as an avenue tree
396	*Melia dubia* Cav.	Meliaceae	Large	Rare	Moist eciduous forests
397	*Meliosma pinnata* (Roxb.) Walp.	Sabiaceae	Medium	Rare	Moist deciduous forests
398	*Meliosma simplicifolia* (Roxb.) Walp.	Sabiaceae	Medium	Rare	Moist deciduous forests
399	*Memecylon angustifolium* Wight	Melastomataceae	Small	Rare	Deciduous forests, along the banks of rivers
400	*Memecylon edule* Roxb.	Melastomataceae	Small	Common	Dry evergreen forests along rocky ravines
401	*Memecylon lushingtonii* Gamble	Melastomataceae	Small	Common	Deciduous forests
402	*Memecylon molestum* (Clarke) Cogn.	Melastomataceae	Small	Rare	Deciduous forests
403	*Memecylon umbellatum* Burm.f.	Melastomataceae	Small	Occasional	Deciduous forests

(continued)

Table 3.2 (continued)

S. No.	Botanical name	Family	Tree height	Status	Remarks
404	*Mesua ferrea* L.	Clusiaceae	Large	Rare	Semievergreen forests
405	*Meyna spinosa* Roxb. ex Link var. *pubescens* Robyns	Rubiaceae	Small	Rare	Deciduous forests
406	*Micromelum pubescense* Blume	Rutaceae	Small	Rare	Deciduous forests
407	*Miliusa montana* Gardn. ex Hook. f. & Thoms.	Annonaceae	Small	Rare	Deciduous forests
408	*Miliusa tomentosa* (Roxb.) Sinclair	Annonaceae	Large	Common	Deciduous forests
409	*Miliusa velutina* Hook. f. & Thoms.	Annonaceae	Medium	Occasional	Deciduous forests
410	*Millingtonia hortensis* L.f.	Bignoniaceae	Medium	Cultivated	Plains, planted
411	*Mimusops elengi* L.	Sapotaceae	Small	Occasional	Plains and deciduous forests
412	*Mitragyna parvifolia* (Roxb.) Korth. Fig. 3.4d	Rubiaceae	Medium	Occasional	Deciduous forests
413	*Mitrephora heyneana* (Hook. f. & Thomson) Thwaites.	Annonaceae	Medium	Rare	Deciduous forests on hill tops
414	*Morinda angustifolia* Roxb.	Rubiaceae	Small	Rare	Deciduous forests
415	*Morinda citrifolia* L.	Rubiaceae	Small	Occasional	Deciduous forests
416	*Morinda pubescens* J. E. Smith var. *stenophylla* (Sperng.) Kumari	Rubiaceae	Small	Occasional	Deciduous forests
417	*Morinda pubescens* J. E. Smith. var. *pubescens* (*M. tinctoria* Roxb.)	Rubiaceae	Small	Common	Deciduous forests
418	*Moringa concanensis* Nimmo ex Gibs	Moringaceae	Small	Occasional	Deciduous forests
419	*Moringa oleiferea* Lam. (*M. pterygosperma* Gaertn.)	Moringaceae	Small	Cultivated	Planted in Plains and hills
420	*Morus alba* L.	Moraceae	Samll	Cultivated	Plains

(continued)

Table 3.2 (continued)

S. No.	Botanical name	Family	Tree height	Status	Remarks
421	*Muntingia calabura* L.	Elaeocarpaceae	Large	Cultivated	Grown as ornamental tree
422	*Murraya koenigii* (L.) Spreng.	Rutaceae	Small	Cultivated	Planted for its aromatic leaves
423	*Murraya paniculata* (L.) Jack (*M. exotica* L.)	Rutaceae	Small	Cultivated	Deciduous forests
424	*Naringi alata* (Wall. ex Wight & Arn.) Ellis (*Limonia alata* Wall. ex Wight & Arn.)	Rutaceae	Small	Occasional	Deciduous forests
425	*Naringi crenulata* (Roxb.) Nicolson	Rutaceae	Small	Common	Deciduous forests
426	*Neolamarckia cadamba* (Roxb.) Bosser (*A. cadamba* (Roxb.) Miq.)	Rubiaceae	Large	Occasional	Deciduous forests
427	*Neolitsea foliosa* (Nees) Gamble	Lauraceae	Small	Occasional	Deciduous forests
428	*Neolitsea zeylanica* (Nees) Merr.	Lauraceae	Medium	Occasional	Deciduous forests
429	*Neonauclea purpurea* (Roxb.) Merr.	Rubiaceae	Medium	Rare	Deciduous forests
430	*Nothopegia beddomei* Gamble	Anacardiaceae	Small	Rare	Semievergreen forest
431	*Nothopegia heyneana*(Hook. f)	Anacardiaceae	Small	Rare	Semievergreen foressts
432	*Nyctanthes arbor-tristis* L.	Nyctanthaceae	Small	Cultivated	Plains
433	*Ochna gamblei* DC.	Ochnaceae	Small	Occasional	Deciduous forests
434	*Ochna lanceolata* Spreng.	Ochnaceae	Small	Rare	Deciduous forests
435	*Ochna obtusata* DC. Fig. 3.4f	Ochnaceae	Small	Occasional	Deciduous forests
436	*Olea dioica* Roxb.	Oleaceae	Medium	Rare	Semievergreen forests
437	*Olea paniculata* R. Br. (*Olea glandulifera* Wall. ex. G. Don)	Oleaceae	Medium	Rare	Deciduous forests

(continued)

Table 3.2 (continued)

S. No.	Botanical name	Family	Tree height	Status	Remarks
438	*Olea polygama* Wight	Oleaceae	Medium	Rare	Deciduous forests
439	*Oroxylum indicum* (L.) Vent.	Bignoniaceae	Small	Occasional	Deciduous forests, along streams and ravines and on cool sides of the hills
440	*Ougeinia oojeinensis* (Roxb.) Hochr. (*Ougeinia dalbergioides* Benth.)	Fabaceae	Medium	Occasional	Deciduous forests
441	*Pamburus missionis* (Wight) Swingle	Rutaceae	Medium	Rare	Deciduous forests
442	*Pandanus fascicularis* Lam. (*P. odoratissimus* L.f.).	Pandanaceae	Small	Occasional	Stream side and ravines in deciduous forests
443	*Parkia biglandulosa* Wight & Arn.,	Mimosaceae	Large	Cultivated	Planted in plains and hills
444	*Pavetta breviflora* DC.var. *ciliolata* Gamble ex Bremek.	Rubiaceae	Small	Occasional	Deciduous forests
445	*Pavetta hispidula* Wight & Arn.	Rubiaceae	Small	Rare	Deciduous forests
446	*Pavetta indica* L.	Rubiaceae	Small	Common	Deciduous forests
447	*Pavetta tomentosa* Roxb. ex Smith	Rubiaceae	Small	Common	Deciduous forests
448	*Peltophorum pterocarpum* (DC.) Backer	Caesalpiniaceae	Small	Planted	Grown as avenue tree
449	*Persea macrantha*(Nees.) Kosterm	Lauraceae	Large	Rare	Semievergreen and deciduous forests
450	*Phoebe wightii* Meissn	Lauraceae	Medium	Occasional	Semievergreen forests
451	*Phoenix lourierii* Kunth	Arecaceae	Small	occasional	Open forest, in sandy soils
452	*Phoenix robusta* (Becc) Becc & Hook. f.	Arecaceae	Large	rare	Open forests
453	*Phoenix sylvestris* Roxb.	Arecaceae	Large	Common	Along streams and waste lands

(continued)

Table 3.2 (continued)

S. No.	Botanical name	Family	Tree height	Status	Remarks
454	*Phyllanthus acidus* L.	Euphorbiaceae	Medium	Cultivated	Planted in gardens and house compounds
455	*Phyllanthus emblica* L.	Euphorbiaceae	Medium	Common	Deciduous forests
456	*Phyllanthus indofischeri* Bennet Fig. 3.5a	Euphorbiaceae	Small	Occasional	Deciduous forests
457	*Phyllanthus polyphyllus* Willd.	Euphorbiaceae	Small	Occasional	Deciduous forests
458	*Picrasma javanica* Blume	Simaroubaceae	Medium	Occasional	Deciduous forests
459	*Pisonia aculeata* L.	Nyctaginaceae	Small	Occasional	Dry deciduous forests
460	*Pithecellobium dulce* (Roxb.) Benth.	Mimosaceae	Medium	Occasional	Fallow lands and road sides. Grows in variety of soils
461	*Pittosporum nepalense* (DC.) Rehder & Wilson	Pittosporaceae	Small	Rare	Deciduous forests
462	*Pleurostylia opposita* (Wall.) Alston	Celastraceae	Small	Common	Deciduous forests
463	*Plumeria alba* L.	Apocynaceae	Small	Cultivated	As an ornamental in gardens and temples
464	*Plumeria obtusa* L.	Apocynaceae	Small	Cultivated	As an ornamental in gardens and temples
465	*Plumeria rubra* L.	Apocynaceae	Small	Cultivated	As an ornamental in gardens and temples
466	*Polyalthia cerasoides* (Roxb.) Bedd.	Annonaceae	Small	Common	Deciduous forests, in hill slopes
467	*Polyalthia korinti* (Dunal) Thwaites	Annonaceae	Small	Rare	Dry evergreen along ravines
468	*Polyalthia longifolia* (Sonner) Thwaites	Annonaceae	Large	Cultivated	Planted in gardens, avenues
469	*Polyalthia suberosa* (Roxb.)Thw.	Annonaceae	Small	Occasional	Deciduous forests near streams
470	*Pongamia pinnata* (L.) Pierre (*P. glabra* Vent.,	Fabaceae	Medium	Common	Plains

(continued)

Table 3.2 (continued)

S. No.	Botanical name	Family	Tree height	Status	Remarks
471	*Premna hamiltonii* J. L. Ellis Fig. 3.5b	Verbenaceae	Small	Rare	Deciduous forests
472	*Premna latifolia* Roxb. var. *latifolia*	Verbenaceae	Small	Common	Deciduous forests
473	*Premna serratifolia* L.	Verbenaceae	Small	Rare	Deciduous forests
474	*Premna tomentosa* Willd.	Verbenaceae	Small	Common	Deciduous forests
475	*Prosopis cineraria* (L.) Druce	Mimosaceae	Small	Common	Dry stony lands and degreded forests.
476	*Prosopis glandulosa* Torrey	Mimosaceae	Small	Introduced	Naturalized in open forests
477	*Prosopis juliflora* (Sw.) DC.	Mimosaceae	Small	Common	Waste lands and road sides
478	*Protium serratum* (Colebr.) Engler	Burseraceae	Large	Common	Deciduous forests
479	*Prunus ceylanica* (Wight) Miq.	Rosaceae	Medium	Common	Moist deciduous forests
480	*Prunus jenkinsii* Hook. f	Rosaceae	Medium	Rare	Moist deciduous forests
481	*Psidim guajava* L.	Myrtaceae	Small	Cultivated	Grown as horticulture plant
482	*Psychotria monticola* Kurz (*P. flava* Hook. f.)	Rubiaceae	Small	Rare	Deciduous forests
483	*Pterocarpus marsupium* Roxb.	Fabaceae	Medium	Common	Deciduous forests
484	*Pterocarpus santalinus* L.f. Fig. 3.5c	Fabaceae	Large	Endemic	Deciduous forests
485	*Pterospermum acerifolium* (L.) Willd.	Sterculiaceae	Medium	Rare	Deciduous forests
486	*Pterospermum canescens* Roxb.	Sterculiaceae	Small	Ocassional	Deciduous forests
487	*Pterospermum xylocarpum* (Gaertn.) Sant.& Wagh	Sterculiaceae	Medium	Common	Deciduous forests
488	*Punica granatum* L.	Punicaceae	Small	Cultivated	Plains and hills
489	*Radermachera xylocarpa* (Roxb.) Schum.	Bignoniaceae	Large	Occasional	Hill slopes of deciduous forests

(continued)

Table 3.2 (continued)

S. No.	Botanical name	Family	Tree height	Status	Remarks
490	*Rhamnus nepalensis* Lawson	Rhamnaceae	Small	Rare	Deciduous forests
491	*Rhamuus virgatus* Roxb.	Rhamnaceae	Small	Rare	Deciduous forests
492	*Rhizophora apiculata* Blume	Rhizophoraceae	Small	Common	Mangrove forests
493	*Rhizophora mucronata* Poir	Rhizophoraceae	Small	Common	Mangrove forest
494	*Rhus paniculata* Wall. ex Hook. f	Anacardiaceae	Small	Rare	Dry hilly regions
495	*Rubus gardnerianus* O. Kuntze	Rosaceae	Medium	Rare	Moist deciduous forests
496	*Sageretia parviflora* (Klein) G. Don	Rhamnaceae	Small	Common	Deciduous forests
497	*Salix tetrasperma* Roxb.	Salicaceae	Medium	Rare	Deciduous forests
498	*Salvadora persica* L.	Salvadoraceae	Small	Common	Deciduous forests
499	*Santalum album* L.	Santalaceae	Small	Occasional	Deciduous forests
500	*Sapindus emarginatus* Vahl Fig. 3.5d	Sapindaceae	Small	Common	plains
501	*Sapium eugeniifolium* Buch.-Ham.	Euphorbiaceae	Small	Rare	Semievergreen forests
502	*Sapium insigne* (Royle) Trimen	Euphorbiaceae	Small	Rare	Deciduous forests
503	*Saraca asoca* (Roxb.) De Wilde	Caesalpiniaceae	Small	Endangered	Deciduous forests
504	*Schleichera* oleosa (Lour) Oken	Sapindaceae	Medium	Common	Deciduous forests
505	*Schrebera swietenioides* Roxb.	Oleaceae	Medium	Occasional	Deciduous forests
506	*Scolopia crenata* (Wight & Arn.) Clos	Flacourtiaceae	Medium	Occasional	Deciduous forests
507	*Scyphiphora hydrophyllacea* Gaertn.	Rubiaceae	Small	Rare	Mangrove forest
508	*Searsia mysorensis* (G. Don) Moffett *(=Rhus mysorensis* G.Don)	Anacardiaceae	Small	Common	Dry stony and scrub forests

(continued)

Table 3.2 (continued)

S. No.	Botanical name	Family	Tree height	Status	Remarks
509	*Semecarpus anacardium* L.f.	Anacardiaceae	Medium	occasional	Dry deciduous forests
510	*Shorea robusta* Roxb. ex Gaertn.f. Fig. 3.5e	Dipterocarpaceae	Large	Rare	Deciduous forests
511	*Shorea roxburghii* G.Don	Dipterocarpaceae	Very large	Occasional	Deciduous forests
512	*Shorea tumbuggaia* Roxb. Fig. 3.5f	Dipterocarpaceae	Large	Occasional	Deciduous forests
513	*Siphondon celastrinus* Griff.	Celastraceae	Small	Occasional	Moist deciduous forests
514	*Sloanea sterculiacea* (Benth.) Rehder Fig. 3.6a	Elaeocarpaceae	Large	Rare	Moist deciduous forests
515	*Solanum erianthum* D. Don	Solanaceae	Small	Occasional	Deciduous forests
516	*Sonneratia apetala* Buch.-Ham	Sonneratiaceae	Medium	Common	Mangrove forests
517	*Sonneratia caseolaris* (L.) Engler	Sonneratiaceae	Medium	Rare	Mangrove forest
518	*Soymida febrifuga* (Roxb.) A.Juss. Fig. 3.6b	Meliaceae	Medium	Common	Deciduous forests
519	*Spathodea campanulata* P.Beauv.	Bignoniaceae	Medium	Cultivated	Plains and hills
520	*Spermadictyon suaveolens* Roxb.	Rubiaceae	Small	Rare	Deciduous forests
521	*Spondias pinnata* (L.f.) Kurz	Anacardiaceae	Medium	Rare	Dry deciduous forests
522	*Sterculia foetida* L.	Sterculiaceae	Large	Planted	As avenue plant
523	*Sterculia villosa* Roxb.	Sterculiaceae	Large	Occasional	Deciduous forests
524	*Stereospermum colais* (Dillw.) Mabb.	Bignoniaceae	Medium	Occasional	Slopes in deciduous forests
525	*Stereospermum personatum* (Hassk.) Chatterjee	Bignoniaceae	Small to medium	Common	Dry deciduous forests
526	*Stereospermum suaveolens* (Roxb.) DC.	Bignoniaceae	Medium	Occasional	Dry deciduous forests and semievergreen forests

(continued)

Table 3.2 (continued)

S. No.	Botanical name	Family	Tree height	Status	Remarks
527	*Streblus asper* Lour.	Moraceae	Small	Common	Deciduous forests
528	*Streblus taxoides* (Roth) Kurz	Moraceae	Small	Occasional	Semievergreen forests
529	*Strychnos nux-vomica* L. Fig. 3.6c	Loganiaceae	Medium	Common	Deciduous forests
530	*Strychnos potatorum* L.f. Fig. 3.6d	Loganiaceae	Medium	Common	Deciduous forests
531	*Suregada angustifolia* (Bail. ex Muell.-Arg.) Airy Shaw (*Gelonium lanceolatum* auct-non Willd.)	Euphorbiaceae	Small	Occasional	Deciduous forests
532	*Suregada multiflora* (Juss.) Bail. (*Gelonium multiflorum* Juss.) Fig. 3.6e	Euphorbiaceae	Small	Occasional	Semievergreen forests
533	*Swietenia macrophylla* King	Meliaceae	Large	Introduced	Planted in hill stations, and also as an avenue tree
534	*Swietenia mahogani* (L.) Jacq	Meliaceae	Large	Introduced	Planted in hill stations, and also as an avenue tree
535	*Symplocos cochinchinensis* (Lour.) S. Moore	Symplocaceae	Medium	Rare	Deciduous forests
536	*Symplocos theaefolia* D. Don	Symplocaceae	Small	Rare	Deciduous forests
537	*Syzygium alternifolium* (Wight) Walp.	Myrtaceae	Medium	Common	Hills of deciduous forests
538	*Syzygium calophyllifolium* Walp.	Myrtaceae	Large	Rare	Dry deciduous forests
539	*Syzygium cuminii* (L.) Skeels	Myrtaceae	Large	Common	Plains and hills
540	*Syzygium heyneanum* (Duthie) Wall. ex Gamble	Myrtaceae	Small	Rare	Dry deciduous forests along streams

(continued)

Table 3.2 (continued)

S. No.	Botanical name	Family	Tree height	Status	Remarks
541	*Syzygium malabaricum* (Bedd.) Gamble	Myrtaceae	Medium	Rare	High hills
542	*Syzygium nervosum* A. Cunn	Myrtaceae	Small	Rare	Moist deciduous forests
543	*Syzygium operculatum* (Roxb.) Nicolezu	Myrtaceae	Large	Occasional	High hills
544	*Syzygium samarangense* (Blume) Merrill & Perry	Myrtaceae	Large	Introduced	Planted in gardens
545	*Tabebuia argentea* (Bur & K. Schum) Britt.	Bignoniaceae	Small	Cultivated	Planted in gardens
546	*Tabebuia rosea* (Bertil.) DC.	Bignoniaceae	Small	Cultivated	Planted in gardens
547	*Tamarindus indica* L.	Caesalpiniaceae	Small	Wild and Cultivated	In fallow lands, road sides and forests
548	*Tamarix aphylla* (L.) Karsten.	Tamaricaceae	Small	Rare	Mangrove forests
549	*Tamilnadia uliginosa* (Retz.) Tirveng. & Sastre (*Randia uliginosa* (Retz.) DC.)	Rubiaceae	Small	Occasional	Deciduous forests
550	*Tarenna asiatica* (L.) O. Kuntze ex K. Schum. (*Chomelia asiatica* (L.) O.Kuntze)	Rubiaceae	Small	Common	Deciduous forests
551	*Tecomella undulata* (Sm.) Seem.	Bignoniaceae	Small	Cultivated	Planted in gardens
552	*Tecoma stans* (L.) Kunth	Bignoniaceae	Small	Cultivated	Planted in gardens
553	*Tectona grandis* L.f.	Verbenaceae	Large	Common	Deciduous forests
554	*Terminalia alata* Heyne ex Roth.	Combretaceae	Large	Occasional	Deciduous forests
555	*Terminalia arjuna* (Roxb ex DC.) Wight & Arn.	Combretaceae	Large	Common	Deciduous forests, frequent along streams, river banks

(continued)

Table 3.2 (continued)

S. No.	Botanical name	Family	Tree height	Status	Remarks
556	*Terminalia bellirica* (Gaertn.) Roxb.	Combretaceae	Large	Common	Deciduous forests
557	*Terminalia catappa* L.	Combretaceae	Medium	Cultivated	Planted in households, gardens
558	*Terminalia chebula* Retz.	Combretaceae	Large	Common	Deciduous forests
559	*Terminalia coriacea* (Roxb.) Wight & Arn.	Combretaceae	Medium	Common	Dry hills of dry deciduous forests
560	*Terminalia crenulata* Roth	Combretaceae	Large	Occasional	Deciduous forests
561	*Terminalia gella* Dalz.	Combretaceae	Medium	Rare	Moist deciduous forests
562	*Terminalia pallida* Brandis Fig. 3.6f	Combretaceae	Small	Occasional	Deciduous forests
563	*Terminalia paniculata* Roth	Combretaceae	Medium	Common	Deciduous forests
564	*Terminalia tomentosa* (Roxb. ex DC.) Wight & Arn.	Combretaceae	Medium	Common	Deciduous forests
565	*Thespesia populnea* (L.) Sol.ex Corr.	Malvaceae	Small	Cultivated	Plains, planted in garden and also as avenue plant
566	*Toddalia asiatica* (L). Lam. var. *floribunda* Gamble	Rutaceae	Small	Common	Deciduous forests
567	*Toona ciliata* Roem.	Meliaceae	Large	Occasional	Moist deciduous forests
568	*Trema orientalis* (L.) Blume	Ulmaceae	Medium	Common	Dry evergreen forests
569	*Trewia nudiflora* L.	Euphorbiaceae	Small	Rare	Deciduous forests along streams
570	*Trichilia connaroides* (Wight & Arn.) Bentvelizen	Meliaceae	Medium	Common	Moist deciduous forests
571	*Vitex altissima* L.f.	Verbenaceae	Medium	Common	Semievergreen and deciduous forests
572	*Vitex leucoxylon* L.f.	Verbenaceae	Medium	Occasional	Deciduous forests
573	*Vitex negundo* L. var. *negundo*	Verbenaceae	Small	Common	Plains and deciduous forests

(continued)

Table 3.2 (continued)

S. No.	Botanical name	Family	Tree height	Status	Remarks
574	*Vitex negundo* var. *pupurpurascens* Sivar. & Moldenke	Verbenaceae	Small	Rare	Deciduous forests
575	*Vitex peduncularis* Wall. ex Schauer	Verbenaceae	Medium	Occasional	Deciduous forests
576	*Vitex pinnata* L.	Verbenaceae	Small	Occasional	Deciduous forests
577	*Vitex quinata* (Lour.) F. N. Willliams	Verbenaceae	Small	Rare	Deciduous forests
578	*Vitex trifolia* L.	Verbenaceae	Small	Occasional	Deciduous forests
579	*Walsura trifolia* (A. Juss.) Harms	Meliaceae	Small	Common	Deciduous forests
580	*Wendlandia angustifolia* (Wight & Hook. f.)	Rubiaceae	Small	Rare	Deciduous forests
581	*Wendlandia bicuspidata* Wight & Arn.	Rubiaceae	Small	Occasional	Deciduous forests
582	*Wendlandia gamblei* Cowan	Rubiaceae	Small	Occasional	Deciduous forests
583	*Wendlandia glabrata* DC.	Rubiaceae	Small	Occasional	Deciduous forests
584	*Wendlandia heynei* (Roem. & Schult.) Santapau & Merchant	Rubiaceae	Small	Rare	Deciduous forests
585	*Wendlandia thyrsoidea* (Roem & Schult.) Steud. (*W. notoniana* Wall. ex Wight & Arn.)	Rubiaceae	Small	Rare	Deciduous forests
586	*Wendlandia tinctoria* DC. ssp. *tinctoria*	Rubiaceae	Small	Common	Deciduous forests
587	*Wrightia arborea* (Dennst.) Mabb. (*W. tomentosa* Roem. & Schult)	Apocynaceae	Small	Occasional	Exposed hilly slopes of deciduous forests
588	*Wrightia tinctoria* R.Br.	Apocynaceae	medium	Common	Foot hills and open dry deciduous forests and scrubs
589	*Xantolis tomentosa* (Roxb.) Rap (*Sideroxylon tomentosum* Roxb.)	Sapotaceae	Medium	Ocassional	Moist evergreen forests

(continued)

Table 3.2 (continued)

S. No.	Botanical name	Family	Tree height	Status	Remarks
590	*Ximenia americana* L.	Olacaceae	Small	Common	Deciduous forests
591	*Xylia xylocarpa* (Roxb.) Taub.	Mimosaceae	Medium	Common	Deciduous forests
592	*Xylocarpus moluccensis* (Lam) M. Roem	Meliaceae	Small	Common	Mangrove forests
593	*Xylosma longifolium* Clos	Flacourtiaceae	Small	Occasional	Moist deciduous forests
594	*Zanthoxylum armatum* DC.	Rutaceae	Small	Occasional	Deciduous forests
595	*Zanthoxylum ovalifolium* Wight	Rutaceae	Small	Occasional	Deciduous forests
596	*Zanthoxylum rhetsa* (Roxb.) DC.	Rutaceae	Large	Occasional	Deciduous forests
597	*Ziziphus glabrata* Heyne ex Roth	Rhamnaceae	Small	Rare	Deciduous forests
598	*Ziziphus horrida* Roth	Rhamnaceae	Small	Rare	Dry deciduous forests
599	*Ziziphus mauritiana* Lam. var *mauritiana*	Rhamnaceae	Small	Occasional	Dry deciduous forests
600	*Ziziphus rugosa* Lam	Rhamnaceae	Small	Common	Dry deciduous forests
601	*Ziziphus xylopyrus* (Retz.) Willd.	Rhamnaceae	Small	Common	Dry deciduous forests

Croton scabiosus Bedd (Euphorbiaceae) (Fig. 3.2f). Small- to medium-sized deciduous tree, leaves simple, lepidote on both surfaces, flowers yellow, capsules globose, golden yellow; common in hills of Kadapa district, rare in Veligonda hill ranges at above 400 m. altitude.

Dimorphocalyx kurnoolensis Venkataraju & Pullaiah (Euphorbiaceae). Dioceous small tree, leaves simple, shiny, flowers white, capsular fruit subglobose; found rare along the sides of the streams at the foot hills of Erramalai hills of middle Eastern Ghats at 500–700 m altitude.

Eriolaena lushingtonii Dunn (Sterculiaceae). Small tree, flowers yellow, fruits capsular, woody; and rare in open slopes of moist deciduous forests of Nallamalais, Northern Circars, between 350 and 900 m.

Glochidion tirupathiense Rasingam et al. (Euphorbiaceae). Small tree, flowers creamy, axillary fascicle; small tree in deciduous forests of Tirupathi hills above 700 m.

Hildegardia populifolia (Roxb.) Schott & Endl (Sterculiaceae) (Fig. 3.3f). Tree, leaves simple, flowers scarlet, fruit follicular, winged, wings flat, inflated; found in middle (few patches of Anantapuramu and Chittoor districts of Andhra Pradesh).

Table 3.3 Medicinally useful tree taxa of Andhra Pradesh and modes of administration

Botanical name/family	Condition treated and mode of administration
Acacia catechu Mimosaceae	Heartwood extract given internally with local liquor made from *Madhuca longifolia* to control bleeding after childbirth in women.
Acacia farnesiana Mimosaceae	Heartwood made into a fine paste applied on the skin to cure leprosy. Fruits of this plant are used to control coughs, 15 ml of decoction of stem bark taken orally, twice a day for three days to cure diarrhoea and dysentery. Paste of bark is used to treat cuts and wounds.
Acacia nilotica Mimosaceae	Stem bark made into powder taken internally twice daily to cure gonorrhoea. Leaves chewed to cure scurvy.
Albizia amara Mimosaceae	Leaf with asphalt (silajittu in Telugu), crude copper sulphate (mayilututtam in Telugu) made into pills, one pill taken daily early in the morning to treat mental illness.
Albizia lebbeck Mimosaceae	Stem bark powder used for diarrhoea. Leaf juice dropped into the eyes to cure night blindness.
Albizia odoratissima Mimosaceae	The paste of leaves is applied in the eye to treat diseases. The decoction of the stem bark is used to relieve body pains. The leaf juice dropped into the eye to cure night blindness.
Alangium salvifolium	Root paste taken internally and externally (bite spot) to cure snakebite. Root juice is used to relieve fever.
Annona squamosa Annonaceae	The root is a drastic purgative and is used in acute dysentery.
Anogeissus latifolia Rutaceae	Stem bark made into paste taken internally and externally (bite spots) As an antidote to treat snakebite. Stem bark paste taken internally after delivery to expel the placenta.
Aegle marmelos Rutaceae	Fruit pulp and stem bark decoction is taken internally with cumin seeds to treat stomach disorders. Fruit pulp mixed with honey or sugar is given for immediate relief of hiccups. The fruit is eaten for curing dysentery and diarrhoea.
Ailanthus excelsa Simaroubaceae	A total of 15 to 20 leaflets are boiled in750 ml of water till it becomes 150 ml and taken 10 ml per time three times a day for three days to cure malaria. Fresh stem bark crushed, made into juice, taken internally for relief from stomachache and chronic fever.
Azadirachta indica Meliaceae	Leaf juice taken externally and internally to cure skin disease. Leaf decoction taken with honey to cure diarrhoea and dysentery. Stem bark decoction is drank twice daily for seven days to cure malarial fever.
Anacardium occidentale Anacardoaceae	Kernel is used as an aphrodisiac.
Bauhinia purpurea Caesalpiniaceae	The root extract is used for haemorrhoids and as a homeostatic agent. The stem bark is boiled in water. This water is used for bathing by mothers to restore health after childbirth.
Bridelia airy-shawii Euphorbiaceae	Stem bark powder is given with water to relieve abdominal pains. Stem bark made into fine powder taken as vapour bath to restore health after childbirth.
Bombax ceiba Bombacaceae	Bark juice, about six teaspoons taken three times a day to cure stomach ache.

(continued)

Table 3.3 (continued)

Botanical name/family	Condition treated and mode of administration
Boswellia serrata Burseraceae	The juice of this plant controls coughs. The paste of stem bark is given orally twice a day for one week to treat skin diseases. The bark of stem is used to treat diabetes. The gum is used to treat fever.
Buchanania axillaris Anacardiaceae	The fruits are edible. Seeds warmed slightly and made into powder taken daily after taking food for easy digestion (dyspepsia).
Butea monosperma Fabaceae	Yellow dye is prepared from flowers and seeds, used as purgative and vermifuge. Resin obtained from the bark taken internally to cure jaundice. Resin is given internally together with milk twice a day for ten days to treat bleeding.
Capparis grandis Capparaceae	The paste of leaf is used to cure skin diseases. Leaves crushed and juice applied to treat insect bite.
Careya arborea Barringtoniaceae	Stem bark mixed with salt made into a fine paste applied externally to cure skin diseases. Juice of leaves, about 15 ml thrice a day is given for about a week to treat fever.
Cochlospermum religiosum Cochlospermaceae	Leaves and flowers are macerated with water and given orally twice a day morning and evening in promoting menstruation. The stem bark decoction is added to the water and taken as a head bath every day for four days to cure jaundice.
Chloroxylon swietenia Flindersiaceae	The powdered bark is put in a thin cloth and soaked in breast milk, and applied in drops to eye injuries. The bark is used as an astringent.
Dichrostachys cinerea Mimosaceae	Decoction of stem bark taken twice a day to relieve fever.
Diospyros chloroxylon Sapotaceae	The paste of stem bark is taken orally thrice a day for two days to treat dyspepsia. Fruits used as diuretic and for constipation.
Diospyros melanoxylon Sapotaceae	Dried flowers crushed, made into juice and taken internally for urinary problems.
Dolichandrone falcata Bignoniaceae	Root powder is mixed with castor oil and applied externally as an antidote for snake and other poisonous bites. Root with goat's urine made into paste taken to treat haemorrhoids.
Erythrina stricta Fabaceae	Stem bark decoction taken internally to cure dysentery.
Ficus hispida Moraceae	The juice of leaves is applied to treat boils. The fruit boiled in goat's milk is used in hepatic hindering. The fruit and root are used to cure diabetes.
Ficus religiosa Moraceae	Stem bark decoction taken internally to cure dysentery. Stem ash with butter applied as an ointment to cure sores on feet.
Erythroxylum monogynum Erythroxylaceae	The leaf paste is applied externally for curing wounds and 10 ml of juice is given twice a day for about a week to treat malarial fever.
Gardenia gummifera Rubiaceae	Gum from the stem dissolved in water and given for constipation and also to kill intestinal worms (anthelmintic).

(continued)

Table 3.3 (continued)

Botanical name/family	Condition treated and mode of administration
Gardenia resinifera Rubiaceae	Juice of leaves is taken internally to cure liver disorders. Gum from the stem dissolved in water and given to kill intestinal worms (anthelmintic).
Gyrocarpus americana Hernandiaceae	Stem bark made into paste taken externally to cure arthritis.
Givotia moluccana Euphorbiaceae	Seed made into fine powder, mixed with *Pongamia pinnata* seed oil and made into paste applied externally as an ointment for curing psoriasis.
Gmelina arborea Verbenaceae	Root paste used internally and externally as an antidote to snakebite.
Gmelina asiatica Verbenaceae	Stem bark is made into paste and appliedon the head to treat dandruff. Root juice is given internally in the treatment of gonorrhoea.
Helicteres isora Sterculiaceae	Root decoction mixed with turmeric powder and applied externally to cure cuts and wounds, fruit paste applied externally to treat skin diseases.
Haldina cordifolia Rubiaceae	Stem bark of the plant is ground into a paste with black pepper and sesame seeds given to women twice a day for three days to treat dysmenorrhoea. Stem bark paste is used to treat sores and galls.
Holarrhena pubescens Apocynaceae	Stem bark powder taken orally with water to cure diarrhoea, dysentery and fever. The powder of stem bark is given orally along with pure honey to the child suffering from cold and coughs. Stem bark paste applied externally to skin diseases.
Hardwickia binata Caesalpiniaceae	Decoction of fresh leaves used as purgative. Gum obtained from the bark mixed with coconut oil applied externally to cure gonorrhoea.
Ixora pavetta Rubiaceae	Resin applied in the eyes for easy delivery in women.
Lannea coromandelica Anacardiaceae	Leaves gently boiled and applied on wounds.
Lagerstroemia parviflora Lythraceae	Root bark made into decoction taken twice daily to reduce fever.
Melia azedarach Meliaceae	Leaves ground with turmeric and made into pills and taken orally for four days after menstruation before baths to treat menstrual disorders. Leaves of *Melia azedarach* and *Aristolochia bracteolata* taken equally and heated. The mixtures taken with hot water to cure haemorrhoids. Leaves with camphor or turmeric made into paste used to cure sores.
Mitragyna parvifolia Rubiaceae	Stem bark mixed with leaf squeezed and inhaled to relieve coughs and cold. Squeezed fruits are applied on forehead to treat headache. Root bark made into paste taken internally and externally (bite spot) as antidote to snakebite.
Madhuca indica Sapotaceae	Juice of stem bark, about three teaspoons twice a day is given for about a week to treat fever. Liquor made from flowers and fruits is given as tonic to relieve dengue fever.

(continued)

Table 3.3 (continued)

Botanical name/family	Condition treated and mode of administration
Mallotus philippensis Verbenaceae	The red powder obtained from the mature fruits is mixed with cow butter and taken internally during menstruation and an abortifacient. Stem bark made into a fine paste and applied externally to cure skin diseases.
Ochna obtusata Ochnaceae	Stem bark made into paste taken externally (bite spot) and internally to treat snakebite.
Oroxylum indicum Bignoniaceae	Seeds are soaked in water for 12 h and mixed with turmeric, made into a fine paste, given internally to relieve abdominal pains and burning sensation while passing urine.
Pongamia pinnata Fabaceae	Seed oil is warmed and applied externally for boils and wounds. Fruits made into necklace and worn around the neck to cure chronic coughs.
Pterocarpus marsupium Fabaceae	Red coloured sap obtained from cut stem ends is stored in bottles and taken two spoons a day with water to treat nervous diseases and weakness.
Pterocarpus santalinus Fabaceae	A paste of the wood is applied externally to give a cooling effect during inflammation and headache.
Premna tomentosa Verbenaceae	The leaf extract is taken orally twice a day to treat diuretic problems. The leaf juice is used internally as well as externally to relieve abdominal pains.
Kavalama urens (Sterculia urens) Sterculiaceae	Stem bark is powdered and made into small pills, two pills twice a day are given for a week in the treatment of rheumatic pains. Small quantity of gum taken internally for dyspepsia.
Soymida febrifuga Meliaceae	Decoction of stem bark is given in chronic cases of diarrhoea and dysentery.
Sapindus emarginatus Sapindaceae	Seeds are taken for blood purification.
Schleichera oleosa Sapindaceae	Stem bark paste applied externally to treat skin diseases.
Schrebera swietenioides Oleaceae	Stem bark and leaf paste applied on cracked lips. Leaf juice used to relieve toothache.
Semecarpus anacardium Anacardiaceae	Seed oil is used externally to relieve rheumatic pains. Seed oil is applied externally to cure cuts.
Strychnos nux-vomica Loganiaceae	Root is made into a fine paste and applied externally (bite spot) and internally taken to treat snakebite (cobra bite).
Strychnos potatorum Loganiaceae	One seed rubbed into a fine paste with buttermilk and given internally for one week to cure chronic diarrhoea. Seed paste used to cure gonorrhoea, to treat scorpion and snakebites and seed powder used for cleaning of muddy water.
Tamarindus indica Caesalpiniaceae	Fresh stem bark decoction is used to cure diarrhoea. Decoction of the leaves used as a vermifuge.
Terminalia alata Combretaceae	Stem bark is chewed as an antidote to snakebite for temporary treatment.

(continued)

Table 3.3 (continued)

Botanical name/family	Condition treated and mode of administration
Terminalia arjuna Combretaceae	Stem bark made into powder and taken with water as an emetic when poison is taken.
Terminalia bellirica Combretaceae	Fruits of *Phyllanthus emblica* and *Terminalia chebula* are mixed and made into decoction, taken internally to cure leucorrhoea. Above fruits (locally known as triphala) mixed in equal parts made into fine powder taken internally twice a day for 15 days to treat haemorrhoids.
Terminalia chebula Combretaceae	Leaf galls (Karakantu in Telugu) and asafoetida resin made into paste, and used for setting bone fracture.
Trema orientalis Ulmaceae	Tips of aerial roots made into paste applied on penis to cure syphilis.
Wrightia arborea Apocynaceae	Roots are made into powder and taken with water, one teaspoon to relieve fever. Stem bark made into powder and taken with water to cure rheumatism.
Wrightia tinctoria Apocynaceae	Leaf juice with milk applied externally to relieve itching (scabies). Stem bark is made into powder and taken with water for epilepsy. Crushed roots given to induce vomiting (emetic).

Lasiococca comberi Haines (Euphorbiaceae). Small tree, often buttressed, leaves simple, panduri-form or obovate, flowers in racemes, capsule 3-lobed, tuberculate; found in northern circars of Andhra Pradesh at an altitudes of 700 m.

Premna hamiltonii (Buch.-Ham.) Ellis (Verbenaceae). Tree, leaves simple, ovate, yellow gland dotted, flowers pale green, drupe globose, black; occurring in Nallamalais of Middle Eastern Ghats at 600–1000 m.

Pterocarpus santalinus L. (Papilionaceae). Moderate-sized tree, leaves 3 rarely 5-foliolate, flowers yellow, fruit samara; common on the hill slopes at 300–700 m altitude, endemic to the southern-Middle Eastern Ghats.

Shorea tumbuggaia Roxb (Dipterocarpaceae). Large resinous tree, flowers creamy white, fruits winged; occurring in Seshachalam hill ranges of Chittoor district at 700–1000 m.

Syzygium alternifolium (Wight) Walp (Myrtaceae). Deciduous middle-sized tree, leaves thick coriaceous, simple, flowers cream or yellowish-white, sweet scented, berry globose, crowned with cup-like persistent calyx tube, dark purple; occurring in Middle Eastern Ghats at 600–850 m altitude.

Terminalia pallida Brandis (Combretaceae) (Fig. 3.6f). Small- to medium-sized tree, leaves simple, thick coriaceous, flowers pale yellow, drupe ovoid, faintly ridged; distribution is restricted to Kadapa and Chittoor districts of Andhra Pradesh at an altitudes of 600–800 m.

Wendlandia gambleii Cowan (Rubiaceae). Small tree, leaves simple, flowers yellowish, in panicles, capsule globose; found in Rampa hills of East Godavari district at 1000–1500 m altitude.

Fig. 3.3 (**a**) *Deccania pubescens* (Roth) Tirveng., (**b**) *Dillenia bracteata* Wight, (**c**) *Diospyros melanoxylon* Roxb., (**d**) *Ficus racemosa* L., (**e**) *Haldina cordifolia* (Roxb.) Ridsd., (**f**) *Hildegardia populifolia* (Roxb.) Schott. & Endl

3.7 Endemic Trees of Peninsular India Occurring in Andhra Pradesh (Source, Nayar et al. 1984)

Anacardiaceae: *Nothopegia heyneana* (Hook. f.) Gamble

Annonaceae: *Alpnosea madraspatana* Bedd., *Miliusa eriocarpa* Dunn ex Gamble, *Miliusa montana* Gard. ex Hook.f. et Thoms.

Aquifoliaceae: *Ilex malabarica* Bedd.

Bignoniaceae: *Dolichandrone arcuata* Clarke, *Dolichandrone atrovirens* (Heyne ex Roth) Sprague

Capparaceae: *Maerua apetala* (Roth) Jacob (Fig. 3.4e)

Celastraceae: *Eunoymus indicus* Heyne ex Roxb., *Maytenus bailadillana* (Swamy & Mooney) Raju & Biswas

Fig. 3.4 (**a**) *Holoptelia integrifolia* (Roxb.) Planchon, (**b**) *Isonandra villosa* Wight, (**c**) *Kavalama urens* (Roxb.) Raf., (**d**) *Mitragyna parvifolia* (Roxb.) Korth, (**e**) *Maerua apetala* (Roth) Jacobs, (**f**) *Ochna obtusata* DC

Cordiaceae: *Cordia evolutior* Gamble

Dilleniaceae: *Dillenia bracteata* Wt (Fig. 3.3b).

Dipterocarpaceae: *Shorea roxburghii* G. Don (Fig. 3.5e)

Ebenaceae: *Diospyros assimilis* Bedd.

Euphorbiaceae: *Bridelia crenulata* Roxb; *Glochidion ellipticum* Wt. *Glochidion neilgherrense* Wight, *Glochidion tomentosum* Dalz, *Mallotus aureopunctatus* Muell.-Arg, *Mallotus muricatus* Bedd, *Mallotus stenanthus* Muell.-Arg., *Trewia polycarpa* Benth. ex Hook.f.

Fig. 3.5 (**a**) *Phyllanthus* indofischeri Bennet, (**b**) *Premna hamiltonii* J.L.Ellis, (**c**) *Pterocarpus santalinus* L.f. (**d**) *Sapindus emarginatus* Vahl, (**e**) *Shorea roxburghii* G.Don, (**f**) *Shorea tumbuggaia* Roxb

Icacinaceae: *Apodytes dimidiata* E. Meyer ex Arn

Lauraceae: *Cinnamomum macrocarpum* Hook.f., *Cinnamomum malabaratum* (Burm. f.) Bl., *Cryptocarya neilgherrensis* Meissn., *Litsea oleoides* Hook.f., *Litsea wightiana* (Nees.) Hok.f., *Neolitsea foliosa* Gamble, *Neolitsea scrobiculata* Gamble

Melastomataceae: *Memecylon lushingtonii* Gamble

Meliaceae: *Aglaia elaeagnoidea* (Juss.) Benth var. *courtallensis* (Gamble) K.K.N. Nair

Mimosaceae: *Acacia campbelli* Arn.

Moraceae: *Ficus beddomei* King, *Ficus dalhousiae* Miq

Myristicaceae: *Knema attenuata* (Hook.f. et Thoms.) Warb.

Myrsinaceae: *Myrsine capitellata* Wall. ex Roxb

Ochnaceae: *Ochna obtusata* DC. var. *gamblei* (King ex Brandi) Kanis (Fig. 3.4f)

Pittosporaceae: *Pittosporum dasycaulon* Miq.

Fig. 3.6 (**a**) *Sloanea sterculiacea* (Benth.) Rehder, (**b**) *Soymida febrifuga* (Roxb.) A.Juss., (**c**) *Strychnos nux-vomica* L., (**d**) *Strychnos potatorum* L.f., (**e**) *Suregada angustifolia* (Baill. ex Muell.-Arg.) Airy Shaw, (**f**) *Terminalia pallida* Brandis

Rubiaceae: *Hymenodictyon obovatum* Wall., *Wendlandia angustifolia* Wt., *Deccania pubescens* (Roth) Tirveng (Fig. 3.3a).

Rutaceae: *Pamburus missionis* (Wt.) Swingle

Sapotaceae: *Isonandra villosa* Wt, *Manilkara roxburghinana* (Wt.) Dubard

Vacciniaceae: *Vaccinium neilgherrense* Wt.

3.8 Overexploited Tree Taxa

The following tree taxa are overexploited by local peoples for their needs, i.e. for timber, cordage, gums, and fuel wood.

Anogessus latofolia (Roxb.) ex DC. Wall ex Guill. & Perr., *Cochlospermum religiosum* (L.) Alston,

*Chloroxylon swietenia*Adr. Juss, *Chukrasia tabularis*Adr. Juss., *Dalbergia latifolia* Roxb., *Diospyros melanoxylon* Roxb. (Fig. 3.3c), *Dolichandrone atrovirens* (Roth). Sprague., *Givotia moluccana* (L.) Sreem., *Gmelina arborea* Roxb., *Haldina cordifolia* (Roxb.) Ridsd., *Kavalama urens* (Roxb.) Raf. (Syn.: *Stercularia urens* Roxb.), *Pterocarpus marsupium* Roxb., *Pterocarpus santalinus* L., *Santalum album* L., *Shorea robusta* Roxb. ex Gaertn., *Soymida febrifuga* (Roxb.) Adr. Juss. (Fig. 3.6b), *Tecotna grandis* L.f., *Terminalia paniculata* Roxb. and *Terminalia coriaeea* (Roxb.) Wt. & Arn.

3.9 Causes for Depletion of Tree Species and Its Conservation

The natural forests all over the Andhra Pradesh are under great pressure. Overgrazing, over-exploitation of trees for timber, fuel wood, fodder, etc., forest fires, encroachment of forest for agriculture, urbanization, and construction of reservoirs, plantation of exotic trees for shade or afforestation has decreased the tree and other biodiversity. Rich biodiversity has been thereby lost in Andhra Pradesh. Plantations favoured only few valuable species; thus, natural species are dwindling and shrinking. The biotic interference has had a marked effect on the vegetation leading to the extinction of valuable and rare species. Due to various anthropogenic factors and natural catastrophes, there is perceptible decline in the population of many valuable trees, making them rare and threatened.

Illegal felling of trees is a problem to reckon with in most reserve forests and the best way to solve it is protecting with the full involvement of the local communities living in the area. Forest fires should be prevented at any cost. Frequent fire has reduced the moist forest into drier forest. Grazing should be restricted to carrying capacity of forests and degraded grasslands should be planted up with improved fodder grass and legumes to increase the productivity of the area and to meet the present fodder demand. Enrichment plantations could be taken in degraded forestlands to introduce valuable species and a mixture of species having multipurpose characters. Appropriate conservation measures will not only stop further depletion of trees but will also help in improving tree diversity in Andhra Pradesh.

Acknowledgements The authors are thankful to the authorities of BSID, BLAT, CAL and MH for the permission to consult their Herbaria and Libraries.

References

Ahmedullah M, Nayar MP (1987) Endemic Plants of the Indian region, vol 1. Peninusular India. Bot. Surv. India, Calcutta

Gamble JS, Fischer CEC (1915–1935) Flora of Presidency of Madras, London Rep ed 1957. Calcutta

Hooker JD (1872–1897) Flora of British India, vol. 7. London

Nayar MP, Ahmed M, Raju DCS (1984) Endemic and rare plants of Eastern Ghats. Indian J For 7:35–42

Pullaiah T (2018) Flora of Andhra Pradesh, vol 5. Scientific Publishers, Jodhpur

Pullaiah T, Karuppusamy S (2018) Flora of Andhra Pradesh, vol 3. Scientific Publishers, Jodhpur

Pullaiah T, Muralidhara Rao D (2002) Flora of Eastern Ghats. In: Ranunculaceae-Moringaceae, vol 1. Regency Publications, New Delhi

Pullaiah T, Sandhya Rani S (1999) Trees of Andhra Pradesh, India. Regency Publications, New Dehli

Pullaiah T, Sri Rama Murthy K (2001) Flora of Eastern Ghats. In: Leguminosae, vol 2. Regency Publications, New Delhi

Pullaiah T, Sri Rama Murthy K (2018) Flora of Andhra Pradesh, vol 2. Scientific Publishers, Jodhpur

Pullaiah T, Sri Rama Murthy K, Karuppusamy S (2007) Flora of Eastern Ghats, India. In: Rosaceae - Asclepiadaceae, vol 3. Regency Publications, New Delhi

Pullaiah T, Sandhya Rani S, Karuppusamy S (2011) Flora of Eastern Ghats. In: Stylidaceae – Plantaginaceae, vol 4. Regency Publications, New Delhi

Pullaiah T, Chennaiah E, Sandhya Rani S (2018a) Flora of Andhra Pradesh, vol 1. Scientific Publishers, Jodhpur

Pullaiah T, Ali Moulali D, Sandhya Rani S (2018b) Flora of Andhra Pradesh, vol 4. Scientific Publishers, Jodhpur

Genetic Diversity and Variability Analysis in Sweet Flag (*Acorus calamus* L.)

4

D. Aparna, M. Ravindra Babu, and P. Rama Devi

Abstract

Twenty-nine genotypes of sweet flag (*Acorus calamus* L.) were evaluated to estimate genetic diversity, variability, heritability and genetic advance. Among the six clusters formed, cluster II and III had constituted a maximum of eleven genotypes each and cluster I had four genotypes. Cluster IV, V and VI had single genotype. Among the six clusters, cluster III recorded the maximum intra-cluster distance (21.38), followed by cluster II (11.89), indicating that the genotypes included in these clusters are closely related and population within the cluster are homozygous. Highest inter-cluster distance was found between cluster IV and VI (169.31), followed by cluster II and VI (111.43). The high genotypic coefficient of variation and phenotypic coefficient of variation were noticed for number of leaves, leaf length, number of scales, intermodal length, number of lets, rhizome length and rhizome weight, suggesting greater phenotypic and genotypic variability among the genotypes and responsiveness of the attributes for making further improvement by selection. Moderate phenotypic coefficient of variation and genotypic coefficient of variation estimates were observed for plant height and leaf width, indicating moderate range of genetic variability and considerable influence of environment in the expression of these traits. Based on the variation present in the genotypes as exhibited by the estimates of various parameters, continuous selection can be adopted for crop improvement in sweet flag.

Keywords

Acorus calamus · Genetic diversity · Variability · Heritability

D. Aparna (✉) · P. R. Devi
All India Co-ordinated Research Project on Medicinal and Aromatic Plants and Betelvine, Horticultural Research Station, Dr. Y S R Horticultural University, Venkataramannagudem, Andhra Pradesh, India

M. R. Babu
Horticultural Research Station, Dr. Y S R Horticultural University, Venkataramannagudem, Andhra Pradesh, India

S. M. Khasim et al. (eds.), *Medicinal Plants: Biodiversity, Sustainable Utilization and Conservation*, https://doi.org/10.1007/978-981-15-1636-8_4

4.1 Introduction

Sweet flag (*Acorus calamus* L.) is a commonly known drug in traditional systems of medicine like Ayurveda and Unani. It is a perennial wetland monocot plant belonging to the family *Acoraceae*. The scented leaves and rhizomes of sweet flag are being traditionally used as a medicine and the dried and powdered rhizome has a spicy flavour and is used as a substitute for ginger, cinnamon and nutmeg for its odour (Balakumbahan et al. 2010). It has been long known for its medicinal value, it is wild or cultivated throughout Himalayas at an altitude ascending up to 6000 ft. The rhizomes of *Acorus calamus* possess a characteristic sweet smell due to the presence of β-asarone. The rhizomes are considered to possess antispasmodic, carminative, anthelmintic, aromatic, expectorant, nauseate, nervine, sedative and stimulant properties and also used for the treatment of epilepsy, chronic diarrhoea and abdominal tumours (Paithankar et al. 2011). Most of the medicinal plants are under threat worldwide due to habitat fragmentation, deforestation, overharvesting and exploitation to meet the demands of raw drug markets (Thriveni et al. 2013). Adaptability to the environmental changes decreases due to reduction of loss of genetic diversity among populations. The level of genetic diversity also affects individual fitness and potential persistence of population. This has resulted in reduced population size, increased genetic drift and decreased population adaptability, resulting in increased risk of extinction of valuable medicinal plants (Lande 1995). Thus, assessment of genetic diversity is the first step in the evaluation of long-term conservation need of any plant species (Astorga and Campos 2004). The herb rarely produces seeds and is mainly propagated by vegetative means. However, for crop conservation and improvement information on genetic diversity and variability are prerequisites. The phenotypic variability among a collection of genetic stock gives an indication of potential genotypic variability. Hence, information on the extent of variability for some of the economic traits and their heritability will be helpful to the breeders to formulate sound breeding programme. Hence, this study was taken up to study the genetic diversity and to find out the extent of variability, heritability and expected genetic advance of the available germ plasm of sweet flag.

This study was conducted at All India Co-ordinated Research Project on Medicinal and Aromatic Plants and Betelvine, Horticulture Research Station, Dr. Y.S.R. Horticultural University, Venkataramannagudem, West Godavari District, Andhra Pradesh, during 2015–2017. The experimental material comprises 29 genotypes of sweet flag which were collected from different parts of India. The germ plasm was grown in Randomized Block Design and the data was recorded for nine characters, viz., plant height, number of leaves, leaf length, leaf width, number of lets, number of scales, intermodal length, rhizome length, rhizome width and rhizome weight. Genetic diversity was studied following Mahalanobis's (1936) generalized distance (D^2) extended by Rao (1952). Based on the D^2 values, the genotypes were grouped into clusters following the method suggested by Tocher (Rao 1952). Intra- and inter-cluster distances were calculated by the methods of (Singh and Chaudhury 1985). Phenotypic and genotypic coefficient of variability, heritability and expected genetic advance were determined according to Burton and Devane

(1953). Heritability in broad sense was calculated according to the formula suggested by Hanson et al. (1963).

4.2 Genetic Diversity Analysis in *Acorus calamus* L.

The use of Mahalanobis D^2 statistics for estimating genetic divergence was emphasized by many workers like Shinde et al. (2012) because it permits precise comparison among all possible pairs of population in any given group affecting actual crosses.

4.2.1 Clustering Pattern

Twenty-nine genotypes of sweet flag were evaluated for ten characters and the analysis revealed the existence of wide genetic diversity as they formed six clusters. Among the six clusters, cluster II and III had constituted a maximum of 11 genotypes each and cluster I had four genotypes. Cluster IV, V and VI had single genotype (Table 4.1, Fig. 4.1). Clusters II and III had maximum number of genotypes indicating that differences between them were less and individuals from similar population formed tight clusters. The clustering pattern revealed that the genotype did not resolve according to their geographical origin. Geographic diversity, though appears as an important factor, it seems that it is not the only factor determining the genetic divergence. It also indicates that the factors other than geographical diversity may also be responsible for such grouping. Similar findings were reported by Mythili Avadhani et al. (2016).

Table 4.1 Distribution of 29 genotypes of sweet flag in different clusters (Tocher's method)

Group	Number of genotypes	Cluster members
Cluster I	4	TNAc-12, TNAc-13, TNAc-8, Gubbi
Cluster II	11	APAc-6, APAc-12, APAc-13, APAc-15, APAc-16, APAc-17, APAc-18, APAc-19, Symbolia, Pandirimamidi and Raipur
Cluster III	11	APAc-1, APAc-2, APAc-3, APAc-4, APAc-5, APAc-8, APAc-9, APAc-10, APAc-11, APAc-14 and Kodaikanal
Cluster IV	1	Bengaluru
Cluster V	1	TNAc-9
Cluster VI	1	APAc-7

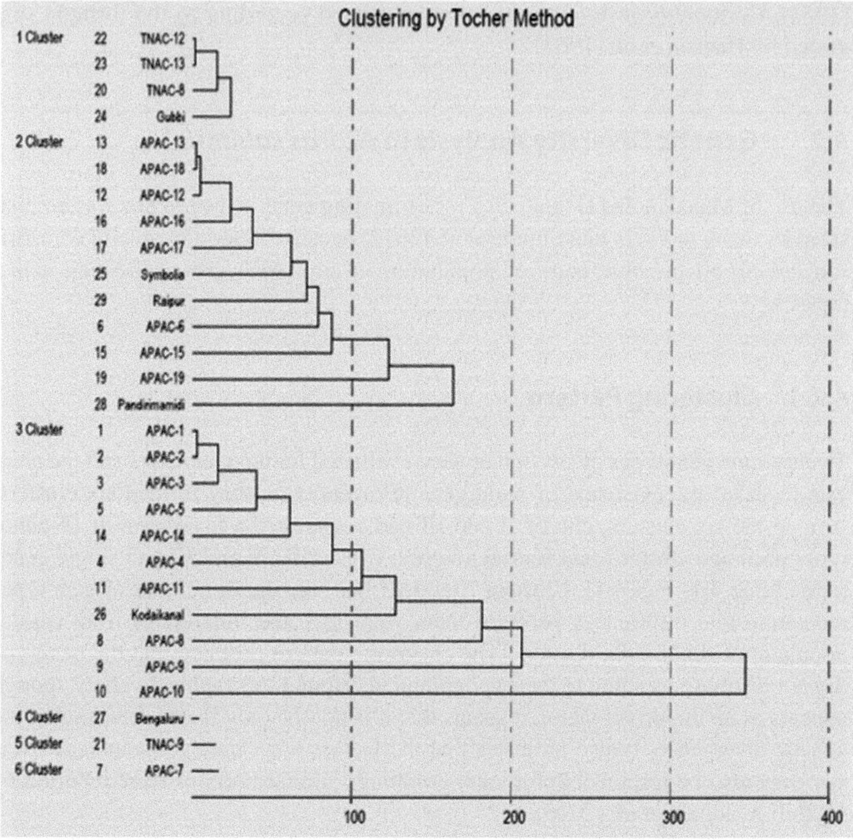

Fig. 4.1 Diagram depicting distribution of different genotypes of sweet flag in six clusters

Table 4.2 Average intra- (bold) and inter-cluster D^2 values for six clusters of sweet flag genotypes (Tocher's method)

	Cluster I	Cluster II	Cluster III	Cluster IV	Cluster V	Cluster VI
Cluster I	**6.70**	20.82	22.22	45.01	38.14	82.01
Cluster II	20.82	**11.89**	26.65	19.77	57.38	111.43
Cluster III	22.22	26.65	**21.38**	57.53	36.95	84.27
Cluster IV	45.01	19.77	57.53	**0.00**	103.69	169.29
Cluster V	38.14	57.38	36.95	103.69	**0.00**	16.31
Cluster VI	82.01	111.43	84.27	169.29	16.30	**0.00**

4.2.2 Intra- and Inter-Cluster Distances

Average intra- and inter-cluster D^2 values of 29 genotypes were presented in Table 4.2 and Fig. 4.2. Among the six clusters, the cluster III recorded the maximum intra-cluster D^2 value (21.38), followed by cluster II (11.89), indicating that the

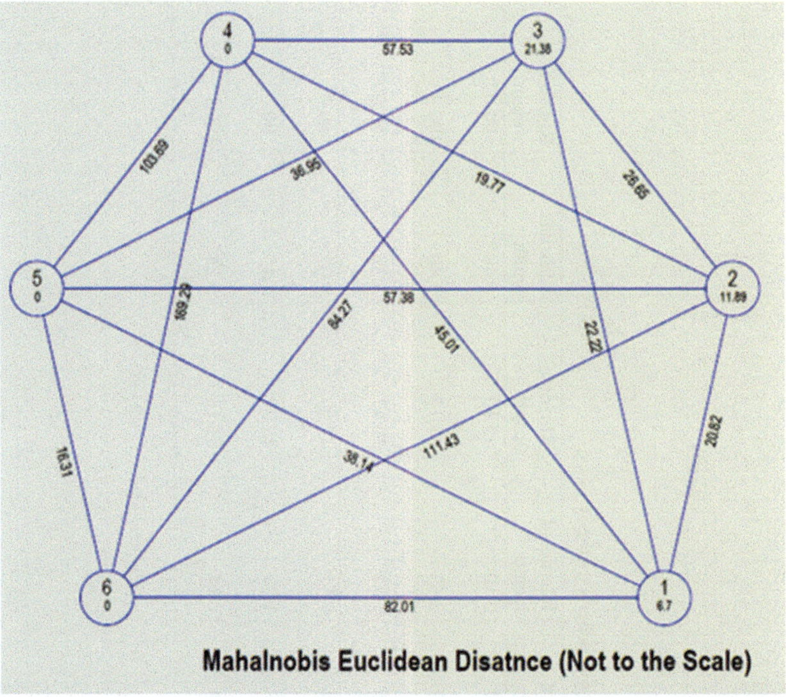

Fig. 4.2 Diagram depicting average intra- and inter-cluster distances for six clusters of different genotypes of sweet flag (Tocher's method)

genotypes included in these clusters are closely related and population within the cluster are homozygous. Highest inter-cluster distance was found between cluster IV and VI (169.31), followed by cluster II and VI (111.43). The genotypes belonging to clusters II, IV and VI had moderate genetic divergence and hence inter-mating between the genotypes belonging to these clusters may yield transgressive segregants in advanced generations. The emergence of homozygosity due to dominance of vegetative propagation may preserve the adaptive gene complexes in stable environment (Pai 2005).

4.2.3 Cluster Mean for Different Characters

The clusters differed with respect to their per se performance (Table 4.3). The results revealed that the clusters had slight variation with respect to different characters. The cluster II recorded maximum plant height, leaf width, number of scales, internodal length, rhizome length and rhizome weight, whereas cluster I recorded highest number of leaves, leaf length and number of lets. Rhizome weight and rhizome length were highest in cluster II followed by cluster VI.

Table 4.3 Mean values of clusters for ten characters in 29 genotypes of sweet flag (Tocher's method)

	Plant height	No. of leaves	Leaf length	Leaf width	No. of lets	No. of scales	Internodal length	Rhizome length	Rhizome width	Rhizome weight
Cluster I	55.33	79.67	44.37	2.07	11.00	27.67	2.10	43.10	5.40	94.67
Cluster II	63.10	56.00	40.20	2.10	8.00	28.67	2.53	67.17	5.40	197.33
Cluster III	48.02	45.72	33.44	1.88	9.33	22.56	1.72	40.47	5.39	63.00
Cluster IV	40.79	35.76	30.72	1.62	3.38	21.14	1.57	42.27	5.07	69.91
Cluster V	54.28	53.06	39.85	2.05	9.39	23.28	1.93	45.77	5.45	90.67
Cluster VI	57.42	47.54	41.03	2.01	6.04	23.54	2.14	51.13	5.23	109.17

4.2.4 Contribution of Characters to Genetic Divergence

In addition to the knowledge on the degree of divergence, the study also revealed the contribution of different characters studied towards total genetic divergence (Table 4.4). The character that appears maximum number of times ranks first and greater is its contribution to genetic divergence. In the present investigation, assessment of the contribution of different characters indicated that, the character rhizome weight (42.36%) was the maximum contributor for genetic divergence, followed by number of lets (23.15%). The relative contribution by intermodal length, number of scales, leaf width and leaf length for divergence among the genotypes was negligible. The study indicated that the characters such as rhizome weight, number of lets and number of leaves should be considered while selecting the individuals for further crop improvement.

4.2.5 Genetic Variability

The extent of variability for the characters in different genotypes measured in terms of variance, phenotypic coefficient of variation (PCV) and genotypic coefficient of variation (GCV) along with the amount of heritability (h^2) and expected genetic advance as per cent of mean are given in Table 4.5 and Fig. 4.3. The estimates of phenotypic coefficient of variation (PCV) were higher than their respective genotypic coefficient of variation (GCV) for all the traits which might be due to the influence of environment to certain extent in expression of these characters. The high GCV and PCV were noticed for number of leaves, leaf length, number of scales, internodal length, number of lets, rhizome length and rhizome weight, suggesting greater phenotypic and genotypic variability among the genotypes and responsiveness of the attributes for making further improvement by selection. Moderate PCV and GCV estimates were observed for plant height and leaf width, indicating moderate range of genetic variability and considerable influence of environment in the expression of these traits.

Table 4.4 Per cent contribution of different characters towards genetic divergence in genotypes of sweet flag

S. No	Source	Contribution %	Times ranked 1st
1	Plant height (cm)	9.85	40
2	Number of leaves	11.58	47
3	Leaf length(cm)	0.99	4
4	Leaf width (cm)	0.74	3
5	Number of lets	23.15	94
6	Number of scales	0.49	2
7	Internodal length (cm)	0.25	1
8	Rhizome length (cm)	7.64	31
9	Rhizome width (cm)	2.96	12
10	Rhizome weight (g)	42.36	172

Table 4.5 Estimates of variability, heritability and genetic advance for ten characters in sweet flag germ plasm

S No	Character	Variance		ECV (%)	PCV (%)	GCV (%)	h^2	Genetic Advance	
		Phenotypic	Genotypic					Standard	As per cent of mean
1	Plant height (cm)	73.78	60.91	7.04	16.86	15.32	0.83	14.61	28.68
2	No of leaves	145.12	105.17	13.49	25.71	21.88	0.72	17.98	38.38
3	Leaf length (cm)	55.27	32.98	12.83	20.20	15.60	0.60	9.14	24.82
4	Leaf width (cm)	0.12	0.05	14.09	18.02	11.23	0.39	0.27	14.43
5	No. of lets	14.79	12.73	20.49	54.85	50.88	0.86	6.82	97.23
6	No. of scales	22.77	10.45	15.25	20.73	14.04	0.46	4.51	19.59
7	Internodal length (cm)	0.15	0.08	14.45	20.79	14.94	0.52	0.42	22.13
8	Rhizome length (cm)	116.78	55.01	17.10	23.52	16.14	0.47	10.49	22.82
9	Rhizome width (cm)	0.12	0.03	5.66	6.57	3.33	0.26	0.18	3.47
10	Rhizome weight (g)	1207.72	1076.22	12.91	39.11	36.92	0.89	63.79	71.80

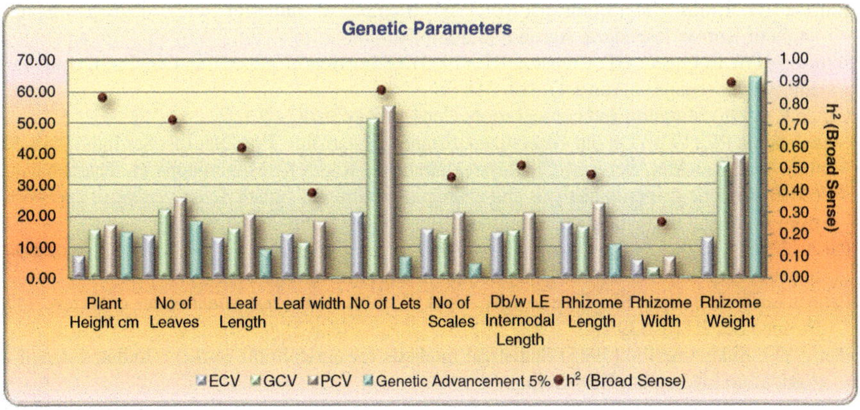

Fig. 4.3 Diagram depicting magnitude of genetic parameters

The heritability estimates will help the breeders in selection based on the phenotypic performance. In this study, high heritability was observed for plant height, number of leaves per plant, leaf length, number of lets, internodal length and rhizome weight, which indicates that selection is effective. But this selection is misleading because Johnson et al. (1955) reported that heritability estimate along with genetic advance is more useful than the heritability value alone for improving a particular trait. The high heritability combined with high genetic advance as per cent of mean was observed for number of leaves, number of lets and rhizome weight, indicating that these characters are under the control of additive gene action (Panse and Sukhatme 1985) and would respond very well to continuous selection.

4.3 Conclusion

Twenty-nine genotypes studied in the present investigation exhibited moderate genetic divergence. The emergence of homozygosity due to dominance of vegetative propagation may preserve the adaptive gene complexes in stable environment. The genotypes have shown wide range of variation for various characters studied. The variation present in the genotypes as exhibited by the estimates of GCV, PCV, broad sense heritability and genetic advance suggested ample scope of crop improvement through selection and inclusion of yield and yield attributing characters in sweet flag.

References

Astorga JG, Campos GC (2004) Genetic variability of the narrow endemic tree Antirhea aromatic in a tropical forest of Mexico. Ann Bot 93:520–528

Balakumbahan R, Rajamani K, Kumanan K (2010) *Acorus calamus*: an overview. J Med Plant Res 4:2740–2745

Burton GW, Devane EW (1953) Estimating heritability in tall fescue (*Festuca arundinaceae*) from replicated clonal material. Agron J 45:478–481

Hanson CH, Rininson HF, Comstock RE (1963) Biometrical studies of yield in segregating population of Korean lespedeza. Agron J 45:268–272

Johnson HW, Robinson HP, Comstoc RE (1955) Estimation of genetic and environmental variability in soybeans. Agron J 47:314–318

Lande R (1995) Mutation and conservation. Conserv Biol 9:782–791

Mahalanobis PC (1936) On the generalized distance in statistics. Proc Nat Inst Sci India 2:49–55

Mythili Avadhani MN, Selvaraj CI, Rajasekharan PE, Rao VK, Munirajappa H, Tharachand C (2016) Genetic diversity analysis and chemical profiling of Indian *Acorus calamus* accessions from south and north-East India. Indian J Biotechnol 15:560–567

Pai A (2005) The population ecologyof a perennial clonal herb (*Acorus calalmus* L.) (Acoraceae) in south East Ohio, USA. Ph. D. Dissertation. Ohio University, USA

Paithankar VV, Belsare SL, Charde RM, Vyas JV (2011) *Acorus calamus*: an overview. Int J Biomed Sci 2:518–529

Panse VG, Sukhatme PV (1985) Statistical methods for agricultural workers. Indian Council of Agricultural Research, New Delhi

Rao CR (1952) Advanced statistical methods in biometrical research. John Wiley and Sons, New York, p 374

Shinde KG, Birajdar UM, Bhalekar MN, Patil BT (2012) Genetic divergence in brinjal (*Solanum melongena* L.). vegetable. Science 39(1):103–104

Singh RK, Chaudhury BD (1985) Biometrical methods in quantitative genetic analysis. Kalyani Publishers, New Delhi

Thriveni HN, Sumangala RC, Shivaprakash KN, Ravikanth G, Vasudeva R (2013) Genetic structure and diversity of Coscinium fenestratum : a critically endangered liana of Western Ghats, India. Plant Syst Evol 300:403–413

5 Flora of Mangrove Species Utilized for Ethnomedicinal Practices in Gautami Godavari Estuary, Andhra Pradesh, India

G. M. Narasimha Rao

Abstract

Mangroves are one of the unique estuarine ecosystems among the aquatic habitats on the earth. Flora that grows in this region is capable of tolerating different fluctuations in environmental, hydrographical, and chemical parameters of the concerned water bodies. These tropical tidal forests play a vital role in fetching the livelihood of coastal inhabitants locally and enhancing the economy of the countries as a whole. Godavari mangroves are the second largest mangrove ecosystems in India. In the present investigation, medicinal importance and application of the mangrove species from Gautami Godavari estuary was documented. Data was gathered from the herbal doctors and local elderly people having knowledge about mangrove species. A total of 27 mangrove species were identified to treat various ailments such as skin diseases, cough, fever, orthopedic problems, constipation, piles, dysentery, and diabetic problems of human beings and to cure several diseases of cattle also.

Keywords

Mangrove flora · Medicinal uses · Gautami Godavari estuary · Andhra Pradesh · India

5.1 Introduction

Mangrove forests are special, peculiar, highly productive, and diversified estuarine ecosystems in their morphology, anatomy, physiology, and reproduction. These are the tidal forests which are daily inundated by high tides of the sea. Besides, salinity,

G. M. N. Rao (✉)
Medicinal Applications of Mangrove Flora, Department of Botany, Andhra University, Visakhapatnam, India

S. M. Khasim et al. (eds.), *Medicinal Plants: Biodiversity, Sustainable Utilization and Conservation*, https://doi.org/10.1007/978-981-15-1636-8_5

silt, clay and sand proportions in the soil play key role for formation of mangrove forests. Mangrove forests support the millions of the people's livelihood across the globe. These beautiful forests act as nursery grounds for the migratory birds and migratory fish. Rich organic matter present in these waters supports the growth of phytoplankton, which in turn facilitates the production of more and more fish. Sunderbans are the second largest mangrove forests in the world, while Godavari mangroves are the second largest mangrove ecosystems in India. Landscape, low-lying regions, and alluvial soils support the development as well as protection of mangrove flora. Gautami Godavari is one of the major tributaries of River Godavari in Andhra Pradesh. Major portions of the mangrove flora are distributed along the mud flats, creeks, and channels of Gautami Godavari estuary. Mangroves represent a unique and ecologically important estuarine habitat in the tropical and subtropical zones of the globe. Mangrove zones are the most productive, highly dynamic, and morphologically complex ecosystems. Saline environment is required for stable mangrove ecosystem. But each mangrove plant has specific requirement for distribution and growth within the ecosystem.

Across the globe lot of work was carried out by several authors (Chapman 1976; Saenger et al. 1983; Bunt 1992; Twilley 1998; Spalding et al. 1997; Valiela et al. 2001). Mangrove ecosystems of Gautami Godavari estuary were studied by Rao (1959), Sidhu (1963), Raju (1968), Blasco (1975), Umamaheswara Rao and Narasimha Rao (1988), Narasimha Rao and Dora (2009), Narasimha Rao and Subba Rangaiah (2010), Narasimha Rao (2012, 2014). In the recent years based on the utility of the mangrove species in various application by human beings lot of research has been conducted and going on for the benefit of the mankind. Still today in the developing countries, majority of the population relay on indigenous medicines prepared from the plant populations (Srinivastava et al. 1996; Saranraj and Sujitha 2015; Dossou-Yovo et al. 2017). Flora of mangroves are physiologically and chemically unique and possess active metabolites (Patra and Thatoi 2011), Nishiyama et al. (1978) reported the antimicrobial activity in mangrove species. Studies on microbial activity of mangrove plant species of Gautami Godavari estuary were conducted by several investigators (Vadlapudi et al. 2009; Vadlapudi and Naidu 2009a, b, c, d; Prasanna Lakshmi and Narasimha Rao 2013) on antibacterial and antifungal activity. Several mangrove species of the Godavari estuary was utilized by the local inhabitants for treating ailments of human beings and livestock. In the present exploration work information was gathered on the applications of different plant parts of the mangrove species used for treating the physical disorders of human beings and cattle.

Gautami Godavari estuary is one of the major branches of the River Godavari which flows south west and opens into the Bay of Bengal at Bhiravaplaem and Kothapalem. It is also connected to Kakinada Bay by two rivers, namely Gaderu and Coringa arising at Bhiravapalem and Yanam, respectively. There are many criss and cross canals which pass through mangrove areas and open into Kakinada Bay. The Gautami branch of Godavari is also connected to Pandi backwaters by a channel called as Peddaru. Mud flats with mangrove forests are more extensive along the estuarine regions of Gautami Godavari estuary. These tidal forests are located between 82° 12′ and 82° 21′ E longitudes and 16° 31′ and 16° 54′ N latitudes and cover an area of 31,622 hectares (Umamaheswara Rao and Narasimha Rao 1988).

Present investigation was carried out for a period of 1 year from October 2016 to September 2017 in the mangrove habitats of the Gautami Godavari estuary. Twelve field trips were made to the study sites to collect the information on ethnomedicinal practices by local inhabitants of mangrove and nearby habitations through interviewing elderly people, educated persons, and medicinal and herbal practitioners. Five types of informants were chosen by random sampling.

1. Vaidhyas and local herbal doctors.
2. Persons involved in fishing in the mangrove forest regions.
3. Locals who are collecting the wood, honey, and other forest products from the mangrove habitats.
4. The interpreters.
5. Persons with more than 50 years age.

   Each medicinal practice was cross-checked with 3 or 4 informants. Methodology for collection of data was followed by Jain (1981, 1984) and Martin (1995). Plant species were collected and identified with the help of Flora of British India (Hooker 1872–1897) and Flora of the Presidency of Madras (Gamble and Fischer 1936). Specimens were deposited in the Department of Botany, Andhra University, Visakhapatnam.

5.2 Mangrove Flora and Ethnomedicinal Purposes

In the present investigation, information on medicinal applications of various mangrove species situated in between Chollangi and Pandi regions of Gautami Godavari estuary was selected for documentation of the medicinal practices prevailing among the herbal doctors, traditional healers, village heads, and elderly persons who are using these mangrove species. During the present exploration study, a total of 44 plant species such as true mangroves, associated mangroves, and halophytes were collected from the mangrove habitats of the Gautami Godavari estuary. Among these species, 27 plant species were identified as medicinal plants which help to cure various ailments of the human beings and cattle. These mangrove species are generally used to treat common health problems such as skin allergy/skin diseases, small pox, constipation, leprosy, to control sugar levels, tooth decay, snake bite, diarrhea, mumps, and bone-related problems. In the study information was presented along with species key information.

Acanthus ilicifolius L. Sp. Pl.639.1753; Clarke in Hook. f. Fl. Brit. India. 4:481. 1984; Gamble, Fl. Pre. Madras.2:712.1957(repr.ed) *Acanthus ilicifolius*(L.) Juss. Gen. Pl. 103. 1789; Wight, Ic.t. 459, 1981.

> Erect, simple branched herbs; leaves elliptic-oblong, toothed, acute or truncate at apex, spiny along margins, Flowers bluish-purple, arranged in spikes; spikes simple or branched, capsules 2–3 cm in diam., ovoid-oblong, compressed, apiculate, oblong, shining green brownish. Seeds uniform. Flowering and fruiting: February and March.

Aegialitis rotundifolia Roxb. Fl.Ind.2:111, 1832; Clarke in Hook. f.Fl.Brit. India 3:479, 1180; Haines 2:529, 1922; V. Steenis Fl. Male. 4:108, 1949.

Evergreen shrubs or small trees 1–7 m tall; conically swollen at the base with stilt roots; bark thin, lenticellate. Leaves 2.5–9.5 × 3–9 cm, alternate, panicles sometimes forked, capsules 8–9 × 0.4–0.5 cm, linear, curved, 5-ribbed, splitting finally along the ribs from the apex, often tipped with stamina tube. Hypocotyl 6.5–8 cm, flowering and fruiting: February and July.

Aegiceras corniculatum (Linn.) Blanco in Pl. Philip. (ed.1) 79. 1837; G. 2: 532. Rhizophora corniculata Linn., in Stickman, Herb. Amb. 13.1754 et.Amoen. Acad.4:123. 1759. Aegiceras majus Gaerth., Fruct. 1:216.f.46.1788; FBI: 3: 515; G.2:532; H.2:535.

An erect, bushy shrub, flowers white, fragrant, in dense umbel, flowering and fruiting: January to May.

Arthrocnemum indicum (Willd.) Moq., Chenop. Monogr. Enum.113.1840(p.p.) et in DC., Prodr. 13:151.1849;FBI 5:12; G.2:828. Salicornia indica Willd., Ges. Naturf. Fr. Neve Scvhr.2:111, t.4.1799.

Much branched herbs; seeds ovoid. Flowering and fruiting: April to June.

Avicennia alba Bl., Bijd.821.1826; G.2:774; H.2:760; Meesue in Blumea 5: 78.1942.*A. officinalis* Linn. var. alba (Bl.) C.B.Cl.in Hook.f.Fl.Brit.India4:604.1885.

Trees, 5–6 m high; flowers yellow, in heads of stout peduncles; capsules conical and beaked when young, ovoid when old, flowering and fruiting: April to August.

A. marina (Forsk.) Vierh., Akad.Wiss.Wien.71:435.1907;G.2:774;M.258; Moldenke in Fl.Ceylon, 4:127.1983.*Sceura marina* Forsk., Fl.Aegypt. – Arab.37.1775. *Avicennia officinalis* sensu Cl.in FBI 4.604.1885 pro parte; Biswas in Notes R.B.G. Edin.18:163.1934, non Linn

Trees 3 m tall, flowers 0.4–0.6 cm across; reddish yellow capsules ovoid, yellow when ripe compressed. Flowering and fruiting: March to October.

A. officinalis Linn., Sp.Pl.110.1753; FBI4:604 p.p; G.2:774; H.2:760; Moldenke in Fl.Ceylon 4:132.1983. *A. tomentosa* sensu Biswas, Notes R.B.G. Edin 0.18:163.1934 non Jacq.(Enum.Pi.25.1760).

Trees ±10 m high, flowers 0.4–0.6 cm diam, yellow, cymes, capsules ovoid, seeds compact. Flowering and fruiting: March to August.

Bruguiera gymnorhiza (Linn.)Savi in Lamk., Encycl.4:696.1798; FBI2:437. *Rhizophora gymnorhiza* Linn. Sp.Pl.443.1753.*Bruguiera conjugata* (Linn.) Merr.in Philipp.Journ.Sci.9:118.1914;G.!: 324;H.2:364.*B. rheedii* Bl.,Enum.Pl.Java1:92.1827.

Trees, 3.5 to 5 m tall, leaves 8–11 × 4–6 cm, lathery, ovate, flowers solitary, drooping, fruits indehiscent, drupes 2–3 cm long, flowering and fruiting: November to April.

Ceriops decandra (Griff.) Ding Hou in Steenis, Fl.Males. (Ser.1) 5:471. 1958. *Bruguiera decandra* Griff., Trans. Phys. Soc. Cal.8: 10.1836. *Ceriops roxburghiana* Arn. In Ann. Nat.Hist. 1: 364.1838; FBI 2:436; G. 1:324; H. 2: 363.

Small trees, 2–4 m tall, leaves obovate, flowers greenish-white, fruits 1.2–1.5 cm long, flowering and fruiting: July to September.

Clerodendrum inerme (Linn.) Gaertn., Fruct. 1:271.t.57.f.1.1788; FBI 4: 589; G.2: 769; H.2:755. *Volkameria inermis* Linn., Sp.Pl.637.1753.

A much branched, scandent shrub; leaves 2.2 × 1.2 cm, flowers blue color, flowering and fruiting: July to April.

Cressa cretica Linn., Sp.Pl.223.1753; FBI 4:225; G.2:647; H. 2:613.

Erect, gray hairy, tender shrub, ±20 cm high; leaves dirty green, 2.5–9 × 1–6 mm, flowers white, seed shiny, flowering and fruiting: October to January.

Derris trifoliata Lour.,Fl. Cochinch. 2: 433.1790; Thothath. In Bull.Bot.Surv. India 3:181.1962 et Fasc. Fl.India 8:301982. D.uliginosa (Roxb.) Benth. In Miq., Pl.Jungh 1: 252.1852; FBI 2: 241; G.1.273; H.2:311. Galedupa uliginosa roxb., Fl.Ind.3: 243.1832.

Large shrubs; leaves 10–25 cm long, flowers rosy-pink, pods oblong, seeds ovate, flowering and fruiting: January to March.

Excoecaria agallocha Linn., Syst. Nat.(ed.10) 1288.1759 et .Sp. Pl(ed.2)1451.1763; FBI 5:472; G.3:941; H.1: 121.

Moderate shrubs; 1–2 m height, with white, highly acrid juice, bark with prominent lenticels; flowers in catkins, yellowish green; fruits 3-lobed capsules; seeds sub-globose; smooth, flowering and fruiting: May to July.

Heliotropium curassavicum Linn., Sp.Pl.130. 1753; G.2: 630.

An erect, drooping herb, about 15 cm long, flowers white, scorpeoid cyme, flowering and fruiting: October.

Lumnitzera racemosa Willd. Ges.Naturf. Berlin Neva Schr.4: 187.1803; FBI 2:452; G. 1:331; H. 2: 312; Exell in van Steenis. Fl. Males. ser. 1, 4: 588.1954.

Large shrubs, 2 m high, leaves 3–5 × 1.2–1.5 cm, flowers whitish-green in 0.8–1 cm long spikes; drupes ovoid, flowering and fruiting: February to March.

Myriostachya wightiana (Nees ex Steud.) Hook.f., FBI 7: 327. 1896; G.3: 1259; Bor 518. Leptochloa wightiana Nees ex Steud., Syn. Pl. Glum. 1:209. 1854.

Tall, perennial, marshy herbs, rootstock thick, panicles narrow, 4–20-flowered, glumes 2, grains obliquely ovoid. Flowering and fruiting: September to December.

Prosopis chilensis (Molina) Stuntz in U.P Dept. Agri. Bur. Pl. Indust. Invent. Seeds 31:85. 1914. *Ceratonia chilensis* Molina, Sagq. Chil. 172. 1782. *Prosopis juliflora* (SW.) DC., Prodr. 2:447. 1825. *Mimosa juliflora* SW., Prodr. 85. 1788.

A shrub to a small tree, with stout, axillary spines; leaves 2-pinnate; flowers pale yellow, in spikes; pods straight or curved. Flowering and fruiting: September to February.

Rhizophora apiculata Bl., Enum.Pl. Java 1:91:1827; Ding Hou in Steenis, Fl. Males. (Ser.1) 5: 452. 1958. *R. candelaria* DC., Prodr. 3: 32. 1828 (non Wt. & Arn. 1834); G.1:323; H.2:362. *R. conjugata* sensu Arn. In Ann. Nat. Hist. 1:363. 1838, non Linn. 1753; FBI 2:436.

Trees 8–12 m high; leaves 5–13 cm, flowers 3.5 m diam, fruits indehiscent drupes, 2–2.5 cm long. Flowering and fruiting: August to February.

Rhizophora mucronata Lam. Syn: R. mangle Roxb

Evergreen small tree, bark smooth, brown, flowers creamy-white, fruits 3–3.5 cm long, crowned by persistent calyx lobes.

Salicornia brachiata Roxb., Fl. Ind. 1:84, 1832; FBI 5:12; G. 2:828; H.2:809.

Much branched annuals 30 cm long, branches jointed, greenish-red, flowers pale yellow, minute, slender cylindric spikes, seeds erect, compressed. Flowering and fruiting: September and October.

Sesuvium portulacastrum (Linn.) Linn., Syst.ed.10, 1058.1759; FBI 2:659: G.1:388; H.246. *Portulaca portulacastrum* Linn., Sp.Pl.446.1953.

Sub-erect, prostrate or creeping herbs, flowers axillary, 7–12 mm long, capsules 3-celled, enclosed in persistent calyx. Flowering and fruiting: November to January.

Uses*:* paste of the whole plant is applied externally for healing of wounds.

Sonneratia apetala Buch.-Ham.in Syme.Emb. Ava.3:313.1800; FBI 2:579; G.1:363; H. 2:392.

Evergreen trees, 15 m high, branches slender, flowers 3, berries indehiscent, 2 cm diam, flowering and fruiting: January to April.

Sonneratia caseolaris (Linn.) Engl. in. Engl. &Prantl, Nat. PFlanzenf. 261.1897. *Rhizophora caseolaris* Linn. In Stickmann, Herb. Amb. 13. 1754(p.p). *Sonneratia acida* Linn.f. Suppl. 252.1781; FBI 2: 579; H.2: 393.

Trees 5 m tall, leaves 16–13 × 2.5 cm, flowers terminal, berries indehiscent, 2.5–5 cm diam, with persistent calyx. Flowering and fruiting: February to August.

Suaeda maritima (Linn.) Dumort, Fl. Beng.22. 1827; FBI 5:14; G.2:829; H. 2:810; M.129; V.170. Chenopodium maritimum Linn, Sp.Pl.221.1753.

Slender erect annuals; leaves green or often tinged with pink; flowers minute, flowering and fruiting: January to June.

Suaeda monoica Forsk.ex Gmel.in.Linn., Syst. Nat.(ed.13) 2:503. 1791; FBI 5:13; G. 2:829; M. 259.

Erect bushy herbs, leaves fleshy, linear, obtuse or subacute, 1.5 × 2.0 cm; flowers polygamous, seeds ovoid, erect. Flowering and fruiting: May to December.

Suaeda nudiflora Moq Syn: *S. maritima* Dum. Eng: Common Indian Saltwort Tel: Ilakura. Fam: Chenopodiaceae.

Shrub 75 cm tall, inflorescence with few flowered clusters.

Xylocarpus granatum Koen. Naturforscher 20:2, 1784; harms in Engl.& Prant. Pfam.ed. 2, 19bI 81, 1940; Parkinson in India For.60: 138, 1934; Mooney in Suppl. Bot. Bihar &Orissa 249, 1950. Carapa obovata Blume Bijdr. 179, 1825; Haines 1:187, 1921.

Trees, 10–15 m tall, 60–80 cm in girth; stems buttressed; bark smooth, yellowish-white with papery flakes. Leaves unijugate or bijugate, flowers 5–7 mm across, white, red glandular within, on short sparingly branched, axillary thyrses; fruits 30–40 cm, septifragal capsules, seeds more than 15; each 4–7 × 3–4 cm pyramidal or triangular; testa corky. Flowering and fruiting: July to August.

Information presented in this communication would be useful for the researchers for selecting the mangrove plant species for antimicrobial studies or extraction of novel compounds from these plants (Table 5.1). Further, this data can aid as baseline information for the scholars and students and supports for enhancing the knowledge in the field of ethnomedicine.

Table 5.1 Ethnomedicinal uses of some important mangrove species of Godavari estuary

S. No	Name of the species	Family	Habit	Ethnobotanical use
1	*Acanthus ilicifolius* L	Acanthance	Shrub	Fruit pulp used as blood purifier, leaf for rheumatic pains, roots along with Ocimum leaves prepared a paste for back pain.
2	*Aegialitis rotundifolia Roxb*	Plumbaginaceae	Shrub	Fresh leaves crushed and made as a paste is applied for joint pains, knee pains, and orthopedic-related problems.
3	*Aegiceras corniculatum* (Linn.) Blanco	Primulaceae	Tree	Bark used for diarrhea, leaves applied after slight warming on the mumps.
4	*Arthrocnemum indicum* (Willd.) Moq	Amaranthaceae	Shrub	Seeds soaked in water for 1 h and applied on wounds for better healing.
5	*Avicennia alba* Bl.	Acanthaceae	Tree	Bark is used for skin diseases. Leaves juice used for treating the tooth ache of the cattle
6	*Avicennia marina* (Forsk.) Vierh.	Acanthaceae	Tree	Bark is used for skin diseases. Fruits are used to control the constipation, paste prepared from leaves and roots used to heal the wounds.
7	*Avicennia officinalis* Linn.	Acanthaceae	Tree	Fruits are made into a poultice and used for boils, bark is made into a powder and used for skin diseases.
8	*Bruguiera gymnorhiza* (Linn.) Savi	Rhizophoraceae	Tree	Boiled decoction of bark used to treat diarrhea. Leaves are made into a paste and applied for healing the wounds and injuries
9	*Ceriops decandra* (Griff.) Ding Hou	Rhizophoraceae	Tree	Decoction of stem bark is administered to cure the intestinal problems of the cattle.
10	*Clerodendrum inerme* (Linn.) Gaertn	Laminaceae	Shrub	Paste of the tender leaves is administered to treat the worms of livestock. Decoction of the root is administered orally to cure indigestion and edema.
11	*Cressa cretica* Linn. Sp	Convolvulaceae	Shrub	Paste of the tender stems used to control the bleeding.
12	*Derris trifoliata* Lour.	Fabaceae	Shrub	Decoction of the leaf is administered daily for 3 to 4 days to control pneumonia.
13	*Excoecaria agallocha* Linn.	Euphorbiaceae	Tree	Milky latex of this plant applied on the swelling parts as well as for orthopedic-related problems
14	*Heliotropium curassavicum* Linn.	Boraginaceae	Herb	Whole plant extract with garlic cloves is administered for 3 to 5 days to cure ephemeral fever.

(continued)

Table 5.1 (continued)

S. No	Name of the species	Family	Habit	Ethnobotanical use
15	*Lumnitzera racemosa* Willd	Combretaceae	Tree	Leaves and stem bark used to treat piles and constipation.
16	*Myriostachya wightiana* (Nees ex Steud.) Hook.f.	Poaceae	Herb	Leaf paste is applied on skin wounds of the cattle for quick relief.
17	*Prosopis chilensis* (Molina) Stuntz	Fabaceae	Shrub	Decoction of leaves administered twice daily for 4 to 6 days to cures rheumatism and arthritis.
18	*Rhizophora apiculata* Bl.	Rhizophoraceae	Tree	Wood is used for heavy construction. Pastes from the leaves and prop roots are used to treat the backache and joint pains.
19	*Rhizophora mucronata* Lam.	Rhizophoraceae	Tree	Timber is good for heavy construction. Bark is powerful astringent and is used in the treatment of hemorrhage, hematuria, and angina. Also used to cure diabetes.
20	*Salicornia brachiata* Roxb	Amaranthaceae	Herb	Powder of the whole plant used for rheumatism and skin disorders.
21	*Sonneratia apetala* Buch.-Ham	Lythraceae	Tree	Bark is mixed with pepper and administered for one week to cure gastric problems. Leaves are used to treat skin diseases
22	*Sonneratia caseolaris* (Linn.) Engl	Lythaceae	Tree	Fruit with combination of jiggery is a administrated for 10 to 15 days period for treating gastrointestinal problems
23	*Suaeda maritima* (Linn.) Dumort	Amaranthaceae	Herb	Decoction of the tender stems is administered to cure pneumonia
24	*Suaeda monoica* Forsk. ex Gmel.	Amaranthaceae	Herb	Seed powder with honey administered once daily for 10 to 15 to 20 days for treating respiratory ailments
25	*Suaeda nudiflora* Moq	Amaranthaceae	Herb	Decoction of leaf and stem is administered once daily for 10 to 15 days to treat infertility.
26	*Xylocarpus granatum* Koen	Meliaceae	Tree	Bark used for diarrhea and dysentery.

Acknowledgements I would like to express my grateful thanks to UGC SAP-CAS I for financial assistance to carry out this investigation. I would like to extend my thanks to various people in the mangrove habitations for their valuable support during the period of this investigation. I wish to thank various forest officers in the Mangrove forests of Godavari estuary for their continuous support and cooperation during the exploration of this study.

References

Blasco F (1975) The mangroves of India, Pondicherry. Fren Sect Sci Technol 14:1–175

Bunt JS (1992) Introduction to mangrove ecosystem. In: Robertson AI, Alongi DM (eds) Tropical mangrove ecosystem. American Geophysical Union, Washington D.C., pp 1–6

Chapman VJ (1976) Mangrove vegetation. J Cramer, Vaduz, pp 1–326

Dossou-Yovo HO, Vodouhe FG, Sinsin B (2017) Ethnobotanical survey of mangrove plant species used as medicine from Ouidah to Grand- Popo districts, Southern Benin. Am J Ethnomed 4

Gamble JS, Fischer CEC (1936) Flora of the presidency of Madras. Adlard & Sons Ltd., London, 1915–1930; reprinted. Edn (Botanical Survey of India, Culcutta), (1-3). 1957

Hooker JD (1872–1897) The Flora of British India, vols. 1–7. L. Reeve & Co., Ashford, Kent. Ethnobotany Oxford & IBH, New Delhi, pp. 193–198

Jain SK (1981) Observations on Ethnobotany of the Tribal of Central India. In: Jain SK (ed). pp. 193–198

Jain SK (1984) A manual of ethnobotany. Scientific Publishers, Jodhpur

Martin G (1995) Ethnobotany: a method manual. Chapman and Hall, London

Narasimha Rao GM (2012) Distribution pattern and present scenario of mangroves and associated flora of Andhra Pradesh. In: Biodiversity of Aquatic Resources, Chapter 3, pp. 29–49

Narasimha Rao GM (2014) Distribution, density and conservation of mangroves in the Godavari estuary, Andhra Pradesh, India. J Biol Chem Res 31(2):614–622

Narasimha Rao GM, Dora SVVSN (2009) An experimental approach to mangroves of Godavari estuary. Indian J For 32(3):263–265

Narasimha Rao GM, Subba Rangaiah G (2010) Distribution of mangroves and associated flora of the Pandi Back waters of Gautami Godavari estuary. ANU J Nat Sci 2(1):22–26

Nishiyama Y, Ryuzo Sanchez PC, Kozaki M (1978) Inhibitory functions of mangrove bark towards cell growth of microorganisms. Hakko, Kogaku, Kaishi 56:712–717

Patra JK, Thatoi HN (2011) Metabolic diversity and bio activity screening of mangrove plants: a review. Acta Physiol Plant 33:1051–1061

Prasanna Lakshmi K, Narasimha Rao GM (2013) Antimicrobial activity of mangrove plant *Avicennia officinalis* (lam. Briqvet) on selected pathogens. Res J Pharma Biol Chem Sci 4(3):335–341

Raju DCS (1968) The vegetation on West Godavari. In: Proceedings National Symposium on Recent Advances in Tropical Ecology, Banaras Hindu University, Varanasi

Rao RS (1959) Observations on the mangrove vegetation of the Godavari estuary. In Proceedings of the mangrove symposium, Calcutta

Saenger P, Hegerl EJ, Davie JDS (1983) Global status of mangrove ecosystems. In: Commission on ecology papers no. 3. Gland. World Conservation Union (IUCN), Switzerland

Saranraj P, Sujitha D (2015) Mangrove medicinal plants: a review. American- Eurasian. J Toxicol Sci 7(3):146–156

Sidhu SA (1963) Studies on the mangroves of India 1. East Godavari regio. Ind Forest 86:337–351

Spalding M, Blasco F, Field C (1997) World Mangrove Atlas. International Society for Mangrove Ecology, World Conservation Monitoring Centre, Cambridge, p 178

Srinivastava J, Lambert J, Vietmeyer N (1996) Medicinal plants: an expanding role in development. The World Bank, Washington, D.C., p 18

Twilley RR (1998) Mangrove wetlands. In: Messina MG, Conner WH (eds) Southern forested wetlands ecology and management. Lewis, Boca Raton, pp 445–473

Umamaheswara Rao M, Narasimha Rao GM (1988) Mangrove population of the Godavari delta complex. Indian J Man Sci 17:326–329

Vadlapudi V, Naidu K (2009a) Bioactivity of marine mangrove plant *Avicennia alba* on selected plant and oral pathogens. Int J ChemTech Res 1(4):1213–1216

Vadlapudi V, Naidu K (2009b) *In vitro* antimicrobial activity *Ceriops decandra* against selected aquatic, human and phytopathogens. Int J Chem Tech Res Vol 1(4):1236–1238

Vadlapudi V, Naidu K (2009c) *In vitro* antimicrobial potentiality of mangrove plant *Myriostachya wigtiana* against selected phytopathogens. Biomed Pharmacol J 2(2):235–238

Vadlapudi V, Naidu K (2009d) *In vitro* bioefficiency of marine mangrove plant activity of *Rhizophora conjugata*. Int J Pharm Tech Res 1(4):1598–1600

Vadlapudi V, Bobbarala V, Naidu K (2009) Comparative screening of selected mangrove plants methanolic extracts against clinical and plant pathogens. J Pharm Res 2(6):1062–1064

Valiela I, Bowen JL, York JK (2001) Mangrove forests: one of the world's threatened major tropical environments. Bio Sci 51:807–815

Tree Diversity Assessment in Sacred Groves of Eastern Ghats, Visakhapatnam District, Andhra Pradesh, India

6

Satyavathi Koda, S. B. Padal, and Sandhya Deepika Devara

Abstract

The phytosociological study provides information on the tree species diversity in sacred groves of Eastern Ghats, Visakhapatnam District, Andhra Pradesh, India. Tree diversity and its relation to habitat was explored using tree dataset of 8 belt transects (5 × 500 m) totaling 2 ha in the study area. A total of 67 plant taxa (≥15 cm gbh) belonging to 59 genera of 34 families with 233 individuals were recorded. The most dominant species were *Anogeissus latifolia*, *Xylia xylocarpa*, *Mangifera indica*, and *Lannea coromandelica*. The stand density ranged from 55 to 62 individuals 0.5 ha^{-1} while basal area varied from 1.50 to 5.01 m^2 0.5 ha^{-1}. Shannon-Weiner index (H′) ranged from 3.09 to 3.40 while Simpson's index from 0.95 to 0.96 and Menhinick's index ranged from 3.04 to 4.67.

Keywords

Tree diversity · IVI · Eastern Ghats · Visakhapatnam

6.1 Introduction

Biodiversity is the very basis of human survival and economic development (Singh et al. 1994) upon which communities, countries, and future generations depended (Elizabeth 1995). Plant biodiversity accounting for 0.40 million of species (Chapman 2009) play the important part in the cycle of nature.

Sacred Groves of India are forest fragments of varying sizes, which are communally protected, and which usually have a significant religious connotation for the protecting community. Hunting and logging are usually strictly prohibited within these patches. Sacred groves did not enjoy protection via federal legislation in India.

S. Koda · S. B. Padal (✉) · S. D. Devara
Department of Botany, Andhra University, Visakhapatnam, Andhra Pradesh, India

S. M. Khasim et al. (eds.), *Medicinal Plants: Biodiversity, Sustainable Utilization and Conservation*, https://doi.org/10.1007/978-981-15-1636-8_6

The growing threat of biodiversity loss in the world receives more attention from ecologists and conservationists who seek effective ways to conserve biodiversity. One of the approaches that have received great attention in the recent past is the role of traditional, cultural practices and beliefs in protecting and managing biodiversity (Byers et al. 2001; Infield 2001; Fabricius 2004; Berkes and Davidson 2006; Garnett et al. 2007; Gao et al. 2013; Kandari et al. 2014).

Tropical plant diversity assessment is a tool for the quantitative studies of regional scale biogeographical patterns (Gordon and Newton 2006). International conservation organizations have conducted a variety of biodiversity assessment in global and regional scales for identifying priority areas where only limited resources are available for conservation (Margules et al. 2002). Tropical deciduous forests occur under varied climatic conditions, but essentially with alternate wet and dry tropics (Naidu and Kumar 2015).

Quantitative information on composition, distribution, and abundance of woody species is of key significance to understanding the form and structure of a forest community and also for planning and implementation of conservation strategy of the community (Singh et al. 2016).

Assessment of forest community composition and structure is very helpful in understanding the status of tree population, regeneration, and diversity for conservation purposes. Therefore, the biodiversity-rich sacred groves are of immense ecological significance. They also play an important role in the conservation of flora and fauna. Keeping in view the role of the sacred groves as treasure of repositories of variety of tree species, the present study is conducted with objectives to find out the tree diversity and to carry out inventory of trees in the study area.

The Eastern Ghats are located along the Peninsular India extending over 1750 km under 11°03′ to 22°03″N latitudes and 77°02′ to 87°02′E longitudes covering 2,24,290 km^2. In Andhra Pradesh, the hilly region of Eastern Ghats is divided into Northern, Central, and Southern Ghats. The Northern Eastern Ghats portion covers the districts of Srikakulam, Vizianagaram, Visakhapatnam, East Godavari, West Godavari, and Khamam. Forests are endowed with rich, varied, and endemic flora. The climate of Eastern Ghats is typically tropical enough to support the most luxuriant type of vegetation and maintain rich biodiversity. Modhakondamma Sacred Grove (MK) is situated in Paderu Mandal, Visakhapatnam District, Andhra Pradesh, which is 111 km away from Visakhapatnam town. It lies between 18° 1′11.91″ North latitude and 82°42′37.96″ East longitude and the elevation of the grove is 484 M. Bodakondamma Sacred Grove (BK) is situated in Chintapalli Mandal, Visakhapatnam District, Andhra Pradesh. It lies between 17°47′7.05″ North latitude and 82°30′52.70″ East longitude. Darakondamma Sacred Grove (DK) is situated in Gudem Mandal, Visakhapatnam District, Andhra Pradesh, which is 170 km away from Visakhapatnam town. Downuru Sacred Grove (DN) is situated in Koyyuru Mandal, Visakhapatnam District, Andhra Pradesh, which is 110 km away from Visakhapatnam town. It lies between 17°44′59.72″ North latitude and 82°31′16.20″ East longitude (Fig. 6.1).

For the determination of biodiversity, methodology was adopted from National Bioresource Development Board, Department of Biotechnology, Government of

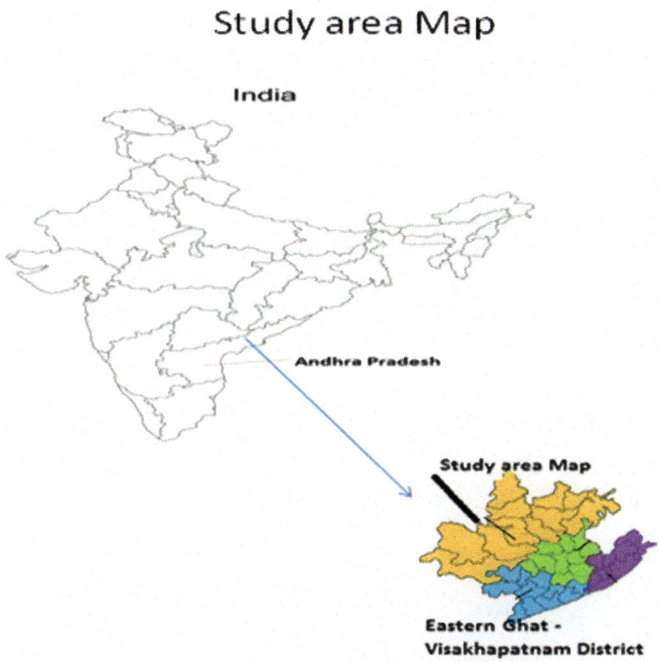

Fig. 6.1 Map of Study area Eastern Ghats of Visakhapatnam District

India, to quantify plant resources of India. In all the four study sites, two belt transects of 5 × 500 m size (totaling 0.5-ha) were laid in each site during 2014–2016 and all live trees with ≥15 cm girth at breast height (gbh) were enumerated. Based on the individuals recorded in the discrete plot samples, vegetation data were quantitatively analyzed for basal area, relative density, relative frequency, and Importance Value Index (IVI). The representative taxa were collected and identified with the help of floras (Pullaiah and Sri Rama Murthy 2001; Pullaiah et al. 2007) and made into herbarium. The voucher specimens were housed in the Botany Department Herbarium (BDH), Department of Botany, Andhra University, Visakhapatnam. Some of the plant photographs were provided for easy identification (Figs. 6.2 and 6.3).

6.2 Tree Diversity in Sacred Groves of Visakhapatnam District, Eastern Ghats of India

6.2.1 Species Richness and Diversity

A total of 67 angiospermous plant taxa of ≥15 cm gbh that belong to 57 genera and 28 families were recorded according to APG III System of Classification. Plot-wise species richness was 35, 30, 28, and 24 in MK, BK, DK, and DN, respectively, with

Plant Photos

Fig. 6.2 Photo plate 1

a major difference between the plots (Table 6.1). The families showing the most diversity in terms of the number of species were the Fabaceae (8), Moraceae (6), Malvaceae and Combretaceae with (5), Rubiaceae, Phyllanthaceae, and Anacardiaceae with (4), Lamiaceae and Meliaceae with (3), Oliaceae, Dilleniaceae, and Burseraceae with (2), and remaining 15 families were represented by single species.

Fig. 6.3 Photo plate 2

The high value of Shannon index in the site MK with 3.408 and lowest in the site BK with 3.206 were obtained. The high value indicates high tree diversity. Among them, site MK is the most species-rich plot with 35 species. Further, the species richness range of 24–35 species indicates the diverse nature of the forests and tree composition in these sacred groves. The Simpson's index was highest in site MK and lowest in site DN. The Evenness index revealed DN as the most diverse site and Menhinick's index was highest in site MK (Table 6.1).

Table 6.1 Summary of tree inventory (≥15 cm) in the four sites of sacred groves

Variable	MK	BK	DK	DN
Species richness ≥15 cm gbh	35	30	28	24
No. of genera	30	27	25	24
No. of families	19	20	16	20
Diversity indices				
Shannon	3.408	3.206	3.225	3.094
Simson_1-D	0.9611	0.9517	0.9567	0.9516
Evenness_e^H/S	0.863	0.8227	0.8988	0.9195
Menhinick's	4.677	3.873	3.776	3.048
Density (stems ha^{-1})	**56**	**60**	**55**	**62**
Basal area (m^2 ha^{-1}) ≥15 cm gbh	**5.01729**	**3.66952**	**4.10343**	**1.50109**
Elevation (m)	484	615	717	240
Latitude (N)	18° 1′11.91″N	17°47′7.05″N	17°58′52.86″N	17°44′59.72″N
Longitude (E)	82°42′37.96″E	82°30′52.70″E	82° 7′56.39″E	82°31′16.20″E

Density (Stems ha^{-ha})

6.2.2 Stand Density and Basal Cover

The total stand density of trees for the four sites of study area was 233 individuals. The highest stand density was observed in DN (62 individuals' ha^{-1}), whereas the lowest stand density was in DK (55 individuals' ha^{-1}). The other two sites had moderate stand densities. The distribution of the basal area across four sites, using gbh interval classes, revealed the dominance of smallest individuals in all the sites. Basal area in all the study sites ranges from 1.50 m^2 ha^{-1} (site DN) to 5.01 m^2 ha^{-1} (Fig. 6.4).

6.2.3 Girth Class Distribution

Tree species richness as well as density decreased with increasing girth class in all the study plots. The girth class distribution has revealed that majority of tree individuals were represented in 31–60 cm class and less abundance in >120 cm girth class. The basal area values ranged between 1.05 and 5.01 m^2 ha^{-1}. In all the four sites except DN, the higher gbh class recorded higher basal values. The distribution of the basal area across the study sites using gbh interval classes reveals the dominance of small stemmed individuals in the sites.

6.2.4 Tree Dominance and Rarity

The density of different species varied widely in the four study sites. Based on their density in 0.5-ha sites, the predominant species group represented with 50% of total individuals. The most dominant species were *Anogeissus latifolia*, *Xylia xylocarpa*,

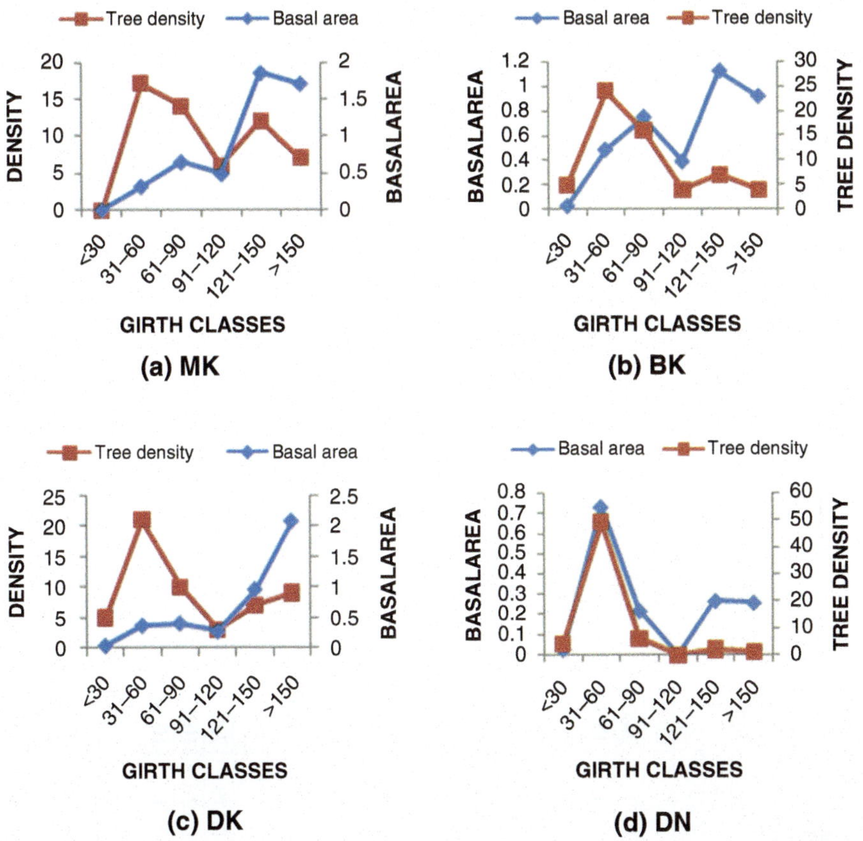

Fig. 6.4 (a–d) Contribution of tree species stands density and basal area based on girth class distribution in sacred groves of Eastern Ghats. *MK* Modhakondamma Sacred Grove, *BK* Bodakondamma Sacred Grove, *DK* Darakondamma Sacred Grove, *DN* Downuru Sacred Grove

Cleistanthus collinus, and *Lannea coromandelica.* Common species represented with 20% individuals. Rare species were such as *Bischofia javanica, Trichilia connaroides, Walsura trifoliolata,* and *Syzygium cumini.*

6.2.5 Importance Value Index (IVI)

Analyzing the IVI for the individuals in each site, a distinct pattern can be observed, which is present in all study sites. Top ten species accounted for 47.9% in MK; *Mangifera indica* was the dominant species with 24.17 IVI. The codominant species were *Anogeissus latifolia, Xylia xylocarpa,* and *Schleichera oleosa.* In BK, top ten species contributed 33% and *Mangifera indica* was the dominant species with 25.98 IVI. The codominant species were *Xylia xylocarpa, Anogeissus latifolia,* and *Haldina cordifolia. Anogeissus latifolia* was the dominant tree in site DK with 21.03

Table 6.2 Density, basal area, and importance value index (IVI) of the tree species in four 0.5-ha sites

S. No.	Name of the plants	Family	MK			BK			DK			DN		
			D	BA	IVI	D	BA	IVI	D	BA	IVI	D	BA	IVI
1	*Aegle marmelos*	Rutaceae	0.5	0.009	4.063	0.5	0.21	4.71	1.5	0.04	10.93	–	–	–
2	*Alangium salvifolium*	Alangiaceae	–	–	–	–	–	–	–	–	–	2	0.04	14.07
3	*Albizia odoratissima*	Fabaceae	–	–	–	0.5	0.22	10.05	0.5	0.3	11.41	2	0.26	25.67
4	*Anogeissus acuminata*	Combretaceae	–	–	–	0.5	0.02	4.66	0.5	0.02	4.62	–	–	–
5	*Anogeissus latifolia*	Combretaceae	1.5	0.72	24.01	2	0.27	18.57	1.5	0.38	21.03	2	0.01	9.17
6	*Artocarpus heterophyllus*	Moraceae	1	0.28	13.34	0.5	0.25	10.8	1.5	0.45	20.99	–	–	–
7	*Atalantia monophylla*	Rutaceae	–	–	–	–	–	–	–	–	–	0.5	0.02	5.6
8	*Bischofia javanica*	Phyllanthaceae	0.5	0.2	7.92	–	–	–	–	–	–	–	–	–
9	*Bombax ceiba*	Malvaceae	0.5	0.3	9.89	0.5	0.17	8.62	0.5	0.25	10.32	0.5	0.25	21.11
10	*Buchanania lanzan*	Anacardiaceae	0.5	0.0127	4.12	–	–	–	–	–	–	1.5	0.03	11.59
11	*Callicarpa tomentosa*	Lamiaceae	–	–	–	0.5	0.06	5.69	–	–	–	–	–	–
12	*Canthium dicoccum*	Rubiaceae	–	–	–	–	–	–	1.5	0.04	10.91	–	–	–
13	*Careya arborea*	Lecythidaceae	–	–	–	–	–	–	–	–	–	1.5	0.09	16.04
14	*Caryota urens*	Arecaceae	–	–	–	0.5	0.16	8.49	–	–	–	–	–	–
15	*Cassia fistula*	Fabaceae	0.5	0.0286	4.43	2.5	0.09	15.44	1.5	0.03	10.82	–	–	–
16	*Chloroxylon swietenia*	Rutaceae	–	–	–	–	–	–	–	–	–	2.5	0.12	21.19
17	*Cleistanthus collinus*	Phyllanthaceae	–	–	–	–	–	–	1	0.03	9	1.5	0.04	12.16
18	*Cochlospermum religiosum*	Bixaceae	–	–	–	–	–	–	–	–	–	–	–	–
19	*Dalbergia lanceolaria*	Fabaceae	0.5	0.179	7.43	–	–	–	0.5	0.13	7.31	–	–	–
20	*Dillenia indica*	Dilleniaceae	1	0.12	10.13	–	–	–	–	–	–	–	–	–
21	*Dillenia pentagyna*	Dilleniaceae	1	0.14	10.54	0.5	0.16	8.49	–	–	–	–	–	–
22	*Diospyros sylvatica*	Ebenaceae	–	–	–	–	–	–	–	–	–	1.5	0.02	10.98
23	*Erythrina suberosa*	Fabaceae	0.5	0.02	4.43	–	–	–	–	–	–	–	–	–
24	*Erythrina variegata*	Fabaceae	–	–		–	–	–	1	0.07	9.99	–	–	--
25	*Ficus auriculata*	Moraceae	1.5	0.04	10.49	–	–	–	–	–		–	–	–
26	*Ficus exasperata*	Moraceae	0.5	0.03	4.64	1	0.21	13.82	1	0.4	18.01	–	–	–

S. No.	Name of the plants	Family	MK			BK			DK			DN		
			D	BA	IVI	D	BA	IVI	D	BA	IVI	D	BA	IVI
27	*Ficus microcarpa*	Moraceae	0.5	0.03	4.53	–	–	–	–	–	–	–	–	–
28	*Ficus semicordata*	Moraceae	1	0.02	8.31	0.5	0.03	4.91	1	0.03	8.84	1	0.03	9.93
29	*Firmiana colorata*	Malvaceae	–	–	–	–	–	–	0.5	0.13	4.91	–	–	–
30	*Gardenia latifolia*	Rubiaceae	–	–	–	--	–	–	0.5	0.03	4.62	–	–	–
31	*Garuga pinnata*	Burseraceae	0.5	0.05	4.88	–	–	–	–	–	–	1.5	0.05	10.9
32	*Glochidion velutinum*	Phyllanthaceae	0.5	0.009	4.06	0.5	0.008	4.17	0.5	0.02	16.86	–	–	–
33	*Gmelina arborea*	Lamiaceae	0.5	0.12	6.34	1	0.04	8.99	1	0.36	10.11	–	–	–
34	*Grewia eriocarpa*	Malvaceae	–	–	–	–	–	–	–	–	–	1.5	0.05	12.95
35	*Haldina cordifolia*	Rubiaceae	1	0.25	12.73	1	0.09	10.57	0.5	0.24	5.13	0.5	0.03	6.38
36	*Holoptelea integrifolia*	Moraceae	–	–	–	–	–		0.5	0.04	8.49	–	–	–
37	*Hymenodictyon orixense*	Rubiaceae	0.5	0.02	4.21	–	–	–	–	–	–	–	–	–
38	*Heynea trijuga*	Meliaceae	2	0.08	9.39	–	–	–	–	–	–	–	–	–
39	*Kydia calycina*	Malvaceae	–	–	–	–	–	–	–	–	–	1	0.01	8.92
40	*Lagerstroemia parviflora*	Lythraceae	0.5	0.01	4.21	0.5	0.01	12.35	–	–	–	–		
41	*Lannea coromandelica*	Anacardiaceae	–	–	–	1.5	0.1	12.35	1	0.01	10.48	1	0.02	9.66
42	*Macaranga peltata*	Euphorbiaceae	1	0.04	8.57	–	–	–	1.5	0.02	11.89	–	–	–
43	*Madhuca longifolia*	Sapotaceae	–	–	–	1.5	0.12	12.93	–	–	–	0.5	0.01	5.01
44	*Mangifera indica*	Anacardiaceae	2	0.64	24.17	1.5	0.6	25.98	–	–	–	–	–	–
45	*Magnolia champaca*	Magnoliaceae	0.5	0.07	5.03	–	–	–	–	–	–	–	–	–
46	*Nyctanthes arbor-tristis*	Oleaceae	–	–	–	1.5	0.01	9.87	–	–	–	–	–	–
47	*Oroxylum indicum*	Bignoniaceae	0.5	0.01	4.21	–	–	–	–	–	–	2	0.05	14.68
48	*Phyllanthus emblica*	Phyllanthaceae	–	–	–	–	–	–	1.5	0.08	5.79	1.5	0.03	11.93
49	*Premna tomentosa*	Lamiaceae	–	–	–	–	–	–	0.5	0.07	10.66	–	–	–
50	*Protium serratum*	Burseraceae	0.5	0.03	4.58	–	–	–	1	0.1	24.14	–	–	–
51	*Pterocarpus marsupium*	Fabaceae	0.5	0.15	6.97	–	–	–	–	–	–	–	–	–

(continued)

Table 6.2 (continued)

S. No.	Name of the plants	Family	MK			BK			DK			DN		
			D	BA	IVI	D	BA	IVI	D	BA	IVI	D	BA	IVI
52	*Pterospermum xylocarpum*	Sterculiaceae	–	–		–	–	–	–	–	–	1.5	0.08	14.93
53	*Schleichera oleosa*	Sapindaceae	1	0.3	13.75	–	–	–	1.5	0.58	9.06	–	–	–
54	*Schrebera swietenioides*	Oleaceae	–	–	–	0.5	0.03	4.85	–	–	–	–	–	–
55	*Semecarpus anacardium*	Anacardiaceae	–	–	–	–	–	–	–	–	–	1.5	0.05	12.98
56	*Sterculia villosa*	Malvaceae	–	–	–	0.5	0.01	4.39	–	–	–	–	–	–
57	*Strychnos nux-vomica*	Loganiaceae	–	–	–	–	–	–	–	–	–	1	0.02	9.79
58	*Syzygium cumini*	Myrtaceae	0.5	0.14	6.71	–	–	–	–	–	–	–	–	–
59	*Tamarindus indica*	Fabaceae	–	–	–	1	0.23	14.38	–	–	–	–	–	–
60	*Terminalia alata*	Combretaceae	1.5	0.11	11.73	0.5	0.04	5.15	1	0.03	9.02	–	–	–
61	*Terminalia arjuna*	Combretaceae		–	–	–	–	–	–	–	–	–	–	–
62	*Terminalia bellirica*	Combretaceae	0.5	0.02	4.36	1	0.11	11.13	0.5	0.02	4.73	–	–	–
63	*Toona ciliata*	Meliaceae	0.5	0.16	7.2	–	–	–	–	–	–	–	–	–
64	*Walsura trifoliolata*	Meliaceae	0.5	0.06	5.06	0.5	0.03	4.97	–	–	–	–	–	–
65	*Wrightia tinctoria*	Apocynaceae	–	–	–	2	0.06	12.93	–	–	–	1	0.01	9.06
66	*Xylia xylocarpa*	Fabaceae	2.5	0.49	22.96	3	0.22	20.75	1.5	0.06	11.54	2	0.06	15.19
67	*Zanthoxylum armatum*	Rutaceae	–	–	–	0.5	0.01	4.28	–	–	–	–	–	–
	Grand total		13	2.3084	300	11.5	1.898	300	16	2.71	300	18	0.96	300

MK Modhakondamma Sacred Grove, *BK* Bodakondamma Sacred Grove, *DK* Darakondamma Sacred Grove, *DN* Downuru Sacred Grove, *D* Density, *BA* Basal area, *IVI* Important Value Index

∗value <0.001

IVI and top ten species accounted for 25.59% of the total IVI. The codominant species were *Artocarpus heterophyllus*, *Dillenia pentagyna*, and *Xylia xylocarpa*. In site DN, the dominant species was *Xylia xylocarpa* (15.19 IVI) and the codominants were *Artocarpus heterophyllus* and *Haldina cordifolia* (Table 6.2).

Tree species richness at defined study sites and in minimum diameter classes gives a reliable instrument to indicate the diversity level of a study site (Premavani et al. 2014). The present data can be compared with the large number of similar plots inventoried in India and elsewhere in the tropics. A total of 135 species belonging to 105 genera of 45 families were recorded in the study area. Species richness ranging from 46 to 82 ha^{-1} with mean value of 62 ha^{-1} was recorded which is very distantly close to 64–82 species ha^{-1} in Sengaltheri-Kakachi (Parthasarathy 2001), 25–61 species ha^{-1} in Saddle peak of North Andaman Islands (Tripathi et al. 2004), and 18–84 ha^{-1} in Little Andaman Island (Rasingam and Parathasarathy 2009).

The mean value in the present study is higher than that of 21 species ha^{-1} in Kolli hills of India (Chittibabu and Parthasarathy 2000), 31 species ha^{-1} in dry deciduous forests of western India (Kumar et al. 2010), 37 species ha^{-1} in Kaan forest of Western Ghats (Gunaga et al. 2013), 38 species ha^{-1} in reserved forests of southern Eastern Ghats of Andhra Pradesh (Rao et al. 2011), and 58–59 species ha^{-1} in Great Andaman (Padalia et al. 2004).

Basal area of a tree is the girth occupied at breast height (gbh) and it is an important attribute to quantify the vegetation structure and site quality (Suthari 2013).

Girth class frequency showed population structure of trees exhibited in the study sites are in conformity with other forest stands (Sahu et al. 2012). Tree density distribution across different girth classes indicates how well the growing forest is utilizing site resources.

The most dominant tree species in the study area were *Anogeissus latifolia*, *Xylia xylocarpa*, *Cleistanthus collinus*, and *Lannea coromandelica*. Pragasan and Parthasarathy (2010) reported that the dominant species in the southern Eastern Ghats were *Albizia amara*, *Euphorbia antiquorum*, *Canthium dicoccum* var. *dicoccum*, *Memecylon edule*, and *Chloroxylon swietenia*, whereas in northern Eastern Ghats, *Shorea robusta*, *Lannea coromandelica*, *Madhuca indica*, and *Diospyros melanoxylon* (Panda et al. 2013).

According to the present study, the number of species were Moraceae, Fabaceae, and Combretaceae with (5), Rubiaceae, Euphorbiaceae, and Anacardiaceae with (4), Verbenaceae, Sterculiaceae, Rutaceae, and Meliaceae with (3), Dilleniaceae, Caesalpiniaceae, and Burseraceae with (2), and 22 families were represented by single species. Interestingly, similar findings were reported by Padalia et al. (2004) where Euphorbiaceae and Rubiaceae were the most dominant family in all forest types, except mangrove. Sandhyarani et al. (2007) resulted that Euphorbiaceae is the dominant family followed by Moraceae and Lauraceae in the Eastern Ghats.

6.3 Conclusion

This study will serve as a primary input towards maintaining and sustaining the phytodiversity and also would help in understanding the threats that are faced by the tropical forests and would help in driving conservation policies. The study revealed that, main driving force behind the disturbance and degradation of the trees occurs due to human activities. Also the local people are encouraged to grow indigenous tree species plantation. There is an urgent need for recognizing these traditionally valued natural systems at various levels and planning for their better management, ultimately aiming to conserve biodiversity.

Acknowledgements Authors are thankful to the Head, Department of Botany, Andhra University, for providing necessary facilities for the study. K. Satyavathi (Author) is also grateful to UGC, New Delhi, for providing financial support under Junior Research Fellowship scheme.

References

Berkes F, Davidson IJH (2006) Biodiversity, traditional management systems, and cultural landscapes: examples from the boreal forest of Canada. Int Soc Sci J 58:35–47

Byers BA, Cunliffe RN, Hudak AT (2001) Linking the conservation of culture and nature: a case study of sacred forests in Zimbabwe. Hum Ecol 29:187–218

Chapman AD (2009) Numbers of living species in Australia and the world, 2nd edn. Australian Biological Resources Study, Canberra

Chittibabu CV, Parthasarathy N (2000) Attenuated tree species diversity in human-impacted tropical evergreen forest sites at Kolli hills, Eastern Ghats, India. Biodiver Conserv 9:1493–1519

Elizabeth D (1995) Global biodiversity assessment. UNEP, CUP, UK

Fabricius C (2004) Rights, resources and rural development: community-based natural resource management in Southern Africa. Earthscan, London

Gao H, Ouyang Z, Chen S, Koppen CSA (2013) Role of culturally protected forests in biodiversity conservation in Southeast China. Biodivers Conserv 22:531–544

Garnett ST, Sayer J, DuToit J (2007) Improving the effectiveness of interventions to balance conservation and development: a conceptual framework. Ecol Soc 12:2

Gordon JE, Newton AC (2006) The potential misapplication of rapid plant diversity assessment in tropical conservation. J Nat Conserv 14:117–126

Gunaga S, Rajeswari N, Vasudeva R (2013) Tree diversity and disturbance of *Kaan* forests: relics of a community protected climax vegetation in the Central Western Ghats. Trop Ecol 54:117–131

Infield M (2001) Cultural values: a forgotten strategy for building community support for protected areas in Africa. Conserv Biol 15:800–802

Kandari LS, Bisht VK, Bhardwaj M, Thakur AK (2014) Conservation and management of sacred groves, myths and beliefs of tribal communities: a case study from North-India. Environ Syst Res 3:16

Kumar JIN, Kumar RN, Bhoi RK, Sajish PR (2010) Tree species diversity and soil nutrient status in three sites of tropical dry deciduous forest of western India. Trop Ecol 51:273–279

Margules CR, Pressey RL, Williams PH (2002) Representing biodiversity: data and procedures for identifying priority areas for conservation. J Biosci 27(4):309–326

Naidu MT, Kumar OA (2015) Tree species diversity in the Eastern Ghats of northern Andhra Pradesh, India. J Threat Taxa 7(8):7443–7459

Padalia H, Chauhan N, Porwal MC, Roy PS (2004) Phytosociological observations on tree species density of Andaman Islands, India. Curr Sci 87(6):799–806

Panda PC, Mahapatra AK, Acharya PK, Debata AK (2013) Plant diversity in tropical deciduous forests of eastern Ghats, India: a landscape level assessment. Int J Biodiv Conserv 5:625–639

Parthasarathy N (2001) Changes in forest composition and structure in three sites of tropical evergreen forests around Sengaltheri. West Ghats Curr Sci 80:389–393

Pragasan LA, Parthasarathy N (2010) Landscape-level tree diversity assessment in tropical forests of southern eastern Ghats, India. Flora Morphol Distri Funct Ecol Plants 205:728–737

Premavani D, Naidu MT, Venkaiah M (2014) Tree species diversity and population structure in the tropical forests of north central Eastern Ghats, India. Notulae Scientia Biologicae 6(4):448–453

Pullaiah T, Sri Rama Murthy K (2001) Flora of eastern Ghats. In: Leguminosae, vol 2. Regency Publications, New Delhi

Pullaiah T, Sri Rama Murthy K, Karuppusamy S (2007) Flora of eastern Ghats, India. In: Rosaceae—Asclepiadaceae, vol 3. Regency Publications, New Delhi

Rao BRP, Babu MVS, Reddy MS, Reddy AM, Rao VS, Sunitha S, Ganeshaiah KN (2011) Sacred groves in southern eastern Ghats, India: are they better managed than forest reserves? Trop Ecol 52:79–90

Rasingam L, Parathasarathy N (2009) Tree species diversity and population structure across major forest formations and disturbance categories in little Andaman Island. India Trop Ecol 50(1):89–102

Sahu SC, Dhal NK, Mohanty RC (2012) Tree species diversity and soil nutrient status in a tropical sacred forest ecosystem on Niyamgiri hill range, eastern Ghats, India. Trop Ecol 53:163–168

Sandhyarani S, Murth KSR, Pullaiah T (2007) Tree flora in eastern Ghats of southern peninsular India. Res J Bot 2(4):176–185

Singh JS, Raghubanshi AS, Varshney CK (1994) Integrated biodiversity research in India. Curr Sci 66(70):109

Singh SS, Malik ZA, Sharma CM (2016) Tree species richness, diversity, and regeneration status in different oak (Quercus spp.) dominated forests of Garhwal Himalaya, India. J Asia Pacif Biodiv 9:293–300

Suthari S (2013) Biodiversity characterization and aboveground vegetation carbon pool assessment in northern Telangana at landscape level using geospatial technique (PhD thesis). Department of Botany, Kakatiya University, Warangal

Tripathi KP, Tripathi S, Selvan T, Kumar K, Singh KK, Mehrotra S, Pushpangadan P (2004) Community structure and species diversity of saddle peak forest in Andaman Island. Trop Ecol 45:241–250

Part II

Ethnobotany and Ethnomedicine

Structure Design and Establishment of Database Application System for Miao Medicinal Plants in Guizhou Province, China

7

Sizhao Liu, Wenqiang Liu, Beixi Zhang, Qiyi Lei, Jiangju Zhou, and Chunlin Long

Abstract

Qiandongnan Miao and Dong Autonomous Prefecture, located in southeast Guizhou, China, harbors rich biodiversity. As one of the indigenous minorities, the Miao people rely on medicinal plants for self-medication. However, during this era most traditional knowledge is facing the danger of disappearing and biodiversity is being lost. This has created a method to realize the information dissemination, the application of computerized databases. Database system establishing and data mining of traditional medicinal plants are also beneficial for ethnomedicinal researches. Thus, on the basis of top-level design, an ethnobotanical research focused on Miao medicinal plants in Guizhou was conducted during 2016 and 2017. Their names, parts used, and values were documented. By using Microsoft Windows 2003 server, B/S system structure, Microsoft SQL Server, and Microsoft Visual Studio 2005 Network search system, an open database system was designed and established. A total of five modules were shown on the home page, including Homepage, Species identification, Data retrieval, Plant lists, and User Center. All users could click on the corresponding module

S. Liu · B. Zhang
College of Life and Environmental Sciences, Minzu University of China, Beijing, China

College of Life and Environmental Sciences, Kaili University, Guizhou, China

W. Liu
College of Information Engineering, Minzu University of China, Beijing, China

Q. Lei · J. Zhou
College of Life and Environmental Sciences, Kaili University, Guizhou, China

C. Long (✉)
College of Life and Environmental Sciences, Minzu University of China, Beijing, China

Key Laboratory of Ethnomedicine, Minzu University of China, Ministry of Education, Beijing, China
e-mail: long.chunlin@muc.edu.cn

S. M. Khasim et al. (eds.), *Medicinal Plants: Biodiversity, Sustainable Utilization and Conservation*, https://doi.org/10.1007/978-981-15-1636-8_7

to get the information according to their own requirements. This new database system was designed to support a project evaluating the biodiversity of Miao medicinal plants. The software used to create the database is open source freely available which will be convenient for inquiring and statistics, and also could potentially be applied to other ethnobotanically driven natural product collection and drug discovery programs.

Keywords
Database · Ethnomedicine · Ethnobotany · Miao medicinal plants

7.1 Introduction

With emerging modern science and technology, the number of database systems based on different purposes is rapidly increasing, which makes it easy for users to access information and expand knowledge (Smith et al. 2000, 2009; Pilon et al. 2017; Torres et al. 2006). Nowadays, information stored in electronic medium has remarkably increased (Obayashi et al. 2018; Thomas 2003) with medicinal plant database emerging as one of the trends. In terms of ethnomedicinal database, of high practicability mainly serving medical sanitary, medicinal plants and other specialized fields (Hrudayanath and Jayanta 2011; Dighe et al. 2010; Ma 2014). In China, "Traditional Chinese Medical Database System," "China Ethnomedicine Information Service System," and "Scientific Database of China Plant Species" are the most common databases in ethnomedicinal fields (Luo et al. 2017).

Ethnomedicine is a study or comparison of the traditional medicine practiced by ethnic groups, and especially by indigenous peoples. Documenting indigenous medical knowledge and scientific study of indigenous medicines and sharing this knowledge are some of the priorities for the current ethnomedicinal research (Tuasha et al. 2018; Heinrich et al. 2009; Leonti 2011). Also, application of computerized databases has supplemented new methods for information dissemination.

Computerized database can provide efficient solution for data management. Database can also contribute towards ethnomedicinal research by facilitating the process of analysis and interpretations (Afendi et al. 2012). Moreover, web-based databases facilitate knowledge transmission and feed the circle of information exchange between the ethnomedicinal studies and public audience (Mohanraj et al. 2018). With the growing popularity of Internet this becomes more significant also for healthcare systems.

Due to this, an ethnobotanical research focused on Miao medicinal plants in Guizhou was conducted during 2016 and 2017. Their names, parts used, values, and other information were documented. By using Microsoft windows 2003 server, B/S system structure, Microsoft SQL Server, and Microsoft visual studio 2005 Network search system, an open database system was designed and established. Thus, the purposes of the present work were: (1) to document and digitize the wealth of information contained within traditional Miao medicinal knowledge at the study area;

(2) to provide a query platform for users to query information easily and accurately, which can achieve resource sharing and inheriting real sense; and (3) to provide valuable clues for future development and also give comprehensive and scientific guidance for local people to consume the medicinal herbs in a safe range.

7.2 Database Structure

7.2.1 Demand Analysis

Qiandongnan Miao and Dong Autonomous Prefecture, located in southeast Guizhou, China, harbors rich biodiversity and has specific karst topography. As one of the indigenous minorities, Miao is mainly living in this area. Most of them rely on medicinal plants for self-medication. These traditional medicinal plants play an important role of protecting people's lives and health, especially in remote and poor areas. Compared with high level of consumption in hospital treatment, it is low-cost for easily accessing raw materials, even in modern times. Therefore, Miao ethnic group has its own medicinal characteristic and has various experiences to use medicinal knowledge.

However, from the perspective of traditional Miao medicine, there have been relatively few efforts to build online databases that include medicinal plants, their therapeutic uses, and value. If the situation of Miao medicinal resources is not yet clear, disorder and overuse will lead to ecological destruction and biodiversity loss. Especially during this era when most traditional knowledge is facing the danger of disappearing, Miao medicine should be taken more care. Thus, based on investigations and literature, the database of Miao medicinal plants was established. It will provide scientific evidence for the research and utilization of Miao medicinal plant, and also provide a medical plant information inquiry service for scientific research, medicine, teaching, and individual.

7.2.2 Data Sources of Medicinal Plants

Based on the approaches of ethnobotany and taxonomy, investigations were carried out during 2016–2017. A total of 385 medicinal plant species used by Miao people were identified, which belong to 101 families and 364 genera. Their names, parts used, and values were documented. In addition, we also reviewed *Flora of China*, *Flora of Guizhou*, Web of Science, VIP database, and Scientific Database of China Plant Species for perfecting and supplementing Miao medicinal information.

7.3 Web Development Platform

7.3.1 Software and Hardware Configuration

To construct this database, the compiled and curated data was integrated using MySQL (https://www.mysql.com/), a relational database management system which serves as a back end for our resource (Afendi et al. 2012). For operating system and service offering Microsoft Windows Server 2003, a Datacenter Edition, was used which supports mission-critical database application. Visual Studio .NET used to provide the reliable development environment includes a set of default device configuration files (Ono et al. 2017).

7.3.2 Development of Technology

In the B/S architecture system (Fig. 7.1), the user through a browser on the network to the distribution of many of the server request, the server on the browser to process the request, the user information needed to return to the browser (Lebo et al. 2017). The application is described with C#, which is a simple, modern, user-oriented, and secure programming language. As a part of the Microsoft Visual Studio 2005, providing the access of the Microsoft. NET platform for users.

7.3.3 Framework Database

In order to build a comprehensive repository for Miao medicinal plants, the data based on traditional information including name, habitat, pesticide effect, medicinal parts, dosage, and administration are shown in Fig. 7.2.

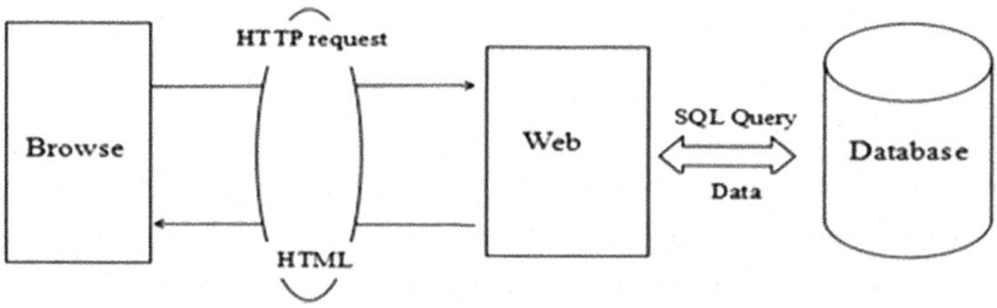

Fig. 7.1 B/S three-tier structure

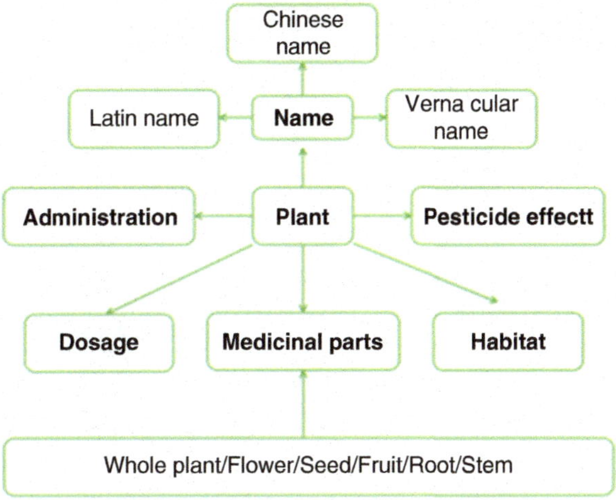

Fig. 7.2 The connection of different types of data

Fig. 7.3 Home page of the database system

7.4 Implementation of the Database System

7.4.1 Homepage

Five modules are seen on the home page (Fig. 7.3), including Homepage, Species identification, Data retrieval, Plant lists, and User center. All users could click on the corresponding module to get the information according to their own requirements. The design of page mainly uses white as background, meanwhile the latest uploaded pictures of Miao medicinal plants are displayed on the home page in a dynamic form.

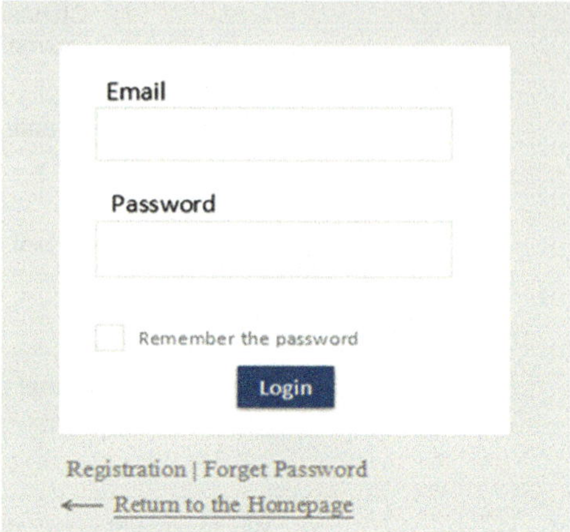

Fig. 7.4 Logon screen of the species identification

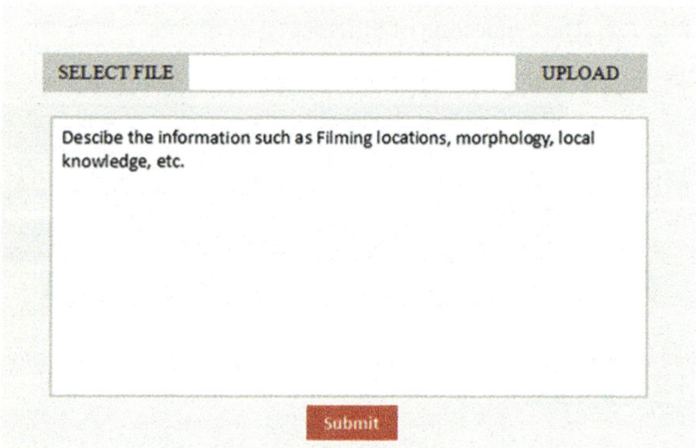

Fig. 7.5 Uploaded screen of the species identification

7.4.2 Species Identification

The function of this module is the user's application for identifying species which is not recognized. When the user accesses the module, the page will jump to the Logon Screen (Fig. 7.4). After the login, users shall identify the species.

The main steps of species identification include: (1) This gives the user a button for browsing the local file system and selecting an image to be uploaded, as shown in Fig. 7.5; (2) Describe the upload image, such as photographing place and morphological characteristics; (3) Click the Submit button, after the image has successfully been uploaded, the order is sent to system for review; (4) The identification results are displayed in the User center-My identification. The catch: images might be no more than 3 M, common supported formats include jpg. and png. Please allow 1–2 business days to see the audit results; if the picture does not show clearly or lack information as listed, the application will not be considered.

7.4.3 Data Retrieval

Information retrieval system is an important part of database. The accuracy and availability of retrieval system are related to the availability of database system (Pathania et al. 2015). Thus, database design with various factors should be adopted, such as the internal structure of the database and the configuration of software and hardware for users.

The web-interface of the database enables users to query for each of these associations using (a) scientific names of plants, (b) Chinese name, (c) therapeutic uses, and (d) medicinal parts. Each search will launch a more detailed range of search prompts by category, which attempts to narrow down the user's choices. The search will populate the middle pane with the results depending on category selected. Clicking on View button leads to more detail towards medicinal plants (Fig. 7.6). In addition, when browsing this page, at the top of right-hand side, "Print" and "Add to favorite" will emerge, medicinal plants can be added to favorite, users could go back to User center and find the medicinal plants they had bookmarked. At the bottom, allowing users to leave a comment below.

7.4.4 Plant Lists

This module offers a summary table for Family (Fig. 7.7). Users should be able to drill down from the summary of Family table into Genus table which belongs to Family selected, ultimately arriving at the table of species (If that is the level of detail required). Then click View button that leads to more detail towards medicinal plants.

***Gastrodia elata* Bl.**

Chinese name: Tian ma

Miao name: yangf wid vud

Orchidaceae Gastrdia

Using part: Tuber

Effect: Dizziness; Numberss of the limbs; Infantile convulsion

Usage: Grinding, decoction

Fig. 7.6 The information of medicinal plants

Fig. 7.7 A summary table of medicinal plants

7.4.5 User Center

User center is a design philosophy in which users, their needs, interests, and behavior define the foundation of website in terms of site structure and obtaining the information. This module supports a variety of features, including My identification, Favorite, My reply, and Account management.

1. After they have uploaded images, users can follow the progress on "My identification." A dialog box displays: your application has been queued for review by site administrators and will be identified after approval. The result might take a couple of days for manager's confirming. When identification has been finished you will see a dialog box showing the results.
2. After the medicinal plants have been added to your list of favorites, select the List of Favorites table. Users can be a convenient one-touch your favorite collection. Similarly click to view the message reply pages, the figures show that the total number of messages.
3. User account management (self-service) that allows users to add a user, change a user's password, log out, and so on (You want to provide this function only for registered users.)

7.5 Conclusion

With the help of web technology development, information database for Miao medicinal plants was established. A total of five modules were seen on the home page, including Homepage, Species identification, Data retrieval, Plant lists, and

User center. All users could click on the corresponding module to get the information according to their own requirements. This system provides a unifying platform for the application of computational approaches to elucidate mechanistic links between traditional knowledge of Miao medicinal plants and their therapeutic action. Through the ministerial dispatching computer management subsystem, data can be shared, information is centralized, working efficiency is raised, and it is convenient for inquiring and statistics offering a query service to the people who take up the correlative research.

In the future, we hope to update this database with the following additional information. Firstly, it will be important to update our database with more detailed information on the relative phytochemical composition of medicinal plants will be crucial for evaluating and developing traditional Miao medicine formulations. Secondly, during the investigation, we also found that lots of precious endangered species were used by Miao ethnic group. In this system, no module was involved. In our future work, the management and conservation of endangered plants will be integrated into the system. Thirdly, in addition to traditional classification, modern classification of disease should be followed for sharing information at global level. The current version of International Classification of Diseases (ICD-10) which is available on WHO website (http://www.who.int/classifications/icd10/) can be used as reference for the ethnopharmacological works.

Acknowledgements We are very grateful to the Miao people who provided valuable information about medicinal plants. We would like to thank College of Information Engineering at Minzu University of China for providing database technique. Without their contribution, this study would have been impossible. This work was financially supported by the National Natural Science Foundation of China (31761143001, 31870316, 31161140345 & 31660083), the Key Laboratory of Ethnomedicine (MinzuUniversity of China) of Ministry of Education of China (KLEM-ZZ201806), Minzu University of China (Collaborative Innovation Center for Ethnic Minority Development, yldxxk201819), Ministry of Education of China and State Administration of Foreign Experts Affairs of China (B08044)

References

Afendi FM, Okada T, Yamazaki M (2012) KNApSAcK family databases: iIntegrated metabolite-plant species databases for multifaceted plant research. Plant Cell Physiol 53(2):e1. https://doi.org/10.1093/pcp/pcr165

Dighe NS, Nirmal SA, Musmade DS (2010) Herbal database management. Syst Rev Pharm 1(2):152

Heinrich M, Edwards S, Moerman DE (2009) Ethnopharmacological field studies: a critical assessment of their conceptual basis and methods. J Ethnopharmacol 124(1):1–17

Hrudayanath T, Jayanta KP (2011) Biotechnology and pharmacological evaluation of medicinal plants: an overview. J Herbs Spices Med Plants 17(3):214–248

Lebo MS, Zakoor KR, Chun K (2017) Data sharing as a national quality improvement program: reporting on BRCA1 and BRCA2 variant-interpretation comparisons through the Canadian Open Genetics Repository (COGR). Genet Med:294–302. https://doi.org/10.1038/gim.2017.80

Leonti M (2011) The future is written: impact of scripts on the cognition, selection, knowledge and transmission of medicinal plant use and its implications for ethnobotany and ethnopharmacology. J Ethnopharmacol 134(3):542. https://doi.org/10.1016/j.jep.2011.01.017

Luo BS, Guo ZY, Chen XY, Gu RH, Liu B, Long CL (2017) Database establishment of ethnomedicinal plants during dragon-boat festival in Jingxi, Guangxi, China. Guihaia 12(1). https://doi.org/10.11931/guihai.gxzw201705021

Ma KP (2014) Rapid development of biodiversity informatics in China. Biodiv Sci 22(3):251–252

Mohanraj K, Karthikeyan BS, Vivekananth RP (2018) IMPPAT: a curated database of Indian medicinal plants, Phytochemistryand therapeutics. Sci Rep 8(4329). https://doi.org/10.1038/s41598-018-22631-z

Obayashi T, Aoki Y, Tadaka S (2018) ATTED-II in 2018: a plant coexpressiondatabase based on investigation of statistical property of the mutual rank index. Plant Cell Physiol 59(1):e3. https://doi.org/10.1093/pcp/pcx209

Ono H, Ogasawara O, Okubo K (2017) RefEx, a reference gene expression dataset as a web tool for the functional analysis of genes. Sci Data 72(1). https://doi.org/10.1038/sdata.2017.1054:170105

Pathania S, Ramakrishnan SM, Bagler G (2015) Phytochemica: a platform to explore phytochemicals of medicinal plants. Database J Biol Data Curat 2015:1–8. https://doi.org/10.1093/database/bav075

Pilon AC, Valli M, Dametto AC (2017) NuBBEDB: an updated database to uncover chemical and biological information from Brazilian biodiversity. Sci Rep 7(1):7215. https://doi.org/10.1038/s41598-017-07451-x

Smith AT, Boitani L, Bibby C (2000) Databases tailored for biodiversity conservation. Science 290(5499):2073. https://doi.org/10.1126/science.290.5499.2073b

Smith VS, Rycroft SD, Harman KT (2009) Scratchpads: a data-publishing framework to build, share and manage information on the diversity of life. BMC Bioinformatics 10(Suppl 14):S6. https://doi.org/10.1186/1471-2105-10-S14-S6

Thomas MB (2003) Emerging synergies between information technology and applied ethnobotanical research. Ethnobot Res Appl 1(1). https://doi.org/10.17348/era.1.0.65-74

Torres RDS, Medeiros CB, Gonçcalves MA (2006) A digital library framework for biodiversity information systems. Int J Digit Lib 6(1):3–17

Tuasha N, Petros B, Asfaw Z (2018) Medicinal plants used by traditional healers to treat malignancies and other human ailments in Dalle District, Sidama zone, Ethiopia. J Ethnobiol Ethnomed 4(1):15. https://doi.org/10.1186/s13002-018-0213-z

Documentation and Protection of Traditional Knowledge

8

Tapan Mukherjee

Abstract

Traditional medicine in India can be classified into codified (*Ayurveda, Unani, Siddha, Amchi*, etc.) and non-codified (folk medicine) systems. Both the systems are contributing equally to the primary healthcare in India. All the types of traditional medicine practices found in India take care of the primary health needs of about 70% of the Indian population. Among them, the non-codified and codified systems of traditional medicine share the equal task in managing primary healthcare. The non-codified system of traditional medicine is diverse and varies with geography, local flora and culture. The system differs from one region to the other and is known by various names like indigenous medicine, ethnomedicine, folk/folklore medicine and many more. This system is diminishing with time due to various factors. Therefore, proper documentation is needed to preserve this ancient, non-codified traditional knowledge.

Keywords

Traditional knowledge · Indian systems of medicine · Documentation

8.1 Introduction

Traditional knowledge refers to knowledge systems, encompassing a wide variety of areas, held by indigenous communities. Traditional knowledge has been passed from generation to generation by word of mouth and confined to certain localities. Traditional knowledge (TK) is a collectively owned property and is integral to the cultural or spiritual identity of the social group in which it operates and is preserved. Conservation of indigenous biodiversity over a long time period has been possible

T. Mukherjee (✉)
Indian Journal of Traditional Knowledge, CSIR-NISCAIR, New Delhi, India

S. M. Khasim et al. (eds.), *Medicinal Plants: Biodiversity, Sustainable Utilization and Conservation*, https://doi.org/10.1007/978-981-15-1636-8_8

because of the application of traditional knowledge (TK) systems by native communities applying complex and location specific practices. Traditional knowledge (TK) is both cumulative and dynamic, building upon the experience of earlier generations and adapting over time to socio-economic changes and new technologies. The importance of TK in the conservation of biodiversity and sustaining natural resources has been recognised. TK has received much attention in resource management regimes and it suggests new paths for ecological research, conservation monitoring and better understanding of ecological processes (Balaraman and Gulati 2005; Uppeandra et al. 2002; Dutfield 2000; Fabricant and Farnsworth 2001; Fokunang et al. 2011; Garí 2000; Hansen 2002; Karki 2001; Kuanpoth 2001; Lai and Roy 2004; Mukherjee 2008, 2011, 2014; Sharma 2000; Singh and Mukherjee 2008; Subcharoen et al. 2000; Tapsell et al. 2006).

8.2 Biodiversity

India is one of the 12 mega biodiversity centres in the world and has two of the world's 18 biodiversity hotspots located in the Western Ghats and in the Eastern Himalayas. The forest cover in these areas is very dense and diverse and of incredible biodiversity. The country is estimated to have over 47,000 plant species and 81,000 animal species representing 7% of the world's flora and 6.5% of its fauna. About 33% of the country's recorded flora is endemic. There are 89 national parks and 504 wildlife sanctuaries in the country. India is bestowed with a treasure of medicinal plants. Medicinal plants are, at present, obtained from both cultivated and wild sources. India hosts about 17,000 species of higher plants, out of which 7500 are known for their therapeutic uses.

8.2.1 Medicinal Plants

Medicinal plants have been identified and used since time immemorial. Medicinal plants and their application in therapy have a long tradition. The World Health Organization (WHO) estimates that 80% of the population of some Asian and African countries presently uses herbal medicine for some aspect of primary healthcare. According to the use, medicinal plants can be classified into three categories: (a) plants used in Codified System or used in Indian Systems of Medicine (over 1500), (b) plants used in ethno or folk medicine based on oral (undocumented) information being passed from generation to generation (3000) and (c) plants which have been investigated pharmacologically and chemically, and their active principles are used in modern medicine or are providing valuable leads for partial or total synthesis of new drugs (700). Medicinal products may be broadly categorised into (a) herbal medicinal products and (b) traditional herbal medicinal products. Herbal medicinal products are the chemical constituents isolated from medicinal plants. For traditional herbal medicinal products, efficacy must be based on their long-standing use instead of being proven by clinical trials. A harmonised scientific evaluation is needed and should be in place. Plants have the ability to synthesise a wide variety of chemical compounds that are used to perform important biological functions. At least 12,000 such compounds have been isolated so far, a number estimated to be less than 10% of the total.

8.2.2 Indian Systems of Medicine

Traditional medicine in India can be classified into codified (Ayurveda, Unani, Siddha, Homoeopathy) and non-codified (indigenous/ folk / tribe medicine) systems. Both the systems contribute equally to the primary healthcare in India. The traditional medicine and practices can be classified broadly into (a) traditional medicine with a systematic codified body of knowledge (*Ayurveda, Siddha, Unani, Tibetan, Amchi, Sowa Rigpa Chinese,* etc.); (b) non-codified system of traditional or folk medicine, which is transmitted by oral means and is mostly acquired through trial-and-error approaches and (c) spiritual or shamanistic medicine, which has a strong religious/spiritual element and is practised only by highly specialised local experts. India has well-developed codified systems of medicine over the centuries. In India, regulation and systematic functioning of these codified traditional systems of medicine, starting with education at university level, registration of qualified practitioners, production, safety and efficacy issues of the drugs are monitored by the Government. However, in the presence of a strong codified traditional medicine, the non-codified system has lost its importance though there is high usage by the public while the non-codified traditional medicine or folk medicine lack precise policy issues in India.

8.2.3 Traditional Medicine

India has the largest total tribal population compared to any other country in the world. The large tribal population of India is not a homogeneous group. It is composed of a number of different tribes. In India there are 461 tribal communities, of which 174 have been identified as subgroups. Our Constitution recognises 212 tribes and these tribes are found in different parts of the county.

Traditional medicine, also known as folk medicine, is a non-codified system of medicine. As defined by WHO, traditional medicine refers to health practices, approaches, knowledge and beliefs incorporating plant, animal and mineral based medicines, spiritual therapies, manual techniques and exercises, applied singularly or in combination to treat, diagnose and prevent illnesses or maintain well-being (Uppeandra et al. 2002). Traditional medicines provide an important healthcare service and have shown great potential to meet a broad spectrum of healthcare needs. Nearly 25% of modern medicines are derived from plants first used in traditional medicine. However, further research, clinical trials, and evaluations are needed. Traditional medicine occupies a very important place in healthcare in the world. Traditional knowledge and the use of plant-based medicines remain important in the prevention and treatment of various diseases and ailments as the traditional medicine is quickly accessible and affordable.

Traditional medicine practices found in India take care of the primary healthcare needs of about 70% of the population. Among them, the non-codified and codified systems of traditional medicine share the equal task in managing primary healthcare. The non-codified system of traditional medicine is diverse and varies with geography, local flora and culture. It was developed in accordance with primary needs and locally available resources of a particular region. The system differs from one region to the other and is known by various names like indigenous medicine, ethnomedicine, tribal medicine, folk medicine, folklore medicine and many more. However, the practice is not formalised in many countries and to a large extent remains in non-codified form. Traditional medicines have links with spiritual and magico-religious rituals.

8.2.4 Some Examples of Traditional Medicine

- Use of *plao-noi* by Thai traditional healers to treat ulcers.
- Use of the *Ayahuasca* vine by Western Amazonian tribes to prepare various medicines.
- Use of *Hoodia* cactus by the San people to stave off hunger while outhunting.
- In Indonesia, phyto-pharmaceutical products developed from ginger and *noni* as the raw material exhibited efficacy in treating tuberculosis. These plants are also used in folk medicine for the treatment of cough, whooping cough, asthma and to increase immunity.
- Use of *Jeevanti* (*Trichopus zeylanicus*) by Kani tribes of Kerala, to stave off hunger and thirst on long hunting trips.

8.2.5 Misappropriation of Traditional Knowledge

Appropriation of traditional medicinal knowledge without the prior consent of the community is biopiracy. In the recent past, there have been several cases of biopiracy of traditional knowledge from India. First, it was the patents on *Haldi* (turmeric), Neem and *Basmati* rice. Now patents have been obtained on compositions comprising of various indigenous plants. Even the misappropriation of TK has been reported globally. In South Africa, the San, hunter-gatherers are the sole preservers of the knowledge of the *Hoodia cactus* which is a viable appetite-suppressant drug. A similar example of biopiracy is of *Maca* plant of the Andes on which a patent has been provided only because the age-old knowledge was not in written form. Bioprospecting is collecting indigenous knowledge after a prior consent by the community and exploiting the biological resources.

Traditional knowledge (TK) has always been an easily accessible treasure and thus has been susceptible to misappropriation. The traditional knowledge, particularly, related to the treatment of various diseases has provided leads for the development of biologically active molecules by the technology-rich countries. Traditional knowledge includes both the codified (documented) and non-codified information (not documented but may be orally transmitted). Bio-piracy of codified traditional knowledge continues, since this information exists in regional languages, and there exists a language barrier due to which the patent offices are unable to search this information as prior art, before granting patents. Formulations used for the treatment of human ailments from traditional knowledge are time-tested since they have been in practice for centuries. The reliability of the traditional medicine systems coupled with the absence of such information with patent offices provides an easy opportunity for interlopers for getting patents on these therapeutic formulations derived from traditional medicine systems. The grant of patents on non-patentable knowledge (related to traditional medicines), which is based on the existing traditional knowledge, has been causing a great concern to the developing world. Some of the examples illustrate the bio-piracy of traditional knowledge and in many of these cases the country had to fight for revocation of the granted patents. Revocation may not be a feasible option possible for all the patents taken on the traditional knowledge since it involves huge costs and time.

8.2.6 Need for Protection

These knowledge systems have significance and relevance not only to its holders but also to the rest of the humanity. The knowledge needs to be documented and preserved to facilitate protection. Efforts are needed to develop an effective system to collect, classify and inventorise traditional knowledge (TK) without losing its essential native content and value. Need has also been felt to protect this from inappropriate use and to provide economic benefits to the holders of such knowledge. The traditional knowledge, which is the wisdom developed over many generations, needs to be scientifically validated and utilised for wider applications. Traditional knowledge and its practices can play a great role in enhancing our understanding for conservation.

The existing forms of intellectual property protection regimes do not adequately recognise the rights of TK holders. National level mechanisms and legal provisions to prevent biopiracy as well as to install informed consent mechanisms to ensure reward to TK holders also require international recognition for their effective implementation and for their enforcement in other countries. In this regard, there is a need for the development of an international mechanism for protecting TK. Issues relating to protecting, recognising and rewarding of TK associated with biological resources are very complex. The modalities for protecting TK are still emerging and evolving. The nature of entitlements and share in benefits is also a grey area. Proper documentation of associated TK could help in checking biopiracy. Documentation would facilitate tracing of indigenous communities with whom benefits of commercialisation of such materials/knowledge has to be shared. Documentation has one clear benefit, it would check patents based on TK in the public domain that is today difficult to prevent due to lack of availability of information with patent examiners.

The need to protect TK has arisen in many instances in the context of claims relating to the unauthorised appropriation of TK based products, processes and the biological resources on which they are based. There is a need to preserve and document traditional healing practices and for more future scientific research on the plants to determine their efficacy and safety. This could improve the traditional recipes and might contribute to a better integration of traditional medicine into the national health system in the future. Preserving and documenting the local traditions and indigenous knowledge could facilitate future research on the safety and efficacy of medicinal plants and provide a starting point for identifying single chemical entities leading to the development of standardised phytomedicine. Traditional medicine now should be used as a discovery engine to provide new targets, leads, botanical drug discovery and development including bioprospecting and bioavailability.

In India, several initiatives have been taken to improve the traditional system of medicine including Acts and Amendments like Biological Diversity Act (2002), Biological Diversity Rules (2004), Scheduled Tribes and other Traditional Forest Dwellers (Recognition of Forest Rights) Act (2006). These policies, instead of being concrete and elaborative, address only a few issues of traditional practitioners and their practice. Moreover, as these laws are nonspecific and covers various other aspects, there is a need to develop *Sui-generis* laws for governing the rights of traditional practitioners.

The establishment of Traditional Knowledge Digital Library (TKDL) has facilitated the registry for traditional knowledge, which is a big step forward towards protection of the knowledge from bio-piracy. TKDL has thousands of Indian traditional formulations from traditional systems of medicine, which is available online for referencing and to deal with IPR issues. Still, the major portion of Indian non-codified traditional medicine remains undocumented and unprotected, paving way to bio-piracy. At the same time, intellectual property rights of non-codified traditional medicine practices are unprotected mainly because of their collective ownership and perpetuity development pattern from generation to generation which made the system "not so novel" commercially.

While implementing any system, recognising ownership of knowledge holders, the community should be taken into confidence so that they are convinced that their knowledge will not be misappropriated. The possibility of economic returns generated from their knowledge will act as an incentive for community to continue with their practices. Communities may be more inclined to preserve and transmit their knowledge to next generations if their rights are respected. The knowledge holders will be willing to disclose their knowledge once they are assured and convinced that the proposed IPR law (*Sui generis*) can give them control over their knowledge. This in turn will encourage disclosure, use and proliferation of traditional knowledge that might otherwise be lost.

The case study of *Jeevani* drug (*Trichopus zeylanicus*) gives an insight into the concept of Access and Benefit Sharing. The importance of traditional knowledge is highlighted in the revocation of patents granted to derivatives of *Neem*, turmeric and *Basmati* on the ground that they were part of the traditional knowledge of our country. Attention has been drawn to the international initiatives at protecting traditional knowledge including the Convention on Biological Diversity (CBD). There are no uniform norms regarding the protection of different types of traditional knowledge owned by local communities. There is an urgent need to enact a *sui generis* or alternative law to protect traditional knowledge.

8.3 Conclusion

Traditional medicines are in use since time immemorial, but due to lack of sufficient scientific evidence required by the scientists for the safety, efficacy and quality, it has not been popular worldwide. Traditional practitioners are integrated and well-respected people in their communities and can play a vital role in the community social health information and healthcare sensitisation.

Traditional medicine is making a comeback. Researchers and scientific organisations increasingly recognise the value of traditional medicine as a source of potential new blockbuster drugs and as alternative providers of primary healthcare. There is a need for global effort to break down the legal, regulatory and conceptual barriers that support the promotion of modern medicine at the expense of traditional practices and Access and Benefit Sharing (ABS) to ensure that the originators and custodians of traditional medicine get the respect, recognition and equitable share of the benefits they deserve. Need is also felt to adapt systems of regulation, testing, training and licensing or certification with methodologies that suit traditional medicine. Integrating modern and traditional medicine is a major challenge.

Stringent intellectual property regime cannot easily cope with the traditional approach to medical knowledge, which is commonly owned and freely handed down through generations. The tightly defined tests for safety and efficacy that are a hallmark of modern drug regulation are another constraint. They have been developed to test standardised drugs at fixed dosages and rely on being able to identify active ingredients and provide easily reproducible results. But traditional medicines are inherently diverse, both in how they are formulated and dispensed.

References

Balaraman R, Gulati OD (2005) Ayurveda through the ages. In: Patil PN, Gulati OD, Balaraman R (eds) Topics in the history of pharmacology. BS Shah Prakashan, Ahmedabad, Gujarat

Dutfield G (2000) The public and private domains: intellectual property rights in traditional knowledge. Sci Commun 21(3):274–295

Fabricant DS, Farnsworth NR (2001) The value of plants used in traditional medicine for drug discovery. Environ Health Prospect 109(1):69–75

Fokunang CN, Ndikum V, Tabi OY, Jiofack RB, Ngameni B, Guedje NM, Tembe-Fokunang EA, Tomkins P, Barkwan S, Kechia F, Asongalem E, Ngoupayou J, Torimiro NJ, Gonsu KH, Sielinou V, Ngadjui BT, Angwafor IIIF, Nkongmeneck A, Abena OM, Ngogang J, Asonganyi T, Colizzi V, Lohoue J, Kamsu-Kom (2011) Traditional medicine: past, present and future research and development prospects and integration in the National Health System of Cameroon. Afr J Trad Complement Altern Med 8(3):284–295

Garí J (2000) Biodiversity conservation and use: local and global considerations, science, technology and development papers, no 7. Center for International Development, Harvard University, Cambridge

Hansen S (2002) Traditional ecological knowledge and prior act. WIPO ConfInt Patent System, Geneva

Karki MMS (2001) Neem-based natural product innovations: analysis of patents. J Intellectual Property Rights 6:27–37

Kuanpoth J (2001) Legal protection of traditional knowledge: the case of Thai traditional medicine, ASEAN Workshop on the TRIPS Agreement & Traditional Medicine. WHO, Jakarta

Lai PK, Roy J (2004) Antimicrobial and chemopreventive properties of herbs and spices. Curr Med Chem 11(11):1451–1460

Mukherjee TK (2008) Protection of Traditional Knowledge for Sustainable Development and Wealth Creation, ProcNtl Seminar on "Community Based Sustainable Natural Resources Management & Development in North-East India", Pasighat, Arunachal Pradesh

Mukherjee TK (2011) Medicinal Plants in India and traditional Medicine, Proc. ICMCM Int. Conf on "Milestones in Chinese Medicine Development", Hong Kong

Mukherjee TK (2014) Traditional medicine: road ahead. In: Mukherjee PK (ed) Traditional medicine and globalization – the future of ancient systems of medicine. Maven Publishers, Kolkata

Sharma SK (2000) "National Measures and experience for Protection of Traditional Indian Medical Knowledge of Ayurveda" - in the Regime of Intellectual property Rights (IPR), Interregional Workshop on intellectual Property Rights in the Context of Traditional Medicine, Bangkok

Singh RK, Mukherjee TK (2008) Ecoliterary tools: methodology and approaches on biocultural knowledge learning and conservation. CSIR, NISCAIR, New Delhi

Subcharoen P, Luanratana O, Kuanpoj J, Assawintrangura S (2000) Indigenous Knowledge and Intellectual Property -Thai Study, Inter-regional Workshop on intellectual property rights in the context of traditional medicines, Bangkok

Tapsell LC, Hemphill I, Cobiac L (2006) Health benefits of herbs and spices: the past, the present, the future. Med J Aust 185(4 Suppl):S4–S24

Uppeandra D, Sumit M, Mitali J, Arvind B, Bisht AK, Meena J (2002) Current status and future strategy for development of medicinal plants sector in Uttaranchal, India. Curr Sci 83(8):956–964

Ethnobotanical Assessment of Medicinal Plants Used by Indigenous People Living Around the Sacred Groves of East Godavari District, Andhra Pradesh, India

9

Venkatesh Rampilla, S. M. Khasim, and K. Thammasiri

Abstract

The present study was conducted for the first time in sacred groves located in East Godavari District, Andhra Pradesh, India. The purpose was to document the indigenous knowledge of the tribal people used in the preparation of herbal medicines. To get the data on traditional uses of ethnomedicinal plants, 25 key informants were interviewed. Quantitative ethnobotanical indices, i.e., fidelity level (FL), data matrix ranking (DMR), priority ranking (PR), and Jaccard index (JI), were calculated for recorded ethnomedicinal plants. A total of 104 ethnomedicinal plants belonging to 89 genera and 53 families used in 17 disease categories were documented. Leaves (32.59%) were the frequently used plant parts, and most of the informants suggested taking herbal medicines orally. In our study, we found highest FL value for *Andrographis paniculata* and *Gymnema sylvestre* (100%) and the lowest value was recorded for *Gloriosa superba* (40%). *Syzygium cumini* was the most multipurpose plant other than medicinal uses. In PR, informants ranked poducultivation (Burn agriculture) (19.25%) as a leading threat to ethnomedicinal plants. The present study provides useful information about traditional uses of ethnomedicinal plants used by indigenous communities in different ailments. Most of the plants having highest use values could be employed in pharmaceutical research in order to achieve adequate revenue. Some of the plants in the study area are facing threats; hence, sustainable harvesting and conservation initiatives are needed in the region.

V. Rampilla
Department of Botany, Dr. L.H.R. Government Degree College,
Mylavaram, Andhra Pradesh, India

S. M. Khasim (✉)
Department of Botany and Microbiology, Acharya Nagarjuna University,
Guntur, Andhra Pradesh, India

K. Thammasiri
Department of Plant Science, Faculty of Science, Mahidol University, Bangkok, Thailand

S. M. Khasim et al. (eds.), *Medicinal Plants: Biodiversity, Sustainable Utilization and Conservation*, https://doi.org/10.1007/978-981-15-1636-8_9

Keywords
Sacred groves · Ethnomedicine · Priority ranking · Fidelity level

9.1 Introduction

Sacred groves are patches of forests conserved through man's spiritual belief and faith. International Union for the Conservation of Nature (IUCN), Sacred Natural Sites (SNS) has been defined more specifically as "areas of land or water bodies having special spiritual significance to ethnic communities" (Wild and McLeod 2008). In India they are extended from the Himalayas to North-East India and Bihar, Orissa, Madhya Pradesh, Andhra Pradesh, Karnataka, Kerala, and Tamil Nadu (Malhotra 1998; Malhotra et al. 2001). Ethnomedicine, a traditional systematic medicine practiced by indigenous communities, is related to cultural interpretation of health, disease, and illness (Suthari et al. 2014). Globally different ethnic groups are using around 70,000 plants species for their medical purposes. It is estimated that 64–84% of world's population depends on traditional medicine (Khan et al. 2004; Augustine and Sivadasan 2004). India is a treasure house for ethnomedicinal knowledge due to the presence of diverse vegetation, ancient lineage, and multi-ethnic groups. All India Coordinated Research Project on Ethnobiology (AICRPE 1992-1998) mentioned that tribals has been using more than 10,000 wild plants for meeting their primary healthcare, food, and other material requirements. The study of ethnobotany through a qualitative survey method is a very old history but the interest in numerical ethnobotany has established progressively in last couple of decades (Amjad and Arshad 2014). A number of earlier authors have explored qualitative ethnobotanical data (Naidu and Khasim 2010; Khasim et al. 2013; Padal et al. 2013; Savithramma et al. 2015; Srinivasa Rao et al. 2015). The sacred groves have a rich wealth in ethnomedicinal plants and this area remained unexplored due to its remoteness. This is the first effort in this region to provide quantitative ethnobotanical data employed by indigenous people. The objectives of the present study were (1) enlisting native ethnomedicinal plants and (2) recording the ethnic formulations of this flora along with their mode of preparation, administration, and dosage (3) to find out the correlation between ethnomedicinal uses and (4) to provide baseline data for pharmaceutical research by the application of quantitative indices.

The present study was carried out on ethnobotany of four sacred groves, viz., Dummakonda, Korukonda, Peddakonda, and Rampa (Fig. 9.1), located in East Godavari district of Northern part of Eastern Ghats of Andhra Pradesh, India. So far, the sacred groves to be studied are unexplored for their floristic diversity and ethnomedicinal wealth. All the groves under study were intensively explored for a period of 4 years during 2011 to 2015. The exploration trips were conducted for the same grove sites covering different seasons.

Ethnomedicinal data documentation on traditional herbal remedies is an important aspect of conservation approach. The present study, therefore, documents the traditional knowledge of local indigenous communities based on questioner of Jain

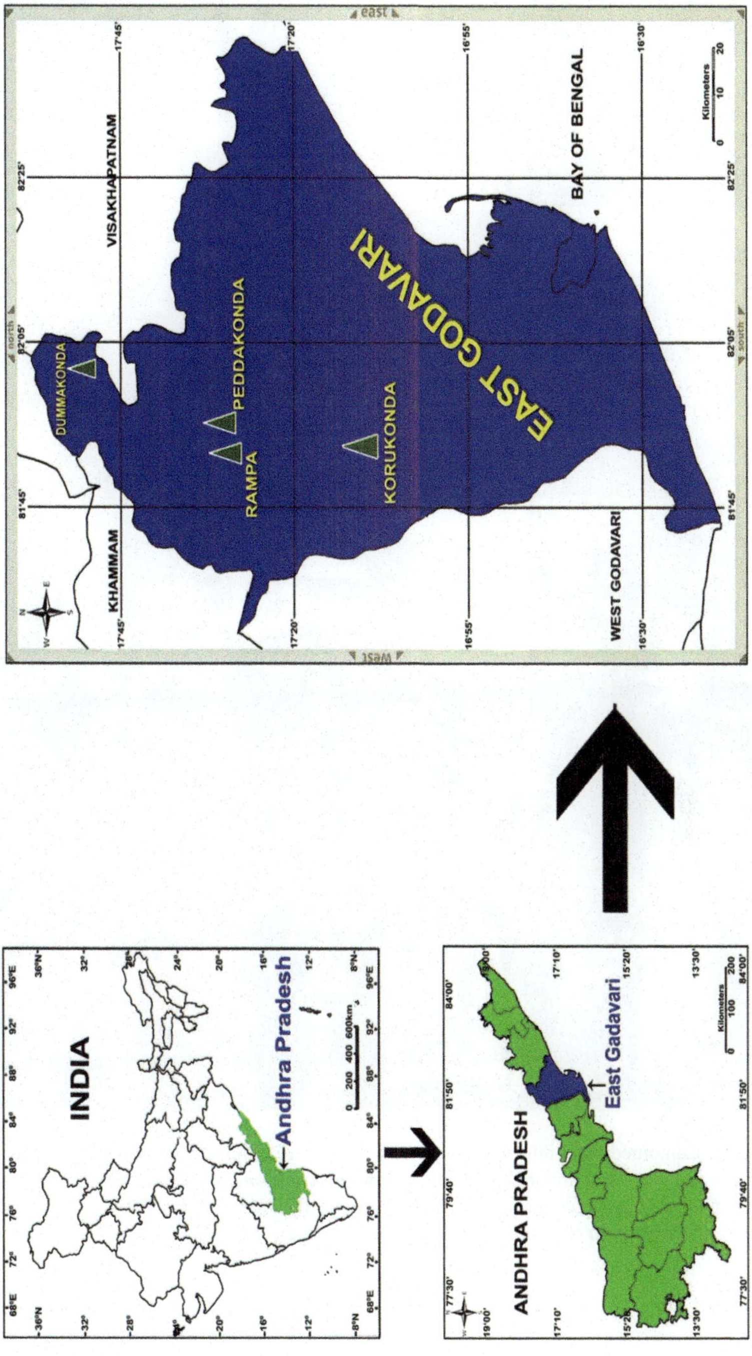

Fig. 9.1 Map of study area (Sacred Groves) in East Godavari District, Andhra Pradesh, India

Fig. 9.2 (**a**) Ethnomedicinal information collection from tribal physicians, (**b**) local market, (**c**) toddy preparation and drinking, (**d**) poducultivation, (**e**) *Argemone Mexicana*, (**f**) *Alangium salvifolium*, (**g**) *Curcuma pseudomontana*, (**h**) *Caryota urens*, (**i**) *Dichrostachys cinerea*, (**j**) *Rauvolfia serpentina*, (**k**) *Terminalia chebula*, (**l**) *Woodfordia fruticosa*

and Mudgal (1999) (Fig. 9.2a). All the questions were asked and categorized into five. They are: (1) demographic characteristics of the informants, (2) botanical investigation (source of plant, local name, numbering, preliminary classification, specimen for herbarium), (3) pharmacological investigation (how did the tribe acquire the ethnomedicinal knowledge and gather plants, name of ailments they have dealt with, dosage, method of use), (4) what are the destructive agencies (poducultivation/burn agriculture, forest fire, fuel wood, etc.), and (5) what type of benefits are obtained from natural resources to the local communities. The interviews and discussions were organized in their native languages (Telugu, Koya) with the help of local translators.

9.1.1 Ethical Issues

Ethnic groups, also known as tribes, have traditional cultural identity, distinct language or dialect of their own and culturally homogeneous and live in seclusion, governed by their own social norms and largely having a self-contained economy. There are 33 tribal or indigenous groups inhabited in Eastern Ghats forests of Andhra Pradesh (AP). The tribals of Eastern Ghats of AP originated from prehistoric era. Several micro and megalithic sites discovered in upper Godavari are the evidences for Stone Age culture of these tribals (Senapathi and Sahu 1966). In the present study area, 5 tribal communities were observed such as Koya Dora, Konda Reddi, Konda Kammara, Valmikies, and Konda Kapu. The houses of these communities are constructed with bamboo, palm, and other timber-yielding trees. The leaves of *Borassus flabellifer* L. (Palmyra) and *Phoenix sylvestris* L. (Etha) are extensively used for thatching. All these tribal communities celebrate festivals at the beginning of any economic activity and before harvesting agricultural produce. "Kothe, Bali panduga and Agam panduga" to please their ancestral spirits are some of the important festivals. The young men carry their bows, axis, arrows, and knives to hunt wild animals like wild goat, wild pig, wild sheep, rabbit, pigeon, small birds, etc. Drinking of toddy (Fig. 9.2c) and arrack is common among these tribes. The forest of this area provides firewood, timber, medicinal plants, and edible plants to the community. In India, the Biological Diversity Bill (No. 93 of 2000), Wildlife Protection Act, 1972, and the Protections of Plant Varieties and Farmers' Rights Act, 2001, support the indigenous people for conservation of sacred groves and their own territories (Anon 2001).

9.1.2 Plant Collection, Identification, and Preservation

The specimens were collected in well-tied thick polythene bags. The collected specimens were made into herbarium following Jain and Rao (1997). The ethnomedicinal plant species used by the indigenous communities of the study area were authenticated using the International Plant Name Index (http://www.ipni.org), The Plant List (www.theplantlist.org), GRIN Taxonomy site (http://www.ars-grin.gov/cgi-bin/npgs/html/queries.pl), "Flora of the Presidency of Madras"

(Gamble and Fischer 1915-1935), "Flora of Andhra Pradesh" (Pullaiah and Chennaiah 1997), "Flora of East Godavari" (Rao et al. 1999), "Flora of Guntur City" (Mohana Rao et al. 2015) and "Genera Plantarum" (Bentham and Hooker 1862–83). All the herbarium specimens were deposited in Botany Department, Acharya Nagarjuna University.

9.2 Data Matrix Ranking and Data Analysis

Data matrix ranking was conducted using ten multiuse plants commonly reported by key informants based on the protocol of Cotton (1996). The selected informants were asked to give value to each character on the basis of benefits obtained from each plant.

Ethnobotanical data obtained during the study were summarized using Microsoft Excel spreadsheet (2007®) used to calculate proportions like plant families, habit, plant parts, frequency of citation, and popularly used plants. The most used medicinal plants against human ailments are calculated using Fidelity Level (FL) value. This value explains relative healing potential of medicinal plant.

$$\mathrm{FL}\ (\%) = \mathrm{Ip} / \mathrm{Iu} \times 100$$

where Ip is the number of informants who independently indicated the use of species for the same major ailment and Iu the total number of informants who gave information about the plant for any major ailment (Alexiades 1996). Before the calculation of FL, all human ailments were grouped into major disease categories, using a similar approach employed elsewhere (Heinrich et al. 1998).

Jaccard index is used to compare study data with that of other ethnobotanical studies conducted in other parts of Andhra Pradesh as well as other states of India, and also among the indigenous communities in the study areas. The formula used to evaluate the JI index (González-Tejero et al. 2008) was JI = c × 100/a + b−c, where "a" is the recorded number of species in our study area , "b" is the documented number of species in neighboring area, and "c" is the common number of species in both areas. In case of indigenous communities, "a" is the number of species reported by an indigenous community "A," "b" is the number of species cited by the indigenous community "B," and c is the number of species reported by both "A" and "B."

9.3 Ethnobotanical Assessment of Medicinal Plants Used by Communities

9.3.1 Demography of Informants

A total of 25 key informants (21 men and 4 women) were interviewed. The age of the respondents ranged from 35 to 65 years (Table 9.1). Most of the respondents were men compared with women of this area. In the study area there are about five

Table 9.1 The key informants of studied sacred groves

SI. No.	Name of the tribal physician	Gender	Age (years)	Community
1	M. Devaiah	Male	60	Konda Kapu
2	T. Subba Rao	Male	65	Konda Kammara
3	M. Sitarame Reddy	Male	57	Konda Reddy
4	N. Raja Rao	Male	45	Konda Reddy
5	K. Bujji Dora	Male	60	Konda Dora
6	K. Chinna Reddy	Male	48	Konda Reddy
7	Kanem Thamma Rao	Male	50	Konda Dora
8	Saramkota Basava Reddi	Male	56	Konda Reddy
9	Gorle Vinasa Rao	Male	53	Valmiki
10	Ganta Chinna Rao	Male	59	Konda Kammara
11	Vetla Lacchi Reddi	Male	55	Konda Reddy
12	Kattula Rami Reddi	Male	75	Konda Reddy
13	Vajjam Baludora	Male	48	Koya Dora
14	Koyyala Muniya	Female	45	Konda Kammara
15	Vasam Satyavathi	Female	50	Valmiki
16	Shambhoji Mysareddy	Male	45	Konda Reddy
17	Jammala Kannayya	Male	52	Konda Reddy
18	Para Bullemma	Female	70	Konda Kapu
19	Vanka Muthayya	Male	50	Konda Kapu
20	Kuram Laxmayya	Male	35	Konda Kammara
21	Madakam Ramanna Dora	Male	36	Koya Dora
22	Mirthivada Latchi Reddy	Male	38	Konda Reddy
23	V. Venkateswara Rao	Male	40	Valmiki
24	Kangala Ramana	Female	36	Koya Dora
25	Kanigiri Lakshmana Rao	Male	45	Konda Kapu

tribal groups residing in surrounding villages. These tribal groups have been professing their original ethnic culture and also dependent on plant resources for their livelihood. Interestingly, literate persons have less knowledge compared to illiterate persons, as they are not interested in this profession and doing other jobs for their livelihood. Similar observations were recorded elsewhere by other workers (Gedif and Hahn 2003; Uniyal et al. 2006).

9.4 Taxonomy of Documented Species

In the present investigation about 104 medicinal plants, belonging to 89 genera and 53 families, have been recorded and enumerated systematically. Out of 104 species, dicotyledons are 88 and monocotyledons are 16 (Table 9.2). Some of the ethnomedicinal plants are shown in Fig. 9.2d–l. Based on plant habit-wise analysis (Fig. 9.3), trees (32.69%) are frequently used by local people, followed by herbs (28.84%), shrubs (23.07%), and climbers (15.38%). The most commonly represented families of medicinal plants were Caesalpiniaceae, Euphorbiaceae (7 species each); Fabaceae

Table 9.2 List of ethnomedicinal plants and their ethnobotanical uses in the study area

Sl. No	Species name/ voucher specimen No.	Local name (in Telugu)	Family	Habit	Parts used	Preparation and mode of use	Dosage	Therapeutic importance
1	*Abrus precatorius* L. ANUBH01101	Gurivinda	Papilionaceae	Climber	Leaves and seeds	*Aristolochia* root and *Abrus precatorius* (white) seeds paste applied on the surface of white spots of the skin.	Once in a day.	Leukoderma
						Seed of *Sapindus* and *Abrus* paste applied externally on forehead.	During headache.	Headache
						Leaf past applied on the surface of the bald head for hair growth.	Once in a day (1 month)	Hair growth on bald head
2	*Acacia sundra* (Roxb.) DC. ANUBH01122	Sundra, chandra	Mimosaceae	Small tree	Bark and leaves	Prepare 1 g size tablets by using bark powder.	Three times per day.	Cough
						Prepare extract by boiling leaves in water.	Gargling of extract 3 to 4 times per day; ½ gm orally once in a day.	Dental diseases.
						Clean the fistula with *Ficus religiosa* bark extract and apply *Acacia* bark powder as a balm on the surface of the fistula area.	Once in a day	Fistula

3	*Acalypha indica* L. ANUBH01111	Murkonda, kumpati	Euphorbiaceae	Herb	Leaves	Prepare paste by using 9 leaves, 9 pepper and make it into 2 g size tablets.	Take two times per day with water.	Jaundice
						Prepare paste by using leaves and black *Cuminum* and apply as a balm.	Apply twice per day.	Scabies.
4	*Achyranthes aspera* L. ANUBH01110	Uttareni	Amaranthaceae	Herb	Whole plant	Prepare ash by burning whole plant.	Take ½ tea spoon daily two times orally.	Bronchitis and long term indigestion.
						Paste prepared by using leaves.	Apply around the sting area as band.	Scorpion sting
5	*Acorus calamus* L. ANUBH01151	Vasa	Araceae	Shrub	Tubers and rhizome	Rhizome is ground with water or arrack to prepare a paste and applied on the surface of nipple before the milk is drunk by baby.	Daily one or two times.	Cough/cold
						Prepare powder by using shade-dried root tubers. ½ g of this powder taken orally with ghee or milk.	Morning and evening about 1 month.	Develop memory

(continued)

Table 9.2 (continued)

Sl. No	Species name/ voucher specimen No.	Local name (in Telugu)	Family	Habit	Parts used	Preparation and mode of use	Dosage	Therapeutic importance
6	*Adhatoda zeylanica* Medic. ANUBH01152	Mandi	Acanthaceae	Shrub	Leaves	Take equal amounts of leaves of *Adhatoda* and *Ocimum*, prepare extract, and add adequate amount of honey and jaggery and form paste.	A spoonful of paste taken orally in both morning and evening every day.	Asthma and cough
7	*Aegle marmelos* (L.) Correa. ANUBH01153	Maredu, Bilva chettu	Rutaceae	Tree	Leaves & Fruit	Make a tablet-like by using leaf paste	Apply on eye surface before sleep (avoid entry of extract into the eye)	Eye diseases
						Prepare juice from ripen fruit pulp and sugar.	½ cup taken orally once in a day.	Mucoid stools & blood motions.
8	*Aerva lanata* (L.) Juss. ANUBH01112	Telagapindikura or kondapindiaku	Amaranthaceae	Herb	Roots, whole plant	25 g of *Aerva lanata* roots and 25 g poppy seeds are macerated, prepared as paste, and taken orally.	Twice in a day.	Leukorrhea
						Prepare extract by using whole plant and take orally.	20 or 40 g three times/day	Renal stones and over motions.

9	*Ailanthus excelsa* Roxb. ANUBH01154	Pedda manu	Simaroubaceae	Tree	Bark	Prepare a thick extract by using wet bark and take orally with 10 g curd.	Two times per day (morning and evening).	Blood and gum motions, diarrhea.
						Prepare powder by using shade-dried bark.	Taken 1 g with glass of butter milk both morning and evening.	Piles
10	*Alangium salvifolium* (L.f.) Wanger ANUBH01155	Uduga	Alangiaceae	Tree	Roots bark	Prepare paste by maceration of roots with gout urine. Apply this paste on bitten area externally.	Twice in a day.	Rat bite
						Prepare oil by mixing bark extract of *Alangium* and Sesamum oil.	Massage with this oil twice in a day.	Relieves from paralytic pains.
11	*Aloe vera* (L.) Burm.f. ANUBH01125	Kalabanda	Liliaceae	Shrub	Stem	Remove the epidermal peel of stem and take the pulp. Wash the pulp with hot water 5-6 times and take orally with sugar.	Once in a day	Body coolant

(continued)

Table 9.2 (continued)

Sl. No	Species name/ voucher specimen No.	Local name (in Telugu)	Family	Habit	Parts used	Preparation and mode of use	Dosage	Therapeutic importance
12	*Ammannia baccifera* L. ANUBH01156	Agnivendramu	Lythraceae	Herb	Leaves and leaf tips	Prepare 2 g sized tablet by mixing leaf tips and pepper.	Taken orally both in morning and evening.	Malaria
						Shade-dried leaves are powdered and made into 2 g sized tablets.	Taken orally with honey both morning and evening.	Whooping cough
13	*Andrographis paniculata* (Brum.f.) Nees. ANUBH01126	Nelavemu	Acanthaceae	Herb	Whole plant	Make 5 g sized tablets by mixing whole plant powder and leaf extract.	Twice/day about 15 days.	Malaria
						Make 10 g sized tablets by using whole plant extract.	One tablet early morning of the day about 2 days.	Worm infection
14	*Annona squamosa* L. ANUBH01127	Setaphalam	Annonaceae	Tree	Leaves	Prepare leaf powder and take orally with honey or water.	1/4 tea spoon both morning & evening.	Diabetes
						Prepare paste by mixing *Annona* leaf and rice grains; apply as band over swelling area.	Twice in a day.	Furuncle

15	*Argemone mexicana* L. ANUBH01157	Balurakkisa	Papavaraceae	Herb	Roots and latex	Take equal parts of *Argemone* root bark, *Cissus* stem, salt, macerate and prepare as tablets.	Twice in a day.	Asthma
						Mixture prepared by mixing root extract of *Argemone* and pepper powder, taken orally.	Once in a day	Gonorrhea
16	*Argyreia nervosa* (Burm.f.) Boj. ANUBH01158	Samudra Pala	Convolvulaceae	Climber	Root and seedling	Prepare juice by maceration of seedlings with water.	Taken ½ tea glass once in a day.	Gum motions & diarrhea.
						The root powder soaked in Asparagus root juice in 7 times, after that shade dried and make it into powder.	Take 2 to 3 g powder dissolved in butter and taken orally for about 1 month.	Semen development and male fertility.

(continued)

Table 9.2 (continued)

Sl. No	Species name/ voucher specimen No.	Local name (in Telugu)	Family	Habit	Parts used	Preparation and mode of use	Dosage	Therapeutic importance
17	*Aristolochia indica* L. ANUBH01128	Nalla eswari	Aristolochiaceae	Climber	Roots	Prepare extract by using root powder with water, apply on snake or scorpion bite area and take 2 or 3 spoons orally.	One time after the bite.	Snake bite or scorpion bite.
						Prepare paste using *Abrus precatorius* white type seeds and *Aristalochia* roots and apply on white spots.	Twice in a day.	Leukoderma
18	*Asparagus racemosus* Willd. ANUBH01129	Pilli teegalu	Liliaceae	Herb	Roots	Root tubers are fried in ghee and added sufficient amount of sugar and taken orally.	Take 3 tubers at morning and evening.	Leukorrhea
						Root tubers macerated in milk and made into extract.	Taken orally once in a day.	Blood motions
19	*Azadirachta indica* A. Juss. ANUBH01113	Vepa	Meliaceae	Tree	Leaves	Take equal parts of *Ocimum* leaves, *Azadirachta,* pepper, patika macerate and prepare mixture.	Taken orally 5 mg tablets three times in a day.	Malaria fever
						Prepare leaf paste by using leaf tips and make into 5 mg tablets.	Taken orally early in morning for 3 days.	Worm infection

20	*Bauhinia racemosa* Lamk. ANUBH01114	Goddukura, Deva kanchanamu	Caesalpiniaceae	Tree	Roots and bark	Mix bark powder of *Bauhinia racemosa* with *Borassus* jaggery and make into 5 g sized tablets.	Take orally with honey once in day.	Menorrhagia and sexually transmitted diseases.
						Prepare powder from roots.	Take orally with butter milk both morning and evening.	Anal diseases.
21	*Boerhavia diffusa* L. ANUBH01115	Atukamamidi	Nyctaginaceae	Herb	Leaves and roots	Prepare juice by using leaves and take orally with curd.	Twice in a day for 3 days.	Jaundice
						Boil the secondary roots in cow milk, dry it, and make into powder, take this powder with ghee and jaggery.	3 g of powder twice in a day	Menstruation
22	*Bombax ceiba* L. ANUBH01130	Burugu, Konda burugu	Bombacaceae	Tree	Bark	The extract of the bark taken orally.	Twice in a day.	Blood motion and tuberculosis
23	*Butea superba Roxb.* ANUBH01159	Teega moduga	Fabaceae	Climbing tree	Bark and seeds	Prepare seed powder and take orally with rice-washed water	Once in a day	Intestinal worms

(continued)

Table 9.2 (continued)

Sl. No	Species name/ voucher specimen No.	Local name (in Telugu)	Family	Habit	Parts used	Preparation and mode of use	Dosage	Therapeutic importance
						Seeds are macerated with *Calotropis* latex and applied on sting area, it takes off poison.	Immediate after scorpion sting	Scorpion sting
						Barks boiled in water, make extract, and add sufficient sugar.	Take ½ tea glass size extract for about 15 days.	Menorrhagia
24	*Caesalpinia bonduc* (L.) Roxb. ANUBH01160	Gacha kaya	Caesalpiniaceae	Shrub	Seed	Prepare macerated paste by using seed dal and apply as lotion on lymph nodes.	Twice in a day.	Inguinal lymph nodes
						Seed dal is macerated with egg yolk and applied as lotion.	Twice in a day.	Hydrocele
25	*Calotropis gigantea* (L.) R.Br. ANUBH01131	Tella Jilledu	Asclepiadaceae	Shrub	Leaves and flowers	Leaf juices prepared from heated leaves by adding *Zinger powder*, applied as bandage over the tonsils, it removes swelling and tonsils.	Apply twice/ day	Tonsils
						White *Calotropis* flower latex applied in the dental caries, it relieves pain.	Once in a day	Dental caries

26	*Canthium parviflorum* Lamk. ANUBH01124	Balusu	Rubiaceae	Shrub	Leaves and bark	Prepare extract by using bark or leaf and take orally.	15 to 25 g once in a day	Blood motions and diarrhea
27	*Caryota urens* L. ANUBHO1197	Jeeluga	Arecaceae	Tree	Inflorescence	The inflorescence is tapped for toddy.	–	Anesthetic tonic
28	*Cassia angustifolia M. Vahl.* ANUBH01117	Sunamukhi	Caesalpiniaceae	Shrub	Leaves	Prepare powder by mixing equal parts of fruit wall of *T. chebula* and *Cassia* leaves and add honey to this powder, make into 5 g tablets, and take orally.	One tablet/day at night only.	Indigestion
29	*C. auriculata* L. ANUBH01132	Thangedu	Caesalpiniaceae	Shrub	Flowers	Take equal amount of shade-dried flower powder and honey, make into 2–3 g sized tablets.	Taken orally both morning and evening about 30 days.	Polyuria
						Prepare extract by taking equal parts of bark and *Ficus hispida* bark. Take ½ tea glass orally.	2–4 times/day	Blood and gum motions.

(continued)

Table 9.2 (continued)

Sl. No	Species name/ voucher specimen No.	Local name (in Telugu)	Family	Habit	Parts used	Preparation and mode of use	Dosage	Therapeutic importance
30	*C. fistula* L. ANUBH01161	Rela or rella	Caesalpiniaceae	Tree	Leaves	Leaves are macerated with rice-washed water and prepared ointment and applied as lotion.	Apply twice/ day	Leprosy
						Make chutney by using leaves and take with rice.	Once in day	Free motion and constipation
31	*C. occidentalis* L. ANUBH01133	Kasintha	Caesalpiniaceae	Herb	Leaves	Prepare paste by grinding leaves, salt, and turmeric powder.	Apply twice/ day	Wound healer
						Macerate the leaves with cow ghee and apply as band over eye, it reduces eye redness.	Apply twice/ day	Conjunctivitis
32	*C. tora* L. ANUBH01116	Tagirasa, tantepu rodda	Caesalpiniaceae	Shrub	Leaves and seeds	Macerated paste of grains applied on forehead as band relieves pain.	Apply twice/ day	Migraine headache.
						Prepare leaf paste by using leaves and apply as lotion.	Apply twice/ day	Honeybee sting.
33	*Centella asiatica* (L.) urban. ANUBH01123	Saraswathi aku	Apiaceae	Herb	Leaves	Extract prepared by using boiling leaves of *Centella asiatica* and Mentha seeds.	Take 10 to 15 g per day orally.	Motion and fever.

						A table spoon of dried leaf powder taken orally with glass of milk.	Table spoon/ day 3 to 4 days.	Jaundice
						A table spoon of dried leaf powder with glass of butter milk.	Table spoon/ day (use in 10 days)	Gonorrhea
34	*Cissampelos pareira* L. ANUBH01163	Chiruboddi, Vishaboddi	Minispermaceae	Climber	Roots	Take equal parts of *C. pareira* roots, *Ailanthus excelsa* bark, garlic cloves, and *Cuminum* seeds, macerate into paste, and make it into 5 g sized tablets.	Taken orally both morning and evening.	Fever
						Prepare mixture by mixing 5 parts of *Cissampelos pareira* root powder, pepper 5 parts, Asafoetida 3parts, *Zinger* 3 parts, and add honey, make into 3 g sized tablets.	Taken orally 3 tablets/day	Indigestion

(continued)

Table 9.2 (continued)

Sl. No	Species name/ voucher specimen No.	Local name (in Telugu)	Family	Habit	Parts used	Preparation and mode of use	Dosage	Therapeutic importance
35	*Cissus quadrangularis* L. ANUBH01164 .	Nalleru	Vitaceae	Climbing shrub	Stem	Take *Cissus* stem boil in the pot and make into juice, add honey to this and take orally.	Once in a day	Hiccups
						Cut small piece from woody twiner and shade dried and make into powder. To this add roasted *Piper longum* powder and take orally with honey.	Once in a day	Bronchitis and cough
						10 g without surface layer of *Cissus* stem and *Phaseolus aureus* seeds are macerated and taken orally.	1 g in both morning and evening.	Removes paralytic pains.
36	*Costus speciosus* (Koenig.) J. E. smith. ANUBH01165	Basica dumpa, Changalava costu	Costaceae	Herb	Root tubers	Grind tubers of *Costus speciosus* boiled in *Sesamum* oil and filter the extract and prepare oil.	Used this oil as lotion twice in a day.	Piles
						Prepare extract by mixing *Costus* tuber and *Ricinus communis* root bark with soured rice extract.	Once in a day	Headache

37	*Curculigo orchioides* Gaertn. ANUBH01166	Nelathadi	Hypoxidaceae	Herb	Tubers	The paste prepared by using roots with gout milk and jaggery and applied as face pack.	Once in week.	Cosmetics
						Powder prepared by using tuber, from this 1.5 g to 3 g combined with milk and sugar and taken orally.	Once in a day.	Leukorrhea and sperm wastage.
						Take equal amounts of tuber and *Psoralea* seeds and prepare powder.	Take 1 g daily morning and evening.	Deafness removal.
38	*Curcuma angustifolia* Roxb. ANUBH01167	Palabittra dumpa	Zingiberaceae	Herb	Tuber	Take boiling milk, add Palagunda and sugar, stir it till it becomes semisolid. To this add elachi powder, make it into small size biscuits (cut the curcuma tuber into small pieces and keep in water and sundry it about 2 days; this is known as palgunda).	Once in a day	Venereal and menstrual diseases.

(continued)

Table 9.2 (continued)

Sl. No	Species name/ voucher specimen No.	Local name (in Telugu)	Family	Habit	Parts used	Preparation and mode of use	Dosage	Therapeutic importance
39	*C. longa* L. ANUBH01168	Pasupu	Zingiberaceae	Herb	Tuber	Prepare oil by mixing turmeric powder and neem oil, apply as lotion.	Apply twice in a day	Skin disease.
						Prepare turmeric water by mixing stem powder with water and use to clean the eyes. It reduces redness of eyes.	Apply twice in a day	Conjunctivitis
40	*C. pseudomontana* Graham. ANUBH01202	Adavi pasupu	Zingiberaceae	Herb	Tuber	Boiled tubers ground with a pinch of salt is administered orally to increase lactation in nursing mother.	Once a day for 7 days	Lactation
41	*Cymbidium aloifolium* (L.) SW. ANUBH01199	Pedda vajanika	Orchidaceae	Epiphytic herb	Leaf	Juice prepared by grinding the leaves and used to cure earache.	2-4 drops of leaf juice poured inside the ear twice per day.	Earache
42	*Datura metal* L. ANUBH01118	Ummetta	Solanaceae	Herb/ under shrub	Leaves and roots.	Prepare extract by maceration of *Datura* leaves, *Piper longum* with cold water and take orally.	Once in a day	Filariasis

						Take north side roots of *Datura* plant, grind them, mix with milk, jiggery, and ghee, and take orally.	Once in a day	Mental recovery (use only in the presence of herbal doctor)
43	*Dichrostachys cinerea* (L.) ANUBH01198	Veluthuru chettu	Mimosaceae	Tree	Root	Extract prepared by using root powder and used to remove or dissolve the renal stones (kidney).	Take a half spoon extract twice a day for 2 weeks.	Renal stones
						Extract prepared by using root powder, to this extract a spoon full of garlic juice and pinch of black pepper powder are added.	A spoonful of extract taken orally once in a week for about 5 weeks.	Paralysis
44	*Dillenia indica* L. ANUBH01193	Kallinga	Dilleniaceae	Tree	Fruits	Fruits are macerated and prepared as juice which has laxative effect.	One time only.	Laxative agent
45	*Dioscorea hispida* Dennst. ANUBH1194	Adavi dumpa.	Dioscoreceae	Climber	Leaf	Leaves are macerated and prepared as juice to cure earache.	2–3 drops twice in a day.	Earache

(continued)

Table 9.2 (continued)

Sl. No	Species name/ voucher specimen No.	Local name (in Telugu)	Family	Habit	Parts used	Preparation and mode of use	Dosage	Therapeutic importance
46	*D. oppositifolia* L. ANUBH01195	Tella gadda	Dioscoreceae	Climber	Tuber	The tubers are ground and prepared as paste.	Apply twice in a day on the surface of the wound.	Wound healer
47	*Diospyros melanoxylon* Roxb. ANUBH01196.	Beedi aku.	Ebenaceae	Tree	Flowers	Decoction prepared by using shade-dried flowers.	Take a full spoon orally twice in a day for about week.	Urinary disorders
48	*Dodonaea viscosa* L. ANUBH01189	Bandam	Sapindaceae	Shrub	Leaves	Prepare leaf juice by using fresh leaves.	Two drops of juice put into the nostrils thrice a day for 15 days.	Fits
49	*Eclipta prostrata* L. ANUBH01119	Guntagalagara-ku	Asteracea	Herb	Whole plant	Prepare tablets by mixing with whole plant powder with whole plant extract.	Taken orally twice in a day.	Blood motions
						Fish egg taken with whole plant extract orally reduces xerophthalmia.	Once in a day for 7 days.	Xerophthalmia

50	*Euphorbia hirta* L. ANUBH01120	Reddyvari nanudubalu	Euphorbiaceae	Herb	Leaf tips	Take 9 leaf tips and add sufficient pepper, mix each other and make into 1 g tablets.	Taken orally early in the morning.	Cold fever
						Prepare extract by mixing 5 g leaf juice, 5 g pepper and take orally.	Once in day after menstruation for only 3 days.	Menstrual pain
51	*E. cattimandoo* W. Elliot. ANUBH01169	Kattimandu	Euphorbiaceae	Shrub	Latex	Take equal parts of milky latex, oil and cow milk, boil the contents till oil evaporated. Massage with this oil on paralytic hand reduces the pain.	Apply twice/ day	Paralysis
						Collect the latex from stem and apply externally on cut parts of the body, it clots the blood.	Apply at the time of cuts.	Wound healer and blood clot.
52	*Ficus religiosa* L. ANUBH01134	Ravi chettu	Moraceae	Tree	Leaves	Leaves are boiled in water for about one hour and applied for healing of wounds.	Twice per day.	Healing of wounds

(continued)

Table 9.2 (continued)

Sl. No	Species name/ voucher specimen No.	Local name (in Telugu)	Family	Habit	Parts used	Preparation and mode of use	Dosage	Therapeutic importance
53	*Gloriosa superba* L. ANUBH01170	Potti dumpa, Adavi nabi	Liliaceae	Herb	Root tuber	Root tuber pastes with kerosene is applied on scorpion bite area.	Twice in a day.	Scorpion bite.
						Prepare root tuber paste by using pomegranate juice and keep in hot air oven until it becomes ash. This ash is mixed with lemon juice and applied on piles area.	Twice in a day.	Piles
54	*Gymnema sylvestre* (Retz.) R.Br.ex.Schultes. ANUBH01171	Podapatri	Asclepiadaceae	Climber	Leaves	Prepare powder by using equal amount of leaf powder and *Syzygium* seed powder.	1 g with glass of water in morning and evening.	Diabetes
						Prepare solution by using leaves and pour 2 to 4 drops into eye of animal to cure watery eye.	Two times per day.	Animal eye diseases.
55	*Habenaria plantaginea* Lindl. ANUBH01202	Kusumagedda	Orchidaceae	Terrestrial herb	Tuber	Tubers ground with black pepper (*Piper nigrum)* and garlic (*Allium sativum*) and the paste made into pills.	Take 1-2 pills orally twice a day till to cure.	Chest-pain and stomach-pain

56	*Hemidesmus indicus* (L.) R.Br. ANUBH01172	Pala sugandi or avu sugandi	Asclepiadaceae	Climber	Root	Prepare extract by mixing 2 parts of roots and 1 part of ghee.	Take orally once in a day.	Rat poisoning
						Prepare extract by using root bark and apply on diseased area. Use this extract to prepare tablets.	Apply lotion two times in a day. Take one tablet with milk.	Herpetic
57	*Holarrhena antidysenterica* (Roth) Wall. ex A. DC. ANUBH01173	Kodisa Pala or Pala chettu.	Apocyanaceae	Tree	Bark and roots.	Make extract by taking equal amounts of stem bark of *Holarrhena, Calotropis gigantea,* and *Nerium odoratum.*	Apply lotion two times.	Wound cleaning
						Prepare extract by using root bark and take orally with curd.	Once in a day.	Dissolves renal stones.
58	*Hybanthus enneaspermus* (L.) J. Muell. ANUBH01102	Nelakobbari, ratnapurusha	Violaceae	Herb	Leaves	Prepare paste by maceration of leaves and stalks. Use this paste as lotion over filarial parts.	Twice in a day.	Filariasis
59	*Ipomoea hederifolia* L. ANUBH01192.	Kasiratnam	Convolvulaceae	Climber	Roots	Prepare paste by maceration of tuberous roots.	Take small amount orally for about 15 days.	Gonorrhea

(continued)

Table 9.2 (continued)

Sl. No	Species name/ voucher specimen No.	Local name (in Telugu)	Family	Habit	Parts used	Preparation and mode of use	Dosage	Therapeutic importance
60	*I. mauritiana* Jacq. ANUBH01174	Palagummadu	Convolvulaceae	Climber	Root tuber	Prepare powder from shade-dried tuber pieces. Taking 2 spoonful powder with fresh toddy (alcoholic product of *Borassus flabellifer*) enhances the milk production in delivery women.	Once in a day	Lactation
61	*Jatropha gossypifolia* L. ANUBH01197.	Seema nepalamu	Euphorbiaceae	Shrub	Leaves	Extract prepared by using leaves and applied as lotion.	Twice in a day.	Eczema
62	*Lannea coromandelica* (Houtt.) Merr. ANUBH01175	Gumpena	Anacardiaceae	Tree	Bark	Bark and 7 pepper seeds ground and prepared as 1 g size tablets.	Take two tablets per day.	Blood motions
63	*Lawsonia inermis* L. ANUBH01103	Gorinta	Lythraceae	Tree	Leaves	A paste of leaves and piece of soap applied for swellings or sprains.	Twice in a day.	Sprains and swellings.
						Prepare oil by boiling leaves in coconut oil or Sesamum oil.	Massage the hair with this oil once in a day.	Hair fall &and blackening hair.
						Leaf paste is applied on the surface of the nail.	Once in a month	With low and herpetic

64	*Madhuca longifolia* (J. Koenig ex L.) J. F. Macbr. ANUBH01176	Vippa, Ippa	Sapotaceae	Tree	Flowers	Prepare extract using flowers; add ¼ part honey, pour into pot. Keep this pot below the soil for 1 month.	Take orally 1-2 ounce in a day to control mucoid stools.	Mucoid stools.
						Prepare flower powder and take one teaspoon with a glass of milk	Twice in a day to enhance male fertility.	Enhances male fertility.
65	*Mimosa pudica* L. ANUBH01104	Attipatti	Mimosaceae	Shrub	Leaves	Take 4 g of *Mimosa* leaves, black pepper, and a cup of water, macerate and filter extract, consume orally early morning. The extract is applied as a bandage around the filarial leg. It removes pain and boils.	Once in a day early morning for about 30 days.	Filariasis
66	*Mirabilis jalapa* L. ANUBH01177	Mogga malle, chandrakantha	Nyctaginaceae	Shrub	Leaves and root tubers	Prepare paste by using root tuber and black pepper in equal amounts and form 1 g size tablets.	Take one tablet at morning and evening daily for 10 days.	All diseases.

(continued)

Table 9.2 (continued)

Sl. No	Species name/ voucher specimen No.	Local name (in Telugu)	Family	Habit	Parts used	Preparation and mode of use	Dosage	Therapeutic importance
						Take castor oil onto the surface of the leaf and heat the leaf, place in gonorrheal wounds.	Once in a day.	Gonorrhea wounds.
67	*Momordica dioica* Roxb. ex. Willd ANUBH01105	Agakara	Cucurbitaceae	Climber	Leaves and root tubers	Tuber is macerated with honey and prepared as lotion.	Apply the lotion twice in a day on forehead to relive from headache.	Headache.
						Leaf juice is poured in *Ricinus* oil and the contents are boiled and used as massage oil for joint pains.	Twice in a day to set relief from arthritis.	Arthritis
68	*Mucuna pruriens* (L.) DC. ANUBH01178	Dulagondi	Fabaceae	Climbing vine	Seeds	Prepare powder by taking equal parts of *Mucuna* grains, *Tribulus* grains, seeds of *Hygrophila, Asparagus* roots, *Curculigo* tuber.	One tea spoon powder taken orally with cow milk once in a day.	Enhances male fertility.
69	*Ocimum basilicum* L. ANUBH01106	Sabja, rudrajeda tulasi	Lamiaceae	Shrub	Leaves and roots	The grains are soaked in water for about 6 hours till they become imbibed and taken orally with sugar.	Once in a day.	Body coolants

						Prepare extract by using roots and take ¼ tea glass orally.	Once in a day.	Constipation
70	*O. sanctum* L. ANUBH01107	Tulasi	Lamiaceae	Shrub	Leaves	Prepare juice by using leaves and take orally with honey.	Both morning and evening	Cough with mucous
71	*Operculina turpethum* (L.) Silva Mansu. ANUBH01179	Tegada	Convolvulaceae	Twiner	Roots	Prepare powder mixture by mixing 3 parts of white *Operculina* roots and one part of *Zingiber*, *Piper* mixture, *Phyllanthus* fruit wall, elachi, *Cinnamon* powder mixture	Take 3 to 5 g orally twice in a day	Piles and constipation
72	*Oroxylum indicum* L. ANUBH01191	Dekki/Pimpini	Bignoniaceae	Tree	Stem bark	Paste prepared by maceration of the stem bark.	Small amounts taken orally for 2-4 days and also apply on infected portions.	Skin diseases

(continued)

Table 9.2 (continued)

Sl. No	Species name/ voucher specimen No.	Local name (in Telugu)	Family	Habit	Parts used	Preparation and mode of use	Dosage	Therapeutic importance
73	*Pergularia daemia* (Forsk.) ANUBH01108	Dustapu teega	Asclepiadaceae	Climber	Roots and latex.	Make extract using twining root of *Pergularia* and sesame seeds.	Take 2 g of extract orally once in a day.	Dysmenorrhea
						Make extract by mixing *Coleous* and *Cuminum* seeds with *Pergularia* latex.	1 g with hot water once in a day	Constipation
74	*Phyllanthus amarus* Schum.& Thonn. ANUBH01109	Nela usiri	Euphorbiaceae	Herb	Whole plant and leaves	Juice prepared from whole plant by mixing with curd or butter milk is taken orally.	Daily early morning one glass.	Jaundice
						Prepare paste by using leaves and make into 3 g size tablets and take orally.	One tablet/day.	Motions
						Prepare paste by using leave of *Phyllanthus* and *Trigonella* seeds and make into 2 g size tablets.	Three times per day with a cup of curd.	Blood motions

75	*P. emblica* L. ANUBH01110	Usiri kaya	Euphorbiaceae	Tree	Fruits	Prepare mixture by mixing 120 g fruit wall of *Phyllanthus emblica*, 120 g *Terminalia. chebula*, 60 g *T. bellarica* powder and 270 g sugar pellets. 10 g of this powder is taken orally with water.	Twice in a day.	Indigestion
						Fruit macerated paste applied as band below the abdomen relieves the urinary tract pain and irritation.	Twice in a day.	Urinary tract irritation and pain.
76	*Piper longum* L. ANUBH01181	Pippali	Piperaceae	Herb	Root and pods	Take equal parts of roots and pepper, make into powder, and ½ g of this powder is taken orally with milk. This promotes lactation.	Once in a day at night time	Galactagogue
						Take equal parts of Asafoetida, rock salt, *Terminalia chebula* fruit wall, Pippali, prepare into powder.	½ tea spoon taken orally both morning and evening.	Stomachache

(continued)

Table 9.2 (continued)

Sl. No	Species name/ voucher specimen No.	Local name (in Telugu)	Family	Habit	Parts used	Preparation and mode of use	Dosage	Therapeutic importance
77	*P. nigrum* L. ANUBH01180	Meriyalu	Piperaceae	Climbing shrub	Seeds	Prepare 2 mg size tablets by maceration of equal parts of pepper and *Holarrhena* root bark.	Taken orally 3 times in a day for 3 days.	Gum motions
78	*Plumbago zeylanica* L. ANUBH01143	Tella chitramulamu	Plumbaginaceae	Shrub	Root bark	Prepare powder by using root and take orally with honey.	1 g in a day.	Obesity
						Mixture prepared by mixing equal amounts of root bark of *Plumbago, Zinger, Peppe*, rocksalt, *Coleous* powder.	4grams twice/ day	Indigestion.
79	*Pongamia pinnata* (L.) Pierre. ANUBH01142	Kanuga chettu	Fabaceae	Tree	Seeds and bark	Prepare juice by mixing bark juice of *Pongamia* and *Phyllanthus* and take orally with sugar.	¼ tea glass taken once in a day.	Small pox
						The seeds are macerated with water and applied as lotion over diseased parts of skin.	Apply twice in a day.	Scabies

80	*Pueraria tuberosa* (Willd.) DC. ANUBH01150	Nelagummadu, Vidari	Fabaceae	Climbing vine	Tubers	Prepare powder by using tubers and take orally with cow ghee.	3-5 g three times/day	Reduces sperm discharge
						Prepare powder by taking equal parts of *Pueraria* tuber and *Asparagus* root, take orally with milk.	Take one spoon both morning & evening	Menstrual and pregnancy problems.
81	*Randia spinosa* Thumb. Poir. ANUBH01149	Manga chettu	Rubiaceae	Tree	Pods	Take equal parts of *Randia* pods and Katukarohini (*Picrorhiza kurroa*), macerate with hot water, apply as band over the stomach.	Twice in a day.	Stomachache
82	*Rauvolfia serpentina* (*L.*) Benth. ex Kurz. ANUBH01148	Sarpagandi	Apocyanaceae	Shrub	Root bark	Make powder by using 100 g root bark of *Rauwolfia*, 200 g *Phyllanthus* powder without seeds, 50 g katukarohini (*Picrorhiza kurroa*). Add sufficient amount of *Cuminum* extract and make into tablets.	Take one tablet daily (morning and evening).	Blood pressure reduces.

(continued)

Table 9.2 (continued)

Sl. No	Species name/ voucher specimen No.	Local name (in Telugu)	Family	Habit	Parts used	Preparation and mode of use	Dosage	Therapeutic importance
83	*Ricinus communis* L. ANUBH01141	Amudamu	Euphorbiaceae	Shrub	Leaves and seeds	Prepare tablets by mixing leaf tips of *Ricinus* and pepper paste.	Take one tablet once in a day for three days.	Jaundice
						Prepare tablets by grinding *Ricinus* seeds and sugar.	One tablet twice/day	Arthritis
84	*Sapindus emarginatus* Vahl. ANUBH01140	Kunkudu	Sapindaceae	Tree	Flowers	Make a powder from shade-dried flowers. Taken orally 1 g of this powder with 1 spoon sugar.	Twice/day both morning and evening.	Menorrhagia
85	*Semecarpus anacardium* L.f. ANUBH01147	Nallajeedi	Anacardiaceae	Tree	Seed	Prepare oil by mixing pod juice of *Semecarpus* and sesamum oil.	Massage the paining part with this oil twice in a day.	Arthritis
						Take pod juice by using pin and apply as lotion over foot cracks and soul pains.	Twice in a day.	Foot and soul pain
86	*Solanum nigrum* L. ANUBH01139	Kamanchi, gajukura	Solanaceae	Herb	Whole plant and leaves	Whole plant dry in shade area and make into powder, take orally one table spoon.	Both morning and evening for about 3 days.	Jaundice
						Prepare leaf juice by using young leaves.	Apply twice in a day	Swelling (due to water accumulation)

87	*Soymida febrifuga* (Roxb.) A. Juss. ANUBH01187	Somita	Meliaceae	Tree	Bark	Prepare extract from 20 g of bark powder and take ½ tea glass orally.	Morning and evening for 4 days.	Menstrual pain.
88	*Sphaeranthus indicus* L. ANUBH01138	Bodatharam, kodipenu mokka	Asteraceae	Herb	Whole plant and leaves	Prepare extract by using whole plant.	Taken orally 20 ml of extract before lunch for 7 days.	Migraine headache
						Prepare oil by mixing leaf juice and coconut oil.	Hair tonic for about 1 month	Blackening of hair
89	*Sterculia urens* Roxb. ANUBH01188	Kovelachettu.	Sterculiaceae	Tree	Stem	Gum is trapped from the stem and used to cure stomach ulcers.	Once in week.	Stomach ulcers
90	*Strychnos nux-vomica* L. ANUBH01162	Mushini, vishamushti	Strychnaceae	Tree	Seeds and fruit pulp	Seeds are heated till it becomes black soot and ground, make it into powder.	½ tea spoon of powder taken orally with milk once in a day.	Leukorrhea.
						Lotion prepared by using fruit pulp and turmeric.	Once in a day and used as cosmetic.	Cosmetics and foot cracks.

(continued)

Table 9.2 (continued)

Sl. No	Species name/ voucher specimen No.	Local name (in Telugu)	Family	Habit	Parts used	Preparation and mode of use	Dosage	Therapeutic importance
91	*Syzygium cumini* (L.) Skeels. ANUBH01186	Neredu	Myrtaceae	Tree	Leaves and seeds	Seeds are heated and then made into powder. Add a spoon of seed powder and tadi *(Borassus flabellifer)* jaggery to a cup of milk and make it coffee like.	Take two times per day.	Diabetes
						Prepare bark extract and take with sugar.	Take two times per day.	Leukorrhea
						2 g of seed powder with glass of water.	Take three times per day	Heavy motions.
						Prepare extract by using leaf tips of *Syzygium* and mango, take with sugar or honey. It reduces vomiting.	Take three times per day.	Vomiting
92	*Terminalia arjuna* (Roxb. ex. Dc.) wt & Arn. ANUBH01144	Tella maddi or yeti maddi	Combretaceae	Tree	Bark and leaves.	Extract of bark is mixed with ghee and applied on breakage portion of the bone.	Once in a day	Bone fracture
						Boil the bark in milk and add little amount of ghee and honey.	Once in a day	Blood motions and heart diseases.

						Leaf extract is applied as a lotion to reduce joint pains.	Once in a day	Arthritis pains.
93	*T. bellirica* (Gaertn.) Roxb. ANUBH01145	Tani, Tademanu	Combretaceae	Tree	Pods and seed	The roasted pods with salt is taken orally.	Once in a day.	Diarrhea
						Seed inner part is taken with thatikallu (alcoholic liquid of *Borassus flabellifer*).	Take 1 gm 3 times in a day.	Dissolve stones in kidney.
						Seed powder	0.5 g taken orally in a day.	Cough and bronchitis.
94	*T. chebula* Retz. ANUBH01146	Karakaya	Combretaceae	Tree	Fruits	Take equal amounts of powder of fruit and Jaggery to prepare paste.	Small balls prepared and taken after meal per day two times.	Blood piles
						Fruit only taken orally.	Single fruit every day two times. Continue up to two months.	Indigestion

(continued)

Table 9.2 (continued)

Sl. No	Species name/ voucher specimen No.	Local name (in Telugu)	Family	Habit	Parts used	Preparation and mode of use	Dosage	Therapeutic importance
95	*Tinospora cordifolia* (Willd) Miers. ex Hook.f & Thoms. ANUBH01182	Kodipudi teega or teppa teega.	Minispermaceae	Climber	Stem and leaves	Prepare extract by using leaf and stem.	A cup of extract in the morning and evening every day.	Fever
						Grind the leaves and prepare paste.	A small amount of past taken orally every day morning with glass of butter milk.	Jaundice
96	*Tribulus terrestris* L. ANUBH01137	Palleru	Zygophyllaceae	Herb	Whole plant and leaves	Leaf juice prepared by grinding leaves with water and taken orally.	½ cup both morning and evening	Dissolve renal stones
						Prepare whole plant extract and add ¼ part sesamum oil to this extract, use as massage oil for arthritis.	Apply twice in a day.	Arthritis
97	*Vanda tessellata* (Roxb.) G. ANUBH01200	Vajanika or Bdanika	Orchidaceae	Epiphytic herb	Leaves	The juice of the leaves used to cure earache and skin diseases.	2-3 drops of juice poured into ear twice in day.	Earache and skin diseases.

						The aerial roots and leaves are ground and paste in plastered for bone fracture	Once in week.	Bone fracture
98	*V. testacea* (Lindl.) ANUBH01201	Vajanika or Bdanika	Orchidaceae	Epiphytic herb	Leaves	The juice of the leaves used to cure earache and skin diseases.	2-3 drops of juice poured into ear twice in day.	Earache and skin diseases.
						The aerial roots and leaves are ground and paste in plastered for bone fracture.	Once in week	Bone fracture
99	*Vitex negundo* L. ANUBH01183	Vavili	Verbenaceae	Tree	Leaves	Prepare extract by using leaves. Take 4 parts of this extract and 1 part of cow ghee/butter, boil it till the extract removal.	2 g of this ghee twice in a day.	Tuberculosis
						Take leaf and sesamum oil into dish and boil and add Saindhava salt (rock salt) and old jaggery. Prepare curry by using leaf tips and ghee.	Clean the ear with cotton and pour two drops per day in 3 to 4 days. Eat a cup of curry with rice once in day per/week.	Ear nerve wound discharge. Blood piles and red discharge(over bleeding)

(continued)

Table 9.2 (continued)

Sl. No	Species name/ voucher specimen No.	Local name (in Telugu)	Family	Habit	Parts used	Preparation and mode of use	Dosage	Therapeutic importance
100	*Withania somnifera* (L.) Dunal ANUBH01184	Penneru	Solanaceae	Shrub	Roots	Prepare powder by taking equal parts of roots and sugar pellets.	Morning and evening for few days.	To overcome anemia
						Prepare root powder and add little amount of sugar pellets powder, ghee and take orally.	Once in a day for 40 days	Increases sperm count.
101	*Woodfordia fruticosa* (L.) Kurz ANUBH01185	Jaji, Arepuvvu	Lythraceae	Shrub	Flowers	2 g sized tablets are prepared by the maceration of flowers and taken orally with rice water.	Once in a day	Menorrhagia
102	*Wrightia tinctoria* R, Br. ANUBH01136	Aku Pala	Apocyanaceae	Tree	Leaves	Prepare leaf extract and take ½ tea glass orally.	Once in a day	Stomachache
103	*Zingiber officinale* roscoe. ANUBH01135	Sonti	Zingiberaceae	Perennial herb	Stem	Dried pieces are soaked in cow butter over night, these are macerated and prepared into 5 g sized tablets.	Two tablets in a day in two days.	Blood motions
104	*Ziziphus xylopyrus* (Retz.) Willd. ANUBH01190	Gotti chettu	Rhamnaceae	Tree	Bark	Bark of the stem and pepper are ground and made into 2 g size tablets.	Taken orally twice a day for 3 days.	Cough

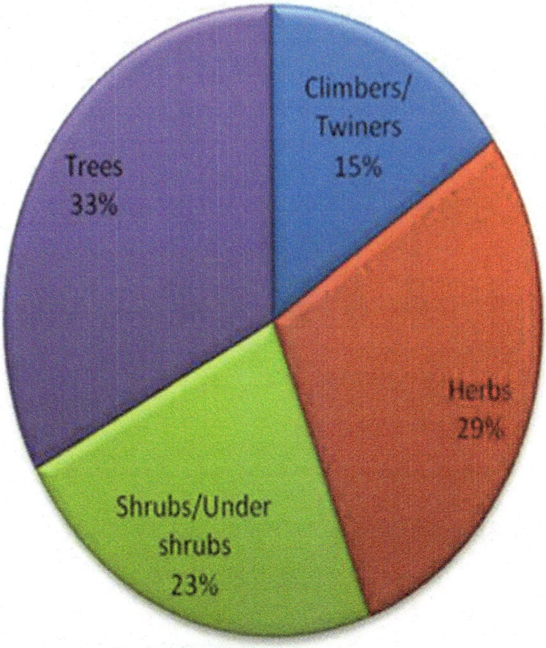

Fig. 9.3 Habit-wise analysis of ethnomedicinal plants

w(5 species); Convolvulaceae, Orchidaceae, Zingiberaceae (4 species each); and Apocyanaceae, Combretaceae, Mimosaceae, Liliaceae, Lythraceae, Solanaceae (3 species each). Fourteen families were represented by 2 species each and 26 families by 1 species each. The present study further indicates that the majority of medicinal plants are dicotyledons rather than monocots. Suthari et al. (2014) and Gottlieb et al. (2002) also observed that the dicotyledonous families possess more medicinal properties than the monocots.

9.5 Mode of Preparation

Traditional practitioners used different parts of the plant to prepare herbal medicine. Depending on the plant part used for medicinal purposes, the leaf constituents showed the highest percentage (32.59%), followed by root (14.81%), bark (11.11%), seed (9.6%), tubers (8.8%), flower and inflorescence (5.9%), whole plant (5.9%), stem (3.7%), latex, fruit, and pods (2.2% each), and rhizome [(0.74%); (Fig. 9.4)]. The use of leaves in the preparation of remedies for various ailments is also very common in other parts of the world (Bonet et al. 1992; Tabuti et al. 2003: Muthu et al. 2006; Teklehaymanot et al. 2007). Most commonly used methods of plant preparations are decoctions (26%) and powders (26%), followed by paste (23%), juice (12%), crushing (9%), chewing (3%), and others (1%). Methods of preparations are presented in Fig. 9.5. Majority of the informants suggested taking herbal

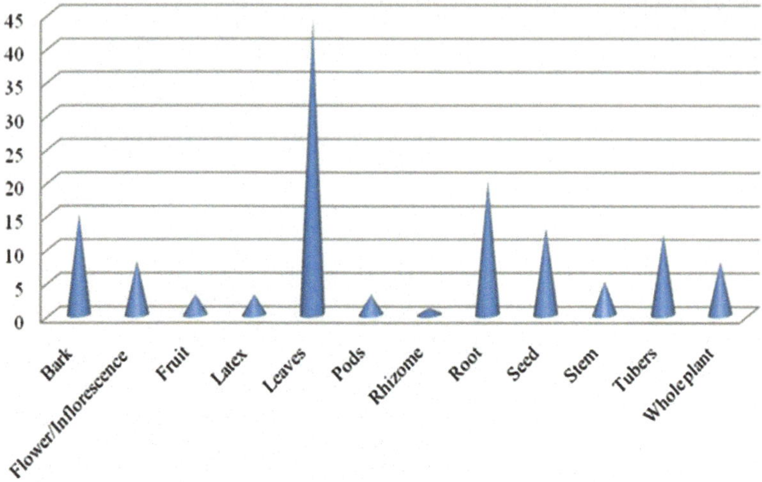

Fig. 9.4 Plant part-wise analysis of crude drug samples

Fig. 9.5 Methods of medicinal plant preparation

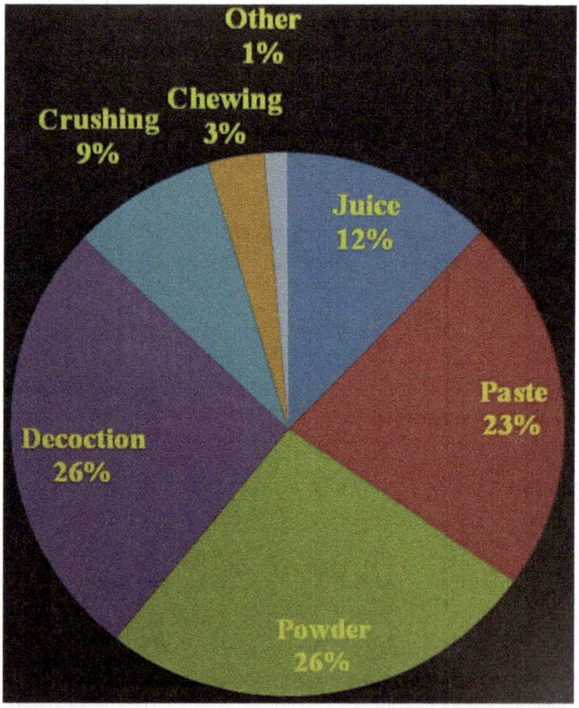

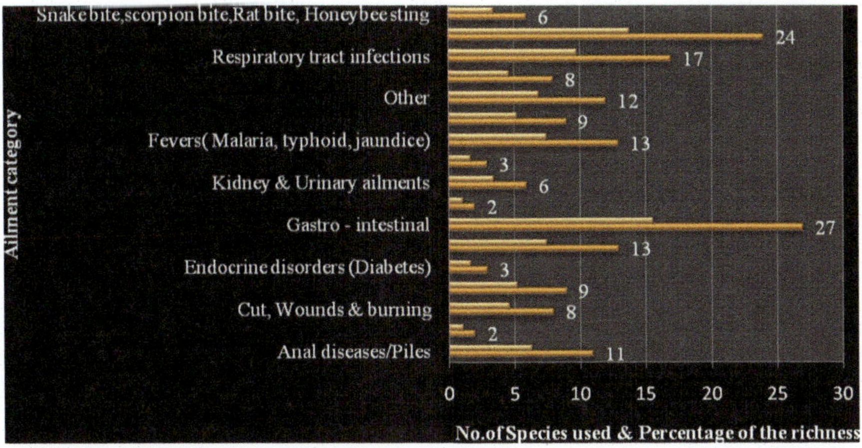

Fig. 9.6 Percentage of ailments treated

medicines orally (111 remedies), externally applying or rubbing (48 remedies) and eye problems (4 remedies), as consistent with comparable investigations (Kayani et al. 2015; Sadat-Hosseini et al. 2017).

About 104 plant species reported in this study have been used in curing 173 different ailments either singly or in combination. The plants were used to treat various diseases grouped under 17 categories (Fig. 9.6). Highest number of species (27) has been used in different formulations to treat gastrointestinal ailments while 24 species have been used against reproductive disorders.

9.6 Fidelity Level of Medicinal Plants

Mesfin et al. (2014) studied the ethnomedicinal plants in Amaro Woreda (Ethiopia) and mentioned that medicinal plants that are known as remedies of single ailment have 100% FL value as compared to those that used as remedies for more than one type of ailment. Medicinal plants with high FL values are those that are widely used by the local practitioners. In our study, we found highest FL value for *Andrographis paniculata* and *Gymnema sylvestre* (100%), and lowest value was recorded for *Gloriosa superba* (40%) (Table 9.3). Similarly, Ugulu (2012) studied the medicinal plants used for therapeutic Turkish bath by local people of various areas in Turkey and reported that *Cucumis sativus* (100%) has the highest FL value and *Thymus vulgaris* (32%) has the lowest value.

Most of the medicinal plants were freely harvested by users from the nature in which they were abundantly found. About 80% of the medicinal plants were wild species. Some species like *Sterculia urens*, *Gymnema sylvestre*, and *Hemidesmus indicus* were reported to be planted by tribals in the surroundings of the sacred groves and home gardens. Plant collection is their part-time activity. Many of the medicinal plants are collected in winter or early spring seasons such as *Gloriosa*

Table 9.3 Fidelity Level of medicinal plants used by tribal people of sacred groves

Plant species	Family	Therapeutic uses	No. of informants (Ip)	Total no. of informants (In)	Fidelity level (%)
Acorus calamus	Acoraceae	Cough and cold	11	14	78
Adhatoda zeylanica	Acanthaceae	Asthma	7	13	54
Andrographis paniculata	Acanthaceae	Malaria	6	6	100
Cassia fistula	Caesalpiniaceae	Anal diseases	7	14	50
Curcuma longa	Zingiberaceae	Skin diseases	9	11	81
Gloriosa superba	Liliaceae	Scorpion sting	4	10	40
Gymnema sylvestre	Asclepiadaceae	Diabetes	7	7	100
Hemidesmus indicus	Asclepiadaceae	Herpetic	4	7	54
Phyllanthus amarus	Euphorbiaceae	Jaundice	10	11	91
Soymida febrifuga	Meliaceae	Menstrual pain	6	13	46
Strychnos nux vomica	Strychnaceae	Leukorrhea	9	13	69
Terminalia arjuna	Combretaceae	Bone fracture	5	7	71
Terminalia chebula	Combretaceae	Blood piles	6	9	66
Woodfordia fruticosa	Lytraceae	Mennoergia	4	9	45

Table 9.4 Direct matrix ranking (DMR) of tree species with different use other than medicinal value in the study area

Use	*A. excelsa*	*A. indica*	*O. indicum*	*C. fistula*	*M. longifolia*	*P. emblica*	*S. cumini*	*T. chebula*
Construction	10	15	10	8	5	5	10	6
Hedge, fencing	8	5	02	4	2	2	6	5
Fire wood	10	5	10	6	8	10	20	10
Cash income	15	15	12	5	25	20	16	25
Fruit, food	4	5	10	5	20	25	20	20
Total	47	45	44	28	60	62	72	66
Rank	**5th**	**6th**	**7th**	**8th**	**4th**	**3rd**	**1st**	**2nd**

superba, *Costus speciosus*, and *Acorus calamus*, when the plants are in dormant state, and species such as *Terminalia chebula* and *Terminalia bellirica* fruits are collected during early summer.

9.6.1 Direct Matrix Ranking (DMR)

Direct matrix ranking (DMR) for tree species showed that *Syzygium cumini* had the most multipurpose use other than medicinal uses, Fruit and Fire wood (20), cash income (16), construction (10), and Hedge, fencing (6) (Table 9.4). Other tree species with high DMR value were *Terminalia chebula* and *Phyllanthus emblica.* The lowest DMR value was recorded for *Cassia fistula* (Table. 9.4). These result indicated that indigenous communities of the study area are highly reliant on the tree species for household furniture, cash income, construction, and wall thatching. Similar observations were reported by Lulekal et al. (2013) in the North Zone of Ethiopia. Ishtiaq et al. (2012) mentioned that multiuse of plants other than medicinal significance plays an important role in the routine life of the native people of eastern medicine.

9.6.2 Priority Ranking (PR)

The priority ranking data based on the degree of threats to ethnomedicinal pants was conducted using key informants (Table 9.5). In PR, informants ranked poducultivation (Burn agriculture) (19.25%) as a leading threat, followed by fuel wood (14%), monocultivation (13.33%), lack of awareness on medicinal plants (11.85%), overexploitation (11.11%), forest fires (9.6%), and drought (4.4%). Hussain et al. (2004) mentioned that overgrazing, habitat destruction, and overexploitation are increasing day by day, which are main factors responsible for species extinction. According to Tesfaye Seifu et al. (2006), globally, agricultural encroachment in natural habitat leads to consequent loss of medicinal plant species.

Table 9.5 Priority ranking (PR) of factors perceived as threats to ethnomedicinal plants based on their level of destructive effects in the study area (R1 = Rampa; R2 = Peddakonda; R3 = Dummakonda; R4 = Korukonda)

	Respondents						
Factor	R1	R2	R3	R4	Total	Percentage	Rank
Drought	2	2	1	1	6	4.4	8
Forest fires	3	5	3	2	13	9.6	7
Fuel wood	4	6	5	4	19	14	2
Lack of awareness in public with regards medicinal plants	3	5	3	3	16	11.85	5
Monocultivation of plants like teak, and eucalyptus, by forest dept.	6	5	3	4	18	13.33	3
Overexploitation	3	3	4	5	15	11.11	6
Poducultivation (burn agriculture)	5	10	6	5	26	19.25	1

Table 9.6 Jaccard index comparing the present study with previous reports at the regional and national scales

Previous study area	References	Total documented species	Total documented species in present study	Similar use of plant	Dissimilar use of plant	Plants common to both areas	Jaccard index (JI)
Nallamalais in Kurnool district, Andhra Pradesh.	Sudhakar reddy et al. (2007)	51	104	3	10	13	11.20
Rudrakod sacred grove in Nallamalais, Andhra Pradesh.	Raviprasad and Sunitha (2011)	69	104	8	21	29	30.52
Diversity of medicinal plants in upper Assam	Saikia and Khan (2011)	96	104	9	15	24	18.75
Ethnomedicinal plants of Adilabad District, Telangana state	Murthy (2012)	96	104	14	22	36	39.13
Ethnobotanical studies Japali hanuman Theertham sacred grove Tirumala Hills, Andhra Pradesh.	Savithramma et al. (2014)	85	104	02	07	09	5.55
Ethnobotanical studies of Kotia hills of Vijayanagaram district, Andhra Pradesh.	Sankara Rao et al. (2015)	30	104	06	07	13	13.68
Sacred groves in Cuddalore and villupuram district, Tamil Nadu	Karthik et al. (2016)	85	104	4	16	20	15.5
Quantitative Ethnobotanical survey of traditional siddha medical practitioners from Thiruvarur District, Tamil Nadu	Vijay kumar et al. (2016)	33	104	07	05	12	11.88

9.6.3 Jaccard Index

Documented ethnobotanical data of the present study compared with other ethnobotanist reports of local and national regions shows great variation, because of the difference in cultural and social behaviors of indigenous communities. Jaccard index was used to find out the degree of similarities. This kind of comparative analysis among regions and communities is very useful to finding new drugs (Leonti 2011). The JI ranged from 5.5 to 39.13 (Table 9.6). The highest degree of similarity was found in the study of Murthy (2012), which revealed the same ethnic values and same type of vegetation of both areas. The minimum JI values were recorded for the work conducted by Savithramma et al. (2014) in Japali Hanuman Theertham, a sacred grove of Tirumala Hills, Andhra Pradesh. The difference in ethnobotanical knowledge may be due to the presence of some ecological barriers that resulted into the diversities of vegetation and habitats. Based on this result it can be concluded that geographical isolation among communities shows great impact on change in ethnobotanical knowledge.

9.7 Conclusion

This study demonstrates that many plant species found in sacred groves play an important role in local healing practices and that knowledge of ethnomedicine is still utilized. The documentation of this ethnomedicinal knowledge has furnished us with novel information that will provide recognition of this undocumented knowledge. In the present study, there were identified 104 plant species of medicinal importance and used for 52 different human ailments. In majority of cases, leaf parts are mostly used for drug preparation. It was observed that the medicinal plants used by traditional practitioners for the preparation of medicine had been collected from the wild, seed generated, and root parts. Overexploitation and poducultivation is one of the biggest threats to conservation of ethnomedicinal plants. Hence sustainable harvesting is an urgent need to conserve ethnomedicinal plants. A planned approach is necessary to conserve medicinal plants and encourage native communities in trading off locally prepared herbal products. This knowledge will also helpful to link ethnobotanical and chemical knowledge to understand the use of ethnic formulations by local communities.

Acknowledgements The authors are grateful to Dr. K. Babu, Post Doctoral Fellow of Department of Botany and Microbiology, Acharya Nagarjuna University, for his help to assist our entire work and use of all laboratory facilities.

References

AICRPE (All India Coordinated Research Project on Ethnobiology) (1992-1998) Final Technical Report, Ministry of Environment and Forest, Govt. of India

Alexiades MN (1996) Collecting ethnobotanical data: an introduction to basic concepts and techniques. In: Alexiades MN (ed) Selected guidelines for Ethnobotanical research: a field manual. The New York Botanical Garden, New York, pp 53–94

Amjad MS, Arshad M (2014) Ethnobotanical inventory and medicinal uses of some important woody plant species of Kotli, Azad Kashmir. Pakistan. Asian Pacific J Tropic Biomed 4:952–958

Anon (2001) Census of India. New Delhi

Augustine J, Sivadasan M (2004) Ethnomedicinal plants of Periyar Tiger Reserve, Kerala, India. Ethnobotany 16:44–49

Bentham G, Hooker JD (1862–83) Genera Plantarum, Vol. III. Reeve & Co, London

Bonet MA, Blanche C, Xirau JV (1992) Ethnobotanical study of the river Tenes Valley (Catalonia, Iberian Peninsula). J Ethnopharmacol 37:205–212

Cotton CM (1996) Ethnobotany: principles and applications, vol 412. John Wiley and Sons, New York

Gamble JS, Fischer CEC (1915-1935) Flora of Presidency of Madras, Vol. 1-3. Adlard and Son Ltd, London; pp. 112-305

Gedif T, Hahn H (2003) The use of medicinal plants in self-care in rural Central Ethiopia. J Ethnopharmacol 87:155–161

González-Tejero MR, Casares-Porcel M, Sánchez-Rojas CP, Ramiro-Gutiérrez JM, Molero-Mesa J, Pieroni A et al (2008) Medicinal plants in the Mediterranean area: synthesis of the results of the project Rubia. J Ethnopharmacol 116:341–357

Gottlieb OR, Borin MR, De MB, Brito NRS (2002) Integration of ethnobotany and phytochemistry: dream or reality? Phytochemistry 60:145–152

Heinrich M, Ankil A, Frel B, Weimann C, Sticher O (1998) Medicinal plants in Mexico: healers' consensus and cultural importance. Soc Sci Med 47:1859–1871

Hussain F, Sher H, Ibrar M (2004) Ethnomedicinal profile of some plants of district swat, Pakistan. Pak J Plant Sci 10:85–104

Ishtiaq M, Mumtaz AS, Hussain T, Ghani A (2012) Medicinal plant diversity in the flora of Leepa Valley, Muzaffarabad (AJK), Pakistan. Afri J Biotechnol 11:3087–3098

Jain SK, Mudgal VA (1999) Hand book of ethno-botany. Bishen Singh Mahendra Pal Singh, Dehra Dun

Jain SK, Rao RR (1997) A handbook of field and herbarium methods. Today and Tomorrow's Printers and Publishers, New Delhi

Karthik S, Subramanian M, Ravikumar S, Dhamotharan R (2016) Medicinal plants and their uses: a study of twelve sacred groves in Cuddalore and Villupuram districts, Tamil nadu, India. Int Edu Res J 2(5):95–102

Kayani S, Ahmad M, Sultana S, Khan Shinwari Z, Zafar M, Yaseen G (2015) Ethnobotany of medicinal plants among the communities of alpine and sub-alpine regions of Pakistan. J Ethnopharmacol 164:186–202

Khan ZS, Khuroo AA, Dar GH (2004) Ethnomedicinal survey of Uri, Kashmir Himalaya. Indian J Trad Knowl 3(4):351

Khasim SM, Naidu KA, Jhansi K (2013) Fourth world botany documentation and preservation. Curr Sci 3:292–293

Leonti M (2011) The future is written: impact of scripts on the cognition, selection, knowledge and transmission of medicinal plant use and its implications for ethnobotany and ethnopharmacology. J Ethnopharmacol 134:542–555

Lulekal E, Asfaw Z, Kelbessa E, Damme PV (2013) Ethnomedicinal study of plants used for human ailments in Ankober District, north Shewa zone, Amhara region, Ethiopia. J Ethnobiol Ethnomed 9:63. https://doi.org/10.1186/1746-4269-9-63

Malhotra KC (1998) Anthropological dimensions of sacred groves in India: an overview. In: Ramakrishnan PS, Saxena KG, Chandrashekara UM (eds) Conserving the sacred for biodiversity management. UNESCO and Oxford & IBH Publishing, New Delhi, pp 423–438

Malhotra KC, Gokhale Y, Chaterjee S, Srivastava S (2001) Cultural and Ecological Dimensions of Sacred Groves in India. Indian National Science Academy, New Delhi, and Indira Gandhi Rashtriya Manav Sangrahalaya, Bhopal

Mesfin F, Seta T, Aseefa A (2014) An ethnobotanical study of medicinal plants in Amaro Woreda, Ethiopia. Ethnobot Res Appl 12:341–354

Mohana Rao PR, Khasim SM, Aasiya B (2015) Flora of Guntur City, Andhra Pradesh. Bishen Singh Mahendra Pal Singh, Dehra Dun

Murthy EN (2012) Ethno medicinal plants used by gonds of Adilabad district, Andhra Pradesh. Ind Int J Pharm Life Sci 3(10):2034–2043

Muthu C, Ayyanar M, Raja N, Ignacimuthu S (2006) Medicinal plants used by traditional healers in Kancheepuram district of Tamil Nadu. Ind J Ethnobiol Ethnomed 2:43. https://doi.org/10.1186/1746-4269-2-43

Naidu KA, Khasim SM (2010) Contribution to the floristic diversity and ethnobotany of eastern Ghats in Andhra Pradesh, India. Ethnobotanical Leaflets 14:920–941

Padal SB, Sathyavathi K, Chandrasekhar P (2013) Ethnomedicinal uses of trees by tribals of munchangiputtu mandalam, Visakhapatnam district, Andhra Pradesh, India. Int J Pharm Biol Sci 3(2):441–449

Pullaiah T, Chennaiah E (1997) Flora of Andhra Pradesh. In: Ranunculaceae-Alangiaceae, vol I. Scientific Publishers, Johdpur

Rao RS, Sudhakar S, Venkanna P (1999) Flora of east Godavari district, Andhra Pradesh, India. INTACH, Hyderabad

Raviprasad B, Sunitha S (2011) Medicinal plant resources of Rudrakod sacred grove in Nallamalais, Andhra Pradesh, India. J Biodiver 2(2):75–89

Sadat-Hosseini M, Farajpour M, Boroomand N, Solaimani-Sardou F (2017) Ethnopharmacological studies of indigenous medicinal plants in the south of Kerman, Iran. J Ethnopharmacol 199:194–204

Saikia P, Khan ML (2011) Diversity of medicinal plants and their uses in home gardens of upper Assam, Northeast India. Asian J Pharma Biol Res 1(3):296–309

Sankara Rao B, Sarada Mani N, Sujatha B, Polumahanthi S (2015) Ethnobotanical investigation of underground plant parts of Kotia hills of Vizianagaram district, Andhra Pradesh, India. J Med Plant Stud 3(2):118–120

Savithramma N, Yugandhar P, Linga Rao M (2014) Ethnobotanical studies on japali hanuman theertham - a sacred grove of tirumala hills, Andhra Pradesh. Ind J Pharm Sci Res 6:83–88

Savithramma N, Yugandhar P, Suhrulatha D (2015) Traditional medicinal plants used by local people of Kailasakona - a sacred grove of Chittoor district, Andhra Pradesh, India. Int J Pharm Pharm Sci 7(3):407–411

Seifu T, Asres K, Gebre-Mariam T (2006) Ethnobotanical and Ethnopharmacological studies on medicinal plants of chifra district, Afar region, Northeast Ethiopia. Ethiop Pharm J 24:41–58

Senapathi N, Sahu NK (1966) Koraput district Gazetteer. Govt. of Orissa, Cuttack

Srinivasa Rao D, Shanmukha Rao V, Prayaga Murthy P, Narasimha Rao GM, Venkateswara Rao Y (2015) Some ethnomedicinal plants of Parnasala sacred grove area eastern Ghats of Khammam district, Telangana, India. J Pharm Sci Res 7(4):210–218

Sudhakar Reddy C, Reddy KN, Thulasi K, Chiranjibi P (2007) Ethnobotanical studies on medicinal plants used by the Chenchus of Nallamalais in Kurnool district, Andhra Pradesh, India. Res J Med Plants 1:128–133

Suthari S, Sreeramulu N, Omkar K, Reddy CS, Raju VS (2014) Intracultural cognizance of medicinal plants of Warangal north Forest division, northern Telangana, India. J Ethnobot Res Appl 12:211–235

Tabuti JRS, Lye KA, Dhillion SS (2003) Traditional herbal drugs of Bulamogi, Uganda: plants, use and administration. J Ethnopharmacol 88:19–44

Teklehaymanot T, Giday M, Medhin G, Mekonnen Y (2007) Knowledge and use of medicinal plants by people around Debre Libanos monstery of Ethiopia. J Ethnopharmacol 111:271–283

Ugulu I (2012) Fidelity level and knowledge of medicinal plants used to make therapeutic Turkish baths. J Ethnomed 6(1):1–9

Uniyal SK, Singh KN, Jamwal P, Lal B (2006) Traditional use of medicinal plants among the tribal communities Chhota, Western Himalaya. J Ethnobiol Ethnomed 2:14. https://doi.org/10.1186/1746-4269-2-14

Vijayakumar S, Harikrishnan JP, Prabhu S, Morvin Yabesh JE, Manogar V (2016) Quantitative Ethnobotanical survey of traditional siddha medical practitioners from Thiruvarur District with Hepatoprotective potentials through in Silico methods. Achiev Life Sci 10:11–16

Wild R, Mcleod C (2008) Sacred Natural Sites. Guidelines for Protected Area Managers. Best Practice Protected Area Guidelines Series No16, IUCN Gland, Switzerland

Ethno-botanico-medicine in the Treatment of Diabetes by the Tribal Groups of Visakhapatnam District, Andhra Pradesh, India

10

M. Tarakeswara Naidu, S. Suthari, O. Aniel Kumar, and M. Venkaiah

Abstract

Diabetes is not a disease; it is a condition and metabolic disorder. An ethnobotanical investigation on antidiabetic plants used by primitive tribes of Visakhapatnam district was conducted through semi-structured questionnaires and personal interaction. The predominant indigenous tribes in the area are Kammara, Kondadora, Khondus, Kotia, Kulia, Malis, Mannedora, Mukhadora, Porja, Reddidora, and Nookadora. A total of 30 plant taxa pertaining to 28 genera and 19 families were recorded for the treatment of diabetes. The routine use of medicinal plants for the treatment of diabetes either separately or in combination with other plant parts or either with honey or cow or goat milk/urine or lemon juice was documented. Mostly the tribal people prepare leaves or whole plant on top priority to cure the diabetes. The present study clearly indicates that the tribal groups have been using traditional knowledge since ancient times to cure or minimize the diabetic condition though hospital facilities were established in the region. The present paper suggests that further chemical and pharmacological investigations are much needed to validate the claims scientifically.

Keywords

Diabetes · Ethnomedicine · Medicinal plants · Primitive tribes

M. T. Naidu (✉) · O. A. Kumar · M. Venkaiah
Department of Botany, Andhra University, Visakhapatnam, Andhra Pradesh, India

S. Suthari
Department of Botany, Vaagdevi Degree & PG College, Hanamkonda, Telangana, India

S. M. Khasim et al. (eds.), *Medicinal Plants: Biodiversity, Sustainable Utilization and Conservation*, https://doi.org/10.1007/978-981-15-1636-8_10

10.1 Introduction

Diabetes mellitus (DM) is a chronic disorder of carbohydrate, fat, and protein metabolisms, which increase the sugar levels in the blood due to defect in insulin secretion. It is one of the major health problems affecting millions of people and causing death globally. It is categorized into Type 1 (insulin dependent: ID) and Type 2 (non-insulin dependent: NID). ID is due to autoimmune destruction of pancreatic β-cells in genetically susceptible persons. NID is often diagnosed in adults and it is the defect of insulin secretion and insulin action (Sridhar et al. 2002). NID is very common and constituting 90–95% of the diabetic population. It affected more than 100 million and 2–3% of the population worldwide. International Diabetes Foundation (IDF) reported that the menace affected well over 336 million people globally, and that by 2030, the figure is predicted to affect 552 million (Whiting et al. 2011). The common symptoms of diabetic condition are high sugar levels in blood, frequent urination, unusual thirst, extreme weight loss, hunger, blurred vision, easy tiredness, weakness, irritation, etc. (Modak et al. 2007). The diabetic condition is a serious threat to mankind health and the control, treatment and its complications mainly depend on the chemical or biochemical agents.

Traditional botanical knowledge (TBK) plays a major role in the formulation of new drug discovery through scientific research. WHO (2000) depicted that more than 80% of global population depends on traditional system of medicine for their primary healthcare demands. In traditional medicine, the diabetes is treated with potential medicinal plants. More than 1200 medicinal plants are being used at global level to control the diabetes and approximately 30% of the traditionally used antidiabetic plants were pharmacologically and chemically investigated (Alarcon-Aguilar et al. 2002; Babu et al. 2006). In India, indigenous remedies have been used for treatment of diabetes since ancient times. Earlier, it was reported that several plant species have the potential phytochemical constituents to treat the diabetes (Wadkar et al. 2008). Therefore, the present study has taken up to document the antidiabetic plants to cure diabetes and their parts used by ethnic tribes of Visakhapatnam district, Andhra Pradesh. The paper reports the information about the scientific name, vernacular (local) name, and family of antidiabetic plants, along with the mode of administration.

The study area, Visakhapatnam district, is located between latitudes 17°12′ to 18°33′ N and longitudes 82°18′ to 83°22′ E. It is bounded by Orissa and Vizianagaram district on the North, by East Godavari district on the South, by Orissa on the West, and by Bay of Bengal on the East (Fig. 10.1). It spreads over an area of 11,161 km^2. The mean annual temperature ranges between 12 °C and 35 °C. The predominant indigenous tribes are Kammara, Konda doras, Khondus, Kotia, Kulia, Malis, Manne dora, Mukha dora, Porja, Reddi doras, and Nooka dora, whose population accounts for 14.55% (Anonymous 2011).

Extensive field surveys for antidiabetic plants were conducted during July 2013 to March 2016 and ethnobotanical data was collected through household surveys, interviews, semi-structured questionnaires, and general discussions with the ethnic tribes. Fifty-five (55) informants of ethnic groups were identified based on medicinal plants knowledge, experience in practicing ethnomedicine, age, gender, and

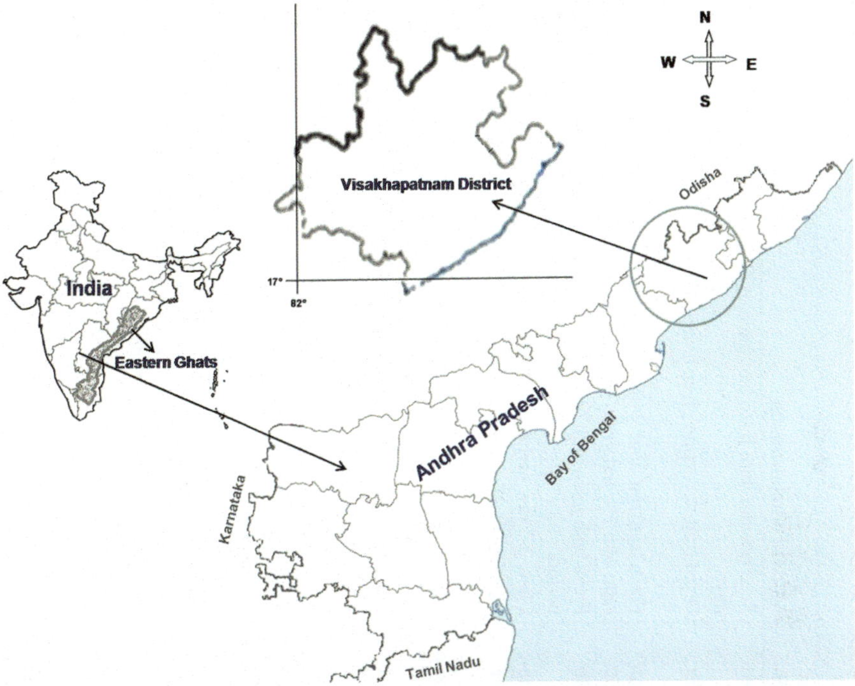

Fig. 10.1 Map showing the high diversified Eastern Ghats zone in India, Andhra Pradesh State, along with boundaries

systematically interviewed. The data was collected for the implicated source plants. Almost all the tribes or traditional healers can easily diagnose the diabetic patient when they direct the patients to urinate in a visible place. If the practitioners observe any sweet smell or a large number of ants hovered around the urinated place, then they confirm that the patient is suffering from diabetic problem and then only they start the treatment. And, they also query the patients about their urination duration and thirst in their daily life. The collected antidiabetic plant materials were photographed at their locality and identified with local floras (Gamble 1915; Subbarao and Kumari 2008). Voucher specimens were housed at Botany Department Herbarium (BDH), Department of Botany, Andhra University, Visakhapatnam.

10.2 Ethno-botanico-medicine in the Treatment of Diabetes

The present study revealed that totally 30 plants distributed in 28 genera belonging to 19 families and different modes of treatment were followed to cure diabetes by tribes inhabiting in Visakhapatnam district, Andhra Pradesh, India. In the following documentation, the species are arranged alphabetically and include information about scientific name, vernacular name, and family names of these medicinal plants along with the plant parts used and ethnomedicinal usage (Table 10.1).

Table 10.1 Medicinal plants as antidiabetic agents by tribes of Visakhapatnam

	Scientific name	Vernacular name	Family	Part useful	Mode of administration
1	*Abelmoschus moschatus* Medik.	Okra	Malvaceae	Seed	Seed powder with fresh leaf juice of *Ocimum* and take 2-3 times a day.
2	*Abrus precatorius* L. (Fig. 10.2a)	Gurivinda	Fabaceae	Fruit	Dried fruit decoction given twice a day.
3	*Aegle marmelos* (L.) Corrêa (Fig. 10.2b)	Maredu	Rutaceae	Leaf/ fruit	Fresh leaf/fruit juice with mustard seed powder and honey in the ratio of 6:3:1 and given daily for 2 months.
4	*Allium cepa* L.	Neerulli	Amaryllidaceae	Bulb	Fresh bulb juice with low concentration of honey for prolonged period.
5	*Allium sativum* L.	Vellulli	Amaryllidaceae	Bulb	Chew 3-5 raw garlic cloves with empty stomach for 10-15 days.
6	*Andrographis paniculata* (Burn.f.) Nees (Fig. 10.2c)	Nelavemu	Acanthaceae	Leaf	Pills from leaf powder at a dose of 2-3 per day for duration of 3 months.
7	*Bacopa monnieri* (L.) Wettst.	Sambrani chettu	Plantaginaceae	Whole plant	Pills (100-200 mg each) whole plant given 15-30 days.
8	*Balanites aegyptica* (L.) Delile	Gara	Zygophyllaceae	Root	Fresh juice from roots with honey and lemon juice in the ratio of 20:1:2 and given for 1-2 weeks.
9	*Bauhinia purpurea* L.	Devakanchana	Fabaceae	Leaf/ bark	Decoction (2-3 leaves/100 ml water or 1 in. bark soaked in 100 ml water) given orally for 1 month.
10	*Beta vulgaris* L.	Ella kura	Amaranthaceae	Whole plant	Whole plant decoction with 1-2 g of garlic paste given twice a day for 1 month.
11	*Bidens pilosa* L.	Aggi mokka	Asteraceae	Whole plant	A crude drug is prepared from whole plant and given once in a week for 6 months.
12	*Caesalpinia bonduc* (L.) Roxb. (Fig. 10.2d)	Gaccha kaya	Fabaceae	Leaf	Fresh leaf juice (1-2 leaves/glass of water) given once in a fortnight for prolonged periods.
13	*Capparis zeylanica* L. (Fig. 10.2e)	Aridonda	Capparaceae	Fruit	Fruits taken as vegetable.
14	*Centella asiatica* (L.) Urb.	Saraswathi aku	Apiaceae	Whole plant	Decoction from whole plant given 50 ml for 1 week.
15	*Coccinia grandis* (L.) Voigt	Kakidonda	Cucurbitaceae	Leaf	Fresh leaves decoction given orally twice a day.

16	*Coix lacryma-jobi* L.	Adavi guriginja	Poaceae	Whole plant	Pills (5–10 mg) from dried leaves given two pills for each day with hot water/milk.
17	*Enicostema axillare* (Poir. ex Lam.) A. Raynal	Nelagolimidi	Gentiaceae	Whole plant	One cup of juice obtained from whole plant and is taken orally before meals.
18	*Ficus racemosa* L.	Atti medi	Moraceae	Leaf/fruit	Pills from dried leaves with honey and given 2 pills for 3 months. Fruit extract with *Ocimum* paste given one spoonful for week.
19	*Finlaysonia wallichii* (Wight) Venter	Sugandhi	Apocynaceae	Whole plant	Juice prepared from crushed plant is taken in the morning for a month.
20	*Gymnema sylvestre* (Retz.) R. Br. ex Sm. (Fig. 10.2f)	Podapatri	Apocynaceae	Leaf	A cup of fresh leaves decoction given twice a day.
21	*Ichnocarpus frutescens* (L.) W. T. Aiton	Muntha gajanamu	Apocynaceae	Root	Juice obtained from crushed roots is taken in the morning before breakfast.
22	*Ipomoea batatas* (L.) Lam.	Yerra dumpa	Convolvulaceae	Leaf	10–15 ml of dried leaves decoction given
23	*Maerua oblongifolia* (Forssk.) A. Rich.	Bhoochakra gadda	Capparaceae	Leaf	A cup of juice from fresh leaves is taken in the morning for 3 months.
24	*Momordica charantia* L.	Kakara	Cucurbitaceae	Leaf/fruit	Fresh decoction from leaves/fruit given twice daily.
25	*Momordica dioica* Roxb. ex Willd. (Fig. 10.2g)	Angakara	Cucurbitaceae	Leaf/fruit	Fresh decoction from leaves/fruit given twice daily.
26	*Pterocarpus marsupium* Roxb.	Yegisa	Fabaceae	Whole plant	Whole plant decoction given 10–15 ml each day along with 1–2 leaves of *Aegle marmelos*.
27	*Rauvolfia serpentina* (L.) Benth. ex Kurz	Patalagaruda	Apocynaceae	Root	Dried root powder (800–1000 mg/day) given along with cow urine.
28	*Scoparia dulcis* L.	Bimbika	Plantaginaceae	Whole plant	Whole plant decoction with 50 mg powder of *Achyranthes aspera* given 15–20 ml each day.
29	*Syzygium cumini* (L.) Skeels	Neredu	Myrtaceae	Fruit	Fruits are taken orally when in season.
30	*Tinospora cordifolia* (Willd.) Miers. (Fig. 10.2h)	Tippa teega	Menispermaceae	Leaf	Juice obtained from crushed leaves and taken daily in the morning.

Fig. 10.2 Noteworthy antidiabetic plants from Visakhapatnam district: (**a**) *Abrus precatorius*, (**b**) *Aegle marmelos*, (**c**) *Andrographis paniculata*, (**d**) *Caesalpinia bonduc*, (**e**) *Capparis zeylanica*, (**f**) *Gymnema sylvestre*, (**g**) *Momordica dioica*, (**h**) *Tinospora cordifolia*

In the study, Fabaceae and Apocynaceae (four species of each) are the most commonly used families for the treatment of diabetes and Cucurbitaceae is with three species. Other families such as Amaryllidaceae, Plantaginaceae, and Capparaceae have two species for each family and one species each for Malvaceae, Acanthaceae, Zygophyllaceae, Amaranthaceae, Asteraceae, Apiaceae, Poaceae, Gentiaceae, Moraceae, Convolvulaceae, Myrtaceae, and Menispermaceae. Of all these plants recorded, 20 grow wild, seven were cultivated, and three were both cultivated and wild. Among all the species, herbs were found to be more (13 species), followed by climbers (nine species), trees (five species), and shrubs (three species).

From the data, it could be inferred that for more number of remedies whole plant was used, followed by leaf materials and fruits. However, plant parts like roots, bark, and seeds were less frequently used by tribal healers in the study area. The plants were used either separately or in combination with other plant parts. Always they prepared medicine to heal diabetes in the combination of more number of plants and they believe that combination of several plant parts cures disease rapidly. Taking the medicine as infusion either with water, lemon juice, milk, or honey is the major mode of treatment of diabetes. Ethnomedicine practices are employed as mostly infusions, powdered, and decoctions. To improve the acceptability of certain remedies that are taken orally some additives are frequently used. Before starting the treatment the condition of the patient is observed deeply and then prepared medicines are given to treat diabetes.

The dicots and monocots among the total of 30 plant species of present survey fall in the ratio of about 9:1 (Fig. 10.2). This kind of unequal distribution pattern suggests that attempts may be made to verify whether any common mechanism of action is in operation in these dicots. The whole plant was reported (9) to be the most used part of the plants, constituting 30% of herbal preparation. This kind of distribution pattern amply suggests that the active compounds in majority of the plant species were localized in whole body. This was followed by the leaves that constituted nearly 20% (6 species) and leaf/fruit that constituted 13.3% (4 spp.), while roots, fruits, bulbs, seeds, and bark (10–6.6%) were rarely used for the preparation of the medicines. Another interesting point was that majority of the plants of the present study were herbs (76.6%) which facilitates easy propagation requiring limited space and labor for their maintenance.

Several herbs have shown antidiabetic activity when assessed with the available experimental techniques (Mukherjee et al. 2006). For the treatment of type 2 diabetes, *Momordica charantia, Pterocarpus marsupium,* and *Trigonella foenum-graecum* are good remedies (Jung et al. 2006). Wide array of plant-derived compounds with consistent antidiabetic activity have proven their possible use in the treatment of diabetes (Farnsworth 1998). Plants like *Allium cepa, Allium sativum, Aloe vera, Brassica juncea, Cajanus cajan, Coccinia indica, Caesalpinia bonducella, Ficus benghalensis, Gymnema sylvestre, Momordica charantia, Murraya koenigii, Ocimum tenuiflorum, Pterocarpus marsupium. Swertia chirayita, Syzygium cumini, Tinospora cordifolia,* and *Trigonella foenum-graecum* have been studied in relation to diabetes and their complications. However, these plants have shown varying degree of hypoglycemic and antihyperglycemic activity (Jain and Goel 1995; Ignacimuthu et al. 2006; Modak et al. 2007; Ramya et al. 2008).

10.3 Conclusion

The present study indicates that Visakhapatnam is blessed with splendid diversity of ethnomedicinal plants. Ethnomedicinal approach for the treatment of diabetes is practical, cost-effective, and biologically safe. Very prominent method of herbal preparation was paste or decoction usually taken orally on daily basis for long period of time. The indigenous knowledge of the tribal communities must be properly documented and preserved so that their knowledge could be passed on to the future generations. Such studies and documents provide important and vital clues for the understanding of the complex heritage of tribal communities and their association with environment and nature.

Acknowledgements The funding for the present study by the UGC-SAP, Department of Botany, Andhra University, Visakhapatnam, is gratefully acknowledged. We thank the traditional practitioners, herbal doctors, plant specialist, and village elders of Visakhapatnam district for sharing their valuable botanical knowledge.

References

Alarcon-Aguilar FJ, Roman-Romos R, Flores saenz JL, Aguirre-Garcia F (2002) Investigation on the Hypoglycaemic effects of extracts of four Mexican medicinal plants in normal and alloxan diabetic mice. Phytother Res 16:383–386

Anonymous (2011) Census of India. Government of India, New Delhi

Babu PA, Suneetha G, Boddepalli R, Lakshmi VV, Rani TS, Rambabu Y, Srinivas K (2006) A database of 389 medicinal plants for diabetes. Bioinfo 1(4):130–171

Farnsworth NR (1998) Screening plants for new medicines. Biodiversity National Academy Press, Washington DC, pp 83–97

Gamble JS (1915) Flora of Presidency of Madras. Adlard and Son Ltd, London

Ignacimuthu S, Ayyanar M, Sivaraman KS (2006) Ethnobotanical investigations among tribes in Madurai district of Tamil Nadu. J Ethnobiol Ethnomed 2:25–30

Jain SK, Goel AK (1995) A manual of ethnobotany. Scientific Publishers, India, pp 142–153

Jung M, Park M, Lee HC, Kang YH, Kang ES, Kim SK (2006) Antidiabetic agents from medicinal plants. Curr Med Chem 13(10):1203–1218

Modak M, Dixit P, Londhe J, Ghaskadbi S, Devasagayam JPA (2007) Indian herbs and herbal drugs used for the treatment of diabetes. J Clin Biochem Nutr 40(3):163–173

Mukherjee PK, Maiti K, Mukherjee K, Houghton PJ (2006) Leads from Indian medicinal plants with hypoglycemic potentials. J Ethnopharmacol 106(1):1–28

Ramya S, Rajasekharan C, Sivaperumal R, Krishnan R, Jayakumararaj R (2008) Ethnomedicinal perspectives of botanicals used by Malayali tribes in Vattal Hills of Dharmapuri (TN), India. Ethnobot Leaflets 12:1054–1060

Sridhar GR, Rao PV, Ahuja MMS (2002) Epidemiology of diabetes and its complications. RSSDI, Textbook of Diabetes Mellitus, Hyderabad, pp 95–112

Subbarao GV, Kumari GR (2008) Flora of Visakhapatnam District. Botanical Survey of India, Andhra Pradesh

Wadkar KA, Magdum CS, Patil SS, Naikwade NS (2008) Anti-diabetic potential and India medicinal plants. J Herbal Med Toxicol 2:45–50

Whiting DR, Guarignata L, Weil C, Shaw J (2011) IDF diabetes atlas: global estimates of the prevalence of diabetes for 2011 and 2030. Diabetes Res Clin Pract 94(3):311–321

WHO (2000) General guidelines for methodologies on research and evaluation of traditional medicine. WHO/EDM/TRM/2000.1, Geneva, pp 1–80

Ethnomedicine from Konda Reddis of High Altitude Agency Tracts of East Godavari District, Andhra Pradesh, India

11

Krishna Rao Mortha and Mutyala Naidu Lagudu

Abstract

The high altitude agency tracts of East Godavari district, Andhra Pradesh, India, were explored during 2015–2017 for ethnomedicinal plants. The information was gathered through conducting interviews and having interactive sessions with Konda Reddi ethnic community at their dwelling places. The ethnomedicinal information pertains to 19 ailments afflicting the local people. The medicinal plants collected were documented with vernacular and scientific name of species and their families, ethnic use, parts used and mode of administration. One hundred species of flowering plants belonging to 88 genera and 50 families are used as phytomedicine. These were analysed for growth forms and plant part-wise usage.

Keywords

Ethno-botanico-medicine · Konda Reddis · Human ailments · Eastern Ghats

11.1 Introduction

Medicinal plants are nature's gift to humankind and are a rich heritage of India. India is known as an "Emporium of medicinal plants". About 70% of the rural folk depend on medicinal plants for their healthcare. India is rich in biodiversity and ranks tenth among plant-rich countries of the world and fourth among the Asian countries. India is the highest producer of medicinal plants and reportedly has

K. R. Mortha (✉)
S.R.V.B.S.J.B. Maharanee College, Peddapuram, Andhra Pradesh, India

M. N. Lagudu
Department of Botany, Adikavi Nannaya University, Rajamahendravaram, Andhra Pradesh, India

S. M. Khasim et al. (eds.), *Medicinal Plants: Biodiversity, Sustainable Utilization and Conservation*, https://doi.org/10.1007/978-981-15-1636-8_11

45,000 plant species of which 15,000–20,000 species are of potent medicinal value while 2500 plant species of 1000 genera are used by traditional healers.

The study area, the upper hilly forest tracts in East Godavari district, Andhra Pradesh, has the fast dwindling forests due to biotic interference and climate change. The major factors are podu cultivation, overgrazing, forest fires, alien weed invasions, etc. In future, the degradation of the area will hasten up due to steep increase in human population and livestock along with human's avaricious activities. As most of the tribal people are migrating to urban areas in search of livelihoods, their precious knowledge of the medicinal plants may not be available to the future generations. And, also as per Rio Earth Summit, it is needed to collect, document and preserve the valuable information on medicinal plants and their uses by the local people from this area. The database made available will benefit the future generations and the researchers in the fields of Botany, Chemistry, Pharmacology, etc. The perusal of ethnobotanical literature reveals that less attention has been paid to the present area of study by the earlier researchers. In the light of these observations, the present investigation was taken up for ethnomedicinal survey in order to evaluate the efficiency of drugs, with the following objectives:

1. The collection, identification and documentation of the plants used by Konda Reddis in high altitude agency tracts of East Godavari district, Andhra Pradesh, Southern India.
2. To record the first-hand information from the tribal practitioners of common human ailments in the study area.
3. The taxonomic analysis and systematic evaluation of drug-yielding plants.
4. To suggest the means of protection, conservation and preservation of medicinal plants for sustainable use for the primary healthcare of tribal communities.

Rampa was the chief place in the small Mutta of the same name. Its ruler was known as Zamindar, Mansabdar or the Raja of Rampa, and he was the resident of the place. The earliest known chief was the Mansabdar Rambhupati Devu, who lived early in the nineteenth century A.D. East Godavari district figures quite prominently in the contribution it made to the liberation from British imperialist rule. During the 1890s, the District Association of Godavari has formed devoted to national activity. When the entire country was under non-violent non-cooperation movement, a youth by name Alluri Sitarama Raju, fired by patriotic zeal and fervour, rose in rebellion against the British on the 22nd of August 1922. He wanted to establish an independent kingdom by defying the British and spread the message of independence to other areas. He moved from place to place very vigilantly in the upper East Godavari district. His itinerary included Addateegala, Rajavommangi and several villages in Yellavaram and Rampachodavaram subtaluks in the district.

The agency area (study area) of East Godavari district is under the administrative control of the Integrated Tribal Development Agency (ITDA), Rampachodavaram. The tribal population is mostly concentrated in 608 villages of 11 *mandals*, which include four *mandals* of sub-plan area having 49 villages only. The villages of the agency are mostly remote and interior. The main tribes of East Godavari district are

Konda Reddi, Konda Dora, Koya Dora, Konda Kammara, Konda Kapu, Manne Dora and *Valmikis*. *Konda Reddis* are the prime Primitive Tribal Group of the area and they constitute 33% of the total population.

Tribal societies are storehouses of accumulated experience and knowledge on indigenous vegetation. The focus of the present study is to register the ethnomedicinal knowledge possessed by ethnic people, the Konda Reddi. Ethnomedicinal surveys were undertaken for 2 years (2014–2015 and 2015–2016). Interviews were conducted with Konda Reddi at their dwellings in tribal Mandals of high altitude East Godavari district (Fig. 11.1). During the oral interviews, specific questions were asked and the information supplied by the informants was noted. The data were verified interviewing the ethnic people in different hamlets/villages showing the same plant sample to different informants. The knowledgeable informants were taken to the field for the collection of plants used by them.

The ethnomedicinal data of 100 plants which cure 19 different ailments/diseases presented was the outcome of the intensive field studies conducted in 20 interior tribal pockets with good forest cover. During the field work, special care was taken to record data on phenology, habit, habitat, and availability of the crude drug. Herbarium specimens were prepared following the conventional methods and deposited in the college herbarium.

11.1.1 Ethnomedicine from Konda Reddis from East Godavari

The study area comprises high altitude area of East Godavari district, covering 11 mandals inhabited by the ethnic tribes like Konda Reddi and Koyas. The ethnic people and 45 local vaidyas were consulted for the documentation of the plant medicines.

During the exploratory trips, medicinal uses have been recorded on 100 plant species belonging to 88 genera of 50 plant families (Table 11.1). Of these, 94 species represent 44 families of Dicotyledonae, five species belong to five families of Monocotyledonae, and one fern Pteridophyte (Table 11.2).

The family-wise analysis of ethnomedicinal data was assessed for the top 10 dominant families which contributed the greater number of medicinal species. Euphorbiaceae are represented by 21 species (20.5%), followed by Leguminosae (17 species: 16.6%), Solanaceae (16 species: 15.6%), Asteraceae (13 species: 12.7%), Amaryllidaceae (11 species: 10.7%), Acanthaceae and Combretaceae (10 species: 9.8%), Cannabinaceae and Zingiberaceae (8 species: 7.8%) and Convolvulaceae (6 species: 5.8%) (Table 11.3).

From the present study, it is evident that the local people use herbs (41.1%), followed by trees (21.5%), shrubs (18.6%), creepers and undershrubs (3.9%), climbing shrubs and small trees (2.9%), and climbing herbs, epiphytes, twiners, vines and woody climbers (0.9%) (Table 11.4). Of the total 100 plant species recorded for ethnomedicinal purposes, 80 species are wild, 19 are cultivated and one plant parts is purchased from local market.

Fig. 11.1 Tribal habit and habitat. (**a**) Deforestation, (**b**) Konda Reddi family, (**c**) Tribe selling *Terminalia chebula*, (**d**) Kommu dance, (**e**) Dhimsa dance, (**f**) Author interaction with tribal doctor, (**g**) Asthma medicine preparation and usage

Table 11.1 Demographic particulars of the ethnic tribes in the study area (2011 Census)

Habitat	Total population	Tribal population	Tribal population (%)
India	121,07,26,932	9,92,35,860	8.19
Andhra Pradesh	8,45,80,777	59,18,073	6.99
East Godavari district	1,54,296	2,13,958	4.15

Table 11.2 Statistical analysis of flora explored in the study area

	Angiosperms				
	Dicots	Monocots	Total	Pteridophytes	Total
Families	44	5	49	1	50
Genera	82	5	87	1	88
Species	94	5	99	1	100

Table 11.3 Top ten families of ethnomedicinal use

Euphorbiaceae	21
Leguminosae	16
Solanaceae	16
Asteraceae	13
Amaryllidaceae	11
Acanthaceae	10
Combretaceae	10
Cannabinaceae	8
Zingiberaceae	8
Convolvulaceae	6

Table 11.4 Richness of crude drug diversity: habit-wise analysis

Herb	42
Tree	22
Shrub	19
Creeper	4
Under shrub	4
Climbing shrub	3
Small tree	3
Climbing herb	1
Epiphyte	1
Twiner	1
Vine	1
Woody climber	1

The plant parts used for ethnomedicinal purposes were classified into root, root bark, rhizome, tuber, bulb, whole plant, stem, heart wood, stem bark, leaf, bud and petiole, flower, fruit, seed, oil, latex and mucilage. Depending upon the plant part used for medicinal purposes, leaf constitutes the highest percentage (17.2%), followed by fruit (16.2%), root (15.5%), seed (13.1%), whole plant (12.1%), root bark (5.7%), rhizome and stem (each with 3.0%), flower (2.7%), stem bark and tuber (2.3%), heart wood (2.0%), bulb (1.3%), oil (1.0%), latex and mucilage (0.6%), and bud and petiole (each with 0.3%) (Table 11.5).

The common diseases prevailing in tribal group habitations are ascertained in consultation with local vaidyas. The common ailments are appetite, aphrodisiac, asthma, boils and tumours, diabetes, dysentery, fever, indigestion, jaundice, joint pains, leucorrhoea, life span, memory power, piles, rheumatism, skin diseases, spermatorrhoea, strength and energy, wounds and warts.

One hundred species reported in the present study are used in curing 19 different ailments/diseases/disorders (Table 11.6). The top ten plant species of multiple uses are *Phyllanthus emblica* and *Withania somnifera* (eight ailments), *Terminalia chebula* (seven ailments), *Andrographis paniculata* and *Sphaeranthus indicus* (six ailments), *Acalypha indica*, *Asparagus racemosus*, *Curculigo orchioides* and *Tinospora cordifolia* (five ailments) and *Abutilon indicum* (four ailments).

The upper hilly agency tracts of East Godavari district is a botanically rich region (Rao et al. 1999). As per the family-wise use index, Euphorbiaceae with 21 species stands first in offering more ethnomedicinal species. The Asteraceae, with over 1200 native and naturalized species in India, are expected to top the list. But, in the present study, this family occupies the fourth place with 13 ethnomedicinal species. Obviously, the preponderance of plant species and their medicinal uses need not correspond. However, it is generally reported that Dicot families (Magnoliopsida) dominate the medicinal plant list (e.g., Gottlieb et al. 2002; Suthari et al. 2013). It is the case with the present study as well.

In the present study, *Tinospora cordifolia* is used in asthma, diabetes, lifespan increase and rheumatism by the Konda Reddis. The same species is used to cure tuberculosis and fevers by *Maldharis* of Kachchh district of Gujarat (Chandrashekar and Avinash 2000) and whole plant for gonorrhoea and diabetes by *Mullu kuruma* tribe of Wayanad district, Kerala (Silja et al. 2008), and stem paste for bone fractures by the natives of Eastern Ghats, Andhra Pradesh (Suneetha et al. 2011). In contrast to the use of the whole plant of *Achyranthes aspera* to cure asthma, piles, boils, tumours and skin diseases in the present study, it is employed for ringworm and eczema problems in eastern India (Sen and Behera 2003), for mouth sores and piles in Uttar Pradesh (Khanna and Kumar 2000), root juice for mental disorders, leaves and root for asthma in Arghakhanchi district of western Nepal (Panthi and Chaudhary 2003), leaf for piles and root paste for headache by *Mullu kuruma* tribe of Wayanad district of Kerala (Silja et al. 2008), and root paste for postpartum treatment by Konda Reddis of Andhra Pradesh (Pandurangaraju et al. 2011b).

Table 11.5 Crude drug samples: plant part-wise analysis

Leaf	51
Fruit	48
Root	46
Seed	39
Whole plant	36
Root bark	17
Rhizome	9
Stem	9
Flower	8
Stem bark	7
Tuber	7
Heart wood	6
Bulb	4
Oil	3
Latex	2
Mucilage	2
Bud	1
Petiole	1

11.1.2 Dysentery

For dysentery and diarrhoea, Raju and Reddy (2005) enumerated 37 ethnomedicinal plant species used by tribes of Khammam district, Telangana, whereas Pragada et al. (2012) reported 40 medicinal plants used for the treatment of dysentery in north coastal Andhra Pradesh. The present study recorded the use 11 species, viz., *Acacia chundra, Aegle marmelos, Argyreia nervosa, Cannabis sativa, Cinnamomum tamala, Holarrhena pubescens, Musa paradisiaca, Nelumbo nucifera, Piper nigrum, Punica granatum* and *Terminalia chebula,* for dysentery. Of these, only two species (namely, *Acacia chundra* and *Aegle marmelos*) were earlier reported (Pandurangaraju et al. 2011a) for the same disease treatment.

11.1.3 Jaundice

The leaves of *Andrographis paniculata* (Fig. 11.2b), *Blumea axillaris, Erythroxylum monogynum, Mitragyna parviflora, Phyllanthus amarus* and *Zaleya decandra*, the stem bark of *Holarrhena pubescence* and *Oroxylum indicum*, roots of *Pavetta indica* and *Tinospora cordifolia*, seed of *Piper nigrum* and rhizome of *Curculigo orchioides* are used to treat jaundice by the tribals of East Godavari district. These results are similar with stem and leaf juice of *Tinospora cordifolia* in Gulbarga district, Karnataka (Ghatapanadi et al. 2011), and root of *Curculigo orchioides* in northern Maharashtra (Badgujar and Patil 2008) are used for curing jaundice.

Table 11.6 Medicinal plants used for 19 different human ailments

Disease/ailment	Plant name	Vernacular	Plant part used	Habit	Family
Joint pains	*Abutilon indicum*	Tutturu benda	Root	Shrub	Malvaceae
	Acampe rigida	Sanna rastramu	Leaf, root	Epiphyte	Orchidaceae
	Allium sativum[b]	Neerulli	Bulb	Herb	Amaryllidaceae
	Aristolochia indica	Eeswari	Root	Herb	Aristolochaceae
	Calotropis gigantea	Jilledu	Leaf	Shrub	Apocyanaceae
	Cissus quadrangularis	Nalleru	Stem	Climbing shrub	Vitaceae
	Moringa oleifera[a]	Munaga chettu	Leaf	Tree	Moringaceae
	Pedalium murex	Pedda palleru	Whole plant	Herb	Pedaliaceae
	Plumbago zeylanica	Tella chitramulam	Root	Herb	Plumbaginaceae
	Strychnos nux-vomica	Mushini	Fruit, bark	Tree	Loganiaceae
	Vitex negundo[a]	Vavili	Leaf	Shrub	Verbenaceae
Aphrodisiac	*Andrographis paniculata*	Nelavemu	Whole plant	Herb	Acanthaceae
	Asparagus racemosus	Pilli tegalu	Root	Vine	Asparagaceae
	Hybanthus enneaspermus	Ratna purusha	Whole plant	Herb	Violaceae
	Moringa oleifera[a]	Munaga chettu	Fruit, seed	Tree	Moringaceae
	Mucuna pruriens	Pedda dulagondi	Seed	Herb	Leguminosae
	Psoralea corylifolia	Bhavanchalu	Seed	Herb	Leguminosae
	Pueraria tuberosa	Nela gummudi	Tuber	Climbing shrub	Leguminosae
	Senna angustifolia	Sunamukhi	Leaf	Shrub	Leguminosae
	Spermacoce hispida	Madanaku	Whole plant	Herb	Rubiaceae
	Sphaeranthus indicus	Bodatharamu	Whole plant	Herb	Asteraceae
	Syzygium cumini	Neredu	Bark	Tree	Myrtaceae
	Withania somnifera[a]	Ashwagandha	Root	Shrub	Solanaceae
Leucorrhoea	*Acacia chundra*	Sandra	Heart wood	Tree	Leguminosae
	Alangium salviifolium	Uduga	Seed, bark	Small tree	Cornaceae
	Aloe vera[a]	Kalabandha	Mucilage	Undershrub	Amaryllidaceae

	Andrographis paniculata	Nelavemu	Whole plant	Herb	Acanthaceae
	Aristolochia bracteolata	Gadidha gadapaku	Leaf	Herb	Aristolochiaceae
	Bauhinia variegata	Deva kanchanam	Root, bark	Tree	Leguminosae
	Ficus benghalensis	Marri chettu	Fruit, root tips	Tree	Moraceae
	Ficus religiosa	Ganga raavi	Fruit	Tree	Moraceae
	Mirabilis jalapa[a]	Nitya malli	Whole plant	Herb	Nyctaginaceae
	Phyllanthus amarus	Nela usiri	Whole plant	Herb	Phyllanthaceae
	Phyllanthus emblica	Usiri	Fruit, seed	Tree	Phyllanthaceae
	Strychnos nux-vomica	Mushini	Seed	Tree	Loganiaceae
Memory power	*Acorus calamus*	Vasa	Rhizome	Herb	Acoraceae
	Argyreia nervosa	Samudra Pala	Leaf	Climber	Convolvulaceae
	Azadirachta indica	Vepa	Leaf, bark	Tree	Miliaceae
	Centella asiatica	Saraswathi aaku	Whole plant	Herb	Apiaceae
	Clitoria ternatea	Shankupushpi	Root	Climbing herb	Leguminosae
	Papaver somniferum[a]	Gasa gasaalu	Fruit	Herb	Papaveraceae
	Piper longum[b]	Pippallu	Fruit	Herb	Piperaceae
Asthma	*Acalypha indica*	Murupinda	Whole plant	Herb	Euphorbiaceae
	Achyranthes aspera	Uthareni	Whole plant	Herb	Amaranthaceae
	Acorus calamus[a]	Vasa	Seed, root	Herb	Acoraceae
	Azadirachta indica	Vepa	Leaf, bark	Tree	Miliaceae
	Datura metel	Ummetha	Seed	Shrub	Solanaceae
	Terminalia bellirica	Tanikaya	Fruit	Tree	Combretaceae
	Justicia adhatoda	Adda saramu	Leaf	Shrub	Acanthaceae
	Senna occidentalis	Kasintha	Seed	Undershrub	Leguminosae
	Solanum surattense	Nela vakudu	Fruit	Herb	Solanaceae
	Tinospora cordifolia	Tippa teega	Leaf, stem	Woody climber	Menispermaceae

(continued)

Table 11.6 (continued)

Disease/ailment	Plant name	Vernacular	Plant part used	Habit	Family
Piles	*Acalypha indica*	Murupinda	Whole plant	Herb	Euphorbiaceae
	Achyranthes aspera	Uthareni	Leaf	Herb	Amaranthaceae
	Aloe vera[a]	Kalabandha	Mucilage	Undershrub	Amaryllidaceae
	Amorphophallus sylvaticus	Kandha	Tuber	Herb	Araceae
	Curcuma longa[b]	Pasupu	Rhizome	Undershrub	Zingiberaceae
	Euphorbia neriifolia	Aakujemudu	Latex	Shrub	Euphorbiaceae
	Holarrhena pubescens	Kodesapala	Seed	Shrub	Apocynaceae
	Holoptelea integrifolia	Nemali chettu	Bark	Tree	Ulmaceae
	Phyllanthus emblica	Usiri	Fruit	Tree	Phyllanthaceae
Boils and tumours	*Achyranthes aspera*	Uthareni	Root, leaf	Herb	Amaranthaceae
	Aegle marmelos	Maredu	Leaf, fruit	Tree	Rutaceae
	Albizia lebbeck	Dirisena	Seed oil	Tree	Leguminosae
	Andrographis paniculata	Nelavemu	Whole plant	Herb	Acanthaceae
	Argemone mexicana	Rakkisa	Latex	Herb	Papaveraceae
	Aristolochia bracteolata	Gadidha gadapaku	Leaf	Herb	Aristolochiaceae
	Aristolochia indica	Eeswari	Root	Shrubby twiner	Aristolochiaceae
	Curcuma longa[b]	Pasupu	Rhizome	Undershrub	Zingiberaceae
	Datura metel	Ummetha	Leaf	Shrub	Solanaceae
	Elytraria acaulis	Adugu thappeta	Root	Herb	Acanthaceae
	Kalanchoe pinnata	Gorrela masala kura	Leaf	Herb	Crassulaceae
Jaundice	*Abutilon indicum*	Thutturu benda	Root	Shrub	Malvaceae
	Andrographis paniculata	Nelavemu	Whole plant	Herb	Acanthaceae
	Asparagus racemosus	Pilli tegalu	Root	Vine	Asparagaceae
	Azadirachta indica	Vepa	Leaf, bark	Tree	Miliaceae
	Blumea axillaris	Kukka pugaku	Leaf	Herb	Asteraceae
	Curculigo orchioides	Nela thadi	Root	Herb	Hypoxidaceae
	Ipomoea digitata	Kasi ratnalu	Root	Twiner	Convolvulaceae

	Ocimum tenuiflorum	Thulasi	Leaf, seed	Shrub	Lamiaceae
	Phyllanthus amarus	Nelausiri	Whole plant	Herb	Phyllanthaceae
	Phyllanthus emblica	Usiri	Fruit	Tree	Phyllanthaceae
	Piper nigrum[b]	Mirialu	Seed	Climber	Piperaceae
	Ricinus communis[a]	Aamudamu	Oil, seed, root	Shrub	Euphorbiaceae
	Sarcostemma acidum	Aku jemudu	Whole plant	Twiner	Apocynaceae
	Solanum americanum	Kamanchi	Leaf	Herb	Solanaceae
Wounds and warts	*Acacia chundra*	Sandra	Heart wood	Tree	Leguminosae
	Acalypha indica	Murupinda	Leaf	Herb	Euphorbiaceae
	Alangium salviifolium	Uduga	Leaf	Tree	Cornaceae
	Aristolochia bracteolata	Gadidha gadapaku	Leaf	Herb	Aristolochiaceae
	Boerhavia diffusa	Atuku mamidi	Root	Herb	Nyctaginaceae
	Celastrus paniculatus	Jyothishmathi	Leaf	Shrub	Celastraceae
	Ficus religiosa	Ganga raavi	Bark	Tree	Moraceae
	Phyla nodiflora	Bokkena	Whole plant	Herb	Verbenaceae
	Phyllanthus amarus	Nelausiri	Whole plant	Herb	Phyllanthaceae
	Santalum album[b]	Manchi gandham	Heart wood	Small tree	Santalaceae
	Trianthema portulacastrum	Galijeru	Root, leaf	Herb	Aizoaceae
	Tridax procumbens	Gaddi chamanthi	Leaf	Herb	Asteraceae
Life span	*Abutilon indicum*	Thutturu benda	Root	Shrub	Malvaceae
	Asparagus racemosus	Pilli tegalu	Root	Vine	Asparagaceae
	Curculigo orchioides	Nela thadi	Root	Herb	Hypoxidaceae
	Ipomoea digitata	Kasiratnalu	Leaf	Herb	Convolvulaceae
	Pueraria tuberosa	Nela thadi	Tuber	Climbing shrub	Leguminosae
	Senna occidentalis	Kasintha	Leaf	Shrub	Leguminosae
	Solanum americanum	Kamanchi	Whole plant	Herb	Solanaceae

(continued)

Table 11.6 (continued)

Disease/ailment	Plant name	Vernacular	Plant part used	Habit	Family
	Tinospora cordifolia	Tippa teega	Whole plant, stem	Woody climber	Menispermaceae
Indigestion	*Allium sativum*[b]	Vellulli	Bulb	Herb	Amaryllidaceae
	Coleus aromaticus[a]	Vamu aaku	Leaf	Herb	Lamiaceae
	Ferula asafoetida[b]	Inguva	Seed	Herb	Apiaceae
	Phyllanthus emblica	Usiri	Fruit	Tree	Phyllanthaceae
	Piper longum[b]	Pippallu	Fruit	Herb	Piperaceae
	Sphaeranthus indicus	Bodatharamu	Whole plant	Herb	Asteraceae
	Tribulus terrestris	Palleru	Root	Herb	Zygophyllaceae
Skin disease	*Acacia chundra*	Sandra	Heart wood	Tree	Leguminosae
	Acalypha indica	Murupinda	Leaf	Herb	Euphorbiaceae
	Achyranthes aspera	Uthareni	Leaf	Herb	Amaranthaceae
	Albizia lebbeck	Dirisena	Oil, bark	Tree	Leguminosae
	Aristolochia bracteolata	Gadidha gadapaku	Leaf	Herb	Aristolochiaceae
	Azadirachta indica	Vepa	Leaf, bark	Tree	Miliaceae
	Blumea axillaris	Kukka pugaku	Leaf	Herb	Asteraceae
	Calotropis gigantea	Jilledu	Leaf	Shrub	Apocyanaceae
	Curcuma longa	Pasupu	Rhizome	Herb	Zingiberaceae
	Datura metel	Ummetha	Leaf, seed	Shrub	Solanaceae
	Eclipta prostrata	Guntakalagara	Whole plant	Herb	Asteraceae
	Plumbago indica	Yerra chitramulam	Root	Herb	Plumbaginaceae
	Solanum anguivi	Chitti mulaga	Seed	Shrub	Solanaceae
	Terminalia bellirica	Tanikaya	Fruit	Tree	Combretaceae
	Withania somnifera[a]	Ashwagandha	Root	Shrub	Solanaceae
Strength and energy	*Abutilon indicum*	Thutturu benda	Root, flower	Shrub	Malvaceae
	Argyreia nervosa	Samudra Pala	Leaf	Climber	Convolvulaceae
	Asparagus racemosus	Pilli tegalu	Root	Vine	Asparagaceae
	Hybanthus enneaspermus	Ratna purusha	Whole plant	Herb	Violaceae

	Phyllanthus emblica	Usiri	Fruit	Tree	Phyllanthaceae
Dysentery	*Acacia chundra*	Sandra	Heart wood	Tree	Leguminosae
	Aegle marmelos	Maredu	Bark, fruit	Tree	Rutaceae
	Argyreia nervosa	Samudrapala	Leaf, stem	Climber	Convolvulaceae
	Cannabis sativa[a]	Ganjaee	Leaf, seed	Herb	Cannabinaceae
	Cinnamomum tamala[b]	Dhalchina chekka	Bark	Tree	Lauraceae
	Holarrhena pubescens	Kodesa Pala	Seed	Tree	Apocynaceae
	Musa ×paradisiaca	Arati	Tuber	Herb	Musaceae
	Nelumbo nucifera	Thamara	Flower, petiole	Herb	Nelumbonaceae
	Piper nigrum[b]	Mirialu	Seed	Climber	Piperaceae
	Punica granatum	Dhanimma	Fruit skin	Tree	Lythraceae
	Terminalia chebula	Karakkaya	Fruit	Tree	Combretaceae
Spermatorrhoea	*Actiniopteris radiata*	Nemali chettu	Whole plant	Herb	Actinopteridaceae
	Asparagus racemosus	Pilli tegalu	Root	Vine	Asparagaceae
	Azadirachta indica	Vepa	Leaf, bark	Tree	Miliaceae
	Curculigo orchioides	Nela thadi	Tuber	Herb	Hypoxidaceae
	Hygrophila auriculata	Kokilaksham	Seed	Tree	Acanthaceae
	Ipomoea digitata	Kasiratnalu	Seed	Herb	Convolvulaceae
	Hybanthus enneaspermus	Ratna purusha	Whole plant	Herb	Violaceae
	Pueraria tuberosa	Nela gummudi	Tuber	Woody climber	Leguminosae
	Senna auriculata	Thangedu	Flower, seed	Shrub	Leguminosae
Diabetes	*Allium sativum*	Vellulli	Bulb	Herb	Amaryllidaceae
	Alpinia galanga	Dumparastram	Rhizome	Herb	Zingiberaceae
	Azadirachta indica	Vepa	Leaf, bark	Tree	Miliaceae

(continued)

Table 11.6 (continued)

Disease/ailment	Plant name	Vernacular	Plant part used	Habit	Family
	Cinnamomum tamala[b]	Dhalchina Chekka	Leaf, bark	Tree	Lauraceae
	Citrullus colocynthis	Chedu putsa	Fruit	Herb	Cucurbitaceae
	Elytraria acaulis	Adugu thappeta	Root	Herb	Acanthaceae
	Psoralea corylifolia	Bhavanchalu	Seed	Herb	Leguminosae
	Senna auriculata	Thangedu	Flower, seed, root	Shrub	Leguminosae
	Senna occidentalis	Kasintha	Seed	Shrub	Leguminosae
	Strychnos potatorum	Induba	Seed	Tree	Loganiaceae
	Syzygium cumini	Neredu	Bark, seed	Tree	Myrtaceae
	Tinospora cordifolia	Thippa teega	Stem	Woody climber	Menispermaceae
Apatite	*Azadirachta indica*	Vepa	Leaf, bark	Tree	Miliaceae
	Cinnamomum tamala[b]	Dalchina chekka	Bark, leaf	Tree	Lauraceae
	Citrus medica	Maadi phalam	Fruit juice	Tree	Rutaceae
	Coriandrum sativum[a]	Dhaniyalu	Seed, whole plant	Herb	Apiaceae
	Phyllanthus emblica	Usiri	Fruit	Tree	Phyllanthaceae
	Plumbago indica	Yerra chitramulam	Root	Herb	Plumbaginaceae
	Syzygium cumini	Neredu	Bark, seed	Tree	Myrtaceae
	Tinospora cordifolia	Thippateega	Leaf, stem	Climber	Menispermaceae
	Trigonella foenum-graecum[a]	Menthulu	Seed	Herb	Leguminosae
Fever	*Acorus calamus*[a]	Vasa	Root	Herb	Acoraceae
	Andrographis paniculata	Nelavemu	Whole plant	Herb	Acanthaceae
	Azadirachta indica	Vepa	Leaf, bark	Tree	Miliaceae
	Cissampelos pareira	Adavi bankateega	Root	Climbing shrub	Menispermaceae
	Eclipta prostrata	Guntakalagara	Whole plant	Herb	Asteraceae
	Gmelina arborea	Tella gummudu	Stem bark	Tree	Verbenaceae
	Justicia adhatoda	Addasaramu	Leaf	Shrub	Acanthaceae

	Ocimum tenuiflorum	Thulasi	Leaf, seed	Shrub	Lamiaceae
	Soymida febrifuga	Somidha	Bark	Tree	Meliaceae
	Zingiber officinale[b]	Allam	Rhizome	Herb	Zingiberaceae
Rheumatism	*Acalypha indica*	Murupinda	Whole plant	Herb	Euphorbiaceae
	Allium sativum[b]	Vellulli	Bulb	Herb	Amaryllidaceae
	Alpinia galanga[b]	Dhumpa rastram	Rhizome	Herb	Zingiberaceae
	Andrographis paniculata	Nelavemu	Whole plant	Herb	Acanthaceae
	Aristolochia indica	Eeswari	Root	Climber	Aristolochiaceae
	Azadirachta indica	Vepa	Leaf, bark	Tree	Miliaceae
	Calotropis gigantea	Jilledu	Leaf	Shrub	Apocyanaceae
	Celastrus paniculatus	Jyothishmathi	Leaf	Shrub	Celastraceae
	Citrullus colocynthis	Chedu putsa	Fruit	Herb	Cucurbitaceae
	Clitoria ternatea[a]	Shankupushpi	Root	Twiner	Leguminosae
	Operculina turpethum	Thegada	Root	Climber	Convolvulaceae
	Pergularia daemia	Dustupu teega	Leaf	Climbing shrub	Apocynaceae
	Piper longum[b]	Pippallu	Fruit	Shrub	Piperaceae
	Senna tora	Thantepu	Bark	Shrub	Leguminosae
	Strychnos nux-vomica	Mushini	Fruit pulp, seed	Tree	Loganiaceae
	Tinospora cordifolia	Thippa teega	Leaf, stem	Woody climber	Menispermaceae
	Zingiber officinale[b]	Allam	Rhizome	Herb	Zingiberaceae

[a]Cultivated
[b]Plant part purchased from market

Fig. 11.2 Medicinal plants, (**a**) *Alangium salvifolium*, (**b**) *Andrographis paniculata*, (**c**) *Actiniopteris radiate*, (**d**) *Elytraria acaulis*, (**e**) *Plumbago indica*, (**f**) *Strychnos nux-vomica*

11.1.4 Rheumatism

Naidu et al. (2008) described 38 plant species for curing rheumatism by tribals of Srikakulam district of Andhra Pradesh whereas the aborigines of the present study area use 21 plant species (Table.11.6).

11.1.5 Leucorrhoea

For leucorrhoea, Hemadri and Rao (1983a, b) reported 11 claims of folklore medicine by local tribals of combined Andhra Pradesh and Odisha and Venkataratnam and Venkataraju (2005) recorded 25 crude drugs used by Adivasis in Eastern Ghats whereas Konda Reddis of the present study used 12 species (Table. 11.6).

11.1.6 Asthma

To treat asthma, the leaf juice of *Justicia adhatoda* is used in Macchegaun, Nepal (Joshi et al. 2011), and in Thaluka Purandhar district, Maharashtra (Desale et al. 2013), rhizome of *Acorus calamus* in wetlands of southern Odisha (Panda and Misra 2011), whereas the Konda Reddis of the present study use these two species besides nine species (Table 11.6) for the same purpose.

11.1.7 Diabetes

Mudgal et al. (2001) reported the use of leaves of *Acalypha indica* by Santals of Bihar, *Azadirachta indica* by Khonds of Odisha, *Gymnema sylvestre* by Saoras and Khonds of Odisha and seeds of *Syzygium cumini* by Oraons and Mundas of Bihar and Odisha and stem bark of *Ficus benghalensis* by Santals of Bihar and Khonds of Odisha. In the present study area, the bulb of *Allium sativum*, the roots of *Elytraria acaulis*, rhizome of *Alpinia galanga*, leaf of *Azadirachta indica*, stem of *Tinospora cordifolia*, bark of *Cinnamomum tamala*, fruits of *Citrullus colocynthis* and *Syzygium cumini*, seeds of *Psoralea corylifolia* and *Senna occidentalis*, the whole plant parts of *Senna auriculata* and fruits of *Strychnos potatorum* are used by the ethnic people.

The study involves collection of data on plant biomass resource utilization by the local inhabitants particularly with medicinal uses. Attempts have been made to observe the ethnomedicine for 19 different ailments which are commonly afflicting the local people. Their knowledge of medicinal uses of plants has not received much attention in the past in Godavari Region. In the present era ethnomedicinal claims of the indigenous groups have been documented and led modern researchers to discover new drugs of the modern system. The uses of folk medicines are widespread particularly in aboriginal communities. In rural India, due to lack of adequate transport facilities, primary healthcare centres, awareness and sufficient financial support to meet the expenses for treatments, a high percentage of tribal and rural people in this region still depends on it. The Konda Reddis of upper hilly area of East Godavari still depend on the age-old indigenous healthcare system.

In the absence of official patronization, ethnomedicine could not receive due support from researchers and thus remained unknown or lesser known for quite a long period to the public. Because of many side effects of the modern system of medicines and non-availability of satisfactory drugs for treatment of different ailments, studies on ethnomedicinal claims have been encouraged globally. Ethnomedicinal study has provided lead in the development of modern drugs like atropine, digitoxin, quinine, serpentine, reserpine, morphine and strychnine.

The present study reveals that the floral wealth of the Konda Reddis of the East Godavari district includes a diverse plant species which serve as valuable source of medicines to the masses. The medicine men or healers of the traditional healthcare system play an important role in the study area. These traditional healers contribute much to the rural healthcare practices. Besides, elderly experienced men and women also render their treatment practices to the rural folk. There are as many as 45 traditional healers in the study area. The crude drugs are prepared from plant parts such as roots, stems, leaves, flowers, fruits, underground parts and plant product such as latex. The traditional healers are very selective in the choice of plant parts while preparing the drugs. This is suggestive of the fact that the lead chemicals are deposited in the specific part of the plant.

The healers are very much particular about the specific time of collection and administration. The method of preparation of drugs also varies. Sometimes, they add various additives such as oil, honey, ghee and milk to make the therapeutic dosages effective. The amount and number of dosages vary according to age, sex and health condition of the patient. Very often single drug is administered. But in few cases more than one type of herbal medicines are given to cure a particular disease. In such cases, it is difficult to determine the more efficacious plant for that particular disease. It can only be confirmed by chemical screening and chemical studies.

Information from the Konda Reddis on traditional herbal medicine had always played a vital role in the discovery of new drugs. In general, the success of medicine depends on standard proportions and intelligent combination of different plants. Therefore, further investigations on compound formulations to check their efficacy and safety are necessary so that they can further be utilized for medicinal purposes. Large numbers of plants or their parts are used singly to treat different ailments given a clue of their use in combination with other suitable plants or plant parts and the resultant experiences can be exploited in future practices by the vaidhyas and medicine men for effective cure of diseases.

11.2 Conclusions

1. The Konda Reddis of East Godavari district are still dependent on indigenous knowledge for healthcare that are being influenced by culture and socio-economic aspects such as high cost pharmaceutical remedies.
2. If the herbal medicine is used properly they do not have any side effects. Other studies are also necessary to preserve the popular medicinal knowledge which is

important to enhance our understanding of the relationship among men, society and the nature.

3. It is suggested that medicinal plant cultivation or farming should be taken up in all the open forest areas by the tribes. This helps in alleviating poverty of tribes as well as conservation and preservation of original germplasm.
4. In order to avert marketing problems with crude drug, processing industry must be established in Rampachodavaram.
5. The expertise of the local vaidhyas, medicine men and ethnomedicinal practitioners must be identified, and training should be imparted to them to enhance their expertise. These trainees are to be given priority in the AYUSH Department appointments as helpers and above.
6. The same medicine knowledge can further be scrutinized, so that it can be exploited by the modern medicine. Therefore, the validation of some of the major claims of the ethnic people of their herbal medicine is called for.

Acknowledgements MKR is obliged to the Principal, S.R.V.B.S.J.B. Maharanee College, Peddapuram, for facilities and the UGC, SERO, Hyderabad, for the funding through a minor research project. LM Naidu is thankful to the Head, Department of Botany, Sri Adikavi Nannaya University, Rajahmundry, for encouragement. We express our gratitude to Dr. Vatsavaya S. Raju, Retd. Professor, Kakatiya University, Warangal, for guidance and critically going through the manuscript.

References

Badgujar SB, Patil MB (2008) Ethnomedicines for jaundice used in tribal areas of north Maharastra. Nat Prod Radian 7(1):79–81

Chandrashekar S, Avinash R (2000) Indigenous knowledge on medicinal plants and their use in Narayan Sarovar sanctuary, Kachchh. Ethnobotany 12:1–7

Desale MK, Bharmare PB, Sawant PS, Patil SR, Kamble SY (2013) Medicinal plants used by the rural people of Taluk Purandhar, district Pune, Maharashtra. Indian J Tradit Knowl 12(2):334–338

Ghatapanadi SR, Johnson N, Rajasab AH (2011) Documentation of folk knowledge on medicinal plants of Gulbarga district, Karnataka. Indian J Trad Knowl 10:349–353

Gottlieb OR, Borin MR, de Brito NR (2002) Integration of ethnobotany and phytochemistry: dream or reality? Phytochemistry 60:145–152

Hemadri K, Rao SS (1983a) Leucorrhoea and menorrhagia: tribal medicine. Ancient Sci Life 3:40–41

Hemadri K, Rao SS (1983b) Antifertility, abortifacient and fertility promoting drugs from Dandakaranya. Ancient Sci Life 3:103–107

Joshi K, Joshi R, Joshi AR (2011) Indigenous knowledge and uses of medicinal plants in Macchegaun, Nepal. Indian J Trad Knowl 10:281–286

Khanna KK, Kumar R (2000) Ethnomedicinal plants used by the Gujjar tribe of Saharanpur district, Uttar Pradesh. Ethnobotany 12:17–23

Mudgal V, Khanna KK, Pal DC, Sahoo AK, Pathak MK (2001) Floristic diversity and conservation strategies in India. Ethnobotany 4:2323–2336

Naidu BVAR, Reddi TVVS, Prasanthi S (2008) Folk herbal remedies for rheumatoid arthritis in Srikakulam district of Andhra Pradesh. Ethnobotany 20:76–79

Panda A, Misra MK (2011) Ethnomedicinal survey of some wetland plants of South Orissa and their conservation. Indian J Tradit Knowl 10(2):296–303

Pandurangaraju M, Prasanthi S, Reddi TVVS (2011a) Folk claims from Konda Reddis for the treatment of diarrhea and dysentery. J Non Timber Forest Prod 48(7):77–80

Pandurangaraju M, Prasanthi S, Reddi TVVS (2011b) Medicinal plants in folk medicine for women's diseases in used by Konda reddis in Andhra Pradesh. Indian J Tradit Knowl 10(3):563–567

Panthi MP, Chaudhary RP (2003) Ethnomedicinal plant resources of Arghakanchi district, West Nepal. Ethnobotany 15:71–86

Pragada PM, Rao DS, Venkaiah M (2012) Study of some ethnomedicinal plants for treatment of dysentery of north coastal Andhra Pradesh, India. Int J Biosci 2(1):18–24

Raju VS, Reddy KN (2005) Ethnomedicine for dysentery and diarrhoea from Khammam district of Andhra Pradesh. Indian J Trad Knowl 4:443–447

Rao RS, Sudhakar S, Venkanna P (1999) Flora of east Godavari district, Andhra Pradesh, India. INTACH, Hyderabad

Sen SK, Behera LM (2003) Ethnomedicinal plants used against skin diseases in Bargarh district in Orissa. Ethnobotany 15:90–96

Silja VP, Samitha Varma K, Mohanan KV (2008) Ethnomedicinal plant knowledge of the Mullu kuruma tribe of Wayanad district, Kerala. Indian J Trad Knowl 7:604–612

Suneetha J, Prasanthi S, Naidu BVAR, Reddi TVVS (2011) Indigenous phytotherapy for bone fractures from eastern Ghats. Indian J Tradit Knowl 10(3):550–553

Suthari S, Sreeramulu N, Omkar K, Reddy CS, Raju VS (2013) Intracultural cognizance of medicinal plants of Warangal North Forest Division, Northern Telangana, India, Ethnobotany Research & Applications 11: (in press)

Venkataratnam K, Venkataraju RR (2005) Folk medicine used for common women ailments by Adivasis in the eastern Ghats of Andhra Pradesh. Indian J Trad Knowl 4:267–270

12 Traditional Use of Plants for the Treatment of Bone Fracture by the Local People of West Sikkim, India

Mahendra Tamang, Krishan Pal, and Santosh Kumar Rai

Abstract

West District of Sikkim is the second largest district among four districts of Sikkim, situated in the south-western corner covering the area of 1166 km^2. Mt. Khangchendzonga, the world's third highest mountain, is situated in the border of West district Sikkim with Eastern part of Nepal. Geographically, most of the areas are of hilly region with diverse topography and climate determining various types of forest cover enriching biodiversity. Lepcha, Bhutia and Nepali are the major communities of Sikkim with distinct cultural diversity and they also possess rich traditional knowledge. Keeping in mind the importance of ethnobotanical knowledge and the present trend of its application, this study was felt very important. In the present investigation, 41 plants have been documented that are used by the local people for the treatment of bone fracture in the West District of Sikkim. Different parts of the plants that are used for the formulation of herbal medicine and their ecological status are thoroughly discussed.

Keywords

Traditional knowledge · Ethnomedicine · Bone fracture · Khangchendzonga National Park · West Sikkim

M. Tamang (✉) · K. Pal
Department of Botany, Sri Venkateshwara University, Gajraula, UP, India

S. K. Rai
Department of Botany, Sikkim University, Gangtok, Sikkim, India

S. M. Khasim et al. (eds.), *Medicinal Plants: Biodiversity, Sustainable Utilization and Conservation*, https://doi.org/10.1007/978-981-15-1636-8_12

12.1 Introduction

Human being has been using plant for his benefit since the time immemorial. Today, innumerable plants and their products are being used to fulfill the ever-increasing human needs. The traditional knowledge about the usefulness of different plants evolved through trial-and-error process and has been transferred from generation to generation through different traditional modes. This valuable knowledge has enabled the human to improve their quality of life and it has been estimated that about 80% of world's population depends on indigenous medicine for their health-care (WHO 2000). Old religious literatures existing in India served the repository of many traditional medicine practices which is being practiced in the country since hundreds of years (Jain 1991). The documentation of age-old practices of utilizing plants for treating various common ailments not only helps in preserving for future prospects but also paves a way to the discovery of many lives saving drugs (Cox and Ballick 1994).

Sikkim is a small mountainous northeastern state of India has a total number of four districts covering just 0.2% area of the country but contributes 26% of flowering plants. It occupies an area of 7069 km^2 in total and has altitudinal range variation from 300 m and 8568 m above sea level. It is wedged by Nepal on the west, Tibetan Autonomous region on the North, Bhutan on the east, and Darjeeling District of West Bengal on the South. Sikkim is very diverse witnessing deep valleys to high steep mountain peaks, tropical to alpine climate and biogeography which have led in having varied ecological zones within a very short stretch. A major portion of the state is covered by forest. Sikkim is an integral part of Eastern Himalayas and known for its rich biodiversity and the indigenous knowledge throughout the world. It harbors rich biological diversity and is known to be one of the biodiversity hotspot of the world. West district of Sikkim is the second largest district and lies between 27°00'46"–28°07'48" N latitude and 88°00'58"–88°55'25" E longitude (Fig. 12.1) covering an area of 1166 km^2 which is 16.43% of the total area of Sikkim. A large area of the West district of Sikkim falls under Khangchendzonga National Park (KNP now it is declared as Khangchendzonga Biosphere Reserve which is inscribed as world heritage site by UNESCO in the year 2016). The West district of Sikkim is inhabited by three major communities viz. Lepcha, Bhutia, and Nepali. The distinctiveness and practices of these three communities have added more values on the rich reservoir of traditional knowledge available around them. These ethnic communities have also learnt the uses of plants for medicine since time immemorial. To know the rich biological and traditional knowledge of the region, several ethnobotanical studies were carried out in Sikkim since the last 135 years notably by Atkinson 1882; Biswas 1956; Hajra and Chakraborty 1981; Bennet 1983, 1985; Rai and Sharma 1994; Jana and Chauhan 2000; Gurung 2002; Singh et al. 2002; Maiti et al. 2003; Panda 2007; Pradhan and Badola 2008, 2013; Dash 2009; Idris et al. 2010; Panda and Misra 2010; Sharma and Sharma 2010; Bharati and Sharma 2012; Das et al. 2012; Sharma et al. 2012. However, ethnobotanical study in the Western part of Sikkim is very limited due to remoteness of the area and most of the studies are in general uses of plants. In the present study

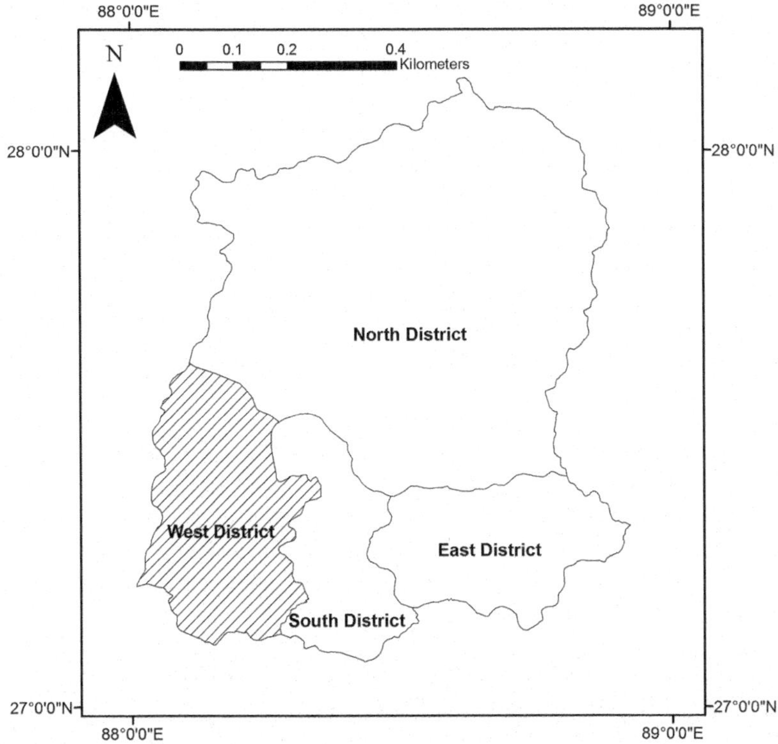

Fig. 12.1 Map of West District Sikkim

an attempt has been made to tap the information about the use of medicinal plants especially for the bone fractures by the local people of West Sikkim.

12.2 Materials and Methods

In the present study, a comprehensive ethno-medico-botanical study was conducted in West District of Sikkim from 2014 to 2017. Before starting the field survey, extensive literatures were surveyed to obtain the information about the healing practices of the local people of this region. In the initial stage, identification of local people having ethno-medicinal knowledge were made that included experienced elderly people, farmers, traditional healers such as *Dhami, Bijuwa*, *Jhakri*, *Bongthing*, *Lama*, and *Vaidyas*. The informants having the broad knowledge about the treatment of bone fractures were selected as the main informants from the rest of the target group. A total of 20 traditional bone fracture healers and 120 local people were interviewed for the authenticity of information and the treatment of bone fractures. It was then followed by field survey to collect the medicinal plants. Semi-structured interviews were conducted and focus group discussions were also organized to derive comprehensive informaton during the course of field survey and the information like plant species, local name, plant parts used, method of preparation, mode of administration, present status of the reported plants, duration of treatment, and reliability of treatment were also recorded. Medicinal plants and the treatment procedures applied by the various healers were also observed and recorded in different seasons. Reports claimed by the informants were cross-checked with each other to authenticate the information. Attempt also has been made to study the distribution of reported medicinal plants in the district in order to assess the current ecological status. The entire field study and the collection of voucher specimens was carried out with the involvement of knowledgeable local medicine men that has the sound information about the location of medicinal plants. It was observed that rural folks do not share information particularly on medicinal plants due to local beliefs that the medicinal plant will be ineffective due to sharing of knowledge, or it may also due to fear of jeopardizing their business interests. Thus, many of them wish to retain such knowledge among their community only.

The collected plant specimens were properly processed in the department of Botany, Sikkim University following the technique suggested by Rao and Sharma (1990) and identified with the help of Flora of India and other regional literatures available in the department. The authentication of plants was done in the Herbaria of Botanical Survey of India, Sikkim Himalayan Circle, Gangtok, Sikkim. The voucher specimens are deposited in the Department of Botany, Sikkim University, for the future references.

12.3 Method of Traditional Treatment

The local herbal practitioners of the study area formulate the locally available medicinal plants in different combinations for the treatment of the bone fractures. The process of making medicine and the usage of plants varies greatly from one healer to another and also from one community to another community. It also depends upon the knowledge about the medicinal plants and availability of these plants during the time of treatment as many of the plants reported are used in fresh conditions only. Generally, the local *Vaidyas* combines at least six to ten different plants for the best result of the treatment. They plants prefered by many are *Acacia pennata, Astilbe rivularis, Bergenia ciliata, Kaempferia rotunda, Lepidium sativum, Pouzolzia hirta, Ulmus lanceifolia and Viscum articulatum*. Even the choice of medicinal plants also rests on the availability of the plants during the treatment time. First of all, the required plants or plant parts are collected fresh from the nearby surroundings, and they are sorted and washed properly. They are then crushed or grounded in mortar with the help of pestle to prepare a fine sticky paste and it is heated on oven for some minutes before applying on the affected area of fractured bones. Secondly, the fractured or dislocated bone of joint of the patient are aligned to fix the correct position of the bones and then the paste is applied thoroughly all around it. Finally, the plastered portion is tightly bandaged with cotton cloths to hold the paste and fractured bones in the right position for some days. The patient is advised to be very careful and any sudden or forceful movement of the body should be avoided during the treatment because this may result into loosening of the grip of paste and the possibility of deformation in the future. The duration of treatment depends much on the condition of the fractures but it usually last for about 7–14 days. At the end of the treatment, the bandage is removed and the affected area is examined thoroughly to see whether the bone setting has occurred properly or not, and in doubtful cases, the plastering and bandaging is repeated further and again kept for another 1 week. The patient is also orally administered the decoction *Astilbe rivularis, Bergenia ciliata, Lepidium sativum and Viscum articulatum* for speedy recovery and reduction of swelling and pain endure during the treatment.

12.4 Traditional Usage of Plants for the Treatment of Bone Fractures

In the present study following medicinal plants that are preferably used by the *Vaidyas* or local practitioners as well as the local communities for the treatment of bone fractures in West District of Sikkim has been documented. Recorded plants are arranged alphabetically in the table (Table 12.1).

In the present study, 41 medicinal plants have been documented that were used for the treatment of bone fracture in West district of Sikkim. These medicinal plants belong to 31 families, out of which the plants belonging to the family Urticaceae have the maximum usage in terms of the treatment of bone fracture followed by the plants belonging to the family Saxifragaceae and Zingiberaceae. Plants belonging to remaining families represent only one species. It is found that the Nepali community use more number of plants for the treatment of bone fracture in comparison to Lepcha and Bhutia communities. This may be due to their larger population and diversity of subgroups within the community. Though the Lepcha community is largely scattered in this district, only a few herbal practitioners or traditional herbal healers with rich knowledge and practices were found in some isolated patches in West district of Sikkim. They acquired this knowledge through centuries of experience while living in the far flung areas as they are thought to be the original inhabitants of Sikkim (Pradhan and Badola 2008). The present study revealed that the stem bark is predominant part used in the preparation of herbal medicine which comprises 15 plants, constituting 36.58% (Fig. 12.2). Subsequently, roots of 12 plants constituting 29.26%, whole plant parts of five plants, constituting 12.19%, aerial parts, rhizome, tuber, and seeds of two plants each is used, constituting 4.87% each, and flower of only one plant is used that constitutes 2.43% in the uses are noted (Fig. 12.2). Root and stem bark are the main parts used in the preparation of herbal paste for plastering the affected areas of bone fractures by all three communities. The medicine men or the herbal practitioners usually do not find these plants in a single place so they need to go a long distances to collect these plants. Due to the increasing intensity of unscrupulous harvesting and preference of plant parts, population of some of these plants is reducing alarmingly in their natural habitat. Presently, out of the above-documented 41 plant species, 18 species (Fig. 12.3) are found in abundance marking no serious concern of threats so far even after

Table 12.1 Plants used for bone fractures in West District Sikkim

Sl. no.	Botanical name	Family	Local name	Habit	Distribution (m)	Parts used	Ethnic communities	Present status
1	*Acacia pennata* (L.) Willd.	Fabaceae	Areri	Climber	300–1000	Roots	NBL	Abundant
2	*Aeschynanthus parviflorus* C. B. Clarke	Gesneriaceae	–	Herb	700–2500	Aerial parts	N	Abundant
3	*Aleuritopteris anceps* (Blanf.) Panigr.	Pteridaceae	Rani Sinka	Herb	300–2500	Rhizome	NBL	Abundant
4	*Ampelocissus sikkimensis* (Lour.) Planch	Vitaceae	Macha Joday	Climber	300–1600	Whole plant	N	Sparse
5	*Artocarpus lacucha* Wall. ex Roxb.	Moraceae	Barhar	Tree	300–1200	Stem bark	N	Abundant
6	*Asparagus racemosa* Willd.	Asparagaceae	Kurilo	Herb	300–2100	Aerial roots	NBL	Sparse
7	*Astilbe rivularis* D. Don	Saxifragaceae	Buro Okhoti	Herb	1800–2600	Roots	NBL	Abundant
8	*Beaumontia grandiflora* Wall.	Apocynaceae	Gothale Lahara	Climber	300–1200	Roots	N	Abundant
9	*Bergenia ciliata* (Haworth.) Sternberg.	Saxifragaceae	Pakhanbet	Herb	1000–3800	Whole plant	NBL	Threatened
10	*Betula alnoides* Buch.-Ham. ex. D. Don.	Betulaceae	Saur	Tree	600–1400	Stem bark	NBL	Abundant
11	*Curcuma longa* L.	Zingiberaceae	Hardi	Herb	300–1800	Rhizome	NBL	Cultivated
12	*Datura metel* L.	Solanaceae	Dhatura	Herb	500–1800	Seeds	NBL	Sparse
13	*Diploknema butyracea* (Roxb.) H. J. Lam.	Sapotaceae	Chiwri	Tree	300–1300	Stem bark	L	Abundant
14	*Engelhardia spicata* Lesch. ex Blume	Juglandaceae	Mauwa	Tree	1200–2200	Stem bark	N	Common
15	*Equisetum debile* Roxb. ex Vaucher	Equisetaceae	Kurkure Jhar	Herb	1000–2200	Roots	NBL	Abundant
16	*Fraxinus floribunda* Wall.	Oleaceae	Lakuri	Tree	700–1900	Stem bark	NBL	Sparse

17	*Garuga pinnata* Roxb.	Burseraceae	Dabdabey	Tree	300–1200	Stem bark	L	Abundant
18	*Girardinia diversifolia* (Link.) Friis.	Urticaceae	Bhangre Sisnu	Herb	300–3000	Roots	NBL	Common
19	*Gymnadenia orchidis* Lindl.	Orchidaceae	Panchamle	Herb	3300–4000	Tuber	NBL	Threatened
20	*Kaempferia rotunda* L.	Zingiberaceae	Bhui Champa	Herb	700–1500	Roots	NBL	Abundant
21	*Laportea bulbifera* Siebold & Zucc.)	Urticaeae	Patle Sisnu	Herb	1300–3300	Roots	NBL	Common
22	*Lepidium sativum* L.	Brassicaceae	Chamsur	Herb	400–1800	Seeds	NBL	Cultivated
23	*Meconopsis horridula* Hook. f. & Thomson	Papaveraceae	Kesar	Herb	3400–5400	Whole plant	B	Sparse
24	*Myrica esculenta* Buch.-Ham. ex D. Don	Myricaceae	Kaphal	Tree	1000–2300	Stem bark	NBL	Abundant
25	*Nyctanthes arbor-tristis* L.	Oleaceae	Parijat	Tree	300–1200	Stem bark	NBL	Sparse
26	*Oxalis corniculata* L.	Oxalidaceae	Chari Amilo	Herb	350–2000	Whole plant	NBL	Common
27	*Plumbago zeylanica* L.	Plumbaginaceae	Seto Chitu	Herb	400–1300	Tubers	N	Abundant
28	*Pouzolzia hirta* Blume ex. Hassk.	Urticaceae	Chipley Jhar	Herb	1000–2500	Roots	NBL	Common
29	*Prunus cerasoides* D. Don.	Rosaceae	Painyung	Tree	800–2100	Stem bark	NBL	Abundant
30	*Rheum australe* D. Don. Thomson	Polygonaceae	Khokim	Herb	3700–4000	Whole plant	NB	Threatened
31	*Rubus ellipticus* Smith.	Rosaceae	Aiselu	Shrub	700–2000	Roots	NBL	Abundant
32	*Schefflera pubigera* (Brongn. ex Planch.) Frodin.	Araliaceae	Kursinglo	Climber	800–1800	Stem bark	NB	Abundant
33	*Schima wallichii* (DC.) Kort.	Theaceae	Chilaune	Tree	300–2100	Stem bark	NBL	Common
34	*Schisandra propinqua* (Wall.) Baillon	Schisandraceae	Singauto	Climber	700–1900	Stem bark	N	Sparse
35	*Smilax elegans* Wall. ex. Kunth.	Smilacaceae	Kukurdaino	Climber	2200–2800	Stem bark	NL	Sparse

(continued)

Table 12.1 (continued)

Sl. no.	Botanical name	Family	Local name	Habit	Distribution (m)	Parts used	Ethnic communities	Present status
36	*Spermadictyon suaveolens* Roxb.	Rubiacea	Ban Champa	Shrub	600–1700	Stem bark	L	Abundant
37	*Tinospora sinensis* (Lour.) Merr.	Menispermaceae	Gurjo Lahara	Climber	300–1000	Root and stem bark	NBL	Sparse
38	*Ulmus lanceifolia* Roxb. ex Wall.	Ulmaceae	Chamlayo	Tree	600–2000	Stem bark	N	Threatened
39	*Urtica dioica* L.	Urticaceae	Gharia Sisnu	Herb	600–2600	Roots	NBL	Common
40	*Viscum articulatum* Burman	Loranthaceae	Harchur	Herb	1500–2300	Aerial parts	NBL	Threatened
41	*Woodfordia fruticosa* (L.) Kurz.	Lythraceae	Dhainyaro	Shrub	300–900	Flower	N	Abundant

Note: *N* Nepali, *B* Bhutia and *L* Lepcha. Local name = Nepali

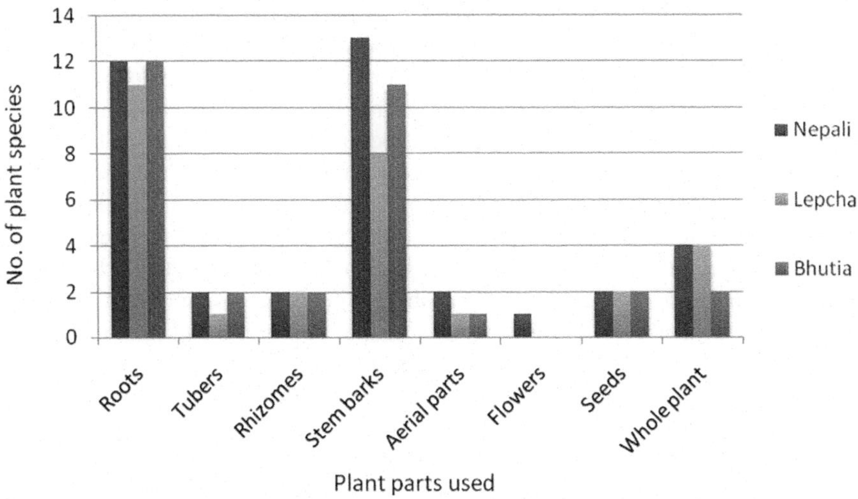

Fig. 12.2 Frequency of parts used for bone fractures in West District Sikkim

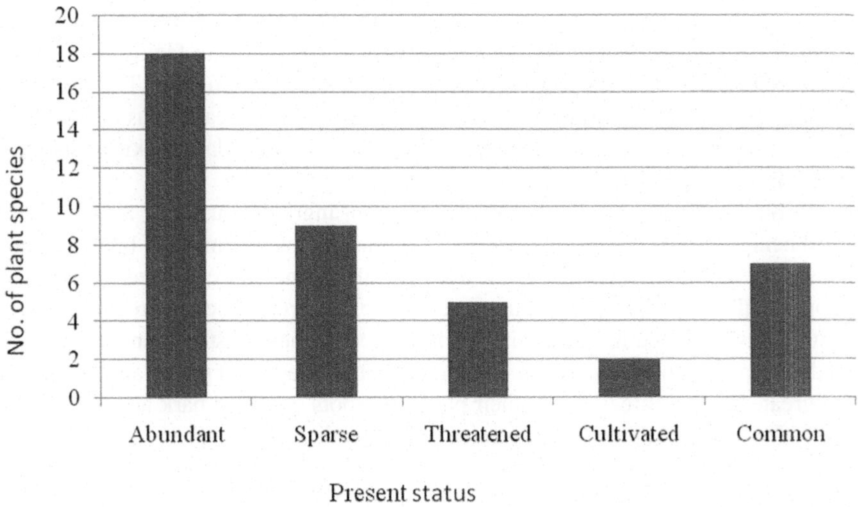

Fig. 12.3 Present status of medicinal plants used for the treatment of bone fractures in West District Sikkim

continuous uses. Remaining ten species are sparsely available, seven species are common and can be spotted in most of the places in West District of Sikkim, two species are cultivated for the purpose of consumption but also finds its application in traditional treatment of certain ailments, and five species are highly exploited and are found only in limited area (Fig. 12.3). Among these five threatened species, though they do not appear in IUCN threat list, *Gymnadenia orchidis* and *Rheum australe* are found in high altitudes 3500 m above sea level whereas human inhabitation is found far below this range in West District of Sikkim. Tuber of *Gymnadenia orchidis* and the whole plant of *Rheum austrule* arc used extensively for the preparation of herbal medicine. This leads to the destruction of whole plants while harvesting them. The local healers meet up their requirement of these plants by visiting these places illegally once every year or two and collect as much as they require. This trend has resulted in the severe pressure on the valuable mountain flora.

Most of the local people still prefer the use of herbal medicine prepared through traditional method rather than the modern system of treatment particularly for bone fracture and dislocation, as they believe that the modern system of treatment may follow painful therapies. Therefore, many natives still prefer and trust upon using traditional healthcare system as the excellent and much effective means to cure their ailments over allopathic drugs (Abebe and Ayehu 1993). As per the response of informants, paste prepared from the roots of *Plumbago zeylanica* is very effective in treating the joint dislocation, bone injuries, and swelling part of injured area if one could tolerate the burning sensation of plant extract. The rhizome of *Bergenia ciliata* and the whole plant of *Rheum australe* are most commonly used in case of bone fracture by the local people who look after the herd of yak, cow, sheep, horses in the high altitude away from their village and similar method of treatment is applied among these animals in case of bone fracture and joint dislocation.

Plants where the stem bark is used to prepare medicine are facing tremendous pressure to meet the need of local people in West district of Sikkim. The stem bark of *Ulmus lanceifolia* is highly preferred by some local healers; as a result the repeated extraction of bark has caused early death of this plant making this plant difficult to survive in natural habitat.

Similarly, *Viscum articulatum* is an important ingredient used by almost all the local healers, *Vaidyas* as well as the local people for the treatment of bone fracture. This plant grows as parasite on tall trees and it is generally collected by the local people without retaining any living parts on the host trees. Such approaches of harvesting have resulted in threatening such species on their existence and further will lead to wipe out their population completely. Other medicinal plants are also facing the greater risk of threats when their parts like roots and stem bark are used excessively for the formulation of traditional herbal medicines.

12.5 Conclusion

Different plants are being extensively used for the treatment of bone fracture in West District Sikkim by the herbal healers of their local communities since time immemorial. Such method of treatment is still prevalent in almost all the rural areas of Sikkim in spite of having the modern medical system of treatment. People of rural areas strongly rely on the traditional method of treatment due to easy accessibility, cost-effective and experience of minimum pain during the process of treatment. However, new generation having exposure to modern medicine and the patients with severe illness and chronic condition visit the hospitals for the treatment. During the field study it is also revealed on interviews that the local people who received treatment from the local herbal healers in the past have no complaint on their health issues. This shows that the local system of treatment is as effective as modern system of treatment. With the advent of technology, young generation people are not attracted to such traditional system of treatment that is experienced by their forefathers and they are giving less importance to the resources available around their nearest surrounding which is likely to ruin the rich heritage of traditional knowledge of this region. Realizing all these conditions, the present investigation was initiated to tap the traditional knowledge and to document such rich traditions of bone fracture treatment for future generations. The herbal treatments are worth appreciating; thus, it becomes imperative for the present generation to conserve it for the future by sustainable practices.

Acknowledgement The authors are immensely grateful to the Forest, Environment and Wildlife Management Department, Govt. of Sikkim, for granting permission to carry out this research in West District of Sikkim and preparation of maps. Authors are also thankful to the scientists of Botanical Survey of India, Sikkim Himalayan Circle, Gangtok, Sikkim, for the identification of plants and other necessary help.

References

Abebe D, Ayehu A (1993) Medicinal plants and enigmatic health practices of northern Ethiopia. B. S. P. E, Addis Ababa

Anonymous (2016) Conservation of natural and cultural values of Khangchendzonga National Park. Prop World Herit Site 8(1)

Atkinson ET (1882) The Himalayan Gazetteer. 2 (Reprint 1973). Cosmos Publication, Delhi

Bennet SSR (1983) Ethnobotanical studies in Sikkim. Indian Forester 109(7):477–481

Bennet SSR (1985) Ethnobotanical studies in West Sikkim. J Econ Tax Bot 7(2):317–321

Bharati KA, Sharma BL (2012) Plants used as ethnoveterinary medicines in Sikkim Himalayas. Ethnobot Res Appl 10:339–356

Biswas K (1956) Common medicinal plants of Darjeeling and Sikkim Himalaya. Bengal Government Press, West Bengal

Census (2011) Census reference tables-total population. Government of India. Accessed 12 May 2016

Cox PA, Ballick MJ (1994) Ethnobotanical approaches to drug discovery. Sci Am 270:82–87

Das T, Mishra SB, Saha D (2012) Ethnobotanical survey of medicinal plants used by ethnic and rural peoples in eastern Sikkim Himalayan region. Afr J Basic Appl Sci 4(1):16–20

Dash SS (2009) Traditional herbal remedies used in Sikkim, India. Nelumbo 51:123–156

Gurung B (2002) The medicinal plants of Sikkim Himalaya. Maples, Chakung

Hajra PK, Chakraborty P (1981) A survey of wild plants sold in the Lal market of Gangtok. Indian J Forestr 4(3):217–220

Idris MS, Badola HK, Singh R (2010) Indigenous knowledge and medicinal use of plants by local communities in Rangit Valley, South Sikkim. India Ne Bio 1(2):34–35

Jain SK (1991) Dictionary of Indian folk medicine and Ethnobotany. Deep Publication, New Delhi

Jana SK, Chauhan AS (2000) Ethnobatical studies on Lepchas of Dzongu, North Sikkim, India. Ann Forestr 8(1):131–144

Maiti D, Pradhan N, Chauhan AS (2003) Folk uses of medicinal plants from North Sikkim. Indian J Trad Know 3(1):66–71

Notlite HJ (1994) Flora of Bhutan. Royal Botanic Garden, Edinburgh, UK and Royal Government of Bhutan 3(1)

Notlite HJ (2000) Flora of Bhutan. Royal Botanic Garden, Edinburgh, UK and Royal Government of Bhutan 3(2)

Panda AK (2007) Medicinal plants of Sikkim in Ayurvedic practice. Regional Research Insitute (Ayurvedic) Gangtok, Sikkim

Panda AK, Misra S (2010) Health traditions of Sikkim Himalaya. J Ayur Integr Med 3:183–189

Pradhan BK, Badola HK (2008) Ethnomedicinal plant use by Lepcha tribe of Dzongu valley, bordering Khangchendzonga Biosphere Reserve, in North Sikkim, India. J Ethnobiol Ethnomed 4(22):4–22

Pradhan, Badola (2013) Plants used in healthcare practices by Limboo tribe in South-West of Khangchendzonga Biosphere Reserve, Sikkim, India. Indian J Tradit Knowl 12(3):355–369

Rai LK, Sharma E (1994) Medicinal plants of Sikkim Himalaya: status, uses and potential. Bishen Singh Mahendra Pal Singh, Dehradun

Rao RR, Sharma BD (1990) A manual for herbarium collections. Botanical Survey of India, Calcutta

Sharma TK, Sharma S (2010) Medicinal plants of Sikkim, Bermiok, Barthang, West District, Sikkim

Sharma TP, Dahal S, Borthakur SK (2012) Documentation of ethno-veterinary practices in Sikkim, India. Pleione 6(2):353–358

Singh BH, Prasad P, Rai LK (2002) Folk medicinal plants in the Sikkim Himalaya of India. Asian Folklore Stud 61:55–62

Tamang M, Pal K, Rai SK, Kalam A, Ahmed SR (2017) Ethnobotanical survey of threatened medicinal plants of. West Sikkim 2(6):116–125

WHO (2000) Turning the tide of malnutrition: responding to the challenge of the 21st century. World Health Organization, Geneva. Accessed 12 Aug 2016. http://who.in/mip2001/files/2232/NHDbrochure.pdf

God's Tree: A Culturally Coded Strategy for Conservation (A Case Study of Gairsain Ecoregion of District Chamoli, Uttarakhand)

13

V. P. Bhatt and Dinesh Singh Rawat

Abstract

This chapter presents a study of the holy trees of Gairsain eco-region of district Chamoli of Uttarakhand. These trees are regarded as god's tree and worshiped by the local inhabitants for propitiating the wishes of the latter. This is an implied technique of conservation and regeneration of ecologically and economically valued plants. A total of 18 tree species belonging to 14 genera of 11 families were recorded as god's trees in the 61 surveyed villages of the region. Nearly six trees per village were found which are considered as god's trees and worshiped during the different occasions. *Quercus leucotrichophora, Prunus cerasoides, Celtis australis* and *Myrica esculenta* are some of the prominent god's trees. All gods' trees are highly valued in all domains of life. The data were collected through extensive and intensive interviews with olden local people and *pujaaris* or priests during 2014–2016. Unfortunately, all these trees and practices are under severe threat and diminishing in number in the process of modernization of culture.

Keywords

God's trees · Gairsain · Conservation · Traditional belief · Modernization · Sacred place · Cultural code

V. P. Bhatt (✉)
Department of Botany, Govt. PG College, Karanprayag, Uttarakhand, India

D. S. Rawat
Department of Botany and Microbiology, HNB Garhwal University, Srinagar Garhwal, Uttarakhand, India

S. M. Khasim et al. (eds.), *Medicinal Plants: Biodiversity, Sustainable Utilization and Conservation*, https://doi.org/10.1007/978-981-15-1636-8_13

13.1 Introduction

Nature has certain symbiotic rules and regulations for proper maintenance and sustainability of its own life. Understanding and following those symbiotic rules in social life is known as religion. The Hindu religion and culture has a deep association with nature, especially with the plants and animals. It is believed that God has bestowed specific power to them (Ranta and Pirta 2007; Robinson and Cush 1997; Hema and Vinaya 2006). Trees have been always as the first temple of God and sacred groves as their first place of worship, and both were held in outmost reverence in the past (Danfi 2006; Pliny 1945; Porteous 1928; Quantz 1898). Many trees are considered sacred and are worshipped in many parts of India (Pande 1964). Trees are not only worshipped for themselves but for what is revealed through them, for what they imply and signify (Eliade 1958; Zahan 1979) and especially for various powers attributed to them (Hamiltone 2002; Milllar et al. 1999). However, it is believed that using one plant as part of ritual is not only good for an individual's health but also helps in the conservation and sustainability of biodiversity (Gam and Nath 2012; Liu et al. 2002; Manilal 1989). Several other researchers also have reported the conservational and religious aspects of sacred trees from different parts of the country (Bhatla et al. 1984; Dhiman 2003; Gadgil 1992; Gam and Nath 2012; Gogoi and Borthakur 1991; Gupta 1980; Jain and Kapoor 2007; Jain 1981; Manilal 1989; Murty 1979; Pande 1964; Sensarma 1995; Ranta and Pirta 2007; Sharma and Joshi 2010; Hema and Vinaya 2006; Sinha 1996; Sinha 1991; Upadhyaya et al. 1997).

Mountainous people of Uttarakhand Himalaya also have dedicated several plants to their respective deities or gods and goddess and have been worshiping these plants on different occasions, for generations together. Unfortunately, no authentic and proper documentation of such plants in the Uttarakhand Himalaya has so far been done. Therefore, the present study aims at identifying and documenting such plants which are conserved through traditional belief and worship system by the local people of the Gairsain block of district Chamoli, Uttarakhand.

Extensive field surveys were made in six different localities (Aadibadri, Dewalkot, Gairsian, Maithaan, Mehalchauri and Rohira) of Gairsain block in district Chamoli, Uttarakhand, during the years 2014–2016 (Fig. 13.1). The area has varied topographic features. As a result of this the region has diverse climatic conditions. The study area (Gairsain) spreads in montane to timberline/alpine type of vegetation which presents a panoramic view of the dense forests. Forests are represented by Oak-Abies mixed forest (2700–3114 m), Oak-mixed forest (1500–2500 m), Oak forest (1800–2000 m) and Pine-mixed forest (1200–1500 m). However, some patches are occupied by pine and scrub forest along with grassy slopes (Dinesh Singh 2010). It is bestowed with three dense forest ranges known as Dudhatoli, Painsar and Angyari while a large portion of the region is occupied by local inhabitants. Randomly ten nearby villages were selected from each locality to cover the ideal representation of all villages (total 232 villages, who's native language is Garhwali) of the block and at least five villagers (mostly *pujaris* or priests, *vaidhyas* or herbalist and aged persons like Amar Singh Belwal (46) of Rohira,

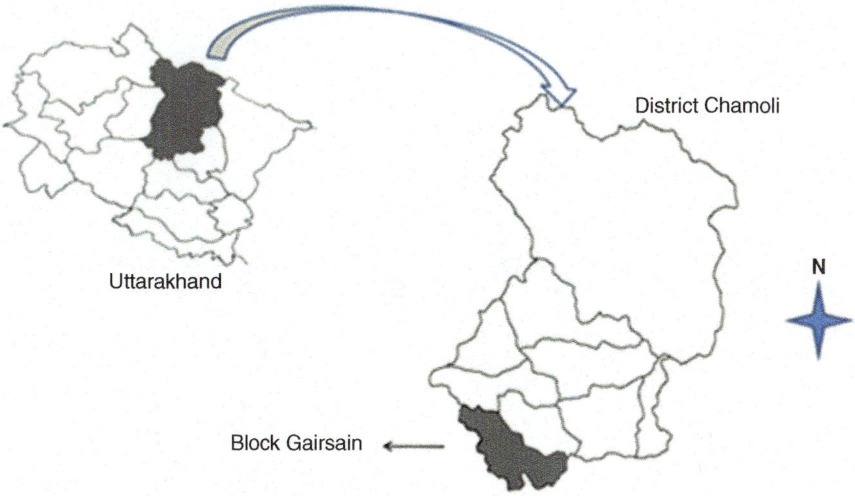

Fig. 13.1 Location map of the study area

Baktawar Singh Negi (54) of Kunigad, Bachan Singh (50) of Thaala, Bhawani Ram (54) of Sarkot, Ganga Singh (42) of Langta, Mohan Singh Rawat (64) of Bukhali, Narandra Singh Nawani (51) of Silanga, Pratap Singh (56) of Kainsua, Prem Singh Negi (60) of Laatu gair, Rai Singh Madhuwal (45) of Jhumakhet, Ramchandra Joshi (56) of Dungri and Ranjeet Singh (50) of Nandasain villages) were interviewed (Fig. 13.2a) from each village to generate the information on different aspects (generation producer, soil binder, wild-life protector-feeder-shelter, medicinal, timber, fruit, oil, resin, gums, fodder, fibre, fire-wood, etc.) of holy trees which are traditionally worshipped as god's trees. Plant specimens were collected and identified by consulting the regional floras (Gaur 1999; Naithani 1984) and finally confirmed by matching the specimens with the Herbarium of Garhwal University, Srinagar.

13.2 God's Tree: A Culturally Coded Strategy for Conservation

Through intensive interviews with locals and surveys conducted, it was known that several gods, goddess, and other deities of the particular region have an importance in material life as well as in sustaining the clean and healthy environment. For this, the local people have since ages developed several myths and rituals for conservation of forests and forest resources. The natural or supernatural beings who fulfil the wishes of locals are prayed and regarded as sacred in the region. *Dev-thaan* (sacred place) and *Dev-daali* (god's tree) are some of them. The god's trees can easily be recognized by seeing their huge virgin canopy with coloured flags (red, yellow or white) and numerous *Ghanties* (bells) hanging on their branches (Fig. 13.2b). For fulfilling individual wishes and wishes of villagers of the region, worship of healthy and huge type of trees is a common custom in the Gairsain region (Fig. 13.2c).

Fig. 13.2 (**a**) Interview with the locals. (**b**) *Rhododendron arboretum.* (**c**) View of worship during *Ashtabali.* (**d–e**) Quercus *leucotrichophora.* (**f**) *Quercus semecarpifolia.* (**g**) *Cupressus torulosa.* (**h**) *Prunus cerasoide.* (**i, j, k**) Close- up of the sacred place under God's tree. (**l**) *Ficus religiosa*

The trees offered worship are called *Dev daali* or *Devta ki daali* (god's tree), and the place where this tree is located is regarded as a sacred place. A place of worship decorated with ritual properties (*Diya, Dhupiyanu, Chimta Trisula, Sankh, Jhanda,* statue, *Maala*) is known as *Thaan* or *Dev-thaan or Dev-sthaan* (sacred place).

Flowers, *dhoop* and edibles are offered during worship to the deity inhabiting that place (Fig. 13.2). The shape and size of the *Thaan* depends upon availability of the area. The *Thaan* observed in the study area were 1 m to 5 m in diameter and rectangle, circular and triangle in shape. Intensive enquiry from the locals also revealed that the selection of *Dev-daali* (god's tree) depends upon the majorly valued and utilized plants in different services. The valuation is made either by the *Pujaari* or the priest and community members. Healthy, well-developed crowns, straight bole and immunity from any kind of diseases are the points which are considered during the selection. These god's trees are always located on the top of the forest or middle slope of the forest. Locals believe that praying god's trees fulfils their wishes, protects them from natural hazards and keeps them away from any kind of mishappening. But at the same time this is also a fact that these trees act as parent trees by dispersing the healthy and maximum number of seeds far and wide so that proper regeneration may occur for future ecological balance.

Entry to the tree or place of trees remains restricted against the touch of the menstruating women, children and even drunken men. It was believed that such persons pollute and deteriorate the quality and sacredness of the tree and the place. It was strongly believed that if anyone made use of any part of such trees to meet personal requirements, such a person would face the wrath of the deity and would also earn punishment for the villagers in the form of some mishappening with animals of the villagers or in the shape of drought, hail storm or excessive rainfall which paralyse the normal life in the region.

13.2.1 God's/Divine Trees in the Region

The present study discovered a total of 18 tree species (*Quercus leucotrichophora* (Fig. 13.2d, e), *Prunus cerasoides, Pyrus pashia, Myrica esculenta, Pinus roxburghii, Ficus palmata, Cupressus torulosa* (Fig. 13.2g), etc.) belonging to 14 genera of 11 families which are considered as god's trees in the 61 surveyed villages of the study area (Table 13.1) and which are worshiped by the locals for fulfilling their wishes. *Quercus leucotrichophora* tree (26%) was found most common in use in the area, followed by *Prunus cerasoides* (13%), *Pyrus pashia* (12%), *Myrica esculenta* (9%), *Pinus roxburghii* (7%), *Ficus palmata* and *Cupressus torulosa* with 5%. Family Fagaceae and Rosaceae together contributed 51% in the total recorded trees (Fig. 13.3).

It is evident from Table 13.1 that these trees are very important to the human beings because of their variety of utility (religious ceremonies, oxygen provider, medicinal, oil, edibles, fodders, ornamental, timber, fire-wood, agricultural implements, bee forages, wildlife, avis shelter and soil binder).

From the above facts examined, it is clear that only the highly valued plants/trees of different climatic and geographic locality were regarded as god's trees. The produces or any part (flower, fruit, leaf, bark, etc.) of the trees is taken as *Prasaad* (grace or blessing) of God so that they may heal their external and internal woes.

Table 13.1 List of the trees with short description conserved as God's trees in the Gairsain region (During 2011–2012)

S.No	Botanical name	Local name	Family	Average height (m)	Average CBH (Cm)	Total	% of total trees	Other uses[a]	Propagation
1.	*Abies pindrow* Royle	Raisul	Pinaceae	40	66	4	1.0	Ag, Md, Tm, Wd	Seed
2.	*Albizia lebbeck* (L.) Benth.	Siras	Mimosaceae	15	20	13	3.2	Ag, Bf, Fd, Wd	Seed
3.	*Bombax ceiba* L.	Semal	Bombacaceae	35	70	9	2.2	Ed, Bf, Fd, Tm, Wd	Seed
4.	*Cedrus deodara* (Roxb. ex D. Don.) G. Don.	Deodar	Pinaceae	24	27	10	2.5	Ag, Fd, Md, Ol, Rc, Tm, Wd	Seed
5.	*Celtis australis* L.	Kharik	Ulmaceae	45	42	5	1.2	Ag, Bf, Ed, Fd, Wd	Seed
6.	*Cupressus torulosa* D. Don	Surai	Cupressaceae	47	23	21	5.2	Ag, Ed, Wd	Seed
7.	*Ficus auriculata* Lour.	Timla	Moraceae	26	41	10	2.5	Ed, Fd, Rc ,Wd	Seed
8.	*Ficus neriifolia* Smith	Khilk	Moraceae	30	51	8	2.0	Ed, Fd, Wd	Seed
9.	*Ficus palmata* Forsk.	Bedu	Moraceae	20	38	20	5.0	Ed, Md, Fd, Wd	Seed
10.	*Ficus religiosa* L.	Peepal	Moraceae	34	35	11	2.7	Wd, Md, Rc	Seed
11.	*Myrica esculenta* Buch.- Ham ex D. Don	Kaphal	Myricaceae	24	52	36	8.9	Ag, Bf, Ed, Wd	Seed
12.	*Pinus roxburghii* Sarjent	Kulain	Pinaceae	40	43	28	6.9	Ag, Ed, Rc, Tm, Wd	Seed
13.	*Prunus cerasoides* D. Don	Payain	Rosaceae	38	48	53	13.1	Ag, Bf, Ed, Fd, Rc, Wd	Seed
14.	*Pyrus pashia* Buch.-Ham. ex D. Don	Melu	Rosaceae	35	41	47	11.6	Ag, Bf, Ed, Fd, Md, Md, Wd	Seed
15.	*Quercus leucotrichophora* A. Camus	Banj	Fagaceae	50	90	105	26.0	Ag, Fd, Wd	Seed
16.	*Quercus semecarpifolia* J. E. Smith	Kheru	Fagaceae	36	80	3	0.7	Ag, Fd, Wd	Seed
17.	*Rhododendron arboreum* Smith	Burans	Ericaceae	19	33	6	1.5	Ag, Bf, Ed, Fd, Md, Tm, Wd	Seed
18.	*Sapium insigne* (Royle) Benth. ex Trimen	Khinna	Euphorbiaceae	12	23	15	3.7	Fd, Wd	Seed and cuttings

[a]*Ag* Agricultural implements, *Bf* Bee forage, *Ed* Edible, *Fd* Fodder, *Md* Medicinal, *Ol* Oil, *Or* Ornamental, *Rc* Religious ceremony, *Tm* Timber, *Wd* Wood

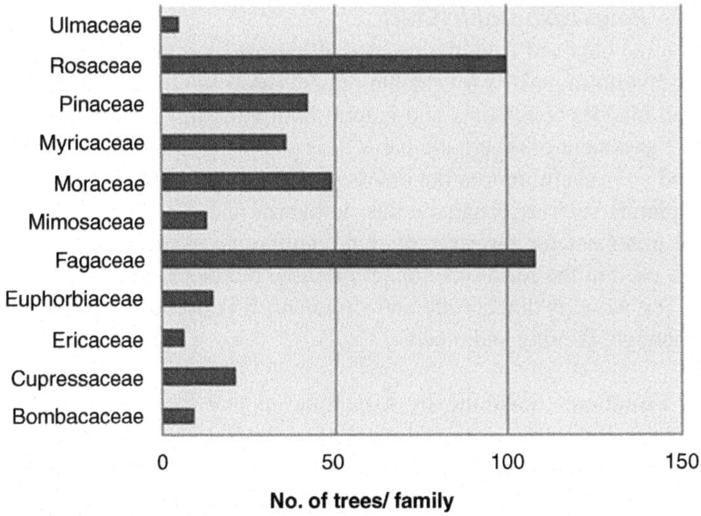

Fig. 13.3 Contribution of different families in the God's tree

13.2.2 Values of Some of the God's Tree

13.2.2.1 *Ficus religiosa* (Peepal)

Peepal has a great significance in ecology, medicine and rituals (Fig. 13.21). Its leaves serve as a wonderful laxative as well as tonic for the body. It is especially useful for patients suffering from jaundice. For constipation and dysentery problem, there can be no better remedy than the consumption of leaves of peepal. Dry the peepal leaves in sun and powder them. Add a solution of jaggery and anise to it. Mix it with water and consume it. This concoction will ensure proper bowel movement. Prepare a mixture of grinded coriander leaves, peepal leaves and sugar and chew it slowly. The leaves of peepal are highly effective in treating heart, various kind of skin disorders and mumps. It helps to control the palpitation of heart and thereby combat the cardiac weakness.

Cultural Functions The tree is described in *Bhagwadgita* as a manifestation of God himself. Thus the tree is the very dwelling place of the spirit of gods and also of the mislead souls like *brahmrakshas*. It is used for ritual property as it is the part of "*panch palaasas*", the five sacred trees. For 12 years daily rituals are offered to the growing plant and at end of the twelfth year a grand wedding ceremony of the tree is held in which the tree as daughter is married off to the mango tree. All wedding rituals are held on the occasion including the *yajna*, the holy sacrifice. Feast is offered to the Brahmins and also to the community members. Peepal tree also functions as *kalpa vriksa* on the occasion of the post funerary rites of a dead person. For 11 days the wandering soul of a dead man is worshipped at the base of the peepal tree where the *preta* rises transformed as *pitra* on the eleventh day. There are thousand other ritual uses of the peepal tree in different rituals of the mountain society in particular and the Hindu society as a whole.

13.2.2.2 *Pinus roxburghii* (Chir)

The turpentine obtained from the resin of all pine trees is antiseptic, diuretic, rubefacient and vermifuge. It is a valuable remedy used internally in the treatment of kidney and bladder complaints and is used both internally and as a rub and steam bath in the treatment of rheumatic pains. It is also very beneficial to the respiratory system and so is useful in treating diseases of the mucous membranes and respiratory complaints such as coughs, colds, influenza and TB. Externally it is a very beneficial treatment for a variety of skin complaints, wounds, sores, burns, boils, etc. and is used in the form of liniment plasters, poultices, herbal steam baths and inhalers. The wood is diaphoretic and stimulant. It is useful in treating burning of the body, cough, fainting and ulcers.

Cultural Functions Scientifically it is believed that the chir or pine tree is an exotic species. But in the lores and practice of the local folks the pine tree is ages old. It functions as the *moru ki daali,* the sacred tree of *Mahur* (Mathura) to be blessed by Krishna at the end of *Pandava lila*. It is believed to have given shelter to Nanda Devi when she was being given a mortal chase by *Mahishasur*. Chir is also used as *bandhan* and *toran* (decorative property) on the occasion of wedding. For all practical purposes pine is one of the five sacred trees used for rituals.

13.2.2.3 *Quercus leucotrichophora* (Banj)

The seeds are astringent and diuretic. They are used in the treatment of gonorrhoea, indigestion, diarrhoea and asthma. Any galls produced on the tree are strongly astringent and can be used in the treatment of haemorrhages, chronic diarrhoea, dysentery, etc. A mulch of the leaves repels slugs, grubs, etc. Oak galls are excrescences that are sometimes produced in great numbers on the tree. When the insect pupates and leaves, the gall can be used as a rich source of tannin, which can also be used as a dyestuff. It is also a good fuel and makes a good charcoal (Fig. 13.2d, e).

Cultural Functions Banj is one of the five sacred trees in the higher Himalayan habitations. It is believed to house the sacred spirits of deities like *Jakh* and *Bhumyaal*. The god of land or *Bhumyaal* loves to live at the base of a Banj tree. Its fuel is used in burning the fire of *yajna*. Its fungus is used as a sacred property in the processional rituals. The fungus is put inside the cabin of a palanquin or ensign and is counted among 64 magic properties put inside such palanquins.

13.2.2.4 *Prunus cerasoides* (Panya)

The fruit is astringent. The juice of the bark is applied externally to treat backaches. A green dye can be obtained from the leaves. A dark grey to green dye can be obtained from the fruit. The seeds are used as beads in necklaces and rosaries. Wood of the tree is durable and aromatic. The branches are used as walking sticks (Fig. 13.2h).

Cultural Functions The tree was brought from the netherworld by Lord Krishna on the challenge of *Kansa*, the villainous maternal uncle who wanted that the poisonous snakes of the netherworld might sting him to death. However, Krishna was victorious and brought up not only the sapling of *Panya* but also roped into captivity the poisonous *Basuki* snake and tossed him at the court of Kansa. Therefore the tree is supposed to be the most sacred of the five trees. It is used in *Styanarayan brat katha*, on the occasion of *ygyopavit rite*, *Naag* ritual, *Nrisingh* ritual and so many rituals held for the well-being of domestic animals. It is the most frequently used tree of all trees.

13.2.2.5 *Bombax cieba* (Semal)

Bombax ceiba (Semal) is used as local says: half a cup of ethanol extract of bark and flower was given for 3 days to both men and women with sexual diseases like hydrocele, leucorrhoea and gonorrhoea and was also used to check menstrual disorders in women. One teaspoon juice of fresh stem bark, one teaspoon juice of fresh root of *Asparagus racemosus* (Jhirna), powder of seven black peppers and one teaspoon of processed sugar or gum are taken orally on an empty stomach two times daily for 21 days to cure gonorrhoea, impotency, spermatorrhea, sterility, nocturnal emission and leucorrhoea. It is also prescribed for increasing sperm in semen and to act as aphrodisiac.

The fresh paste prepared from the bark mixed with cow dung is applied over back muscle of leg at night to treat hotness and inflammation. Root powder is used as a tonic to treat impotency; 10 g of root powder was advised daily with a glass of milk. A powder of stem prickles is used to treat asthma; about 10 g (one spoonful) powders was taken with a glass of cow's milk/fresh water in the morning for 3–4 months. Seed paste prepared in water was applied on small-pox boils. Flowers (25–50 g as feedstuff) were fed to the animal as anthelmintics. Preparation of about 30 g of seed powder of *B. ceiba* and about 10 g Hing *(Ferula foeitida)* are used as an abortifacient.

Cultural Functions The flowers and branches are used in *tantras* and Psycho-medicine.

13.2.2.6 *Pyrus pashia* (Mehal)

The Mehal helps reduce the risks of colon cancer and can prevent the growth of polyps. The astringent juice is used medicinally to stop diarrhoea, but little else seems to be known about the ways locals use the fruit. Most of the vitamin C content is in the skin of the fruit.

Cultural Functions Mehal is a sacred tree and one of the five sacred trees of the communities. Its twigs are planted in the paddy fields for warding off the evil impact of pests insects and bacterial blast. Mehal has been the subject of folk songs and the tree of Mehal is supposed to be the dearest one of the out married daughters who often feels nostalgic for the tree and its ambiance.

13.3 Discussion

The study revealed that the entire Development Block (232 villages) has 1392 trees which are regarded as god's trees. This is a handsome number of trees protected through this method in the region and ultimately having the good regeneration potential for furtherance. This is culturally coded custom of the region which has maintained ecology of the region for a long time. But as a result of devastating modernization, the present generation has altogether ignored it and its value. The new mindset has resulted in severe deterioration of the environment. Therefore, we have to create a new body of knowledge which is scientifically convincing to the degree of faith, the significance of such trees and their cult of worship to the modern generation which is afflicted with of over scienticism and complete loss of spiritual inheritance.

Acknowledgements The authors are grateful to the inhabitants of the Gairsain region for their kind cooperation and sharing the valuable information during the study.

References

Bhatla N, Mukherjee T, Singh G (1984) Plants: traditional worshiping. Ind J Hist Sci 19(1):37–42
Danfi A (2006) On the Typhology and worship status of sacred trees with a special reference to the Middle East. J Ethnobiol Ethnomed 2:26
Dhiman AK (2003) Sacred plants and their medicinal uses. Daya Publishing House, New Delhi
Eliade M (1958) Patterns in comparative religion. Sheed and Ward, New York
Gadgil M (1992) Conserving biodiversity as if people matter: a case study from India. Ambio 21:266–270
Gam NK, Nath PC (2012) Conservation of plant diversity through traditional beliefs and religious practices for rural Mishing tribes in Majuli River Island, Assam, India. Ind J Fund Appl Life Sci 2(2):62–68
Gaur RD (1999) Flora of the District Garhwal: North West Himalaya (with ethnobotanical notes). Transmedia, Srinagar (Garhwal)
Gogoi P, Borthakur SK (1991) Plants in religion-cultural beliefs of the Tai Khamatis of Assam (India). Ethnobotany 3:89–95
Gupta SS (1980) Sacred trees across cultures and nations. Indian Publications, Calcutta, p 27
Hamiltone LS (2002) Forest tree conservation through metaphysical constraints. The George Write Forum 19:57–78
Hema S, Vinaya G (2006) Sacred conservation practices at species level through tree worship. Ethnobotany 18:46–52
Jain SK (1981) Glimpses of Indian ethnobotany. Oxford and IBH Pub, New Delhi
Jain SK, Kapoor SL (2007) Divine botany - universal and useful but under explored traditions. Ind J Trad Knowl 6(3):534–539
Liu H, Xu Z, Xu Y, Wang J (2002) Practice of conserving plant diversity through traditional belief: a case study in Xishuangbanna, Southwest China. Biodivers Conserv 11:705–713
Manilal KS (1989) Linkage of ethnobotany with other sciences and discipline. Ethnobotany 1:15–24
Milllar D, Aniah R, Atoyure P (1999) Shrine and groves. COMPAS News Letter, pp 34–36
Murty SK (1979) Significance behind the worship of plants. Sachitra Ayurveda 32(4):180
Naithani BD (1984) Flora of Chamoli, vol 1-2. Botanical Survey of India, Howrah, New Delhi
Pande T (1964) Tree worshiped in ancient India. Folklore 5(6):213–218

Pliny (1945) Natural history (trasl. by H. Rackham). Havard University Press, Loeb Classical Library, London & Cambridge, Mssachusetts

Porteous A (1928) Forest folkore, mythology and romance. George Allen & Unwin Ltd, London

Quantz JO (1898) Dendro-psychosis. Am J Psychol 9(4):449–506

Ranta RS, Pirta RS (2007) Socio-ecology religious affiliation in three Himalayan villages: implication for mental health. J Ind Acad Appl Psychol 23(1):23–30

Robinson C, Cush D (1997) The sacred cow: Hinduism and ecology. J Beliefs Value 18(1):25–37

Sensarma P (1995) The sacred and religious plants in tantra sastra. Ethnobotany 7:51–61

Sharma V, Joshi BD (2010) Role of sacred plants in religion and health care system of local people of Almora district of Uttarakhand state (India). Acad Arena 2(6):19–22

Dinesh Singh (2010) Flora of lohba range of Kedarnath Forest Division. M.Sc Thesis HNB Garhwal University Srinagar Garhwal; pp. 5–6

Sinha RK (1991) Ecosystem preservation through faith and tradition in India. J Human Ecol 2(1):21–24

Sinha RK (1996) Conservation of cultural diversity of indigenous people essential for protection of biological diversity. In: Jain SK (ed) Ethnobotany in HumanWelfare. Deep Publications, New Delhi, pp 280–283p

Upadhyaya AS, Kumbhojkar MS, Kulkarni DK (1997) Ethno-medico botany of some sacred plants of western Maharashthra. Ethnobotany 9:65–68

Zahan D (1979) The religion, sprituality thought of traditional Africa. The University of Chicago Press, Chicago and London

Ethnomedicinal Plants Used by Ethnic People in Eastern Ghats of Visakhapatnam District, Andhra Pradesh, India

14

S. B. Padal, K. Satyavathi, and J. Prakasa Rao

Abstract

An ethnomedicinal survey was carried out among the ethnic people of Khodu, Mali and Gadaba tribals inhabited in Visakhapatnam District, Andhra Pradesh, India. A total of 55 ethnomedicinal plants belonging to 55 genera under 35 families were recorded. The maximum contribution was recorded for trees, followed by herbs, shrubs and climbers. Among the plant parts used in different formulations, stem and bark are profusely used, which is followed by root, leaves and latex. A total of 35 diseases are known to be cured by using 55 medicinal plants species.

Keywords

Ethnomedicinal plants · Ethnic people · Eastern Ghats · Visakhapatnam district

14.1 Introduction

Medicinal plants are nature's gift to mankind and are rich heritage of India. India is well known as an "Emporium of medicinal plants". About 70% of the rural folk depend on medicinal plants for their health care. India is a treasure of biodiversity which hosts a large variety of plants and ranks tenth among plant-rich countries of the world and fourth among the Asian countries.

According to the World Health Organization (2008), the term "traditional medicine" is to be understood as the sum total of the knowledge, skills and practices based on theories, beliefs and experiences indigenous to different cultures that are used to maintain and improve health, as well as to prevent, diagnose and treat

S. B. Padal (✉) · K. Satyavathi · J. P. Rao
Department of Botany, Andhra University, Visakhapatnam, Andhra Pradesh, India

S. M. Khasim et al. (eds.), *Medicinal Plants: Biodiversity, Sustainable Utilization and Conservation*, https://doi.org/10.1007/978-981-15-1636-8_14

physical and mental illnesses. The World Health Organization has a keen interest in documenting the use of medicinal plants by native peoples from different parts of the world (Buragohain 2011). Across the world and throughout the ages plants have traditionally played a major role in the treatment of human diseases (Thirumalai et al. 2009).

Plants are an integral part of life in many indigenous communities (Sidigia et al. 1990). Besides, being the source of food, fodder, fuel, etc., the use of plants as herbal medicines in curing several ailments goes parallel to the human civilization. India is endowed with a rich wealth of medicinal plants, being perhaps the largest producer and rightly acclaimed as the Botanical garden of the world (Dubey et al. 2004).

In India, the use of plants for medicinal treatment dates back to 5000 years. It was officially recognized that 2500 plant species have medicinal value while over 6000 plants are estimated to be explored in traditional, folk and herbal medicine (Huxley 1984).

Ethnobotanical investigation has led to the documentation of a large number of wild plants used by tribal for meeting their multifarious requirements (Anonymous 1990). Studies on ethnobotany were initiated by Janaki-Ammal as an official programme in the Economic Botany Section of Botanical Survey of India (Howrah) in 1954. From 1960, Jain started intensive field studies among tribal areas of central India (Jain 1963a, b, c, d, e; 1964a, b, c; 1965).

Although different workers have documented the use of various medicinal plants from different parts of Andhra Pradesh (Hemadri et al. 1987a, b; Prayaga Murty and Venkaiah 2010; Prayaga Murty and Mohana Narasinha Rao 2010; Padal et al. 2010), detail information on ethnomedicinal plants is unavailable. Very little literature was available on herbal folk medicine of Visakhapatnam District.

14.2 Materials and Methods

14.2.1 Medicinal Plant Survey and Data Collection

Ethnobotanical exploration trips were carried out in primitive tribal people dominated villages during 2016-2017. The area under study was thoroughly covered and the people were interrogated for information. After selecting the people, knowledge about their interests and skills in identification and utilization were obtained through informal interviews and discussion was made with the informants in their local language. A total of 50 informants (Bagata 8, Konda Dora 8, Mali 12, Gadaba 6, Porja 7 and Valmiki 9) between the age group 30 and 75 years were interviewed with a questionnaire. Out of 50 informants, 35 are male and 15 are female. At the end of each interview, specimens of plants mentioned for medicinal uses were collected and identified. Identification of species was made with the help of Floras (Gamble and Fischer 1915; Subba Rao and Kumari 2003). The species and family names are mentioned according to APG III Systems of Classification. These specimens have been housed in the Herbarium of the Department of Botany, Andhra University, Visakhapatnam, Andhra Pradesh, for further reference.

14.3 Study Area

14.3.1 Eastern Ghats

The Eastern Ghats are located along the Peninsular India extending over 1750 km under 11°03′ to 22°03′ N latitudes and 77°02′ to 87°02′ E longitudes covering 224,290 km^2. The Eastern Ghats are delimited in the north by Similipal hills of Odisha State. The middle section extends from river Krishna (Andhra Pradesh) to near about Chennai city (Tamil Nadu) and includes the Nallamalais, Nigidi, Seshachalam and Veligonda hills. The last section runs in S-SW direction meeting the Western Ghats in the Nilgiris (Reddy et al. 2006).

In Andhra Pradesh the hilly region of Eastern Ghats is divided into Northern, Central and Southern Ghats. The Northern Eastern Ghats portion covers the districts of Srikakulam, Vizianagaram, Visakhapatnam, East Godavari, West Godavari and Khamam. Forests are endowed with rich, varied and endemic flora. Climate of Eastern Ghats typically tropical is enough to support the most luxuriant type of vegetation and maintain rich biodiversity.

Visakhapatnam district is situated in north-eastern part of Andhra Pradesh. The district is bounded by Vizianagaram district on the North, East Godavari district on the South, Koraput of Orissa on the West and Bay of Bengal on the East (Ramesh 1979). Visakhapatnam is a home of 14 tribes, namely Bagata, Gadaba, Gond, Konda Kammara, Konda Dora, Kotiya, Khond, Kulia, Mali, Manne Dora, Nooka Dora, Porja, Reddi Dora and Valmiki. Percentage of ST population in Visakhapatnam district is 13.74%. The geographical area of Visakhapatnam district is 11,161 km^2 and covers the forest area of 3435 km^2 (41.40%).

14.4 Results and Discussion

A total number of 55 plants belonging to 35 families were recorded (Table 14.1). Fabaceae is the most dominated family with 6 species, followed by Apocynaceae with 4 species, Rutaceae and Lamiaceae with 3 species, Verbenaceae, Schrophulariaceae, Rutaceae, Rubiaceae, Nyctaginaceae, Lythraceae, Asteraceae and Liliaceae with 2 species, Caesalpiniaceae with 4 species, Araceae, Cucurbitaceae and Lamiaceae with 3 species in each, Rubiaceae, Plantaginaceae, Malvaceae, Lythraceae, Euphorbiaceae, Asteraceae and Acanthaceae with 2 species in and rest of the 24 families contain only one species. Among the total plant species, trees are highest in number (22), followed by herbs (19), shrubs (9) and climbers and stragglers (5). In this the number of plants used for each ailment is given in brackets, i.e. abortion (3), anaemia (2), anthelmintic (1), antifertility (1), asthama (5), blisters (2), boils (1), breast pain (1), cold (1), cough (1), cuts (1), dandruff (1), diabetes (2), diarrhoea (2), dysentery (3), eczema (2), fractures (1), fever (2), gonorrhoea (1), head ache (1), hydrocele (2), leucorrhoea (5), menstrual problem (1), mosquito repellant (1), muscle pains (1), obesity (1), paralysis (1), piles (1), scurvy (1),

Table 14.1 Ethnomedicinal plants used by tribal people in Eastern Ghats in Visakhapatnam District, Andhra Pradesh

S. No	Plant name	Family name	Local name	Habit	Disease and mode of uses
1	*Abrus precatorius* L.	Fabaceae	Guruvinda	Straggler	Cough: a spoonful of leaf juice is administered daily twice for 3 days (Valmiki).
2	*Acacia sinuata* (Lour.) Merr.	Fabaceae	Shikaya	Shrub	Dandruff: seed paste is massaged on the scalp once a day for a week (Bagata).
3	*Acalypha indica* L.	Euphorbiaceae	Kuppinta	Herb	Scorpion sting: leaf paste is applied on the affected areas and 20 g of this paste is also administered orally (Bagata).
4	*Achyranthes aspera* L	Amaranthaceae	Duchini	Herb	Cuts: leaf paste is applied on the affected areas daily once for 3 days (Mali).
5	*Acorus calamus* L.	Acoraceae	Vasa	Herb	Rheumatoid arthritis: root poultice is applied over the affected area till cure (Valmiki).
6	*Adiantum lunulatum* Burm	Adiantaceae.	Gatumandu	Herb	Abortifacient: a decoction of the fern is used by the tribal people as an abortifacient (Porja).
7	*Aegle marmelos* L.	Rutaceae	Maredu	Tree	Cold: leaf is poured into nostrils once a day for 5 days (Porja).
8	*Alangium salvifolium* (L.f.) Wang.	Alangiaceae	Uduga	Tree	Paralysis: stem bark ground to past with *Piper nigrum* and the past is administered two spoonfuls per day for 7 days (Valmiki).

(continued)

Table 14.1 (continued)

S. No	Plant name	Family name	Local name	Habit	Disease and mode of uses
9	*Aloe vera* (L.) Burm. f.	Asphodelaceae	Kalabanda	Herb	Leucorrhoea: 5 ml of juice extracted from the crushed leaves is given once a day for 7 days (Bagata).
10	*Alstonia venenata* R.Br.	Apocynaceae	Edakulapala	Tree	Anthelmintic: stem bark along with *Piper longum* is made into an extract and is administered in doses of 5 spoonfuls twice a day for 3 days Valmiki).
11	*Artocarpus heterophyllus* Lam.	Moraceae	Panasa	Tree	Skin diseases: tender leaf paste is applied on the affected area till cure (Porja).
12	*Bacopa monnieri* (L.) Penn.	Plantaginaceae	Neeti brahmi.		Diabetes: herbalists prepared decoction for the whole plant given daily for a month to control diabetes or its related symptoms (Konda Dora).
13	*Bauhinia racemosa* Lam.	Fabaceae	Ari	Tree	Asthma: the paste of the stem bark is eaten till cure (Konda Dora) (Fig. 14.1)
14	*Boerhavia diffusa* L.	Nyctaginaceae	Punarnava	Herb	Leucorrhoea: 15 ml of decoction of plant is taken orally once a day for 3 days (Valmiki).
15	*Bombax ceiba* L.	Malvaceae	Buruga	Tree	Skin disease: seed oil is applied to remove the scars (Bagata) (Fig. 14.1)
6	*Buchanania lanzan* Spreng.	Anacardiaceae	Chinna murli	Tree	Boils: stem bark paste with *Ricinus communis* oil is applied on boils (Bagata).

(continued)

Table 14.1 (continued)

S. No	Plant name	Family name	Local name	Habit	Disease and mode of uses
17	*Butea monosperma* (Lam.) Taub.	Fabaceae	Moduga chettu	Tree	Antifertility: stem bark extracts with *Sesamum indicum* oil one spoonful twice a day is given from fourth day of menstruation to 11th day (Bagata) (Fig. 14.1)
18	*Calotropis gigantea* (L.) R.Br.	Apocynaceae	Jilledu	Shrub	Leucorrhoea: root decoction with paste of long peppers (3:1) is taken orally (Porja).
19	*Capparis zeylanica* L.	Capparaceae	Aridonda,	Climber	Tuberculosis: root bark ground with *Piper nigrum* and made into pills, two pills are administered twice a day for 15 days (Valmiki).
20	*Caryota urens* L.	Arecaceae	Jeeluga	Tree	Dandruff: nut powder made into paste applied to the head and bath is taken after 1 h for twice a week (Porja).
21	*Chamaecrista absus* L.	Fabaceae	Chanupala vittulu	Herb	Asthma: 3 ml decoction of flowers in combination with the decoction of leaves is administered internally till cure (Valmiki).
22	*Cassytha filiformis* L	Lauraceae	Seeta savaram	Climber	Hydrocele: the whole plant crushed with sand and tied tightly on testicles for 3 days (Porja).

(continued)

Table 14.1 (continued)

S. No	Plant name	Family name	Local name	Habit	Disease and mode of uses
23	*Celastrus paniculatus* Wild.	Celastraceae	Dati chettu	Climber	Leucorrhoea: root bark ground with black pepper grains, 3 ml of extract taken orally once a day for about 2 weeks (Konda Dora).
24	*Chloroxylon swietenia* DC.	Rutaceae	Billachettu.	Tree	Mosquito repellant: the smoke of burnt leaves is used as mosquito repellant (Porja).
25	*Cheilocostus speciosus* (Koenig) Smith	Costaceae	Adavidumpa	Herb	Abortion: 10 g of rhizome paste is administered twice a day for 5 to 7 days (Valmiki).
26	*Curculigo orchioides* Gaertn.	Hypoxidaceae	Nelatadi	Herb	Piles: two spoonfuls of tuber extract is administered twice a day till cure (Porja) (Fig. 14.1)
27	*Cuscuta reflexa* Roxb.	Convolvulaceae	Bangaru theega	Herb	Tongue ulcer: plant paste is applied for tongue ulcers (Konda Dora).
28	*Dendrophthoe falcata* (L.f.) Ett.	Loranthaceae	Badanika	Shrub	Asthma: 10-12 g of stem bark powder is administered daily twice for 3 days (Konda Dora) (Fig. 14.1)
29	*Diospyros chloroxylon* Roxb.	Ebenaceae	Ellinda	Tree	Diarrhoea: two spoonfuls of leaf juice is given twice a day for 3 days (Valmiki).
30	*Eclipta prostrata* (L.) L. Mant	Asteraceae	Guntakalagaraku	Herb	Fever: 5 ml of plant extract is given orally twice or thrice a day for 7 days.

(continued)

Table 14.1 (continued)

S. No	Plant name	Family name	Local name	Habit	Disease and mode of uses
31	*Elephantopus scaber* L.	Asteraceae	Nelamarri	Herb	Eczema: dried root paste mixed with mustard oil is applied on the affected parts (Konda Dora) (Fig. 14.1)
32	*Elytraria acaulis* (L.f.)	Acanthaceae	Kukkapan	Herb	Menstrual disorders: two spoonfuls of the leaf juice is administered twice a day for 9 days (Mali).
33	*Gloriosa superba* L	Colchicaceae	Nabhi	Herb	Asthma: leaf paste is heated and applied on the forehead and neck for 7 days (Mali).
34	*Gmelina arborea* Roxb	Lamiaceae	Gummidi	Tree	Cough: 5 ml of leaf juice is administered twice a day for 7 days (Mali).
35	*Grewia tiliaefolia* Vahl	Tiliaceae	Tadachettu	Tree	Fracture: root bark paste is applied as plaster on dislocated joints of cattle (Mali).
36	*Haldinia cordifolia* (Roxb.) Ridsd	Rubiaceae	Bandaru	Tree	Wounds: leaves made into paste and applied on the affected parts (Bagata).
37	*Holarrhena pubescens* Wall. ex G. Don.	Apocynaceae	Palakodisa	Tree	Asthma: one teaspoonful of bark powder is given orally till cure (Bagata).
38	*Justicia adhatoda* Medik.	Acanthaceae	Addasaramu	Shrub	Diarrhoea: one spoonful of the leaf extract is taken twice a day till cure (Mali).
39	*Lawsonia inermis* L.	Lythraceae	Gorinta	Shrub	Syphilis: half spoon of leaf paste is administered daily for 16 days.

(continued)

Table 14.1 (continued)

S. No	Plant name	Family name	Local name	Habit	Disease and mode of uses
40	*Leonotis nepetifolia* (L.)	Lamiaceae	Ranabheri	Herb	Breast pain: ash of inflorescence mixed with mustard oil is applied on breast for postnatal breast pain (Bagata) (Fig. 14.2)
41	*Lygodium flexuosum* (L.) Sw.	Lygodiaceae	Khorothi	Herb	Anaemia: 1 pinch of root powder once a day for 15-21 days (Mali).
42	*Mallotus philippensis* (Lam.) Muell.	Euphorbiaceae	Sinduram	Tree	Wounds: seed paste is applied on the wounds of cattle to kill worms.
43	*Nyctanthus arbor-tristis* L	Nyctanthaceae	Parijatam	Tree	Muscle pains: 5 ml of leaf decoction is taken orally once a day for 2 days (Bagata).
44	*Pavetta indica* L.	Rubiaceae	Papidi	Shrub	Blisters: Warm leaf paste applied over the affected parts (Valmiki).
45	*Phyllanthus emblica* L.	Phyllanthaceae	Peddausiri	Tree	Scurvy: 1 green fruit is administered twice a daily for about 2-3 months (Mali) (Fig. 14.2)
46	*Plumbago zeylanica* L.	Plumbaginaceae	Chitra mulamu	Shrub	Abortion: root paste made into pills 2 pills are administered orally twice a day for 5 days (Bagata) (Fig. 14.2)
47	*Pterocarpus marsupium* Roxb.	Fabaceae	Yegisa	Tree	Dysentery: 5 ml of root bark extract mixed with curd and administered orally once a day for 3 days (Valmiki)
48	*Scoparia dulcis* L.	Plantaginaceae	Ghod Thulasi.	Herb	Fever: 2 spoonfuls of plant juice is administered thrice a day for 3 days (Porja).

(continued)

Table 14.1 (continued)

S. No	Plant name	Family name	Local name	Habit	Disease and mode of uses
49	*Solanum nigrum* L.,	Solanaceae	Kamanchi	Herb	Gonorrhoea: 5 ml juice of whole plant is given thrice a day for 15 days (Bagata).
50	*Sterculia urens* Roxb.	Malvaceae	Konda thamara	Tree	Dysentery: 1 spoonful of stem bark extract is administered with honey twice a day for 2 days (Valmiki).
51	*Terminalia chebula* Retz.	Combretaceae	Karakkaya	Tree	Cough: fruit paste mixed with breast milk and administered orally to infants (Mali).
52	*Toddalia asiatica* (L.) Lam.	Rutaceae	Kondakasinda	Straggler	Dysentery: 5 ml of root juice is given twice a day for 3 days (Gadaba).
53	*Vitex negundo* L	Lamiaceae	Vavilli	Shrub	Headache: leaves are made into paste and the paste is applied over the head (Gadaba).
54	*Woodfordia fruticosa* (L.) Kurz.	Lythraceae	Jeguru	Shrub	Leucorrhoea: a spoonful of powder of dried flower mixed in half glass of hot water is administered daily once for 3 days (Mali) (Fig. 14.2)
55	*Wrightia tinctoria* (Roxb.) R. Br.	Apocynaceae	Ankudu	Tree	Obesity: bark along with *Cuminum cyminum* and garlic is used to reduce weight (Valmiki).

rheumatic pains (1), skin disease (2), tongue ulcers (1), tuberculosis (1) and wounds (2). With regard to the frequency of plant parts used in preparations, stem bark was most often used, followed by roots, leaves, whole plant, roots, seeds and tubers. The total 55 plants are used for ethnomedicine to cure 35 ailments. For women problems like leucorrhoea curing *Grewia tiliaefolia* Vahl, *Haldinia cordifolia* (Roxb.) Ridsd, *Holarrhena pubescens* Wall. ex G. Don., *Justicia adhatoda* Medik, and *Lawsonia inermis* L plants are used. An ethnomedicinal study was carried out on the tribes of

Plant Photos

Fig. 14.1 Photo plate 1

Leonotis nepetiifolia

Phyllanthus emblica

Plumbago zeylanica

Woodfordia fruticosa

Terminalia chebula

Vitex negundo

Fig. 14.2 Photo plate 2

Srikakulam district for leucorrhoea problems; a total of 31 species belonging to 27 families were recorded (Satyavathi et al. 2015). From Visakhapatnam District 455 ethnomedicinal plants were reported for various diseases used by local tribes (Padal et al. 2010). Data was compared with the available literature and found that many of the usages listed are not recorded earlier (Rout et al. 2009; Nagariya et al. 2010; Jeeva et al. 2007).

In recent years many attempts were made to study the medicinal plants used by various aboriginal tribes and other people in India, particularly in Western Ghats and Eastern Ghats. So many research institutes and universities showed much interest in studying the Ethnobotany in the Eastern Ghats. Visakhapatnam is one of the rich districts of Andhra Pradesh with good forests and rich biodiversity.

14.5 Conclusion

The rapid denudation of natural vegetation in general and medicinal plants in particular has caused much concern among vegetation managers, botanists, ethnobotanists, ecologists and environmentalists. It is therefore necessary to document the first-hand information of the region and stress on their conservation to the future generations before the species become extinct. There is a need to support indigenous practices about medicinal plants with a vision to conservation and community development. The medico-botanical survey of the area revealed that the people of the area are possessing good knowledge of herbal drugs, but as the people are in progressive exposure to modernization, their knowledge of traditional uses of plants may be lost in due course. So it is important to study and record the uses of plants by different tribes and sub-tribes for futures study. Such studies may also provide some information to biochemists and pharmacologists in screening of individual species and in rapid assessing of phytoconstituents for the treatment of various diseases.

Acknowledgements The authors are thankful to the notified and denotified adivasi groups, their vaids, ojhas, bhopas, etc. and forest officials who provided valuable information on this subject.

References

Anonymous (1990) Ethnobiology in India: a status report. Ministry of Environment & Forests, Govt. of India, New Delhi, pp 1–68

Buragohain J (2011) Ethnomedical plants used by the ethnic communities of Tinsukia District of Assam, India. Recent Res Sci Technol 3(9):31–42

Dubey NK, Kumar R, Tripathi P (2004) Global promotion of herbal medicine: India0 s opportunity. Curr Sci 86(1):37–41

Gamble JS, Fischer CEC (1915) Flora of the Presidency of Madras. vol. 1-3 (vol. 3 by C.E.C Fischer). Adlard & Sons Ltd., London

Hemadri K, Sarma CRR, Rao SS (1987a) Medicinal plant wealth of Andhra Pradesh, part 1. Ancient Sci Lfe 6:167–186

Hemadri K, Sarma CRR, Rao SS (1987b) Medicinal plant wealth of Andhra Pradesh, part II. Ancient Sci Life 7:55–64
Huxley A (1984) Green inheritance: the world wildlife fund book of India. Collins/Harvel, London
Jain SK (1963a) The origin and utility of some vernacular plant names. Proc Nation Acad Sci India Sec B 33:525–530
Jain SK (1963b) Studies in Indian Ethnobotany – II. Plants used in medicine tribals of Madhya Pradesh. Bull Reg Res Lab Jammu 1:126–128
Jain SK (1963c) Magico-religious beliefs about plants among the adivasis of Bastar. Q J Myth Soc 4:73–94
Jain SK (1963d) Observations on ethnobotany of tribals of Madhya Pradesh. Vanyajati 11:177–183
Jain SK (1963e) Studies on Indian Ethnobotany – less known uses of 50 common plants from tribal areas of Madhya Pradesh. Bull Bot Surv India 5:223–226
Jain SK (1964a) The role of botanist in folklore research. Folklore 5:145–150
Jain SK (1964b) Wild plant foods of the tribals of Bastar. Khadi Gramodyog 10:557–561
Jain SK (1964c) Native plant remedies for snake-bite among adivasis of Central India. Indian Med J 57:307–369
Jain SK (1965) Wooden musical instruments of the Gonds of Central India. Ethnomusicology 9:39–42
Jeeva GM, Jeeva S, Kingston C (2007) Traditional treatment of skin diseases in South Travancore, southern peninsular India. Indian J Tradit Knowl 6(3):498–501
Nagariya AK, Meena AK, Jain D, Gupta BP, Yadav AK, Gupta MR (2010) Medicinal plants used in the healing of skin diseases in different regions of India: a review. Int J Chem Anal Sci 1(5):110–113
Padal SB, Prayaga Murty P, Srinivasa Rao D, Venkaiah M (2010) Ethnomedicinal plants from paderu division of Visakhapatnam District, A.P, India. J Phytology 2(8):70–91
Prayaga Murty P, Mohana Narasinha Rao G (2010) Unique etnomedicinal uses some plant species of Andhra Pradesh, India. J Phytology 2(4)
Prayaga Murty P, Venkaiah M (2010) Some abortifacient plants used by the tribal people of Andhra Pradesh. J Phytology 2(4):07–12
Ramesh N (1979) Andhra Pradesh district gazetteers. Govt. of. Andhra Pradesh, Visakhapatnam
Reddy CS, Brahmam M, Raju VS (2006) Conservation prioritization of endemic plants of Eastern Ghats, India. J Econ Tax Bot 30:755–772
Rout SD, Panda T, Mishra N (2009) Ethno-medicinal plants used to cure different diseases by tribals of Mayurbhanj district of North Orissa. Ethno Med 3(1):27–22
Satyavathi K, Sandhya Deepika D, Padal SB, Prakasa Rao J (2015) Ethnomedicinal plants used for leucorrhoea by tribes of Srikakulam District, Andhra Pradesh, India. Malaya J Biosci 2(4):194–197
Sidigia I, Nyaigotti-Chacha C, Kanunah MP (1990) Traditional medicine in Africa. East African Educational Publishers, Nairobi
Subba Rao GV, Kumari GR (2003) Flora of Visakhapatnam district. Andhra Pradesh, vol 1. Botanical Survey of India, Kolkata
Thirumalai T, Kelumalai E, Senthilkumar B, David E (2009) Ethnobotanical study of medicinal plants used by the local people in Vellore District, Tamilnadu, India. Ethnobotanical Leaflets 13:1302–1311
World Health Organization (2008) Traditional medicine. Fact sheet No.134. December, 2008. Accessed 06 Mar 2013. http://www.who.int/mediacentre/factsheets/fs123/en/

Ethnobotany of Medicinal Plants of Eastern Ghats of Andhra Pradesh for the Identification of Plants with Antitumour and Antimicrobial Potential

15

M. Santosh Kumari and Kandru Ammani

Abstract

Plants have profoundly influenced the culture and civilization of man in many countries. Plants have been an integral part of Indian life and culture. The real and indigenous culture of any nation is reflected in its folklore. In folk religion and folklore, trees are often said to be the homes of tree spirits or deities curing their physical and mental problems. "Man ever desirous of knowledge, has already explored many things, but more and greater still remains concealed, perhaps reserved for far distant generations, who shall prosecute the examination of their creator's work in remote countries, and make many discoveries for the pleasure and convenience of life…". This quotation of Linnaeus suggests for the anthropological approach of the traditional knowledge left unexplored. With the advancing civilization there is urgency in the exploration of the field resources, because the tribal pockets in the country are fast dwelling. So preservation and perpetuation of this knowledge should be emphasized for the benefit of mankind. This chapter deals with some of the important less-known uses of folklore medicinal plants used by the tribes of Eastern Ghats for treating wounds and tumours. Eastern Ghats are an abode of rich biodiversity with mixed deciduous forests inhabiting diverse tribal communities and ethnic groups like Chenchus, Sugalis, Yerukulas, Yanadis and Koyadoras living in small hamlets called Gudems or Tandas. The traditional practices and ideas of body physiology and health preservation known to some, transmitted informally as general knowledge, and practiced or applied by anyone in the ethnic group having prior experience. The main aim of the study was to collect data of the plants with antitumour, wound healing and antimicrobial properties. Analysis of the data revealed a number of less-known uses of plant taxa used by the tribes of Eastern Ghats for various ailments. The plant part, disease and ethnic tribe wise data of plant taxa are

M. S. Kumari · K. Ammani (✉)
Department of Botany and Microbiology, Acharya Nagarjuna University, Guntur, Andhra Pradesh, India

S. M. Khasim et al. (eds.), *Medicinal Plants: Biodiversity, Sustainable Utilization and Conservation*, https://doi.org/10.1007/978-981-15-1636-8_15

presented that aims at preservation and perpetuation of this knowledge of the local plants possessing antitumorous and antimicrobial properties and it may lead to the discovery of anticancerous novel drugs for the benefit of future generations of mankind.

Keywords

Indigenous culture · Anthropological approach · Traditional knowledge · Folklore medicinal plants · Eastern Ghats · Antitumour · Wound healing · Antimicrobial properties

15.1 Introduction

Plants have been an integral part of human life and culture through ages. As per the report of All India Ethnobiological Survey by the Ministry of Environment and Forests (MOEF), Government of India, there are over 8000 plant species being used by the ethnic people and folklore who live in and around the forests. Large number of folk medicine remained endemic to certain regions of people in the country, and due to cultural diversity and lack of communication, many of these earlier remedies survived orally only, from generation to generation. Ethnobotanical research acts as a bridge between botany and medicinal aspects of plants. It gives clues for new materials for pharmacological and chemical research and provides new vistas for application, improvement and popularization of traditional medicines.

The ethnobotanical studies show that much valuable information has still remained unverified and unutilized. There is an urgency for documentation of this information gathered from the tribals, traditional healers and herbal medicine practitioners and from various previous works since these habitats which are indigenous knowledge centres are shrinking in size and fast dwelling as they are being subjected to anthropogenic influences due to civilization (Fig. 15.1a). Eastern Ghats is an abode of rich biodiversity with mixed deciduous forests inhabiting diverse tribal communities and ethnic groups with traditional knowledge of plants. The Eastern Ghats is a discontinuous range of mountains along the east coast of India located between 10° 05′ and 22° 30′ N latitude and 76° 23′ and 86° 50′ E longitude in north-east to southwest covering a total area of 75,000 km^2 extending over Orissa, Telangana, Andhra Pradesh, Tamil Nadu and Karnataka states.

The main aim of the study was to collect data of the plants with antitumour, wound healing and antimicrobial properties from Eastern Ghats of Andhra Pradesh. The earliest reference about the usage of plants in this region is seen in the work by Roxburgh (1795–1820 & 1832). An intensive folklore investigation on 32 medicinal plants used by tribals of Andhra Pradesh and Orissa was made by Jain et al. (1973 & 1976). Banerjee (1977) made ethnobotanical observations of Araku Valley in Visakhapatnam district. Harasreeramulu (1980) reported the medicinal importance of 40 plants used by the tribals of Srikakulam dist. Sudhakar (1985) mentioned about 30 medicinal plant species used by Konda Reddis and Koyas of East Godavari

Fig. 15.1 (**a**) Mining in Eastern Ghats, shrinking of tribal areas; (**b**) Lambadi woman with herbs; (**c**) Yanadi woman showing herbal medicine; (**d**) Tribals collecting medicinal plants to the researchers

district. Venkaiah (1980) enumerated 35 medicinal plants used by the local tribes of Vijayanagaram district.

15.2 Methods

15.2.1 Study Sites

The methodologies of ethnobotanical studies have been explained by various scholars (Jain 1964, 1967; Schultes 1960, 1962; Reis 1962). They have mentioned various tools for ethnobotanical research. The methods recognized for this purpose can be divided into two categories: (1) Ethnobotany of the present-through extensive field study among the aboriginals and (2) Ethnobotany of the past. Glimpses of the ethnobotany of the past can be obtained by the study of literature, herbarium, museum material and archaeological remains. The present study includes the folklore medicinal plants used for tumours, wounds and ulcers by the tribals inhabiting the three regions of Eastern Ghats of Andhra Pradesh. It includes, namely, northeastern region consisting of Srikakulam, Vijayanagaram and Visakhapatnam districts, Godavari region consisting of East and west Godavari districts and Nallamalai region consisting of Guntur, Prakasam, Kurnool, Cuddapah and Chittoor districts.

Out of the 27 tribal communities of the state, Bhagatas, Chenchus, Jatapas, Khonds, Kondadora, Konda Reddis, Koyas, Lambadis or Sugalis (Fig.15.1b), Nookadoras, Porjas, Savaras, Valmikis, Yerukulas and Yanadis (Fig.15.1c) are the predominant tribal communities in the above districts who have their own traditional practices of collecting and using the medicinal plants (Fig.15.1d) for curing various disorders in human and cattle.

A database on traditional uses of medicinal plants of Eastern Ghats of Andhra Pradesh for treating wounds, tumours, boils, blisters, ulcers, etc. was collected through field visits (Fig. 15.2a) and using available literature on medicinal plants published in scientific journals, books, reports from national and international organizations, theses and conference papers. From this a list of about 50 plants have been identified as used by tribals of Eastern Ghats of Andhra Pradesh in treating wounds and tumours and they can be predicted to have antitumour and antimicrobial potential.

Fig. 15.2 (**a**) Field visit to Nallamalai forest; (**b**) *Abrus precatorius*; (**c**) tribal woman processing herbs; (**d**) Traditional medicine practitioner

15.3 Results and Discussion

15.3.1 Enumeration of Ethnobotanical Data

In the following enumeration, medicinal plants used for treating wounds, ulcers, boils, blisters, warts and sores are mentioned. The species with details of scientific name, local names, habit, ethnic tribes and medicinal uses along with methods of administration are recorded. The plant species are arranged in alphabetical order. The following abbreviations have been used for the ethnic groups

Bhagatas—B	Konda Reddis—KR	Nookadoras—ND	Yerukulas—YR
Chenchus—CH	Khonds—KH	Porjas—P	Yanadis—Y
Jatapas—J	Kondadora—KD	Savaras—S	
Lambadis (Sugalis)—L	Koyas—K	Valmikis—V	

The following compilation includes plant taxa, local names, families, habit, the ethnic groups and medicinal uses of the plants.

1. *Abrus precatorius* (Gurivinda) Papilionaceae; Twining shrub—K—Root and leaf paste applied on wounds and ulcers (Fig.15.2b).
2. *Acacia chundra* (Sundra) Mimosaceae; Tree—CH—Leaf paste mixed with castor oil applied on boils and blisters.
3. *Acacia torta* (Gachhakorinda) Mimosaceae; Prickly climbing shrub; CH—Root bark ground with turmeric and the paste applied on ulcers and wounds.
4. *Achyranthes aspera* (Vuttareni, Duchhini) Acanthaceae; Herb S—Leaf paste along with Shorea robusta applied on body for smallpox.
5. *Acorus calamus* (vasakommu) Arecaceae; Herb; CH, K—Rhizome paste used to treat tumours.
6. *Alocasia montana* (Roxb) Arecaceae; Herb S—Rhizome paste applied on ulcers and wounds.
7. *Aloe vera* (Kalabandha) Liliaceae; Herb—KR—Fleshy portion of leaves mildly heated and applied on boils, blisters and wounds.
8. *Anacardium occidentale* (Jeedi Chettu) Anacardiaceae; Tree—CH, K—Seed coat oil used for treating tumours and wounds.
9. *Andrographis paniculata* (Nelavemu) Acanthaceae: Herb; CH, K, V, S—Leaves ground with turmeric applied for cuts, wounds and skin diseases.
10. *Annona reticulata* (Ramaphalam) Annonaceae; Small tree; CH, J, KR, —Leaf paste applied on wounds of cattle and human to remove germs.
11. *Annona squamosa* (Seetha Phalam) Annonaceae; Small tree; CH, KR—Leaves ground with tobacco and paste applied on wounds.
12. *Argemone mexicana* (Mundla Rakkasi) Papaveraceae; Herb; KR, V—Roots paste /seeds paste, latex applied on the syphilis sores.
13. *Argyreia daltonii* (Paruma Kodi, Tharpa) Convolvulaceae; climbing shrub; K, KR, V—Leaves ground and paste applied on body for small pox.

14. *Argyreia nervosa* (Chandrapala Teega) Convolvulaceae; climbing shrub; B, KH, V—Leaves dipped in castor oil mildly heated and applied on boils and blisters.
15. *Aristolochia indica* (Nallaeswari, Nagasaramu) Aristlochaceae; Herb; B, KH, V—Leaf paste mildly heated and applied on boils and blisters.
16. *Artocarpus heterophyllus* (Panasa) Moraceae; Tree; J, S—Tender leaf paste applied for skin diseases.
17. *Barleria strigosa* (Adavi Kanakambaram) Acanthaceae; Shrub; CH, KR—Roots crushed with black pepper and filtrate used orally for TB and lung infections.
18. *Bidens pilosa* (Rekkalaraju) Asteraceae; Herb; Whole plant used to treat tumours.
19. *Biophytum sensitivum* (Marugu Mokka) Oxalidaceae; Herb; K, KR—Leaves ground with garlic used for treating wounds and ulcers.
20. *Bombax ceiba* L. (Burugu) Malvaceae; Tree; K, KR, V—Stem bark ground with urine of infant and the paste mildly heated and applied on blisters, wounds and ulcers.
21. *Buchanania lanzan* (Jerri manu) Anacardiaceae; Tree; J, KH, ND—Stem bark paste with castor oil and applied on boils and ulcers.
22. *Calycopteris floribunda* (Yerrateega) Combretaceae; Shrub; CH—Root bark paste applied on ulcers.
23. *Capparis grandis* (Gulimi) Capparaceae; small tree; CH—Stem bark made into paste and applied on wounds and ulcers.
24. *Carissa spinarum* (Vaka) Apocynaceae; shrub; KD, S—Root paste applied on wounds of cattle.
25. *Casearia tomentosa* (Jidumu/ Giridi) Flacourtiaceae; tree; K, S, V—Stem bark powder applied on wounds and ulcers.
26. *Centella asiatica* (Saraswathi Aaku) Apiaceae; Herb; KH, KK, ND, S—The whole plant crushed with turmeric and black pepper and the extract gargled for mouth ulcers.
27. *Chloroxylon swietenia* (Billudu) Flinderciaceae; Tree; KH, KK, ND—Leaves ground with turmeric and paste applied on wounds and ulcers.
28. *Cissus pallida* (Adavi Gummadi) Vitaceae; Shrub—K—Tuber ground with urine of an infant and the paste mildly heated and applied on wounds and ulcers.
29. *Costus speciosus* (Besuka) Zingiberaceae; Herb; KR, ND, S—Rhizome paste is applied on body for chicken-pox and also for treating boils.
30. *Dalbergia latifolia* (Jattregi) Fabaceae; Tree; KH—The bark or leaf extract administered orally for leprosy—1 spoonful once a day for 3 months, and the paste applied on the wounds.
31. *Ehretia aspera* (Mukka Chettu) Boraginaceae; Tree; KR—Dried stem bark powder applied on boils and ulcerous wounds.
32. *Euphorbia banhart* (Adavi Jamuku) Euphorbiaceae; shrub; CH, L—Latex applied for muscle pain, wounds and ulcers.
33. *Ficus benghalensis* L. (Marri) Moraceae; Tree; CH, K—Latex applied on boils.

34. *Ficus hispida* (Budda Medi) Moraceae; Tree; KR—Latex applied on warts, boils and blisters.
35. *Firmiana colorata* (Mundla Mothuka) Sterculiaceae; Tree; KD—Tender leaf juice for eye infections and wound infections.
36. *Glochidion zeylanicum* (Neeru mamidi, pageri) Euphorbiaceae; Small tree; CH, K—Whole leaf used to treat tumours.
37. *Holoptelea integrifolia* (Nevili) Ulmaceae; Tree; K—Leaf paste applied on boils and blisters.
38. *Maytenus emarginata* (Danthi) celastraceae; CH—Shrub; Whole plant and leaves used to treat tumours.
39. *Mitragyna parvifolia* (Batta Ganapu) Rubiaceae; Tree; CH—Stem bark used for Skin disease, wound healing.
40. *Nyctanthes arbortristis* (Parijatham, pagada malli) Nyctaginaceae; Small tree; CH, L—Leaf paste used to treat tumours.
41. *Ipomoea eriocarpa* (Golla Jiddi) Convolvulaceae; Twining herb; K, L—Leaf paste mixed with castor oil, applied over ulcers and wounds.
42. *Paracalyx scariosa (Roxb)* (Adavi Kandi) Papilionaceae; Climbing shrub; L—Leaf paste applied for boils and blisters.
43. *Parkinsonia aculeata* (chinna thumma) Caesalpiniaceae; Tree; CH, L—; Whole plant parts used for curing tumours.
44. *Plumeria rubra* L. (Champagochh, tella Deva ganneru) Apocyanaceae; Tree; KH—Stem bark paste applied on ulcerous wounds.
45. *Rhinacanthus nasutus* (Naga malli) Acanthaceae; under shrub; Y—Leaf paste used to treat tumours and wounds.
46. *Sarcostemma secamone* (Chitti Pala) Asclepediaceae; Shrub; K—Boiled Leaf paste applied for tumours and boils.
47. *Sophora interrupta* (Adavi billu) Fabaceae; Shrub; CH, K, Y—Whole plant used to treat tumours.
48. *Spilanthes calva* (Akkala karra) Asteraceae; Herb; K, Y—Whole plant used to treat wounds and tumours.
49. *Taxillus tomentosus* (Tella noogu, gilludu badanika) Loranthaceae; Small parasitic shrub; CH, Y—Whole plant used to treat wounds and tumours.
50. *Zingiber roseum* (Adavi allam) Zingeberaceae; Herb; CH, YR, Y, K, KD—Rhizome used to treat tumours.

A brief perusal of the literature and the ethnobotanical data on these plants has yielded some interesting findings. The Present analysis of the data clearly reveals that out of 50 species used for treating tumours, wounds and ulcers by the different ethnic groups of Eastern Ghats of Andhra Pradesh majority of the species are wild and mostly they are trees, climbing shrubs and some herbs. Wild habitats were found to be major pools of medicinal plants. More than one ethnic group used the plant for the same ailment. Majority of the species belong to Acanthaceae and Moraceae. Many of them are latex producing. Mostly leaf paste and stem bark are used. The broad utilization of these parts might be attributed to the occurrence of effective bioactive ingredients for antitumour and antimicrobial activities. Some herbs are administered in combination with other plants, so the actual medicine acting should be identified in these combinations. The processing of the crude drugs by the ethnic groups (Fig.15.2c) should also be studied for the evaluation of the potential of the drug. Some plants are endemic like *Alocasia Montana* and *Sophora interrupta*, and few like *Zingiber roseum* and *Acorus calamus* are also rare and endangered.

15.4 Conclusion

The present study supports the belief that this traditional knowledge gathered from the tribals, traditional healers and herbal medicine practitioners (Fig.15.2d) and from the previous works of Eastern Ghats could be exploited for phytochemical and pharmacological studies. Phytochemical and pharmacological investigation of these plant species with antitumour and wound healing potential may result in the discovery of novel drug compounds with anticancerous and antimicrobial activity. Since some plants among these are endemic and few are also rare and endangered, over-exploitation and use of entire plant, particularly root and bark parts, for medicinal preparations could pose a long-term threat on the survival of these plants. Effective measures for both in situ and ex situ conservation of these plants should be administered besides the exploration of the bioactive compounds for the benefit of future generations of mankind.

References

Banerjee DK (1977) Observations on the ethnobotany of Araku valley, Visakhapatnam district, Andhra Pradesh. J Sci Club 31:14–21

Jain SK (1964) The role of a botanist in folklore research. Folklore 5:145–150

Jain SK (1967) Plants in Indian medicine and folklore associated with healing of bones. Indian J Orthop 1:95–104

Jain SK, Benerjee DK, Pal DC (1973) Medicinal plants among certain Adibasis in India. Bull Bot Sur India 15:85–91

Jain S, Kand R, Rao R (1976) Field and Herbarium methods. Today and Tomorrow's Printers and Publishers, New Delhi

Roxburgh W (1795-1820) Plants of the Coast of Coramandel. W. Bulmer & Co., London

Roxburgh W (1820) Flora indica, vol 1, 1st edn. Mission press, Serampore

Roxburgh W (1832) Flora indica, or, descriptions of Indian plants, vol 3. W. Thacker and co Calcutta, Serampore

Schultes RE (1960) Tapping our heritage of ethnobotanicallore. Econ Bot 14:257–262

Schultes RE (1962) The role of ethnobotanist in the search for new medicinal plants. Llyodia 25(4):257–266

Sudhakar S, Rao RS (1985) Medicinal plants of East Godavari district Andhra Pradesh. J Econ Tax Bot 7:399–406

Venkaiah M (1980) Some medicinal plants of Vizianagaram district, Andhra Pradesh. J Indian Bot Soc 59(suppl):168

Von Reis S (1962) Herbaria: sources of medicinal folklore. Econ Bot 16:283–287

Controlling Biological Infestations in Museums by Medicinal Plants

16

Soumoni De

Abstract

The paper comprises of the chemistry and the application of some medicinal plant products which have been used as insecticides and insect repellents also. Plants play an important role in every aspect of human activity even in the conservation of our heritage. Over the past few years, there has been growing worldwide concern about the adverse effects of certain chemical pesticides, natural substances being integral part of biosphere are less disruptive to ecological balance; hence, several plants and their products have been experimented against museum insects. At present there are no dearth of modern chemical pesticides and repellents for the safe upkeep of museum objects. The advent of technology has also given rise to greater concerns of preservation of museum objects by adopting modern technologies. Still the uses of herbal remedies for preservation are in vogue, as these have their own merits. Unfortunately, these recipes are being lost as advertising to use modern chemicals increases. It is not old fashioned to use natural pesticides—rather it is a smart use of the natural resources that are readily available. In this context an attempt has been made to summarize the effectiveness of various Indian herbal pesticides and insect repellents which are being used by different organizations or could be used by the organizations to seize the growth of insect infestation in the different museums. This present investigation deals with the application of leaves, bark, seeds, wood and roots of different medicinal plants having toxic effects on different museum pests. Natural pesticides usually take longer to work than synthetic pesticides. Thus, it is important to apply them as soon as the objects show evidence of pests. Insects are killed either by contact or by ingestion of the insecticides. Some insecticides only repel the insects by a strong odour, they are known as insect repellent. The effectiveness of insecticides and repellents varies

S. De (✉)
Department of Museology, Rabindra Bharati University, Kolkata, India

S. M. Khasim et al. (eds.), *Medicinal Plants: Biodiversity, Sustainable Utilization and Conservation*, https://doi.org/10.1007/978-981-15-1636-8_16

with the active ingredient(s), the concentration and the target species. Finally, the paper spells out the preparation processes of pesticides from plants.

Keywords
Biodeteriogens · Biodegradation · Pesticides · Herbal remedies · Preservation

16.1 Introduction

There are thousands of essential oil-bearing aromatic plants which can be used as effective insect repellents across the world. India is also rich with varieties of these types of plants either cultivated or in wild state. Their efficacy of essential oils as insect repellent has been recognized for thousands of years.

The aroma is the main property of plant origin products that they possess for the eradication of insects. Essential oil, the complex mixture of natural substances (ketone, terpene, ester, alcohol, etc.) present in different plant parts like leaves, seeds, barks, fruits, flowers, roots and rhizomes, acts as pesticides either to repel or to kill certain biological agents. These essential oils can be used after extraction from plant parts by distillation process, such as neem oil, or the plant part can be used as the product, such as dried neem leaves.

Instead of using chemicals there are several positive reasons of using herbal remedies to repel insect pests in museums. Most of these herbal products also have other uses like medicinal applications. These herbal remedies are non-hazardous for human health. They reduce the use of chemicals in museums. The application of herbal remedies does not have any adverse effect on the treated materials. The methods do not require much expertise, equipment and money. These eco-friendly herbal remedies even do not pollute the environment. Above all, these can play an important role in integrated pest management programme.

16.2 Medicinal Plants and Germicidal Properties

Some of the plants and their products, which have been recognized since ancient times for their germicidal properties and insect repellency potentialities, have been discussed here:

16.2.1 Sweet Flag

Scientific name—*Acorus calamus*; Family: Acoraceae

It is a semi-aquatic perennial herb with sword-shaped leaves and small, yellow/green flowers, a creeping and munched branched aromatic rhizome, lives near

Fig. 16.1 Dried sweet flag rhizomes

swamps and banks in Europe, Asia and North America. Mainly rhizome and its mature leaves can be dried and used to repel insects (Sing and Maurya 2005; Mandavgane et al. 2005). The essential oil is contained mainly in the outer skin of the rhizomes. Dry rhizomes contain aromatic oil having mellow odour and it possesses the insecticidal properties. The main constituents of this oil are b-asarone, calamene, calamenol and methylisoeugenol (Marongiu et al. 2005; Raina et al. 2003; Lokesh 2004) (Figs. 16.1 and 16.2). It is useful to protect textiles and other materials like documents from bedbugs, moths, lice, etc. in museums. Small cotton pouch, filled with powdered roots of dried sweet flag, can be kept in cupboards of manuscripts and textiles (Fig. 16.3).

Fig. 16.2 Dried sweet flag rhizome powder

Fig. 16.3 Insect repellent (sweet flag powder) is applied in cloth pouches within manuscript storage almirah

16.2.2 Turmeric

Scientific Name—*Curcuma longa*; Family: Zingiberaceae

Turmeric is a perennial herb, 2-3 ft. high and short steam and tufted leaves. The rhizome constitutes the main part. The plant is a native of South Asia and is cultivated extensively on a large scale in India and China. Rhizome (yellow colour) is the only usable plant part having insect repellent, insecticidal and antifungal properties (Sing and Maurya 2005; Sing et al. 2007). The main constituents are curcumin, α-phellandrene, turmerones, zingiberene and arturmerone (Kharbade et al. 2008). Turmeric is very effective to prevent damages of mainly palm leaf manuscripts and textiles and paper-related objects from different insect and fungus pests (Vaishnavi et al. 2006; Bhatia and Kharbade 2001). Chopped turmeric rhizomes are soaked overnight and then diluted into water; white cotton cloths are dipped in the solution; after drying the cloths get yellow colour and it can be used to wrap manuscripts in storage; before wrapping with this treated yellow cloth the manuscripts should be wrapped with acid-free handmade or another piece of clean cloth, otherwise the manuscripts can get yellow stain from the treated yellow cloth. Palm leaves are seasoned with turmeric paste to make them insect repellent.

16.2.3 Neem

Scientific Name—*Azadirachta indica*; Family: Meliaceae

Neem is a fast-growing tree. It is evergreen, but in severe drought it may shed most or nearly all of its leaves. Major producing countries are India, Sri Lanka, Burma, Pakistan, Tropical Australia and Africa. Seeds, leaves and wood are the mainly used parts. Neem has insecticidal, repellent, antifeedant, growth-inhibiting, fungicidal and antiviral properties (Ranaware et al. 2010; Nandagopal and Ghewande 2004). The seed oil is acrid and bitter in test and has a disagreeable odour. The active ingredients are bitter limonoid compounds like azadirachtin, nimbin, nimbidin and salannin (Sing and Maurya 2005). The leaves are also bitter and have faint but characteristics unpleasant smell. In museums neem can be used in different ways to protect textiles, book, manuscripts, and other organic objects from termites, moths, silverfish, and other insects causing harm to museum objects; traditionally it has been used in rural areas of India for the same purpose. The shade dry leaves can be kept in between acid-free handmade paper or in muslin cloth pouch which will be placed in showcases, storage boxes or in storage cupboards housing the organic museum objects. The planks and museum furniture can be made of neem wood which can ward off termite. Cloths and cloth bags which will be used for storage can be pre-soaked with aqueous solution of neem extract; after drying these can be used for storage purpose.

16.2.4 Red Chilli

Scientific Name—*Capsicum annum*; Family: Nightshade

It is grown in many parts of India, especially in Tamil Nadu, Bihar, Andhra Pradesh, Maharashtra and the plain of North India. Dry chilli is extensively used as a spice in India. The only usable part is fruit. Chilli acts as insecticide, insect repellent, antifeedant and fumigant (Gurnani et al. 2016; Verma et al. 2008). Capsaicin is an active component of red chilli. Dry red chilli is useful to protect textiles especially woollen and silk cloths from cloth moths, carpet beetle, etc. in museums. A layer of dry chillies can be laid in the box and over it tissue papers or clean pieces of cloth are spread; the textiles to be stored will be put over this.

16.2.5 Camphor

Scientific Name—*Cinnamomum camphora*; Family: Lauraceae

Camphor is also known as Kapur, a waxy, white or transparent solid with a strong, aromatic odour. It is found in the wood of the camphor laurel, a large evergreen tree found in Asia. It is used for its scent, as an ingredient in cooking (mainly in India), as an embalming fluid, for medicinal purposes and in religious ceremonies. Bark contains aromatic oil, the oil distilled mainly from chips and refuses, leftover after the collection of cinnamon quills as spice. The active ingredients are α-pinene, bisalboline, cineol, terpineol and caryophyllene. Camphor sublimates, meaning they evaporate from a solid state to a gas; this gas is toxic to the moths and also acts as a fungicide. In museums it can be used for protecting textiles, book, manuscripts and other organic objects from cloth moths, silverfish, etc. and from fungus also. Camphor oil is also used to protect palm leaf manuscripts against insect attack. Small cloth pouches, filled with camphor, can be kept inside the storage of manuscripts and textiles. The camphor when placed on a warm surface evaporates, acts with air quickly to repel insects away (Verma et al. 2008) (Figs. 16.4).

16.2.6 Custard Apple

Scientific Name—*Annona squamosa, A. muricata*; Family: Annonaceae

The tree occur wild and is also cultivated all over India. The fruit is used for eating, the pulp mainly contains glucose and saccharose and vitamin C. Unripe fruit, seed, leaf and root have insect-controlling property. Powdered custard

Fig. 16.4 Placing of camphor in readymade moth ball cases

apple seeds (de-oiled) act as contact poison for flies. It has insecticidal, antifeedant and repellent properties (Er Vaidya A et al. 2016; Ignacimuthu 2004). Cotton bollworms shoot and fruit borers, hairy caterpillars, lice, moths, flies and other destroying museum insects are the target pest. De-oiled seed powder can be placed in a cloth sachet in between the boxes and cupboards housing the materials like textiles, paper and other organic objects. Dilute solution of aqueous extract of custard apple seeds can be sprayed periodically over the plants of the museum premises.

16.2.7 Cumin

Scientific Name—*Cuminum cyminum*; Family: Apiaceae

Cumin is commonly known as Jeera, it is cultivated in almost all the states in India. It is an annual herb with bluish green leaves, small white or rose-coloured flowers and greyish fruits. The seeds have an aromatic odour. Cumin is available in white or black colours. Its seeds contain volatile oil (thiamine), which is responsible for its test and smell. Cumin also acts as an insect repellent (Verma et al. 2008). The main chemical components of cumin oil are cuminic, cymene, dipentene, limonene, phellandrene, α–pinene, linalool and α-terpineol and cumaldehyde. Cumin is useful to repel insects causing damage to the textiles and paper-related objects (Sing et al. 2007; Vaishnavi et al. 2006). Seeds can be placed in a cloth sachet in between the boxes and cupboards housing the materials like textiles, paper and other organic objects.

16.2.8 Kalongi

Scientific Name—*Nigella sativa*; Family: Ranunculaceae

Kalongi (Black Cumin), commonly known as *Kala Jeera*, is an annual flowering plant, native to south and southwest Asia. The fruit is a large and inflated capsule composed of three to seven united follicles, each containing numerous seeds. The seed is used as a spice. The seeds have a pungent bitter taste and a faint smell (Paarakh 2010). The seeds contain a volatile oil which probably accounts for its insect repellent property (Gholamnezhad et al. 2016). The main important compound of this oil is toxic glycoside, melanthin (Gupta 2007; Kalidasu et al. 2017). In museums to protect textiles especially linen and woollen materials and paper-related objects from insect and microbes attack Kalongi can be used (Salman et al. 2008; Bhatia and Kharbade 2001). Seeds can be scattered between folds of linen and woollen clothes and at the manuscript storage.

16.2.9 Black Pepper

Scientific Name—*Piper nigrum*; Family: Piperaceae

Black pepper is a flowering vine, cultivated for its fruit, which is usually dried and used as a spice and seasoning. Black pepper is native to India, cultivated particularly in Kerala, Tami Nadu and Karnataka. Dried fully developed unripe fruit is known as pepper which owes its characteristic pungency and aroma. It has insect repellent property (Verma et al. 2008; Karsha and Lakshmi 2010). The main constituents are linalool, α-phellandrene, limonene, myrcene and α-pinene (Gupta 2007). It is used for preserving textiles and paper-related objects from pest infestations (Vaishnavi et al. 2006; Verma et al. 2011). A mixture of kalongi, sweet flag, cinnamon, clove and black pepper can be used as an insect repellent. Small cloth pouches, filled with this mixture, can be kept within the storage boxes or within the storage cupboards (Fig. 16.5).

16.2.10 Ajowan/Ajwain

Scientific Name—*Trachyspermum ammi*; Family: Apiaceae

Ajowan/Ajwain, also known as ajowan caraway or bishop's weed or carom, is an annual herb. The active ingredients of essential oil of ajowan are primarily thymol, a-pinene, γ-terpinene and p-cymene as well as more than 20 trace compounds (Singh et al. 2004; Gupta 2007). Among these, thymol has fungicidal, insecticidal and insect repellent properties (Mandavgane et al. 2005; Vaishnavi et al. 2006). So the dried ajowan seed can be used as fungicide, insecticide and insect repellent (Sing et al. 2007) and can be kept in storage cabinets and in showcases in small cotton sachet to control pest problem in museums.

Fig. 16.5 Preparation of cloth pouches filled with insect repellents (a mixture of clove, cinnamon and cumin seeds)

16.2.11 Clove

Scientific Name—*Syzygium aromaticum*; Family: *Myrtaceae*

Cloves are the aromatic flower buds of a tree in the family Myrtaceae. They are commonly used as a spice. Cloves are available throughout the year. The useable part is the dried flower buds. The major component of clove is eugenol (toxic) and eugenol acetate (Kamatou et al. 2012*)*. Clove acts as insecticide and insect and fungus repellent (Soonwera and Sinthusiri 2014). To eradicate the insect pests (Jairoce 2016) of manuscripts, clove can be placed in cloth pouch and placed within showcases and storage cupboards.

16.2.12 Karanja

Scientific Name—*Pongamia glabra*; Family: Fabaceae

The karanja tree is of medium size. Major producing countries are East Indies, Philippines and India. Karanja seeds' oil has toxic flavonoids having insecticidal and bactericidal properties (Bhatia and Kharbade 2001). The main constituents are kharanjin and pongamol alkaloids (Majumdar 2008; Majumdar et al. 2004). To protect textiles and paper objects from insect damage karanja seeds are very useful (Shareen 1995). De-oiled seed powder can be placed in a cloth sachet in between the boxes and cupboards housing the materials like textiles, paper and other organic objects.

16.2.13 Tobacco

Scientific Name—*Nicotiana* tabacum; Family: Solanaceae

Tobacco is an agricultural product processed from the leaves of herbs. The dried leaves are also used as snuff or are smoked. It is mainly used to make cigarettes, cigars and other smokable tobacco preparations. Leaves and stalk are the most useful plant parts. Only real tobacco contains nicotine acting as an insecticide. The volatile nicotine possesses insecticidal, repellent and fungicidal properties. The main constituent is nicotine. Tobacco has been used in rural India for protecting textiles, especially woollens, and paper objects from different damaging insect and fungus species. Cloth bags, filled with dry tobacco leaves, can be kept in cupboard or boxes where the cloths are stored; the leaves remain effective for 6 months after drying. Clean cotton cloths are dipped in the tobacco dust juice solution and dried; after drying, the cloths can be used to wrap manuscripts and textiles in storage; this treated cloth will act as insect and fungus repellent. Tobacco juice can be used to make insecticidal paper. Tobacco juice can also be painted on the storage and display showcases and cupboards to repel out insect pests.

16.2.14 Tulsi

Scientific Name—*Ocimum* sanctum; Family: Lamiaceae

Tulsi is an erect sweet scented pubescent herb. The active compounds that are present in tulsi leaves include eugenol, apigenin, luteolin, oleanolic acid and ursolic acid (Sundaram et al. 2012; Padalia and Verma 2011). The essential oil of tulsi has antiviral, antifungal and antibacterial properties (Sethi et al. 2013). The leaves of tulsi also have insect repellent property. So, dried leaves can be kept between the folios of manuscripts or in storage cupboards as insect repellent. Plantation of tulsi tree in the surroundings of museum can be found effective to keep out insects from museum buildings.

16.2.15 Citronella

Scientific Name—*Cymbopogon* sp.; Family: Poaceae

Citronella oil is one of the essential oils obtained from the leaves and stems of different species of *Cymbopogon*. The active ingredients of this aerometric oil are citronellol, citronellal, geraniol, geranyl acetate and limonene (Yaneff 2017). The oil possesses antifungal and insect repellent properties (Kim 2005). It is highly volatile in nature, so to reduce the rate of its evaporation the oil can be mixed with some other inert oils. This oil is generally applied directly on the surface of palm leaf manuscripts to regain flexibility, at the same time they are protected from insect and fungus attack also (Figs. 16.6).

Fig. 16.6 Application of citronella oil on old dried palm leaf manuscript folios for returning their flexibility

16.3 Controlling Biological Infestations in Museums by Medicinal Plants

The plants (or plant products) used as botanical pesticide, should have the wide distribution in nature. They should be grown easily even within a little space and time for cultivation and procurement. These plants pose no hazard to humans or environment. When preparing a pesticide from plants, always let them dry in the shade, because direct sunlight could break down the active ingredients.

The plant origin chemicals act in different ways on various stages of insect life. These may act as fumigant / vaporizer, growth regulator, insect deterrents, feeding deterrents, repellents or confusants (Gupta 2007). The plant products, having these properties, are mostly effective for 6 months, after that they should be replaced with some freshly prepared products.

Besides benefits, the botanical pesticides or herbal remedies have some limitations too for applying them in pest management programme. For the application of botanical pesticides the seed or essential oil extraction methods are not standardized. These botanical pesticides degrade rapidly when compared with chemical one. Compared with chemicals, these pesticides and repellents are less effective. The appropriate pesticide formulas are also not available easily at the time of application of herbal remedies.

There are many herbal remedies which were used in ancient time to preserve art objects which are now forgotten. Today there is a need of intensive survey of such materials so that they are not lost completely in the middle of modern materials and techniques. India's rich floral treasure occupies tenth position in the world and fourth in Asia in plant diversity but only a few of the plant resources were explored and exploited for the preservation purpose, a vast number of plants may be of use if they too are explored. More researches are needed to revert to the use of herbal remedies in museums which are both cost effective and harmless.

References

Bhatia SK, Kharbade BV (2001) Traditional herbal remedies in the preservation of museum objects in India. Indian Sci Cruiser 15(2):31–37

Er Vaidya A et al (2016) Chemical composition, physicochemical and functional properties of custard apple (*Annona squmosa*) seed flours and protein isolate. Int J Sci Eng Technol 5(4):205–209

Gholamnezhad Z, Havakhah S, Boskabady MH (2016) Preclinical and clinical effects of *Nigella sativa* and its constituent, thymoquinone: a review. J Ethnopharmacol 190:372–386

Gupta KK (2007) An introduction to the traditional practices for the control of biodeterioration of manuscripts. In: Sah A (ed) Indigenous methods and manuscript preservation, Samraksika series: 1. National Mission for Manuscripts, Delhi, pp 35–57

Gurnani N, Gupta M, Shrivastava R, Mehta D, Mehta BK (2016) Effect of extraction methods on yield, phytochemical constituents, antibacterial and antifungal activity of *Capsicum frutescens* L. Indian J Nat Prod Resour 7(1):32–39

Ignacimuthu (2004) Botanical pesticides. Curr Sci 86(8):1059–1060

Jairoce CF (2016) Insecticide activity of clove essential oil on bean weevil and maize weevil. AGRIAMBI 20(1):72–77

Kalidasu G, Reddy GS, Kumari SS, Kumari AL, Sivasankar A (2017) Secondary volatiles and metabolites from *Nigella sativa* L. seed. Indian J Nat Prod Resour 8(2):151–158

Kamatou GP, Vermaak I, Viljoen AM (2012) Eugenol-from the remote Maluku islands to the international market place: a review of a remarkable and versatile molecule. Molecules 17(6):6953–6981

Karsha PV, Lakshmi OB (2010) Antibacterial activity of black pepper (*Piper nigrum* Linn.) with special reference to its mode of action on bacteria. Indian J Nat Prod Resour 1(2):213–215

Kharbade BV, Rajmalvar S, Manjunathachari RC (2008) The use of turmeric, a traditional Indian material in the preservation of old manuscripts. In: ICOMCC, 15th Triennial Conference, New Delhi, 22-26 September 2008. ICOMCC, New Delhi, pp 270-275

Kim JK (2005) Evaluation of repellency effect of two natural aroma mosquito repellent compounds, citronella and citronellal. Entomol Res 35(2):117–120

Lokesh GB (2004) Sweet flag (*Acorus calamus*)- cultivation and economic aspects. Nat Prod Rad 3(1):19–20

Majumdar D (2008) Unexploited botanical nitrification inhibitors prepared from karanja plant. Nat Prod Rad 7(1):58–67

Majumdar D, Pandya B, Arora A, Dhara S (2004) Potential use of karanjin (3-methoxy furano-2′,3′,7,8-flavone) as a nitrification inhibitor in different soil type. Arch Agron Soil Sci 50(4):455

Mandavgane SA, Pattalwar VV, Kalambe (2005) Development of cow dung based herbal mosquito repellent. Nat Prod Rad 4(4):270–272

Marongiu B, Piras A, Porcedda S, Scorciapino A (2005) Chemical composition of the essential oil and supercritical CO_2 extract of *Commiphora myrrha* (Nees) Engl. And of *Acorus calamus* L. J Agric Food Chem 53(20):7939–7943

Nandagopal V, Ghewande MP (2004) Use of neem products in ground nut pest management in India. Nat Prod Rad 3(3):150–155

Paarakh PM (2010) *Nigella sativa* Linn.- a comprehensive review. Indian J Nat Prod Resour 1(4):409–429

Padalia RC, Verma RS (2011) Comparative volatile oil composition of four *Ocimum* species from northern India. Nat Prod Res 25(6):569–575

Raina VK, Srivastava SK, Syamasunder KV (2003) Essential oil composition of *Acorus calamus* L. from the lower region of the Himalayas. Flavour Fragr J 18(1):18–20

Ranaware A, Sing V, Nimbkar N (2010) In vitro antifungal study of the efficacy of some plant extract for inhibition of *Alternaria carthami* fungus. Indian J Nat Prod Resour 1(3):384–386

Salman MT, Khan RA, Shukla I (2008) Antimicrobial activity of *Nigella sativa* Linn. seed oil against multi-drug resistant bacteria from clinical isolates. Nat Prod Rad 7(1):10–14

Sethi S, Prakash O, Chandra M, Punetha H, Pant AK (2013) Antifungal activity of 59 essential oils of some *Ocimum* species collected from different locations of Uttarakhand. Indian J Nat Prod Resour 4(4):392–397

Shareen F (1995) Application of kharanja seeds (*Pongamia glabra* vent) for the control of museum insects. Biodeter Cult Prop 3:567–570

Sing S, Maurya S (2005) Antimicrobial, antifungal and insecticidal investigations on essential oils: an overview. Nat Prod Rad 4(3):179–192

Sing G, Maurya S, Marimuthu P, Murali HS, Bawa AS (2007) Antioxidant and antimicrobial investigations on essential oils and acetone extracts of some spices. Nat Prod Rad 6(2):114–121

Singh G, Maurya S, Catalan C, Lampasona MP (2004) Chemical constituents, antifungal and antioxidative effects of ajwain essential oil and its acetone extract. J Agric Food Chem 52(11):3292–3296

Soonwera M, Sinthusiri J (2014) Thai essential oils as botanical insecticide against house fly (*Musca domestica* L.). In: International Conference on Agricultural, Ecological and Medical Sciences (AEMS-2014) Feb. 6-7, 2014 Bali (Indonesia), pp 67- 69. Accessed on 07.09.2015. http://iicbe.org/upload/4602C0214046.pdf

Sundaram RS, Ramanathan M, Rajesh R, Satheesh B, Saravanan D (2012) LC-MS quantificadtion of rosmarinic acid and ursolic acid in *Theocimum sanctum* Linn. Leaf extract (holi basil, Tulsi). J Liq Chromatogr Relat Technol 35(5):634–650

Vaishnavi C, Kaur S, Kaur M (2006) Bacterial activity of kitchen spices and condiments on enteropathogens. Nat Prod Rad 6(1):40–45

Verma RK, Chaurasia L, Katiyar S (2008) Potential antifungal plants for controlling building fungi. Nat Prod Rad 7(4):374–387

Verma RK, Chaurasia L, Kumar M (2011) Antifungal activity of essential oils against selected building fungi. Indian J Nat Prod Resour 2(4):448–451

Yaneff J (2017) Citronella essential oil: facts, health benefits, uses, and recipes. Doctors Health Press, Boston

Hitherto Unexplored Aspects of Medicinal Plants from Ayurveda and Vrikshayurveda

17

M. N. Shubhashree, V. Rama Rao, S. H. Doddamani, and Amit Kumar Dixit

Abstract

Ayurveda is the traditional, ancient Indian system of health science. Vrikshayurveda is the ancient Indian science of plant life which deals with various trees and plant species and ensures the healthy growth and productivity. One of its chapters "Chitrikarana" depicts astounding techniques to make a plant bloom throughout the year irrespective of the season and other horticultural wonders to obtain different varieties of yield. The proper interpretation and availability of Vrikshayurveda can also play an important role in the field of intercropping and put forward for the use of organic fertilizers and can play a crucial role to build the ecofriendly environment. Kunapajala, a liquid biofertilizer explained in Vrikshayurveda is of great relevance in agriculture and horticulture. It can be prepared in a cost-effective way by using the raw materials like flesh and bone of animals, husk, oil cakes, dung, and urine of cattle which are easily available around us. The procedure of preparation is easy and economical too, which is an added advantage. A detailed study of Vrikshayurveda would also provide us with information on pest management based on the ecological principles. This paper intends to understand the traditional knowledge, particularly in the light of contemporary research, and demonstrate the value of traditional knowledge for cultivation and harvesting with special reference to medicinal plants. It is hoped that the ancient wisdom coupled with modern technology would benefit the mankind.

Keywords

Traditional conservation · Medicinal plants · Ayurveda · Vrikshayurveda · Kunapajala

M. N. Shubhashree (✉) · V. Rama Rao · S. H. Doddamani · A. K. Dixit
Regional Ayurveda Research Institute for Metabolic Disorders, (CCRAS, Ministry of AYUSH), Bengaluru, India

S. M. Khasim et al. (eds.), *Medicinal Plants: Biodiversity, Sustainable Utilization and Conservation*, https://doi.org/10.1007/978-981-15-1636-8_17

17.1 Introduction

Medicinal plants have been known for millennia and are highly esteemed all over the world as a rich source of therapeutic agents for the prevention of diseases and ailments (Sharma et al. 2008). Demand for herbal products is surging in India and abroad. India is the second largest exporter of AYUSH (Ayurveda, Yoga, Unani, Siddha, and Homeopathy) drugs as per Pharmaceutical Export Promotion Council of India (Pharmaexil). There is global resurgence in traditional and alternative health care systems resulting in world herbal trade which stands at US$ 120 billion and is expected to reach US$ 7 trillion by 2050 (Ganesan et al. 2016). Cultivation of medicinal plants has become the need of the hour to meet the growing demand of herbal drug requirement in the field of medicine. It is a well-established fact that the pesticide and fertilizers in the form of chemicals may have undesirable impact on the quality of the plants as well as human health (Bhat Savitha et al. 2012). Present scenario is very suitable for the development of the ancient sciences as there is a huge demand for the medicinal plant sciences. Traditional knowledge is vital for sustainability of natural resources including medicinal plants. Of late, there is a revival of interest with herbal-based medicine due to the increasing realization of the health hazards associated with the indiscriminate use of modern medicine. The herbal drug industries are now very fast growing sector in the international market. The focus is on preventive health and especially on medicinal plants. The demand for plant-based medicines, health products, pharmaceuticals, food supplement, cosmetics, etc. are increasing in both developing and developed countries (Sharma et al. 2008). Of late, there has been ruthless exploitation of natural resources and erratic collection practices. For medicinal plants with limited abundance and slow growth, destructive harvesting has led to species extinction. Hence, the collection of drugs with highest pharmacotherapeutic activity is to be planned so that the number of trees chopped off for medicinal purpose is reduced (Chen et al. 2016). Consequently, there has been a resurgence of interest in traditional health sciences.

Ancient India had achieved a level of agricultural technology which is equal to the knowledge in modern times. The seed selection, selection of soil, classification of diseases, management for various diseases, techniques like bonsai, production of seedless fruits, and alteration in fruits and flowers clearly indicate that level of knowledge was high and equivalent to modern agriculture (Shaik Nisar et al. 2017).

The concept of sustainable harvesting and conservation was deeply embedded in the Indian culture. For the people of India, environmental conservation is not a new concept. Historically, the protection of nature and wildlife was an ardent article of faith, reflected in the daily lives of people, enshrined in myths, folklore, religion, arts, and culture. The fundamental principles of ecology—the interrelationship and interdependence of all life—are conceptualized in the Indian ethos and reflected in the ancient scriptural text. Sustainability was ingrained in the thought processes of early Indians as evident from the teachings of Vedas. Sacred groves designated to particular God is an example of traditional practice for plant conservation (Chaithra and Thomas 2017; Kasagana and Karumuri 2011). Many useful plants have been mentioned to have special proximity to a particular God. For instance, Durva

[*Cynodon dactylon* (L.) Pers] is offered to Lord Ganesha and Parijata (*Nyctanthes arbor-tristis* L.) is offered to Lord Krishna. These plants are considered as pious, hence nurtured.

17.2 Method

Ayurveda is perceived as one of the most ancient and well-documented systems of medicine equally relevant in modern times. *Charaka Samhita, Sushruta Samhita, Bhavaprakasha Nighantu, Raja nighantu,* and *Vrikshayurveda* were some of the texts referred for this work. *Acharya Charaka* in *Charaka Samhitha* describes an excellent design of drug research and has given much importance for season of collection along with the place and method of collection (Swagata and Nishteswar 2014).

17.2.1 Drug Research (Ayurvedic Viewpoints)

Bhumipariksha (Selection of land).
Sangrahaniya dravyas (Selection of drug).
Sangrahaniya Vidhi (Method of cultivating).
Sangrahaniya Kala (Time for collection).

Under the heading of *Desha Vichara* (Details pertaining to place) *Kalavichara* (Season and time factor), Ayurveda has emphasized about the collection of raw materials at specific time of plant's life cycle or collection from specific places (Narahari, RajaNighantu) as given in the Table 17.1.

Collection of herbs at right maturity determines the efficacy of medicinal plants. Traditional medical knowledge is bioactivity oriented and informs about best time of collection for certain medicinal species, as observed in case of *Vidari* (*Ipomoea mauritiana* Jacq.). It is interesting to note that the traditionally recommended mature tubers

Table 17.1 Showing the parts of plants and collecting seasons

Sl. No.	Part of the plant specified for therapeutic use	Harvesting season
1.	Flowers	Spring (Vasantha) /Yatharutu (whenever it blooms)
2.	Fruits	Summer/yatharutu
3.	Roots	Summer or in the late winter greeshma, shishira
4.	Leaves and spring (early summer)	Rainy season and spring (early summer)
5.	Branches	Rainy season (varsha)
6.	Bark	Sharad (early winter)
7.	Stem & tuber	Sharad (early winter)
8.	Latex	Sharad (early winter)

of the plant *Ipomoea mauritiana* Jacq. are richer in terms of both their biodiversity and in phytoconstituents than their immature counterparts (Sajjad Khan et al. 2009). Similar case involves the difference between the bioactivity of fresh and dry Brahmi [*Bacopa monnieri* (L.) Wettst.]. Another example of collection as mentioned in the classical texts of Ayurveda is about turmeric (*Curcuma longa* L.) whose rhizomes for therapeutic purposes are collected at night time. Preliminary studies indicate that the turmeric collected at night was significantly more bioactive than that collected during the day (Venkatasubramaniam et al. 2007). In contrast to the above, unripe Bilwa [*Aegle marmelos* (L.) Correa] is to be collected as unripe fruit as it has higher tannin content and acts as *stambhaka* (binding agent), whereas ripened *Bilwa* fruit has higher amount of mucilage and sugar and acts as a mild laxative (Dwivedi 2012).

There are numerous examples to substantiate this view point. In the immature plant of *Datura stramonium* L., the ratio of hyoscine and hyoscyamine is 80:20 and in mature plants it is 20:80 (Miraldi et al. 2001). Similarly, morphine is highest in opium plant 2–3 weeks after flowering. The total alkaloidal contents of Guduchi [*Tinospora cordifolia* (Willd.) Miers.] satva are slightly higher in rainy season and spring while the minimum yield was obtained in Grishmarutu (Summer) (Sharma et al. 2013). The leaves of Parijata (*Nyctanthes arbor-trisis* L.) should be collected in VarshaRutu (rainy season). The leaves show that loss on drying was minimum when they are fully grown. Daily fluctuations were also seen in the essential oil of wild basil herb, or *Ocimum gratissimum* L. (Lamiaceae), where levels of eugenol in the essential oil were observed to drop from 98% at 12 a.m. to 11% at 5 p.m. (Shah and Goyal 2017).

In yet another study on the antidiabetic and hypolipidemic effects of *Momordica charantia* L. fruit extract, it is observed that the highest activity was with spring sample, followed by the summer sample (Kolawole and Ayankunle 2012).

17.2.2 Relevance of Traditional Advice

Pippali (*Piper longum* L.) is a widely used Ayurvedic medicinal plant. For therapeutic purposes Pippali has been used as Ksheerapaka (Milk decoction). It has been observed that the brine shrimp bioactivity studies indicate that the milk decoction is 27 times more effective than the aqueous extract (Venkatasubramaniam et al. 2007) which justifies the practice of traditional methods of preparation. Similarly, the drug Kupilu (*Strychnos nux-vomica* L.) which is a poisonous drug is purified by pretreatment with milk. There has been a reduction in the toxicity by tenfolds upon detoxification of seeds and this is well correlated with the fact that the alkaloid content reduces with detoxification, which further influences the toxicity (Kolawole and Ayankunle 2012).

Yet another interesting fact is about the recommendation of the use of Musta (*Cyperus rotundus* L.) instead of Ativisha (*Aconitum heterophyllum* Wall. ex Royle) in Ayurvedic classical texts. Ancient texts have mentioned them as substitutes in spite of having no botanical similarities. However, it is interesting to know that high performance liquid chromatography (HPLC) has revealed that they have similar chemical profiles (Venkatasubramaniam et al. 2007).

17.2.3 Desha Vichara

Acharya Sushrutha has emphasized on "*Bhumi Päriksha*" (Examination of the cultivated land) in the context of collection of medicinal plants. The effect of ecological conditions on properties of plants has been subtly conveyed by *Acharya Charaka* who states that plants from Himalayas are qualitatively better than those of Vindhya Mountains. The concept of *Vanya* (wild) and *Gramya* (cultivated) varieties of some medicinal plants like *Masha* (Vignamungo (L.) Hepper) mentioned in *Nighantus* (lexicons) possess different medicinal properties and it reveals the effect of ecosphere on the plants) (Agnivesha 2014). There are innumerable practical examples to support this view. For instance, the Neem plant (*Azadirachta indica* A. Juss) collected from dry, hot, and arid habitat is considered to be more potent. Tinnevally Senna from Southern parts of Tamil Nadu (India) has higher concentration of alkaloids than others. Yashtimadhu (*Glycyrrhiza glabra* L.) cultivated in Afghanistan is better than North Indian sample (Kokate 2008).

17.3 Relevance of Vrikshayurveda

Ancient texts contain many descriptions of the uses and management of forests and highlight sustainability as an implicit theme. Treatise called Vrikshayurveda mentions in depth about the plants, its importance, diseases suffered by them, treatment, protection from external factors, increasing the yield, conservation techniques like protection of plants from mist, pests, etc.

Vrikshayurveda is an ancient science of plant life consisting of the details of about 170 different plant species including herbs, shrubs, and trees. The different chapters of *Vrikshayurveda* deal with agri horticulture, home gardening, intercropping, and storage and conservation (Srikanth et al. 2015).

The chapter "Chitrikarana" depicts astounding techniques such as to make a plant boom throughout the year irrespective of the season brings forth premature maturity to plants and fruits and change the shape and form of fruits. For instance, the seeds of pumpkin (*Cucurbita maxima* Duchesne) when processed with fat of animals and planted yield big-sized fruits without seeds (Surapala 2010).

Sl. No	Yield	Recipes
1.	Beejarahitaphala (seedless variety)	Application of paste of *Yastimadhu* (*Glycerrhiza glabra* L.) and *Kusta* [*Cheilocostus speciosus* (J. Koenig) C. D. S pecht], *Madhu pushpa* (*Madhuca indica* J. F. Gmel.) and Sita (sugar) to the roots of the plant)
2.	AkaleKusuma (to obtain unseasonal flowering)	Application of ripened sugarcane juice to the root of *Vidarikhanda* [*Pueraria tuberosa* (Willd.) DC.]
3.	To obtain fragrant flowers	Application of sandalwood (*Santalum album* L.) and fumigation with ghee to the roots
4.	To obtain flowers and fruits early	Fish, fat and meat of pig is triturated with kshara (alkali) of *Chanaka* (*Cicer arietinum* L.)

Sl. No	Yield	Recipes
5.	To have perennial fruits	Mango (*Mangifera indica* L.) seed is triturated with rabbits blood and washed with milk
6.	To obtain big-sized radish	Bones of cow and pig along with the cow dung are burnt and filled into a pit and radish is grown there
7.	To obtain sweet neem fruits	Application of the paste of *Vidanga* (*Embelia ribes* Burm. f.) *Yastimadhu* (*Glycerrhiza glabra* L.) honey, milk and jaggery to neem tree (*Azadirachta indica* A. Juss) and wash with milk and water

Drumaraksa is the chapter which deals with several advices to save plants and trees from the weather and other conditions like winds and storms. It also tells about the medicinal plants used on the broken branch to protect the whole tree from dying. Use of powders of *Solanum indicum* L., *Sesamum indicum* L., *Embelia ribes* Burm. f., and *Brassica juncea* (L.) Czern., milk, ghee, and cow dung has been mentioned in almost all the texts for protection during storage (Surapala 2010; Geetha Suresh 2013). In addition to pretreatments applicable to all seeds in general, treatments specific to particular plants also have been described. Various seed priming processes have been carefully designed in Vrikshayurveda to allow early germination, to obtain good quality of seedlings by following the classical techniques. A study conducted to compare the effects of Vrikshayurveda and Modern cultivation techniques on germination of *Bakuchi* [*Cullen corylifolia* (L.) Medik.] has revalidated the germination behavior of dormant seeds when treated with milk overnight and shade dried, kept in paste of *Brihati* (*Solanum indicum* L.), *Tila* (*Sesamum indicum* L.), *Kamala nala* (*Nelumbo nucifera* Gaertn.), and ghee as compared to control (treated with water) and standard groups (treated with sulfuric acid) (Gangadhar et al. 2016).

For nourishment of plants, use of a biofertilizer called "*Kunapajala*" has been mentioned. Kunapajala is a natural organic product derived from animal and plant products containing a significant quantity of one or more of the primary nutrients like nitrogen, phosphorus, and potassium which are necessary for plant growth. The literary meaning of the Sanskrit word Kunapa is "smelling like dead or stinking" and the name is apt for the liquid manure which is prepared using excreta, bones, body, flesh and marrow of animals, fish, decayed plant products, etc. Kunapajala has some plant growth regulatory actions through which it enhances the overall growth of plants. Being a liquid biofertilizer it is more suitable form of manure and can be beneficial in growth of medicinal plants with probably minimal toxic effects on human body when compared to chemical fertilizer (Mridula Chaturvedi and Chaturvedi 2017).

The other important micronutrients are magnesium, calcium, zinc, manganese, copper, iron, and selenium which are also supplemented by the organic compost Kunapajala (Bhat Savitha et al. 2012). Researchers suggest that application of the principles of Vrikshayurveda like Kunapajala does produce phenomenal and interesting results. Since few research works have been carried out, this discipline of science needs to be developed through concerted research efforts to ascertain its utility.

Though, chemical fertilizers increase the yield, they pose certain serious health threats to human beings, especially infants and pregnant and nursing mothers. Another concern for health is contamination of medicinal plants with toxic heavy metals like mercury, lead, and cadmium, through fertilizers, harmful industrial wastes contaminating the water sources, etc. In contrast, organic manures are considered to be safe and yielding good product by improving water penetration, water holding capacity, improvement in soil structure, microbial biomass, nutrient availability, drought, and heat stress resistance. It also helps in improving the soil pH which has an impact on plant growth and soil microbial activity (Vermeer et al. 1998). Studies using Kunapajala for growing Senna have shown that the total Sennoside content per plant was more. Similarly, for *Langali* (*Gloriosa superba* L.) the active principle (methanol extract) Colchicine was found in higher amount. When Kunapajala was used for *Vrntaki* (*Solanum melongena* L.), it produced large number of branches, higher yield, fruits with lesser seeds, and lower susceptibility to diseases, when compared with plants grown with artificial fertilizer. Similarly for mango, coconut, chilly, paddy, vegetables, etc. similar results have been found (Bhat Savitha et al. 2012). Thus, Kunapajala by virtue of its behavior as growth regulator has been effective in increasing the leaf area, yield of flowers and fruits as well as phytoconstituents.

Some major centers carrying out Vrikshayurveda-related work are Centre for Indian Knowledge Systems (CIKS, Chennai) (Balasubramanian et al. 2003), Asian Agri-History Foundation (AAHF, Secunderabad), and National Institute of Vrikshayurveda; Jhansi. Prof. Nene and his group at the AAHF are promoting Vrikshayurveda in a big way. CIKS, Chennai, is involved in promoting organic farming and works along with farmers belonging to various villages in Tamil Nadu. They are also involved in testing and validation of indigenous knowledge of agriculture by rapid assessment of traditional agricultural practices (Suresh et al. 2017). As a result of their experiments, as well as that of Indian Council of Agricultural Research, using the modern research procedures, it has been proved that the traditional knowledge is valid beyond doubt (Balasubramanian et al. 2009).

17.4 Road Map for Future

There are numerous recipes in Vrikshayurveda which awaits authenticity and validity. Some of them are listed below

- Combination of Sesame, Turmeric, and Yava (*Hordeum vulgare* L.) are to be added to the roots of Cotton plant, and decoction prepared from these drugs is added to obtain red colored cotton.
- Similarly, the drugs like *Shalmali* (*Bombax ceiba* L.), *Nisha* (*Curcuma longa* L.), *Neeli* (*Indigofera tinctoria* L.), *Triphala* (Combination of three drugs—*Terminalia chebula* Retz., *Terminalia bellirica* Wall., *Phyllanthus emblica* L.), and *Kushta* [*Saussurea costus* (Falc.) Lipsch.] when added along with medicated wine to the cotton plant root, it yields green cotton.

- By addition of sugarcane juice to the roots of any tree and smearing sugarcane juice to the cut shoots, it flowers and bears fruits in off season too.
- If the bones of the pig ones are nailed into the shoot of any tree, the tree does not die.
- If the seeds of pumpkin (*Cucurbita pepo* L.), brinjal (*Solanum melongena* L.), etc. are processed with fat of animals and planted, one can expect big-sized fruits without seeds. (Similarly many such recipes are mentioned for increasing the yield or improving the size of fruit which can be explored with the help of allied sciences like biotechnology and genetic engineering (Surapala 2010).

17.5 Conclusion

The use of *Pañcagavyam* (combination of cow dung, cows urine, milk, ghee, and curds), *Kuṇapajala,* and other procedures mentioned in the various texts can be studied further for efficacy, and if found to be suitable, they can be adopted for the various steps involved in development of organic nursery protocol for medicinal plants. Chemical fertilizers show dramatic short-term benefits, but, in the longer run they adversely impact the soil, water, and perhaps the nutritional quality of the plants. Hence, there is great scope to integrate traditional practices for better productivity of quality planting materials. Ayurvedic literature recognized the role of multidisciplinary approach and emphasized on integration of diverse fields as potential tool for development of medical science. It is hoped that ancient wisdom coupled with modern technology would benefit the mankind.

Acknowledgements We thank Director General and Deputy Director General, Central Council for Research in Ayurvedic Sciences, Ministry of AYUSH, and Government of India for their valuable support and encouragement. We are also indebted to Dr. S. M Khasim and his team, Department of Botany, Acharya Nagarjuna University, India, for encouragement and support.

References

Agnivesha (2014) CharakaSamhita, Ayurveda Deepika Commentary by Chakrapanidatta. In: Acharya VYT, Prakashan CS (Ed.), Varanasi, Vimanasthana, chapter:8 verse:101, p 278

Aralelimath GT, Nayak SU, Ankad GM, Hegde HV, Hegde L (2016) Comparative study on effects of Vrikshayurveda and modern techniques on germination of bakuchi (*Psoraleacorylifolia* L.) seeds. Nat Acad Sci Lett 39(4):241–244

Balasubramanian AV (2003) Knowledge and belief systems in the Indian subcontinent. In: Haverkort B, Hooft KV, Hiemstra W (eds) Ancient roots, new shoots - endogenous development in practice. Zed Books, London, pp 37–48

Balasubramanian AV, Vijayalakshmi K, Sridhar S, Arumugaswamy S (2003) Modern dilemmas and traditional insights. In: Haverkort B, Hooft KV, Hiemstra W (eds) Ancient roots, new shoots - endogenous development in practice. Zed Books, London, pp 74–83

Balasubramanian AV, Nirmala TD, Franco M (2009) Use of animal products in traditional agriculture, a pilot project in southern India. Centre for Indian Knowledge Systems, Chennai, p 60

Bhat Savitha D, Ashok BK, Rabinarayan A, Ravishankar B (2012) Importance of Kunapajala (traditional liquid organic manure) of Vrikshayurveda in medicinal plant cultivation. GJRMI 1(7):272–279

Chaithra M, Thomas B (2017) Traditional worshiping plants from selected sacred groves of Kozhikode. Res J Recent Sci 6(4):7–13

Chaturvedi M, Chaturvedi AK (2017) Concept of plant tissue culture in ancient science W.S.R Vriksha-Ayurveda. Imp J Interdiscip Res 3(1):1495–1498

Chen S-L, Hua Y, Luo H-M, Qiong W, Li C-F, Steinmetz A (2016) Conservation and sustainable use of medicinal plants: problems, progress, and prospects. Chin Med 11:37. https://doi.org/10.1186/s13020-016-008-7

Dwivedi DV (2012) Bilva in Indian tradition. Indian J Hist Sci 47.1:37–61

Ganesan A, Marichamy K, Thagamariappan J (2016) Development of medicinal plants sector in India–an empirical study. Int J Eng Technol Manag Appl Sci 4(1):82–88

Kasagana VN, Karumuri SS (2011) Conservation of medicinal plants (past, present & future trends). J Pharma Sci Res 3(8):1378–1386

Kokate CK (2008) Text book of pharmacognosy. Nirali Prakashan, Nirali

Kolawole OT, Ayankunle AA (2012) Seasonal variation in the anti-diabetic and hypolipidemic effects of *Momordica charantia* L. fruit extract in rats. Eur J Med Plants 2(2):177–185

Miraldi E, Masti A, Ferri S, Barni Comparini I (2001) Distribution of hyoscyamine and scopolamine in Datura stramonium L. Fitoterapia 72(6):644–648

Sajjad Khan M, Nema N, Kharya MD, Khanam S (2009) Chromatographic estimation of maturity based phytochemical profiling of *Ipomoea mauritiana* Jacq. Int J Phytomed 1(1):22–30

Shah CS, Goyal R (2017) A textbook of Pharmacognosy. BS Shah publication Pankorenaka, Ahmedabad

Shaik Nisar VH, Joshi S, Ravi S, Rao S (2017) Vrikshayurveda-a review. Int J Ayur Herb Med 7(5):2912–2914

Sharma A, Shanker C, Tyagi LK, Singh M, Rao CV (2008) Herbal medicine for market potential in India: an overview. Acad J Plant Sci 1(2):26–36

Sharma R, Amin H, Galib R, Prajapati PK (2013) Seasonal variations in physicochemical profiles of Guduchi Satva {starchy substance from *Tinospora cordifolia* (Willd.)Miers}. J Ayur Integr Med 4(4):193–197

Srikanth N, Tewari D, Mangal AK (2015) The science of plant life (Vrikshaayurveda) in archaic literature: an insight on botanical, agricultural and horticultural aspects of ancient India. World J Pharm Pharm Sci 4(6):388

Surapala (2010) In: Shrikrishanajugnu (ed) Vrkshayurveda: ancient treatise on plant life. Chaukhamba Sanskrit Series Office, Varanasi

Suresh G, Haridasan K, Krishnamurthy KV (2013) Relevance of Vṛkṣāyurveda and other traditional methods for organic production of nursery seedlings of useful plants. Anc Sci Life 33(1):60–70

Suresh G, Haridasan K, Pandala RC (2017) Comparison between traditional and conventional methods of seed storage and pretreatment in Embeliaribes - a threatened medicinal plant. J Trad Folk Prac 02, 03 & 04(1). https://doi.org/10.25173/jtfp.2016.2.0.50

Swagata DT, Nishteswar K (2014) Collection practices of medicinal plants - Vedic, Ayurvedic and modern perspectives. Int J Pharm Biol Arch 5(5):54–61

Venkatasubramaniam P, Shankar D, Unnikrishnan PM (2007) Need to develop intercultural standards for safety efficacy of traditional Indian system of medicine. Curr Sci 92(11):1499–1505

Vermeer IT, Pachen DM, Dallinga JW, Kleinjans JC, van Maanen JM (1998) Volatile N-nitrosamine formation after intake of nitrate at the ADI level in combination with an amine-rich diet. Environ Health Perspect 106(8):459–463

Ethnoveterinary Medicinal Plants and Practices in Andaman and Nicobar Islands, India

18

Sujatha Tamilvanan and Jai Sunder

Abstract

A majority of population living in the villages of Andaman & Nicobar Islands depends not only on the farming but a major role of their livelihood is animal rearing. Due to remote locality, the cure for their cattle from diseases is not easier; therefore, they develop their own methods by using locally available plants in their backyard, nearby forest, and home remedies to cure their animals and poultry. Many of these practices are quite effective and being used by them since long past. In most of the cases plants or plant parts are given as such, seldom in combination with some other plants, plant parts, or home remedies. The old-aged farmers and livestock owners gained the ethnoveterinary knowledge from their own experience or from their ancestors through the words of mouth. Livestock are an important and integrated component of the agricultural production system in developing countries. Antibiotics are indiscriminately used in livestock and poultry production and the phenomenon of the problem of multidrug resistance has evolved with a great concern for an organic livestock production system. The importance of traditional and alternative medicine out of medicinal plants has been realized for sustainable livestock production by the veterinary practitioners and medicinal plants have comprised a significant proportion of veterinary research. Due to isolated spread of 572 islands in remote locations of A&N Islands, rural farmers are dependent on indigenous knowledge for the treatment of their livestock and poultry rather than immediate animal husbandry and veterinary services. A&N Islands are the hot spot of a large number of medicinal plants. A field survey was done to study and document these ethnoveterinary medicinal plants. A total of 41 plants from 27 families were identified which are of ethnoveterinary importance from the different villages of South and North

S. Tamilvanan (✉) · J. Sunder
Division of Animal Science, ICAR-Central Island Research Institute,
Port Blair, A&N Islands, India

S. M. Khasim et al. (eds.), *Medicinal Plants: Biodiversity, Sustainable Utilization and Conservation*, https://doi.org/10.1007/978-981-15-1636-8_18

Andaman districts of Andaman & Nicobar Islands. Extensive research works are being carried out to explore medicinal plants as growth promoter and immunomodulator in poultry.

Keywords

Andamans · Ethnoveterinary practices and plants · *Morinda citrifolia* · *Andrographis paniculata* Nees · *Aloe vera* · *Azadirachta indica*

18.1 Introduction

Animal husbandry in Andaman & Nicobar Islands has its origin in 1858 when some cattle were transported from mainland. The planned activity and expansion of these sectors took place with the establishment of Animal Husbandry Department in 1942 and the postcolonial settlement of refugees from Bangladesh, Sri Lanka, Burma, and some other parts of India. The animal population of these islands consists of cows, buffalos, goats, pigs, and poultry birds of mostly indigenous type. Livestock is a profitable and integral component in agriculture and an important activity for accelerating the rural livelihood and nutritional security of rural farmers of these islands; however, livestock farming is not an organized/scientific farming except few farmers. A majority of population living in Andaman villages is settlers who came here before independence and their main livelihood support is farming and animal husbandry. Livestock and poultry of these islands in general were free from most of the dreaded diseases. Over the years due to entry of livestock, meat, meat products, egg, poultry, and feed from mainland the diseases like chronic respiratory disease, infectious bursal disease, and Ranikhet disease in poultry have been reported. The climatic condition of the islands, high rainfall and high humidity favors the prevalence of parasitic and gastrointestinal parasitic diseases in livestock. Due to isolated spread of 572 islands in remote locations, rural farmers are dependent on indigenous knowledge for the treatment of their livestock and poultry rather than immediate animal husbandry and veterinary services. Ethnoveterinary medicine (EVM) system is indigenous knowledge on animal health (Mc Corkle 1989) and refers to holistic and interdisciplinary study of traditional knowledge, skills, methods, practices, and folk beliefs of the people about animal health care found among the local community (Mc Corkle et al. 1996). Traditional practices cover diseases and their control including remedies and clinical practices for treatment and prevention to keep their animals healthy and productive. The medicines that livestock keepers are using at their level, other than modern synthetic drugs, are a part of the indigenous knowledge system of people all over the world (Mathius-Mundy and McCorkle 1989; Mathias 2004). The traditional knowledge differs not only from region to region but also among and within communities. Traditional knowledge has been developed through trial and error and deliberate experimentation; therefore, it is less systematic, less formalized, and not universally recognized as a valid method of disease control in animals; further traditional healers have less to offer in the treatment and control of epidemic and endemic infectious diseases like Rinderpest, septicemia, anthrax, and acute life-threatening bacterial diseases though they can cope with a reasonable spectrum of common diseases such as diarrhea, cuts and wounds, cold and cough, worms, retention of placenta, sickness, and stomach problems. With such successes, very little of this traditional knowledge has been documented in developing countries. In recent years, however, increasing attention has been paid to ethnoveterinary knowledge and local veterinary practices (Martin et al. 2001) as antibiotics are indiscriminately used in livestock and poultry production and the phenomenon of the problem of multidrug resistance has evolved with a great concern for an organic livestock production system. Antibiotics are common feed additives in livestock feed at suboptimum level as growth promoters and are mandatory to control and treat the infection. A&N Islands are the hot spot for medicinal plants; however, ethnoveterinary knowledge has had no place in mainstream of veterinary medicine. The indigenous knowledge and the natural resources are depleting day by day due to lack of proper records and documentation. Several ethnoveterinary studies have been conducted in many parts of India but still no work has been done in A&N Islands. Therefore, a study was carried out to document the information on ethnoveterinary practices being followed in pristine Andaman Islands. Further, research works were carried out in broad to explore medicinal plants as enhancers for gut health and its development, immunity, and growth in poultry.

18.2 Methodology

18.2.1 The Study Area and Survey

In Andaman & Nicobar Islands, 18 villages from South Andaman (Fig. 18.1) and 1 village from North Andaman (Fig. 18.1) were selected in this study. 18 villages of South Andaman are Maymyo, Manjeri, Indira Nagar, Guptapara, Manpur, Collinpur, Chouldari, Sippighat, Calicut, Cattleganj, Bamboo flat, Stewartganj, Wimberliganj, Malapuram, Wrightmyo, Manarghat, Shoalbay, and Kodiyaghat and one village from North Andaman is Diglipur. The information on indigenous/traditional/local knowledge, practices, and beliefs of the rural people were collected and the data were gathered from the knowledgeable persons, mostly old farmers, who use these practices with their livestock or had experience about animal husbandry and ethnoveterinary medicines. Adopting the method of Jain (1991), ethnoveterinary practices were collected through personal interview, observations, and group discussion from the selected households. The questionnaires were used to obtain information on medicinal plants with their local names, parts used, mode of preparation, and administration. A total of ten informants, comprising eight males and two females, were identified between the ages of 48 and 74 from each village. They were selected based on their knowledge of medicinal plants either for self-medication or for treating others. The information thus collected are presented in simple descriptive manner.

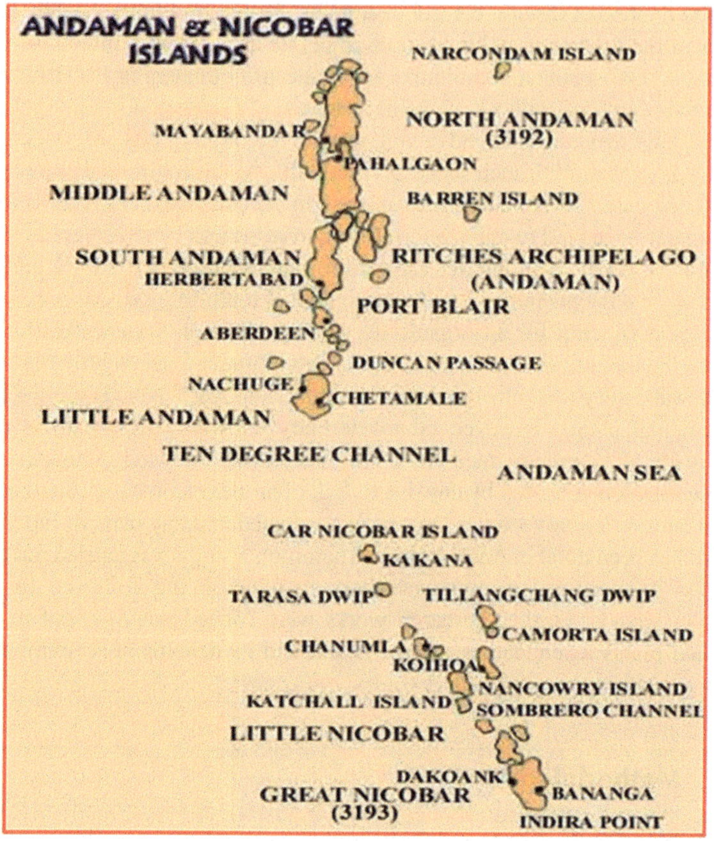

Fig. 18.1 Survey area: South and North Andaman

18.3 Ethnoveterinary Medicinal Plants and Practices

The results of the survey are presented in Table 18.1. There were various practices, conditions, and ailments used by the local people to cure their animals. Many of these practices are quite effective and being used by them since long past. A total of 78 different plants including kitchen spices like garlic, onion, ajowain, cardamom, clove, and cinnamon were found to be of ethnoveterinary use. Some selected plants that are known and frequently used in veterinary practices are enumerated with botanical name, family, common names, parts used, and uses. Results of the experiments involving *Morinda citrifolia, Andrographis panniculata, Aloe vera,* and *Azadirachta indica* (neem) have been extensively reviewed in this paper.

Table 18.1 Ethnoveterinary practices used by local people of Andaman Islands

Botanical name and family	Local name	Parts used	Method of preparation and administration	Medicinal uses
Acacia catechu (Mimosaceae)	Supari, betel-nut, cutch tree, heartwood	Leaves	The leaves are soaked well in water.	Applied for 7 days to cure foot diseases
Achras sapota (Sapotaceae)	Sapota, Chiku, tree potato, sapodilla plum	Fruits	One fruit is crushed and fed daily.	To cure diarrhea in cattle.
Acorus calamus (Acoraceae)	Vacha	Whole plant	Crushed whole plant with ash of burnt wood, two buds of pipli, salt, and turmeric powder and fed twice a day.	To cure diarrhea in goat.
Adhatoda vasica (Acanthaceae)	Vasak, Malabar nut	Leaves	Leaves are fed to the cattle twice a day for 2–3 days.	To cure diarrhea.
Aegle marmelos (Rutaceae)	Bael, Sirphal, Bengal quince, Golden apple, stone apple	Fruit pulp	The pulp of a green fruit is boiled, cooked the tender fruit in low heat and taken out the pulp without seeds, prepared juice of pulp, given once in a week.	To cure diarrhea. To increase the body weight and milk production.
Arachis hypogaea (Fabaceae)	Peanut, ground nut, monkey nut	Leaves	Leaves are given with feed.	To increase the milk production and body weight.
Azadirachta indica (Meliaceae)	Neem	Leaves	50 g neem leaves crushed in water. Neem leaf paste and oil. Neem oil	To cure stomachache. Applied in cut and wounds, abscess and broken horn To remove insects from the eyes and to control flies nuisance.

(continued)

Table 18.1 (continued)

Botanical name and family	Local name	Parts used	Method of preparation and administration	Medicinal uses
Bambusa spinosa (Gramineae)	Bamboo	Leaves	Feed a handful of bamboo leaves once in a day for 2–3 days. Bamboo leaves with rice husk/100 g dried bottle gourd leaves or 1 kg bamboo leaves thrice a day. Raw bamboo leaves Crushed bamboo leaves with one spoon black salt and a small amount of heeng (*Asafoetida*)	To kill ecto-parasites. Diarrhea Retention of placenta, tongue sore Appetizer/to improve the digestion/to cure the stomach problem and fever.
Bauhinia racemosa (Fabaceae)	Kachnar, Jhinjheri, Sirhatta, Asimlous	Leaves	Leaf juice	Applied over forehead to heal redness of eye.
Bauhinia variegate (Fabaceae)	Orchid tree, Camel's foot tree and mountain-ebony.	Bark and root	Root paste using cow urine. Bark decoction Two cups of root decoction twice daily.	Applied twice a day on the eyelids to treat blindness. To wash and cure wounds in foot and mouth. To expel the placenta
Bryophyllum pinnatum (Crassulaceae)	Pathar chutti, Pather chat, Patharchur	Leaves	Leaves are fed	To cure indigestion in goat.
Calotropis procera (Asclepiadaceae)	Madar, aak	Latex of plant	1/2 spoonful of latex extract with an equal amount of mustard oil. A loop made of thick thread is put around the neck of the cow. Now the medicine is applied well on the loop and kept it moving around the neck.	In sickness
Carica papaya (Caricaceae)	Papaya, papaw	Fruit, seed and root	Juice	To cure fever in goat
Centella asiatica (Apiaceae)	Brahmi, Indian pennywort, Brahma manduki, Medhak bhaji	Whole plant	Fed with pad patti (*Phyllanthus* spp.)	Diarrhea

(continued)

Table 18.1 (continued)

Botanical name and family	Local name	Parts used	Method of preparation and administration	Medicinal uses
Cissus quadrangularis (Vitaceae)	Harjor, Parande	Stem	Crushed stem with egg white. One teaspoon of salt is mixed in hot water and hot compression is given to the broken area with this solution. A paste is prepared by crushing 2 pieces of "harjor" stem (each 6 in. long), 100 g raw turmeric and 25 g alum. And slightly cooking in 100 ml mustard oil. Juice with turmeric and jeera.	Applied on broken bones and bandage is supported by bamboo splinters to avoid dislocation. Application of hot compression and medicines are repeated until healing. To increase the body weight and milk production and as dewormer.
Cocos nucifera (Arecaceae)	Coconut	Endosperm of the fruit, oil	Fed one coconut daily. Naphthalene powder with coconut oil Turmeric powder with coconut oil.	To increase the milk production and body weight To cure wounds. Applied over the wound area twice a day wound.
Colocasia esculenta (Araceae)	Auram	Stem	2–3 spoonful salt with the stem is rubbed on the tongue Crushed well matured leaves and tablets are prepared and are fed 2 times for 7 days	To cure sore/ hardening. Diarrhea.
Cynodon dactylon (Poaceae)	Doob, Doorva, Dhub, Harialil	Whole plant	Paste of the flowers with ginger and covered the wound with a tight cloth-bandage.	To cure wound
Datura metel (Solanaceae)	Dhatoora, black Datura, thorn apple	Whole plant	Seed powder Crushed leaf Soak tobacco leaves in water for 12 h and then crushed well. Crushed raw fruit and leaf paste	Skin diseases Insect bite Tick infection. Mastitis
Eclipta alba (Asteraceae)	Bhringaraj, Bhangra, kala makka	Whole plant	Mixed with feed	To improve body weight and milk production.

(continued)

Table 18.1 (continued)

Botanical name and family	Local name	Parts used	Method of preparation and administration	Medicinal uses
Enhydra fluctuans (Asteraceae)	Helencha	Leaves	Fed for 3–4 days	Deworming
Entada phaseoloides (Mimosaceae)	Gila	Fruit and seed	Seed powder with water One crushed fruit with some neem leaves and 50 g salt Fruit paste and wrapped with banana leaf and fed.	Diarrhea and Dysentery Appetizer
Ipomoea batatas (Convolvulaceae)	Sweet potato	Leaves	Mixed with feed	To improve milk production and body weight.
Jatropha curcas (Euphorbiaceae)	Arandi	Seeds	Crushed one or two seeds with water and drenched.	To treat constipation in cattle and goats.
Lagenaria leucantha (Cucurbitaceae)	Bottle gourd	Fruits and leaves	100 g leaves thrice a day Boil one fruit and mixed with feed. Boiled fruit with molasses and fed for 7 days. Four leaves and 30 g salt—paste is applied thrice for two days.	Diarrhea. In retention of placenta When animal takes its own placenta. Hump sore
Luffa acutangula (Cucurbitaceae)	Turia, ridge gourd	Fruit	Dried ridge gourd is burnt on one side and smoke is applied to nose to inhale the smoke	Cold and coughing.
Mangifera indica (Anacardiaceae)	Mango	Fruits, seed and leaves	Seeds. Leaves Mango fruit and ginger are macerated and fed with water.	Dysentery To expel placenta In flatted stomach.
Mimosa pudica (Fabaceae)	Touch me not, Chuimui	Whole plant	Shade dried whole plant and fed.	To improve weight and milk production
Momordica charantia (Cucurbitaceae)	Bitter gourd	Fruits	Crushed inner white mass and paste is given with banana leaves or 100 g leaves alone or with green bamboo leaves fed thrice a day.	In diarrhea

(continued)

Table 18.1 (continued)

Botanical name and family	Local name	Parts used	Method of preparation and administration	Medicinal uses
Moringa oleifera (Moringaceae)	Drumstick, Saijhan, Sajna	Leaves and stem bark	Paste of the stem. Leaves	In external wound For good health and better lactation after delivery.
Nicotiana tobacum (Solanaceae)	Tobacco	Leaves	Paste of crushed leaves and lime/paste of one tobacco leaf, 1–2 scales of garlic, one spoon salt and lime./ paste of one tobacco leaf and 15–20 g of sugar. Paste of 1 dry tobacco leaf, 3–4 spoonful white lime and 3–4 spoonful neem oil. Leaves are soaked in water for 12 h, crushed well and applied on the affected areas.	Applied on the broken horn and covered with a bandage. Applied on wound. In tick infection. Within 24–48 h of application of this mixture, all the ticks get removed.
Musa paradisiacal (Musaceae)	Banana	Whole plant	Fed directly the inner most white portion of banana flower to the animal once in a day for 2–3 days. Dried and crushed roots with water and allowed to swallow. Paste of fruit and sugar candy in water twice a day.	To cure diarrhea. In cases of snake bites and worms. To cure the blisters and hoof sore.
Oryza sativa (Poaceae)	Rice	Grains	Boiled dhan (rice) 4 kg of boiled crushed paddy mix with half kg of wheat flour. Paddy grains with salt are rubbed on the tongue.	In retention of placenta. To let down milk. In tongue sore.
Phyllanthus niruri (Euphorbiaceae)	Niruri, Bhui amla, Jungli amla	Whole plant	Whole plant mixed with feed.	To improve the body weight and milk.
Phyllanthus spp. (Euphorbiaceae)	Pad patti	Leaves	Leaves mixed with feed Mix the *Phyllanthus* spp. with *Centella asiatica* (medhak bhaji)	To control loose motion. To cure diarrhea in poultry.
Piper nigrum (Piperaceae)	Black pepper, Gol mirch	Fruit	25 g black pepper is powdered and mixed with 1/2 kg hot ghee and fed.	In case of snake bite.

(continued)

Table 18.1 (continued)

Botanical name and family	Local name	Parts used	Method of preparation and administration	Medicinal uses
Swertia chirata (Gentianaceae)	Chirota	Fruit	Soaked in water overnight and the water is drenched.	*Deworming*
Tamarindus indica (Caesalpiniaceae)	Tamarind, Imli	Fruit and leaves	Leaf paste Mix tamarind pulp with salt and rubbed over the tongue. A 200 g leaf is boiled in water and with 10 g of salt. 250 g leave is boiled in water to form a thick mass and drenched. 1/2 cup juice of each old tamarind and tender guava leaves. Boiled leaves in water and given the hot fomentation over the swollen area by dipping a cloth in the water. Mix leaves with turmeric powder and salt and rub on the affected portion.	Applied to reduce wound swelling. Anorexia Flatulence In muscle sprain/injury. Recovery occurs within 3–4 days/ mastitis In mouth ulcer.
Vitex trifolia (Verbenaceae)	Samalu	Tubers and leaves	Leaves are made hot and tied around injured part. Boiled neem leaves and mango tree stem bark with tubers of the *V. trifolia*, and applied. Boiled tubers in water and water is applied and applied turmeric powder for 2–3 days.	For muscle sprain/injury. In mouth ulcer To wash the wart in poultry
Weedelia biflora (Compositae)	Burma booti	Whole plant	Leaves are crushed, macerated, and applied on the wound 2–3 times a day. Paste of booty leaves. Plant is crushed and fed with water. Paste of the grinded booty leaves is applied on the muscles.	Broken horn Applied for muscle sprain/ injury. In snake bite

18.3.1 Documentation of Ethnoveterinary Medicinal Plants and Practices

A total of 41 genera from 27 families, 1 from each Acanthaceae, Acoraceae, Anareadeaceae, Apiaceae, Asclepiadaceae, Araceae, Caesalpiniaceae, Cariaceae, Convolvulaceae, Compositae, Crassulaceae, Gentianaceae, Gramineae, Meliaceae, Moringaceae, Musaceae, Piperaceae, Rutaceae, Sapotaceae, Verbenaceae, and Vitaceae; 2 from each Astaraceae, Solanaceae, and Mimosacaae; 3 from each Cucurbetaceae and Euphorbeaceae; and 4 from each Poaceae and Fabaceae were documented as ethnoveterinary use. The study surveyed 24 types of diseases in cattle and poultry. The information obtained is comparable favorably with the result of similar studies conducted in other states of India (Tiwari and Pande 2010; Bish et al. 2004; Shah et al. 2008; Pande et al. 2007; Kulkarni et al. 2007; Kiruba et al. 2006; Biren et al. 2007; Gaur et al. 2010; Khan 2009; Deshmukh et al. 2011; Sathish et al. 2010).

From this study it was revealed that the local people use the medicinal plants not only to cure various diseases but also for increasing body weight and milk production. There was 11 plants found to be used in diarrhea, i.e., *Achras sapota, Acorus calamus, Adhatoda vasica, Aegle marmelos, Bambusa spinosa, Centella asiatica, Colocasia esculenta, Entada phaseoloides, Lagenaria leucantha, Momordica charantia, Musa paradisiacal*, and *Phyllanthus* spp. while other ten plants such as *Aegle marmelos, Arachis hypogea, Cissus quadrangularis, Cocos nucifera, Eclipta alba, Ipomoea batatas, Mimosa pudica, Oryza sativa, phyllanthus niruri,* and *Zea mays* were found to be used for increasing body weight and milk production. The four plants *Bambusa spinosa*, *Cissus quadrangularis*, *Lagenaria leucantha,* and *Nicotina tabacum* reported various ethnoveterinary uses.

These traditional ethnoveterinary information suggested that local people of these islands have some unique knowledge of disease causes and patterns and their treatment using available resources inside and outside the home. As most of these methods show very effective medicinal properties, local people use them very frequently. Documentation of these methodologies has reiterated the strong base for conservation of medicinal plants and underlined the importance and crucial role of medicinal plants in livestock farming of these islands. In addition to this, the traditional medicinal system gives better result in drug resistance diseases with zero side effects (Davis 1994). In future, detailed chemical and pharmacological investigations of these traditional formulations and conversion of medicinal plants to develop local herbal products will help local farmers to access to herbal therapy at their hand extended.

18.3.2 *Morinda citrifolia* (Noni)

Morinda citrifolia L. is commonly known as Noni and belongs to the family Rubiaceae (Nelson 2006). It grows widely throughout the coastal regions of many countries including the Andaman & Nicobar group of Islands. In these islands it is commonly known as *Lorang, Burmaphal, Pongee phal,* and *Surangi* by the tribal. *Morinda citrifolia* has a rich history in India, where it has been used as Ayurveda medicine. However, many animals do not consume the product and avoid contact with its fruit and seeds due to its strong smell and taste. The residents of South Pacific islands have noted health benefits for themselves and their animals by ingesting the *Morinda citrifolia* fruit (Whistler 1985). Some animals such as pigs consume the fruit in its natural state (Fugh-Berman 2003). Most animals have difficulty in consuming and digesting whole fruit. Over the year's research on the use of fruit and leaf extract, literatures are creaming upon anthelmintic (Morton 1992), anti-inflammatory (Fletcher et al. 2013), antitumor (Jinhua et al., 2013), and hepatic stimulatory properties (Sunder et al. 2013a). Sunder et al. (2012) have studied the wide-spectrum antibacterial and antifungal activity of various parts of the *Morinda citrifolia* extracts. They have found that the methanol, ethanol, ethyl acetate, chloroform, and acetone extracts of leaf, stem bark, fruit, and seed showed broad-spectrum antibacterial and antifungal activity in vitro.

1. **Growth-enhancing properties of Noni:** *Morinda citrifolia* juice has high nutritive value containing all the essential amino acids, minerals, vitamins, and other nutrients which are required for the growing cells. It is very rich in proxeronine which is believed to be a precursor to xeronine which helps in activation of xeroninase. Morinda fruit contains several amino acids, vitamins, minerals, coenzyme carbohydrates, and alkaloids which directly or indirectly help in metabolism of the nutrients and overall growth of the cell. More than 200 nutraceutical compounds have been identified from the plant (Singh 2012). The juice of *Morinda citrifolia* fruit at 1.5 ml/bird/day is a potent growth promoter in the native Nicobari fowl of Andaman & Nicobar Islands, India, and broiler (Sunder et al. 2011a). The overall mean 300 g higher body weight gain and better feed conversion efficiency could be possible by supplementing this juice. The fruit extract at 5% per day accelerates the growth of Japanese quail that showed better body weight gain (10 g) and feed conversion ratio (0.18) and it supports for better hen-day egg production (Sunder et al. 2011c). Leaf and fruit of this plant are reported to be used as feed for livestock and poultry (Fugh-Berman 2003). However, further investigation on several bioactive compounds present in the *M. citrifolia* will help in understanding the actual mechanism in detail and its use in livestock and poultry as a source of vitamins and mineral supplement for higher growth, production, and immunity.
2. **Aniticholesterolemic properties and other blood biochemical profile:** Reports suggest that the Noni has photochemicals and beta-sitosterol, a plant sterol with potential for anticholesterol activity (Wang et al. 2006). *Morinda citrifolia* could effectively produce low cholesterol eggs on the platform of enriched egg production. *Morinda citrifolia* fruit and leaf extract at 5% reduces blood cholesterol level by 15% depositing low cholesterol in meat of broilers and egg of Nicobari fowl (Sunder et al. 2011b). Concentration of serum protein is useful in monitoring various disease status. It increases during dehydration, multiple myeloma, and chronic liver diseases. Creatinine is a waste product formed

in the muscle from the high energy storage compound creatinine phosphate and an indicator of renal functions and increases in various renal diseases. Albumin is a plasma protein synthesized in liver from amino acids absorbed from ileum. The extract of *Morinda* lowers serum total protein, serum creatinine, and albumin (Sunder et al. 2011b) indicating its boosting effect on kidney and liver function.

3. ***Morinda citrifolia* as feed supplement:** *Morinda citrifolia* fruit has been exploited as potent feed ingredient in the feed ration for Nicobari fowl and Japanese quail (Sunder et al. 2013b, c). The first report of its kind is the use of Morinda citrifolia fruit granules as feed supplement in the poultry by Sunder et al. (2013b) though some reports are available on the use of leaf and fruit of this plant as feed for livestock and poultry (Fugh-Berman 2003). The processed *Morinda citrifolia* fruit granules replaces maize by 15% and rice bran and wheat by 5% each in the quail ration supporting to produce higher body (19%) and hen-day egg production (24%). *Morinda citrifolia* can also be used as feed ingredient in Japanese quail ration as replacer of concentrate feed up to 20% without sparing the body weight at market age of 5 weeks and hen-housed egg production. However, egg production cost of Rs. 0.45 with *Morinda* is economically least cost effective.
4. ***Morinda citrifolia* as Immunomodulator:** An immunomodulatory polysaccharide-rich substance (Noni-ppt) from the fruit juice of *Morinda citrifolia* has been found to possess both prophylactic and therapeutic potentials against the immunomodulator sensitive Sarcoma 180 tumor system (Furusawa et al. 2003). The anti-tumor activity of Noni-ppt produced a cure rate of 25–45% in allogeneic mice and its activity was completely abolished by the concomitant administration of specific inhibitors of macrophages (2-chloroadenosine), T cells (cyclosporine), or natural killer (NK) cells (anti-asialo GM1 antibody). Noni was also capable of stimulating the release of several mediators from murine effector cells, interleukin-1beta (IL-1), IL-10, IL-12, interferon-gamma (IFNs), and nitric oxide (NO) (Hirazumi and Furusawa 1999). The first report of its immunomodulatory properties in poultry was by Sunder et al. (2007) who has reported that the crude extract of leaf has potent immune-modulating property in Nicobari fowl when fed at 1.5 ml/bird/day mediating the B cell immune response; however it is not stimulating T cell response. Unlike Nicobari fowl, its fruit juice at 5% enhanced both humoral (B cell mediated) and cellular (T cell mediated) immunity in broilers. The protective response was sustained till third week post inoculation. The direct challenge test of infectious bovine disease virus in the *Morinda*-fed birds also showed protection against the infection, indicating that Noni increases the defenses and reinforces the immune system of the body, neutralize its function in all the cells and regenerates the affected cells, thus helping in preventing the development of diseases. The available data suggest for feeding of *Morinda citrifolia* at 1.5 ml/bird/day as immune enhancer in Nicobari fowl.
5. **Synergistic effect of Morinda with *Lactobacillus* spp. on gut health and development:** The use of lactic acid bacteria in livestock poultry feed as probiotic has been well studied by many workers as feed supplements for growth, to

enhance production performance and immune response (Salarmoini and Fooladi 2011; Zamanzad-Ghavidel et al. 2011). This wonder *Morinda* has been scientifically to have synergistic effect in combination with *lactobacillus* spp. The compound terpenoids and phenolic compounds such as aucubin, alizarin, scopoletin, and other anthraquinones (Lavanya and Brahmaprakash 2011) present in the fruit extract being responsible for the antimicrobial activity (Narimani-Rad et al. 2011) significantly reduced gut coliform load (Sunder et al. 2012) as it was also reported by Salarmoini and Fooladi (2011). The *Lactobacillus* metaboilizes the feed to produce lactic acid which in turn lowers the pH of gut intestinal flora and inhibit the pathogenic organisms, mainly the Salmonella and *E. coli.* Further, supplementation of Morinda juice supported the bird to convert feed efficiently so that its supplementation saved 10% of feed and in turn feed cost of 5.39 per bird could be saved. The juice of Morinda fruit has potency to accelerate gut development. There were significant microscopical changes in crypt depth and villi height at the level of duodenum which is the primary site for the development of immune response and where nutrient uptake takes place. The intestinal mucous villi and crypt depth of lactobacillus-fed broilers were significantly very high, approximately 439.64 μ and 51.39 μ, respectively, with the base having ridges. The *Morinda* juice improves the histological indexes of the intestinal mucosa helping the function of gut. Feeding of *Morinda citrifolia* fruit juice (10 ml), Kalmegh (3 g/bird/day), and *Lactobacillus acidophilus* (*1X10*^{8} *cfu/ml)* significantly improved the villi height and crypt depth in broiler duodenum.

6. **Grommune and Morical supplements:** Herbal based feed supplement (Morical) and tonic (Grommune) have been prepared from *Morinda citrifolia* fruit. These preparations have been scientifically evaluated for their growth-promoting, production-enhancing, and immune-modulating properties (Sunder et al. 2013b, c, c). The dose of Grommune is recommended at the rate of 15 ml per bird up to 4 weeks and 30 ml per bird up to eighth week. Similarly, the morical is added as feed supplement at the rate of 8% in feed.

18.3.3 *Andrographis paniculata* Nees (Kalmegh)

1. ***In ovo effect* on post-hatch gut development:** Kalmegh *(Andrographis paniculata)* is a promising medicinal plant commonly used in humans as an immune system booster. Main bioactive compounds are andrographolide and diterpenoid lactone. Its immunomodulatory and growth-promoting activities have been scientifically validated (Mathivanan and Kalaiarasi 2007). The progeny from breeders fed with 3 g of Kalmegh powder produced significantly ($P < 0.05$) higher antibody titer at both 12 (0.55) and 16 (0.38) weeks. This significant humoral response was present till third week of post inoculation in progeny. The in ovo exposure provides the foundation for immunomodulation through tissue deposition of bioactive components that can affect post-hatch chick immune responses. The mechanism for embryonic effects later in the chick's life may be related to developmental effects in the post-hatch chick. In this way enrichment of immune

cells with herbal immune active principles from dietary or embryonic origin might signal the inflammatory immune response.

Structure of intestinal mucosa can reflect the health condition of intestine (Xu et al. 2003). Significant microscopical changes also have taken place in crypt depth and villi height at the level of duodenum in the gut of progeny from breeders fed with Kalmegh feed additive. The control group mucosa contained villi with a height of approximately 269.28 ± 18.48 μm that was statistically ($P < 0.05$) lower than the dietary inclusion levels of 1 g (323.11 ± 16.48) and 3 g (365.06 ± 16.0). The crypt depth of progeny from 3 g (54.42 ± 4.81 μm) and 1 g (58.15 ± 3.42 μm) were significantly ($P < 0.05$) lower than the progeny of control group (63.79 ± 1.72 μm). The histological changes that were brought in the progeny of breeding fowl fed with Kalmegh has provided in-depth base information of Kalmegh for its growth-promoting and immune-enhancing properties that can be utilized as an alternative to antibiotic growth promoters and to produce quality chicks. Further, this study has provided the base to strengthen the yolk sac at in ovo stage itself since yolk sac is a source of critical nutrition for the immunity development and gut function in post-hatch period.

2. **Synergistic effect on Expression of Toll-Like Receptors in broilers:** *Andrographis paniculata* has been proven to inhibit lipid peroxidation and free radical activities. Toll-like receptors (TLRs) are innate immune receptors and induce fast and appropriate host defense reaction against pathogens. TLRs recognize the conserved microbial patterns such as flagellin, LPS, and peptidoglycan in an efficient and non-self-reactive manner to initiate pro-inflammatory and cytokine. Supplementation of Noni and Kalmegh influenced the expression levels of TLR2, TLR3, TLR4, TLR5, TLR15, and TLR21 significantly ($P < 0.05$). The observed increased TLR-3, TLR-4, and TLR-5 gene expression and decreased TLR-7 gene expression in gut associated cecal tonsil in chickens fed dietary Noni and Kalmegh indicated that combination of herbal extracts have better immunomodulatory properties than commercial tonic. The increased expression of TLR-3, TLR-4, and TLR-5 may be due to the effects of phytochemicals on these TLR signal transduction pathways. This is supported by the fact that some phytochemicals like quercetin which is present in the Noni fruit stimulate the production of interferon-γ (IFN-γ), an antiviral product (Park et al. 2009). IFN-γ then acts in an autocrine manner and provides positive feedback for the expression of TLR-3, TLR-4, and TLR-5 (Tohyama et al. 2005). The selectively increased level of TLR-3, TLR-4, and TLR-5 and decreased TLR7 gene expression indicated that supplementing Noni fruit and Kalmegh induces antiviral and antibacterial responses in chicken.

   In another study, the *M. citrifolia* fruit was fed to broiler and the expression of TLR, immunoglobulins, and α1-acid glycoprotein was measured. The results showed that Noni at 6% concentration increased the expression of Toll-like receptor-4 (TLR-4) and TLR-5, chemokine (IL-8), and interleukin-12 (IL-12), and decreased the expression of IL-6 and TLR-7. Increased expression of TLR-3, TLR-4, and TLR-5 indicates antiviral and antibacterial properties of Noni fruits in chickens that are further strengthened by increased expression of IL-12 and IL-8.

3. **In ovo effect on expression of Toll-Like Receptors in Indigenous Nicobari Fowl:** Kalmegh and Noni has in ovo effect to modulate the immunity of chicks when breeders are treated with these herbal plants (Sujatha et al. 2015). In ovo exposure to bioactive components present in herbal plants upregulated the expression of TLR3, TLR4, and TLR5 by 5.4, 9.7, and 1.6 times, respectively, in chicks that is responsive for their antiviral inducing property at gene level.

18.3.4 *Aloe vera* and *Azadirachta indica* (Neem)

Aloe vera is among the most well-known herbs. It is similar to cactus in appearance and mostly grows in and around area of Asia. *Aloe vera* (Aloe barbadensis miller) belongs to Asphodelaceae (Liliaceae) family. An important *Aloe vera* property which has received attention from researchers is the polysaccharide acemannan—a manno polymer. Several studies revealed that properties of *A. vera*, including immunomodulatory and antibacterial properties, may stem from acemannan (Thiruppathi et al. 2010; Jothi Karumari et al. 2014). Similarly *Azadirachta indica* (neem) is used in traditional medicine as a source of many therapeutic agents in the Indian culture and grows well in the tropical countries. *Azadirachta indica* leaves also contain compounds with proven antimicrobial activity. The antimicrobial activity of extracts of neem leaves against such microorganisms as *Staphylococcus* spp., *Streptococcus* spp., *Pseudomonas* spp., *and Escherichia coli* and some fungal strains have been reported (Koona and Budida 2011). Studies on the effects of neem on poultry production especially of broilers and laying hens also exist.

1. ***Aloe vera* and *Azadirachta indica* (Neem) as growth promoters:** *A. vera* has been extensively reported in poultry feed for enhanced growth effect in broiler and better egg production in laying hens (Guo et al. 2004). *A. vera* has potential to be a growth promoter in broiler chicks and its growth-promoting effects are comparable to that of antibiotic growth promoter (AGP). *A. vera* did not show any adverse effects on the health status of birds as evident from their hematological and biochemical studies (Singh et al. 2013). The improvement in live body weight in broilers may be due to antibacterial related to flavonoids in *Aloe barbadensis* that led to maintaining normal intestine microflora by competitive exclusion and antagonism, altering metabolism and increased liver and muscle glycogen contents (Gomez et al. 1998). There are many number of literatures available for *Aloe vera* as growth promoters in birds. In recent past, supplementation of *A. vera* leaves at 5% in cockerals (Odo et al. 2010) and at 0.4% in broilers (Salary et al. 2014) improved body weights and feed efficiency.

   Neem plays an important role as a growth promoter due to its antibacterial and hepatoprotective properties (Landy et al. 2011). It improves performance (Nayaka et al. 2013) and immune response (Jawad et al. 2013) in broilers. Aqueous extract of neem leaves influence the body weight positively. These results may be due to antimicrobial and anti-protozoal properties of neem leaves, which help to reduce the microbial load of birds. Sultan et al. 2015 also recommended the extracts of *Azadirachta indica* leaf in broiler as inexpensive and efficient growth-promoting agent without residual effects like antibiotic growth promoter. *Aloe vera* and neem at 3 ml per day per chick are highly effective as growth promoters if chicks are fed in immediate post-hatch period (Abinaya 2016).

2. ***Aloe vera* and *Azadirachta indica* (Neem) in health management:** Broiler fed with *aloe vera* has improved the hematological indices and biochemical values (Singh et al. 2013; Tariq et al. 2014). Similarly, huge literatures give evidence for the use of *A. indica* safely as growth promoters in poultry. The neem leaf meal has been extensively researched in the chickens for hematology parameters (Esonu et al. 2006; Jawad et al. 2013). In all of these studies in chicken, the neem leaf meal has been reported to have positive impacts on the studied parameters at varying dietary inclusion levels. The neem leaf extracts act as a growth promoter (Landy et al. 2011) and improve immune response (Jawad et al. 2013) in broilers. Neem leaves contain biological active components that affect the nutrient utilization and associated with tissue damage and serum enzymes alteration (Akpan et al. 2008). These bioactive compounds may also alter the hematological and serum biochemical parameters of animals (Sharma et al. 2015). Opender et al. (2004) and Ogbuewu et al. (2009) described that neem leaf meal contained anti-nutritional factors which might affect blood profiles, utilization of nutrients, and growth of animals. *A. indica* at limited dose rate might be used as hepatoprotector in commercial poultry without any toxic effects (Jawad et al. 2014) as it decreased serum alkaline phosphatase (ALP) and aspartate aminotransferase (AST), serum creatinine, and especially serum uric acid values with increased level of neem leaf meal in broiler chickens. Neem also has anti-coccidial effect in broilers and is used as pesticide (Tipu et al. 2002; Esonu et al. 2006). However, severe depressive effects in the blood constituents of broiler chicks and laying birds fed NM at 10% and 15% have been reported by Obikaonu et al. (2012). Abinaya (2016) recommended for supplementing the early chick feeding with Aloe and neem as all hematological indices fall within the normal range.
3. ***Aloe vera* and *Azadirachta indica* (Neem) as immune enhancers:** *A. vera* polysaccharides, anthraquinones and lectins have immunomodulatory effects (Leung et al. 2004; Akev et al. 2007; liu and Wang 2007). The most important component of *A. vera* gel is acemannan and has potential to act against viral infections, which reduced opportunistic infections and stimulate wound healing (Ni et al. 2004). Its immunomodulatory effects had been reported in different animal models (Krishnan 2006). *A. vera* (0.1 and 0.2%) improved titer values against Newcastle disease in broiler chickens (Mehala and Moorthy) (2008). *A. vera* could be supplemented either as powder or in drinking water (Durrani et al. 2007). Neem fruit, also a potent immune booster in broilers (Nasir Landy et al. 2011), is used either as feed additive (7 g/kg) or as neem oil. It causes an elevation of both immune-reactive and bioactive TNF-alpha and gamma-interferon in serum (Ray et al. 1996). Hence, these herbs could be beneficial in immunosuppressant diseases of poultry (Renu Sridhar Rakha et al. 2003).

4. **Early chick nutrition with *Aloe vera* and *Azadirachta indica* (Neem) for gut health and its development:** The morphology of intestinal villi and crypts has been associated in chickens with intestinal function and growth. Adverse changes in the content of the digesta, such as high population of pathogenic bacteria, parasites, or damaging substances, could lead to changes in the surface of intestinal mucosa, because of their close proximity. The microbial population in the gastrointestinal tract of poultry plays an important role in normal digestive processes and health maintenance (Montagne et al. 2003). The reports of Lin et al. (2005) and Talwar et al. (1997) evidences for effective modulating property of *Aloe vera* gel on gut microbes and they increased *Lactobacillus* spp. and reduced *E. coli* count in gut of poultry. The microbial shift could be possible with *Aloe vera* and neem as water additives reducing and increasing the number of gut *E. coli* and *lactobacillus* spp. colonies, respectively (Abinaya 2016), due to the presence of huge numbers of chemically diverse and biologically active ingredients (Devakumar and Suktt 1993) in neem and acemannan, the polysaccharides of *Aloe vera* (Lin et al. 2005).

The microbial population in the gastrointestinal tract of poultry plays an important role in normal digestive processes and health maintenance (Montagne et al. 2003). *Aloe vera* and neem in water significantly reduces gut *E. coli* and increases the number of *Lactobacillus* spp. colonies (Abinaya 2016). This microbial shift might be due to the presence of huge numbers of chemically diverse and biologically active ingredients (Devakumar and Suktt 1993) in neem and acemannan, the polysaccharides of *Aloe vera* (Lin et al. 2005). In general, medicinal plants as alternatives to antibiotics exhibit the direct or indirect effects on intestinal microflora in poultry.

Changes in intestinal morphology, such as reduced villus height or increased crypt depth, can indicate the presence of toxins. According to Cera et al. (1988) maximum absorption and digestion capacity is given by a large luminal area with high villi and is essential to animal development. The longer villi would result in an increased surface area and higher absorption of available nutrients and is essential to animal development through increase in capable surface area and absorption of nutrients (Caspary 1992). Significantly taller and broader villi, significantly lower crypt depth and higher villus height:crypt depth ratio were reported on 14 days of age across the duodenum, jejunum, and ileum of chicks fed with Aloe and neem supplements. This post-hatch supplementation of plant extracts of Aloe and neem in poultry greatly improves gut health and its gut development (Abinaya 2016). Accordingly, the significantly higher villus height:crypt depth ratio with herbal supplements indicated the status of gut environment that is devoid of microbial toxins. This might be due to modified harmful microbial population and establishment of beneficial *Lactobacillus* spp. in that intestine that are responsible for favorable change in intestinal morphology. Lactic acid, the short-chain fatty acid as the final fermentation product by *lactobacillus* spp., might have made the gut environment unfavorable for pathogens. Further these short-chain fatty acids might have stimulated the proliferation of epithelial cells of the bowel. In addition, lower crypt depth indicated for slow tissue turnover by preventing the pathogens from tissue destruction in the gut.

In summary, an attempt has been made to document and describe the knowledge, practices, and belief of the rural community in veterinary practices in South and North Andaman. Based on this exploratory survey, it is inferred that rural livestock farmers of Andaman Islands have vast and rich knowledge and understanding of ethnoveterinary practices. Medicinal plants such as *Morinda citrifolia* L. (Noni), *Andrographis paniculata* (Kalmegh), *Aloe vera,* and *Azadirachta indica* (neem) could be the promising alternatives for antibiotic growth promoters and commercial immune boosters to improve and sustain the production of residue-free poultry produce. This lead paper could be the comprehensive base for further in-depth research on medicinal plants to develop promising replacers for antibiotic growth promoters in the organic poultry production.

References

Abinaya S (2016) Effect of early chick supplementation of *Aloe vera* and *Azadirachta indica* in Indigenous Nicobari fowl and Vanaraja birds. M.Sc Dissertation submitted to the Alagappa University in partial fulfillment of the requirement for the award of the degree of Master of Science in Biotechnology

Akpan MJ, Enyenihi GE, Obasi OL, Solomon IP, Udedibie ABI (2008) Effects of dietary neem leaf extract on the performance of laying hens. In Proceedings of 33rd Annual Conference Nigerian Society Animal Production March 20–25

Biren NS, Nayak BS, Bhatt SP, Jalapure SS, Seth AK (2007) The anti-inflammatory activity of the leaves of *Colocasia esculenta* Linn. Saudi Pharm J15:228–232

Bish N, Pande PC, Tiwari L (2004) Ethnoveterinary practices of Kapkot block of Bageshwar district, Uttaranchal. Asian Agri Hist 8:209–314

Caspary WF (1992) Physiology and pathphysiology of intestinal absorption. Am J Clin Nutr 55:299S–308S

Davis J (1994) Inactivation of antibiotics and the dissemination of resistance gene. Science 264:375–382

Deshmukh RR, Rathod VN, Pardeshi VN (2011) Ethnoveterinary medicine from Jalna district of Maharashtra state. Indain J Trad Know 10:344–348

Devakumar C, Suktt DV (1993) Chemistry. In: Randhawa NS, Parmar BS (eds) Neem research and development, publication no. 3. Society of pesticide science, India, pp 63–96

Esonu BO, Opara MN, Okoli IC, Obikaonu HO, Udedibie C, Iheshiulor OOM (2006) Physiological response of laying birds to neem (Azadirachta Indica) leaf meal-based diets: body weight organ characteristics and haematology. J OJHAS 5:972–997

Fletcher HM, Dawkins J, Rattray C, Wharfe G, Reid M, Gordon-Strachan G (2013) *Morinda citrifolia* (Noni) as an anti-inflammatory treatment in women with primary dysmenorrhoea: a randomised double-blind placebo-controlled trial. Obs Gyne Inter. https://doi.org/10.1155/2013/195454

Fugh-Berman A (2003) The 5 minute herbal and dietary supplement consult. Lippincott, Philadelphia, pp 236–237

Furusawa E, Hirazumi A, Story SJ, Jensen J (2003) Antitumour potential of a polysaccharide-rich substance from the fruit juice of *Morinda citrifolia* (Noni) on sarcoma 180 ascites tumour in mice. Phytother Res 17:1158–1164

Gaur RD, Sharma J, Painuli RM (2010) Plant used in traditional health care of livestock by Gujjar community of Sub-Himalayan tracts, Uttarakhand, India. Indain J Nat Prdts Res 1:243–248

Gomez MP, Geetha B, Aasker G (1998) Antidiabetic effects of fenugreek extract (Trigonella foenum-graecum L.) on domestic animals with special reference to carbohydrate metabolism. J Ecotoxi. Environ Monit 8:103–108

Guo FC, Kwakkel RP, Williams BA, Li WK, Li HS (2004) Effects of mushroom and herb polysaccharides, as alternatives for an antibiotic, on growth performance of broilers. Br Poult Sci 45:684–694

Hirazumi A, Furusawa E (1999) An immunomodulatory polysaccharide rich substance from the fruit juice of *Morinda citrifolia* (noni) with antitumour activity. Phytother Res 13:380–387

Jain SK (1991) Dictionary of Indian folk medicine and ethnobotany. Deep Publications, Paschim Vihar, New Delhi

Jawad Z, Younus M, Rehman M, Maqbool A, Munir R, Muhammad K, Ali Korejo K, Qazi IH (2013) Effect of neem leaves (Azadirachta indica) on immunity of commercial broilers against new castle disease and infectious bursal disease. Afr J Agric Res 8:4596–4603

Jothi Karumari R, Vijayalakshmi K, Tamilarasi L, Ezhilarasi B (2014) Antibacterial activity of leaf extracts of *Aloe vera, Ocimum sanctum* and *Sesbania grandiflora* against the gram positive bacteria. Asian J Biomed Pharma Sci 04:60–63

Khan FM (2009) Ethno-veterinary medicinal usage of flora of greater cholistan desert (Pakistan). Pak Vet J 29:75–80

Kiruba S, Jeeva S, Dhas SSM (2006) Enumeration of ethnoveterinary plants of Cape Comorin, Tamil Nadu. Indain J Tradi Know 5:576–578

Koona SJ, Budida S (2011) Antimicrobial potential of the extracts of the leaves of Azadirachta indica Linn. Nat Sci Biol 3:65–69

Kulkarni AP, Policegoudra RS, Aradhya SM (2007) Chemical composition and antioxidant activity of sapota (*Achras sapota* linn.). Fruit J Food Biochem 31:399–414

Landy N, Ghalamkari G, Toghiani M, Yazdi FF (2011) Humoral immune responses of broiler chickens fed with antibiotic and neem fruit powder (*Azadirachta indica*) as feed additive supplemented diet. IPCBEE 3:153–155

Lin SY, Yang JH, Hsia TC, Lee JH, Chiu TH, Wei YH, Chung JG (2005) Effect of inhibition of aloe-emodin on N-acetyltransferase activity and gene expression in human malignant melanoma cells (A375.S2). Melanoma Res 15:489–494

Martin M, Mathias E, McCorkle CM (2001) Ethnoveterinary medicine: an annotated bibliography of community animal healthcare. ITDG Publishing, Essex

Mathias E (2004) Ethnoveterinary medicine: harnessing its potential. Vet Bullet 74:27–37

Mathius-Mundy E, McCorkle CM (1989) Ethnoveterinary medicine: an annotated bibliography. Bibliographies in Technology and Social Change, Technology and Social Change Program, Iowa State University, Ames, Iowa 50011, USA, p 199

Mathivanan R, Kalaiarasi K (2007) Panchagavya and Andrographis Panniculata as alternatives to antibiotic growth promoters on haematological serum biochemical parameters and immune status of broilers. Indain J Poult Sci 44:198–204

Mc Corkle CM (1989) Veterinary anthropology. Hum Org 48:156–162

Mc Corkle CM, Mathius-Mundy E, Schillhorn Van Veen TW (1996) Ethnovet Res Develop, IT Publication

Montagne L, Pluske JR, Hampson DJ (2003) A review of interactions between dietary fibre and the intestinal mucosa, and their consequences on digestive health in young non-ruminant animals. Ani Feed Sci Tech 108:95–117

Morton JF (1992) The ocean-going Noni, or Indian mulberry (*Morinda citrifolia*, Rubiaceae) and some of its 'colorful' relatives. Econ Bot 46:241–256

Nayaka HBS, Umakantha B, Ruban SW, Murthy HNN, Narayanaswamy HD (2013) Performance and hematological parameters of broilers fed Neem, turmeric, vitamin e and their combinations. Emi J Food Agric 25:483–488

Nelson SC (2006) *Morinda citrifolia* (noni), ver. 4. In: Elevitch CR (ed) Species profiles for Pacific Island agroforestry. Permanent Agri Resou (PAR), Hôlualoa, Hawai

Odo BI, Ekenyem BU, Nwamo AC (2010) Effects of Aloe vera as leaf protein concentrate on growth performance of cockerels. Int J Poult Sci 9:426–428

Pande PC, Tiwari L, Pande HC (2007) Ethnoveterinary plant of Uttaranchal- a review. Indain J Tradi Know 6:444–458

Park HJ, Lee CM, Jung ID, Lee JS, Jeong YI, Chang JH, Chun SH, Kim MJ, Choi W, Ahn SC, ShinYK YSR, Park YM (2009) Quercetin regulates Th1/Th2 balance in a murine model of asthma. Intern Immuno 9:261–267

Salarmoini M, Fooladi MH (2011) Efficacy of lactobacillus acidophilus as probiotic to improve broiler chicks performance. J Agri Sci Tech 13:165–172

Salary J, Kalantar M, Sahebi ala M, Ranjbar K, Hemati Matin HR (2014) Drinking water supplementation of licorice and Aloe vera extracts in broiler chickens. Sci J Ani Sci 3:41–48

Sathish KD, Sharathnath KV, Yogeswaran P, Harani A, Sudhakar K, Sudha P, Banji D (2010) A medicinal potency of *Momordica charantia*. Int J Pharm 1:95–100

Shah R, Pande PC, Tiwari L (2008) Folklore biomedicine for some veterinary diseases in western part of Almora district. Indain J Trad Know 7:355–359

Singh DR (2012) *Morinda citrifolia* L. (Noni): a review of the scientific validation for its nutritional and therapeutic properties. J Dia Endo 6:77–91

Singh J, Koley KM, Chandrakar K, Pagrut NS (2013) Effects of aloe vera on dressing percentage and haemato-biochemidal parameters of broiler chickens. Vet World 6:803–806

Sujatha T, Sunder J, Kundu A (2015) Effect of *Andrographis Panniculata* on immunity and duodenal morphology of progeny of Nicobari fowl. Indain J Vet Ani Sci Res 44:305–310

Sunder J, Rai RB, Yasmeen J, Kundu A, Jeyakumar S (2007) Immunomodulator effect of Morinda citrifolia in poultry. Indain J Ani Sci 77(11):1126

Sunder J, Kundu A, Singh DR, Jeyakumar S, Srivastava RC (2011a) Effect of feeding of *Morinda citrifolia* fruit juice on growth, production and immune response in Nicobari fowl. Indain J Ani Sci 81:68–71

Sunder J, Singh DR, Kundu A, Jeyakumar S (2011b) Effect of feeding of Morinda citriffolia extract on blood biochemical profile of Nicobari fowl and broiler. Ani Sci Rep 5:141–146

Sunder J, Singh DR, Kundu A, Jeyakumar S, Verma SK (2011c) Supplementation of *Morinda citrifolia* extract on growth, production and immune response of Japanese quail. Indain J Vet Res 20:61–65

Sunder J, Singh DR, Jeyakumar S, Kundu A, Srivastava RC (2012) Antimicrobial activity of *Morinda citrifolia* solvent extracts. Indain Vet J 89:9–11

Sunder J, De AK, Jeyakumar S, Kundu A (2013a) Effect of feeding of *Morinda citrifolia* fruit juice on the biophysical parameters of healthy as well as mastitis-affected cow milk. J Appl Ani Res 41:29–33

Sunder J, Jeyakumar S, Sujatha T, Kundu A (2013b) Effect of feeding of Morical:a herbal based supplement on production and egg quality in Japanese quails. Adv Ani Vet Sci 1:157–160

Sunder J, Jeyakumar S, Sujatha T, Kundu A (2013c) Effect of feeding of morical (M.citrifolia) based herbal supplement on production and egg quality in Nicobari fowl. J Medi Plant Res 7:2999–3002

Tariq H, Raman Rao PV, Mondal BC, Malla BA (2014) Effect of Aloe Vera (Aloe barbadensis) and clove (Syzigium aromaticum) supplementation on immune status, haematological and serum biochemical parameters in Japanese quails. Indain J Ani Nutr 31:293–296

Thiruppathi S, Ramasubramanian V, Sivakumar T, Thirumalai Arasu V (2010) Antimicrobial activity of Aloe vera (L.) Burm.f. against pathogenic microorganisms. J Bio Sci Res 1:251–258

Tiwari L, Pande P (2010) Ethnoveterinary medicines in India perspective: reference to Uttarakhand, Himalaya. Indain J Trad Know 9:611–617

Tohyama M, Dai X, Sayama K, Yamasaki K, Shirakata Y, Hanakawa Y, Tokumaru S, Yahata Y, Yang L, Nagai H, Takashima A, Hashimoto K (2005) dsRNA-mediated innate immunity of epidermal keratinocytes. Biochem Biophy Res Commun 335:505–511

Wang GJ, Chu LY, Zhou MY, Chen WM (2006) Effects of preparation conditions on the microstructure of porous microcapsule membrane with straight open pores. J Memb Sci 284:301–312

Whistler WA (1985) Traditional and herbal medicine in the Cook Islands. J Ethnopharmacol 13:239–280

Xu ZR, Hu CH, Xia MS, Zhan XA, Wang MQ (2003) Effects of dietary fructo-oligosaccharide on digestive enzyme activities, intestinal microflora and morphology of male broilers. Poult Sci 82:648–654

Ethnobotanical Trees of Sri Lankamalleswara Wildlife Sanctuary, Eastern Ghats, Andhra Pradesh

19

S. K. M. Basha and P. Siva Kumar Reddy

Abstract

Sri Lankamalleswara Wildlife Sanctuary is one among the 13 sanctuaries present in Andhra Pradesh established in the year 1988. It extends over 464.42 km^2 (46,400 ha). Although the sanctuary was declared with the sole purpose of protecting the habitat of the Jerdon's courser, many other species are also benefited, perhaps more than the target species. Leopard *Panthera pardus*, dhole or Indian wild dog *Cuon alpinus*, sloth bear *Melursus ursinus*, chital *Axis axis*, sambar *Cervus unicolor*, chinkara *Gazella bennettii* and wild boar *Sus scrofa* are seen in this sanctuary. Even the wolf *Canis lupus* has been sighted a couple of times. The present paper deals with nearly 140 tree species belonging to 42 families of their ethnobotanical uses.

Keywords

Ethnobotanical trees · Wildlife Sanctuary · Jerdon's courser

19.1 Introduction

Medicinal plants are being used in India either directly as folk remedies or as medicaments of different ancient system of medicine like Ayurveda, Siddha and Unani. About 40,000 plant species are there in India and have different medicinal uses and have attributed for about 7500 species (Anonymous 1994). India hosting about

S. K. M. Basha (✉)
Vikrama Simhapuri University PG Centre, Kavali, Andhra Pradesh, India

P. S. K. Reddy
Research and Development Centre, Bharathiyar University, Coimbatore, Tamil Nadu, India

S. M. Khasim et al. (eds.), *Medicinal Plants: Biodiversity, Sustainable Utilization and Conservation*, https://doi.org/10.1007/978-981-15-1636-8_19

4,00,000 practitioners of herbal medicine, whose services could not be adequately, utilized in the healthcare delivery systems. About 80% of raw materials for drugs used in the Indian systems of medicine and homoeopathy are based on plant products. Nearly 80% of world's population does not have access to modern medicine. Most of the money for healthcare in the developing world goes to the remaining 20% of the population. Healthcare expenses is predicted to double in the next 10 years. The complementary and alternative medicine with low-cost intervention such as life style changes, diet and nature therapy can be delivered as substitute for high-cost drugs and technological intervention. The credibility of these systems of medicine, therefore, depends on the availability of authentic raw material in sufficient quantities. Drugs are derived from different sources in different ways. The earliest drugs were plant extracts followed by pure natural compounds of known structures. The detection of biologically active compound in extracts through bioassay directed fractionation has gained much importance. In addition, after identifying lead compound the attempt to study the mechanism of action of biosynthetic pathway lead to the drug discovery.

Macroanalysis of the distribution of medicinal plants shows that they are distributed across diverse habitats and landscape elements. Around 70% of India's medicinal plants are found in tropical areas mostly in the various forest types spread across the Western and Eastern Ghats, the Vindhyas, Chotta Nagpur plateau, Aravallis and Himalayas. Although less than 30% of the medicinal plants are found in the temperate and alpine areas and higher altitudes, they include species of high medicinal value. Macro studies show that a larger percentage of the known medicinal plant occur in the dry and most deciduous vegetation as compared to the evergreen or temperate habitats. Analysis of habits of medicinal plants indicates that they are distributed across various habitats. One-third are trees and equal portion shrubs and the remaining one-third herbs, grasses and climbers. A very small proportion of the meidicinal plants are lower plants like lichens and fern algae. Majority of the medicinal plants are higher flowering plants.

Sri Lankamalleswara Wildlife Sanctuary is one among the 13 sanctuaries present in Andhra Pradesh established in 1988. It is situated in the Lankamallai hill ranges and is about 30 km from Kadapa. These hills are often called as Island hills; it is very difficult to distinguish the Lankamallias from the Nallamalias in the North and also from the Palakondas in the South. A depression separates the Lankamallias from the Nallamalias in the North and the Penna river separates from the Palakondas in the South. The Sanctuary was declared mainly for the Critically Endangered Jerdon's courser *Rhinoptilus bitorquatus* (Kalivikodi) after its rediscovery in 1986 (Bhushan 1986). In 1988 the Indian Postal Service department released a stamp to commemorate the rediscovery.

The Government of Andhra Pradesh has declared the Lankamalleswara forest as a wildlife sanctuary in October 17, 1988. On this day Go. Ms. No. 285 was issued by the government of AP. This sanctuary lies between 78° 45′-79° 10 E and 14° 25′-14° 45 N and spread over Proddatur, Badvel and Siddavatam ranges covering five mandals, i.e. Badvel, Siddavatam, Mydhukur, Chennur and Khajipeta. A total of 46,442.82 hectares of land belonging to the Kadapa and Proddatur forest divisions are declared as a wildlife sanctuary. It extends over 464.42 km^2 (46,400 ha).

Lankamalla forest in kadapa district was declared as wildlife sanctuary only to protect the habitat of Jerdon's courser, an endangered bird which was rediscovered in Reddipalli village, but now this sanctuary is also the habitat of tigers. The tiger was first seen in the Rollabodu beat of Sidhavatam range. Later based on the foot-marks tigers are also found in Sidhavatam, Kondur and Madhur beat. The Forest officials based on the foot marks described that there are four tigers present in this sanctuary. The forest officials have arranged cameras in Sabbavi and Madhuru in the sanctuary for discovering the tigers. On 3 December 2015 the camera arranged in Sabbavi has captured the tiger. So, the government has declared that these forests have "Tiger Corridor."

The major types of forest in this sanctuary are southern tropical thorn and southern tropical dry deciduous (Champion and Seth 1968). The sanctuary bears dry deciduous forest in the higher elevations to scrub forest in the plains. Major floristic components of the scrub forest are thorny species of Accacia, Zizipus and Carissa, and non-thorny species of Cassia, Hardwikia and Anogeisus. The Red Sandal *Pterocarpus santalinus* (Fig. 19.1d) endemic to Andhra Pradesh is present here. Sri Lankamalleswara Wildlife Sanctuary has five types of vegetations: (1) moist deciduous forest, (2) dry deciduous forest, (3) riverine forest, (4) scrub forest and (5) barren/sparse vegetation.

Nearly 200 bird species are found in this sanctuary. The critically endangered Jerdon's courser was restricted to a small patch of scrub jungle within the sanctuary. Recent studies reveal that this bird is found in three more localities within the sanctuary (Jeganathan et al. 2002). The Lankamalleswara Wildlife Sanctuary has been selected on the basis of the presence of single species. It is a secondary area where a single restricted range species is found (Stattersfield et al. 1998).

Ethnobotanical data were collected with prescribed pro forma. Enquiries were made on the food habits, occupation, health practices, medicine, trade, beliefs, rituals, ceremonies, traditions and customs of tribals included in the present study, namely Sugalis, Yanadies, Yerukalas and very little chenchus. The original information given by the tribals was recorded in the field book. Besides tribals, local people (villagers, housewives and traditional healers and natuvaidyas) were also interviewed and information on their traditional knowledge was recorded. The enquiries were made repeatedly to understand their knowledge, methods of diagnosis and treatment of diseases. Data were collected on the specific part(s) of the plants used, its collection, processing, preparation of the drugs and its dosage and mode of administration. Each collected individual species were labelled with field numbers in quadruplets after making a critical observation on the habit, habitat, frequency of occurrence, association, colour and scent of flowers and other macroscopic characters, which cannot be observed from the dried herbarium sheets. The details were recorded in the field book. Flowering and fruiting seasons were also recorded for each species.

The collected species were poisoned by dipping the whole twig in saturated mercuric chloride in ethyl alcohol solution and immediately placed between the blotting papers with forceps. These specimens were tied properly in iron pressures with the help of a rope. The blotting papers were changed after 12 hours and specimens were

a *Cassia fistula* L **b** *Ficus racemosa* L. **c** *Mitragyna parviflora* (Roxb.) **d** *Pterocarpus santalinus* L. **e** *Santalum album* L. **f** *Soymida fabrifuga* (Roxb.)

Fig. 19.1 (**a**–**h**) Plant photographs

spread properly. Some leaves were placed facing upwards and some leaves were placed facing downwards. These spread specimens were once again placed in the other dry blotting sheets and tied again. These processes were continued till the specimen dried properly.

The poisoned, dried and pressed specimens were pasted with glue on the thick herbarium sheets measuring 42 × 28 cm and after that stitched. Labels containing all relevant information, viz. name of the plant, family, locality, altitude, date of collection, notes as in field book and collector's name were affixed on the right-hand bottom corner of the herbarium sheets.

The identification of the specimens was made with the help of Gamble's "Flora of Presidency of Madras" (1915-35). The identifications were confirmed with the help of latest monographs, floras and by comparisons with authenticated specimens in the S. K. University Herbarium (SKU), Anantapur. The specimens were deposited in the Department of Botany, Vikrama Simhapuri University P. G Center, Kavali, Nellore (dt), Andhra Pradesh.

19.2 Enumeration of Ethnobotanical Trees of Sri Lankamalleswara Wildlife Sanctuary

Accacia chundra (Rottler) Willd, *Mimosa chundra* Roxb., *Mimosa sundra* Roxb., *Accacia sundra* (Roxb.) DC.

Moderate-sized tree, leaves pinnate, pinnate 10-20 pairs, leaflet elliptic, entire, subacute, rachis with glands at the basal and the two upper most pinnate, flowers white in axillary spikes or 1-3, in a cluster, calyx tube 5 lobed, petals 5, stamens indefinite, pods flat, thin depressed between seeds. Fl. & Frt.: April-December, Near Lankamalla temple, PSKR: 675.

Local name: Sundra	Family: Mimosaceae
Information source: Yanadi	Plant part used: Stem bark

Use description:

Stem bark decoction given orally for weak cattle; wood is used as a fire wood and in making tool-handles.

Acacia leucophloea (Roxb.) Willd. F. B. I. ii. 294; W. & A. 277; Bedd. Fl. t. 48; Brand. For. Fl. t. 27. *Mimosa leucophloea*, Roxb. Cor. Pl. t. 150.

A moderate-sized or large tree with pale yellowish-white flowers heads in large terminal panicles. Bark grey and smooth when young, dark brown and rough when old; wood reddish brown, hard and tough, a good fuel. The bark is used in distilling and gives a good rough fibre. Fl. & Frt.: Most of the year, Ranibavi, PSKR: 682.

Local name: Tellathumma	Family: Mimosaceae
Information source: Yanadi	Plant part used: Root bark, Stem bark

Use description:

Allexiteric (Cold & Bronchites), root bark powder mixed with bulb of *Allium cepa,* make it as a paste and give orally twice a day till cured; Stem bark decoction is used to prevent antenathal complications; wood is used for hut making; wood is used as timber.

Aegle marmelos Corr.; F. B. I. i. 516; W. & A. 96; Wt. Ic. t. 16.; Roxb. Cor. PI. ii. t. 143; Bedd. Fl. t. 161.

A small deciduous throny tree, the leaves usually glabrous, but grey pubescent in the more arid localities. Wood yellowish-white, hard. The fruit is valued for its aromatic pulp. Fl. & Frt.: October-March, Besides Lankamalla temple, PSKR: 626.

Local name: Maredu	Family: Rutaceae
Information source: Yerukala	Plant part used: Leaf & Fruit

Use description:

Leaves decoction taken orally for curing asthma thrice a day for one week; raw fruits decoction is taken orally for curing dysentery; fruits are edible; wood is used as a timber.

Alangium Salvifolium Wang. *A. Lamarckii*, Thw., F. B. I. ii. 741; Bedd. Fl. t. 215, *A. deeapeta lum*, Lamk.: W. & A. 325.

A deciduous small tree, sometimes straggling, sometimes spinous, with pretty white-scented flowers. Bark grey, orange-yellow when young; wood olive-brown, hard and close-grained, accented, useful for ornamental works and a good fuel. Fl. & Frt.: February-July, Siddavatam, PSKR: 165.

Local name: Ooduga	Family: Alangiaceae
Information source: Yanadi	Plant part used: Seed & Root bark.

Use description:

Seeds are powdered and mixed with honey, given orally for curing aphrodisiac; root bark ground, made as a paste and given orally for snakebite; fruits are edible.
Wood is used as timber.

Albizia amara Boivin; F. B. I. ii. 301; Bedd. Fl. t. 61. *Accacia amara*, Willd. W. & A. 274, *A. Wightii*, Grab.; W. & A. 274. *Mimosa amara*, Roxb. Cor. Pl. t. 122.

A moderate-sized deciduous tree with pinkish-white flowers and leaves with many very small leaflets, the young shoot yellow-pubescent. The wood is purplish-brown with lighter bands, very hard and strong. Fl. & Frt.: Throughout the year, on the way to Nithyapoojakona, PSKR: 193.

Local name: Seekireni	Family: Mimosaceae
Information source: Yanadi	Plant part used: Leaf

Use description:

Leaves as insecticide; leaves are dried in shade and powdered; the powder is mixed with water and applied to hair for dandruff.

Albizia lebbeck Benth.; F. B. I. ii. 298; Bedd. Fl. t. 53., *Accacia speciosa*, Willd., W. & A. 275.

A large deciduous tree, with rather large white flowers and very long greenish stamens, the pods characteristic and long remaining on the tree when ripe. Bark brownish-grey. Fl. & Frt.: Throughout the year, Siddavatam, PSKR: 195.

Local name: Dirisana	Family: Mimosaceae
Information source: Yerukula	Plant part used: Seed

Use description:

Seeds are powdered and boiled in water, taken orally for curing piles; seeds are powdered and mixed with water, made as a paste and given orally for curing diarrhoea; wood is used for making furniture.

Ailanthus excelsa Roxb. Cor. Pl. i. 24, t. 23; F. B. I. i. 518; W. & A. 150; Wt. III. i. t. 67.

Densely foliaceous, deciduous trees with persistent leaf scars, leaves imparipinnate, leaflets 8-14 pairs, alternate and subopposite, elliptic to ovate-lanceolate, unequal at base, flowers polygamous, yellow in large axillary or terminal panicles, sepals and petals 5 each, seed 1, in the middle of the samara compressed. Fl. & Frt.: January-May, Near Lankamalla temple, PSKR: 1200.

Local name: Peddamanu	Family: Simaroubaceae
Information source: Yanadi	Plant part used: Stem bark

Use description:

Stem bark as antipyretic. The decoction prepared from the stem bark taken orally twice a day for three days for curing fever; stem bark decoction given orally thrice a day for three days for curing cough; wood is very soft and used in matchbox industry for making matchsticks.

Allophylus serratus (Roxb.) Kurz. Radik. *A. Cobbe*. Bl.; F. B. I. i. 673 *in part*, *Oraitrophe serrata*, Roxb. Cor. Pl. i. t. 61. *Schmidelia serrata* var. a; W. & A. 110.

Tree, leaves alternate, 1-or 3 foliolate; leaflets entire or serrate; stipules 0, flowers small, irregular polygamodioecious. Pedicelled, thymes shorter than the leaves, sometimes shorter than the petiole, not branched. Bark of branchlets whitish; leaflets about 2-5 in. long, serrate, crenate-serrate or dentate, usually hoary-tomentose beneath, drupes small. Fl. & Frt.: August-January, Sabbavi, PSKR: 1250.

Local name: Guvvaguti	Salikunkudu; family: Sapindaceae
Information source: Yanadi	Plant part used: Root

Use description:

Root decoction given orally for curing diarrhoea.

Alstonia scholaris R. Br.; F. B. I. iii. 642; Wt. Ic. t. 422; Bedd. Fl. t. 242.

A large evergreen tree with greenish-white flowers and long very narrow pendulous follicles. Bark dark grey, rough, lenticellate; wood white, soft, even grained, used for boxes and temporary purposes, but not durable. Fl. & Frt.: December-September, Nithyapoojakona, PSKR: 750.

Local name: Eda-kula pala	Family: Bignoniaceae
Information source: Yanadi	Plant part used: Bark

Use description:

Bark decoction is used for curing fever.

Annona squamosa Linn.; F. B. I. i. 78; W. & A. 7.

Small tree, leaves oblong–lanceolate, elliptic, entire, acute or obtuse, flowers solitary or 2-4inclusters, stamens numerous, anthers hooded, carpels more than 50, fruits ovoid, seeds many, indehiscent, connate into a fleshy syncarp. Fl. & Frt. March-July, On the way to Lankamalla temple from Ranibavi, PSKR. 1036.

Local name: Seethapalam	Family: Annonaceae
Information source: Yanadi	Plant part used: Bark & Seed

Use description:

Seed paste is applied externally on the head for curing lice; bark juice used as antidote for snakebite; fruits are edible.

Annona reticulata Linn., F. B. I. i. 78; W. & A. 7.

Leaves acuminate, 5-8 in. long, green beneath, nerves divergent, flowers pale green; fruit orange, subglobose, with flat, 5-cornered areoles.

Local name: Ramapalam	Family: Annonaceae
Information source: Yanadi	Plant part used: Fruit

Use description:

Fruits are edible.

Anogeissus latifolia (Roxb. ex. DC.) Wall. ex. Guill. & Perr, *Conocarpus latifolia* Roxb. ex. DC.

Large deciduous tree; branchlets grey-pubescent. Leave chartaceous, elliptic, oblong-elliptic, entire, obtuse-rotund. Flowers small, numerous in axillary pedunculate cymose heads. Seed solitary. Fl. & Frt. Most of the year, Kapilthertheswarakona, PSKR. 1046.

Local name: Chirumanu velama	Family: Combretaceae
Information source: Yanadi	Plant part used: Stembark

Use description:

Stem bark decoction given orally for snakebite.

Atalantia monophylla Corres.; F. B. I. i. 511; W. & A. 91, *A. floribunda*, Wt. Ic. t. 1611, *Limonia monophylla*, Linn.; Roxb. Cor. Pl. i. t. 82.

A small throny tree with yellowish very hard close-grained wood, leaves alternate, 1 foliolate, leaflet coriaceous, entire, stipules like scales often present representing undeveloped leaf-beds. Flowers in axillary, seeds ovoid or oblong. Fl. & Frt. April-October, Near Lankamalla temple, PSKR 428.

Local name: Adavinimma	Family: Rutaceae
Information source: Yanadi	Plant part used: Fruit

Use description:

Oils extracted from berries applied externally for curing rheumatism and paralysis; wood is used for making furniture.

Azadirachta indica A. Juss.; W. & A. 118; Wt. Ic. t. 17. *Melia azadirachta*, Linn. F. B. I. 544; Bedd. Fl. t. 14.

Tree, leaves alternate, imparipinnate; leaflets subopposite, serrate, very unequal at base, flowers hermaphrodite, in axillary panicles, fruit 1 seeded drupe, endocarp woody, seed ellipsoid. Fl. & Frt. March-July, Jyothi, PSKR. 842.

Local name: Yepa	Family: Meliaceae
Information source: Yanadi	Plant part used: Leaf; Seed & Bark

Use description:

Seed paste mixed with flowers paste of *Ricinus communis* L. and neem leaf juice inserted into the vagina up to cervix to abort two- to three-month pregnancy; leaf juice mixed in honey applied in the ear for curing ear ache; a spoonful of roasted flowers mixed with sugar given three times a day for curing jaundice; stem bark juice mixed with sugar taken orally for curing leucorrhoea.

Bauhinia racemosa Lamk.; F. B. I. ii. 276; W. & A. 295; Bedd. Fl. t. 182.
A small deciduous rather crooked tree with pale yellow flowers. Flowers zygomorphic, in terminal or rarely axillary racemes, pod indehiscent, thick, falcate, flowers in lax racemes, small leaflets cornate for two-thirds of their length, 1-2 in. long, rounded at apex, filaments 5 in. long slender; pod 6-12 in. long, 7-1 in. broad. Bark rough, nearly black; wood brown, hard, a good fuel. Fl. & Frt: February-March, On the way to Lankamalla from Ranibavi, PSKR. 354.

Local name: Are	Family: Caesalpiniaceae
Information source: Yanadi	Plant part used: Stem bark

Use description:

Stem bark decoction given orally for curing diarrhoea; stem bark decoction with jaggery given orally for curing dysentery; wood is used for making furniture.

Boswellia serrata Roxb. ex. Colebr., F. B. I. i. 528, *excl, var. glabra, B. thurifera,* Colebr., W. & A. 174.

- Trees with peppery bark. Leaves deciduous, alternate, imparipinnate, crowded at the ends of the branches; leaflets opposite, flowers hermaphrodite, small, white or pink, in axillary racemes or panicles, leaflets deeply broadly crenate, oblong lanceolate, pubescent; racemes usually panicled; drupes ovoid. Fl. & Frt: March-August, Sabbavi, PSKR. 749.

Local name: Sambrani, Andhuga	Family: Bursaraceae
Information source: Yanadi	Plant part used: Leaf, Stem bark & Stem (Resin)

Use description:

Resin extracted from stem applied for curing skin eruptions.
Stem bark decoction given orally for snakebite.
Tender leaves powdered and mixed with turmeric and made as a paste, applied externally for skin diseases.

Bridelia cinerascens Gegrm.; *B. retusa* (L.) *Spr. var. glauca* Hook. f.
Dioecious trees, branches puberulous, leaves coriaceous, elliptic-oblong, glaucous above, cinerous-glaucous below, crenulate, obtuse-acute, flowers greenish-yellow,

unisexual, axillary in clusters, in the leaf axils, ovary half inferior, bilocular, ovules 4 per locules, styles 2, forked, drupes globose, purple black.. Fl. & Frt: June- October, Sabbavi, PSKR. 678.

Local name: Bontha	Family: Euphorbiaceae
Information source: Yanadi	Plant part used: Bark

Use description:

Bark powder mixed with gingelly oil applied externally for curing rheumatism; wood is used for making furniture.

Bridelia retusa Spreng.; F. B. I. v. 268 inpart; Bedd. Fl. t. 260; Brand. For. Fl. t. 55. *Cluytia spinosa,* Roxb. Cor. Pl. t. 172.

A small- or moderate-sized tree, with conical thorns on the stems and branches when young. Bark grey or brown; wood grey to olive-brown, with a pretty silver grain, useful and durable. **Fl. & Frt:** June-October, Sabbavi, PSKR. 679.

Local name: Korrameddi	Family: Euphorbiaceae
Information source: Yanadi	Plant part used: Bark

Use description:

Bark decoction given orally for curing rheumatism.

Buchanania axillaris Desr. et. Raman., *Mangifera axillaries* Desr, *Buchanani angustifolia* Roxb.

Trees, leaves scarcely coriaceous, narrowly oblong, entire, retuse, flowers green-white, in axillary panicles, sepals and petals 5 each, stamens 10, ovary of 5 carpels, free, ovule 1 per locule, pendulous, style short, lateral, drupes globose, compressed, black, when ripe. Fl. & Frt: June- October, Nithyapoojakona, PSKR. 1026.

Local name: Sarapappu	Family: Anacardiaceae
Information source: Yerukula	Plant part used: Gum

Use description:

1. Gum mixed with hot water given orally for curing diarrhoea.
2. Fruits edible.

Buchanania lanzan Spr.; *B. latifolia* Roxb.

Trees, branchlets clothed with silky hairs, leaves broadly oblong-ovate, villous beneath, entire, retuse or obtuse; flowers cream, in axillary and terminal villous panicles, sepals and petals 5 each, stamens 10, ovary 5-carpellary, free, ovule. Perlocule, pendulous; style lateral, drupes globose, oblique, black. Fl. & Frt: June-October, Nithyapoojakona, PSKR. 1028.

Local name: Chinnasara	Family: Anacardiaceae
Information source: Yanadi	Plant part used: Gum; Leaf

Use description:

1. Leaf paste applied externally for curing glandular swellings.
2. Fruits are edible.

Butea monosperma (Lam.) Taub.; *Erythrina monosperma* Lam.; *Butea Frondosa* Koen. ex. Roxb. Cor. Pl. t. 21; F. B. I. ii. 194; W. & A. 261; Bedd. Fl. t. 176.

Deciduous trees; branchlets densely tomentose, leaves 3-foliolate, leaflets coriaceous, sericeous below, entire, obtuse, retuse, flowers salmon or flame-coloured, clustered at the nodes of rachis, pedicles twice as long as calyx, brown-sheath curved, pods oblong, compressed, base flat wing like, indehiscent, stacked, velvety brown. Fl. & Frt: April-May, Near Lankamalla temple PSKR. 128.

Local name: Modhuga	Family: Fabaceae
Information source: Yanadi	Plant part used: Root

Use description:

Root extract given orally for curing gastric troubles.

Calophyllum inophyllum Linn.; F. B. I. i. 273; W. & A. 103; Wt. III. i. t. 77.

Trees with milky latex, brachlets glabrous; leaf scars persistent, leaves decussate, oblong-obovate, thick-coriaceous, glossy, glabrous, dark green above, entire, rounded, flowers 10-20, in axillary subcorymbose, racemes, sepals 4, petals 4, white, stamens numerous, ovary unilocular; ovule-1, axile or basal. Drupes globose, in massive clusters with thick pericarp; seed solitary. Fl. & Frt: Throughout the year, Sabbavi, PSKR. 1010.

Local name: Ponnachettu	Family: Clusiaceae
Information source: Yerukala	Plant part used: Seed

Use description:

Body Swellings: Seed powder mixed with water, made as a paste and applied externally for curing body swellings.

Capparis apetala Dunn.; *Capparis apetala* Roth.; *Niebuhria linearis* DC. *Niebuhria apetala Dunn.*

Unarmed trees, leaves 3-5 foliolate, flowers dark, greenish purple, in axillary and terminal racemes, sepels 4, petals 0, stamens numerous, berries ovoid, I-seeded. Fl. & Frt: February-August; Near Gopalaswamy temple, PSKR. 1687.

Local name: Pilliaadugu	Family: Capparaceae
Information source: Yanadi	Plant part used: Stem bark

Use description:

Stem bark decoction mixed with black pepper and garlic given orally for curing asthma.

***Capparis divaricata*,** Hook. f. & Thoms.; *Capparis stylosa*, DC.; W. & A. 25, F. B. I. i. 174; Wt. Ic. t. 889 (not of Lam.)

Small tree, leaves elliptic or lanceolate, coriaceous, thorns recurved, flowers yellow, axillary, solitary, sepals and petals 4 each, stamens numerous, berries woody, ribbed, apiculate. Fl. & Frt: March-August; Near Gopalaswamy temple, PSKR. 1689.

Local name: Uppi	Family: Capparaceae
Information source: Yanadi	Plant part used: Leaf

Use description:

1. Leaves are used as antiseptic. Leaf paste applied externally on wounds until cured.
2. Tender fruits are used as vegetables.

Capparis grandis Linn. f; F. B. I. i. 176; W. & A. 27., *C. bisperma*, Roxb.; W. & A. 26.

A small tree with white durable wood. Base unbranched, branchlets valvately, tomentose, throns 0, bark deeply furrowed, leaves ovate, entire, acute, flowers creamy, in terminal corymbs, sepals and petals 4 each, stamens 45-50. Berries globose; 2 seeded, red when ripe. **Fl.** & Frt: April- June; Near Gopalaswamy temple, PSKR. 1690.

Local name: Guti; Ragota	Family: Capparaceae
Information source: Yanadi	Plant part used: Stem bark & Leaf

Use description:

1. Dried leaves and stem bark powdered and applied externally for inflammation.
2. Tuberous roots are cooked and taken with honey in the morning and evening for 15 days for curing the male sterility.

Careya arborea Roxb. Cor. PI. t. 218; F. B. I. ii. 511; W. & A. 334; Wt. III. tt. 99 and 100; Bedd Fl. t. 205.

A large deciduous tree with yellowish-white flowers, large obovate-serrate leaves and large globular fruit. Bark thick, dark grey, branchlets with persistent leaf scars, leaves coriaceous, broadly oblong-obovate, crenulate to denticulate, acuminate, flowers white, in terminal spikes, calyx tube campanulate, lobes 4, revolute along margins, stamens numerous, seeds embedded in fleshy pulp. Fl. & Frt: March-August; Reedybavi, PSKR, 1130.

Local name: Buddadarimi	Family: Baringtoniaceae
Information source: Yanadi	Plant part used: Stem bark

Use description:

1. Stem bark decoction mixed with rice water given orally for curing stomach ache.
2. Wood is used for making furniture.

Carissa carandas Linn., F. B. I. iii. 630; Roxb. Cor. Pl. t. 77; Wt. Ic. t. 426.

A large throny shrub or small tree with white or pale rose-coloured flowers and red berry turning purple or black on ripening and good to eat, especially in tarts and preserves, bark yellowish-brown; wood white with an irregular grey or orange-yellow streaked heartwood, used to make comb and spoons and as fuel. The thorny branches are used for fencing. Fl. & Frt: March-June; Reedybavi, PSKR, 1132.

Local name: Kalekayalu	Family: Apocynaceae
Information source: Sugali	Plant part used: Fruits

Use description:

Fruits are edible. Unripened fruits are used in pickle preparation.

Casearia elliptica Willd. *Casearia tomentosa,* Roxb. F. B. I. ii. 593., Brand. For. Fl. t. 31; Wt. Ic. t. 1849.

A bushy shrub or small deciduous tree in open lands, a much larger tree in forest, the oblong leaves generally tomentose, sometimes nearly glabrous. Bark thick, pale; wood yellowish-white, close-grained. Fl. & Frt: February-August; Nithyapoojakona, PSKR, 1375.

Local name: Girigudu	Family: Flacourtiaceae
Information source: Yanadi	Plant part used: Stem bark

Use description:

1. Stem bark decoction taken orally for curing dropsy.
2. Wood is used for making combs.

Casearia esculenta, Roxb.; F. B. I. ii. 592.

A small tree with many fascicles of flowers from the leaf axils, evergreen. Fl. & Frt: February-August; Nithyapoojakona, PSKR, 1376.

Local name: Kunda jungara	Family: Flacourtiaceae
Information source: Yanadi	Plant part used: Leaf

Use description:

Leaves are used as leafy vegetables.

Cassia fistula Linn. (Figure 19.1a); F. B. I. ii. 261; W. & A. 285. *C. rhombifolia* Roxb.; Wt. Ic. t. 269.

A moderate-sized tree with pale smooth bark when young, darker and rough when old, at once recognized by its long racemes of bright yellow flowers and long cylindrical pod. Leaves paripinnate, leaflets 4-5 pairs, elliptic, ovate or ovate-lanceolate, flowers bright yellow,, in drooping, terminal racemes, calyx lobes 5, reflexed, Petals 5, obovate, stamens 10, all fertile, pods terete, cylindrical, pendulous, brown. Fl. & Frt: March-August; Ranibavi, PSKR, 342 (Fig. 19.1a).

Local name: Rela	Family: Caesalpiniacaeae
Information source: Yanadi	Plant part used: Leaf, Root & Stem bark

Use description:

1. Root and stem bark decoction given orally for curing rheumatism.
2. Leaf juice administrated externally for curing skin diseases.
3. Leaf paste applied externally for curing skin eruptions.

Cassia tora Linn.; F. B. I. ii. 263 *in part*; W. & A. 290 *in part.*

An annual weed, herbs, leaflets 3 pairs, obovate, pubescent below, entire, obtuse, rachis with 2 glands between two lowest pair of leaflets, flowers golden yellow, in terminal corymbose racemes, sepals 5, ovate, pubescent, petals 5, obovate, stamens 10, upper 3 staminodes, ovary subsessile, curved, pubescent, ovules numerous, pods oblong. Fl. & Frt: October-February; Ranibavi, PSKR, 344.

Local name: Pedda Kasindha	Family: Caesalpiniacaeae
Information source: Yanadi	Plant part used: Leaf

Use description:

Leaf juice mixed with lemon juice taken orally for curing stomach ache.

Cassine glauca (Rottb.) Kuntz.; *Mangifera glauca* Rottb.; *Elaeodendron glaucum* (Rottb.) Pers.; *E. roxburghii* Wt. & Arn.; *Celastrus glaucus* Vahl.

Evergreen trees, branchlets glabrous, leaves decusate, ovate-oblong, thin coraceous, flowers green, in axillary corynbose cymes. Fl. & Frt: March-July; Ranibavi, PSKR, 446.

Local name: Neridi	Family: Celastraceae
Information source: Yanadi	Plant part used: Leaf

Use description:

Leaf juice mixed with lemon juice taken orally for curing stomach ache.

Catunaregam spinosa (Thumb.) Triven.; *Gardenia spinosa* Thumb. *Randia dumentorum* (Retz.); Poir, *Gardenia dumentorum* Retz.; *Randia dimentosum* (Retz.) Poir. var. *floribunda* (DC.) Gamb.; *R. floribunda* DC.; *R. longispina* (Roxb.), DC. *Posoqueria longispina* Roxb.; *Randia brandisii* Gamb.

Throny shrubs, leaves clustered, spathulate, entire, obtusely apiculate, flowers axillary, solitary or in fascicles, sepals 5, imbricate, obovate, leafy, petals 5 white (turning yellow), stamens 5, ovary 2to 4 locular; ovules numerous per locules, axile, berries globose with a crown of enlarged calyx lobes.. Fl. & Frt: Throughtout the year; Near Lankamalla temple, PSKR, 1346.

Local name: Manga (or) Chinnamanga	Family: Rubiaceae
Information source: Yanadi	Plant part used: Fruit & Stem bark

Use description:

1. Fruit decoction given orally for curing piles.
2. Stem bark paste (with Sesamum oil) applied externally for curing rheumatism.

Ceiba pentandra (L.) Gaertn. *Bombax pentandrum* L.; *Eleodendron pentandrum* (L). Kurz.; *E. anfractuosum* DC.

Deciduous unarmed trees; prickly when young, leaves 5-9 foliolate, oblanceolate or elliptic, acuminate, entire, 12-15 nerved. Flowers yellow white in axillary clusters or grouped at the ends of leafless branchlets, calyx tube campanulate, lobes 4 or 5, petals5, obovate, stamina tube short, stamens 5, ovary 5-locular, ovules numerous perlocule, axile, fruits capsule indehiscent, green in young becomes brown when mature. Fl. & Frt: December-April, Reedibavi, PSKR, 1146.

Local name: Tellaburuga	Family: Bombacaceae
Information source: Yerukula	Plant part used: Stem bark

Use description:

1. Stem bark decoction given orally for curing diarrhoea.
2. Stem decoction mixed with jaggery, pepper, garlic, kasturi given orally for curing epilepsy.

Chloroxylon swietenia, DC.; F. B. I. i. 569; W. & A. 123; Wt. III. i. t. 56 bis; Bedd. Fl. t. II *Swietenia chloroxylon* Roxb. Cor. Pl. i. t. 64.

A deciduous tree, leaves pinnate, leaflets numerous, very oblique, entire, gland-dotted. Flowers small, in terminal and axillary, pubescent panicles. Calyx short, 5-lobed petals 5, clawed, imbricate. Fruit oblong, 3-celled, loculicidal capsule, seeds about 4 in each cell, imbricate, compressed, winged especially above their margin. Fl. & Frt: September-February, Near Lankamalli, PSKR, 1461.

Local name: Billudu	Family: Flinderciaceae
Information source: Yerukala	Plant part used: Stem bark

Use description:

1. Stem bark is used as antidote. The paste prepared from stem bark applied externally for scorpion sting.
2. Wood is used for making furniture.

Cipadessa baccifera (Roth.) Miq. *C. fruticosa*, Bl.; F. B. I. i. 545. *Mallea rothii*, W. & A. 118.

Small tree, leaves alternate or subopposite, imparipinnate, leaflets opposite, or entire or coarsely serrate; stipules 0, flowers small, hermaphrodite, in axillary peduncled panicles. Fruit drupe, seeds angled; albumen fleshy; embryo curved, the cotyledons oblong; radical superior. Fl. & Frt: Throughout the year, Vaneswaram, PSKR, 1361.

Local name: Ranaberi	Family: Meliaceae
Information source: Yerukala	Plant part used: Leaf

Use description:

Leaves as stimulant. Leaves are dried in shade. The dried leaves are burnt and the smoke was exposed to the body of the children for 5 minutes twice a day for weakness in the children.

Cleistanthus collinus (Roxb.) Benth. ex. Hook. f.; F. B. I. v. 274. *Cluytia collina*, Roxb. Cor. PI. t. 169.

A small deciduous tree. Bark dark brown, rough, wood dark reddish-brown, hard and strong. The outer crust of the capsule is poisonous. Leaves alternate, entire, flowers monoecious, small in axillary clusters or spikes of clusters, capsule large, 75-1 in. in. diam., hard and woody, dark brown, shinning, seed globose, ovary glabrous. Fl. & Frt: February-October, Gollapalli, PSKR, 734.

Local name: Wodisha, Kodisa	Family: Euphorbiaceae
Information source: Yanadi	Plant part used: Leaf

Use description:

1. Leaf roasted in castor oil and used as poultice.
2. Shoot tips with jaggery eaten results in death-poison.
3. Wood is used as timber.

Cochlospermum religiosum (L.) Alston.; *Bombax religiosum* L.; *Cochlospermum gossypium* DC.

Deciduous trees with yellow or red latex, branches crooked, leaves palmately lobed, glabrous above, tomentose below, flowers golden yellow, in terminal panicles, seeds numerous, covered with white cottony hairs. Fl. & Frt: February-March, Rollabodu, PSKR, 1435.

Local name: Kondaburuga	Family: Cochlospermaceae
Information source: Yerukula	Plant part used: Leaf

Use description:

1. Leaf decoction is used for hair wash.
2. Gum powder mixed in milk is used for curing ulcers.

Commiphora caudata, Engl. Protium caudatum, W. & A. 176; F, B. I. i. 530; Bedd. Fl. t. 125.

Deciduous trees with mango like aroma, bark greenish grey, smooth, papery, branchlets glabrous, leaves imparipinnate, flowers polygamous, cream to reddish, axillary paniculate dichasial cymes. A deciduous tree with papery bark and greyish softwood. Fl. & Frt: April-December, Lankamalli, PSKR, 1220.

Local name: Kondamamidi	Family: Burseraceae
Information source: Yanadi	Plant part used: Stem bark

Use description:

Stem bark mixed with *Schleichera oleosa* boiled in water and the decoction given orally for curing dropsy.

Cordia dichotoma Frost. f.; *C. obliqua* Willd.; *C. myxa* auct. Non L.; *C. latifolia.* Roxb.

Deciduous trees, branchlets pendulous, bark thick, grey, wood moderately hard, leaves elliptic-oblong, orbicular, entire, or dentate or sinuate, flowers white, fragrant in axillary and terminal cymes. Fl. & Frt: February-April; September, November, Vaneswaram, PSKR, 330.

Local name: Nakkera	Family: Cordiaceae
Information source: Yanadi	Plant part used: Stem Bark& Fruit

Use description:

1. Fresh stem bark boiled in water, made as decoction. The decoction is taken in orally twice a day for 7 days for curing excessive menstruation.
2. Stem bark paste is used for bone fractures.
3. Fruits are edible.
4. Wood is used for making cots.

Cordia wallichii G. Don; Bedd. Fl. t. 245, C. obliqua, Willd., var. Wallichii, C. B. Clarke; F. B. I. iv. 137. C. oblique, Wt. Ic. t. 1378.

A moderate size tree with a useful wood. Leaves softly fulvous or white tomentose, beneath, ovate-orbicular, truncate or cordate at base, obtuse or slightly acute at apex, flowers rather large in lax cymes, usually long-peduncled. Fl. & Frt: February-April, Rollabodu, PSKR, 330.

Local name: Peddarikhera	Family: Cordiaceae
Information source: Yanadi	Plant part used: Fruit

Use description:

1. Unripe fruits are boiled in water to get decoction. A cup of decoction is taken orally thrice a day for fever until cured.
2. Wood is used for making cots and huts.

Crataeva magna (Lour.) DC., *Capparis magna* Lour.; *Crataeva nurvala* Buch. Ham., *C. religiosa* Forst. f. var. *nurvala* (Buch-Ham) Hook. f. & Thoms. *C. religiosa* sensu Gamb.

Tree, leaves 3-foliolate; leaflets ovate-lanceolate, entire, caudate, flowers white, in terminal racemes, berries irregularly rugose, yellowish-grey with powdery tomentum. Fl. & Frt: April-August, Madhuru, PSKR, 1425.

Local name: Magalingamu	Family: Capparaceae
Information source: Yanadi	Plant part used: Stem bark

Use description:

1. Stem bark as diuretic. Stem bark is broken into pieces, boiled in water to get decoction. A cup of decoction taken orally twice a day for three days for curing urinary complications.
2. Wood is used for making furniture.

Croton scabiosus, Bedd. Fl. t. 283; F. B. I. v. 386.

A small often gregarious tree, the leaf-glands sessile. Bark dark brown, rough; wood yellowish white, hard and close-grained. Fl. & Frt: Most part of the year, Nithyapoojakona, PSKR, 152.

Local name: Yerri chilla	Family: Euphorbiaceae
Information source: Sugali	Plant part used: Stem bark

Use description:

Stem bark is dried in the shade, powdered, boiled in water to get decoction. The cup of decoction taken orally for curing leucorrhoea.

Dalbergia latifolia Roxb. Cor. Pl. t. 113; Fl. Br. Ind. ii. 231; W. & A. 264; Prain Ann Calc. x. 80. t. 62.

Large trees, leaves odd pinnate, leaflets alternate, obovate-orbicular, coriaceous, entire, retuse, flowers white, on old wood or in axillary panicles. Fl. & Frt: Throughout the year, Reddibavi, PSKR, 324.

Local name: Chittegi	Family: Fabaceae
Information source: Yanadi	Plant part used: Stem bark

Use description:

Stem bark powder is applied on affected parts for curing leucoderma.

Dalbergia paniculata, Roxb. Cor. PI. t. 114; F. B. I. ii. 236; W. & A. 265; Prain Ann. Calc. x. 87. t. 68.

Large deciduous trees, leaves imparipinnate, leaflets 9-13 alternate, oblong, entire, obtuse, turn black when drying, flowers white, in axillary or terminal panicles, sepals rusty-tomentose, seeds 1 or 2. Fl. & Frt: April-January, Gollapalli, PSKR, 326.

Local name: Pachari	Family: Fabaceae
Information source: Yanadi	Plant part used: Leaf

Use description:

1. Tender leaves are warmed, placed on swelling and bandaged twice a day till cured for rheumatic swellings.
2. Leaves are warmed and placed on foot cracks till cured.

Dalbergia sisso, Roxb.

Semi-evergreen tree, leaves odd pinnate, leaflets 3-5, alternate, broadly elliptic or ovate, pubescent when young, entire, cuspidate-acuminate. Flowers yellowish white, in axillary panicles. Pods strap shaped. Fl. & Frt: April-January, Rollabodu, PSKR, 328.

Local name: Jittegi	Family: Fabaceae
Information source: Yanadi	Plant part used: Leaf

Use description:

Leaf juice used as eye drops for curing eye irritation.

Deccania pubescens (Roth.) Tirven; *Randia candolleana* Wt. & Arm. *R. deccanensis* Bedd.; *Aldia candolleana* (Wt. & A.) Swamin.

Small tree, branches with elevated leaves scars, leaves orbicular, entire, obtuse, flowers white, in short axillary cymes, berries globose, clustered, leathery. Fl. & Frt: May-October, Sabbavi, PSKR, 1202.

Local name: Yerrabikki	Family: Rubiaceae
Information source: Yanadi	Plant part used: Leaf.

Use description:

Leaf paste applied externally for curing sores.

Dichrostachys cinerea, W. & A. 271; F. B. I. ii. 288; Roxb. Cor. Pl. t. 174; Bedd. Fl. t. 185.

A thorny shrub or small tree with throny branches pretty tasselled flowers, twisted pods and very small pubescent leaflets. Bark grey or light brown, wood red streaked with black, very hard and tough and a good fuel. Fl. & Frt: Throughout the year, On the way to Gopalswamy temple, PSKR, 1259.

Local name: Velthuru	Family: Fabaceae
Information source: Yanadi	Plant part used: Fibres & leaf

Use description:

1. Fibres from stem bark used as antipyretic.
2. Tender leaflets are ground with water, made as a paste and applied externally for curing scabies.

Diospyros chloroxylon, Roxb. Cor, Pl. t. 49; F. B. I. iii. 560, *D. capitulata*, Wt. Ic. tt. 1224, 1588 bis.

A large much branched tree, often throny, in dry localities with small leaves, in better soils, small tree with larger leaves, flowers especially very small, fruit globose, 2-3 in. diam. edible. Bark dark brown, peeling off in small scales; wood yellowish, in grey, moderately hard a good fuel. Fl. & Frt: March-June, Madhuru, PSKR, 152.

Local name: Illinda	Family: Ebenaceae
Information source: Yanadi	Plant part used: Fruit

Use description:

1. Fruit edible.
2. Wood is used for making furniture.

Diospyros ferrea (willd.) Bakh. *Pisonia buxifolia* Rottb.; *Maba buxifolia* (Rotth.) Juss.; *Ehretia ferra* Willd.

Evergreen bushy shrubs or small trees, branchlets glabrous, bark grey to black, Leaves obovate, spathulate or elliptic, entire, obtuse, base acute to alternate, glabrous, Flowers white or yellow, berries globose, orange when ripe.. Fl. & Frt: April-August, Rollabodu, PSKR, 2571.

Local name: Uti, Utipallu	Family: Ebenaceae
Information source: Yerukula	Plant part used: Fruit

Use description:

1. Fruits are edible.
2. Wood is used in hut making and also for making agricultural implements.

***Diospyros melanoxylon*,** Roxb. Cor. Pl. t. 46; F. B. I. iii. 564, *D. dubia*, Wall.; Wt. Ic. t 223. *D. exsculpta*, Bedd. Fl. t. 66, *D. Wightiana*, Wall; Bedd. Fl. t. 67, *D. trupa*, Buch-Ham.; F. B. I. iii. 563.

A moderate-sized deciduous tree, the young leaves and branchlets and the inflorescence tawny–tomentose. Bark greyish-black, cleft in rectangular plates showing the black inner bark in the clefta; wood hard reddish-brown with an irregular black heartwood. This ebony is however not much used, as pieces of it are not obtainable of any large size. The fruit is edible. Fl. & Frt: December-June, Rollabodu, PSKR, 1277.

Local name: Tunki	Family: Ebenaceae
Information source: Yanadi	Plant part used: Flower & Leaf

Use description:

1. Flower powdered mixed with milk and used for curing urinary disorders.
2. Dried leaf is used for making beedi.

Dolichandrone atrovirens, Sprague in Kew Bull, 1919, 304. *D. cripsa*, Seem., F. B. I. iv. 379.

A moderate-sized tree with pretty white flowers, bark brown, rough, wood yellowish-brown, useful. Young branches velvety pubescent, leaves imparipinnate, rachis-tomentose, leaflets tomentose, 3-7 elliptic, ovate, entire, acuminate, flowers white, in terminal corymbs. Fl. & Frt: July-September, on the way to Gopalaswamy temple, PSKR, 1123.

Local name: Neeruddi	Family: Bignoniaceae
Information source: Yanadi	Plant part used: Fruit

Use description:

1. The decoction prepared from the fruits taken orally thrice a day for 7 days to improve disease resistance in the children.
2. Wood is used for making furniture.

***Ehretia pubescens*,** Benth, in Royle III. 306, *Ehretia laevis*, Roxb., var. *pubescens,* C. B. Clarke; F. B. I. iv. 142.

Trees or shrubs, leaves alternate, flowers small, usually white, in terminal or axillary corymbose or panicled cymes, rarely solitary, leaves ovate, scabrous pubescent, as are the branchets and inflorescence, obtuse at apex with small apiculus, narrowed or rounded at base, up to 2 in. long, 1 in. broad. Fl. & Frt: June-March, Near Balakunta, PSKR, 208.

Local name:	Family: Boraginaceae
Information source: Yanadi	Plant part used: Leaf

Use description:

1. Leaves are boiled in water, made as a decoction and the decoction applied on the muscular pains.
2. The tender leaf paste applied to treat eczema.

Euphorbia nivulia Buch-Ham.

Armed, deciduous, succulent trees, trunk straight, branches whorled, branchlets cylindrical with sharp spikes, leaves obovate to oblanceolate, entire, obtuse, glabrous, cyathia paired in subterminal lax twice forked cymes, involucral glands 5. Fl. & Frt: January-May, Jyothi, PSKR, 1604.

Local name: Akujamudu	Family: Euphorbiaceae
Information source: Yanadi	Plant part used: Leaf.

Use description:

1. Leaf sap mixed with neem oil applied externally for curing rheumatic pains twice a day until cured.
2. One or two spoon of leaf sap taken orally at bed time for constipation.

Flacourtia indica (Burm. f.) Merr.; *Gmelina indica* Burm. f. *Flacoutia sepiaria* Roxb. Cor. Pl. i. t. 69; W. & A. 29.

Armed shrubs, branches with sharp, elongate thorns, leaves ovate, glabrous, entire. Flowers unisexual, cream, in axillary racemes on thorns, sepals 4-5, petals 0, stamens numerous. Ovary 3-locular; ovule 2 per locule, axile; styles 3, drupes fleshy, globose, red when ripe. Fl. & Frt: October-June, Lothuvanka, PSKR, 1086.

Local name: Pulicarapandlu	Family: Flacourtiaceae
Information source: Yanadi	Plant part used: Stem bark & Root

Use description:

1. Stem bark extract given orally for dog bite.
2. Root powder mixed with water, made as paste and applied externally for curing eczema.
3. Fruits are edible.

Ficus benghalensis, Linn.; F. B. I. v. 499; King Ann, Calc. i. t. 13 and 81 c. *Urostigma benghalense* Gasp, Wt. Ic. t. 1989.

Monoecious large, much branched evergreen tree throwing out numerous large aerial roots from the main trunk and large branches, which descend to the soil and form supports, and are then capable of separate existence when served from the parent tree. Fl. & Frt: November-April, Lankamalli, PSKR, 1908.

Local name: Marri.	Family: Moraceae.
Information source: Yanadi	Plant part used: Leaf.

Use description:

1. Leaves are warmed and bandaged tightly on the wounds and removed after curing.
2. Latex is applied externally for curing fissures in the feet.
3. Latex is used for curing ring worm.

Ficus hispida, Linn. f. F. B. I. v. 522; King Ann. Calc. i. t. 154 and 155., *Ficus oppositifolia,* Roxb. Cor, Pl. ii. t. 124; Wt. Ic. t. 638. *Ficus doemonum*, Koen., Wt. Ic. t. 641.

Dioecious large shrubs or small tree, without aerial roots, trunk straight with root suckers, branchlets hispid, internodes hollow, leaves opposite, membranous, ovate-oblong or obovate, toothed, apiculate, base rounded or subcordate, scabrid above, hispid below, 3-nerved, lateral nerves 6-12 pairs, receptacle unisexual, green at first, turn yellow on ripening, ovoid, hispid, fascicled on old wood; bracts triangular, achenes globose, keeled. Fl. & Frt: Throughout the year, Balakunta, PSKR, 1914.

Local name: Bemmedi	Family: Moraceae.
Information source: Yanadi	Plant part used: Root bark & Leaf.

Use description:

Figs roasted in urine, as poultice to serotum for curing hydrocele.

Ficus microcarpa L. f.; *Ficus retusa*, Linn.; F. B. I. v. 511; King Ann. Calc. i. t. 61 and 84p.

Moenaceous tall tree with a few prop roots, branchlets glabrous, leaves coriaceous, elliptic-obovate or rhomboid, entire, acute, retuse or rounded, base narrow or cuneate, glabrous, sometimes epiphyte, bark brown, fairly smooth, wood light reddish-grey, moderately hard, one of the best of the fig woods. Fl. & Frt: February-September, Panchalingalagudi, PSKR, 1916.

Local name: Juvvi, Errajuvvi.	Family: Moraceae.
Information source: Yanadi	Plant part used: Stem bark.

Use description:

1. The decoction prepared from the stem bark taken orally thrice a day for 3 days for curing scabies.
2. Wood is used for making huts.

Ficus racemosa L. (Figure 4); *Ficus glomerata*, Roxb. Cor. Pl. ii. t. 123; F. B. I. v. 535; King Ann. Calc. i. t. 218A; Wt. Ic. t. 667; Brand. For. Fl. t. 49.

A large deciduous tree with few short aerial roots or no aerial roots, branchlets minutely pubescent, reddish brown, leaves coriaceous, elliptic, oblong or lanceolate, entire, acute or obtuse, base oblique, 3-nerved, bark grey to reddish brown, wood greyish brown, soft. Fl. & Frt: Most part of the year, Gopalaswamy temple, PSKR, 1918 (Fig. 19.1b).

Local name: Atti, Medi	Family: Moraceae.
Information source: Yanadi	Plant part used: Bark & Fruit.

Use description:

1. The decoction prepared from the bark taken in the early morning for curing diabetes.
2. Fruit powder mixed with honey taken for curing diarrhoea.
3. The paste obtained from the fruits mixed with curd taken orally for curing dysentery.
4. Fruits are edible.

Ficus religiosa, Linn. F. B. I. v. 513; King Ann. Cale. i. t. 67Aand 84u; Bedd. Fl. t. 314, *Urostigma religiosum*, Gasp. Wt. Ic. t. 1967.

Monoecious large trees, epiphytic, when young, aerial roots absent, branchlets yellowish-grey, glabrous, leaves coriaceous, broadly ovate, wavy, long acuminate, caudate, base truncate or cordate, shinning, prominently veined, receptical unisexual, sessile, paired, smooth, depressed globose, dark purple when ripe. Fl. & Frt: March-August, Lankamallai, PSKR, 1920.

Local name: Ravi.	Family: Moraceae.
Information source: Yanadi	Plant part used: Leaf.

Use description:

1. Fresh tender leaves with honey consumed daily for easy fertilization in women.
2. Fruits are edible.

Ficus tinctoria Forst. f. subsp.

Dioeceous epiphytic, climbing, large shrubs or small trees, leaves tightly coriaceous, obliquely elliptic-ovate, entire, acute, base oblique, cuneate, scabrid on both the sides, Fl. & Frt: February-May, Lankamallai, PSKR, 1922.

Local name: Kondajivvi	Family: Moraceae.
Information source: Yanadi	Plant part used: Leaf.

Use description:

Leaf paste used as dressing as broken bones.

Gardenia gummifera Linn. f. F. B. I. iii. 116; W. & A. 395; Wt. Ic. t. 576.

Small deciduous trees, leaves obovate-oblong, entire, obtuse or bluntly acute, flowers white, turning yellow, 1-3 in axils, calyx lobes 5, triangular, corolla lobes 5, obovate, stamens 5, inclided, ovary 5 angular, unilocular, ovules numerous on 2-4 parietal placentae, berries oblong or ellipsoid with longitudinal elevated lines; seeds rugose. Fl. & Frt: January-June, Sabhavi, PSKR, 2120.

Local name: Bikki, Manchi bikki.	Family: Rubiaceae.
Information source: Yanadi	Plant part used: Gum & Fruit.

Use description:

1. Gum with water given orally for constipation.
2. Fruits are edible.

Gardenia latifolia, Ait., F. B. I. iii. 116; Roxb. Cor. Pl. t. 134; W. & A. 395; Wt. Ic. t. 759. *G. enneandra*, Koen.; F. B. I. iii. 119; W. & A. 395; Wt. Ic. t. 574.

Deciduous trees, leaves broadly elliptic, entire, obtuse to subacute, calyx campanulate, lobes 7(9), corolla lobes 7(9), imbricate, oblong-obovate, stamens 7(9), ovary unilocular, ovules numerous on 3 parietal placentae, berries globose, with a crown of calyx; seeds rugose. Fl. & Frt: February-August, Nithyapoojakona, PSKR, 2210.

Local name: Pedda Bikki.	Family: Rubiaceae.
Information source: Yanadi	Plant part used: Fruit.

Use description:

Fruits are edible.

Gardenia resinifera Roth. Nov. PI. SP. 150, 1821., *G. lucida* Roxb. Fl. Ind 2553.1824.

Small tree, leaves elliptic-ovate, flowers pale yellow, solitary, axillary, berries ellipsoid. Fl. & Frt: May–August, Nithyapoojakona, PSKR, 2102.

Local name:	Family: Rubiaceae
Information source: Yanadi	Plant part used: Gum

Use description:

Gum mixed with water and given orally for curing worms.

Givotia moluccana (L.) Sreem.; *Croton moluccanum* L.; *Givotia rottleriformis*, Griff.; F. B. I. v. 395; Wt. Ic. t. 1889; Bedd. Fl. t. 285.

A moderate-sized tree conspicuous for its large leaves densely white tomentose beneath reaching a diameter of up to 10 in. The flower panicles are usually more fulvous tomentose, with distinct clusters, the drupes 1 in long, bark brown, smooth, peeling off in circular scales; wood white, very light and soft used for carved images, toys, lacquered articles and catamarans. The seeds give an oil. Fl. & Frt: April-August, Vaneswaram, PSKR, 2110.

Local name: Tella Poliki	Family: Euphorbiaceae
Information source: Yanadi	Plant part used: Bark

Use description:

Bark decoction given orally for curing stomach ache.

Givotia rottleriformis Griff. F. B. I. v. 395; Wt. Ic. t. 1889; Bedd. Fl. t. 285.

A moderate-sized tree conspicuous for its large leaves, densely white tomentose, the flowers–panicles are usually more fulvous-tomentose, with distinct clusters. Fl. & Frt: April-August, Madhru, PSKR, 1120.

Local name: Tella Puliki	Family: Euphorbiaceae
Information source: Yanadi	Plant part used: Seed

Use description:

Seed powder applied externally for hairs curing dandruff.

Gmelina arborea, Roxb. Cor. Pl. t. 246; F. B. I. iv. 581; Wt. Ic. t. 1470; Bedd. Fl. t. 253.

A large- or moderate-sized deciduous tree with handsome panicles of brownish-yellow flowers and a large yellow fleshy drupe. Bark smooth, white or whitish grey; wood greyish-white, glossy, smooth and even grained, useful for planking. Furniture and many other purposes and not liable to wrap or contract. Fl. & Frt: February-June, Rajulacheruvu, PSKR, 1065.

Local name: Gummer tek	Family: Verbanaeae
Information source: Yanadi	Plant part used: Root & Leaf.

Use description:

1. Root decoction mixed in honey and taken orally for lactation in nourishing mothers as galactogogue.
2. Leaf paste applied on the sting part for scorpion bite.

Grewia flavescens, Juss. *G. carpinifolia*, Mast. in. F. B. I. i. 387, *G. pilosa*, W. & A. 78.

Small tree, leaves oblong or obolanceolate, 3-nerved, stipules filiform, flowers yellow, in axillary, 3-flowered cymes, sepals-5, petals-5, stamens numerous, ovary globose, bilocular, ovule 1 perlocule, axile, drugs obscurely 2-lobed, wrinkled. Fl. & Frt: June-May, Gopalaswamy temple, PSKR, 1528.

Local name: Jana	Family: Tiliaceae
Information source: Yerukula	Plant part used: Root

Use description:

Root paste applied externally for curing wounds and boils.

Grewia tiliaefolia Vahl; F. B. I. i. 386 (Partly); W. & A. 80, Bedd. Fl. t. 108.

A medium-sized tree, usually with large leaves, wood brown, strong and flexible, used for shafts of carts, oars and other implements. Leaves ovate or elliptic, ovate, crenate, acute or rounded, flowers yellow, in 3-4 axillary cymes, sepals and petals 5 each, stamens numerous, ovary 2-locular, ovules 2 per locule, axile, stigma lobed, Drupes globose, black. Fl. & Frt: June-September, Madhuru, PSKR, 1532.

Local name: Pedda Jana	Family: Tiliaceae
Information source: Yanadi	Plant part used: Stem bark

Use description:

Stem bark decoction given orally for treating dysentery.

Gyrocarpus americanus, Jacq. G. jaequine, Roxb. Cor. Pl. t. 1; Bedd. Fl. t. 196; F. B. I. ii. 461.

A large deciduous evergreen tree with large soft leaves often deeply lobed. Bark thin, greyish-white with a silvery lustre; wood soft, grey, in considerable demand for making catamarans, also boxes, trays and toys. Young branches tomentose, leaves alternate, clustered at the ends of the branches, broadly ovate, entire, or lobed acute, glabrous above, flowers unisexual, small, white or pale yellow, in clustered terminal cymes, males numerous, females few, seed solitary. Fl. & Frt: December-March, Sabbavi, PSKR, 1325.

Local name: Kummarapoliki (helicopter tree)	Family: Hernandiaceae
Information source: Sugali	Plant part used: Root

Use description:

Roots are powdered, mixed with water, made as a paste, applied externally for curing wounds.

Haldinia cordifolia (Roxb.) Ridsd.; *Nauclea cordifolia* Roxb.; *Adina Cordifolia* (Roxb.) Hook. f. ex. Brand.

Deciduous trees, trunk base often buttressed, wood yellow, leaves orbicular, – ovate, densely pubescent, entire, acuminate, flowers creamish-yellow, in axillary globose heads, calyx copular, lobes 5, valvate, corolla funnel shaped, lobes 5, recurved, stamens 5, anthers oblong, ovary bilocular; ovules 3-5 per locule, apical, capsules cuneate; seeds elongate, trailed above. Fl. & Frt: Throughout the year, Near Lankamalla temple, PSKR, 1253.

Local name: Rudragadapa	Family: Rubiaceae
Information source: Yanadi	Plant part used: Stem bark

Use description:

1. Stem bark decoction given orally for curing fever.
2. The decoction prepared from the stem bark taken orally thrice a day for 3 days curing post-natal complaints.
3. Wood is used for making huts, cots, and furniture.

Hardwickia binata Roxb. Cor. Pl. iii. 6, t. 209; F. B. I. ii. 270; W. & A. 284; Bedd. Fl. t. 26.

A large deciduous tree with small yellowish, flowers and *Bauhinia* like leaflets, bark dark grey, rough, wood extremely hard and heavy, dark red, streaked with black or purple, used for agricultural and many other purposes. The leaves are cut for manure and the branchlets for the fibre of their bark. Fl. & Frt: September-July, PSKR, 503.

Local name: Yepi	Family: Caesalpiniaceae
Information source: Yanadi	Plant part used: Stem & Leaf

Use description:

1. Fibres obtained from the stem used for preparing robes.
2. Leaves are used as green manure.
3. Wood is used for making agricultural implements and furniture.

Hibiscus platanifolius, Sweet. *H. collinus*, Roxb.; F. B. I. i. 338, *H. eriocarpus*, DC.; W. & A. 51.

A large shrub, branchlets stellate-pubescent, leaves entire, 3-lobed, cordate, dentate, acuminate, 5-nerved at base pubescent, flowers white, fading to purple by noon, axillary, solitary, calyx tubular, lobes 5. Petals 5. Staminal column antheriferous throughout. Ovary 5 locular, capsules brown with woolly hairs. Fl. & Frt: September-May, PSKR, Reddibavi, 635.

Local name: Kondagogu	Family: Malvaceae
Information source: Yanadi	Plant part used: Gum

Use description:

Gum used as refrigerant. The gum obtained from the stem is mixed with sugar and taken orally twice a day for 3 days.

Holarrhena pubescens (Buch-Ham). Wall. ex. G. Don. *Echites pubescens* Buch-Ham, *Nerium antidysentericum* L.; *Echites antidysentericum* Roxb. ex. Flem.; *Holarrhena antidysentericum* Wall. ex. A. DC.; *Chonemorpha antidysenterica* (Roth) Don. *Holarrhena codaga* Don, *H, malaccensis* Wt.

Deciduous, small trees, leaves chartaceous, ovate or elliptic-oblong, entire, acuminate, base rounded, glabrous above, pubescent below, flowers white to yellowish, in terminal and axillary paniculate or corymbose cymes, seeds oblong. Fl. & Frt: April-August, Lankamallai, PSKR, 135.

Local name: Kodaga, Kolamukipala	Family: Apocynaceae
Information source: Yanadi	Plant part used: Leaf & Stem bark

Use description:

1. Leaf paste (with water) applied externally for curing leprocy and leucoderma.
2. Stem bark decoction given orally for curing snakebite.

Holarrhena antidysenterica, wall.; F. B. I. iii. 644; Brand. For. Fl. t. 40. *H. codaga,* G. Don; Wt. Ic. t. 1297. *Chonemorpha antidysenterica*, G. Don; Wt. Ic. t. 439.

A small tree with white flowers, very long narrow pendulous follicles and large membranous, glabrous or pubescent, ovate to ovate-oblong or elliptic leaves. Bark thick, brown rough; wood white, soft, even grained, used for carving and turning in N. India, rarely in the South. Fl. & Frt: April-August, Panchalingalagudi, PSKR, 1035.

Local name: Kolamukki	Family: Apocynaceae
Information source: Yerukala	Plant part used: Stem Bark & Leaf

Use description:

1. Stem bark mixed with bael fruit, boiled in water, made as decoction and one to two teaspoons are taken orally three to four times for curing dysentery.
2. Leaf juice mixed with pinch of line applied externally on joint pains.

Holostemma ada-kodein Schult, *Sarchostemma annulare* Roth; *Holostemma annulare* (Roxb.) Schumn.; *Asclepias annularia* Roxb.; *Holostemma rheedi* Wall.

Perennial large, glabrous climbers, leaves ovate to oblong, narrowly cordiform, entire, acute, base cordate to hastate, chartaceous, puberulous beneath penninerved, lower 2 pairs arising from base, flowers purple or pinkish-red, in few flowered axillary cymes, calyx lobes 5, equal, ovate, concave, bent in the middle at right angle, twisted to right, connate below, follicles usually appressed, boat shaped. Fl. & Frt: July-Januarys, Rajulacheruvu, PSKR, 1235.

Local name: Kodisipala	Family: Apocynaceae
Information source: Yanadi	Plant part used: Stem bark & Leaf

Use description:

1. Stem bark mixed with bael fruit, made as decoction and one teaspoonful is administrated three to four times a day till cured.
2. The decoction prepared from the leaves mixed with dried ginger powder taken orally for curing piles.

Ixora arborea Roxb, *Ixora pavetta* Andr, *I. parviflora* Vahl.

Small trees, leaves elliptic-oblong to oblanceolate, entire, obtuse-acute, flowers white, calyx truncate, 4-toothed, corolla tubular, lobes-4, obtuse, stamens-4, anthers basally tailed, drupes didymous. Fl. & Frt: Throughout the year, Reddibavi, PSKR, 1250.

Local name: Koya	Family: Convolulaceae
Information source: Yanadi	Plant part used: Root bark

Use description:

Root bark as antidote. Root bark is cut into pieces, ground well with water to get paste, applied at the site of the bite for centipede bite.

Ixora pavetta Andr. *I. parviflora* Vahl. *I. arborea* Roxb.

Small trees or evergreen shrubs, leaves elliptic-oblong to oblanceolate, entire, obtuse-acute, flowers white, calyx truncate, 4-toothed, corolla tubular, lobes-4, obtuse, stamens-4, anthers basally tailed, drupes didymous. Fl. & Frt: Throughout the year, Reddibavi, PSKR, 1252.

Local name: Korivi Chettu	Family: Rubiaceae
Information source: Yanadi	Plant part used: Leaf

Use description:

Leaf decoction taken orally twice a day for constipation.

Lannia coromandalica (Houtt.) Mann. *Dialium coromandelicum* Houtt.; *Odina wodier* Roxb.

Deciduous trees, branchlets rustly-tomentose, leaflets 4-pairs, ovate-lanceolate, shiny, entire, caudate, flowers uni-bisexual, pale yellow, in drooping racemes, sepals-4, triangular, petals-4, lanceolate, ovary unilocular, stamens and drupes ovoid or obovoid, subfleshy, compressed. Fl. & Frt: March-July, Madhuru, PSKR, 1423.

Local name: Gumphenna	Family: Anacardiaceae
Information source: Yanadi	Plant part used: Stem bark & Leaf

Use description:

1. The decoction prepared from the stembark and leaves taken orally for curing in the morning on an empty stomach for week.
2. Wood is used for making cots.

Limonia acidissima W. & A. 92.; F. B. I. i. 507, *Limonia crenulata*, Roxb.

A small throny tree, in favourable places reaching a good size. Wood yellowish white, very hard like box wood. Deciduous trees, bark deeply fissured, leaves 1-3 cluster, leaflets1-4 pairs, opposite, oblong, petiole and rachis flattened, narrowly winged. Flowers polygamous, greenish, in lateral or terminal panicles. Seeds numerous. Fl. & Frt: March-December; Gollapalli, PSKR, 1883.

Local name: Velaga	Family: Rutaceae
Information source: Yanadi	Plant part used: Fruit, Gum & Leaf

Use description:

1. Fruits are edible.
2. When stem cut or broken, a transparent gummy substances exudes from the stem. These gum is reduced to powder and mixed with honey. This mixture were taken orally twice a day for three days for curing diarrhoea and dysentery.
3. Tender leaves are crushed to sap and one teaspoon of sap is taken orally in the morning for three days for curing stomach ache in the children.

Listea glutinosa (Lour.) D. C. Robbinson in Philipp.; *Sebifera glutinosa* Lour.; *Listea chinensis* Lam., L. *sebhifera* Pers.

Evergreen deciduous trees, branchlets fulvus-tomentose, leaves coriaceous, crowded towards the ends of branches, elliptic, ovate to obovate, entire, acute, glabrous above, tomentose below, Seed globose. Fl. & Frt: January-December; Lankamallai, PSKR, 1389.

Local name: Pulusumamidi	Family: Lauraceae
Information source: Yanadi	Plant part used: Root & Stem bark

Use description:

Root and stem bark fermented in curd for two days and given orally for curing dysentery.

Listea deccanensis, Gamble n. comb. *L. tomentosa*, Heyne; F. B. I. v. 157 *Tetranthera tomentosa*, Roxb.; Wt. Ic. t. 1834.

A moderate-sized tree with soft grey or tawny branchelets and leaves. The wood is said to be yellowish. Fl. & Frt: January-December; Lankamallai, PSKR, 1390.

Local name: Naramamidi	Family: Lauraceae
Information source: Yanadi	Plant part used: Stem bark

Use description:

Stem bark was selling in Girijan Corporation.

Madhuca indica J. Gmelin., *Madhuca longifolia* (Koen.) Macbr. Var *longifolia* (Roxb.) A. Cheval.; *Bassia latifolia* Roxb.

Deciduous trees, branchlets glabrous, spreading, leaves clustered near the ends of branchlets, thick, elliptic, entire, acuminate, base rounded or acute, flowers pale yellow, scented in dense fascicles on drooping densely tomentose pedicels. Fl. & Frt: February-May; Sabhavi, PSKR, 1923.

Local name: Ippa	Family: Sapotaceae
Information source: Yanadi	Plant part used: Seed & Stem bark

Use description:

1. Germinated seeds are washed, crushed on a stone with few drops of water squeezed through a thin cloth to get the juice. Two or three drops of the juice are put into each nostril once in the morning for headache.
2. Fruit edible.
3. Flowers and stem bark is used for making NATU SARA.

Madhuca longifolia (Koen.) Macbr. Var *longifolia; Bassia longifolia J. Koein. ex. Mant.*

Large deciduous tree, with heavy horizontal branches that go to form a crown of equal spread, branchlets brownish-yellow, glabrous, leaves chartaceous, linear-elliptic, entire, acute, base narrowed, glabrous, green above, flowers creamish-white, in long pedicellate umbels. Fl. & Frt: March-June; Panchalingalagudi, PSKR, 1925.

Local name: Natu Ippa	Family: Sapotaceae
Information source: Yanadi	Plant part used: Stem bark

Use description:

1. The bark is cut into pieces and mixed with juice of sugarcane boiled and prepared as natusara.
2. The bark cut into pieces and dipped in the water for ten minutes. The infusion is taken thrice a day for 3 days in the morning for headache.

Mallotus philippensis, Muell. Arg.; F. B. I. v. 442; Bedd. Fl. t. 289., *Rottlera tinctoria*, Roxb. Cor. Pl. t. 168.

A small tree, much branching lowdown, bark grey, wood grey to light red, smooth and close-grained, a good fuel, dioecious evergreen trees, branchlets straight, tender parts rusty pubescent, leaves ovate or ovate-lanceolate, entire, acuminate, scarlet-glandular below. The chief product is the bright orange dye. Fl. & Frt: October-March; Nithyapoojakona temple, PSKR, 1632.

Local name: Sindhura, Kumkuma	Family: Euphorbiaceae
Information source: Yanadi	Plant part used: Stem Bark

Use description:

Stem bark, fruit, and seed made into paste (with water) and applied externally for curing skin diseases.

Mangifera indica Linn.; F. B. I. ii. 13.; W. & A. 170; Bedd. Fl. t. 162.

A large spreading evergreen tree reaching 50 ft. in height. The oblong-lanceolate, shinning leaves crowded at the ends of the branches, the flowers in dense terminal panicles, cultivated for its edible, very important fruit, which is eaten fresh or made into preserves or pickles. Bark rough, dark grey, wood grey, often streaked, moderately hard, used for planking, packing cases, boats and other purposes. Fl. & Frt: February-July; Lankamallai, PSKR, 1326.

Local name: Mamidi	Family: Anacardiaceae
Information source: Yanadi	Plant part used: Flower

Use description

1. The decoction of dried flowers (hot water) given orally for curing diarrhoea and chronic.
2. Fruits are edible.
3. Wood is used for making furniture, etc.

Maninkara hexandra (Roxb.) Dubard; *Mimusops hexandra* Roxb.; *M. indica*A. DC.

Evergreen, small trees, latex milky, leaves clustered at the ends of branches, coriaceous, shining, broadly ovate or oblong, entire, emarginate, base obtuse or acute, glabrous, dark-green above, flowers white, axillary, solitary or in clusters, berries ellipsoid, reddish-yellow when ripe; seed one, ovoid. Fl. & Frt: April-July; Gopalaswamy temple, PSKR, 1650.

Local name: Palachettu	Family: Sapotaceae
Information source: Yanadi	Plant part used: Fruit

Use description:

1. Fruits are edible.
2. Wood is used for making furniture and agricultural implements.

Margaritaria indica (Dalz.) Airy-Shaw; *Prosorus indicus* Dalz.; *Phyllanthus indicus* Muell. Arg.

Dioecious evergreen trees, branchlets glabrous, lenticillate, crenate, acute, base rounded, flowers unisexual, white in axillary racemes, bracts imbricate, perianth lobes 3, equal ovate, stamens 3-5 free, exserted. Fl. & Frt: March-November, Rajula cheruvu., PSKR, 1127.

Local name: Magaira	Family: Euphorbiaceae
Information source: Sugali	Plant part used: Wood

Use description:

Wood is used for hut making.

Mimusops elongi Linn.; F. B. I. iii. 548; Roxb. Cor. Pl. t. 14; Wt. Ic. t. 1586; Bedd. Fl. t. 40.

A small tree with rather small leaves scarcely 3 in. long in dry forests, a large one with much larger leaves in damp localities. Bark dark grey, rough, deeply fissured, wood dark red, very hard and close-grained, used for building, carts and implements. The scented white corolla which fall off in showers are made into garlands. Seeds give on oil. Fl. & Frt: February-May, Gopalaswamy temple, PSKR, 1568.

Local name: Pogada	Family: Sapotaceae
Information source: Yanadi	Plant part used: Leaf

Use description:

Leaf sap is used for curing eye diseases.

Mitragyna parviflora, Korth., *Stephegyne parcifolia*, Korth, F. B. I. iii. 25. *Nauclea parvifolia,* Roxb. Cor. Pl. t. 52; W. & A. 396; Bedd. Fl. t. 34.

A large deciduous tree with leaves very variable in shape and size. Bark light grey, smooth; wood light pinkish-brown, even-grained, used for furniture, implements and utensils. Flowers fragrant, pale yellow or cream, in globose heads. Seeds numerous. Fl. & Frt: Throughout the year, Lankamallai, PSKR, 1685 (Fig. 19.1c).

Local name: Battaganapa	Family: Rubiaceae
Information source: Yanadi	Plant part used: Bark, Flower bud

Use description:

1. Bark extract given orally for curing snakebite.
2. Dried flower buds powdered and applied for curing toothache.

Morinda pubescens J. E. Smith; *Morinda tinctoria* Roxb.; *Morinda tinctoria* Roxb. var. *tomentosa* (Heyne ex Roth) Hook. f., *M. tomentosa* Heyne ex Roth.

Deciduous trees, bark deeply cracked, leaves elliptic or lanceolate, −ovate, slightly undulate, acuminate, base narrow into petiole. Flowers white, in leaf opposed or terminal globose heads, sepals, petals, stamens 5 each, ovary bilocular, fruits subglobose, syncarpium formed by succulent. Accresent calyx. Fl. & Frt: July-September, Madhuru, PSKR, 1434.

Local name: Maddi	Family: Rubiaceae
Information source: Yanadi	Plant part used: Stem bark

Use description:

Stem bark decoction given orally for curing mad dog bite.

Morinda tinctoria, Roxb.; F. B. I. iii. 156; W. & A. 419. *M. exserta*, Roxb; W. & A. 419. *M. citrifolia*, Bedd. Fl. t. 220, not for Linn.

A moderate-sized deciduous tree. Leaves narrowly linear-oblong, tomentose, stipules bifid, flowers white, in axillary or terminal, umbellate cymes, fruits syncarpium. Fl. & Frt: July- November, Sabhavi, PSKR, 1344.

Local name: Togaru	Family: Rubiaceae
Information source: Yanadi	Plant part used: Leaf

Use description:

Leaf decoction given orally for curing loose motions.

Moringa concanensis, Nimmo; F. B. I. ii. 45; Hook. Ic. t. 2596.

A tree with thick corky bark and white soft wood. Leaves usually two pinnate; leaflets broadly elliptic or orbicular, emerginate at apex, 1-1.5 in. long, flowers yellow streaked with red.; seed wings elongate. Fl. & Frt: July-November, Panchalingalagudi, PSKR, 1444.

Local name: Konda munaga	Family: Moringaceae
Information source: Yanadi	Plant part used: Root

Use description:

Root decoction given orally for curing fever.

Murraya paniculata (L.) Jack.; *Chalcas paniculata* L.; *Murraya exotica* L.

A large shrub or small tree with scented flowers. Wood very hard, resembling box wood. Leaves 3-5, rarely 7 foliolate, leaflets ovate-elliptic, flowers white, in terminal and axillary paniculate cymes, fruits berries ovoid, 1-2 seeded, apex acuminoid. Fl. & Frt: March-September, Nithyapoojakona, PSKR, 445.

Local name: Pulavelagu	Family: Rutaceae
Information source: Yanadi	Plant part used: Root bark

Use description:

1. Root bark is rubbed on a flattened stone with water to get the paste. The paste is applied at the site of the bite.
2. Flowers are used for decorations during morning and festivals.

Naringi crenulata (Roxb.) Nicol.; *Limonia crenulata* Roxb.; *Hesperethusa crenulata* (Roxb.) Roem.; *Limonia acidissima* auct. Non. L.

Deciduous throny trees, spines axillary, straight, leaves in clusters, leaflets 3 pairs, ovate-elliptic, glabrous, crenulate, subacute, petiole and rachis winged, jointed. Flowers 4 or 5 merous, white, in axillary racemes, berries globose, seeds 3 or 4, ovoid. Fl. & Frt: April-September, Bangarlapodu, PSKR, 159.

Local name: Kukka velaga	Family: Rutaceae
Information source: Yanadi	Plant part used: Leaf

Use description:

The decoction prepared from the leaf taken orally for curing epilepsy.

Naringi alata (Wall ex. Wight & Arn.) Eliias. In. Bull. Bot. Surv. Ind. 22; 193.1980. Limonia alata Wall. ex. Wight. & Arn. Prodr. Fl. Ind. Orient. 92.1834; FBI 1: 908.1874.138.

Small tree up to 8 m tall; branchlets spiny, pubescent, bark grey, lenticellate, leaves trifoliate, leaflets subsessile, terminal Fl. & Frt: March-September; Bangarlapodu, PSKR, 161.

Local name: Budidapandlu	Family: Rutaceae
Information source: Yanadi	Plant part used: Fruit

Use description:

Fruits are edible.

Ochna obtusata DC., *Ochna squarrosa*, Linn.; F. B. I. i. 523; Roxb. Cor. Pl. i. t. 89; Wt. III. i. t. 69.

A small tree with thin brown bark and reddish-brown wood of pretty grain, conspicuous for its yellow flowers with persistent sepals. Leaves alternate or apically clustered, obovate or lanceolate, glabrous, crenate or serrate, obtuse. Flowers golden yellow, in terminal or axillary racemes or panicles, seed solitary per cell. Fl. & Frt: March-July, Bangalrlapodu, PSKR, 1123.

Local name: Erravuddi. Kukkamogi	Family: Rutaceae
Information source: Yanadi	Plant part used: Root

Use description:

Root decoction given orally for snakebite.

Oroxylum indicum, Vent.; F. B. I. iv. 378. *Calosanthes indica*, Bl.; Wt. Ic. t. 1337.

A small conspicuous but ungainly tree, at once known by the 2-3 pinnate large leaves and the flat capsule, which may reach 3 ft. in length. Bark light brownish-grey, greenish when cut; wood yellowish-white, soft. Fl. & Frt: July-February, Nagasanipalli, PSKR, 1233.

Local name: Dundilam	Family: Bignoniaceaee
Information source: Yanadi	Plant part used: Root & Bark

Use description:

1. Root and bark decoction given orally for curing cough and fever.
2. Root bark decoction taken orally twice a day for curing dysentery.

***Pavetta indica* L.**

Small trees, leaves oblong, obovate-lanceolate, entire, subacute. Flowers in axillary or terminal, trichotomously branched corymbose cymes, sepals and petals 5 each, stamens 4. Berries globose, fleshy, black when ripe. Fl. & Frt: May-October, Rollabodu, PSKR, 1278.

Local name: Papatikaya	Family: Rubiaceae
Information source: Yerukula	Plant part used: Root

Use description:

Root powdered mixed with ginger and taken orally for curing dropsy.

Phyllanthus emblica, Linn.; F. B. I. v. 288; Bedd. Fl. t. 258; Brand. For. Fl. t. 52. *Emblica officinalis*, Gaertn; Wt. Ic. t. 1896.

A small- or moderate-sized deciduous tree, the branchlets and rachises villous, bark light grey, exfoliating; wood red, hard and close-grained, useful for poles, implements and furniture. Flowers unisexual, greenish-yellow, in axillary clusters Fl. & Frt: March-October, Gollaalli, PSKR, 1404.

Local name: Usirikaya	Family: Euphorbiaceae
Information source: Yanadi	Plant part used: Fruit

Use description:

1. Fermented liquor prepared from the fruit given orally for curing jaundice.
2. The decoction prepared from the fruit mixing with the rhizome of *Curcuma domestica* is taken orally to reduce the levels of blood glucose.
3. Fruits are edible.

Pithacalobium dulce, Benth.; F. B. I. ii. 302.; Bedd. Fl. t. 188. *Mimosa dulcis* Roxb. Cor. Pl. t. 99. *Inga dulcis*, Willd. W. & A. 268; Wt. Ic. t. 198.

Armed tree, branchlets glabrous, leaves bipinnate, pinnate 2 pairs, leaflets paired, oblong-oblanceolate, entire, seeds black, shiny, covered by reddish-white aril. A tree which, when allowed to grow, reaches a good size and gives a useful strong reddish-white wood, a good fuel, pods which are eaten by cattle and foliage which is eaten by goats. Fl. & Frt: November-June, Nithyapoojakona, PSKR, 1708.

Local name: Seemachinta	Family: Mimosaceae
Information source: Yanadi	Plant part used: Stem bark & Leaf

Use description:

1. Stem bark boiled in water prepared decoction. A cup of decoction taken orally thrice a day for 3 days for curing fever.
2. The paste prepared from the leaves directly applied on the wounds and cuts till the cured.
3. Arils of the seeds are edible.

Pleurostylia opposita (Wall.) Alston, *Celastrus opposita* Wall.; *Pleurostylia wightii* Wt. & Arn.: *P. heynei* Wt. & Arn.

Densely foliaceous, semi-evergreen trees, crown compact, branchlets glabrous, leaves opposite, decussate, thin coriaceous, elliptic, obovate, glabrous, entire, obtuse, flowers minute, greenish on lateral cymes, drupes yellowish. Fl. & Frt: October-March, Lothuvanka, PSKR, 1318.

Local name: Nalla Nakkana	Family: Celastraceae
Information source: Sugali	Plant part used: Leaf

Use description:

Leaf paste applied externally for curing wounds.

***Plumaria alba* L.**

Trees, latex milky, bark greyish, smooth, leaves alternate, obovate, oblong, spathulate, oblanceolate, entire, acute, base tapering, glabrous to pubescent below, flowers red or purple, with yellow centre, fragrant, seeds flat, winged, coma absent. Frt: April-July, Ranibavi, PSKR, 1128.

Local name: Telladevaganneru	Family: Apocynaceae
Information source: Sugali	Plant part used: Latex

Use description:

Latex is used for curing sprains.

Polyalthia cerasoides, Hook. f. & Thoms. F. B. I. i. 68; Bedd. Fl. t. 1; King Ann. Calc. iv. 65, t. 86B, *Uvaria cerasoides*, Roxb. Cor. Pl. i. t. 38. *Guatteria cerasoudes*, W. & A. 10.

Evergreen tree, branchlets tomentose, leaves lanceolate or elliptic, pubescent below, acute or acuminate, flowers green, axillary, solitary, berries 30-40, in a cluster, red when ripe, 1-seeded. Fl. & Frt: April-August, Madhuru, PSKR, 1489.

Local name: Gutti	Family: Annonaceae
Information source: Yanadi	Plant part used: Leaf

Use description:

Dried leaves powdered and mixed with water applied as poultice.

Pongamia pinnata (L.) Pier.; *Pongamia glabra*, Vent.; F. B. I. ii. 240; W. & A. 262; Wt. Ic. t. 59; Bedd. Fl. t. 177.

A moderate-sized nearly evergreen tree with 5 or more rather large ovate acuminate leaflets and pinkish-white flowers, bark thick, greyish-brown, tubercled; wood white, moderately hard, used for cart-wheels and other purposes. The seeds give an oil. Fl. & Frt: February-October, Siddavatam, PSKR, 189.

Local name: Kanuga	Family: Fabaceae
Information source: Yanadi	Plant part used: Seed and Stem bark

Use description:

1. Seed paste is applied externally on the boils.
2. Stem bark and bulb of *Allium sativum* L. taken in equal quantities, made into paste and taken with curd for fertility.

Premna latifolia, Roxb.; F. B. I. iv. 577; Wt. Ic. t. 869.

A small tree, the leaves usually drying blue or black, the flowers greenish-white, leaves ovate to sub-orbucular, entire, obtuse, acute, flowers white, in terminal corymbs, drupes depressed globose, with persistent calyx, seeds oblong. Fl. & Frt: Most part of the year, Lankamallai, PSKR, 189.

Local name: Nelli	Family: Verbanaceae
Information source: Yanadi	Plant part used: Leaf & Stem bark

Use description:

1. Leaf extract given orally for curing dropsy.
2. Stem bark paste with cow ghee given orally for curing ulcers.
3. Dry leaf powder with coconut oil applied for dandruff.

Premna tomentosa, Willd. F. B. I. iv. 576; Wt. Ic. t. 1468; Bedd. Fl. t. 251.

A moderate-sized deciduous tree with greenish-yellow flowers. Bark light greyish-brown; wood light brown, smooth and close grained. Leaves ovate-cordate, entire or minutely toothed, acuminate, base rounded to truncate, tawny yellowish-tomentose below, flowers greenish-yellow, in axillary and terminal cymose panicles. Fl. & Frt: January-May, Rollabodu, PSKR, 198.

Local name: Pedda narava	Family: Verbanaceae
Information source: Yanadi	Plant part used: Leaf and Stem Bark

Use description:

Leaf and stem bark decoction given orally for curing Biliousness.

Prosopis cineraria (L.) Druce., *Mimosa cineraria* L. *Prosopis spicigera* L. Mant.

Evergreen tree, leaves bipinnate, pinnate 2 pairs, leaflets oblong, appressed hairy, entire, rounded, flowers yellow, small, in under spikes, pods pendent, cylindrical, coriaceous, with dry pulp, seeds oblong. Fl. & Frt: October-July, Lankamallai, PSKR, 856.

Local name: Jammi	Family: Mimosaceae
Information source: Yanadi	Plant part used: Stem bark & Fruit

Use description:

1. Stem bark is crushed into small pieces and strained in a cup of water for 30 minutes. The infusion is given in orally twice a day for three days for curing jaundice.
2. Stem bark is made into pieces, boiled in water for making decoction. A cup of decoction is given orally for twice a day for three days for curing dysentery.

Pterocarpus marsupium, Roxb. Cor. Pl. t. 116; F. B. I. ii. 239; W. & A. 266; Bedd. Fl. t. 21.

A large deciduous tree, bark thick, grey, with vertical cracks, wood very hard, yellowish, brown with darker streaks, used for building, furniture, agricultural and railway purposes. It gives a red gum-resin "Kino". Leaves imparipinnate, leaflets coriaceous, shining, entire, obtuse, flowers golden yellow, in axillary panicles,. Fl. & Frt: Jane-April, Lothuvanka, PSKR, 985.

Local name: Yepi	Family: Fabaceae
Information source: Yanadi	Plant part used: Gum, Leaf & Wood

Use description:

1. Gum applied externally for curing toothache.
2. Leaves covered on skin for curing skin diseases.
3. Wood extracts taken orally twice a day for curing diabetes.

Pterocarpus santalinus, Linn. f., F. B. I. ii. 239, W. & A. 266; Bedd; Fl. t. 22.

A very pretty and valuable moderate-sized tree, bark blackish-brown deeply cleft into rectangular plates, wood extreamely hard, dark claret red to nearly black. Leaves 3-foliate, leaflets very rarely 2 pairs, coriaceous, entire, obtuse, emarginated. Flowers golden yellow, in axillary panicles, pods flat, orbicular with broad wings, reddish brown. Fl. & Frt: Jan-April, Lankamallai, PSKR, 988 (Fig. 19.1d).

Local name: Errachandanamu	Family: Fabaceae
Information source: Yanadi	Plant part used: Wood & Stem Bark

Use description:

1. Wood powder mixed in hot water given orally for curing diabetes.
2. Stem bark decoction given orally for curing diarrhoea.

Pterospermum canescens Roxb.; *P. Suberifolium* (L.) Lam.; *Pentapetes suberifolia* L.

Trees, leaves coriaceous, obovate-oblong, glabrous above, cream tomentum, irregularly lobed at apex, flowers yellowish-white, fragrant, seeds brown. Fl. & Frt: June-April, Sabhsvi, PSKR, 868.

Local name:	Family: Sterculiaceae
Information source: Yanadi	Plant part used: Wood

Use description:

Wood is used for making cots.

Pterospermum xylocarpum (Gaertn.) Sant. & Wagh.; (*Velaga xylocarpa* Gaerntn.)., *P. heyneanum* Wall. ex. Wt. & Arn.

Evergreen trees, leaves variable, oblong, slightly obovate, entire, dentate, acuminate, flowers pale white, in axillary or terminal clusters, seeds orbicular. Fl. & Frt: October-August, Panchalingalagudi, PSKR, 868.

Local name: Lolugu, Thada	Family: Sterculiaceae
Information source: Yanadi	Plant part used: Stem & Leaves

Use description:

1. Wood is used for making local cots and furniture.
2. Leaves are used as meals plates.
3. Fibre from the stem bark used for making ropes.

Santalum album, Linn; F. B. I. v. 231; Bedd. Fl. t. 256. *Sirium myrbifolium*, Roxb. Cor. Pl. t. 2.

A small evergreen usually semi-parasitic glabrous tree with elliptic-ovate, leaves up to 3 in. long, glacous beneath, the flowers brownish-purple, the drupe black when ripe. Bark dark grey, rough, wood hard and close-grained, sap wood white, scentless, heartwood yellowish-brown, strongly scented. Fl. & Frt: November-April, Nithyapoojakona, PSKR, 741 (Fig. 19.1e).

Local name: Srigandamu	Family: Santalaceae
Information source: Yanadi	Plant part used: Wood

Use description:

Wood paste (with water) applied externally for curing headache and skin eruptions.

Sapindus emarginatus, Vahl; W. & A. III; Bedd. Fl. t. 154. *S. trifoliatus*, Hiern in F. B. I. i. 682 *in part. Not of Linn.*

Densely foliaceous, deciduous trees, branchlets tomentose, leaves even pinnate, leaflets 3-pairs, thick, coriaceous, broadly elliptic, oblong-obovate, glabrous above, pubescent beneath, entire, emarginate, obtuse, flowers brownish, white, in long rusting tomentose panicles,. A stout shady tree. Fl. & Frt: January-April, Panchalingalagudi, PSKR, 748.

Local name: Kunkudu	Family: Sapindaceae
Information source: Yanadi	Plant part used: Fruit

Use description:

1. Fruits extract applied externally for headache.
2. Fruits are used like that of the next to soap.
3. Fruits are edible.

Schefflera stellata Harms. *Heptapleuram stellatum*, Gaertn.; F. B. I. ii. 730. *Hedera aborata,* Wt. Ic. tt. 1011, 1012.

A large straggling or climbing shrub with small flowers, the petals usually falling as an operculam. Branchlets with persistent leaf scars, leaves digitately 5-7 foliolate, leaflets ovate-obovate, retuse, flowers polygamous, white or pale yellow, in terminal panicles, drupes 5-6 angled, orange when ripe. Fl. & Frt: April-July, Panchalingalagudi, PSKR, 847.

Local name: Marrimamidi	Family: Araliaceae
Information source: Yanadi	Plant part used: Flower

Use description:

Flowers are used for decoration in functions.

Schleichera oleosa (Lour.) Oken.; Pistacia oleosa Lour, Schleichera trijuga Willd.

Deciduous trees, branches grey, young shoots silky, leaves paripinnate, leaflets 2-3 pairs, young leaflets bright red, sub-opposite, coriaceous, oblong, elliptic, flowers greenish yellow, in much branched axillary racemes, seed long, brown, smooth. Fl. & Frt: March-July, Panchalingalagudi, PSKR, 478.

Local name: Pulasaram	Family: Sapindaceae
Information source: Yanadi	Plant part used: Stem bark & seed

Use description:

1. Stem bark paste mixed with sesamum oil applied externally for curing itches.
2. Oil obtained from the seeds used as hair tonic.

Soymida febrifuga, Adr. Juss.; F. B. I. i. 567; W. & A. 122; Bedd. Fl. t. 8., *Swietinia febrifuga*, Roxb. Cor. Pl. i. t. 17.

A large tree with greyish-green leaves, thick bluish-grey or brown bitter bark and very hard dark red strong and hand some valuable food. Branchlets wrinkled with persistent leaves scars, leaves paripinnate, leaflets 6 pairs, opposite, ovate, oblong, entire, obtuse or retuse. Flowers greenish-white, in axillary or terminal panicles, seeds oblong winged at both ends. Fl. & Frt: February-July, Sabhavi, PSKR, 574 (Fig. 19.1f).

Local name: Somi	Family: Meliaceae
Information source: Yanadi	Plant part used: Stem bark

Use description

1. The decoction prepared from the stem bark filtrated and cooled. A cup of decoction is taken orally thrice a day for 3 days for curing fever.
2. The decoction of stem bark is taken thrice a day for 3 days for curing diarrhoea.
3. Wood is used for making furniture.

Sterculia urens, Roxb. Cor. Pl. i. t. 24; F. B. I. i. 355; W. & A. 63.

A large deciduous tree, very conspicuous from its pale smooth bark which peels off in flakes. Wood soft and used only for toys. Leaves palmately 3-lobed, entire, acuminate, basically 7-nerved, crowded at the ends of branches, scabrid above, tomentose below, flowers greenish yellow, polygamous in panicles, covered with a dense silky tomentose hairs. Fl. & Frt: December-June; Gopalaswamy temple, PSKR, 774.

Local name: Yerrapoliki	Family: Sterculiaceae
Information source: Sugali	Plant part used: Bark & gum

Use description:

1. Bark paste applied externally for easy delivery.
2. Gum dissolved in water given orally for curing twice a day for reducing Body heat.

Strychnos nux vomica, Linn.; F. B. I. iv. 90; Roxb. Cor. Pl. t. 4; Bedd. Fl. t. 243.

A moderate-sized or large deciduous tree with greenish-white flowers having the scent of fenugreek and orange red berries up to 2 in. in diam. with3-4 flat seeds. Bark thin, dark-grey, hard, close-grained. Leaves coriaceous, orbicular, entire, obtuse or acute, base rounded, glabrous, dark green and shinning above. Fl. & Frt: March-October; Madhuru, PSKR, 908 (Fig. 19.1g).

Local name: Musti	Family: Loganiaceae
Information source: Yanadi	Plant part used: Seed & root bark

Use description:

Seed ash mixed with neem oil applied externally for curing leucoderma. The decoction prepared from the root bark taken immediately after the bite for snake-bite. The paste prepared from the root bark applied externally on the site of bite.

Strychnos potatorum, Linn. f. F. B. I. iv. 90; Roxb. Cor. Pl. t. 5; Wt. III. t. 156.
A moderate-sized tree. Bark brownish-black, deeply cracked or vertical; branchlets fluted, leaves chartaceous, ovate, elliptic, entire, acute or subacuminate, glabrous, flowers white, in axillary cymes, berries globose, thin-shelled, deep blue when ripe. Fl. & Frt: November-June; Sabhavi, PSKR, 912.

Local name: Chilla	Family: Loganiaceae
Information source: Yanadi	Plant part used: Seed & Stem bark

Use description:

1. Seed applied on beaten spot for scorpion sting.
2. The decoction prepared from the stem bark taken orally immediately after the bite.

Suregada angustifolia Baill. ex. Muell. -Arg.; *Gelonium angustifolium* Bail. ex. Muell-Arg.; *G. lanceolatum* auct. Non. Willd.
Dioecious, evergreen shrubs or small trees, leaves alternate, oblong-obovate or lanceolate, entire, subacute, coriaceous, flowers unisexual, greenish-yellow, in sub-umbellate axillary clusters, seeds globose, testa crustaceous. Fl. & Frt: June-December Madhuru, PSKR, 1012.

Local name: Sapranji	Family: Euphorbiaceae
Information source: Yanadi	Plant part used: Root

Use description:

The decoction of root taken orally thrice a day for curing fever.

Syzigium alternifolium Walp., *Eugenia alternifolia*, Wt.; Wt. Ic. t. 537; F. B. I. ii. 497; Bedd. Fl. t. 198.
Deciduos trees, wood dark red; branchlets pale, glabrous, leaves thick, coriaceous, ovate-elliptic, entire, acute, flowers yellowish white or cream, in axillary cymes, scented, berries dark purple, crowned with calyx tube, sweet. Fl. & Frt: March-June, Lankamallai, PSKR, 1122.

Local name: Mogi. Kondaneradu	Family: Myrtaceae
Information source: Yanadi	Plant part used: Fruit

Use description:

1. Fruits are edible.
2. Wood is used in hut making.

Syzigium cumini (L.) Skeels.; *Eugenia jambolana* Lam, *Syzigium Jambolanum* (Lam) DC.

Evergreen trees, branchlets with gall formations, pendulous, leaves opposite-deccusate, elliptic or ovate-lanceolate, entire, acuminate, thick-coriaceous, flowers creamy white or greenish white, in panicles, scented, berries purplish red, black when ripe, seed solitary. Fl. & Frt: March-August, Nithyapoojakona, PSKR, 1125.

Local name: Neradu	Family: Myrtaceae
Information source: Yanadi	Plant part used: Seed

Use description:

1. Seeds are powdered boiled in water make it as a decoction. The decoction is taken orally thrice a day for 15 days for curing diabetes.
2. Seed are dried and powdered. One teaspoon of powder taken orally twice a day for curing gonorrhoea.
3. Wood is used as furniture.
4. Fruits are edible.

Tamarindus indica, Linn.; F. B. I. ii. 273; W. & A. 285; Bedd. Fl. t. 184.

A large evergreen tree of great importance, having rather small pink yellow-striped flowers. Small acid leaflets. The bark is dark grey and the wood, very hard and heavy, is of a dark purplish-brown colour and used as furniture. Fl. & Frt: April-February, Madhuru, PSKR, 1258.

Local name: Chintha	Family: Caesalpiniaceae
Information source: Yanadi	Plant part used: Fruit

Use description:

Mesocarp and young shoots of *Cassia fistula* are mixed powdered made into pills and given orally for curing ascariasis.

Tarenna asiatica (L.) Kuntz. ex. K. Schum. var. *asiatica*, *Rondeletia asiatica* L. *Webera corymbosa* Willd.; *Chomelia asiatica* (L.) Kuntz; *Stylocoryne webera.* A. Rich.

Large shrubs, leaves elliptic-oblong or oblanceolate, entire, acute, base truncate, dark glossy green, black when dry, flowers cream, in terminal trichotomous, n corymbose cymes, berries black, globose with a crown of calyx-lobes. Fl. & Frt: Most part of the year, Rollabodu, PSKR, 1378.

Local name: Kommi	Family: Rubiaceae
Information source: Yanadi	Plant part used: Fruit

Use description:

Fruits are edible.

Terminalia alata Hyene ex Roth; *T. tomentosa* (Roxb. ex. DC.) Wt. & Arn.; *Pentaptera tomentosa* Roxb. ex. DC.

Deciduous trees, bark black, with much vertical fissures and transverse cracks; branchlets villous. Leaves thick, coriaceous, oblong, tomentose, beneath, crenulate, flowers pale yellow in dense panicles. Drupes ovoid, glabrous, 5-winged, wings equal, with horizontal lines and apex obtuse. Fl. & Frt: May-February, Madhuru, PSKR, 1693.

Local name: Nallamaddi. Enumoddu	Family: Combretaceae
Information source: Yanadi	Plant part used: Seed

Use description:

1. Seeds powdered and mixed with gum, turmeric and Bengal gram used as a bath powder.
2. Wood is used for making furniture, doors, windows and agricultural implements.

Terminalia arjuna, W. & A. 314(in note); F. B. I. ii. 447; Bedd. Fl. t. 28. *T. Berryi*, W. & A. 314.

A large and handsome deciduous tree usually with narrow oblong leaves, the stems often buttressed. Bark pinkish-grey, smooth, wood brown, variegated, with darker streaks, very hard. Flowers pale yellow, in axillary panicles, drupes ellipsoid, reddish brown. Fl. & Frt: April-September, Sabhavi, PSKR, 1703.

Local name: Tellama ddi	Family: Combretaceae
Information source: Yanadi	Plant part used: Gum

Use description:

1. Gum is collected from the stem.
2. Wood is used for making agricultural implements.

Terminalia bellerica, Roxb. Cor. Pl. t. 198; F. B. I. ii. 444; W. & A. 313; Wt. Ic. t. 172; Bedd. Fl. t. 18.

A large deciduos tree with long-petioled leaves, pubescent flowers and tomentose fruit. Bark bluish-grey with vertical cracks, wood yellowish-grey, hard, not durable, branchlets white-pubescent with persistent leaf scares, leaves alternate, ovate-obovate or broadly elliptic, dencly clustered at the ends, flowers cream, in axillary, solitary or paired spikes, drupes globular. Fl. & Frt: March-June, Rollabodu, PSKR, 1705.

Local name: Thandra	Family: Combretaceae
Information source: Yanadi	Plant part used: Fruit

Use description:

Fruits along with those of *Terminalia chebula* Retz. and *Emblica officinalis* L. dried, powdered and given orally with honey for curing dropcy.

Terminalia chebula Retz. Obs. Bot. 5: 31, 1789.

Deciduous trees. Leaves elliptic–oblong, coriaceous. Flowers yellow, in axillary spikes. Drupes ellipsoid–obovoid; Fl & Fru:. May–October. Madhuru, 1646.

Local name: Nallakaraka	Family: Combretaceae
Information source: Yanadi	Plant part used: Stem bark

Use description:

Stem bark extract applied externally for curing bone fractures.

Terminalia pallida, Brand. Ind. Trees, 308.

A small subevergreen tree with glacous leaves, glabrous flowers. Leaves thick, coriaceous, broadly oblong, entire, obtuse, flowers pale yellow, in simple terminal or axillary spikes, drupes obovoid when dry. Fl. & Frt: March-June, Madhuru, PSKR, 1710.

Local name: Tella karaka	Family: Combretaceae
Information source: Yanadi	Plant part used: Stem bark & Fruit

Use description:

The decoction prepared from the stem bark is taken orally for curing diarrhoea.

Thespesia populnea, Cav., F. B. I. i. 345; W. & A. 54; Wt. Ic. t. 8; Bedd. Fl. t. 63.
A fairy large evergreen tree with cordate entire long-petioled leaves, long peduncled flowers, the petals yellow fading to purplish pink. Fl. & Frt: Throughout the year Rollabodu, PSKR, 1707.

Local name: Gangaravi	Family: Malvaceae
Information source: Yanadi	Plant part used: Seed & bark

Use description:

1. The capsule is filled with a yellow pigment like liquid gamboges, which is a good external application in scabies and other skin diseases. The juice simply applied to the parts affected thrice a day for three days.
2. The bark boiled in water used as a wash for curing scabies.

Trema orientalis (L.) Blume; *Celtis orientalis* L.
Dioecious, evergreen trees, branchlets terete, sparsely pubescent, leaves chartaceous, broadly lanceolate or ovate, serrulate, acuminate, base uneaual, rounded, rough above, flowers unisexual, light green, in axillary cymes, the male cymes more compact than the female and usually on different trees. Fl. & Frt: May-November, Nithyapoojakona, PSKR, 1117.

Local name: Kaka Musti	Family: Ulmaceae
Information source: Yanadi	Plant part used: Stem bark

Use description:

1. Stem bark boiled in milk, ground and made into pills given orally for curing fever.
2. Fruits are edible.

Vitex altissima, Linn. f.; F. B. I. iv. 584; Wt. Ic. t. 1466; Bedd. Fl. t. 252. *V. alata*, Heyne; F. B. I. iv. 584.
A large and lofty tree with white flowers tinged with blue or violet. Bark yellowish grey; wood grey to yellowish, brown, hard and close grained, valuable for building furniture, carte, etc. The winged petioles are most conspicuous in young plants and coppice shoots. Fl. & Frt: June–October, Lankamallai, PSKR, 1725.

Local name: Nemaliadugu	Family: Verbanaceae
Information source: Yanadi	Plant part used: Stem bark

Use description:

1. Stem bark ground and mixed with goat milk given orally for galactogogue.
2. Bark decoction given as washing agent for toothache.

Vitex leucoxylon, Linn. f.; F. B. I. iv. 587, *Wallrothia leucoxylon*, Roxb.; Wt. Ic. t. 1467.

A deciduous tree with spreading head and white rather large flowers with purple hairs on the long palate, the drupe large and dark purple. Bark grey, smooth; wood light greyish-brown, moderately hard and useful for furniture and carts. Fl. & Frt: February-April, PSKR, Sabhavi, 1735 (Fig. 19.1h).

Local name: Kondavavili	Family: Verbanaceae.
Information source: Yanadi	Plant part used: Root

Use description:

Root decoction given orally for curing fever.

Walsura trifoliate (A. Juss) Harm.; *Heynea trifoliate* A. Juss.; *Walsura ternate Roxb.; Walsura piscida Roxb.*

Evergreen trees, leaves thick coriaceous, 3-foliate, leaflets elliptic or oblong or ovate-lanceolate, shiny, entire rounded, flowers creame, in axillary or terminal panicles, sepals and petals 5 each, stamens 10, berries obovoid, velvety tomentose, bright orange-yellow when mature. Fl. & Frt: March-August, Rollabodu, PSKR, 1554.

Local name: Yerravaludu	Family: Meliaceae
Information source: Yanadi	Plant part used: Stem bark

Use description:

1. Stem bark is powdered and applied externally on the symptoms for curing syphilis.
2. The decoction prepared from the stem bark taken orally twice a day until cured for syphilis.

Wrightia arborea (Dennst.) Mabber.; *Periploca arborea* Dennst.; *Wrightia tomentosa* Roem. & Schult.; *W. mollissima* Wall.; *W. wallichi* A. Dc.

Small deciduous trees, branchlets tomentose, leaves elliptic-oblong, entire, acuminate, base rounded, pubescent on the nerves above, tomentose below, drying, dark green, flowers pale yellow or greenish, in terminal corymbose cymes, seeds slender. Fl. & Frt: April-February, Panchalingalagudi, PSKR, 1542.

Local name: Palvareni	Family: Apocynaceae
Information source: Yanadi	Plant part used: Bark

Use description:

Bark decoction given orally for curing renal complaints.

Wrightia tinctoria R. Br.; F. B. I. iii. 653; Wt. Ic. t. 444; Bedd. Fl. t. 241.

A small deciduous tree with yellowish flowers, the coronal scales orange, connate follicles and ovate or obovate tomentose leaves, the white coma of the seeds at base, bark grey, corky; wood white moderately hard, even grained, useful for carving. Fl. & Frt: October-June, Rajulacheruvu, PSKR, 1540.

Local name: Pala. Ankudu	Family: Apocynaceae
Information source: Yanadi	Plant part used: Stem bark

Use description:

1. Stem bark decoction mixed with garlic and pepper given orally for curing stomach ache.
2. Bark powder used as blisters.

***Xylia xylocarpa*,** Taub. *X. dolabriformis*, Benth.; F. B. I. ii. 286; Bedd. Fl. t. 186. *Inga xylocarpa,* DC.; W. & A. 269.

A deciduous tree with 2 pairs of oblong lanceolate leaflets, the end pair up to 8 in. long, white flowers in globose heads and large woody dolabriform pods, bark reddish-grey; wood hard, reddish brown, durable and useful for sleepers and building purposes but difficult to work. Fl. & Frt: Nithyapoojakona PSKR, 1402.

Local name: Konda tangedu	Family: Leguminosae
Information source: Yanadi	Plant part used: Stem

Use description:

Wood is used for making furniture.

Zizyphus mauritiana Lam.; *Z. jujuba* (L.) Gaertn.; *Rhamnus jijuba* L.

Deciduos trees, branchlets drooping, throny, leaves sub-orbicular, ovate-elliptic, white tomentose below, dentate, flowers yellowish green, in axillary clusters, sepals 5, ovate, stamens 5, disc 10lobed, grooved. Drupes globose, reddish brown. Fl. & Frt: September-February, Near Reggayalakunta, PSKR, 1204.

Local name: Regi chettu	Family: Rhamnaceae
Information source: Yanadi	Plant part used: Stem bark & fruit

Use description:

1. Stem bark powder mixed with water make as a paste applied site of the bite for snakebite.
2. Fruits are edible.

Zizyphus xylopyrus Willd.; F. B. I. i. 634; W. & A. 162.

A large straggling shrubs or small tree with globose velvety, sometimes nearly glabrous, fruit and 3 celled hard nut. The leaves are very white woolly when young but nearly glabrous when old. Wood reddish brown, hard, similar to but not quite so good as that of *Z. jujuba*. Fl. & Frt: October-March, Lankamallai, PSKR, 1216.

Local name: Gotika	Family: Rhamnaceae
Information source: Yanadi	Plant part used: Root bark

Use description:

Root bark paste applied externally for curing wounds.

19.3 Tree Diversity in Eastern Ghats of India

Trees are a major component of the most biologically diverse ecosystems on earth, including woodlands and forests (FAO 2014). During the climate change negotiations in 2015 at Paris, the major issues realized were the *conservation* of trees, the *cessation* of deforestation and the *sustainable management* of forests as the activities that address climate change. Tree conservation was the key agenda item at the fifth Global Botanic Gardens Congress held in New Zealand in October 2013. In India, the native people, regardless of their religious faiths, worship a number of tree species with deep respect, e.g. *Ficus religiosa* and *Aegle marmelses*.

It is estimated that the number of tree species on our planet ranges from 45,000 to 100,000 (Fine and Ree 2006). The number of tree species found in the tropics is slightly more than the double to the temperate region (Hunt 1996; Slik et al. 2015). As per the recent census by GTSG, there are 60,065 tree species recorded worldwide. About 45% of these tree species are members of just 10 angiosperm families; the most dominant being Leguminosae with 5405 species, followed by Rubiaceae (4827) and Myrtaceae (4330). The country with the most diverse tree flora is Brazil with 8715 tree species, followed by Colombia (5776 spp.) and Indonesia (5142 spp.). Nearly 58% (34,575) of tree species are single country endemics. The country with the most *endemic* trees is Brazil (4333 spp.). Ten percent of world's total (*c.* 8000) tree species are globally threatened with extinction. As per the IUCN Red

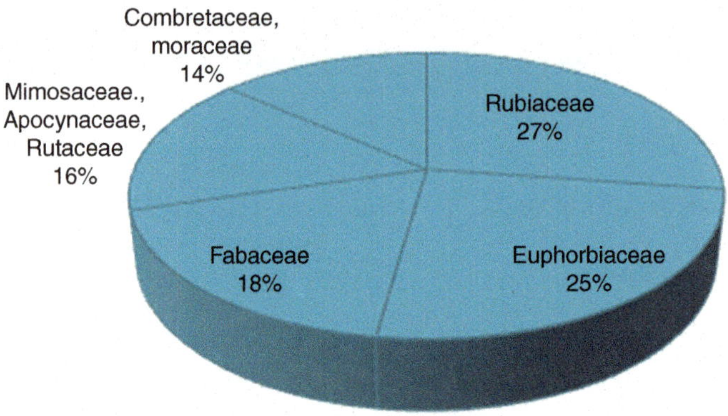

Fig. 19.2 Top five dominant tree families of the study area

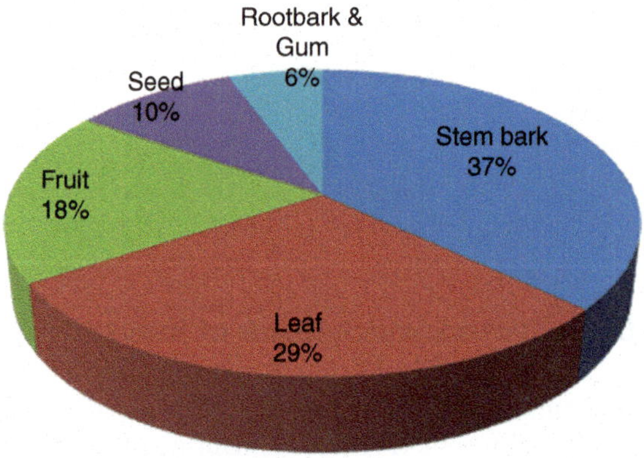

Fig. 19.3 Analysis based on plant part used

List, there are more than 9500 tree species assessed. Of these, about 6400 (67.37%) fall under globally threatened (CR, EN, VU) category; among them are 1100 tree species, which are critically endangered (CR). As a nation, India has 2544 tree species predominated by Lauraceae and Rubiaceae with 180 tree species each, followed by Leguminosae (131 spp.), Euphorbiaceae (113 spp.; excluding the 85 spp. of Phyllanthaceae) and Mytaceae (94 spp.).

Nearly 140 tree species belonging to 42 families of their ethnobotanical uses were recorded (Figs. 19.2 and 19.3). Among 42 families Rubiaceae occupies first with 12 species (27%), Euphorbiaceae occupies second with 11 species (25%), Fabaceae occupies third with 8 species (18%), Mimosaceae, Apocynaceae and Rutaceae occupies fourth with 7 species (16%) and Combretaceae and Moraceae occupies fifth with 6 species (14%) (Fig. 19.2). Based on the plant part used stem bark occupies first place with 52(37%), leaf occupies second with 41 (29%), Fruit occupies third with 26 (18%), seed occupies fourth with 14 (10%) and root bark and gum occupies fifth with 8 (6%) (Fig. 19.3). These 140 tree species curing 71 almities. Some species like *Cassia fistula, Ficus racemosa, Mitragyna parviflora,*

Pterocarpus santalinus, Santalum album, Soymida fabrifuga Strychnos nux vomica and *Vitex leucoxylon* are endemic and threatened are frequently used (Fig. 19.1a–h). The major threat to the Sanctuary is that the Government has rehabilated the villages which have submerged in the Somasila dam backwater here in the Sanctuary. Many anthropogenic activities like widening of roads, construction of high-tension electrical wires, construction of religious monuments like temples, church etc. is also causing damage to the sanctuary. The economic value estimate (oxygen production alone) of a healthy tree by Delhi Greens is Rs. 24 lakhs per year. The revised economic estimate of the University of Calcutta experts is Rs. 3,55,13,000, as the total cost of services rendered by a tree in its lifetime. So, protecting trees is every citizen's responsibility in terms of his own health, habitat health, forest ecosystem'shealth, and ultimately the overall economic well-being of local people.

References

Anonymous (1994) Little known but with great potential. International Agricultural Development, pp 13–14

Bhushan B (1986) Rediscovery of the Jourdon's double –banded courser cursorious bitorquatus (Blyth). J Bombay Nat Hist Sco 83:1–14

Champion HG, Seth SK (1968) A revised survey of forest types of India. Government of India, New Delhi

FAO (2014) The state of the world's forest genetic resources. FAO, Rome

Fine PV, Ree RH (2006) Evidence for a time-integrated species-area effect on the latitudinal gradient in tree diversity. Am Nat 168(6):796–804

Gamble JS (1915) Flora of presidency of Madras, vol 1, 2 & 3. Adlard and Son Lrs Lonswn, Calcutta

Hunt DR (1996) The genera of temperate broadleaved trees. Broadleaves 2:4–5

Jeganathan P, Green RE, Bowden CGR, Norris K, Pain D, Rahmani A (2002) Use of tracking strips and automatic cameras for detecting critically endangered Jerdon's coursers Rhinoptilus bitorquatus in scrub jungle in Andhra Pradesh, India. Oryx 36:182–188

Slik JWF, Arroyo-Rodríguez V, Aiba SI, Alvarez-Loayza P, Alves LF, Ashton P, Venticinque EM (2015) An estimate of the number of tropical tree species. Proc Natl Acad Sci U S A 112(24):7472–7747

Stattersfield A, Crosby MJ, Long AJ, Wege DC (1998) Endemic bird areas of the world: priorities for biodiversity conservation. Bird Life International, Cambridge

A Study on Medical Systems for Dengue Fever

20

B. Swapna

Abstract

Incidence of dengue in India increased drastically. Dengue fever is caused by arthropod-borne flavivirus which have four serotypes that spread by the bite of mosquito *Aedes aegypti*. The objective of this study is to explore the available methods to prevent and control dengue fever in allopathy, homeopathy, Ayurvedic, and traditional medical systems. Primary and secondary sources of data are used in this study. Secondary source of data is Internet. Secondary data is used to analyze allopathy, homeopathy, and Ayurvedic medical systems. Primary data is used to analyze traditional medical system. This is collected by interviewing two traditional medical practitioners. Traditional system is cheaper compared to other systems and it is easy to administer and has no side effects. There is a need to conduct more studies, observations, and investigations to explore safe and effective ways to tackle epidemic diseases like dengue fever and its complications.

Keywords

Dengue fever · Alternative medicine · Traditional medicine

20.1 Introduction

Dengue fever is caused by arthropod-borne flavivirus which have four serotypes that spread by the bite of mosquito *Aedes aegypti*. Dengue fever is a severe, flu-like illness that affects infants, young children, and adults, but seldom causes death. Dengue should be suspected when a high fever (40 °C/104 °F) is accompanied by 2 of the

B. Swapna (✉)
Department of Botany, Vikrama Simhapuri University Post-Graduate Centre, Kavali, Andhra Pradesh, India

S. M. Khasim et al. (eds.), *Medicinal Plants: Biodiversity, Sustainable Utilization and Conservation*, https://doi.org/10.1007/978-981-15-1636-8_20

following symptoms: severe headache, pain behind the eyes, muscle and joint pains, nausea, vomiting, swollen glands, or rash. The temperature in dengue fever is called Saddleback temperature. Symptoms usually last for 2–7 days, after an incubation period of 4–10 days after the bite from an infected mosquito. Severe dengue is a potentially deadly complication due to plasma leaking, fluid accumulation, respiratory distress, severe bleeding, or organ impairment. Severe joint pain is caused in dengue which is also known as break-bone fever. Warning signs occur 3–7 days after the first symptoms in conjunction with a decrease in temperature (below 38 °C/100 °F) and include severe abdominal pain, persistent vomiting, rapid breathing, bleeding gums, fatigue, restlessness, and blood in vomit. The next 24–48 h of the critical stage can be lethal; proper medical care is needed to avoid complications and risk of death.

Dengue fever became a serious global problem. In almost all countries we find the occurrence of this disease. India is no exception. It is estimated that annually above 390 million infections occur globally. During the period between 1996 and 2015, a massive increase of more than 500% has been recorded in number of dengue cases reported in India (Singh and Rawat 2017). Many studies have been reported on incidence of dengue (Mutheneni et al. 2017) and pathophysiology of dengue (Balmaseda et al. 2005).

The objective of this study is to analyze available methods in allopathy, homeopathy, Ayurveda, and traditional system to prevent, control, and cure dangerous dengue fever.

Required information is collected from primary and secondary sources. The secondary source is Internet. Secondary source is utilized for allopathy, homeopathy, and Ayurveda systems. The primary data is obtained by interviewing two traditional herbal practitioners. The two persons are selected purposively for reliable data. The plants were collected from Ghatika Siddheswaram and Narasimha Konda. The collected information is analyzed and presented in Table 20.1.

20.2 Medical Systems for Dengue Fever

20.2.1 Preventive Measures for Dengue Fever

The following are the preventive measures:

(1) Ensure improved water storage, (2) Ensure proper waste disposal, (3) Strictly avoid water stagnancy, (4) Use mosquito repellents containing 10% DEET, (5) Avoid dark colored clothing as mosquitoes are attracted to darker shades, and (6) Wear full-sleeved clothes.

20.2.2 Allopathy Medical System

No specific antiviral medication is currently available to treat dengue. The treatment of dengue fever is symptomatic and supportive in nature. Bed rest and mild analgesic-antipyretic therapy are often helpful in relieving lethargy, malaise, and fever

Table 20.1 List of medicines used by two traditional practitioners for dengue fever

Name of the practitioner	Stage of the dengue fever	Name of the medicinal plant		Nature of medicine	Administration of medicine
		Scientific name	Common name		
1. Sri Peddanna, Ghatika Siddeswaram, SPSR Nellore district, A.P	Prevention	*Croton bonplandianum*	Galivanamokka	Repels mosquitoes	Leaf paste diluted with water sprayed in room
	Control and cure	*Euphorbia hirta*	Reddivarinanubralu	Whole plant decoction	Two tablespoons for every 3 h
	Control and cure	*Ocimum sanctum*	Tulasi	Leaf paste	Leaf paste mixed with lime juice and camphor applied on fore head
2. Sri Magesh Anand Swamy, Narasimhakonda, SPSR Nellore district, A.P	Prevention	*Anisomeles malabarica*	Magabeera	Repels mosquitoes	Fresh leaf lit with castor oil
	Prevention	*Celosia polygonoides*	Elukauttareni	Leaf decoction	Thrice daily
	Control and cure	*Achyranthes aspera*	Uttareni	Leaf paste	Leaves with garlic made into paste given once daily
	Control and cure	*Pseudarthriavicida*	Muyyakuponna	Leaf decoction	Leaf decoction with pepper powder twice daily

associated with the disease. Acetaminophen (paracetamol) is recommended for the treatment of pain and fever. Aspirin, other salicylates, and nonsteroidal anti-inflammatory drugs (NSAIDs) should be avoided. Patients with dengue hemorrhagic fever or dengue shock syndrome may require intravenous volume replacement. Plasma volume expanders can be used in patients who do not respond to isotonic fluids (Gan 2014).

Dengvaxia, vaccine for dengue, was developed and it was approved by the WHO (2016). But, India has not yet introduced the vaccine.

20.2.3 Homeopathy Medical System

Homeopathic literature mentions several medicines for the treatment of dengue fever. *Eupatorium perfoliatum* is an excellent remedy for dengue fever. *Eupatorium Perfoliatum* is also popularly known as "bone set" as it brings about the quickest relief in severe bone and joint pains in fever. It is also very beneficial in providing relief from severe pain in the eyeballs in dengue fever. *Eupatorium Perfoliatum* was announced as the preventive drug for dengue outbreak by the council. *Gelsemium sempervirens*, *Bryonia alba*, and *Rhus toxicodendron* are the other medicines used to treat dengue (Jacobs et al. 2007; Hassan et al. 2013; Mahesh et al. 2018; Novaes 2015).

20.2.4 Ayurvedic Medical System

The other names for dengue fever in Ayurveda are dandakajwara and pittajajwara. Although there is no specific treatment for disease, Ayurveda stresses to strengthen immune system of the body and keeping a control on hyperthermia. Shunthi Churna, Hinguleshwar, tulasi leaves, Amrutharistam, Amruthottaramkashayam, sudarsanamgulika, Panchamruthamkashayam, and Datura are the commonly used medicines for dengue (National Health Portal of India (n.d.), https://www.nhp.gov.in).

20.2.4.1 Diet During Dengue Fever

- Rice gruel (Kanji), rice porridge (khichari) cooked with ginger, and lemon is useful.
- Tea prepared with tulasi, ginger, and cardamom may be taken at two hourly intervals.
- Spicy and oily food should be strictly avoided.

20.2.5 Traditional System

Plants have been a source for traditional medicine. Identification and use of medicinal plants for the treatment of various diseases is an age-old practice in human history. A field visit was conducted to Ghatika Siddeswaram and Narasimhakonda in SPSR Nellore district of Andhra Pradesh and interviewed two traditional medical practitioners, Sri Peddanna and Sri Magesh Anandswamy, respectively. Information about the use of traditional medicinal plants to prevent dengue by repelling mosquitoes and medicine to control dengue fever was collected through the questionnaire. According to the traditional healers *Croton bonplandianum* and *Anisomeles malabarica* are used to repel mosquitoes. *Euphorbia hirta*, *Ocimum sanctum*, *Achyranthes aspera*, and *Pseudarthria viscida* are used to treat dengue fever and *Celosia polygonoides* is used as preventive medicine. The details of the plants are given in Table 20.1. Further studies are required to know the potential of these plants and discover a promising antiviral drug from these medicinal plants and validate it to treat dengue fever. Very few plants have been studied and reported in the literature for their efficacy against dengue virus (Jain et al. 2008; Kala 2012; Kadir et al. 2013; Huang et al. 2017). Medicine in traditional system is cheaper compared to allopathy, homeopathy, and Ayurvedic systems. It is easy to administer and has no side effects. In recent times traditional system is gaining importance in India. The system is to be formalized.

20.3 Conclusion

In India dengue fever is on rise. Since it is transmitted through mosquito bite, preventive measures are to be strengthened. Efforts at individual as well as community level to prevent dengue also help to decrease the prevalence of dengue fever. There is a need to create awareness among people about precautionary measures. Alternative and complementary medicines are to be employed. Government has to consider the introduction of Vaccine approved by WHO.

References

Balmaseda A, Hammond SN, Pérez MA (2005) Assessment of the World Health Organization scheme for classification of dengue severity in Nicaragua. Am J Trop Med Hyg 73:1059–1062

Gan VC (2014) Dengue: moving from current status to the state of-the-art treatment. Curr Treat Opt Infect Dis 6:208–226

Hassan S, Tariq I, Khalid A, Karim S (2013) Comparative clinical study on the effectiveness of homeopathic combination remedy with standard maintenance therapy for dengue fever. Trop J Pharm Res 12:767–770

Huang N, Hung W, Tsai W, Lai F, Lin Y, Huang M, Chen J, Lin W, Weng J, Chang T (2017) Ficus septica plant extracts for treating dengue virus in vitro. Peer J 5:3448. https://doi.org/10.7717/peerj.3448

Jacobs J, Fernandez EA, Merizalde B, Avila-Montes GA, Crothers D (2007) The use of homeopathic combination remedy for dengue fever symptoms: a pilot RCT in Honduras. Homeopathy:22–26

Jain M, Ganju L, Katiyal A, Padwad Y, Mishra KP, Chanda S (2008) Effect of *Hippophae rhamnoides* leaf extract against dengue virus infection in human blood-derived macrophages. Phytomedicine 15(10):793–799

Kadir SLA, Yaakob H, Zulkifli RM (2013) Potential anti-dengue medicinal plants: a review. J Nat Med 67:677–689

Kala CP (2012) Leaf juice of *Carica papaya* L.: a remedy of dengue fever. Med Aromat Plants 1:109
Mahesh S, Mahesh M, Vithoulkas G (2018) Could homeopathy become an alternative therapy in dengue fever? An example of 10 case studies. J Med Life 11(1):75–82
Mutheneni SR, Morse AP, Caminade C, Upadhyayula SM (2017) Dengue burden in India: recent trends and importance of climatic parameters. Emerg Microb Infect 6(8):e70. https://doi.org/10.1038/emi.2017.57
National Health Portal of India (n.d.): Dengue Fever. nhp.gov.in/disease/musculo-skeletal-bone-joints–/dengue-fever. Accessed 06 Jan 2018
Novaes AR (2015) Homeopathic intervention in users treatment network public advised with Dengue in Victory, Brazil. Proceedings of 70th Congress of Liga Medicorum Homoeopathica Internationalis. http://www.lmhi2015.org
Singh PK, Rawat P (2017) Evolving herbal formulations in management of dengue fever. J Ayur Integr Med 8(3):207–210
WHO (2016) Dengue vaccine: WHO position paper – July 2016 30(91):349–364

Part III

Bioactive Compounds from Plants and Microbes

21 Development of Immunoassays for Ginsenosides in Ginseng

Hiroyuki Tanaka

Abstract

Ginsenosides separated by silica gel thin-layer chromatography (TLC) blotted onto a PVDF membrane that was treated with a $NaIO_4$ solution and bovine serum albumin (BSA) resulted in a ginsenoside–BSA conjugate. The blotted spots were stained with anti-ginsenoside Rb1 (G-Rb1) and Rg1 (G-Rg1) monoclonal antibodies (MAbs). A newly established immunostaining method, namely, eastern blotting, was applied to determine whether ginsenosides, which are used in traditional Chinese medicine (TCM), contain protopanaxadiol and/or protopanaxatriol. This is a new method of separating the ginsenoside molecule into two functional parts using a simple and well-known chemical reaction. The sugar parts are oxidized by $NaIO_4$ to produce dialdehydes, which react with amino groups of the protein and covalently bind to the adsorbent PVDF membrane. The MAb binds to the aglycon part of the ginsenoside molecule for immunostaining. Double staining of ginsenosides using anti-G-Rb1 and anti-G-Rg1 MAbs in eastern blotting allows for complete identification of ginsenosides in the *Panax* species. The immunoaffinity concentration of G-Rb1 was determined using an immunoaffinity column conjugated with anti-G-Rb1 MAb that produced the knockout extract, which may be useful for pharmacological investigations. To concentrate and determine the amount of G-Rb1 in *P. japonicus*, the crude extract of *P. japonicus* was fractionated using an immunoaffinity column conjugated with anti-G-Rb1 MAb. Although G-Rb1 was expected to be a component of *P. japonicus* through enzyme-linked immunosorbent assay (ELISA) analysis, two ginsenosides, namely, chikusetsusaponins III and IV, which have protopanaxadiol as an aglycon, were identified using eastern blotting.

H. Tanaka (✉)
Faculty of Pharmaceutical Sciences, Sanyo-Onoda City University, Sanyo-Onoda, Yamaguchi, Japan
e-mail: htanaka@rs.socu.ac.jp

S. M. Khasim et al. (eds.), *Medicinal Plants: Biodiversity, Sustainable Utilization and Conservation*, https://doi.org/10.1007/978-981-15-1636-8_21

Keywords
Ginsenoside · ELISA · Immunostaining · Monoclonal antibody · Ginseng

21.1 Introduction

With the rapid development of the molecular biosciences and their biotechnological applications, immunoassays using monoclonal antibodies (MAbs) against drugs and low-molecular-weight bioactive compounds have become an important tool due to their specificity for receptor binding analyses, enzyme assays, and quantitative and qualitative analytical techniques in both animals and plants. The immunoblotting method is based on a western blotting technique that utilizes antigen–antibody binding properties and has provided a specific and sensitive detection method for higher molecule analytes, such as peptides and proteins. Previously, we prepared many MAbs against naturally occurring bioactive compounds such as forskolin (Sakata et al. 1994), crocin (Xuan et al. 1999), solamargine (Ishiyama et al. 1996), opium alkaloids (Shoyama et al. 1996), marijuana compounds (Tanaka et al. 1996), ginsenosides (Tanaka et al. 1996; Fukuda et al. 2000a), saikosaponin A (Zhu et al. 2004), paeoniflorin (Lu et al. 2003), sennosides (Morinaga et al. 2000; Morinaga et al. 2001), ginkgolic acid (Loungratana et al. 2004), glycyrrhizin (Shan et al. 2001), and berberine (Kim et al. 2004) and established individual competitive enzyme-linked immunosorbent assays (ELISAs) as highly sensitive, specific, and simple methodologies. As an extension of this approach, an immunostaining method using anti-solamargine MAb was established by us (Tanaka et al. 1997).

Ginseng, the crude drug of the *Panax ginseng* root, is one of the most important components of traditional Chinese medicine (TCM). It has been used to enhance stamina and the capacity to cope with fatigue and physical stress; it has been used as a tonic against cancers, disturbances of the central nervous system, and hypothermia; and it has been used to boost carbohydrate and lipid metabolism, immune function, the cardiovascular system, and radioprotections. Its major active components are the ginsenosides, which consist of protopanaxatriol and/or protopanaxadiol that has a dammarane skeleton in their molecules. It is well-known that the concentrations of ginsenosides vary in the ginseng root or the root extracts depending on the method of extraction, subsequent treatment (Kitagawa et al. 1989), or even the season of its collection (Tanaka 1989). Therefore, standardization of quality is required in TCM. Moreover, *Panax japonicus* is morphologically different from *P. ginseng*. It is evident that major components of ginsenosides are oleanane-type saponins such as chikusetsusaponins. Therefore, the content of ginsenosides might be low.

To control the quality of ginseng prescribed in TCM, we previously prepared anti-G-Rb1, anti-G-Rg1, and anti-G-Re MAbs (Tanaka et al. 1999; Fukuda et al. 2000a), We also prepared an ELISA, a new staining method for ginsenosides (eastern blot), and an immunoaffinity concentration method for quantitative analysis of samples containing low concentrations of G-Rb1 (Fukuda et al. 2000b).

As part of our ongoing studies on MAbs against naturally occurring bioactive compounds, we review here a new eastern blotting method for ginsenosides in TCM-prescribed ginseng and the double staining for ginsenosides in the crude drug of the *Panax* species using anti-G-Rb1 and anti-G-Rg1 MAbs. The immunoaffinity concentration of G-Rb1 using an immunoaffinity column is also discussed.

21.2 Eastern Blotting (Tanaka et al. 1997)

Ginsenosides were applied to a silica gel TLC plate and developed with *n*-BuOH/EtOAc/H_2O (15:1:4). The developed TLC plate was dried and sprayed with a blotting solution mixture of *i*-PrOH/MeOH/H_2O (1:4:20 by volume). It was placed on a stainless steel plate and covered with a PVDF membrane. Next, a glass microfiber filter sheet was placed over the plate, and the whole assembly was pressed evenly for 50 s with a 120 °C hot plate as previously described (Tanaka et al. 1997) with some modifications. The PVDF membrane was separated from the TLC plate and dried.

The blotted PVDF membrane was dipped in water containing $NaIO_4$ and stirred at room temperature for 1 h. After washing with water, 50 mM carbonate buffer solution (pH 9.6) containing BSA was added and stirred at room temperature for 3 h. After the PVDF membrane was washed with PBS, it was treated with PBS containing 5% skim milk for 3 h to reduce nonspecific adsorption. The PVDF membrane was immersed in anti-G-Rb1 MAb and stirred at room temperature for 1 h. After the PVDF membrane was washed twice with PBS containing 0.05% Tween 20 and water, a 1:1000 dilution of peroxidase-labeled goat anti-mouse IgG in PBS containing 0.2% gelatin was added and stirred at room temperature for 1 h. The PVDF membrane was washed twice with TPBS and water and then exposed to freshly prepared 1 mg/mL 4-chloro-1-naphthol/0.03% H_2O_2 in PBS for 10 min at room temperature. The reaction was stopped by washing with water, and the immunostained PVDF membrane was allowed to dry.

For staining by anti-G-Rg1 MAb, the blotted PVDF membrane was treated in the same way as was the anti-G-Rb1 MAb except it was exposed to 0.2 mg/mL of 3-amino-9-ethylcarbazole/0.03% H_2O_2 in acetate buffer (0.05 M, pH 5.0) containing 5% *N*,*N*-dimethylformamide.

21.3 Preparation of Immunoaffinity Column Using Anti-G-Rb1 MAb and Immunoaffinity Concentration of G-Rb1 (Fukuda et al. 2000b)

Purified anti-G-Rb1 MAb in diluted Bio-Rad Affi-Gel Hz coupling buffer was dialyzed against the coupling buffer two times. $NaIO_4$ solution was added to the MAb solution and stirred gently at room temperature. After the reaction, glycerol was added to the reaction mixture and stirred to inactivate $NaIO_4$ before being dialyzed. Affi-Gel Hz hydrazide gel was added to the above reaction mixture, resulting in a hydrazone gel and an immunoaffinity gel, which were packed into a plastic minicolumn.

The ginseng root extracts were redissolved in PBS and then filtered with a 0.45-μm MILEX-HV filter (Millipore) to remove insoluble portions. The filtrate was loaded on the immunoaffinity column and allowed to stand overnight at 4 °C. The column was washed with the washing buffer solution (40 mL) and then eluted with 100 mM AcOH buffer containing 0.5 M KSCN and 20% MeOH (pH 4.0). The fraction containing G-Rb1 was used for ELISA to determine its concentration and then subjected to TLC with $CHCl_3$/MeOH/H_2O (7:4:1) and *n*-BuOH/AcOH/H_2O (15:1:4), used as developing solvents, followed by eastern blotting.

Concentration of ginsenoside from the crude extracts of the *P. japonicus* root by immunoaffinity column was carried out in the same way as was for *P. ginseng* as described above. Individual fractions containing ginsenosides were performed for eastern blotting as indicated above.

21.4 Eastern Blotting of Ginsenosides

Although western blotting is a common assay for separating substances of high molecular weight, it has not been used for small molecules, as direct immunostaining of such compounds on a TLC plate has not been performed to date. Therefore, a new method for small molecular compounds is needed. Moreover, fixing them onto the membrane also requires a new methodology. Previously, we successfully separated the functional groups of small molecular compounds such as solasodine glycosides into a part of an epitope and fixed them onto a membrane as follows (Tanaka et al. 1997). The blotted PVDF membrane was treated with $NaIO_4$ solution. This reaction enhanced the fixing of solasodine glycoside via solasodine glycoside–BSA conjugates on the PVDF membrane. The PVDF membrane incubated in the absence of $NaIO_4$ was essentially free of staining for solasodine glycoside. We have applied this new methodology to various glycosides such as glycyrrhizin (Shan et al. 2001) and saikosaponins (Morinaga et al. 2006a). In this paper, we investigate the eastern blotting of ginsenosides.

The H_2SO_4 staining and eastern blotting of ginsenoside standards and the TCM using anti-G-Rb1 MAb are shown in Fig. 21.1. It is impossible to determine the ginsenosides by TLC stained with H_2SO_4 because of the complicated profile, shown in Fig. 21.1. However, clear staining of G-Rb1 was obtained using eastern blotting. Although H_2SO_4 staining detected all standard compounds, eastern blotting showed only limited staining of G-Rb1 and two other ginsenosides, G-Rc and G-Rd, cross-reactivities of which were 0.02% (Fig. 21.1). The eastern blotting method was considerably more sensitive than H_2SO_4 staining. Furthermore, Kikyoto and Daiokanzoto prescriptions, which do not contain ginseng, showed no spot of G-Rb1. This finding suggests that aglycon, protopanaxadiol, and a part of the sugars may be of importance to immunization and may function as an epitope in the structure of ginsenosides. In addition, the specific reactivity of the sugar moiety in the ginsenoside molecule against anti-G-Rb1 MAb may be modified by $NaIO_4$ treatment of ginsenosides on the PVDF membrane, causing G-Rc and G-Rd to become detectable by eastern blotting.

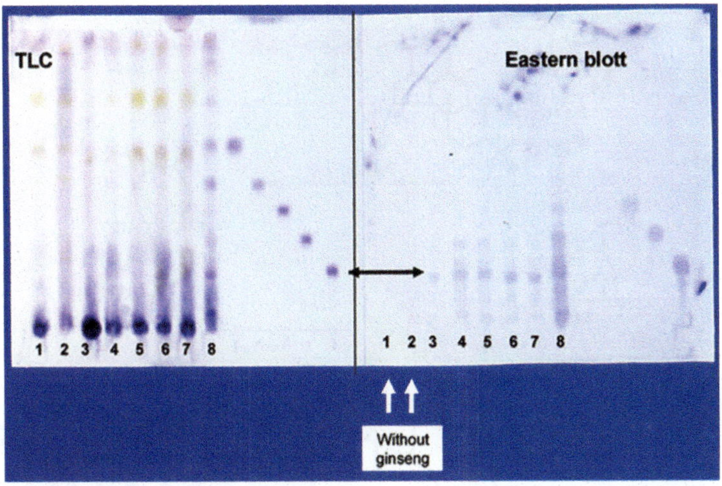

Fig. 21.1 Eastern blotting of ginsenosides in traditional Chinese medicine (TCM) prescriptions by anti-ginsenoside Rb1 MAb. Samples: *1*, Kikyoto; *2*, Daiokanzoto; *3*, Ninjin'yoeito; *4*, Shikunshito; *5*, Ninjinto; *6*, Hangeshashinto; *7*, Sho-saiko-to; *8*, crude extract of ginseng. Samples 1 and 2 do not contain ginseng. Standard of ginsenosides indicated G-Rg1, G-Re, G-Rd, G-Rc, and G-Rb1 from upper

We report here a new methodology to separate the G-Rb1 molecule into two functional parts. The sugar parts are oxidized to give dialdehydes, which react with lysine and/or arginine amino groups of proteins that can bind strongly to the adsorbent membrane, PVDF. The aglycon part of the G-Rb1 molecule is bound and can be visualized by using the anti-G-Rb1 MAb. The method is shown diagrammatically in Fig. 21.2.

Although purple staining of G-Rg1 was expected because of the 3-amino-9-ethylcarbazole substrate, all ginsenosides in the mixture of anti-G-Rg1 and anti-G-Rb1 MAbs and the pair of substrates, including G-Rg1, G-Re, G-Rd, G-Rc, and G-Rb1, were stained blue (data not shown). This finding suggested that the sensitivities of the 3-amino-9-ethylcarbazole and 4-chloro-1-naphthol substrates might be different. Therefore, we performed successive staining of the membrane using anti-G-Rg1 and anti-G-Rb1. Finally, we succeeded in the double staining of the ginsenosides, in which G-Rg1 and G-Re were stained purple and the others were stained blue, as shown in Fig. 21.3. These results indicate that both antibodies can distinguish among individual aglycons, protopanaxatriol, and protopanaxadiol. For this application, the crude extracts of various *Panax* species were analyzed using the double-staining system (Fukuda et al. 2001), allowing all ginsenosides to be determined (Fig. 21.3).

Interestingly, the staining color may indicate pharmacological activity. For example, the purple spots show ginsenosides that have the stimulation activity as the central nervous system. The blue color shows ginsenosides that have the depression effect for the central nervous system. Moreover, the R_f value of the ginsenosides is

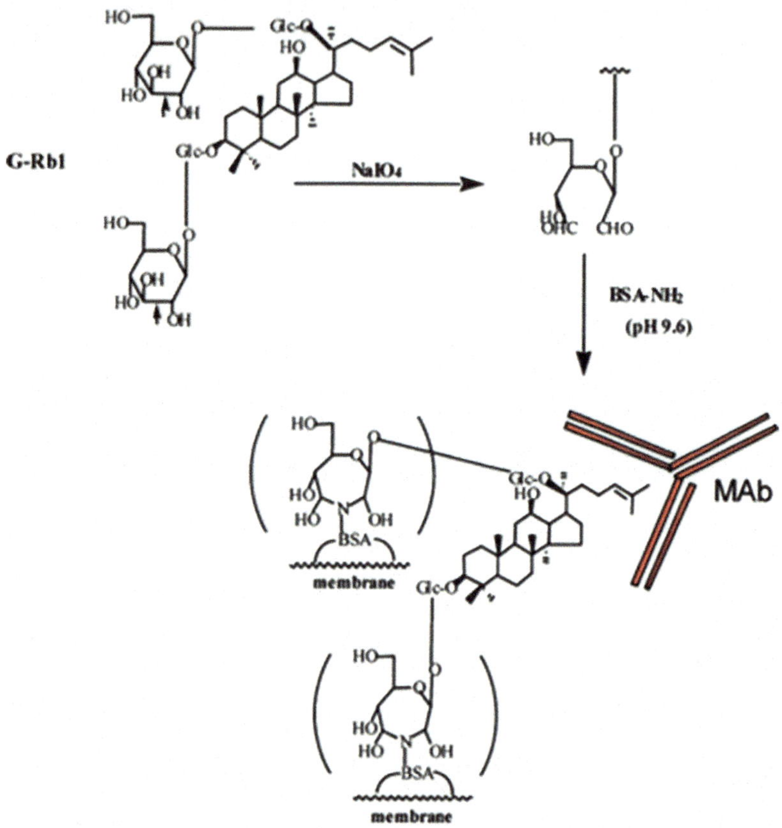

Fig. 21.2 Mechanism of eastern blotting

roughly the same as the number of sugars attached to the aglycon. Therefore, both analyses indicate that it is possible that the aglycon and the number of sugars elucidate the structure of ginsenosides.

We also investigated the Araliaceae species by eastern blotting using anti-G-Rb1 MAb (Fig. 21.4). ELISA analysis and the eastern blotting profile of *Kalopanax pictus* Nakai (Fig. 21.4b, line 13, as indicated by an arrow) suggest that this species may contain G-Rb1. Using this information, we successfully isolated G-Rb1 from the bark of *K. pictus* even though the concentration is 0.0009% dry weight as reported previously (Tanaka et al. 2005).

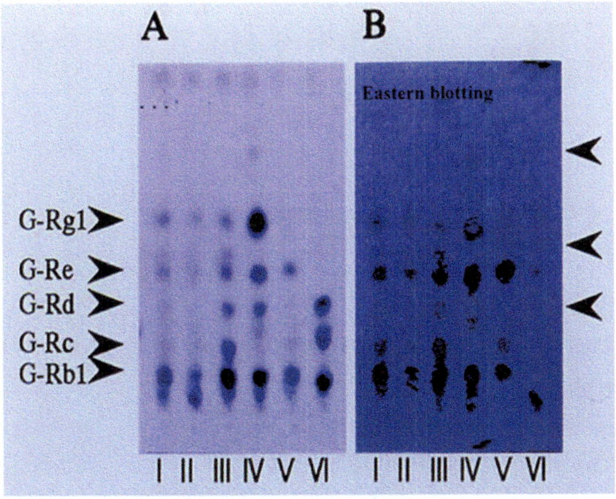

Fig. 21.3 Double staining of eastern blotting for ginsenosides contained in various ginseng samples using anti-G-Rb1 and anti-G-Rg1 monoclonal antibodies: (**a**) TLC profile stained by sulfuric acid; (**b**) eastern blotting by anti-G-Rb1 and anti-G-Rg1 monoclonal antibodies I, II, III, IV, V, and VI indicated white ginseng, red ginseng, fibrous ginseng (*Panax ginseng*), *Panax notoginseng*, *Panax quinquefolius*, and *Panax japonicus*, respectively. Upper purple color spots and lower blue color spots were stained by anti-G-Rg1 and anti-G-Rb1 monoclonal antibodies, respectively

21.5 Immunoaffinity Concentration by Immunoaffinity Column Conjugated with MAb for the Determination of Ginsenosides (Fukuda et al. 2000b)

To confirm the concentration for G-Rb1 by immunoaffinity column conjugated with anti-G-Rb1 MAb, a crude extract of *P. ginseng* roots was loaded onto the immunoaffinity column and washed with the washing solvent. The fractions 1–8 (0–16 mL) containing overcharged G-Rb1 were determined by ELISA. G-Rc, G-Rd, G-Re, and G-Rg1 were also detected in these fractions by the eastern blotting procedure (Fig. 21.5). A sharp peak appeared around fractions 19–25 (40–52 mL), which contained G-Rb1. However, G-Rb1 purified by the immunoaffinity column was still contaminated by a small amount of malonyl-G-Rb1 as detected by eastern blotting. This compound has almost the same cross-reactivity as G-Rb1 (data not shown). Therefore, the mixture was treated with a mild alkaline solution at room temperature for 1 h to give pure G-Rb1. Overcharged G-Rb1, eluted with washing solution, was repeatedly loaded and finally isolated in pure form. These results confirmed that the immunoaffinity column can concentrate G-Rb1 from the ginsenoside mixture.

After washing, fractions were deionized, and the solvent was lyophilized. Figure 21.6 shows the TLC profile of the purification step. Lines 1, 2, and 3 were the crude extract, the washing fraction, and the eluted fraction, respectively. Interestingly, the washing fraction contained all of the compounds in the ginseng crude extract except G-Rb1. It is evident that the G-Rb1 molecule can be eliminated using an immunoaffinity column conjugated with anti-G-Rb1 MAb and that the washing

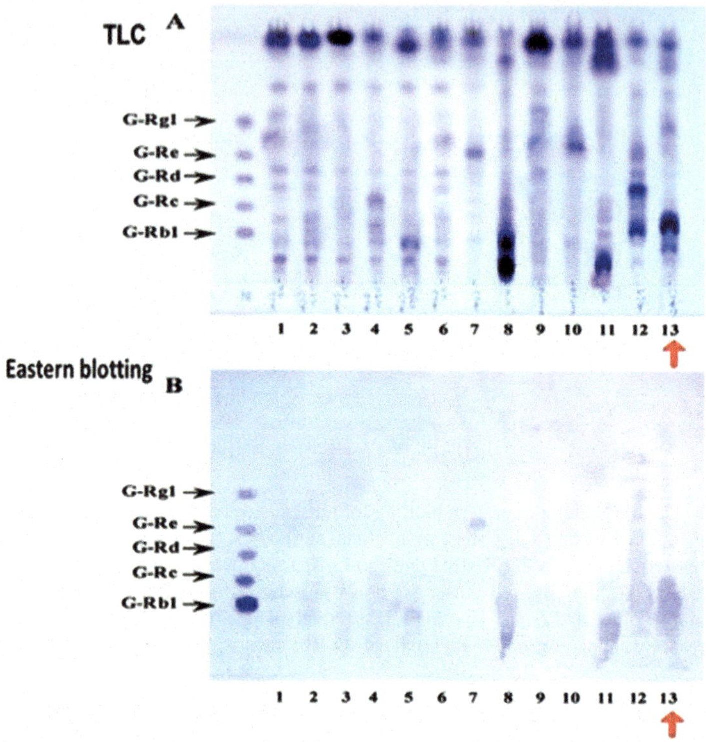

Fig. 21.4 Ginsenosides in Araliaceae plants analyzed by eastern blotting. Arrow indicates *Kalopanax pictus*

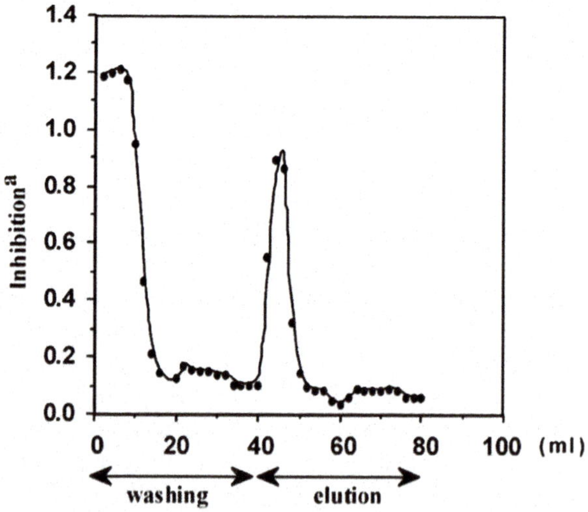

Fig. 21.5 Elution profile of *Panax ginseng* crude extract used immunoaffinity column monitoring by ELISA using anti-G-Rb1 MAb

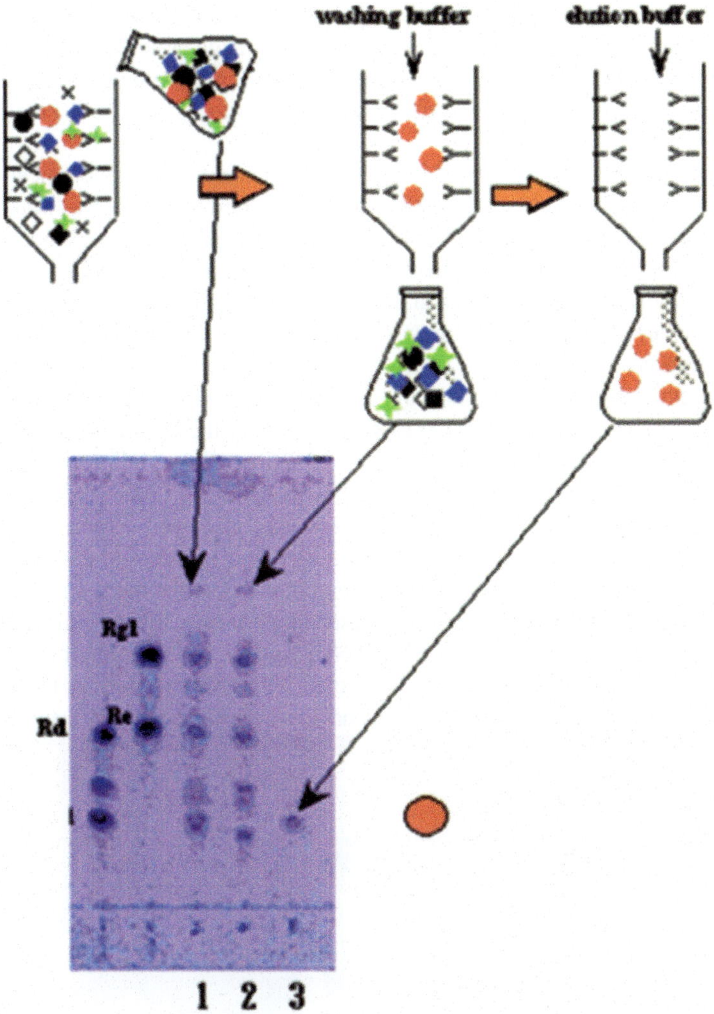

Fig. 21.6 Preparation of knockout extract eliminated G-Rb1 from *Panax ginseng* crude extract using immunoaffinity column conjugated with anti-G-Rb1 MAb. Lines 1, 2, and 3 indicate crude extract, knockout extract, and purified G-Rb1, respectively. Red spot shows G-Rb1

fraction is knocked out only by the G-Rb1antigen. Therefore, we named this washing fraction a knockout extract. This knockout extract may be useful for the determination of real pharmacologically active principles in TCM. The antibody was stable when exposed to the eluent and the immunoaffinity column, indicating almost no decrease in capacity (20 μg/mL gel) after repeated use of more than 10 cycles under the same conditions, as reported for a single-step separation of forskolin from a crude extract of *Coleus forskohlii* root (Yanagihara et al. 1996). Furthermore, because we succeeded in the preparation of MAbs having a wide cross-reactivity with molecules such as anti-solamargine MAb (Ishiyama et al. 1996), anti-saikosaponin A (Zhu et al. 2006), and G-Re (Morinaga et al. 2006b), related total saponins can be concentrated using an immunoaffinity column conjugated with Mab, as reported previously in the case of solasodine glycosides (Putalun et al. 1999).

Panax japonicus is distributed in Japan and China and is morphologically different from the other *Panax* species. Yahara et al. reported that no G-Rb1 was found in *P. japonicus* (Yahara et al. 1977) but instead isolated oleanane-type saponins called chikusetsusaponins and determined their structure. Morita et al. examined the varieties of *P. japonicus* by chemical analysis of saponins (Morita et al. 1985) and found that the concentration of G-Rb1 may exist at trace levels. However, we found higher concentrations of G-Rb1 using ELISA compared to previous reports (Fukuda et al. 2000c), although approximately half the concentration of G-Rb1 was found using HPLC analysis as was found using ELISA. To clarify these differences, we used an immunoaffinity column for concentration of G-Rb1. The crude root extract of *P. japonicus* was loaded onto the immunoaffinity column and washed with the washing solvent and an elution solvent, as previously discussed. Figures showed the H_2SO_4 staining (21.7a) and the eastern blotting (21.7b) profiles of the two fractions separated by the immunoaffinity column. Fraction 1 eluted with the washing solvent showed many spots, including chikusetsusaponins, similar to the original extract of *P. japonicus*. However, fraction 2 contained a higher concentration of compound 1, although two other bands were still detected with eastern blotting. Compound 1 clearly indicated a dammarane saponin having a protopanaxadiol framework and

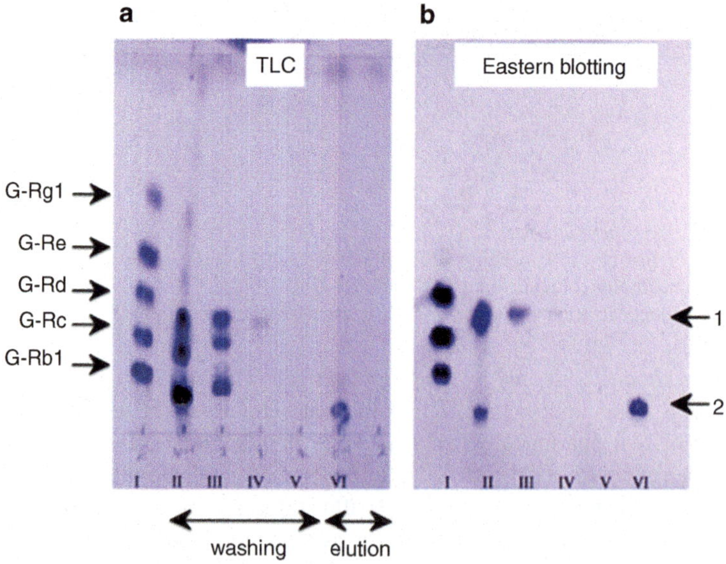

Fig. 21.7 Purification and determination of ginsenosides of *P. japonicus* by immunoaffinity column and eastern blotting

three sugars, a number similar to the R_f value of G-Rd, suggesting that compound 1 is chikusetsusaponin III. We were ultimately able to identify compound 1 as chikusetsusaponin III in a direct comparison with an authentic sample.

A clear unknown band of compound 2 appeared in fraction 5 eluted with the elution solvent. G-Rb1 was not detected by eastern blotting (Fig. 21.7b), although it was detected by TLC (Fig. 21.7a). This finding suggests that compound 2 has a molecular structure and cross-reactivity similar to those of G-Rb1 and seems to be related to the ginseng saponin that has a protopanaxadiol aglycon. Moreover, compound 2 might have the same sugar fragments but possesses five sugar moieties, as indicated by their R_f values. This evidence suggests that compound 2 might be chikusetsusaponin III-20-*O*-gentiobiose, chikusetsusaponin IV, which was identified by direct comparison with an authentic sample (Fukuda et al. 2000c). Therefore, we concluded that *P. japonicus* did not contain G-Rb1 but did contain chikusetsusaponin IV and that it has the same aglycon and the same sugar component as chikusetsusaponin.

Acknowledgments This work was supported by JSPS KAKEN Grant Numbers 11470470, 17750074, and 16K08296.

References

Fukuda N, Tanaka H, Shoyama Y (2000a) Formation of monoclonal antibody against a major ginseng component, ginsenoside Rg1 and its characterization. Cytotechnology 34:197–204

Fukuda N, Tanaka H, Shoyama Y (2000b) Isolation of the pharmacologically active saponin ginsenoside Rb1 from ginseng by immunoaffinity column chromatography. J Nat Prod 63:283–285

Fukuda N, Tanaka H, Shoyama Y (2000c) Application of ELISA, Western blotting and immunoaffinity concentration for survey of ginsenosides in crude drugs of Panax species and traditional Chinese herbal medicines. Analyst 125:1425–1429

Fukuda N, Tanaka H, Shoyama Y (2001) Double staining of ginsenosides by Western blotting using anti-ginsenoside Rb1 and Rg1 monoclonal antibodies. Biol Pharm Bull 24:1157–1160

Ishiyama M, Shoyama Y, Murakami H, Shinohara H (1996) Production of monoclonal antibodies and development of an ELISA for solamargine. Cytotechnology 18:153–158

Kim JS, Tanaka H, Shoyama Y (2004) Immunoquantitative analysis for berberine and its related compounds using monoclonal antibody in herbal medicines. Analyst 129:87–91

Kitagawa I, Taniyama T, Yoshikawa M, Ikenishi Y, Nakagawa Y (1989) Chemical studies on crude drug processing. VI, Chemical structures of malonyl-ginsenosides Rb1, Rb2, Rc and Rd isolated from the root of Panax ginseng C.A. Meyer. Chem Pharm Bull 37:2961–2966

Loungratana P, Tanaka H, Shoyama Y (2004) Production of monoclonal antibody against ginkgolic acids in Ginkgo biloba Linn. Am J Chin Med 32:33–48

Lu Z, Morinaga O, Tanaka H, Shoyama Y (2003) A quantitative ELISA using monoclonal antibody to survey paeoniflorin and albiflorin in crude drugs and traditional Chinese herbal medicines. Biol Pharm Bull 26:862–866

Morinaga O, Tanaka H, Shoyama Y (2000) Production of monoclonal antibody against a major purgative component, sennoside A, its characterization and ELISA. Analyst 125:1109–1113

Morinaga O, Nakajima S, Tanaka H, Shoyama Y (2001) Production of monoclonal antibodies against a major purgative component, sennoside B, their characterization and use in ELISA. Analyst 126:1372–1376

Morinaga O, Zhu S, Tanaka H, Shoyama Y (2006a) Visual detection of saikosaponins by on-membrane immunoassay and estimation of traditional Chinese medicines containing Bupleuri radix. Biochem Biophys Res Commun 346:687–692

Morinaga O, Tanaka H, Shoyama Y (2006b) Detection and quantification of ginsenoside Re in ginseng samples by a chromatographic immunostaining method using monoclonal antibody against ginsenoside Re. J Chromatogr B 830:100–104

Morita T, Tanaka O, Kohda H (1985) Saponin composition of rhizomes of Panax japonicus collected in South Kyushu, Japan and its significance in oriental traditional medicine. Chem Pharm Bull 33:3852–3858

Putalun W, Tanaka H, Shoyama Y (1999) Rapid separation of solasodine glycosides by an immunoaffinity column using anti-solamargine monoclonal antibody. Cytotechnology 31:151–156

Sakata R, Shoyama Y, Murakami H (1994) Production of monoclonal antibodies and enzyme immunoassay for typical adenylate cyclase activator, forskolin. Cytotechnology 16:101–108

Shan SJ, Tanaka H, Shoyama Y (2001) Enzyme-linked immunosorbent assay for glycyrrhizin using anti-glycyrrhizin monoclonal antibody and a new eastern blotting for glucuronides of glycyrrhetinic acid. Anal Chem 73:5784–5790

Shoyama Y, Fukada T, Murakami H (1996) Production of monoclonal antibodies and ELISA for thebaine and codeine. Cytotechnology 19:55–61

Tanaka O (1989) Saponin-composition of Panax species. In: Shibata S, Ohtsuka Y, Saito H (eds) Recent advances in ginseng studies. Hirokawa Publishing, Tokyo, pp 43–47

Tanaka H, Goto Y, Shoyama Y (1996) Monoclonal antibody based enzyme immunoassay for marihuana (cannabinoid) compounds. J Immunoassay 17:321–342

Tanaka H, Putalun W, Tsuzaki C, Shoyama Y (1997) A simple determination of steroidal alkaloid glycosides by thin-layer chromatography immunostaining using monoclonal antibody against solamargine. FEBS Lett 404:279–282

Tanaka H, Fukuda N, Shoyama Y (1999) Formation of monoclonal antibody against a major ginseng component, ginsenoside Rb1 and its characterization. Cytotechnology 29:115–120

Tanaka H, Fukuda N, Yahara S, Isoda S, Yuan CS, Shoyama Y (2005) Isolation of ginsenoside Rb1 from Kalopanax pictus by Eastern blotting using anti-ginsenoside Rb1 monoclonal antibody. Phytother Res 19:255–258

Xuan L, Tanaka H, Xu Y, Shoyama Y (1999) Preparation of monoclonal antibody against crocin and its characterization. Cytotechnology 29:65–70

Yahara S, Kasai R, Tanaka O (1977) New dammarane type saponins of leaves of Panax japonicus CA Meyer 1. Chikusetsusaponins-L5, Chikusetsusaponins-L9A and Chikusetsusaponins-L10. Chem Pharm Bull 25:2041–2047

Yanagihara H, Sakata R, Minami H, Tanaka H, Shoyama Y, Murakami H (1996) Immunoaffinity column chromatography against forskolin using an anti-forskolin monoclonal antibody and its applications. Anal Chim Acta 335:63–70

Zhu S, Shimokawa S, Tanaka H, Shoyama Y (2004) Development of an assay system for saikosaponin a using anti-saikosaponin a monoclonal antibodies. Biol Pharm Bull 27:66–71

Zhu S, Shimokawa S, Shoyama Y, Tanaka H (2006) A novel analytical ELISA-based methodology for pharmacologically active saikosaponins. Fitoterapia 77:100–108

Elicitation of Flavonoids in *Kalanchoe pinnata* by *Agrobacterium rhizogenes*-Mediated Transformation and UV-B Radiation

22

Amalia Fkiara, Gregorio Barba-Espín, Rime El-Houri, Renate Müller, Lars Porskjær Christensen, and Henrik Lütken

Abstract

Kalanchoe pinnata from the stonecrop family (Crassulaceae) is a medicinal plant with high content of bioactive compounds. The plant is known for its anti-allergic, antioxidant, anti-inflammatory, antimicrobial and antibacterial activities, which are mainly attributed to flavonoids. Low yields of bioactive compounds in medicinal plants have led to new strategies for enhancing their biosynthetic capacity in order to increase the content of specialized metabolites. Herein, the accumulation of specialized metabolites in plants can be triggered by elicitation methods. In the current study, different elicitation strategies were conducted towards the enhancement of bioactive compounds in *K. pinnata* leaves. We investigated the effect of natural transformation with the *Agrobacterium rhizogenes* strain A4 as means of biological elicitation on the total content of flavonoids in the leaves of *K. pinnata.* Furthermore, the effect of supplemental UV-B radiation, as physical elicitor, was assessed on the total flavonoid content of both wild-type (WT) and *rol*-transformed plants. The combined effect of the two mentioned elicitation methods was also examined. The data showed that presence of *rol* genes resulted in an increase of 24% in the total flavonoid content compared to WT plants. The supplemental UV-B radiation increased the total content of flavonoids with 95% and 89% in the WT and the *rol*+ plants, respectively. Collectively, a synergistic

A. Fkiara · G. Barba-Espín · R. Müller · H. Lütken (✉)
Section for Crop Sciences, Department of Plant and Environmental Sciences, Faculty of Science, University of Copenhagen, Taastrup, Denmark
e-mail: hlm@plen.ku.dk

R. El-Houri · L. P. Christensen
Section of Biotechnology, Department of Chemical Engineering, Biotechnology and Environmental Technology, University of Southern Denmark, Odense, Denmark

S. M. Khasim et al. (eds.), *Medicinal Plants: Biodiversity, Sustainable Utilization and Conservation*, https://doi.org/10.1007/978-981-15-1636-8_22

effect was shown as the combination of the two factors dramatically increased (133%) the total flavonoid content in *K. pinnata* leaves.

Keywords
A. rhizogenes · Elicitation · Flavonoids · *rol*+ · UV-B radiation

22.1 Introduction

In recent years, medicinal plants have been receiving increasing attention from the pharmaceutical industries as the interest in alternative therapies in the worldwide population is growing (Newman and Crag 2007). With the insecurity that emerged from the danger and toxicity of using some synthetic drugs and antibiotics, there has been a general increase in the perception that naturally derived products are safer than synthetic (Husain 2010). *Kalanchoe pinnata* (also known as *Bryophyllum pinnatum*), from the stonecrop family (Crassulaceae), is a popular plant both as ornamental and in folk medicine in many regions of the world. In the past, *K. pinnata* was used for alleviation or prevention of respiratory, stomach and hepatic problems, against tumour, kidney and spleen disorders, diabetes and vaginal diseases (Rahmatullah et al. 2010). Nowadays, *K. pinnata* is used for the anti-tumourous (Supratman et al. 2001), antihypertensive (Bopda et al. 2014), anti-allergic (Cruz et al. 2012), antimicrobial, antioxidant (Tatsimo et al. 2012) and anti-inflammatory (Nayak et al. 2010; Chibli et al. 2014) properties of its leaf extracts. These medicinal properties are attributed to the major classes of bioactive compounds in the plant, i.e. flavonoids, alkaloids and terpenoids, which constitute the active components of numerous herbal drugs (Kabera et al. 2014).

Agrobacterium rhizogenes-mediated transformation is a promising strategy for increasing the content of bioactive compounds in plants. *Agrobacterium rhizogenes* is a soilborne pathogenic bacterium causing the hairy root disease by infecting and inserting specific genes into the plant host's genome from wounded sites (White et al. 1985). The root-inducing (Ri) phenotype is the result of transfer, integration and expression of bacterial transfer DNA (T-DNA) in the transformed plant. In agropine strains of *A. rhizogenes*, the T-DNA is split into left (T_L) and right (T_R) fragments. The T_L-DNA comprises 18 open reading frames (ORFs) of which four *root oncogenic loci* (*rol* genes), termed *rol*A, *rol*B, *rol*C and *rol*D, are the major determinants for the development of characteristic 'hairy roots' at the infection site and are sufficient for obtaining the *A. rhizogenes*-induced phenotypes (Casanova et al. 2005; Christey 2001; Christensen and Müller 2009). Based on the naturally occurring *A. rhizogenes rol* genes, transformation without the use of recombinant DNA can be termed 'natural transformation', and plants derived from this platform are considered as non-GMO in several countries in Europe (European Union 2001). The expression of the *rol* genes, alone or combined, often leads to profound metabolic alterations that included increased accumulation of bioactive compounds. Moreover, transformed hairy roots often grow faster and are considered to be genetically stable

(Zhou et al. 2007). It has been reported that *Vitis amurensis* plant cells transformed with the *rol*B gene had a 100-fold increased production of the stilbenoid resveratrol (Kiselev et al. 2007), and *rol*C transformation of root cultures of *Atropa belladonna* increased the production of the alkaloids hyoscyamine and scopolamine 12-fold (Bonhomme et al. 2000). Moreover, *Panax ginseng* root cultures transformed with *rol*C displayed a production of ginsenosides of more than 6% dry weight, which is close to the maximum biosynthesis capacity (Bulgakov 2008).

As plants use sunlight for photosynthesis, they are as a consequence exposed to the ultraviolet (UV) radiation that is present in sunlight. Although the shortwave band of the terrestrial solar spectrum (UV-B radiation) accounts for less than 0.5% of the total solar energy reaching the surface of the earth, its high energy causes damages to RNA, DNA, proteins and lipids (Heisler et al. 2003). Therefore, plants have developed mechanisms to protect themselves from the harmful effect of UV-B radiation. One of the protective mechanisms is the accumulation of phenolic pigments in epidermal layers of leaves and stems, acting as filters that absorb UV-B light (Matsuura et al. 2012; Nascimento et al. 2015). Of all classes of phenolic compounds, flavonoids are regarded as the most relevant for UV protection.

In the present study, the role of *A. rhizogenes*-mediated transformation and UV-B radiation as enhancers of flavonoid contents was explored. Total flavonoids were determined by HPLC-DAD in wild-type (WT) and *rol*+ plants, prior and after exposure to supplemental UV-B radiation to determine potential synergistic effects.

22.2 Plant Material and Propagation

WT and *rol*+ *K. pinnata* plants were provided by Knud Jepsen A/S and used as maternal plants for further propagation. The *rol*+ plants were generated by *A. rhizogenes* (strain ATCC43057 containing plasmid pRiA4)-mediated transformation following Christensen et al. (2008). The maternal plants were grown in greenhouse for 9 months, in 2 L pots in a substrate mixture composed of peat supplemented with clay and silica (Weibulls Horto AB, Hammenhög, Sweden), under a 16 h day/8 h night photoperiod (23 °C/20 °C).

For propagation, 10-cm nodal cuttings were excised from the mother plants, dipped in auxin powder (Floramon A 1%, Novo Trade ApS, Odense, Denmark), planted in pots with the same substrate mixture and placed into the greenhouse at the above conditions. Eight-week-old propagated plants of both *K. pinnata* WT and *rol*+ were used in the experiments.

22.3 Climate Chamber Light Settings and UV-B Elicitation

The experiment was set up in a climate chamber (VEPHQ 5/2000, Heraeus Vötsch GmbH, Balingen, Germany) (16 h photoperiod, 28 °C during the day and 20 °C during the night). LED lamps (FL300 SUNLIGHT fixture, Fiona Lighting, Senmatic A/S, Søndersø, Denmark) were used as source of white light, supplying a photosynthetic active radiation (PAR) of 180 μmol s^{-1} m^{-2} at plant height. The chamber was divided into two compartments by non-reflective screens. The first compartment included exclusively the use of white light from 7 to 23 h. In the second compartment, supplemental UV-B radiation was applied. Plants were placed at a distance of 2 m above the UV-B lamps (broadband lamps, Philips TL 40W/12 RS SLV) and irradiated with a UV-B intensity of 3 W m^{-2} during 5 h per day (from 10 to 15) for a week. UV-B light intensity was measured with a RM-12 Ultraviolet Light Meter equipped with a UV-B sensor (Opsytec Dr. Gröbel GmbH, Ettlingen, Germany). The irrigation was conducted manually (150 mL every second day). Two independent experiment repetitions displaced in time were conducted.

22.4 Sample Preparation

The pool of leaves from three plants was considered a biological replicate. After sampling, each biological replicate was ground into a powder in a Mortar Grinder Type RM 100 (Retsch GmBh, Haan, Germany), which was constantly kept cold by adding liquid N_2. One gram of produced powder was extracted twice with 5 mL methanol for 24 h at 5 °C in the dark, mixed frequently and centrifuged at 1000 rpm for 10 min. The supernatant was evaporated to dryness and redissolved in 1 ml methanol and filtered by a Q-Max syringe filter (13 mm Ø, PTFE membrane pore size 0.22 μm, Frisenette APS, Knebel, Denmark) prior to HPLC-DAD analyses.

22.5 Determination of Total Flavonoid Content (TFC) by High-Performance Liquid Chromatography-Diode Array Detection (HPLC-DAD)

The HPLC separations were carried out on a Luna C18(2) column (150 × 4.6 mm, 5 μm, 100 Å, Phenomenex, Allerod, Denmark) at room temperature. The mobile phase consisted of solvent A (aqueous with 0.1% formic acid) and solvent B (acetonitrile with 0.1% formic acid), and the following gradient system was used: 10–99% B (40 min), 99% B (34 min), 99–1% B (1 min) and 10% B (5 min). Flow rate was 0.8 mL min^{-1} and the injection volume 10 μL. Quercetin was used as standard. The retention time and ultraviolet spectra were obtained for chromatogram peaks at 254 nm. A quercetin calibration curve ($R^2 = 0.9974$) was used to quantify the TFC in the samples, constructed by plotting the integrated peak area at 254 nm against the calibration curve.

22.6 Statistical Analysis

All analyses were performed with six biological replicates. Data were subjected to statistical analysis using Microsoft Excel statistical package (version 15.16). Two-tailed Student's *t*-test with 95% and 99.9% degrees of confidence ($p \leq 0.05$ and

$p \leq 0.001$, respectively) was performed to indicate significant differences. Data were expressed as mean ± standard error (SE).

22.7 Elicitation of Flavonoids in *Kalanchoe pinnata*

In this study, the accumulation of flavonoids in *K. pinnata* plants was evaluated upon elicitation. Eight-week-old propagated *K. pinnata* plants were utilized in the experiment (Fig. 22.1). The TFC was calculated in leaves in WT and *rol*+ plants after 7 days of incubation with or without supplemental UV-B radiation (3 J m^{-2} during 5 h per day) (Figs. 22.2 and 22.3). In the WT plants, TFC was 232 ± 12 μg g^{-1}. In comparison, the corresponding content in *rol*+ plants was 287 ± 29 μg g^{-1}, representing a significant increase of 24% compared to the WT plants (Fig. 22.3). These results are supported by studies in which plants derived from *A. rhizogenes* transformation have the ability to produce higher amounts of valuable secondary metabolites in comparison with the wild-type counterparts (Giri and Narasu 2000; Oksman-Caldentey and Hiltunen 1996; Sevon and Oksman-Caldentey 2002). Integration of *A. rhizogenes* T-DNA in the plant's genome and corresponding expression of *rol* genes often alter the plant morphology (Chandra 2011). However, the *rol*+ lines used in this study did not show differences in terms of fresh weight and morphology of leaves compared to control lines (data not shown).

UV-B radiation induces photobiological stress in plants, which, among other effects, may lead to increased production of secondary metabolites (Schreiner et al. 2014). In the present work, for both *K. pinnata* plants, WT and *rol*+, the TFC increased upon UV-B light treatment compared to untreated plants. UV-B-treated WT plants displayed values of 452 ± 33 μg g^{-1}, representing an increase of 95% compared with the values of untreated WT plants. Similarly, the TFC of UV-B-treated *rol*+ plants was 541 ± 21 μg g^{-1}, which was 89% higher than that of the untreated *rol*+ plants (Fig. 22.3). When treated plants were compared with untreated plants, significant differences were observed at *p* values of ≤0.001 for both plant types. These results are in

Fig. 22.1 Representative 8-week propagated WT (**a**) and *rol*+ (**b**) plants utilized in the UV-B elicitation experiment

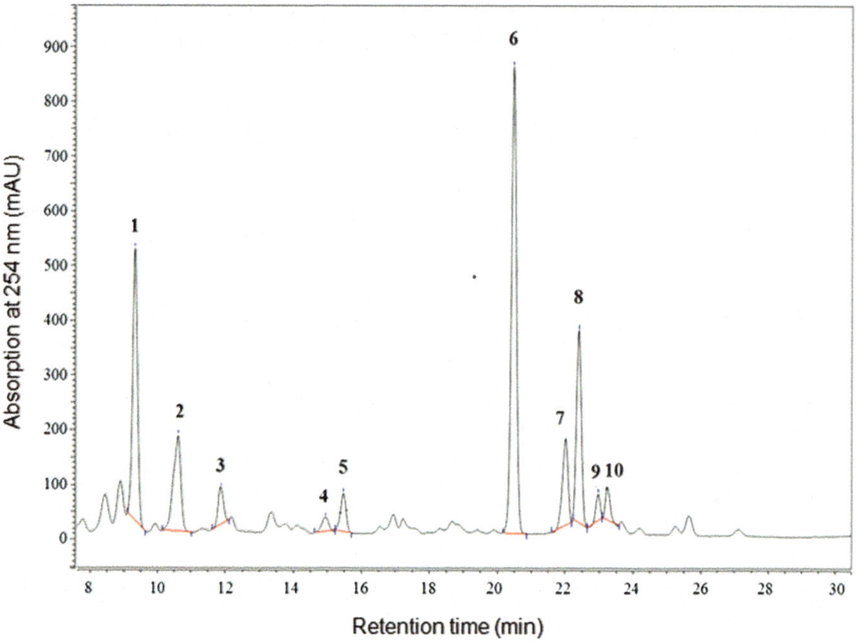

Fig. 22.2 Characteristic HPLC chromatogram of flavonoids in leaf extracts of *K. pinnata* recorded at 254 nm. Peak quantification was performed by plotting the integrated peak areas (1–10) against a quercetin calibration curve

Fig. 22.3 Total flavonoid content (TFC) in leaves of wild-type and *rol+* *K. Pinnata* plants exposed to supplemental UV-B radiation (3 J m^{-2} for 5 h per day) or not (control) for 7 days. Data represent the mean ± SE, $n = 6$. (∗) and (∗∗) denote statistical significance according to two-tailed t-test at $p \leq 0.05$ and $p \leq 0.001$, respectively

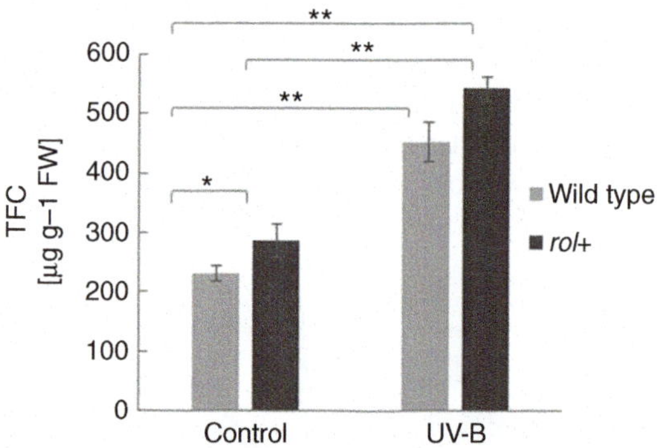

agreement with previous studies that reported induction of flavonoids upon UV-B radiation. UV-B treatment increased total flavonol content in young petunia plants (Ryan et al. 1998) and the concentrations of two surface flavonols (calycopterin and 3′-methoxycalycopterin) in *Gnaphalium luteoalbum* plants (Cuadra et al. 1997). Mahdavian et al. (2008) reported that treatment with UV-B radiation increased proline, quercetin, rutin and anthocyanin concentrations in leaves of *Capsicum annuum*. Synthesis of phenolic substances such as anthocyanin and flavonoids was also observed in UV-B-treated *Arabidopsis thaliana* seedlings (Bieza 2001). In another study in soybean (Middleton and Teramura 1993), it was indicated that UV-B light tolerance was positively correlated with the content of flavonoids. The effect of UV-B radiation was also investigated in vegetables and fruits during the vegetative and post-harvest period. Broccoli treated with supplementary UV-B radiation during the vegetative period increased the content of ascorbic acid, flavonoids and other phenolic compounds (Topcu et al. 2015). Ripe black currant fruits treated with short-term UV-B radiation after harvest showed an increase in the total content of phenolics and an altered phenolic composition (flavonols, anthocyanins, hydroxycinnamic and hydroxybenzoic acids) (Huyskens-Keil et al. 2007).

Additionally, the combination of *rol* transformation and UV-B light treatment resulted in the highest increase in the TFC (133%), when compared with untreated WT plants. To our knowledge, this represents the first reported synergy between the presence of *rol* genes and the use of UV radiation in the elicitation of secondary metabolites in in vivo plants. Hence, we propose that natural transformation followed by UV light elicitation can be a successful strategy for increasing the content of secondary metabolites in plants.

22.8 Conclusion

Agrobacterium rhizogenes-mediated transformation resulted in substantial increase of flavonoid content in *K. pinnata* leaves (24%). Furthermore, supplemental UV-B radiation for 7 days increased the content of total flavonoids in both WT and *rol*+ plants by 95% and 89%, respectively. Taken together, the synergistic effect of the two factors resulted in the highest increase of the total flavonoid content (133%).

Acknowledgements The authors would like to thank Knud Jepsen A/S, Hinnerup, Denmark, for providing plant material. Hanne Hasselager is acknowledged for the *A. rhizogenes*-mediated transformation of *K. pinnata*. This research was a part of the innovation consortium BioFactory funded by The Danish Agency for Science, Technology and Innovation.

References

Bieza K (2001) An *Arabidopsis* mutant tolerant to lethal ultraviolet-b levels shows constitutively elevated accumulation of Flavonoids and other Phenolics. Plant Physiol 126(3):1105–1115

Bonhomme V, Laurain-Mattar D, Lacoux J, Fliniaux MA, Jacquin-Dubreuil A (2000) Tropane alkaloid production by hairy roots of *Atropa belladonna* obtained after transformation with *Agrobacterium rhizogenes* 15834 and *Agrobacterium tumefaciens* containing *rol*A, B, C genes only. J Biotechnol 81(2–3):151–158

Bopda OSM, Longo F, Bella TN, Edzah PMO, Tawe GS, Bilanda DC, Tom ENL, Kamtchouing P, Dimo T (2014) Antihypertensive activities of the aqueous extract of *Kalanchoe pinnata* (Crassulaceae) in high salt-loaded rats. J Ethnopharmacol 153(2):400–407

Bulgakov VP (2008) Functions of *rol* genes in plant secondary metabolism. Biotechnol Adv 26:318–324

Casanova E, Trillas MI, Moysset L, Vainstein A (2005) Influence of *rol* genes in floriculture. Biotechnol Adv 23:3):3–3)39

Chandra S (2011) Natural plant genetic engineer *Agrobacterium rhizogenes*: role of T-DNA in plant secondary metabolism. Biotechnol Lett 34(3):407–415

Chibli LA, Rodrigues KCM, Gasparetto CM, Pinto NCC, Fabri RL, Scio E, Alves MS, Del-Vechio-Vieira G, Sousa OV (2014) Anti-inflammatory effects of *Bryophyllum pinnatum* (Lam.) Oken ethanol extract in acute and chronic cutaneous inflammation. J Ethnopharmacol 154(2):330–338

Christensen B, Müller R (2009) The use of *Agrobacterium rhizogenes* and its *rol*-genes for quality improvement in ornamentals. Eur J Hortic Sci 74(6):275–287

Christensen B, Sriskandarajah S, Serek M, Müller R (2008) Transformation of *Kalanchoe blossfeldiana* with *rol*-genes is useful in molecular breeding towards compact growth. Plant Cell Rep 27(9):1485–1495

Christey MC (2001) Use of ri-mediated transformation for production of transgenic plants. In Vitro Cell Dev Biol Plant 37(6):687–700

Cruz EA, Reuter S, Martin H, Dehzad N, Muzitano MF, Costa SS, Rossi-Bergmann B, Buhl R, Stassen M, Taube C (2012) *Kalanchoe pinnata* inhibits mast cell activation and prevents allergic airway disease. Phytomedicine 19(2):115–121

Cuadra P, Harborne JB, Waterman PG (1997) Increases in surface flavonols and photosynthetic pigments in *Gnaphalium luteo-album* in response to UV-B radiation. Phytochemistry 45(7):1377–1383

European Union (2001) Directive 2001/18/EC of the European parliament and of the council of 12 March 2001 on the deliberate release into the environment of genetically modified organisms and repealing council directive 90/220/EEC-commission declaration

Giri A, Narasu ML (2000) Transgenic hairy roots: recent trends and applications. Biotechnol Adv 18(1):1–22

Heisler GM, Grant RH, Gao W, Slusser JR (2003) Ultraviolet radiation and its impacts on agriculture and forests. Agric For Meteorol 120(3):120–133

Husain A (2010) Economic aspects of exploitation of medicinal plants. In: Akerele O, Heywood V, Synge H (eds) Conservation of medicinal plants. Cambridge University Press (Virtual Publishing), Cambridge, pp 125–140

Huyskens-Keil S, Eichholz I, Kroh LW, Rohn S (2007) UV-B induced changes of phenol composition and antioxidant activity in blackcurrant fruit (*Ribes nigrum* L.). J Appl Bot Food Qual 81:140–144

Kabera J, Semana E, Mussa A, He X (2014) Plant secondary metabolites: biosynthesis, classification, function and pharmacological properties. J Pharm Pharmacol 2:377–392

Kiselev KV, Dubrovina AS, Veselova MV, Bulgakov VP, FedoreyevSA ZYN (2007) The *rolB* gene-induced overproduction of resveratrol in *Vitis amurensis* transformed cells. J Biotechnol 128(3):681–692

Mahdavian K, Ghorbanli M, Kalantari KM (2008) The effects of ultraviolet radiation on the contents of chlorophyll, flavonoid, anthocyanin and proline in *Capsicum annuum* L. Turk J Bot 32:25–33

Matsuura HN, de Costa F, Yendo ACA, Fett-Neto AG (2012) Photoelicitation of bioactive secondary metabolites by ultraviolet radiation: mechanisms, strategies, and applications. In: Chandra S, Lata H, Varma A (eds) Biotechnology for medicinal plants. Springer, New York, pp 171–190

Middleton EM, Teramura AH (1993) The role of flavonol glycosides and carotenoids in protecting soybean from ultraviolet-B damage. Plant Physiol 103:741–752

Nascimento LBDS, Leal-Costa MV, Menezes EA, Lopes VR, Muzitano MF, Costa SS, Tavares ES (2015) Ultraviolet-B radiation effects on phenolic profile and flavonoid content of *Kalanchoe pinnata*. J Photochem Photobiol B 148:73–81

Nayak S, Marshall JR, Isitor G (2010) Wound healing potential of ethanolic extract of *Kalanchoe pinnata* Lam. Leaf—a preliminary study. Indian J Exp Biol 48:572–576

Newman DJ, Crag GM (2007) Natural products as sources of new drugs over the last 25 years. J Nat Prod 70:461–477

Oksman-Caldentey KM, Hiltunen R (1996) Transgenic crops for improved pharmaceutical products. Field Crop Res 45(1-3):57–69

Rahmatullah M, Mollik MAH, Islam MK, Islam MR, Jahan FI, Zubaida K, Seraj S, Chowdhury MH, Islam F, Miajee ZUME, Rownak J (2010) A survey of medicinal and functional food plants used by the folk medicinal practitioners of three villages in Sreepur Upazila, Magura District, Bangladesh. Am-Eurasian J Sustain Agric 4(3):363–373

Ryan KG, Markham KR, Bloor SJ, Bradley JM, Mitchell KA, Jordan BR (1998) UV-B radiation induced increase in quercetin: kaempferol ratio in wild-type and transgenic lines of *Petunia*. Photochem Photobiol 68(3):323–330

Schreiner M, Martínez-Abaigar J, Glaab J, Jansen M (2014) UV-B induced secondary plant metabolites. OPJ 9(2):34–37

Sevon N, Oksman-Caldentey KM (2002) *Agrobacterium rhizogenes*-mediated transformation: root cultures as a source of alkaloids. Planta Med 68(10):859–868

Supratman U, Fujita T, Akiyama K, Hayashi H, Murakami A, Sakai H, Koshimizu K, Ohigashi H (2001) Anti-tumor promoting activity of Bufadienolides from *Kalanchoe pinnata* and *K. Daigremontiana* × *butiflora*. Biosci Biotechnol Biochem 65(4):947–949

Tatsimo S, Tamokou J, Havyarimana L, Csupor D, Forgo P, Hohmann J, Kuiate JR, Tane P (2012) Antimicrobial and antioxidant activity of kaempferol rhamnoside derivatives from *Bryophyllum pinnatum*. BMC Res Notes 5(1):158

Topcu Y, Dogan A, Kasimoglu Z, Sahin-Nadeem H, Polat E, Erkan M (2015) The effects of UV radiation during the vegetative period on antioxidant compounds and postharvest quality of broccoli (*Brassica oleracea* L.). Plant Physiol Biochem 93:56–65

White FF, Taylor BH, Huffmann GA, Gordon MP, Nester EW (1985) Molecular and genetic analysis of the transferred DNA regions of the root-inducing plasmid of *Agrobacterium rhizogenes*. J Bacteriol 164(1):33–44

Zhou X, Wu Y, Wang X, Liu B, Xu H (2007) Salidroside production by hairy roots of *Rhodiola sachalinensis* obtained after transformation with *Agrobacterium rhizogenes*. Biol Pharm Bull 30(3):439–442

Biogenic Silver Nanoparticles from *Trametes ljubarskyi* (White Rot Fungus): Efficient and Effective Anticandidal Activity

23

Krishna Gudikandula, Shaik Jaffar, and
Maringanti Alha Singara Charya

Abstract

Today there is an enormous interest in developing safe, cost-effective and environmentally friendly technologies for nanoparticle synthesis. In the present study, extracellular synthesis of silver nanoparticles was carried by *Trametes ljubarskyi* KU382503.1, white rot fungus isolated from the decayed wood log. The reduction of the silver nanoparticles was monitored by UV–visible spectrophotometry, and the characterization of the AgNPs was carried out by FT-IR and transmission electron microscopy (TEM). The synthesized silver nanoparticles are very stable. Furthermore, the anticandidal activity of the AgNPs was assessed using agar well diffusion method. The biosynthesized AgNPs showed considerable activity against the *Candida albicans*. The present research opens a new path for the biological synthesis of AgNPs, and the process is easy and safe to scale up for biomedical and industrial applications.

Keywords

AgNPs · *Trametes ljubarskyi* · Biological synthesis · *Candida albicans*

K. Gudikandula (✉) · M. A. S. Charya
Department of Microbiology, Kakatiya University, Warangal, Telangana, India

S. Jaffar
Department of Biochemistry, Acharya Nagarjuna University, Guntur, Andhra Pradesh, India

S. M. Khasim et al. (eds.), *Medicinal Plants: Biodiversity, Sustainable Utilization and Conservation*, https://doi.org/10.1007/978-981-15-1636-8_23

23.1 Introduction

Candida species belongs to the normal microbiota of an individual's gastrointestinal tract, mucosal oral cavity, and vagina (Shao et al. 2007), and this is responsible for many clinical manifestations from mucocutaneous overgrowth to bloodstream infections (Eggimann et al. 2003). These yeasts are forming symbiotic relationship in healthy humans and may cause systemic infection in immunocompromised state due to their great adaptability to different host niches (Sardi et al. 2013). Presently, a sudden increase in the number of yeasts that are impervious to antifungal drugs is recognized worldwide; therefore, the use of in vitro laboratory tests may fetch the doctor in selecting an appropriate therapy. Nowadays most of the commercially available effective antifungal agents are designed based upon polyenes (amphotericin B), triazoles (fluconazole, itraconazole, voriconazole, posaconazole) or echinocandins (caspofungin, micafungin and anidulafungin). In any case, administration of these antifungals is often assisted by various complications such as amphotericin B toxicity and adverse effects of some azoles including toxicity and drug interactions (Levin et al. 2007; Taxvig et al. 2007; Worth et al. 2008; Venkatakrishnan et al. 2000) and yeast resistance to antifungal therapy. Due to this, possibility for effective antifungal therapy must be found to avoid the mentioned adverse effects (White et al. 2002; Perea et al. 2001; Panáček et al. 2009).

Now, nanotechnology is emerging as expeditiously growing area with its applications in the field of science and technology for the purpose of manufacturing new material at the nanoscale level (Albrecht et al. 2006). Nanoparticle has multifunctional properties and received special attention due to their potential use in biomedical applications such as infection, prevention, and wound healing. Silver nanoparticles are known to show strong antimicrobial activity against bacteria (Morones et al. 2005; Panáček et al. 2006). Although the effects of silver nanoparticles against fungal pathogens have sustained only marginal attention from the researchers, AgNPs exhibited significant antifungal activity against clinical isolate strains of *Candida* spp. causing undesirably life-threatening fungal infections (Panáček et al. 2009). According to Kim et al. (2008), the AgNPs showed potent fungistatic activity against clinical isolates and ATCC strains of *Trichophyton mentagrophytes and Candida spp*. Consequently, AgNPs are proving to be an alternative for the development of new antifungal drugs; nevertheless, much effort is needed to understand the effects against other fungal pathogens (Qian et al. 2013). The main drawback with the chemical and physical methods of silver nanoparticle formation is that they are extremely costly and also involve the use of toxic, hazardous chemicals and they contain potential environmental and biological stakes (Ingham et al. 2012). Fungi may be considered as good source for the synthesis of silver nanoparticles and ideal in the synthesis of metal nanoparticles, due to their capacity to accumulate large amount of enzymes. Comparatively with the other microbes, fungal mycelia mesh can withstand flow pressure and agitation and other conditions in the bioreactors compared to plant materials and bacteria (Gudikandula et al. 2015). White rot fungi are commonly used in bioremediation; it was reported that these fungi can also serve as a platform for the bioreduction of Ag nanoparticles (AgNPs) (Vigneshwaran et al. 2006).

In this study, the extracellular synthesis and characterization of AgNPs generated by the reduction of white rot fungi were reported. The biologically synthesized *Trametes ljubarskyi* were analysed and tested against *Candida albicans*. The results reported here cover the biological synthesis of AgNPs and their anticandidal activity.

All the chemicals were obtained from Hi-Media Pvt. Limited, Mumbai, India. Distilled water was used for the experiment. *Candida albicans* (ATCC 10231) was procured from the microbial-type culture collection centre Microbiologics, 200 Copper Avenue North, St. Cloud, MN 56303. This culture was grown on potato dextrose agar (PDA) medium at 25 °C for 3 days and maintained at 4 °C in a refrigerator. *Trametes ljubarskyi* KU382503.1 used in this study was isolated from Eturnagaram Forest, Warangal District, Telangana, India (18°20′20″N, 80°25′45″E). It was grown and maintained at 32 °C on malt extract agar medium (MEA: malt extract, 15 g; NH_4Cl, 1 g; KH_2PO_4, 1 g; citric acid, 15N per 1000 mL of distilled water).

23.2 Synthesis of Silver Nanoparticles from White Rot Fungi

The strain *Trametes ljubarskyi* KU382503.1 (Fig. 23.1a, b) was cultured in a 250 mL Erlenmeyer flask containing 100 mL MEA broth at 30 °C and 150 revolutions per minute (rpm) for 5 days. After incubation, biomass was separated by double filtration and then properly washed with sterile distilled water to remove traces of medium components. The obtained culture filtrate was challenged with 1 mM silver nitrate in equal volume and incubated at 30 °C under shaking conditions at 150 rpm. The control (without addition of $AgNO_3$) was also kept simultaneously under the same conditions.

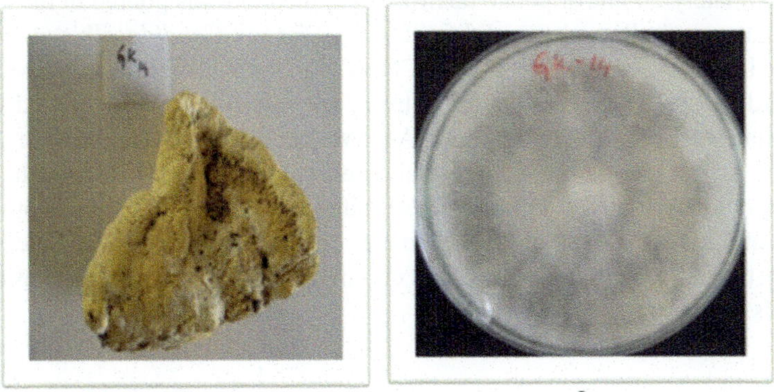

a b

Fig. 23.1 Fruit body and morphology of *Trametes ljubarskyi*. (**a**) View of the fruit body. (**b**) Top view of the colony (7 days)

23.3 Characterization of Silver Nanoparticles

23.3.1 UV–Vis Absorption

Silver nanoparticle synthesis was confirmed by sampling the reaction mixture at regular intervals, and the absorption maxima was scanned by UV–Vis spectra at the wavelength of 350–470 nm (ELICO SL-159 Spectrophotometer).

23.3.2 TEM

Transmission electron microscopy (TEM) was obtained using JEOL (JEM-1010) instrument with an accelerating voltage of 80 kV. AgNPs were loaded on carbon-coated copper grids, and solvent was allowed to evaporate under the infrared light for 30 min. The particle size and surface morphology of nanoparticles were evaluated using ImageJ 1.45s software.

23.3.3 FT-IR

Fourier transform infrared spectroscopy (FT-IR) of silver nanoparticles was carried out by using Digital Excalibur 3000 series, Japan, in diffuse reflectance mode (DRS-800) in the range of 400–4000 cm^{-1} at a resolution of 4 cm^{-1}. FT-IR reveals the biomolecules responsible for the bioreduction of silver ions and stabilization of silver nanoparticles in the solution.

23.3.4 Anticandidal Activity

Anticandidal activity was performed using agar well diffusion method. Approximately 20 mL of potato dextrose agar medium was poured in the sterilized petri plates and allowed to solidify. The test fungus *Candida albicans* was suspended in a saline solution (0.85%) NaCl and adjusted to a turbidity of 0.5 McFarland standards (108 CFU/mL). One millilitre of the test culture was speeded over the PDA plates using sterilized spreader. Agar wells of 8 mm diameter were prepared with the help of sterilized stainless steel borer. The wells were loaded with different concentrations of AgNP solution (40, 60 and 80 μL). The plates were kept for incubation at 32 °C for 48 h; later the diameter of zone of inhibition was measured in millimetres. Silver nitrate solution and culture filtrate extract were used as negative controls.

Fig. 23.2 Conical flasks of extracellular filtrate of *Trametes ljubarskyi* after 12 h at 30 °C in dark condition. (**a**) Cell filtrate without silver ion (control). (**b**) Cell filtrate with silver ion

23.4 Biogenic Silver Nanoparticles and Anticandidal Activity

23.4.1 Biosynthesis of Silver Nanoparticles

The nanoparticle synthesis reaction was started after the culture filtrate extract was introduced right into 1 mM liquid ($AgNO_3$) solution. After 12 h of incubation in shaking condition, the reaction blend was turned into dark brown colour solution (Fig. 23.2). The colour modification was taking place due to the fact that the energetic metabolites present in the culture filtrate essence reduce the silver metal ions right into silver nanoparticles. The intensity of the colour change was increased in direct proportion to the incubation period of nanoparticle synthesis. It could be due to the excitation of surface area plasmon vibration (SPR) and also reduction of $AgNO_3$ (Mulvaney 1996).

23.4.2 Characterization of Silver Nanoparticles

UV–Vis absorption spectra is the most prominent method to observe the forming and stabilization of silver nanoparticles in the aqueous solution, as the spectral response of silver nanoparticles is mainly based on the diameter. The plasmon peak resonance moves to longer wavelengths and broadens as the diameter increases (Ninganagouda et al. 2014). So the aqueous bioreduction of silver nanoparticles could be easily observed by UV–Vis, which is sensitive to several factors such as particles shape, particle–particle interaction with the medium and size (Khan et al. 2013). The synthesized sample reaction mixture exhibits the strong peak at 420 nm (Fig. 23.3) at the end of 12 h and remained closed at 418 nm. This reveals that the reaction was very rapid, and the particles are well distributed in the solution. *Trametes ljubarskyi* has never been reported to biosynthesize AgNPs.

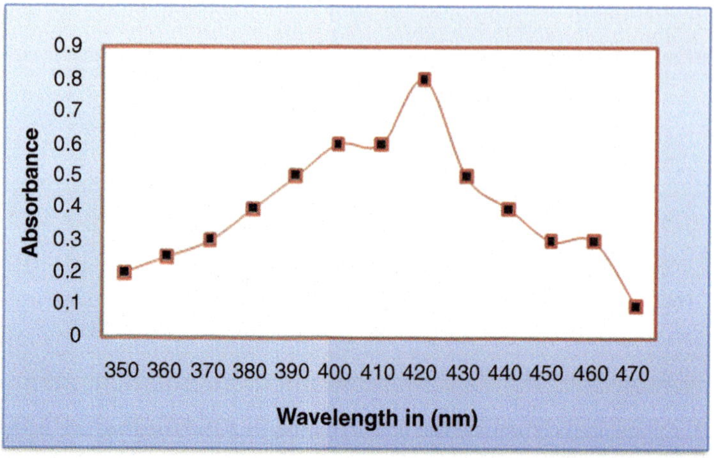

Fig. 23.3 Surface plasmon resonance analysis of synthesized SNPs with UV–Vis spectroscopy shows a typical broad peak at 420 nm

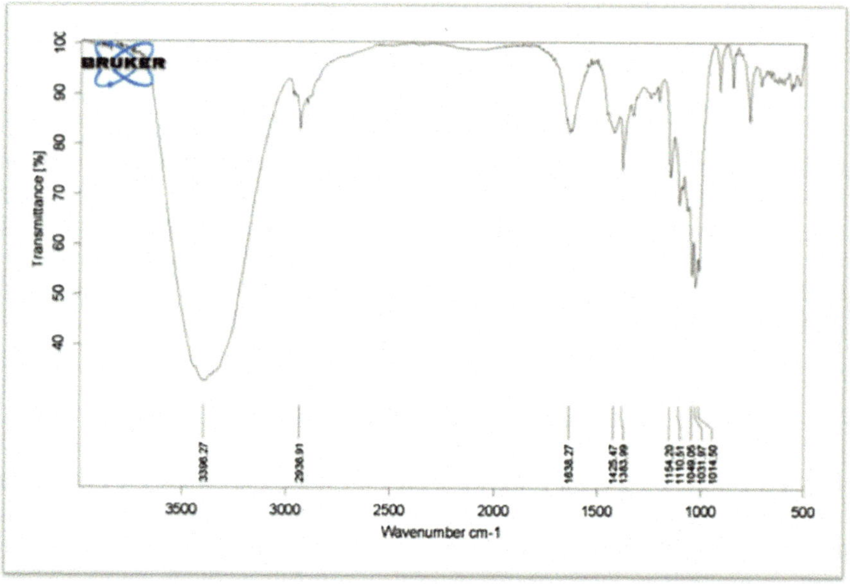

Fig. 23.4 FT-IR spectra of AgNPs synthesized by *Trametes ljubarskyi*

23.4.2.1 FT-IR

FT-IR spectrum of synthesized SNPs was carried out to understand the possible biomolecules responsible for the capping and also stabilization of nanoparticles. For this, the sample was analysed in the scan array from 3500 to 500 cm^{-1} of close to IR spectra by FT-IR. The peaks at 3396 cm^{-1} and also 1638 cm^{-1} were assigned for O–H bond of phenols and also N–H bond of key amines, specifically (Fig. 23.4).

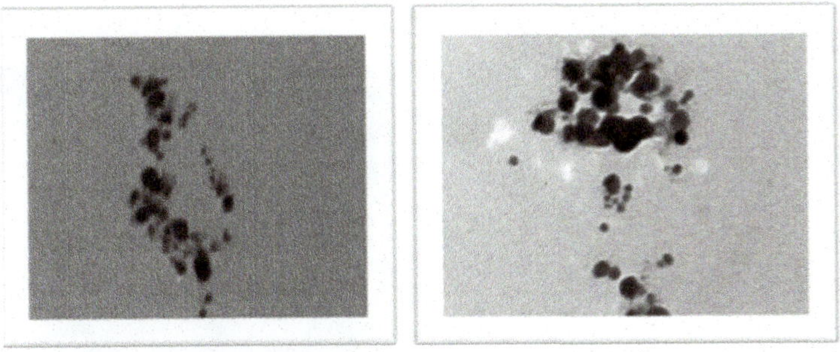

Fig. 23.5 Transmission electron microscopy images of AgNPs

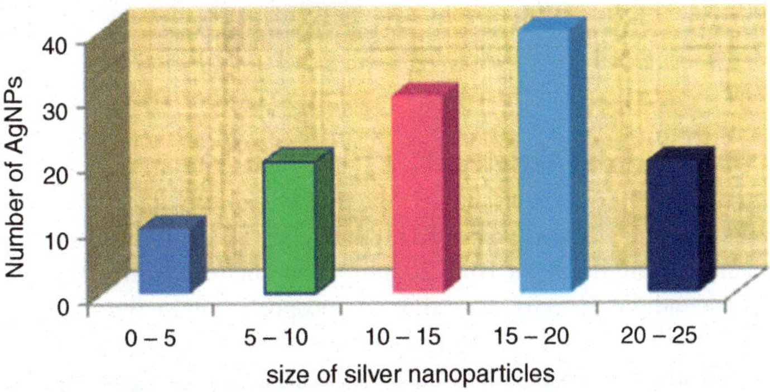

Fig. 23.6 Particle size histogram of AgNPs

This recommends that the hydroxyl groups of phenols as well as amide groups of proteins forming a layer of the nanoparticles act as capping agents to prevent agglomeration and provide stability to the reaction medium. Similar type of results was found in *Syzygium alternifolium* stem bark extract-mediated synthesis of silver nanoparticles (Yugandhar et al. 2015).

23.4.2.2 TEM

A transmission electron microscopy (TEM) picture of the AgNPs is shown in Fig. 23.5. The shapes of AgNPs were either round or near to spherical. A total amount of 120 particles were counted to determine their dimensions. The results showed that the particles size varied from 1 to 25 nm. The particle dimension histogram of AgNPs is shown in Fig. 23.6.

The results of UV–visible absorption spectroscopy as well as TEM recommended that the AgNPs were polydispersed spherical or near to spherical shapes varying from 1 to 25 nm and also were exceptionally similar to the reports of Qian et al. (2013) as well as Verma et al. (2010). We used the extracellular filtrate to synthesize AgNPs, and the procedure was devoid of any type of toxic chemicals or solvents. This would not only result in "natural" AgNPs but also simplify the downstream handling. Provided these benefits, the *Trametes ljubarskyi* KU382503.1 represents a promising prospect for large-scale production of AgNPs.

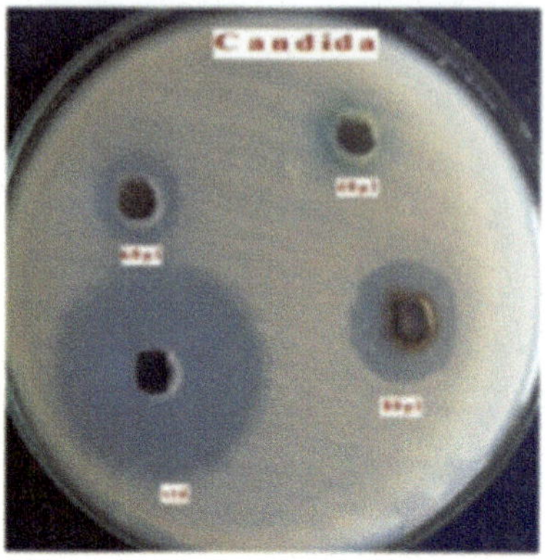

Fig. 23.7 Anticandidal activity of biologically synthesized silver nanoparticles

23.4.3 Anticandidal Activity

The AgNPs of *Trametes ljubarskyi* KU382503.1 at 80 μL/well showed good anticandidal activity ranging from 10 to 20 mm. From the results, it was clearly known that the anticandidal activity was directly proportional to the concentration of AgNPs. A negative control, i.e. fungal culture filtrates, did not show any activity against tested strain. Fluconazole used as standard against fungi showed the inhibition zones of 30 mm (Fig. 23.7). This is the first report on biosynthesis of silver nanoparticles from *Trametes ljubarskyi* and their anticandidal activity. Nanoscale silver has been shown to be among the potential antifungal agents, and also specifically biologically synthesized silver nanoparticles could be a lot more reliable in the medicinal applications as a result of their varied surface coatings. Hwang et al. (2012) established the mechanism of action of AgNPs on *C. albicans*. It was documented that the cells revealed to AgNPs showed increased reactive oxygen species and also hydroxyl radical production leading to the apoptotic features such as phosphatidylserine externalization, DNA and nuclear fragmentation, activation of metacaspases and mitochondrial disorder.

San Chan and Don (2013), in their study, synthesized silver nanoparticles from white rot fungi *Schizophyllum commune* and reported that AgNPs have great promise as antimicrobial agent against *S. aureus*, *S. epidermidis*, *E. coli* and *C. albicans*. Very few are reported on the silver nanoparticle synthesis by white rot fungi, but many reports are available on plant- and chemical-mediated synthesis of silver nanoparticles and their activity against *Candida* sp. In Kim et al.'s (2008) [14] study, nano-Ag was synthesized chemically, and its antifungal effects on clinical isolates and ATCC strains of *Trichophyton mentagrophytes* and *Candida* species were investigated, and they revealed that silver nanoparticles may have considerable antifungal activity.

23.5 Conclusion

In the present study, the focus was on biological synthesis of silver nanoparticles using culture filtrate of *Trametes ljubarskyi*. The physical property of synthesized nanoparticle was characterized using relevant techniques. The synthesized nanoparticles are active against *Candida albicans*. This is the first report on the anticandidal activity of AgNPs synthesized using white rot fungi. The synthesized nanoparticles have the potential to be exploited in the preparation of antifungal drugs, and there is a wide scope for detailed investigation in the future for the application of AgNPs in the field of medicine for controlling the pathogen. Based on these results, it is concluded that *Trametes ljubarskyi* is an efficient and effective source for the synthesis of silver nanoparticles and good source against *Candida albicans*.

Acknowledgement The authors are thankful to Ministry of Environment and Forest, New Delhi, for providing financial assistance.

References

Albrecht MA, Evans CW, Raston CL (2006) Green chemistry and the health implications of nanoparticles. Green Chem 8:417–432

Eggimann P, Garbino J, Pittet D (2003) Epidemiology of Candida species infections in critically ill non-immunosuppressed patients. Lancet Infect Dis 23:685–702

Gudikandula K, Metuku RP, Pabba SK, Laxmi SV, Nidadavolu HB, Burra S, Alha SC (2015) Fungus-mediated synthesis of silver nanoparticles and their activity against gram positive and gram negative bacteria in combination with antibiotics. Walailak J Sci Technol 12:613–622

Hwang IS, Lee J, Hwang JH, Kim KJ, Lee DG (2012) Silver nanoparticles induce apoptotic cell death in Candida albicans through the increase of hydroxyl radicals. FEBS J 279:1327–1338

Ingham CJ, Boonstra S, Levels S, De Lange M, Meis JF, Schneeberger PM (2012) Rapid susceptibility testing and microcolony analysis of Candida spp. cultured and imaged on porous aluminum oxide. PLoS One 7:e33818

Khan M, Khan M, Adil SF, Tahir MN, Tremel W, Alkhathlan HZ, Al-Warthan A, Siddiqui MR (2013) Green synthesis of silver nanoparticles mediated by Pulicaria glutinosa extract. Int J Nanomedicine 2:1507–1506

Kim KJ, Sung WS, Moon SK, Choi JS, Kim JG, Lee DG (2008) Antifungal effect of silver nanoparticles on dermatophytes. J Microbiol Biotechnol 18:1482–1484

Levin MD, den Hollander JG, Van Der Holt B, Rijnders BJ, Van Vliet M, Sonneveld P, Van Schaik RH (2007) Hepatotoxicity of oral and intravenous voriconazole in relation to cytochrome P450 polymorphisms. J Antimicrob Chemother 60(5):1104–1107

Morones JR, Elechiguerra JL, Camacho A, Holt K, Kouri JB, Ramírez JT, Yacaman MJ (2005) The bactericidal effect of silver nanoparticles. Nanotechnology 16:2346

Mulvaney P (1996) Surface plasmon spectroscopy of nanosized metal particles. Langmuir 12:788–800

Ninganagouda S, Rathod V, Singh D (2014) Characterization and biosynthesis of Silver nanoparticles using a fungus Aspergillus niger. Int Lett Nat Sci 10:49–57

Panáček A, Kolář M, Večeřová R, Prucek R, Soukupová J, Kryštof V, Hamal P, Zbořil R, Kvítek L (2009) Antifungal activity of silver nanoparticles against *Candida* spp. Biomaterials 30:6333–6340

Panáček A, Kvítek L, Prucek R, Kolar M, Vecerova R, Pizurova N, Sharma VK, Nevečná TJ, Zboril R (2006) Silver colloid nanoparticles: synthesis, characterization, and their antibacterial activity. J Phys Chem B 110:16248–16253

Perea S, López-Ribot JL, Kirkpatrick WR, McAtee RK, Santillán RA, Martínez M, Calabrese D, Sanglard D, Patterson TF (2001) Prevalence of molecular mechanisms of resistance to azole antifungal agents in Candida albicans strains displaying high-level fluconazole resistance isolated from human immunodeficiency virus-infected patients. Antimicrob Agents Chemother 45:2676–2684

Qian Y, Yu H, He D, Yang H, Wang W, Wan X, Wang L (2013) Biosynthesis of silver nanoparticles by the endophytic fungus Epicoccum nigrum and their activity against pathogenic fungi. Bioprocess Biosyst Eng 36:1613–1619

San Chan Y, Don MM (2013) Optimization of process variables for the synthesis of silver nanoparticles by *Pycnoporus sanguineus* using statistical experimental design. J Korean Soc Appl Biol Chem 56:11–20

Sardi JC, Scorzoni L, Bernardi T, Fusco-Almeida AM, Giannini MM (2013) Candida species: current epidemiology, pathogenicity, biofilm formation, natural antifungal products and new therapeutic options. J Med Microbiol 62:10–24

Shao LC, Sheng CQ, Zhang WN (2007) Recent advances in the study of antifungal lead compounds with new chemical scaffolds. Yao Xue Xue Bao 42:1129–1136. (Article in Chinese)

Taxvig C, Hass U, Axelstad M, Dalgaard M, Boberg J, Andeasen HR, Vinggaard AM (2007) Endocrine-disrupting activities in vivo of the fungicides tebuconazole and epoxiconazole. Toxicol Sci 100:464–473

Venkatakrishnan K, von Moltke LL, Greenblatt DJ (2000) Effects of the antifungal agents on oxidative drug metabolism. Clin Pharmacokinet 38:111–180

Verma VC, Kharwar RN, Gange AC (2010) Biosynthesis of antimicrobial silver nanoparticles by the endophytic fungus *Aspergillus clavatus*. Nanomedicine 5:33–40

Vigneshwaran N, Kathe AA, Varadarajan PV, Nachane RP, Balasubramanya RH (2006) Biomimetics of silver nanoparticles by white rot fungus, Phanerochaete chrysosporium. Colloids Surf B Biointerfaces 53:55–59

White TC, Holleman S, Dy F, Mirels LF, Stevens DA (2002) Resistance mechanisms in clinical isolates of Candida albicans. Antimicrob Agents Chemother 46:1704–1713

Worth LJ, Blyth CC, Booth DL, Kong DC, Marriott D, Cassumbhoy M, Ray J, Slavin MA, Wilkes JR (2008) Optimizing antifungal drug dosing and monitoring to avoid toxicity and improve outcomes in patients with haematological disorders. Intern Med J 38:521–537

Yugandhar P, Haribabu R, Savithramma N (2015) Synthesis, characterization and antimicrobial properties of green-synthesised silver nanoparticles from stem bark extract of *Syzygium alternifolium* (Wt.) Walp. 3 Biotech 5:1031–1039

Herbal Medicinal Markets in China: An Ethnobotanical Survey

24

Yuanyuan Ji, Qiong Fang, Sizhao Liu, Beixi Zhang, and Chunlin Long

Abstract

There are numerous herbal medicinal markets in China where people trade fresh, dried, or processed herbs. The Chinese government has recognized 17 huge herbal medicinal marketplaces throughout the country. Three markets, Anguo, Kunming, and Yulin, were investigated through ethnobotanical methods. In total, 210 vendors and 135 buyers had been interviewed. The results revealed that rich species diversity of medicinal plants existed in the targeted markets. Common and massive medicinal plants traded in the three markets had been presented, including 35 medicinal materials from more than 40 plant species. There is rich traditional knowledge of medicinal plants in the investigated markets. The vendors in Kunming market maintained the richest traditional knowledge. The rich market-based traditional knowledge of herbal medicines is probably related to the biological and cultural diversity in the region where local

Y. Ji · Q. Fang
College of Life and Environmental Sciences, Minzu University of China, Beijing, China

Key Laboratory of Ethnomedicine, Minzu University of China, Ministry of Education, Beijing, China

S. Liu · B. Zhang
College of Life and Environmental Sciences, Minzu University of China, Beijing, China

Key Laboratory of Ethnomedicine, Minzu University of China, Ministry of Education, Beijing, China

College of Life and Environmental Sciences, Kaili University, Guizhou, China

C. Long (✉)
College of Life and Environmental Sciences, Minzu University of China, Beijing, China

Key Laboratory of Ethnomedicine, Minzu University of China, Ministry of Education, Beijing, China

Kunming Institute of Botany, Chinese Academy of Sciences, Kunming, China
e-mail: Long.chunlin@muc.edu.cn

S. M. Khasim et al. (eds.), *Medicinal Plants: Biodiversity, Sustainable Utilization and Conservation*, https://doi.org/10.1007/978-981-15-1636-8_24

market is located. Medicinal plants and associated traditional knowledge from local communities with different linguistic groups will highly enrich herbal diversity in the markets.

Keywords
Diversity of herbs · Herbal medicinal market · Ethnobotany · Traditional knowledge

24.1 Introduction

The traditional markets or marketplaces in the developing countries are locations where people regularly gather for the purchase and sale of food and cereals, livestock and poultry, clothing, accessories, local artifacts, souvenirs and a wide assortment of personal goods, and other animal or plant products. Some markets are permanent, while others are held once a week or on less frequent specified days such as festival days called periodic markets. The form that a market adopts depends on its locality's population, culture, transportation, and ambient and geographic conditions.

Markets are not only public spaces in which many kinds of products are sold but also places in which cultural information are exchanged as well. These spaces are distinctive for a given culture or society because they present small-scale reproductions of the given region's cultural and biological diversity. It is common to find specific locations in these markets where plants and animals are sold for medicinal purposes and locations that serve to unite, maintain, and diffuse empirical knowledge from different regions and of different origins (Albuquerque et al. 2007).

The specialized markets for herbal medicines have developed very well in some developing countries with rich biological and cultural diversity. In recent years, studies on medicinal plants traded in local markets revealed their significances. Ethnobotanical researches conducted in traditional markets became attractive to the public worldwide. Bussmann and collaborators, for example, have spent a dozen of years to study medicinal plants traded in local markets through ethnobotanical approaches in Eastern Nepal (Kunwar et al. 2009, 2010, 2013), Northern and Central Peru (Bussmann et al. 2007, 2010, 2013; Bussmann and Glenn 2010), and Southeast Asia (Hidayati et al. 2015). They recorded and documented the species of medicinal plants, collected local people's traditional medicinal knowledge, and analyzed their values. Their findings revealed that local people, particularly the local healers, possess rich traditional knowledge about herbal medicinal plants which have been used for treating various ailments among indigenous communities. The traditional botanical knowledge and local genetic resources have contributed to the livelihood, income raising, and future development such as pharmaceutical industry.

Medicinal plants traded in traditional markets showed rich diversity in the megadiversity countries. In South America, diversity of plants commercialized in some traditional marketplaces has been described based on ethnobotanical investigations (Nicholson and Arzeni 1993; Narvaez and Stauffer 1999; Marcia et al. 2005; Albuquerque et al. 2007). In South Africa, a study discussed quantitative analyses in ethnobotanical research in the traditional markets of Johannesburg (Williams et al. 2005).

Probably China is the world's largest producer and consumer of herbal medicinal products for its huge population, rich biological and cultural diversity, and long history of herbal medicine uses. China's total population is estimated to be 1.39 billion in 2018, with an annual growth rate at 0.52% (www.chyxx.com), ranked number one in the world. Most people depend on herbal medicine for healthcare. China has been identified as one of the megadiversity countries by Conservation International (Mittermeier et al. 1997). It harbors great plant and animal biodiversity. About 34,000 species of higher plants have been recorded in China according to a recent statistics (Long 2015), in which more than 12,000 species are medicinal plants. There are 56 linguistic groups living in China. Each group has developed its own medicinal system and conducted traditional medicinal practice for many centuries. The traditional Chinese medicine (TCM), for example, has become a special resource in terms of healthcare, an economic resource with great potential, a scientific and technological resource with originality advantages, an outstanding cultural resource, and an ecological resource of great importance of China. The total value of pharmaceutical products reached 786.6 billion RMB (or $123.4 billion USD) in 2015, in which 3.72 billion RMB were obtained from export of herbal medicinal materials (The State Council Information Office 2016).

There are innumerable traditional markets in China. Most of these marketplaces can be found to sell herbal medicines. In Southwest China, in particular, it is very common that fresh herbal medicinal plants are commercialized in periodic markets. Almost in every province there is at least one huge herbal medicinal marketplace with thousands of vendors. Some studies have addressed the importance of these huge herbal medicinal marketplaces for economic development, but few were linked to the ethnobotanical information.

The purpose of the present study was to exploit the ethnobotanical knowledge hiding in the huge marketplaces. Vendors in huge herbal medicinal markets are mostly medicinal businessmen who never collect or grow medicinal plants. A few masters and their students become professionals who have kept traditional knowledge of medicinal plants. This study focused on the ethnobotanical knowledge existing in the huge and permanent markets where hundreds of vendors traded medicinal plants. Medicines derived from animals, animal products, and other materials were not taken into consideration in the present paper.

Three huge herbal medicinal markets, Anguo, Yulin, and Kunming, had been visited from 2012 to 2017, while many smaller markets such as Jingxi (a Zhuang people's periodic herbal market in Guangxi), Gongcheng (a Yao people's periodic herbal market in Guangxi), and Jianghua (a Yao market in Hunan) had been investigated (Yang et al. 2009; Lin et al. 2016; Luo et al. 2018). Three markets are listed in the 17 huge medicinal marketplaces recognized by the Chinese government (Fig. 24.1).

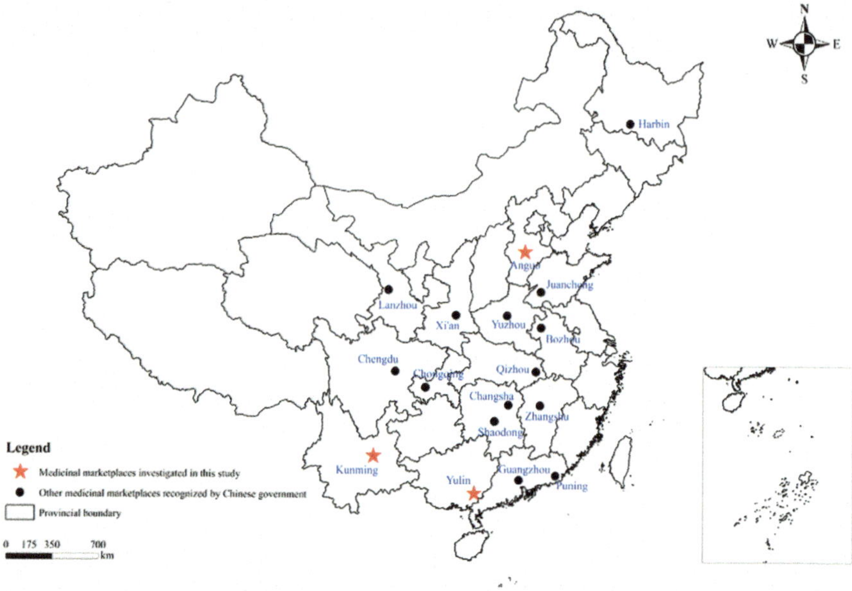

Fig. 24.1 Location of 17 huge medicinal marketplaces recognized by the Chinese government

The Anguo Dongfang Herbal Medicinal Market, located in Anguo, Hebei Province, is regarded as the biggest medicinal market in Northern China. Over 1500 vendors trade herbal medicines in Anguo market which covers 2 hectares. More than 2800 varieties of herbal medicines traded in the market produced about 1.1 billion USD yearly.

The Yulin Herbal Medicinal Market is located in Yulin, Guangxi Zhuang Autonomous Region. The facility is a two-layered building with 812 departments/vendors, occupying an area of 1.9 hectares. More than 1000 species of herbal medicinal plants are traded in this market, and the daily transaction volume reaches 500,000 kg.

The Kunming Juhuayuan Herbal Medicinal Market was moved from the downtown to eastern suburban of Kunming, the capital city of Yunnan Province, which is regarded as the richest center of biological and cultural diversity in China. Over 300 vendors are trading about 4000 varieties of herbal medicines.

The Chinese databases such as VIP and Wanfang and other electronic tools including Google Scholar, PubMed, Scopus, and Web of Science were used for literature investigations. The methods of pharmacognosy, ethnobotany, and participatory rural appraisal had been adopted in investigating the huge markets. Surveys were conducted in Anguo market (2012–2013), Yulin market (2015–2016), and Kunming market (2012–2017). During the surveys, 210 vendors had been visited, and 135 buyers had been interviewed.

24.2 Major Marketplaces of Herbal Medicine in China

In 1950s–1980s, more than 100 big markets, together with smaller markets at county or township levels, traded herbal medicine in China. In 1997, the Chinese government officially recognized 17 huge herbal medicinal markets that passed the evaluations organized by the State Administration of Traditional Chinese Medicine, China Food and Drug Administration, Ministry of Health, and State Administration for Industry and Commerce. The non-recognized big markets had been banned which might be turned to other business. The profiles of 17 huge herbal medicinal markets that remained after recognition in 1996–1997 are presented in Table 24.1.

From Fig. 24.1 and Table 24.1, the huge herbal medicinal markets recognized by the Chinese government are scattered in the country. The distribution pattern is reasonable based on geographical, historical, cultural, and resource dimensions. In the western part of China, including Xinjiang, Tibet, Qinghai, and Inner Mongolia, there were small and periodic markets for trading herbal medicines, but permanent and large-scaled medicinal markets did not exist in history. Another reason might be less or difficult access to the medicinal resources in the region because of harsh biophysical environment. Other provinces such as Shanxi, Jilin, Liaoning, Zhejiang, Jiangsu, Fujian, Guizhou, Taiwan, and Hainan have smaller or periodic herbal medicinal markets but lack super herbal markets as shown in Fig. 24.1.

It was difficult to select marketplaces for collecting medicinal information in China from countless markets. According to literature reviews and our investigations nationwide in the past 20 years, Southwest China (covering Yunnan, Sichuan, Guizhou, Chongqing, Guangxi, and Tibet geographically) is the diversity center of herbal medicines and ethnic cultures in the country. More than 20,000 higher plant species grow in the region, and over 30 ethnic groups are native to Southwest China. There are numerous traditional markets almost in every township. Most of them are periodic markets with herbal medicines sold by different linguistic groups, which imply that rich traditional knowledge about medicinal plants has been maintained and exchanged in Southwest China. It is also why two huge herbal medicinal marketplaces, Kunming and Yulin markets, had been selected from this region in our study. The third marketplace was presented as Anguo, a huge medicinal market in Northern China, where less ethnic culture existed but the traditional Chinese medicinal culture mainstreamed.

24.3 Common and Massive Medicinal Plants Traded in the Targeted Markets

In three herbal medicinal markets we investigated, we found that 35 medicinal materials from more than 40 plant species appeared commonly in most stalls. Over 50% of vendors sold these medicinal herbs during our investigations. These medicinal plants include *Angelica sinensis*, *Arisaema erubescens*, *A. heterophyllum*, *Astragalus membranaceus*, *A. mongholicus*, *Atractylodes macrocephala*, *Aucklandia lappa*, *Cistanche deserticola*, *C. tubulosa*, *Citrus medica* var. *sarcodactylis*, *Coptis chinensis*, *Corydalis*

Table 24.1 Information of 17 huge herbal medicinal marketplaces in China

Market name	Province	Area (m^2)	Vendor number	Species number	Representative medicinal plants
Anguo	Hebei	1,500,000	4000	2800	*Angelica dahurica* (dry root), *Trichosanthes kirilowii* (dry fruit and root), *Atractylodes macrocephala* (root and stem), *Glehnia littoralis* (dry root), *Salvia miltiorrhiza* (root), *Saposhnikovia divaricata* (root), *Platycodon grandiflorus* (root), *Achyranthes bidentata* (root), *Dioscorea opposita* (dry root), *Belamcanda chinensis* (rhizome)
Bozhou	Anhui	1,060,000	6000	2600	*Saposhnikovia divaricata* (root), *Dendranthema morifolium* (flower), *Paeonia lactiflora* (dry root), *Morus alba* (dry root bark), *Trichosanthes kirilowii* (dry root), *Prunella vulgaris* (dry inflorescence), *Lycium chinense* (fruit), *Phellodendron chinense* var. *glabriusculum* (dry bark), *Fraxinus chinensis* (dry bark), *Platycodon grandifloras* (root)
Changsha Gaoqiao	Hunan	5000	85	709	*Gastrodia elata* (tuber), *Codonopsis pilosula* (root), *Bupleurum* spp. (root), *Angelica sinensis* (root), *Panax notoginseng* (leaves, fruit, roots, and rhizomes), *Poncirus trifoliata* (dry fruit), *Polygonatum sibiricum* (rhizomes), *Polygonatum odoratum* (rhizomes), *Lonicera japonica* (dry flower), *Eucommia ulmoides* (bark)

(continued)

Table 24.1 (continued)

Market name	Province	Area (m^2)	Vendor number	Species number	Representative medicinal plants
Chengdu Hehuachi	Sichuan	94,660	2000	2000	*Ligusticum chuanxiong* (root and stem), *Gastrodia elata* (tuber), *Codonopsis pilosula* (root), *Astragalus membranaceus* (root), *Ganoderma lucidum* (fruit body), *Coptis chinensis* (rhizomes), *Citrus limon* (fruit), *Fritillaria cirrhosa* (dry bulb), *Crocus sativus* (dry style and stigma), *Aconitum carmichaelii* (dry root)
Chongqing Caiyuanba	Chongqing	2500	200	800	*Coptis chinensis* (rhizomes), *Poncirus trifoliata* (dry fruit), *Gardenia jasminoides* (dry fruit), *Saussurea costus* (root), *Scrophularia ningpoensis* (root), *Paeonia suffruticosa* (dry root bark), *Pinellia ternata* (tuber), *Eucommia ulmoides* (bark), *Fritillaria cirrhosa* (dry bulb), *Crocus sativus* (dry style and stigma)
Guangzhou Qingping	Guangdong	11,000	1500		*Dendrobium* spp. (dry stem), *Salvia bowleyana* (root and rhizome), *Pseudostellaria heterophylla* (dry root), *Ganoderma lucidum* (fruit body), *Codonopsis pilosula* (root), *Ophiopogon japonicus* (tuber), *Lycium chinense* (fruit), *Sanvitalia procumbens* (whole plant), *Citrus reticulata* (dry peel)

(continued)

Table 24.1 (continued)

Market name	Province	Area (m^2)	Vendor number	Species number	Representative medicinal plants
Harbin Sankeshu	Heilongjiang	6000	1000	1000	*Panax ginseng* (root), *Lonicera japonica* (dry fruit), *Forsythia suspense* (fruit), *Strobilanthes cusia* (dry root and leaf), *Atractylodes lance* (rhizome), *Cyrtomium fortunei* (rhizome and petiole), *Schisandra chinensis* (fruit), *Arctium lappa* (fruit and root), *Bupleurum chinense* (root), *Dictamnus dasycarpus* (root bark)
Juancheng Shunwangcheng	Shandong	140,000	1000	1100	*Paeonia suffruticosa* (dry root bark), *P. lactiflora* (dry root), *Angelica dahurica* (root), *Strobilanthes cusia* (dry root and leaf), *Carthamus tinctorius* (flower), *Astragalus membranaceus* (root), *Pinellia ternate* (tuber), *Rehmannia glutinosa* (rhizome), *Trichosanthes kirilowii* (dry root), *Platycodon grandifloras* (root)
Kunming Juhuayuan	Yunnan	93,500	500	4000	*Panax notoginseng* (dry root and flower), *Amomum tsao-ko* (fruit), *Paris polyphylla* var. *chinensis* (rhizome), *Gastrodia elata* (tuber), *Rhodiola rosea* (root and rhizome), *Saussurea involucrata* (whole herb), *Angelica sinensis* (root), *Dipsacus asperoides* (root), *Nardostachys chinensis* (root and rhizome)

(continued)

Table 24.1 (continued)

Market name	Province	Area (m^2)	Vendor number	Species number	Representative medicinal plants
Lanzhou Huanghe	Gansu	40,000		800	*Codonopsis pilosula* (root), *Glycyrrhiza uralensis* (root and rhizome), *Angelica sinensis* (root), *Rehmannia glutinosa* (rhizome), *Strobilanthes cusia* (dry root and leaf), *Tussilago farfara* (flower bud and leaves), *Lycium chinense* (fruit), *Dipsacus asperoides* (root), *Epimedium brevicornu* (whole plant)
Puning	Guangdong	60,000	405	1000	*Dendrobium officinale* (dry stem), *Amomum villosum* (fruit), *Citrus reticulata* (dry peel), *Citrus medica* var. *sarcodactylis* (fruit), *Evodia lepta* (root and branch), *Polygonum chinense* (rhizome), *Ilex asprella* (dry root), *Abrus cantoniensis* (whole herb and fruit), *Strophanthus divaricatus* (root, stem, and leave), *Ficus simplicissima* (root)
Qichun	Hubei	66,670	328	1000	*Artemisia argyi* (whole herb), *Coix lacryma-jobi* (seed), *Prunella vulgaris* (whole plant), *Dioscorea opposita* (tuber), *Origanum vulgare* (whole plant), *Coptis chinensis* (whole plant), *Ephedra* spp. (whole plant), *Codonopsis pilosula* (root), *Crocus sativus* (dry style and stigma)

(continued)

Table 24.1 (continued)

Market name	Province	Area (m^2)	Vendor number	Species number	Representative medicinal plants
Shaodong Lianqiao	Hunan	13,340	800	1000	*Lonicera japonica* (flower), *Poria cocos* (fruit body), *Polygonatum odoratum* (rhizome), *Siraitia grosvenorii* (fruit), *Dictyophora indusiata* (fruit body), *Nepeta cataria* (whole plant), *Poncirus trifoliata* (dry fruit), *Bupleurum* spp. (root), *Paeonia suffruticosa* (dry root bark), *P. lactiflora* (dry root bark)
Xi'an Tiansheng	Shaanxi	450,000	1500	1600	*Scutellaria baicalensis* (rhizome), *Dendranthema morifolium* (flower), *Panax ginseng* (root), *Lilium brownii* var. *viridulum* (dry bulb), *Lycium chinense* (fruit), *Aconitum carmichaelii* (lateral root), *Astragalus membranaceus* (root), *Salvia miltiorrhiza* (root), *Eucommia ulmoides* (bark), *Polyporus umbellatus* (dry sclerotia), *Gastrodia elata* (tuber)
Yulin	Guangxi	18,700	809	1000	*Siraitia grosvenorii* (fruit), *Illicium verum* (fruit), *Asparagus cochinchinensis* (dry root), *Kaempferia galanga* (rhizome), *Cinnamomum cassia* (bark, branch, leaves, and flowers), *Abrus cantoniensis* (whole herb and seed), *Brucea javanica* (seed), *Pueraria lobata* (root), *Pogostemon cablin* (peduncle and leaves), *Piper nigrum* (fruit)

(continued)

Table 24.1 (continued)

Market name	Province	Area (m^2)	Vendor number	Species number	Representative medicinal plants
Yuzhou	Henan	266,640	2500	1000	*Chaenomeles sinensis* (fruit), *Evodia lepta* (root and leaves), *Forsythia suspense* (fruit), *Atractylodes macrocephala* (rhizome), *Ginkgo biloba* (fruit and leaves), *Saposhnikovia divaricata* (root), *Dioscorea opposita* (dry tuber), *Arisaema erubescens* (tuber), *Typhonium giganteum* (tuber)
Zhangshu	Jiangxi	266,640	2000	1000	*Evodia rutaecarpa* (fruit), *Rosa laevigata* (root, leaves, and fruit), *Corydalis decumbens* (tuber), *Polygonatum sibiricum* (rhizome), *Poncirus trifoliata* (dry fruit), *Bletilla striata* (tuber), *Lilium brownii* var. *viridulum* (dry bulb), *Stephania tetrandra* (fleshy root), *Plantago asiatica* (seed)

yanhusuo, *Chrysanthemum morifolium*, *Cynanchum glaucescens*, *Cyperus rotundus*, *Eucommia ulmoides*, *Forsythia suspensa*, *Fritillaria cirrhosa*, *Gastrodia elata*, *Glycyrrhiza uralensis*, *Ligusticum chuanxiong*, *Liquidambar orientalis*, *Lonicera japonica* (and relatives), *Lycium chinense*, *L. barbarum*, *Magnolia officinalis*, *M. biloba*, *Morinda officinalis*, *Myristica fragrans*, *Ophiopogon japonicus*, *Paeonia lactiflora*, *Panax ginseng*, *P. notoginseng* (both roots and inflorescence), *Piper longum*, *Poria cocos*, *Platycodon grandiflorus*, *Rehmannia glutinosa*, *Sophora flavescens*, *Trichosanthes kirilowii*, and *Vitex trifolia*.

The frequency and quantity of above listed medicinal plants vary from one market to another. For example, *Angelica sinensis*, *Astragalus membranaceus*, *Chrysanthemum morifolium*, *Gastrodia elata*, *Glycyrrhiza uralensis*, *Lycium chinense*, and *Panax notoginseng* are the most common species that appeared in 70% stalls. They are massively consumed by people from both Northern and Southern China. However, the quantity of some species was limited. Only a few kilograms (or less) of *Arisaema erubescens*, *Coptis chinensis*, *Corydalis yanhusuo*, and *Panax ginseng* would be available from each stall. Medicinal herbs from the tropics such as *Morinda officinalis* and *Myristica fragrans* possessed less quantity, although they are used frequently.

The native medicines were commonly sold in the targeted markets. *Achyranthes bidentata*, *Atractylodes macrocephala*, *Belamcanda chinensis*, *Dioscorea opposita*, *Glehnia littoralis*, *Platycodon grandiflorus*, *Salvia miltiorrhiza*, and *Saposhnikovia divaricata*, are local products with large plantation in Northern China. Therefore, in Anguo marketplace, the largest herbal medicinal market in Northern China, these medicines are traded in a large scale. Guangxi has the largest plantation area of *Siraitia grosvenorii* and *Illicium verum*; thus in Yulin market, a lot of vendors sell products of these two species. Yunnan is the native production area of *Panax notoginseng*, *Amomum tsao-ko*, and other herbs. The medicines from these species are commonly sold in Kunming marketplace.

24.4 Market-Based Traditional Knowledge of Medicinal Plants

The herbal medicinal markets are ideal places for people to exchange knowledge of medicinal plants. Although almost all herbal medicines have been recorded in Chinese Pharmacopoeia or other literatures, rich traditional knowledge has still been maintained by vendors in the herbal markets.

Traditional knowledge of identification and authentication of medicinal plants and materials is probably the most important for the vendors and buyers. The experienced vendors study traditional knowledge to identify and authenticate herbal medicines, although they accept training and obtain instructions from books. A lot of herbal medicines in three targeted markets had been processed by cutting into slices or grinding into powder, which destroyed most of morphological characteristics. Thus people cannot identify them easily. Traditional knowledge now provides strong power to support identification and authentication of medicines in the markets. The experienced vendors and buyers use their knowledge to identify or authenticate medicines by colors, tastes, smells, and feels. For instance, the sliced roots of *Astragalus membranaceus* can be easily confused with other medicinal roots. An experienced vendor will take a small piece to chew and tell the taste to confirm the right medicine.

Traditional knowledge of processing medicinal plants and materials can improve the quality and efficacy of herbal medicines. There are plentiful methods and skills to process herbal medicine, such as washing, drying, peeling, cutting, grinding, cooking, heating, selecting, grading, and storing, from fresh materials to patent drugs. Most vendors in the three markets did not have such traditional knowledge because they bought herbal medicines from the suppliers directly. In Kunming market, 27 vendors showed different processing tools from ethnic groups including Bai, Dai, Miao, Tibetan, and Yi but mostly from the Han Chinese. In Yulin market, eight vendors told they had processing tools, but in Anguo market, only two vendors had these tools. Traditional knowledge of processing medicines is normally inherited from the vendors' masters/teachers or parents. A few learnt from local communities in Yunnan Province.

Herbs traded in 17 huge herbal medicinal marketplaces are monitored and supervised by national, provincial, and local authorities. The Chinese Pharmacopoeia proposed guidelines of uses, dosages, taking methods, and course of treatment. The uses of herbal medicines, however, vary case by case. The experienced vendors in the targeted markets sometimes shared with buyers their traditional knowledge of uses of medicinal plants and materials. In Kunming market, for example, some vendors provided the buyers with traditional knowledge of uses of *Panax notoginseng* (sanqi), a native and important medicinal root from Southeast Yunnan. The different methods to prepare sanqi roots can be used for different purposes. A vendor from Yi ethnic group in Kunming market shared her traditional knowledge of uses of *Cynanchum bungei* roots, which is different from the Chinese Pharmacopoeia or other existing literature records.

24.5 Herbal Markets and Ethnobotany

There is rich species diversity of herbal medicinal plants in 17 huge herbal medical marketplaces of China (Table 24.1). Herbal medicinal diversity is closely related to local biodiversity. For instance, the Kunming market is located in the richest province of China's biodiversity (Guo and Long 1998). The herbal diversity in the market reflected Yunnan's rich biological diversity. However, the species diversity in the markets is probably related to ethnic-cultural diversity in the region too, based on our investigations in the three markets, Anguo, Kunming, and Yulin. Taking Kunming market as an example, over 4000 species of medicinal plants are traded in the market, which is the biggest number in all 17 huge herbal medicinal marketplaces recognized by the Chinese government. As the capital city of Yunnan Province, Kunming attracted vendors from different parts of Yunnan. The province has the largest number of linguistic groups in the country, with 25 different ethnic groups (Long et al. 2003). Many of these medicinal plants have been used by indigenous peoples. Examples include *Erigeron breviscapus* used by Miao people, *Veratrum taliense* by Bai people, *Helicia nilagirica* by Dai and Hani people, *Thamnolia vermicularis* by Bai and Naxi people, and *Cynanchum otophyllum* by Naxi people. These medicinal plants are mostly or only traded in Kunming market, not in other huge marketplaces in China.

Since ancient times in China, the theory of homology of medicine and food has widely been accepted by the Chinese people. Thus herbal medicines for dietary uses are warmly welcomed by all linguistic groups. In recent years, they are becoming more and more popular. It is probably because people obtained more income. The most common medicinal plants used as medicinal food or dietary supplements include *Angelica sinensis*, *Astragalus membranaceus*, *Chrysanthemum morifolium*, *Gastrodia elata*, *Glycyrrhiza uralensis*, *Lycium chinense*, and *L. barbarum*. As the most famous but expensive representative of homology of medicine and food, *Panax ginseng*, which has been regarded as the best medicine for tonic, is usually consumed by richer people, although more and more people are able to afford it in recent decades. The herbal medicines for dietary uses in all markets become very popular, and the volumes of transactions reach very high. The goji berries, *Lycium chinense* and *L. barbarum* fruits, for example, are products of Northwest China particularly in Ningxia Region (Yao et al. 2018). They are sold in three targeted markets and other huge herbal medicinal marketplaces. The total volume of goji berries consumed in China reached 70,000 tons annually.

Acknowledgments We are very grateful to the vendors and buyers in the three marketplaces, Anguo, Kunming, and Yulin, who provided valuable information about herbal medicines. This work was supported by the National Natural Science Foundation of China (31761143001, 31870316, and 31161140345), the Key Laboratory of Ethnomedicine (Minzu University of China) of Ministry of Education of China (KLEM-ZZ201806), Minzu University of China (Collaborative Innovation Center for Ethnic Minority Development, yldxxk201819), Ministry of Education of China, and State Administration of Foreign Experts Affairs of China (B08044).

References

Albuquerque UP, Monteiro JM, Ramos MA, de Amorim ELC (2007) Medicinal and magic plants from a public market in northeastern Brazil. J Ethnopharmacol 110:76–91

Bussmann RW, Glenn A (2010) Medicinal plants used in Northern Peru for reproductive problems and female health. J Ethnobiol Ethnomed 6:30. https://doi.org/10.1186/1746-4269-6-30

Bussmann RW, Glenn A, Meyer K, Meyer K, Kuhlman A, Townesmith A (2010) Herbal mixtures in traditional medicine in Northern Peru. J Ethnobiol Ethnomed 6:10. https://doi.org/10.1186/1746-4269-6-10

Bussmann RW, Paniagua-Zambrana N, Chamorro MR, Moreira NM, Negri MLDRC, Olivera J (2013) Peril in the market-classification and dosage of species used as anti-diabetics in Lima, Peru. J Ethnobiol Ethnomed 9:37. https://doi.org/10.1186/1746-4269-9-37

Bussmann RW, Sharon D, Vandebroek I, Jones A, Revene Z (2007) Health for sale, the medicinal plant markets in Trujillo and Chiclayo, Northern Peru. J Ethnobiol Ethnomed 3:37. https://doi.org/10.1186/1746-4269-3-37

Guo HJ, Long CL (1998) Biodiversity of Yunnan, SW China. Yunnan Sci & Tech Press, Kunming, China

Hidayati S, Franco F, Bussmann RW (2015) Ready for phase 5—current status of ethnobiology in Southeast Asia. J Ethnobiol Ethnomed 11:17. https://doi.org/10.1186/s13002-015-0005-7

Kunwar RM, Mahat L, Acharya RP, Bussmann RW (2013) Medicinal plants, traditional medicine, markets and management in far-west Nepal. J Ethnobiol Ethnomed 9:24. https://doi.org/10.1186/1746-4269-9-24

Kunwar RM, Shrestha KP, Bussmann RW (2010) Traditional herbal medicine in Far-west Nepal, a pharmacological appraisal. J Ethnobiol Ethnomed 6:35. https://doi.org/10.1186/1746-4269-6-35

Kunwar RM, Uprety Y, Burlakoti C, Chowdhary CL, Bussmann RW (2009) Indigenous use and ethnopharmacology of medicinal plants in far-west Nepal. Ethnobot Res Appl 7:5–28. https://doi.org/10.17348/era.7.0.5-28

Lin CR, Xu WB, Huang YS, Yu LY, Liu Y (2016) The important medicinal plants of Yao traditional medicinal market during Dragon-boat Festival in Gongcheng Guangxi. Guangxi Science & Technology Publishing Press, Nanning, China

Long CL (2015) Wild plant resources. In: Liu X (ed) The science report on biological germplasm resources in China. Science Press, Beijing, pp 36–74

Long CL, Li H, Ouyang ZQ, Yang XY, Li Q, Trangmar B (2003) Strategies for agrobiodiversity conservation and promotion: a case from Yunnan, China. Biodivers Conserv 12(6):1145–1156. https://doi.org/10.1023/A:1023085922265

Luo BS, Liu YJ, Liu B, Liu SZ, Zhang BX, Zhang LH, Lin CR, Liu Y, Long CL (2018) Herbal medicinal market on Dragon-Boat Festival in Jianghua County, China: a special platform for selling health. J Ethnobiol Ethnomed 14:61

Marcia MJ, Garcia E, Vidaurre PJ (2005) An ethnobotanical survey of medicinal plants commercialized in the markets of La Paz and El Alto, Bolivia. J Ethnopharmacol 97:337–350. https://doi.org/10.1016/j.jep.2004.11.022

Mittermeier RA, Gil PR, Mittermeier CG (1997) Megadiversity earth's biologically wealthiest nations. Conservation International (CI), Washington, DC

Narvaez A, Stauffer F (1999) Products derived from palms at the Puerto Ayacucho Markets in Amazonas State, Venezuela. Palms 43:122–129

Nicholson MS, Arzeni CB (1993) The market of medicinal plants of Monterrey, Nuevo Leon, Mexico. Econ Bot 47:184–192. https://doi.org/10.1007/BF02862021

The State Council Information Office of the People's Republic of China (2016) Traditional Chinese Medicine in China. December 2016

Williams VL, Witkowski ETF, Balkwill K (2005) Application of diversity indices to appraise plant availability in the traditional medicinal markets of Johannesburg, South Africa. Biodivers Conserv 14:2971–3001

Yang CY, Long CL, Shi YN, Wang YH, Wang HS (2009) Ethnobotanical study on medicinal market during Dragon-boat Festival in Jingxi County, South western Guangxi Region. J CUN (Nat Sci Edit) 18(2):16–26

Yao RY, Heinrich M, Weckerle CS (2018) The genus *Lycium* as food and medicine: a botanical, ethnobotanical and historical review. J Ethnopharmacol 212:50–66

Phytochemical Constituents and Pharmacological Activities of a Traditional Medicinal Plant, *Glochidion eriocarpum* (Phyllanthaceae)

25

Beixi Zhang, Sizhao Liu, Qiyi Lei, Jiangju Zhou, and Chunlin Long

Abstract

Glochidion eriocarpum is a shrub in the family Phyllanthaceae. The whole plant, stem bark, roots, and leaves are used for medicinal purposes by ethnic people in China to treat lacquer poison, convergence, diarrhea, dampness, and itching. As the intense research on chemical constituents, pharmacological activity and the active principles of *G. eriocarpum* have attracted the interest to the discovery of new drugs. This review summarizes the traditional uses of *G. eriocarpum* and its chemical constituents, pharmacological effects, and clinical applications with some suggestions for future research.

Keywords

Glochidion eriocarpum · Ethnomedicine · Chemical constituents · Pharmacological activity

B. Zhang · S. Liu · Q. Lei
College of Life and Environmental Sciences, Minzu University of China, Beijing, China

School of Life and Health Science, Kaili University, Kaili, Guizhou, China

Key Laboratory of Ethnomedicine (Minzu University of China), Ministry of Education, Beijing, China

J. Zhou
School of Life and Health Science, Kaili University, Kaili, Guizhou, China

C. Long (✉)
College of Life and Environmental Sciences, Minzu University of China, Beijing, China

Key Laboratory of Ethnomedicine (Minzu University of China), Ministry of Education, Beijing, China

Kunming Institute of Botany, Chinese Academy of Sciences, Kunming, China
e-mail: long.chunlin@muc.edu.cn

S. M. Khasim et al. (eds.), *Medicinal Plants: Biodiversity, Sustainable Utilization and Conservation*, https://doi.org/10.1007/978-981-15-1636-8_25

25.1 Introduction

Glochidion eriocarpum Champ. ex Benth. is a shrub in the family Phyllanthaceae (earlier included in Euphorbiaceae). The medicinal value of the genus *Glochidion* has attracted the interest of many scholars at home and abroad, and many patents have reported compounds from this genus for the treatment of certain diseases.

China is a rich country with abundant plant resources in the world. There are 37,500 plant species; 11,146 species of plants are used by people to treat diseases. *Glochidion eriocarpum* is a national plant in southwestern China and is widely used by many ethnic groups. The species is mainly distributed in China (Jiangsu, Fujian, Taiwan, Hunan, Guangdong, Hainan, Guangxi, Guizhou, Yunnan, etc.), Laos, Thailand, and Vietnam, growing on hillsides, valleys, or forest edges at an altitude of 130–1600 m (Luo and Sun 2013). Especially in China, the whole plant, root, and leaf of *Glochidion eriocarpum* are used for medicinal purposes. And it is commonly used to address hepatitis, intestinal diseases, diarrhea, urticaria, mastitis, toothache, menorrhagia, dysentery, skin eczema, and other diseases (Liu 2012). Thus, the medicinal value of the genus *Glochidion* has aroused the interest of many scholars at home and abroad. Some patents have reported that compounds from this genus were used to treat chronic and acute osteomyelitis, cirrhosis, chronic gastritis, hepatitis B, liver ascites, and malignant tumors. However, the scientific evidence of the abundant use value of this plant is still lacking. This paper summarizes various methods of utilizing this plant by Chinese people and chemical structure and pharmacological activities of *Glochidion eriocarpum* in the recent 10 years. We hope by systematically summarizing the research of the *Glochidion eriocarpum* we can further broaden the development and utilization prospects of the plant and provide clues and scientific theoretical basis for further research and development of the resource.

25.2 Folk Value of *Glochidion eriocarpum*

In China, *Glochidion eriocarpum* is a kind of important ethnic medicinal plant resource widely used by many people (Table 25.1). There are three ethnic medicinal formulas that record *Glochidion eriocarpum* for the treatment of gynecological inflammation, urinary stones, and detoxification. It can also be used as a dietary fiber in diet therapy. Studies have shown that *Glochidion eriocarpum* has anti-*Helicobacter pylori* (Zhang and Wang 2008), which can be scientifically validated as a mechanism for treating stomach pain and gastric ulcer in Zhuang medicine. At the same time, they are used in other countries such as Vietnam for the treatment of asthma, enteritis, cholera, indigestion, rheumatism, etc. (Nhiem et al. 2012). In Laos, they were treated to treat postpartum recovery, anemia (dizziness, headache), mild puerperal fever, and abdominal pain (Vichith et al. 2011).

Table 25.1 Traditional medicinal knowledge of *Glochidion eriocarpum* in China

Linguistic group	Use and value	Part used	References
Han	Clearing heat and dampness, detoxifying, and relieving itching. Enteritis, dysentery. Allergy to raw lacquer, dermatoses, pruritus, urticaria, eczema, exfoliative dermatitis	Root; leaf	"National Compilation of Chinese Herbal Medicine"
Yi	Lacquer poisoning, dermatitis, and eczema; radical bowel disease, dysentery, rectal prolapse	Whole plant	"The Ailao"
She	Flashback, abdominal pain, food bite	Root; leaf	"She Nationality's Medicine"
Lisu	Rheumatic bone pain, bruises, and swelling. Anal prolapse, uterine prolapse, vaginal discharge, diarrhea, hepatitis	Root; leaf	"Nujiang Medicine"
Dai	Acute gastroenteritis, dysentery, rheumatoid arthritis, bruises, traumatic bleeding, sore, eczema, dermatitis	Branch; leaf	"Yunnan Provincial Records"
Dong	Diarrhea, dysentery	Root; leaf	"Guangxi Medical Records"
Yao	Enteritis, diarrhea, dysentery, epistaxis, treatment of burns, allergic dermatitis, lacquer tree allergy, skin eczema, skin irritation, acute gastroenteritis, rheumatic arthralgia, bruises, traumatic bleeding, lacquer sores	Root, leaf, whole plant	"Guangxi Medical Records"
Zhuang	Stomachache, icteric hepatitis, maternal bleeding, menorrhagia, measles, topical allergic dermatitis, lacquer tree allergy, pruritus, skin eczema, acne ulcers	Root, branch, leaf	"Guangxi Medical Records," "Yunnan Provincial Records"
Jinuo	Enteritis, itchy skin, rash	Root; leaf	"Jinuo Nationality's Medicine"

25.2.1 Chemical Constituents of *Glochidion eriocarpum*

There are 50 compounds isolated and identified from *Glochidion eriocarpum*, mainly quinones and glycosides extracted from the roots, stems, and leaves in addition to containing aromatic compounds, steroids, and other compounds.

25.2.1.1 Terpenoids

Terpenoids are the main components of the *Glochidion eriocarpum*, and they are ubiquitous in the plant kingdom which is the generic term for all isoprene polymers and their derivatives (Shi et al. 2012). Terpenoids have many biological activities, such as anticancer, anti-allergy, and anti-HIV (Chen and Li 2016), with important medicinal and economic value (Fu et al. 2003). Twenty-eight terpenoids, the highest proportion of *Glochidion eriocarpum* chemical constituents, and each part of the plant have contained terpenoids with good biological activities (Table 25.2, Fig. 25.1).

Table 25.2 Terpenoids distributed in different parts of *Glochidion eriocarpum*

No.	Compound	Chemical formula	Part	References
1	(6*R*,9*S*)-3-oxo-α-ionol-β-D-Glucopyranoside	$C_{19}H_{30}O_7$	Root	Wang (2014)
2	(6*S*,9*S*)-Roseoside	$C_{19}H_{30}O_8$	Root	Wang (2014)
3	(*Z*)-4-[3′-(β-D-Glucopyranosyloxy) butylidene]-3,5,5-trimethyl-2-cyclohexen-1-one	$C_{19}H_{30}O_7$	Root	Wang (2014)
4	3-Epilupeol	$C_{30}H_{50}O$	Shoots	Puapairoj et al. (2005); Nhiem et al. (2012); Kiem et al. (2009)
5	Blumenol C glucoside	$C_{19}H_{32}O_7$	Root	Wang (2014)
6	Cannabiside D	$C_{19}H_{30}O_9$	Root	Wang (2014)
7	Epifriendelanol	$C_{30}H_{52}O$	Stem bark	Wei (2001)
8	Friedelin	$C_{30}H_{50}O$	Stem bark	Wei (2001)
9	Glochidiol	$C_{30}H_{50}O_2$	Entire plants	Wei (2001); Puapairoj et al. (2005); Nhiem et al. (2012); Hui and Li (1976); Kiem et al. (2009)
10	Glochidionionoside A	$C_{19}H_{30}O_9$	Root	Wang (2014)
11	Glochidionionoside C	$C_{19}H_{30}O_9$	Root	Wang (2014)
12	Glochidionionoside E	$C_{19}H_{30}O_9$	Root	Wang (2014)
13	Glochidioside N	$C_{41}H_{66}O_{12}$	Stem bark	Wei (2001)
14	Glochidol	$C_{30}H_{48}O$	Stems	(Hui and Li 1976)
15	Glochidone	$C_{30}H_46O$	Entire plants	Wei (2001); Puapairoj et al. (2005); Nhiem et al. (2012); Hui and Li (1976); Kiem et al. (2009)
16	Glochidonol	$C_{30}H_{48}O_2$	Entire plants	Wei (2001); Puapairoj et al. (2005); Nhiem et al. (2012); Hui and Li (1976); Kiem et al. (2009); Kiem et al. (2009)
17	Glochierioside A	$C_{48}H_{72}O_{14}$	Shoots	Nhiem et al. (2012)
18	Glochierioside B	$C_{48}H_{72}O_{14}$	Shoots	Nhiem et al. (2012)
19	Glochierioside C	$C_{48}H_{72}O_{14}$	Shoots	Nhiem et al. (2012)
20	Glochierioside D	$C_{48}H_{72}O_{14}$	Shoots	Nhiem et al. (2012)
21	Glochierioside E	$C_{48}H_{72}O_{14}$	Shoots	Nhiem et al. (2012)
22	Glochierioside F	$C_{48}H_{72}O_{14}$	Shoots	Kiem et al. (2009)
23	Glochierioside G	$C_{48}H_{72}O_{14}$	Shoots	Kiem et al. (2009)
24	Glochilocudiol	$C_{30}H_{50}O_2$	Entire plants	Hui and Li (1976)
25	Lup-20(29)-en-3β,23-diol	$C_{30}H_{48}O_2$	Shoots part	Kiem et al. (2009)
26	Lup-20(29)-ene-1β,3β-diol	$C_{30}H_{50}O_2$	Roots	Puapairoj et al. (2005); Kiem et al. (2009)

(continued)

Table 25.2 (continued)

No.	Compound	Chemical formula	Part	References
27	Lupenone	$C_{30}H_{48}O$	Shoots, roots	Puapairoj et al. (2005); Nhiem et al. (2012); Hui and Li (1976); Kiem et al. (2009)
28	Lupeol	$C_{30}H_{50}O$	Stem bark	Wei (2001)

25.2.1.2 Glycoside Compounds

Glycosides are widely distributed in the roots, stems, leaves, flowers, and fruits of plants. It has the efficacy of eliminating phlegm and relieving cough (Liu et al. 2013), nourishing, treating rheumatism (Su 2016), lowering cholesterol, being anti-inflammatory and antibacterial (Peng et al. 2014), improving immune regulation, being antitumor (Wang et al. 2014), and other effects. *Glochidion eriocarpum* contains 10 kinds of saponins and has good activity in improving human immunity (Table 25.3, Fig. 25.2).

25.2.1.3 Other Compounds

In addition to terpenoids and glycosides, the chemical components of *Glochidion eriocarpum* were also found: bergenin, steroids (β-sitosterol, stigmasterol), octadecanoic acid, tetradecyl ester, octanol, myristic acid, lauric acid, etc.

25.3 Pharmacological Activity

25.3.1 Antitumor Activity and Cytotoxic Activity

So far, the research on *Glochidion eriocarpum* in China and abroad has focused on its cytotoxic activity and antitumor activity, but there are few studies on new compounds and pharmacological activities. In 2009, two triterpenoid glochieriosides A and B from *Glochidion eriocarpum* were reported (Otsuka et al. 2000). The results showed that these two compounds have cytotoxic activity against four human cancer cells such as HL-60, HT-29, MCF-7, and SK-OV-3. Studies have shown that glochidonol exhibits strong cytotoxic activity on human breast cancer (MCF-7), lung cancer (NCI-H460), and prostate cancer (DU-145) (Bagalkotkar et al. 2011). Glochidone and lupeol have significant anti-malignant tumor growth effects and are very sensitive and have inhibitory effects on cancerous lung cells (Sakkrom et al. 2010). At the same time, glochidiol, lup-20(29)-ene-1α,23-diol, glochidone, and 3-epi-lupeol have cytotoxic effects on human hepatocellular carcinomas (Xiao et al. 2008). Further experiments showed that glochidonol and glochidiol have the effect of preventing the proliferation of malignant tumors, but lup-20(29)ene-3α,23-diol does not (Puapairoj et al. 2005). In vivo experiments have shown that glochidiol has a strong inhibitory effect on mouse skin tumors (Tanaka et al. 2004).

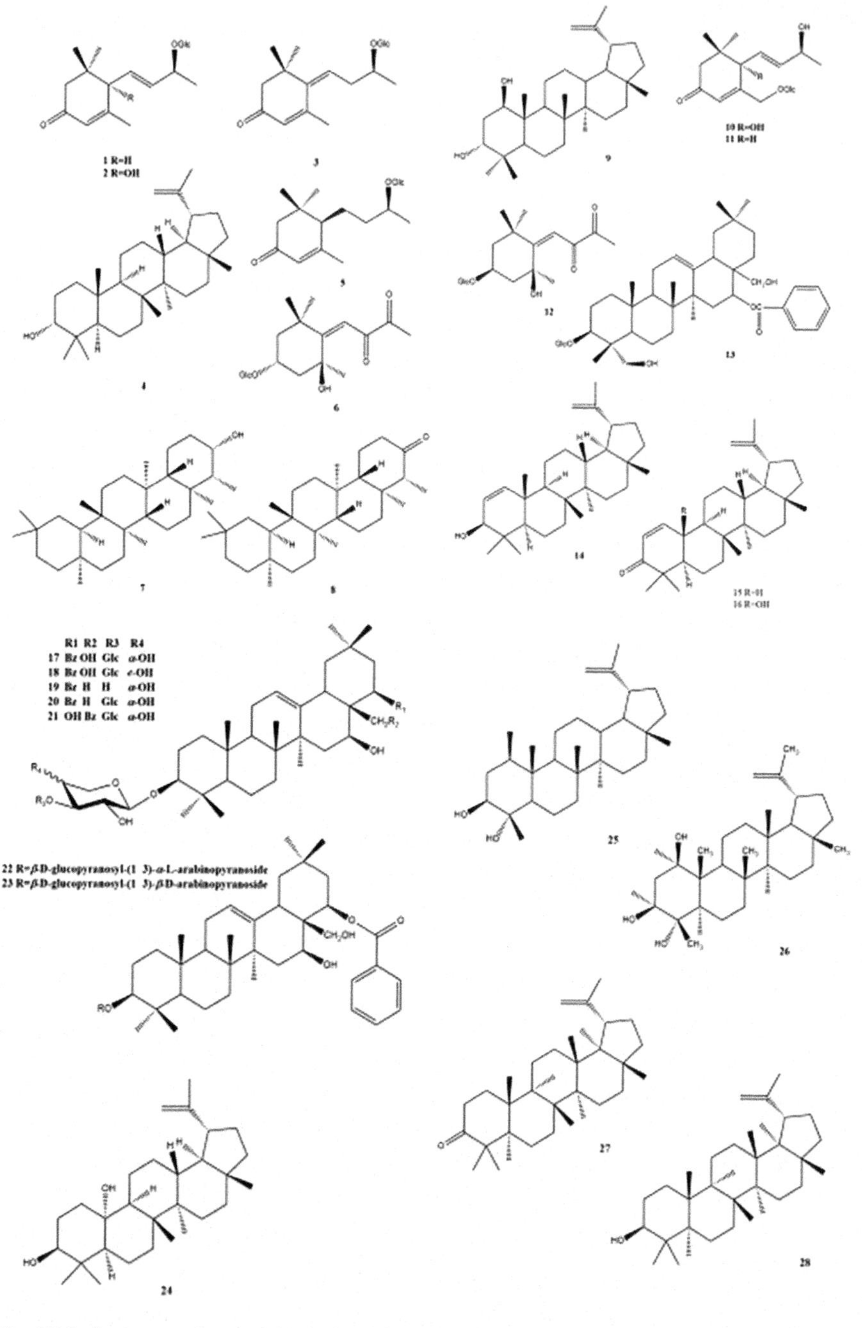

Fig. 25.1 Structures of terpenoids isolated from *Glochidion eriocarpum*

Table 25.3 The distribution of glycosides in different parts of *Glochidion eriocarpum*

No.	Compound	Chemical formula	Part	References
29	1,2-Dimethoxyphenyl-4-*O*-β-D-glucoside	$C_{36}H_{36}N_2O_8$	Root	Wang (2014)
30	1-β-D-Glucopyranosyloxy-3-methoxy-5-hydroxybenzene	$C_{13}H_{18}O_8$	Root	Wang (2014)
31	2-β-D-Glucopyranosyloxy-4-methoxy-6-hydroxy-isovalero-phenone	$C_{18}H_{26}O_9$	Root	Wang (2014)
32	3,4-Dihydroxy-1-naphthoic acid	$C_{11}H_8O_4$	Root	Wang (2014)
33	7*R*,8*S*-Dihydrodehydrodiconiferyl alcohol 4-*O*-β-D-glucopyranoside	$C_{26}H_{34}O_{11}$	Root	Wang (2014)
34	7*S*,8*R*-Dihydrodehydrodiconiferyl alcohol 4-*O*-β-D-glucopyranoside	$C_{26}H_{34}O_{11}$	Root	Wang (2014)
35	Adenosine	$C_{10}H_{13}N_5O_4$	Root	Wang (2014)
36	Koaburaside	$C_{14}H_{20}O_9$	Root	Wang (2014)
37	Koaburaside monomethyl ether	$C_{15}H_{22}O_9$	Root	Wang (2014)
38	Syringin	$C_{17}H_{24}O_9$	Root	Wang (2014)

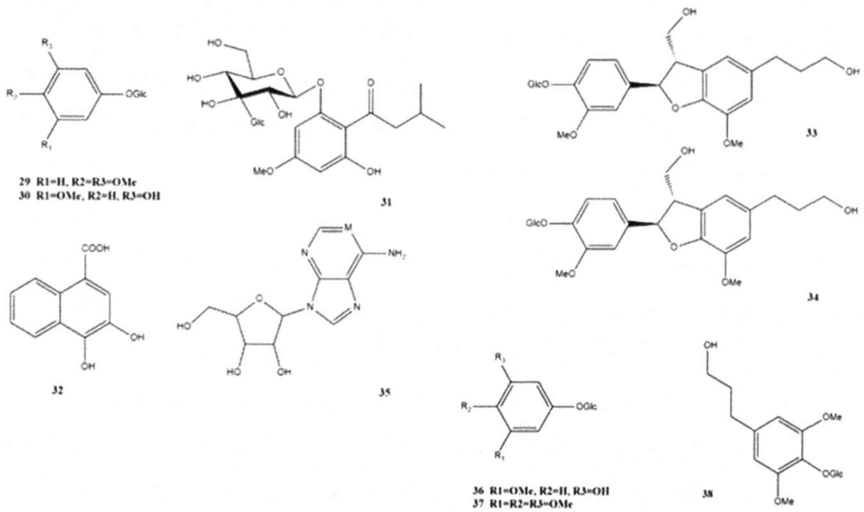

Fig. 25.2 Structures of glycosides isolated from *Glochidion eriocarpum*

25.3.2 Anti-inflammatory and Analgesic Effects

Phyllanthus urinaria was studied by Brazilians to show that its extract has analgesic effect. Chemical composition studies of *P. urinaria* extracts by chromatography and GC indicated that β-sitosterol, glochidonol, and glochidone were present in the extract (Catapan et al. 2001). Glochidone has significant antinociceptive effects on writhing and formalin tests (Krogh et al. 1999) suggesting that these compounds may have analgesic effects. *Glochidion puberum* extracts have significant anti-inflammatory and analgesic effects, and their effects may be related to the reduction of histamine content in the inflammation sites (Huang 2010). *Glochidion eriocarpum* may have the same effect.

25.3.3 Antioxidant Activity

Each extracted part of *Glochidion puberum* has a different in vitro antioxidant capacity (Hu et al. 2014). Pharmacological studies have shown that most of *Glochidion* species exhibit antibacterial, anti-inflammatory and antioxidative effects. The dominant ingredient, lupin-type triterpenoids, has potential antitumor activity and antiviral effect. There are few studies on its chemical components and active ingredients in China. It is necessary to conduct in-depth research on the substance basis of its efficacy (Ganguly et al. 1968; Puapairoj et al. 2004).

25.3.4 Others

Some patents have reported that some plants and their compounds of *Glochidion* can treat some diseases. For example, *Glochidion puberum* and its compound can treat chronic and acute osteomyelitis, chronic gastritis, hepatitis B, liver cirrhosis, liver ascites, and malignancy and assist in the treatment of hepatitis, intestinal diseases, diarrhea, and acute and chronic enteritis (Li 2007). *Glochidion eriocarpum* may contain the same chemical constituents as the *Glochidion zylanicum* root extract. The experiment shows that the water extract has inhibitory effect on *Flavobacterium tegecticola* (Sharma et al. 2010).

25.4 Conclusion

Glochidion eriocarpum is an ethnomedicinal plant in southwestern China. The study on its chemical composition and pharmacological effects can promote the standardization of *Glochidion eriocarpum* materials and provide theoretical guidance for its use, so as to further develop this ethnomedicine. So far, most of the 50 compounds extracted from *Glochidion eriocarpum* have good activity, and a small number of them are cytotoxic. They can be strengthened by structural modification and structure-activity relationship to discover new active lead compounds. However, the current research on the chemical composition of *Glochidion eriocarpum* mainly focuses on the component extraction and activity research; its activity, mechanism, and biosynthetic route and the development of ethnic medicine products need to be further explored; research on these can better develop the plant resources of *Glochidion eriocarpum*.

Acknowledgments This study was supported by the Key Laboratory of Ethnomedicine (Minzu University of China) of Ministry of Education of China (KLEM-ZZ201906), the National Natural Science Foundation of China (31761143001, 31870316, 31161140345, and 31660083), Minzu University of China (Collaborative Innovation Center for Ethnic Minority Development, yldxxk201819), Ministry of Education of China, and State Administration of Foreign Experts Affairs of China (B08044).

References

Bagalkotkar G, Tang SC, Khalivulla SI, Hamzah AS, Shaari K, Lajis NH, Stanslas J (2011) Isolation and cytotoxicity of triterpenes from the roots of *Phyllanthus pulcher* Wall. ex Müll.-Arg. (Euphorbiaceae). Afr J Pharm Pharmacol 5(2):183–188

Catapan E, Otuki M, Am Yunes R, Bresciani L, Ferreira J, Santos A, Cechinel-Filho V (2001) Pharmacological activity and chemical composition of callus culture extracts from selected species of *Phyllanthus*. Pharmazie 55(12):945–946

Chen JZ, Li CY (2016) Research progress on anti-HIV steroids in natural products. Asia-Pac Trad Med 12(7):54–59

Fu J, Wang Y, Qi XF (2003) Ecophysiological functions and economic value of terpenoids. J Northeast For Univ 31(6):59–62

Ganguly AK, Govindachari TR, Mohamed PA (1968) Chemical constituents of *Glochidion hohenackeri*. Tetrahedron 22(4):1513–1519

Hu DQ, Han ZX, Hu JH (2014) The screening of antioxidative active parts of plate leaves by hairy fruit. J Hubei Univ Trad Chin Med 16(4):51–54

Huang AJ (2010) An experimental study on the anti-inflammatory and analgesic effects of the extract from *Glochidion puberum*. J Hubei Inst Natl Med Sci 27(4):17–19

Hui WH, Li MM (1976) Lupene triterpenoids from *Glochidion eriocarpum*. Phytochemistry 15(4):23–26

Kiem PV, Thu VK, Yen PH, Nhiem NX, Tung NH, Cuong NX, Kang HK (2009) New triterpenoid saponins from *Glochidion eriocarpum* and their cytotoxic activity. Chem Pharm Bull 57(1):102–103

Krogh R, Kroth R, Berti C, Madeira AO, Souza MM, Cechinelfilho V, Yunes RA (1999) Isolation and identification of compounds with antinociceptive action from *Ipomoea pes-caprae* (L.) R. Br. Pharmazie 54(6):464–466

Li GY (2007) Two treatments of single Chinese medicine. Chin Naturopathy 15(1):22–23. https://doi.org/10.3969/j.issn.1007-5798.2007.01.027

Liu N (2012) Progress in studies on chemical constituents and pharmacological activities of *Glochidion*, and the chemical constituents of *Glochidion lanceolarium*. Dissertation, Northwest Normal University

Liu Q, Li W, Zheng YN, Wu HF (2013) Research progress of three terpenoid saponins and their pharmacological activities in *Platycodon grandiflorus*. J Jilin Agric Univ 35(2):221–228

Luo YC, Sun QW (2013) Commonly used natural medicines in Guizhou Ethnic Groups, vol 2. Guizhou Sci Technol Press, Guiyang

Nhiem NX, Thu VK, Van KP, Van MC, Tai BH, Quang TH, Kang JI (2012) Cytotoxic oleane-type triterpene saponins from *Glochidion eriocarpum*. Arch Pharm Res 35(1):19–23

Otsuka H, Hirata E, Shinzato T, Takeda Y (2000) Isolation of lignan glucosides and neolignan sulfate from the leaves of *Glochidion zeylanicum* (Gaertn) A. Juss. Chem Pharm Bull 48(7):1084–1086

Peng SL, Zhang W, Li Y (2014) Research progress in the function and mechanism of *Bupleurum* saponin. J Jinlin Med Coll 35(1):66–68

Puapairoj P, Naengchomnong W, Kijjoa A, Pinto MM, Pedro M, Nascimento MS, Herz W (2005) Cytotoxic activity of lupane-type triterpenes from *Glochidion sphaerogynum* and *Glochidion eriocarpum* two of which induce apoptosis. Planta Med 71(3):208–213

Puapairoj P, Naengchomong W, Kijjoa A et al (2004) Cytotoxic activity of lupane-type triterpenes from *Glochidion sphaerogynum* and *Glochidion eriocarpum*. Planta Med 70(9):1234–1236

Sakkrom P, Pompimon W, Meepowpan P, Nuntasaen N, Loetchutinat C (2010) The effect of *Phyllanthus taxodiifolius* Beille extracts and its triterpenoids studying on cellular energetic stage of cancer cells. Am J Pharm Toxicol 5(3):139–144

Sharma JVC, Malleswari G, Rao JV, Muralibalaram V, Sangeetha R (2010) Antimicrobial activity of root extracts of *Glochidion zylanicum*. Int J Chem Sci 3(1):35–37

Shi RB, Jiang YY, Wang Z, Wang YY (2012) On the recent development of Chinese medicinal chemistry. J Beijing Univ Trad Chin Med 35(3):153–159

Su XF (2016) Extraction, purification and antioxidant activity of total flavonoids from *Sambucus williamsii* leaves. Shanxi Norm Univ 2(2):219–228

Tanaka R, Kinouchi Y, Wada S, Tokuda H (2004) Potential anti-tumor promoting activity of lupane-type triterpenoids from the stem bark of *Glochidion zeylanicum* and *Phyllanthus flexuosus*. Planta Med 70(12):1234–1235

Vichith L, De BHJ, Lars B (2011) Traditions and plant use during pregnancy, childbirth and postpartum recovery by the Kry ethnic group in Lao PDR. J Ethnobiol Ethnomed 7(1):14–15

Wang Y, Shao RG, Jiang JD, Wang Z (2014) Progress in research on mechanisms of antitumor effects of cardiac glycosides. Chin New Drugs J 27(6):677–681

Wang YM (2014) Studies on the chemical constituents and bioactivities of *Gentiana dahurica* and *Glochidion eriocarpum*. Kunming Bot 24(5):324–325

Wei XM (2001) Research on chemical constituents of *Clinopodium urticifolium*, *Clerodendrum serratum* and *Glochidion eriocarpum*. Lanzhou Univ 51(5):1043–1049

Xiao HT, Wang YH, Hao XY, Yang XS, Hao XJ (2008) Triterpenes from *Glochidion coccineum* and their cytotoxicity *in vitro*. Lishizhen Med Mater Med Res 57(1):102–103

Zhang Y, Wang Y (2008) Screening study on the role of Chinese herbal medicine and Zhuang medicine in the treatment of *Helicobacter pylori* in Guangxi. Chin J Ethnomed Ethnopharm 17(10):19–20

Endophytic Fungi and Their Impact on Agroecosystems

26

Ahmed Abdul Haleem Khan

Abstract

Plants are known to inhabit a variety of habitats that are exposed to extreme climatic conditions. The mutualistic associations of plants with microorganisms benefit them to coexist in harsh environments. The areas of plant such as rhizosphere (root), rhizoplane, and phyllosphere (apoplast of leaf) will serve microbes to the tissues of roots, shoots, leaves, flowers, and seeds. These microbes enter the plant through lenticels, stomata, wounds, ruptures, and nodules and become endophytes. There are two modes of endophyte transmission—horizontal (through airborne spores) and vertical (seeds) into the plants. The beneficial activities of endophytes are through release of antibiotics, proteases, chitinases, bacteriocins, siderophores, lipopeptides, and volatiles that control phytopathogens and improve plant growth and health. The plant endophytes are evaluated quantitatively by qPCR and culture methods (CFU/g tissue) and qualitatively by FISH, GUS, GFP, and PCR. In this discussion, endophytic fungi are represented to evaluate the studies carried out to present an overview of plant-fungi interactions that could be an alternative for improving agroecosystems by their applications.

Keywords

Endophytic fungi · Bioactive metabolites · Antimicrobials · Biocontrol · Phytopathogens · Stress

A. A. H. Khan (✉)
Department of Botany, Telangana University, Nizamabad, Telangana, India

S. M. Khasim et al. (eds.), *Medicinal Plants: Biodiversity, Sustainable Utilization and Conservation*, https://doi.org/10.1007/978-981-15-1636-8_26

26.1 Introduction

Terrestrial plants exist in variety of habits and habitats that are exposed to different situations. To overcome the extreme climatic conditions, these plants form mutualistic or symbiotic association with microorganisms. The association of microorganisms could be in root (rhizosphere), leaf (phyllosphere), stem, and seed (spermosphere) that improve the various physiological aspects of these plants. The association in root will enhance the availability of nutrients and plant growth in adverse conditions (Rho et al. 2018).

Endophytes are microorganisms (fungi/bacteria/actinomycetes) that occur inside the tissues and organs of healthy plants throughout their life cycle without causing any visible sign of disease. The endophytic fungi are grouped into clavicipitaceous (inhabit shoots and rhizomes of grasses) and non-clavicipitaceous (inhabit nonvascular plants, ferns, conifers, and angiosperms). The fungi that exist as endophytes are predominantly reported from Ascomycota and Basidiomycota. The endophytes were reported to involve in constant interactions within its host that evolve specific ecological niche. The mutualistic interactions between host and endophyte favor the production of secondary metabolites as bioactive natural compounds with several bioactivities for human welfare (Petrini 1991). The abundance and community composition of endophyte in host depend on environmental factors (temperature and humidity), anatomy, chemical composition, maturity, and type of host tissues. The plant habits that exist in all the habitats/environments are known to exhibit the endophyte colonization.

The endophytic fungi are ubiquitous and are reported to colonize approximately 300,000 plant species that exist in the biosphere. The colonization of endophytes varies from tissues to tissues in underground and aerial parts of host plants (Strobel 2003). There is high diversity of endophytes that could be culturable or non-culturable from the plant parts sampled (Table 26.1). Endophytes inside the plants were reported to promote plant growth by facilitating soil nutrient uptake, limit pathogen, and play role in specific defense that protects against biotic or abiotic stresses (Table 26.2). These organisms exist in both free living and association with plants and are involved in detoxification of synthetic recalcitrant chemicals (herbicides) through their ability to release hydrolytic enzymes (Tetrad-Jones and Edwards 2015).

The liquid cultures of endophytic fungi (*Penicillium* sp., *P. brasilianum*, *P. griseoroseum*, *Trichoderma* sp., and *Xylaria* sp.) isolated from different tropical plant species were reported for evaluation of biofuel precursor production through acid-catalyzed transesterification reactions. Among the endophytes, *Xylaria* sp. was found to be the highest producer of methyl esters (Santos-Fo et al. 2011). Three hundred healthy leaves from seven mangrove plant species, *Aegiceras corniculatum, Avicennia marina, A. officinalis, Bruguiera cylindrica, Ceriops decandra, Excoecaria agallocha,* and *Lumnitzera racemosa*, were studied for foliar endophytes. The leaves of these plants were found to contain mitosporic fungi and ascomycetes that were common to more than one host (Kumaresan and Suryanarayanan 2001).

Table 26.1 Distribution of endophytes

Plant used for isolation	Plant part	Endophytic fungi isolated	Reference
Azadirachta indica, Holarrhena antidysenterica, Terminalia arjuna, T. chebula	Bark, roots, twigs	30 Endophytic *Pestalotiopsis* strains	Tejesvi et al. (2009)
Aegle marmelos Correae (Rutaceae)	Bark, leaves, root	*Fusarium* sp., *Alternaria* sp., *Drechslera* sp., *Rhizoctonia* sp., *Curvularia* sp., *Nigrospora* sp., *Stenella* sp., *Chaetomium globosum, Emericella* sp.	Gond et al. (2007)
Santalum album Linn., *Kuhnia rosmarinifolia* Vent.	Roots	*Penicillium* sp., *Fusarium* sp.	Sun et al. (2014b)
Paris polyphylla var. *yunnanensis*	Rhizome	*Gliocladiopsis irregularis, Plectosphaerella cucumerina, Podospora* spp., *Gliomastix murorum, Aspergillus fumigatus, Pichia guilliermondii, Neonectria radicicola*	Li et al. (2008)
Taxus chinensis var. *mairei*	Barks, branches, leaves	145 Fungal taxa	Wu et al. (2013b)
Ceriops tagal, Rhizophora apiculata, R. stylosa, Bruguiera sexangula var. *rhynchopetala*	Stems, roots	*Pestalotiopsis, Phomopsis*	Xing and Guo (2011)
Cereus jamacaru DC (Cactaceae)	Cladodes	*Cladosporium cladosporioides, Fusarium oxysporum, Acremonium implicatum, Aureobasidium pullulans, Trichoderma viride, Chrysonilia sitophila, Aspergillus flavus*	Bezerra et al. (2013)
Luma apiculata, Myrceugenia ovata var. *nannophylla, Eugenia neomyrtifolia* (Myrtaceae)	Leaves	*Amphilogia* sp., *Coniochaeta velutina, Diaporthe* sp., *Diaporthe stewartii, D. phaseolorum, D. helianthi, Greeneria* sp., *Colletotrichum boninense, Colletotrichum* sp., *Cephalosporium* sp., *Annulohypoxylon* sp., *Biscogniauxia* sp., *Nemania* sp., *Xylaria berteri, Xylaria castorea, Xylaria* sp., *Xylaria enteroleuca, Guignardia* sp., *Guignardia mangiferae, Dothiora cannabinae, Cladosporium subtilissimum, Cladosporium colombiae, Mycosphaerella* sp., *Pseudocercospora basintrucata, Camarosporium brabeji, Didymella* sp., *Lewia infectoria, Microsphaeropsis olivacea, Paraconiothyrium* sp., *Cryptosporiopsis actinidiae, Mollisia cinerea, Pezicula corylina, Penicillium restrictum, Peziza* sp., *Mortierella sclerotiella, Trametes hirsuta*	Vaz et al. (2014)

(continued)

Table 26.1 (continued)

Plant used for isolation	Plant part	Endophytic fungi isolated	Reference
Fagus sylvatica	Leaves	*Acremonium strictum*, *Alternaria* sp., *Anthostomella tomicoides*, *Apiognomonia errabunda*, *Arthrinium phaeospermum*, *Aspergillus* sp., *Aureobasidium pullulans*, *Bertia moriformis*, *Bisporella citrina*, *Botrytis cinerea*, *Chaetomium funicola*, Chrysosporium sp., *Cladosporium cladosporioides*, *Cladosporium oxysporum*, *Cladosporium* sp., *Colletotrichum acutatum*, *Coniochaeta ligniaria*, *Coniochaeta velutina*, *Coniothyrium* sp., *Cryptococcus* sp., *Cylindrocarpon didymium*, *Diaporthe* sp., *Didymostilbe* sp., *Diplodina* sp., *Epicoccum nigrum*, *Exophiala* sp., *Fusarium lateritium*, *Fusarium* sp., *Ganoderma* sp., *Geniculosporium* sp., *Helotiales* sp., *Hormonema dematioides*, Hypocreales sp., *Hypoxylon serpens*, *Hypoxylon unitum*, *Isaria farinosa*, *Mycosphaerella punctiformis*, *Mycosphaerella* sp., *Nigrospora oryzae*, *Nodulisporium* sp., *Oedocephalum argillaceum*, *Penicillium* sp., *Pezicula livida*, *Peziza echinospora*, *Phialea fumosella*, *Phialophora* sp., *Phomopsis* sp., *Podospora anserine*, *Rhinotrichiella globulifera*, *Rhizoctonia* sp., *Rhodotorula* sp., *Rosellinia nectrioides*, *Sirococcus* sp., *Sporormiella australis*, *Sporormiella intermedia*, *Sporormiella minima*, *Strattonia minor*, *Thysanophora penicillioides*, *Virgariella* sp., *Xylaria* sp.	Unterseher and Schnittler (2010)
Senecio olgae, *Matthiola integrifolia*, *Asyneuma attenuatum*, *Clementsia semenovii*, *Lathyrus mulkak*, *Iris stolonifera*, *Eremostachys sarawschanica*, *Salvia komarovii*, *Scutellaria iscanderi*, *Scutellaria glabrata*, *Stipa richteriana*, *Aquilegia vicaria*, *Spiraea baldschuanica*, *Asperula pamirica*, *Viola alaica*	Roots	Dark septate endophytes	Zubek et al. (2011)

(continued)

Table 26.1 (continued)

Plant used for isolation	Plant part	Endophytic fungi isolated	Reference
Holcoglossum sinicum, H. flavescens, H. kimballianum, H. lingulatum, H. nujiangense, H. rupestre, H. subulifolium, H. wangii, H. weixiense	Roots	*Alternaria, Cladosporium, Clonostachys, Colletotrichum, Cosmospora, Cryptosporiopsis, Cylindrocarpon, Didymella, Epulorhiza, Fusarium, Myrmecridium, Leptosphaeria, Paraconiothyrium, Phomopsis, Pyrenochaeta, Stephanonectria*	Tan et al. (2012)
Catharanthus roseus	Leaves, stems, roots	*Endophytic fungi hyphae*	Lakra et al. (2013)
Catharanthus roseus	Leaves, stem, root	*Alternaria, Cladosporium, Aspergillus, Drechslera, Curvularia, Bipolaris, Chaetomium*	Kharwar et al. (2008)
Catharanthus roseus	Leaves	*Alternaria, Aspergillus, Curvularia, Penicillium, Trichoderma, Helminthosporium, Phoma, Verticillium, Paecilomyces*	Momsia and Momsia (2013)
Eucommia ulmoides, Forsythia suspense, F. giraldiana, F. ovata, Berberis poiretii, Rhus potanini	Branches, leaves	*Alternaria alternata, Microsphaeropsis conielloides*	Qiu et al. (2008)
Echinacea purpurea	Leaves, lateral shoots, roots	*Ceratobasidium, Cladosporium Colletotrichum, Fusarium, Glomerella, Mycoleptodiscus*	Rosa et al. (2012)
Silybum marianum	Stems, leaves, roots, and seeds	*Diaporthe cotoneastri, Thielavia terricola, Penicillium hispanicum, Cladosporium colombiae, Penicillium copticola, Myrothecium verrucaria, Chaetomidium arxii, Daldinia loculata, Zopfiella longicaudata, Microascus nidicola, Trichophaea hybrida, Biscogniauxia mediterranea, Biscogniauxia atropunctata, Penicillium aculeatum, Talaromyces minioluteus, Nemania serpens, Chaetomium rectangulare, Penicillium restrictum, Alternaria metachromatica*	Raja et al. (2015)
Festuca rubra	Leaves	18 Different taxa	Zabalgogeazcoa et al. (2013)

(continued)

Table 26.1 (continued)

Plant used for isolation	Plant part	Endophytic fungi isolated	Reference
Withania somnifera	Leaves, stem, roots	*Chaetomium bostrycodes, Eurotium rubrum, Melanospora fusispora, Aspergillus awamori, A. terreus, A. auricomus, A. flavus, A. niger, A. pulvinus, A. terricola, A. thomii, Alternaria alternata, Cladosporium cladosporioides, Curvularia oryzae, Drechslera australiensis, Fusarium moniliforme, F. semitectum, Myrothecium roridum, Penicillium corylophilum, Phoma*	Khan et al. (2010)
Diospyros crassiflora	Leaves, roots, fruit, seeds	*Xylaria, Colletotrichum*	Meli and Langer (2012)

The root-associated endophytes were investigated in 12 plant species in temperate forest with dominant oak species (*Quercus serrata*) by pyrosequencing. The colonized endophytic fungal communities observed in the root samples were from ascomycetes in 345 operational taxonomic units (OTUs) (Toju et al. 2013).

The roots of epiphytic orchid *Dendrobium sinense* (family: Orchidaceae) were evaluated for endophytic fungi and correlated with their host tree species (*Syzygium buxifolium, Rhododendron mounlmainense, Podocarpus neriifolius,* and *Cyclobalanopsis disciformis*). The results of the study showed richness and diversity for fungal endophytes from *D. sinense* roots on *S. buxifolium* than other host trees (Wang et al. 2017). The rhizomes of *Paris polyphylla* (family: Liliaceae) medicinal plant that produce steroidal saponins of 4, 6, and 8 years were investigated for fungal endophytes by culture-based and culture-independent methods. The endophytes were characterized by morphological and molecular approaches. Among the endophyte isolates are *Alternaria* sp., *Cylindrocarpon* sp., *Chaetomium globosum, Fusarium oxysporum, F. redolens, Leptodontidium* sp., *Plectosphaerella cucumerina, Paraphaeosphaeria sporulosa, Pyrenochaeta* sp., *Penicillium chrysogenum, P. swiecickii, Trichoderma viride, T. ovalisporum, T. gamsii, Truncatella angustata,* and *Trichocladium opacum* (Liu et al. 2017).

Chenopodium quinoa roots were evaluated for isolation of endophytic fungi; the plant was growing near salt lake from the Atacama Desert, Chile. The isolates were identified as *Alternaria alternata, Bartalinia robillardoides, Cadophora* sp., *Coniochaeta* sp., *Embellisia* sp., *Fusarium acuminatum, F. avenaceum, F. oxysporum, F. sambucinum, F. tricinctum, Neonectria macrodidyma, Penicillium brevicompactum, P. minioluteum, P. murcianum, Phoma* sp., *Plectosphaerella* sp., *Rhinocladiella similis,* and *Sarocladium spinificis* (Gonzalez-Teuber et al. 2017). The foliar endophytic fungi from *Cornus controversa* and *Prunus grayana* were investigated. The total endophytes isolated were 27 and 24 fungal taxa from host plants, and *Colletotrichum* sp. was dominant among the isolates (Kato et al. 2017). The fungal diversity was assessed in pulses (chickpea, lentil, pea) and wheat-based rotation. The fungi were characterized by ITS, 18S rRNA gene markers, and 454 pyrosequencing.

Table 26.2 Plant defense

Plant used for isolation	Plant part	Endophytic fungi isolated	Activity	Reference
Oryza sativa L.	Leaves, roots	*Chaetomium globosum, Cladosporium cladosporioides, Fusarium oxysporum, Penicillium chrysogenum*	Antagonists against fungal pathogens	Naik et al. (2009)
Cirsium arvense	Leaf	*Chaetomium cochliodes*	Plant defense	Hartley et al. (2015)
Pinguicula vulgaris	Roots	*Trichoderma* spp., *Cladosporium macrocarpum*	Protect from stress and nutrient-poor conditions	Quilliam and Jones (2012)
Laguncularia racemosa	Branches, leaves	*Diaporthe phaseolorum*	3-Hydroxypropionic acid (3-HPA)—antibacterial agent	Sebastianes et al. (2012)
Picea abies (Norway spruce)	Seedlings	*Phialocephala subalpina*	Protect against oomycete pathogens	Tellenbach and Sieber (2012)
Picea abies (Norway spruce)	Root	*Phialocephala subalpina, P. fortinii*	Protect against oomycete pathogens	Tellenbach et al. (2011)
Betula pendula Roth (silver birch)	Leaves	*Fusicladium betulae, Gnomonia setacea, Venturia ditricha*	Protect against abiotic and biotic stress	Helander et al. (2006)
Merremia umbellata (Convolvulaceae) *Cordia alliodora* (Boraginaceae)	Leaves	*Colletotrichum tropicale*	Altered the foraging behavior of leaf-cutting ants (*Atta colombica*), and decrease the productivity of this herbivore	Estrada et al. (2014)
Nyctanthes arbor-tristis	Leaf, stem	*Alternaria alternate, Cladosporium cladosporioides, Nigrospora oryzae, Colletotrichum dematium, Chaetomium globosum,*	Antibacterial and antifungal activities	Gond et al. (2012)
Dendrobium loddigesii Rolfe	Leaves, stems, roots	*Fusarium, Acremonium*	Antimicrobial activity	Chen et al. (2010b)

(continued)

Table 26.2 (continued)

Plant used for isolation	Plant part	Endophytic fungi isolated	Activity	Reference
Bacopa monnieri (L.) Pennell (Scrophulariaceae)	Branches, leaves	*Fusarium* spp., *Trichoderma* spp.	Cytotoxic activity, antimicrobial activity	Katoch et al. (2014)
Oryza granulata	Stem	*Muscodor* spp.	Antifungal activity	Yuan et al. (2011)
Picea abies	Roots	*Phialocephala fortinii*	Inhibitory effect on root rot pathogen (*Heterobasidion parviporum*)	Terhonen et al. (2014)
Adhatoda vasica, Cannabis sativa, Ocimum sanctum, Viola odorata, Withania somnifera	Leaves	*Aspergillus niger, A. flavus, A. clavatus, A. variecolor, Penicillium chrysogenum, Alternaria alternata, Curvularia lunata, Haplosporidium* sp., *Phoma* sp., *Nigrospora* sp., *Colletotrichum* sp., *Cladosporium* sp., *Stemphylium* sp., *Fusarium* sp., *Geotrichum* sp., *Phomopsis* sp., *Trichoderma* sp., *Rhizopus* sp.	Establish defense mechanism in plants	Gautam (2014)
Phoenix dactylifera L.	Leaves, roots	*Alternaria, Cladosporium, Davidiella, Pythium, Curvularia,*	Survival in harsh environment	Chobba et al. (2013)
Hordeum vulgare L.	Roots	*Fusarium equiseti, Pochonia chlamydosporia*	Protection against abiotic and biotic stress	Macia-Vicente et al. (2009)
Cinnamomum camphora (L.) Presl.	Leaf, stem, and petiole	*Chaetomium* sp., *Pestalotiopsis* sp., *Phyllosticta nobilis, Phomopsis* sp., *Phacidium* sp., *Alternaria cinerariae, Arthrobotrys, Arthrinium* sp., *Aspergillus fumigatus, Aspergillus niger, Chaetophoma, Curvularia lunata, Cladosporium tenuissimum, Drechslera, Gliomastix, Humicola grisea, Nigrospora oryzae, Penicillium* sp., *Periconia, Stachybotrys, Trichoderma harzianum*	Antimicrobial activity	Kharwar et al. (2012)

Calotropis procera (Apocynaceae)	Leaves	*Acremonium kiliense, Acremonium strictum, Cercospora pulcherrima, Cladosporium cladosporioides, Cladosporium oxysporum, Colletotrichum gloeosporioides, Curvularia pallescens, Diplodina microsperma, Glomerella cingulata, Guignardia bidwellii, Phaeoramularia calotropidis, Rhodotorula glutinis, Xylaria* sp.	Antimicrobial activity	Nascimento et al. (2015)
Madhuca indica Gmel.	Stem, bark, leaves	*Alternaria alternata, Fusarium* sp., *Trichoderma* sp., *Aschersonia* sp., *Colletotrichum coccodes, Botryosphaeria rhodina, Aspergillus flavus, A. niger, A. terreus, A. tubingensis, A. lentulus, Fusarium* sp., *Cladosporium* sp., *Fusarium graminearum, Penicillium* sp., *Taeniolella* sp., *Phlyctaena* sp., *Pleospora* sp., *Emericella* sp., *Xylaria* sp., *Gliocladium* sp. *Macrophoma* sp., *Chaetomium* sp., *Preussia* sp., *Corynespora* sp., *Monodictys* sp., *Botryodiplodia* sp., *Phomopsis liquidambari, Phomopsis* sp., and *Fusicoccum* sp.	Antibacterial activity	Verma et al. (2014b)
Theobroma cacao	Leaves	*Colletotrichum tropicale*	Pathogen resistance	Mejia et al. (2014)
Smallanthus sonchifolius (Poepp.) H. Rob. (Asteraceae)	Rhizophores	*Coniochaeta ligniaria*	Antifungal activity	Rosa et al. (2013)
Convolvulus arvensis, Cirsium arvense, Heracleum sosnowskyi, H. sibiricum, Galinsoga parviflora, Calystegia sepium, Cannabis sativa, Chenopodium spp., *Ch. album, Ch. urbicum, Artemisia vulgaris, Arctium* spp., *A. tomentosum, Atriplex* spp., *Lepidotheca suaveolens, Abutilon theophrasti, Papaver rhoeas*	Leaves	*Acremonium* sp., *Epicoccum nigrum, Vermicularia* spp., *Botrytis cinerea, Sclerotinia sclerotiorum, Ascochyta chenopodiicola, Bipolaris sorokiniana, Chaetopyrena* sp., *Colletotrichum gloeosporioides, Coniothyrium* sp., *Passalora dubia, Phoma* sp., *Septoria atriplicis, S. calystegiae, Stagonospora convolvuli, Phomopsis asteriscus, Ph. albicans, Ph. malvacearum, Ph. morphaea*	Antimicrobial, phytotoxic, insecticidal activities	Berestetskiy et al. (2014)

(continued)

Table 26.2 (continued)

Plant used for isolation	Plant part	Endophytic fungi isolated	Activity	Reference
Donax grandis, *Angiopteris evecta*, *Clidemia hirta*, *Palmae* spp., *Amischotolype mollissima*, *Cnestis palala*, *Sindora coriacea*, *Antiaris toxicaria*, *Phyllagathis rotundifolia*, *Catunaregam spinosa*, *Tacca integrifolia*, *Ixora grandiflora*, *Ampelocissus cinnamomea*, *Tetracera indica*, *Chroesthes longifolia*, *Ancistrocladus tectorius*, *Ardisia colorata*, *Dendropanax laurifolius*, *Zingiber* spp., *Clerodendrum deflexum*, *Cleistanthus* spp., *Koompassia excelsa*, *Anacolosa frutescens*, *Justicia* spp., *Psychotria condensa*, *Costus speciosus*, *Brassaiopsis polyacantha*, *Eurycoma longifolia*, *Leptonychia caudate*, *Dioscorea hispida*, *Phyllanthus pulcher*, *Mimosa* spp., *Thottea* spp., *Molineria latifolia*, *Caesalpinia parviflora*, *Strychnos ignatii*, *Centotheca lappacea*, *Rotheca serrata*	Leaves, stems, roots, rhizomes, flowers, fruits, and bark	300 isolates	Cytotoxic, antibacterial activities	Hazalin et al. (2009)
Pinellia ternata, *P. pedatisecta*, *Lycium chinense*, *Digitalis purpurea*, *Leonurus heterophyllus*, *Bletilla striata*, *Belamcanda chinensis*, *Taxus yunnanensis*	Branch, leaf, root		Cytotoxic, antibacterial, and antifungal activities	Miller et al. (2012)
Quercus variabilis		*Aspergillus* sp., *Alternaria* sp., *Cladosporium* sp., *Penicillium* sp.	Antimicrobial activity	Wang et al. (2007)
Adenocalymma alliaceum Miers	Stems, leaves, petioles	*Alternaria alternata*, *Aspergillus niger*, *Stenella agalis*, *Fusarium oxysporum*, *F. roseum*, *Curvularia lunata*	Antibacterial activity	Kharwar et al. (2011)
Aquilaria sinensis	Wood	28 Isolates	Antimicrobial, antitumor activities	Cui et al. (2011)
Taxus baccata L.	Bark	*Aspergillus*, *Penicillium*, *Fusarium*	Antimicrobial activity	Tayung and Jha (2010)

Ipomoea carnea	Stem	*Quambalaria cyanescens*	Antimicrobial activity	Padhi and Tayung (2013)
Dendrobium devonianum, D. thyrsiflorum	Stem, root	*Phoma* sp., *Epicoccum nigrum, Fusarium*	Antibacterial/antifungal activity	Xing et al. (2011)
Cynodon dactylon	Leaves	*Aspergillus* sp.	Antimicrobial activity	Li et al. (2005b)
Ophiopogon japonicus	Stem, root	*Aspergillus* sp., *Fusarium* sp., *Cylindrocarpon*	Antimicrobial activity	Liang et al. (2012)
Eucalyptus citriodora Hook.	Leaf	*Alternaria alternata, Aspergillus fumigatus, A. terreus, Cladosporium cladosporioides, Drechslera rostrata, Humicola grisea, Nigrospora oryzae, Penicillium cristata, Pestalotia* sp.	Antifungal activity	Kharwar et al. (2010)
Pinus wallichiana	Stem, needle	*Alternaria* sp., *Pestalotiopsis* sp., *Preussia* sp., *Sclerostagonospora* sp., *Coniothyrium carteri, Thielavia subthermophila, Truncatella betulae, Cochliobolus australiensis, Tritirachium oryzae*	Antimicrobial activity	Qadri et al. (2014)
Atractylodes lancea	Roots, leaves	*Cunninghamella* sp., *Gilmaniella* sp.	Symbiosis between endophyte and plant, drought stress	Yang et al. (2013, 2014)
Atractylodes lancea	Stems	*Gilmaniella* sp.	Defense and metabolic responses	Wang et al. (2012b)
Atractylodes lancea	Leaves	*Acremonium strictum*	Antagonistic interactions	Wang et al. (2013b)
Tinospora cordifolia Miers.	Stem, Leaves, petioles, roots	*Guignardia, Acremonium, Penicillium, Colletotrichum, C. linicola, C. crassipes, Cladosporium, Chaetomium globosum, Curvularia, Alternaria alternata, Aspergillus tubingensis, A. sydowii, Pseudofusicoccum violaceum, Botryosphaeria rhodina*	Antibacterial activity	Mishra et al. (2012)

(continued)

Table 26.2 (continued)

Plant used for isolation	Plant part	Endophytic fungi isolated	Activity	Reference
Calotropis gigantea, Datura stramonium, Jatropha curcas, Aconitum carmichaelii, Arisaema erubescens, Beaumontia brevituba, Ervatamia sp., *Hedyotis diffusa, Illicium macranthum, Paris polyphylla, Rehderodendron* sp., *Rhoiptelea chiliantha*	Plant tissues	*Alternaria* sp., *Marssonina* sp., *Pithomyces* sp., *Pestalotiopsis* sp., *Hainesia* sp., *Torula* sp.	Antifungal and antitumor activities	Li et al. (2005a)
Stryphnodendron adstringens (Mart.) Coville (Fabaceae)	Leaves, bark	*Diaporthe phaseolorum, Guignardia camelliae, Preussia pseudominima*	Antimicrobial activity, anticancer activities	Carvalho et al. (2012)
Azadirachta indica A. Juss	Fruits, roots	*Fusarium solani, Chaetomium globosum, Humicola, Drechslera, Colletotrichum, Scytalidium* sp.	Antimicrobial activity	Verma et al. (2011)
Ottelia acuminata, Myriophyllum verticillatum, Equisetum arvense, Cardamine multijuga, Impatiens chinensis	Leaves, stems	*Cladosporium, Geotrichum, Fusarium*	Antifungal activity	Li et al. (2010)
Combretum leprosum	Leaf	*Fusarium solani, Hypocrea koningii, Aspergillus oryzae, Fusarium oxysporum*	Bioactivities, i.e., cytotoxicity and antiproliferative, antifungal, antivascular	Santos et al. (2012)
Centella asiatica	Leaves	*Colletotrichum higginsianum, Guignardia mangiferae, Glomerella cingulata*	Antagonism	Rakotoniriana et al. (2008)
Orthosiphon stamineus Benth		*Phomopsis* sp.	Anti-candidal activity	Yenn et al. (2012)
Lippia sidoides Cham. (Verbenaceae)	Leaves, stem	*Alternaria alternata, Guignardia bidwellii, Phomopsis archeri, Curvularia pallescens, Drechslera dematioidea, Microascus desmosporus, Paecilomyces variotii, Periconia byssoides, Ulocladium oudemansii, Fusarium lateritium, Phoma tracheiphila*	Antimicrobial metabolites against bacteria and fungi	de Siqueira et al. (2011)

Populus angustifolia	Leaf	*Penicillium* sp., *Truncatella*	*Drepanopeziza populi* disease severity without any endophyte	Busby et al. (2013)
Ocimum sanctum, Sapindus detergens	Stem, leaves	*Alternaria* sp., *Phoma sojicola, Exserohilum* sp.	Characterize fungi for their ability to produce antibacterial and anticancer biomolecules	Bhagat et al. (2012)
Picrorhiza kurroa, Cannabis sativa, Withania somnifera, Rauwolfia serpentina, Cedrus deodara, Abies pindrow, Pinus roxburghii, Nothapodytes nimmoniana, Platanus orientalis, Artemisia annua	Stems/twigs	*Chaetomium globosum, Valsa sordida, Thielavia subthermophila, Diaporthe phaseolorum, D. helianthi, Alternaria alternata, A. citri, A. tenuissima, A. arborescens, A. brassicae, Alternaria* sp., *Schizophyllum commune, Gibberella* spp., *Gibberella moniliformis, G. avenacea, Petriella setifera, Cochliobolus lunatus, Fusarium* spp., *F. equiseti, F. proliferatum, F. redolens, F. tricinctum, F. solani, F. flocciferum, Hypoxylon fragiforme, Nigrospora sphaerica, Cercophora caudate, Cladosporium cladosporioides, Lasiodiplodia theobromae, Glomerella acutata, Sordaria humana, S. superba, Talaromyces trachyspermus, Cochliobolus spicifer, Scleroconidioma sphagnicola, Daldinia fissa, Penicillium oxalicum, P. expansum, Ulocladium, Phomopsis, Paraphoma* spp., *Polyporus arcularius, Apiosordaria otanii, Petriella, Bipolaris tetramera, Trichopaea abundans*	Antimicrobial, immunomodulatory activities	Qadri et al. (2013)
Coffea arabica, C. robusta		*Trichoderma harzianum, Guignardia, Phomopsis, Fusarium oxysporum, Aspergillus versicolor, Glomerella, Cladosporium*	Antimicrobial activity	Sette et al. (2006)

(continued)

Table 26.2 (continued)

Plant used for isolation	Plant part	Endophytic fungi isolated	Activity	Reference
Decalepis arayalpathra	Tubers	*Aspergillus, Fusarium, Penicillium, Trichoderma, Mucor*	Endophytic communities against root rot disease	Premalatha et al. (2014)
Nothapodytes foetida	Leaves	*Bionectria ochroleuca*	Antimicrobial	Samaga et al. (2014)
81 species (40 families and 23 orders) of Thai medicinal plants	Leaves, stems	582 Isolates	Antimicrobial, anticancer, antimalarial activities	Wiyakrutta et al. (2004)
Jatropha curcas	Leaf	*Colletotrichum truncatum, Nigrospora oryzae, Fusarium proliferatum, Guignardia cammillae, Alternaria destruens, Chaetomium* sp.	Antifungal activity	Kumar and Kaushik (2013)
Cinnamomum insularimontanum, Cin. kanehirae, Cin. osmophloeum, Lindera aggregata, Lin. communis, Litsea cubeba, Machilus thunbergii, Phellodendron amurense, Tetradium ruticarpum, Zanthoxylum wutaiense, Citrus aurantium, Cit. kotokan, Cit. medica, Cit. medica var. *sarcodactylis, Cit. madurensis, Cit. sinensis, Citrus* sp.*, Fortunella japonica, Murraya euchrestifolia, Toddalia asiatica, Zanthoxylum nitidum, Cinnamomum cassia, Cinnamomum* sp.	Twigs	*Colletotrichum, Guignardia, Hypoxylon, Nigrospora, Phomopsis,* and *Xylaria*	Antimicrobial activity	Ho et al. (2012)
Catharanthus roseus	Shoot	*Epicoccum nigrum*	Host defense response	Musetti et al. (2007, 2011)

Azadirachta indica, Aegle marmelos, Catharanthus roseus, Eucalyptus citriodora, Nyctanthes arbor-tristis, Adenocalymma alliaceum, Tinospora cordifolia, Cinnamomum camphora		*Cladosporium cladosporioides, Alternaria alternata, Curvularia lunata, Aspergillus niger, Chaetomium globosum, Nigrospora oryzae, Phoma glomerata*	Antimicrobial, antimalarial, antioxidant activities, extracellular enzyme production	Kharwar et al. (2014)
Calotropis procera	Leaves	*Alternaria alternata, Aspergillus flavus, Aspergillus ochraceus, Aspergillus terreus, Aspergillus versicolor, Boeremia crinicola, Chaetomium bostrychodes, Chaetomium carinthiacum, Chaetomium elatum, Chaetomium gelasinosporum, Chaetomium globosum, Chaetomium gracile, Cladosporium cladosporioides, Cladosporium sphaerospermum, Cochliobolus lunatus, Eurotium repens, Fusarium pseudoanthophilum, Fusarium dimerum, Fusarium merismoides, Fusarium oxysporum, Fusarium solani, Gibberella pulicaris, Glomerella tucumanensis, Microascus manginii, Monilinia fructigena, Myrothecium verrucaria, Nectria inventa, Nigrospora oryzae, Paecilomyces variotii, Penicillium canescens, Penicillium chrysogenum, Penicillium corylophilum, Sarocladium strictum*	Antifungal activity	Gherbawy and Gashgari (2014)
Juniperus procera	Twigs	*Aspergillus fumigatus, Penicillium oxalicum, Preussia* sp., *Peyronellaea eucalyptica, Peyronellaea sancta, Alternaria tenuissima*	Antimicrobial activities	Gherbawy and Elhariry (2014)
Capsicum annuum		*Penicillium resedanum*	Abiotic stress	Khan et al. (2015a)

(continued)

Table 26.2 (continued)

Plant used for isolation	Plant part	Endophytic fungi isolated	Activity	Reference
Helianthus annuus L.	Roots	*Penicillium citrinum, P. sumatrense, P. steckii, P. sizovae, P. tropicum, Aspergillus terreus, A. niger, A. tubingensis, A. candidus, A. fumigatus*	Biotic stress	Waqas et al. (2015)
Atractylodes lancea	Leaves	*Acremonium strictum*	Drought resistance	Yang et al. (2014)
Dioscorea zingiberensis	Rhizomes	*Fusarium oxysporum*	Defense mechanisms	Li et al. (2014)
Salsola collina, Suaeda salsa, Corispermum declinatum	Stems, leaves	*Alternaria alternata, A. franseriae, Fusarium solani, Phoma pomorum*	Protect from stress	Sun et al. (2012)
Carapa guianensis Aublet (Meliaceae)	Leaves	*Aspergillus, Beltrania, Botryosphaeria, Colletotrichum, Diaporthe, Endomelanconiopsis, Fusarium, Guignardia, Pestalotiopsis, Phomopsis, Pilidiella, Trichoderma*, and *Xylaria*	Chemical defense and adaptive response	Ferreira et al. (2015)

The results showed fungal endophytes such as *Fusarium tricinctum, F. redolens, Clonostachys rosea,* and *Cryptococcus* sp. from pulses and species of *Mortierella, Cryptococcus,* and *Paraglomus* were AM fungi from wheat (Borrell et al. 2017).

The roots of terrestrial photosynthetic orchid *Ophrys bertolonii* Mor. were examined for fungal diversity by using molecular methods (ITS regions) and found the fungal taxa from Basidiomycota and Ascomycota (*Fusarium* spp., *F. oxysporum, F. redolens, Gibberella moniliformis, Epulorhiza* spp., *Alternaria* spp., *A. arborescens, Rhizoctonia* spp.) as symbionts that could influence the growth and metabolism of orchid plant (Pecoraro et al. 2015). The leaf segments from medicinal herbs were examined for endophytes during monsoon, winter, and summer seasons. *Chaetomium globosum, Aspergillus niger, Aureobasidium pullulans, Curvularia lunata, Fusarium* spp., *Penicillium* spp., *Pestalotiopsis* spp., *Trichoderma viride,* and *Cladosporium cladosporioides* were isolated from more than one host plant (Naik et al. 2014).

The pollen, seeds, and seedlings from plants *Centaurea cyanus* L. (Asteraceae), *Papaver rhoeas* L. (Papaveraceae), *Senecio vulgaris* L. (Asteraceae), *Centaurea nigra* L. (Asteraceae), *Plantago lanceolata* L. (Plantaginaceae), and *Rumex acetosa* L. (Polygonaceae) were evaluated for presence of endophytic fungi. The fungal endophytes identified from test plants based on morphology were *Acremonium strictum, Alternaria alternata, Aspergillus niger, Aureobasidium pullulans, Botrytis cinerea, Chaetomium cochliodes, Cladosporium cladosporioides, Cladosporium oxysporum, Cladosporium sphaerospermum, Colletotrichum coccodes, Colletotrichum dematium, Fusarium avenaceum, Epicoccum nigrum, Fusarium equiseti, Fusarium merismoides, Fusarium* sp., *Fusarium tricinctum, Geotrichum candidum, Mucor hiemalis, Penicillium* sp., *Phialophora verrucosa, Rhabdospora coriacea,* and *Trichothecium roseum.* The results showed transmission of mutualistic association between these fungi and their hosts provides trans-generation resistance in plants (Hodgson et al. 2014).

The leaves, stem, and roots of soybean (*Glycine max* L. Merr.) and corn (*Zea mays* L.) were evaluated for diversity and infection frequencies by endophytic fungi. The results showed high isolation of endophytes from stem tissues than leaves and root. From soybean (*Alternaria alternata, Arthrinium phaeospermum, A. niger, Clonostachys rosea, Curvularia lunata, F. graminearum, F. equiseti, F. oxysporum, Macrophomina phaseolina, Trichoderma saturnisporum*), the most frequently isolated was *Fusarium graminearum* and least was *Scopulariopsis brevicaulis. Aspergillus terreus* was frequent and *A. flavus* was least isolated endophyte from corn plants (*Bipolaris tetramera, C. rosea, F. proliferatum, F. graminearum, Mucor circinelloides*) (Russo et al. 2016).

The endophytic fungi were studied in avocado (*Persea americana* Mill., family: Lauraceae) growing in conventional and organic supplements. The isolates were identified as *Alternaria, Cladosporium, Colletotrichum, Corynespora, Diaporthe, Lasiodiplodia, Neofusicoccum, Neopestalotiopsis, Phyllosticta,* and *Strelitziana.* The results proved the difference in endophyte communities from organic and conventional agroecosystems (Shetty et al. 2016).

The leaves of 167 tree species from three sites, subtropical, cool temperate and subboreal forests, were compared for endophytes based on 28S ribosomal DNA and classified into 42 operational taxonomic units (OTUs). The diversity, composition, and host recurrence of endophytic fungi showed 610 isolates from xylariaceous foliar endophytes (Ikeda et al. 2014). The phyllosphere of Mediterranean tree *Olea europaea* L. was studied for structure and fungal composition (epiphytic and endophytic) in the phyllosphere in climate change. The results of this study proved rainfall and temperature influenced the endophytic fungal communities in the olive species (Gomes et al. 2018). The leaves of *Eucalyptus microcorys* were evaluated for structure and endophytic fungi by using two isolation methods (particle filtration and leaf fragment plating). The isolates were identified based on morphology and DNA sequencing. Among the 709 isolates (59 taxa), *Castanediella eucalypticola* and *Neophaeomoniella eucalypti* were newly reported in this study (Lacerda et al. 2018).

Vitis vinifera stems were evaluated for endophytic fungi by culture-dependent and culture-independent molecular approaches (next-generation sequencing, NGS). Endophytes were isolated among the 19 genera (*Acremonium, Alternaria, Arthrinium, Ascorhizoctonia, Aspergillus, Aureobasidium, Bipolaris, Botryosphaeria, Botrytis, Chaetomium, Cladosporium, Curvularia, Hypoxylon, Lasiodiplodia, Mycosphaerella, Nigrospora, Penicillium, Phoma,* and *Scopulariopsis*). Culture-independent approaches were high in endophytic fungal communities than culture dependent (Dissanayake et al. 2018). The three plant species *Begonia fischeri, B. olsoniae,* and *B. venosa* were evaluated for the fungal endophytic communities in leaves. The total endophytes were 426 isolates from 19 genera (*Colletotrichum, Diaporthe, Phyllosticta, Neopestalotiopsis, Stagonospora,* and *Nigrospora*). *Begonia fischeri* was rich and diverse in endophytes than *B. olsoniae* and *B. venosa* (Correia et al. 2018).

There is an ever-increasing demand for crops that yield food grains to feed the human population that grows rapidly. There were multiple strategies developed to meet the solution of food supply crisis in such stream; crop improvement (plant breeding, genetically modified crops) strategies supported with inputs of variety of agrochemicals filled the gap with adverse impacts on the ecosystem. The factors influencing crop yield include biotic and abiotic (Cocq et al. 2017). The biotic factors are insect pests and pathogens that are controlled through alternative approach such as biocontrol methods (Lugtenberg et al. 2016).

26.2 Abiotic Stress

The plants are sedentary living forms that depend on its environment to fulfill its autotrophy via photochemical conversions through sunlight, water-carbon dioxide, and inorganic mineral nutrients. Changes in surroundings are common in plant life that lead to deficiency of primary requirements; it could result in wide range of stress. The stress is extreme environment that effects growth, development, and reproduction. High salt, low/high temperature, low pH, heavy metals, and high/low water content could result in altered physiology of plants (Lata et al. 2018).

26.2.1 Heavy Metals

The anthropogenic activities resulted in pollution of variety of ecosystems, among which mangrove ecosystem was reported for heavy metal pollution and toxicity of plants. *Kandelia candel* (family: Rhizophoraceae) mangrove plant was explored for presence of endophytic fungi and their role in in vitro copper stress. Chlorophyll A and B, relative water content, and water saturation deficit were parameters checked in control and endophyte (*Purpureocillium* sp.)-treated plants. The results under copper-enhanced condition showed that the endophyte protected the *K. candel* plant and proved efficiency of the organism in remediation of heavy metal in polluted mangrove ecosystem (Gong et al. 2017).

The mining waste was used in laboratory condition for growth of *Verbascum lychnitis* (family: Scrophulariaceae) along with inoculation with arbuscular mycorrhizal fungi (AMF) and endophytic fungi (*Cochliobolus sativus, Diaporthe* sp., *Phoma exigua*). The results of the study showed the increased survival and biomass yield in test plant with endophyte inoculation. The endophytic fungi could be significant in phytoremediation of mining wastes (Wezowicz et al. 2017). The roots, rhizome, and leaves of *Typha latifolia* (L.) (family: Typhaceae) were investigated for fungal endophytes, and lead absorption efficiency by isolates was screened. Among the different plant parts, 20 endophyte roots (*Aspergillus niger, Myrothecium* sp., *Phoma* sp., *Penicillium* sp., *Acremonium* sp., *Fusarium* sp.), rhizome (*Penicillium* sp., *A. fumigatus, A. tamarii, Myrothecium* sp., *Fusarium* sp.), and leaves (*Drechslera* sp., *A. terreus, Myrothecium* sp.) were recovered. *Aspergillus niger* was found to be efficient in metal absorption and could remove the soil pollutants with Pb (Sharma et al. 2018). The endophytic fungus *Penicillium oxalicum* isolate from *Artemisia annua* L. was screened for degradation of triclosan (TCS). The endophyte degraded the TCS into 2-chlorohydroquinone, 2,4-dichlorophenol, and hydroquinone. The end products showed no toxicity to *E. coli*, and the endophyte was effective to degrade the contaminant (Tian et al. 2018).

The study was reported to investigate the effects of single and co-inoculation of AM fungus (*Rhizoglomus intraradices*) and fungal endophytes (*Mucor* sp., *Trichoderma asperellum*) on plant growth, vitality, toxic metal accumulation, and production of sesquiterpene lactone and flavonoid by cultivation in polluted substrate. The inoculation of fungi resulted in increase of biomass yield in plants grown in polluted substrate, and *Mucor* sp. enhances the Zn content in the leaves and roots (Wazny et al. 2018). The ryegrass (*Lolium perenne*) was reported to study the co-inoculation of dark septate endophyte and arbuscular mycorrhizae (*Cadophora* sp., *Funneliformis mosseae*) grown in Cd/Zn/Pb-polluted soil. The plant biomass and shoot Na, P, K, and Mg content were enhanced that decreased trace element effects in inoculated plants (Berthelot et al. 2018).

The fungal endophytes were isolated from roots and leaves of *Miscanthus* × *giganteus* and screened in the laboratory, greenhouse, and field conditions for supporting the growth of host plants in polluted and non-polluted soils. The isolates of endophytes from roots belonged to the Pezizomycotina: *Fusarium oxysporum, Periconia, Exophiala, Microdochium,* and *Leptodontidium.* These endophytes were reported to produce phytohormones that improved plant growth in heavy metal-polluted soils (Schmidt et al. 2018). The tobacco roots were investigated for dark septate endophytic fungi and screened for their application to reduce the heavy metal content in leaves of host plant by inoculation. The results of the study showed that among 65 strains of fungal endophytes, two isolates of *Acrocalymma vagum* were effective to decrease the heavy metal content in tobacco leaves (Jin et al. 2018).

26.2.2 Drought

The fungal endophyte *Piriformospora indica* was studied to investigate the effect on inoculation in maize (*Zea mays* L.) for plant water status, physiological traits, and root morphology under mechanical and drought conditions. The results of inoculated plants in terms of leaf water potential, relative water content, root and shoot fresh weights, root volume and diameter, leaf proline content, leaf area, and enzymes (CAT, APX) proved better adaptation to individual and combined stress environments compared to uninoculated plants (Hosseini et al. 2018). The fungal endophyte *Penicillium minioluteum* isolate from naturally growing quinoa at the Atacama Desert was used for *Chenopodium quinoa* under water-deficit conditions. The plants with endophyte (E+) and uninoculated (E−) plants were assessed in water deficit and water availability for plant performance, photosynthesis, water-use efficiency, and photochemical efficiency. The results of mutualistic association with endophyte increased the root biomass and benefitted the host to cope water deficit (Gonzalez-Teuber et al. 2018).

The endophytic yeast *Rhodotorula graminis* was evaluated for drought tolerance in poplar (*Populus deltoids* × *P. nigra*) by monitoring parameters like phytohormones, endophyte colonization on Fv/Fm, chlorophyll and stomatal conductance, photochemical efficiency, and ROS assay. The endophyte-inoculated plants at harvest showed increase in total biomass, IAA, GA3, SA, ABA, JA, and BRS and decrease in ROS (Khan et al. 2016b).

The arid ecosystems are associated with scarcity of water, and the plants are exposed to drought stress. The plants from this environment were reported for root endophytes—*Gomphrena colosacana, G. pulchella* (family: Amaranthaceae), *Flaveria haumanii, Parthenium hysterophorus, Senecio hualtaranensis, Thymophylla pentachaeta, Trixis cacalioides* (family: Asteraceae), *Ehretia cortesia* (family: Boraginaceae), *Bromelia urbaniana, Deuterocohnia longipetala, Dyckia velascana, Tillandsia angulosa, T. xiphioides* (family: Bromeliaceae), *Opuntia sulphurea, Tephrocactus articulatus* (family: Cactaceae), *Atriplex argentina, A. lampa, A. lithophila, A. quixadensis, A. spegazzini* (family: Chenopodiaceae), *Cressa nudicaulis, Evolvulus arizonicus* (family: Convolvulaceae), *Ephedra* sp. (family: Ephedraceae), *Adesmia cordobensis, A.* aff. *trijuga* (family: Fabaceae), *Halophyton ameghinoi* (family: Halophytaceae), *Tricomaria usillo* (family: Malpighiaceae), *Sida argentina* (family: Malvaceae) *Allionia incarnata, Boerhavia pulchella* (family: Nyctaginaceae), *Aristida mendocina, Chloris castilloniana, Pappophorum caespitosum, Setaria cordobensis, Sporobolus phleoides* (family: Poaceae), *Monnina dictyocarpa* (family: Polygalaceae), *Grahamia bracteata, Portulaca confertifolia* (family: Portulacaceae), *Xeroaloysia ovatifolia* (family: Verbenaceae), *Bulnesia retama, Larrea cuneifolia,* and *Plectocarpa tetracantha* (family: Zygophyllaceae). The roots from these plants for dark septate endophytes (DSE) were found to benefit in water stress (Lugo et al. 2015).

The effect of systemic seed-borne fungal endophyte—*Epichloë gansuensis*—in the drunken horse grass (*Achnatherum inebrians*) infected with *Blumeria graminis* (causative agent: powdery mildew) by growing test grass in (15%, 30%, 45% and 60%) water holding capacity (WHC). The endophyte-infected (E+) and endophyte-noninfected (E−) plants were compared for infection incidence, disease lesions, disease index, biomass production, and growth of grass in soil water holding capacities (15%, 30%, 45%, 60% WHC). The E+ plants reduced the ability of pathogen to colonize the host plant and enhanced the plant growth in all soil water conditions (Xia et al. 2015).

The roots of 24 plant species under water and salt stress from 12 sites were surveyed for fungal endophytes, and 180 fungal isolates were identified by morphological and molecular techniques (internal transcribed spacer (ITS), translation elongation factor-1α gene region sequencing). The most frequent genera identified among the endophytes were *Acremonium, Alternaria, Aspergillus, Fusarium,* and *Phoma* (Macia-Vicente et al. 2008).

26.2.3 Salinity

The ectomycorrhizal fungi (EMF) and endophytic fungi associated with *Alnus glutinosa* (family: Betulaceae) growing in natural saline conditions were reported to evaluate the physicochemical parameters, salinity (soil), diversity, and species richness. The fungi belonging to class of Leotiomycetes and Sordariomycetes were found as fungal endophyte species in sites with high phosphorus, and the endophytes were supporting the ectomycorrhizae in the roots of *A. glutinosa* growing in saline conditions (Thiem et al. 2018).

26.3 Biotic Stress

26.3.1 Pest Control

Insect pathogenic fungi (entomopathogens) were reported for alternative biocontrol of insect pests as the overuse of synthetic chemical pesticides killed insect forms and reduced the burden of harmful crop invaders with increase in threats to the ecosystem. The entomopathogenic fungi were introduced on the plants through variety of application approaches such as foliar spray, stem injection, root and seed coating, and soaking (drenching) of soil. The effective entomopathogenic fungi include *Beauveria bassiana* and *Metarhizium anisopliae* that were reported as biocontrol and plant growth promoters (increase root length, plant height, foliar area, dry and wet weight, plant yield, nutritional status, and seed germination) (Jaber and Ownley 2018).

The entomopathogenic fungus *Tolypocladium cylindrosporum* that is an endophyte was isolated from perennial grasses *Holcus lanatus* and *Festuca rubra.* This endophyte was examined for fumonisin B (FB) mycotoxins and cyclosporine A (CyA, immunosuppressive agent) and virus (TcV) infection and temperature on fumonisin production. The endophyte with virus infection increased FB2 but not CyA production (Zabalgogeazcoa et al. 2018).

The study was reported on entomopathogenic fungi *B. bassiana*, *M. anisopliae,* and *M. robertsii* inoculated to soybean (*Glycine max*) plants to assess the effect on the growth and yield in laboratory and field conditions. The foliar spray was found to colonize the test plants with *B. bassiana* and increase in growth and yield (Russo et al. 2018). The common bean (*Phaseolus vulgaris*) seeds were inoculated with entomopathogenic fungi *B. bassiana* and *M. anisopliae* to evaluate them as endophytes in sterile sand/peat and sterile vermiculite. The endophytic colonization was higher in sand/peat than in vermiculite (Parsa et al. 2018).

The symbiotic relationship between cool-season grass *Triticum* (wheat) species—*T. dichasians, T. tripsacoides, T. columnare, T. cylindricum, T. monococcum, T. neglecta, T. recta, T. triunciale, T. turgidum, T. umbellulatum, T. kotschyi, T. ovatum, T. peregrinum, T. speltoides*, and *T. tauschii*—was evaluated for fungal endophytes, and the fungi belonging to genera *Neotyphodium* and *Acremonium* were found in wheat species. These endophytes could serve as biological control agents, i.e., pests and abiotic stress in wheat. The pest deterrence in grasses by endophytes is contributed due to alkaloid production and reported to cause toxicity in cattle, sheep, and livestock that feed on these pastures (Marshall et al. 1999).

The sterile seeds of cotton plant (*Gossypium hirsutum*) were used to isolate endophytic entomopathogens. The isolates were identified based on PCR analysis. The endophytic entomopathogens *Beauveria bassiana* and *Purpureocillium lilacinum* (*Paecilomyces lilacinus*) were inoculated at the seed stage to establish endophytes in cotton seedlings, and the endophytes were used on cotton aphid (*Aphis gossypii* Glover) reproduction in both greenhouse and field environments using in planta feeding trial effects. The results showed the role of endophytes in biological control of pest population (Lopez et al. 2014).

26.3.2 Herbivory

The mutualistic association between cool-season ryegrass (*Lolium perenne*) and endophytic fungus (*Epichloë festucae*) is well known. This fungus belongs to family Clavicipitaceae and transmits into grass sp. through vertical transmission. The association of endophyte with grass affects negatively on the vertebrate and invertebrate herbivores upon consuming the host plant as feed. The endophyte associated with grass releases alkaloids that cause defensive mutualism against herbivores. The common alkaloids by fungi are pyrrolizidine (peramine) that exhibits toxicity for insect pest—Argentine stem weevil (invertebrate herbivore)—and indole-diterpene (lolitrem B) and ergot (ergovaline) resulting in diseases: ryegrass staggers and fescue toxicosis in vertebrate herbivores. The alkaloid content in grass plant was found to be dependent on endophytic fungi association and abiotic factors (plant nutrition, temperature). The young grass plants (6 weeks) were reported to be less toxic than old (2 years) as the quantity of alkaloid and endophytic fungi was increasing with time period. The plant age and season were found to play role in intoxications by *E. festucae* (Fuchs et al. 2017).

The perennial grass tall fescue (*Festuca arundinacea* Schreber cv. KY-31) was infected with the mutualistic fungal endophyte *Neotyphodium coenophialum*. The parameters like rate of photosynthesis, number of tillers, dry matter (DM), organic matter (OM), neutral detergent fiber (NDF), acid detergent fiber (ADF), lignin, crude protein (CP), and in vitro neutral detergent fiber digestibility were tested in infected plants. The results showed significant effects on herbivore feeding of plants infected with endophytes (Newman et al. 2003).

The leaf metabolic profile of creeping thistle (*Cirsium arvense*) infected with foliar endophyte, *Chaetomium cochliodes*, was analyzed after 2 weeks using ultra-high-performance liquid chromatography (UPLC)-linked quadrupole time-of-flight mass spectrometry (QTOFMS). The fungal endophytes were re-isolated from infected leaves of test plant that showed *Acremonium strictum, Chaetomium cochliodes, C. fimeti,* and *Periconia minutissima*. The results showed chemically mediated interactions between host plants and their endophytes that developed responses through oxylipin metabolites associated with attack by pathogens or wounds by herbivores protected the plants (Hartley et al. 2015).

The leaves of *Cirsium arvense* (family: Asteraceae) were infected with endophytic fungi *Chaetomium cochliodes*, *Cladosporium cladosporioides*, and *Trichoderma viride*. The infected leaf material was fed to *Mamestra brassicae* (larva) and *Cassida rubiginosa* (adult). The endophytic fungi effected foliar feeding in insects (herbivory) (Gange et al. 2012).

The leaves of perennial ryegrass (*Lolium perenne* L.) by quantitative trait loci (QTL) analysis were investigated for influence in planta levels of mycelial mass (ELISA) and rapid metabolite profiling of alkaloids (peramine, ergovaline, and *N*-formylloline) in F1 mapping populations with different fungal endophyte—*Epichloë* sp. The symbiotic fungal endophyte association with host plant resulted in production of bioactive alkaloids that conferred protection from invertebrate herbivores (Faville et al. 2015).

26.3.3 Antimicrobial

The endophytic fungus *Trichoderma citrinoviride* isolated from mountain-cultivated ginseng (*Panax ginseng* Meyer) plant was reported for mycoparasitism assay against six fungal pathogens (*Alternaria panax, Botrytis cinerea, Cylindrocarpon destructans, Phytophthora cactorum, Pythium* spp., *Rhizoctonia solani*), dual culture of *T. citrinoviride* and *B. cinerea*, cell wall-degrading enzyme assay, inoculation of ginseng plants with endophyte, qPCR, plant hormone analysis (ginseng hormones, zeatin, abscisic acid, and jasmonic acid), and ginsenoside analysis. The results of this study proved that *Trichoderma citrinoviride* formulated in dustable powder and granules was effective to control the test pathogenic fungi and elicitor for ginsenoside production (Park et al. 2019).

The endophytic yeast-like fungus *Geotrichum candidum* that was isolate from fruit of *Solanum melongena* (eggplant/Indian brinjal) was studied for profiling of volatile organic compounds (VOCs) and antifungal activity against *Rhizoctonia solani* (rice pathogenic fungi). The results proved VOCs from endophyte were enhanced through exogenous addition of naphthalene and utilization of *G. candidum* as mycofumigant for biocontrol of phytopathogen (Mookherjee et al. 2018). *Edgeworthia chrysantha* Lindl. medicinal plant known to use for treatment of eye disorders was studied to isolate endophytic fungi and its antimicrobial activity (*Escherichia coli, Staphylococcus aureus,* and *Candida albicans*). The endophyte *Aspergillus fumigatus* was found to produce heterocyclic alkaloids (NMR, ESI-MS) in fermentation broth with antimicrobial activities (Zhang et al. 2018c).

Zingiber officinale Rosc. rhizomes were investigated for isolation of endophytic fungi and biocontrol of pythium rot (*Pythium myriotylum*), clinical pathogens (*Bacillus subtilis, Escherichia coli, Staphylococcus aureus, Salmonella enterica Typhimurium, Klebsiella pneumoniae*), and phytopathogens (*Pythium myriotylum, Sclerotium rolfsii, Colletotrichum acutatum, Corynespora cassiicola, Fusarium oxysporum, Phytophthora infestans,* and *Rhizoctonia solani*). Among the isolates, *Paraconiothyrium* sp. was the fungal endophyte reported to produce danthron (anthraquinone derivative) with broad range of biological activity (Anisha et al. 2018).

The fungal community in rhizosphere and bulk soil with tree species—barley (*Populus euramericana* and *Taxodium distichum*)—was evaluated with culture-independent 18S rDNA-based polymerase chain reaction-denaturing gradient gel electrophoresis (PCR-DGGE) method. The results of sequencing showed biocontrol fungus as *Chaetomium* and cellulolytic fungus as *Acremonium alcalophilum* (Zhang et al. 2018a).

The study was reported to investigate the leaves of *Pachystachys lutea* for isolation of endophytic fungi and evaluate their antagonistic activity and cellulose production. The endophytes were characterized by molecular taxonomic analysis (ITS, EF 1-α, β-tubulin, MLSA). Among the isolates, *Diaporthe, Colletotrichum, Phyllosticta, Xylaria, Xylaria, Nemania,* and *Alternaria* were identified. The endophytes were effective as in vitro biocontrol of plant pathogens (*Colletotrichum* sp., *Fusarium oxysporum*) and cellulose production (0.87–1.60 μmol/min) (Ribeiro et al. 2018). The rhizome of turmeric (*Curcuma longa* L., family: Zingiberaceae) was evaluated for fungal endophytes, antagonistic activity against *Pythium aphanidermatum* and *Rhizoctonia solani* (rhizome rot disease and leaf blight disease), and plant growth-promoting activities (IAA, HCN, cellulase, and phosphate solubilization) and assay for rhizome colonization. The isolates were identified as *Trichoderma harzianum, T. asperellum, T. atroviride, T. hamatum, F. solani, F. proliferatum, F. fujikuroi,* and *P. microspora*. The isolate *T. harzianum* was found to be biocontrol agent against the disease causative agents, positive for multiple plant growth-promoting activities (Vinayarani and Prakash 2018).

The segments of leaf, twig, and stem of mature healthy plant (*Cupressus sempervirens* L. var. *cereiformis* Rehd.) were inoculated to isolate fungal endophytes during autumn, spring, and summer. The identification of isolates was done by morphology and internal transcribed region sequences (ITS) of nuclear ribosomal DNA (ITS1, 5.8S ITS2). The identified fungal endophytes were *Pyrenochaeta* sp., *Ascorhizoctonia* sp., *Phoma* sp., *P. herbarum, Talaromyces* sp., *Thielavia* sp., *T. microspore, Alternaria multiformis, Penicillium brevicompactum,* and *Didymella* sp. The bioassays studied were intracellular and extracellular metabolites, antiproliferative activity, and antifungal assay (test pathogens: *Pyricularia oryzae*, cypress pathogens: *Diplodia seriata, Spencermartinsia viticola,* and *Phaeobotryon cupressi*). The results showed that the host with endophytes contributes biotic stress tolerance and protection (Soltani and Moghaddam 2015).

The endophyte (*Penicillium* sp., *Truncatella angustata*)-inoculated plant leaf disc of *Populus angustifolia* was studied for presence of endophytic fungi. The identification of endophyte was based on morphology and ITS (ITS1 and ITS4) and was reported to be *Penicillium* sp. The interaction between endophyte and *Drepanopeziza populi* (necrotrophic foliar pathogen) was studied for defense gene expression by qRT-PCR. The results showed that the endophyte conferred defense against pathogen to the host plant (Busby et al. 2013).

The endophytic fungus *Chaetomium globosum* isolated from *Houttuynia cordata* Thunb. was screened for antifungal activity (*Exserohilum turcicum, Botrytis cinerea*). The results proved the endophyte as potential biocontrol agent (Pan et al. 2016).

The fungal endophyte (*Epichloë festucae*) and the cool-season grass (*Festuca rubra*) were studied for in planta expression of salicylate hydroxylase enzyme that degrades salicylic acid. The enzyme confers insect, drought, and disease tolerance to host grass plant. The results proved that the endophyte plays significant role in host plant defense (Ambrose et al. 2015).

The surface-sterilized leaf segments of *Strychnos nux-vomica* L. from forest areas were investigated for endophytes, and 23 different fungal species (*Aspergillus niger, Aureobasidium pullulans, Cephalosporium acremonium, Coniella diplodiella, Microsphaeropsis* sp., *Nigrospora sphaerica, Pestalotiopsis* sp., *Phoma* sp., *Phomopsis* sp., *Pestalotiopsis* sp., *Rhizoctonia solani, Septonema bombayensis, Sordaria fimicola, Torula herbarum, Xylaria* sp., *Pyrenochaeta* sp., *Chaetomium globosum, Phomopsis obscurans, Colletotrichum dematium*) were isolated belonging to Ascomycetes, Coelomycetes, and Hyphomycetes which showed potential to control some pathogenic fungi in vitro (Naik et al. 2015). The roots of tomato (*Lycopersicon esculentum* Mill.) were reported to characterize the impacts of varied agroecosystem practices with survival, infectivity, and occurrence of fungal endophytes. The plants from organic plots were found to colonize with endophyte *Microdochium bolleyi* that reduced infection potentials (Rasmann et al. 2009).

The leaf and root segments of rice (*Oryza sativa* L.) were inoculated on PDA with streptomycin (250 mg/l) to isolate fungal endophytes. The isolates were identified as *Acremonium* sp., *Aspergillus flavus, A. ochraceous, Chaetomium globosum,*

Chlamydomyces palmarum, Cladosporium cladosporioides, Coniothyrium fuckelli, Fusarium oxysporum, Humicola fuscoatra, Nigrospora oryzae, Paecilomyces variotii, Penicillium chrysogenum, P. decumbens, Phialophora verrucosa, Rhizoctonia solani, Speiropsis pedatospora, Stemphylium botryosum, and *Trichoderma viride.* The endophyte isolates *Chaetomium globosum* and *Penicillium chrysogenum* were studied for antagonistic ability by dual culture method against test fungal pathogens *Alternaria alternata, Macrophomina phaseolina, Nigrospora oryzae, Phoma sorghina,* and *Rhizoctonia solani.* The results of the study revealed that fungal endophytes from rice were antagonistic against test pathogens (Naik et al. 2009).

The endophytic fungi (*Penicillium simplicissimum, Leptosphaeria* sp., *Talaromyces flavus, Acremonium* sp.) from roots of cotton plant (*Gossypium hirsutum*) were tested for biocontrol of pathogenic fungi, i.e., *Verticillium dahliae.* Verticillium wilt is a disease caused by *V. dahliae* in cotton that results in severe loss to the crop around the globe. The treatment of seed with endophytic fungi showed effective reduction of disease incidence, disease index, and control efficacy. The endophytes were proved biocontrol agents and eradicate disease from cotton (Yuan et al. 2017).

The oil palm ramets were tested with inoculation of endophytic fungi (*Diaporthe phaseolorum, Trichoderma asperellum, Penicillium citrinum*) and pathogen (*Ganoderma boninense*). The efficacy of endophytes to control pathogen was evaluated with plate assay, polymerase chain reaction, and ergosterol assay. The results showed endophytes could play a role in biocontrol of pathogen effectively (Chow et al. 2017). The larvae of *Spodoptera litura* and pathogenic fungi like *Fusarium oxysporum, Sclerotinia sclerotiorum, Rhizoctonia solani,* and *Botrytis cinerea* were tested for antagonistic activity by using endophytic fungi (*Trichoderma longibrachiatum, Syncephalastrum racemosum*). The endophytic fungi were isolated from leaves of *Markhamia tomentosa.* The extracts from endophyte were proved effective to control the test larvae and pathogens (Ibrahim et al. 2017). The study was reported on field colonies of leaf-cutting ants (*Atta colombica*) to evaluate the role of fungal endophytes. The results proved that fungal endophytes play a defensive role against the leaf-cutting ants in field and laboratory conditions (Coblentz and Van Bael 2013; Estrada et al. 2013).

Pistacia vera was evaluated for the presence of endophytic fungi, and the extracts from these fungi (*Trichoderma harzianum, T. atroviride, Quambalaria cyanescens, Byssochlamys nivea*) were tested for antifungal (*Aspergillus flavus, Rhizoctonia solani, Sclerotinia sclerotiorum*) and plant growth-promoting activities. The results of the study proved that extracts from the endophytic fungi were potent to control the test fungi and plant growth promoters by release of siderophore, chitinase, and phosphate solubilization (Dolatabad et al. 2017).

The endophytic fungi were evaluated from tree species *Protium heptaphyllum* and *Trattinnickia rhoifolia* (family: Burseraceae) and their antagonistic activity against phytopathogen *Fusarium oxysporum* (direct bioautography). Among the 355 isolates, *Chaetomium globosum* and *Meyerozyma* were effective against phytopathogen as they were found to produce cladosporin, chaetoatrosin, and chaetoviridin (RP-HPLC-DAD-ESI-MS) (Fierro-Cruz et al. 2017).

26.3.4 Nematicidal

The nematode parasites of plants are known to cause heavy loss through their parasitic action. The biocontrol of these parasites was reported by use of nematophagous fungus *Pochonia chlamydosporia* in food grain crop plant *Hordeum vulgare* (barley). The test fungus could be colonized in both monocot and dicot plants and was found to infect eggs of cyst and root knot nematodes. The results of plants inoculated with fungus showed plant growth promotion through biosynthesis of plant hormones (auxin, jasmonic acid, and ethylene) and effector and pattern-triggered immunity (ETI and PTI). The study proved *P. chlamydosporia* colonizes as endophyte that eradicates the effects of biotic and abiotic stress in crop plants (Larriba et al. 2015).

The fungal endophyte *Fusarium oxysporum* was evaluated for metabolites for phytohormone and nematicidal compounds. The results showed 11 compounds were found; among them, three possess nematode antagonism and one IAA. This study proved fungal endophyte serves environment-friendly nematocides and plant resistance (Bogner et al. 2017).

26.4 Plant Growth-Promoting Activity

The leaves of *Teucrium polium* L. were used to isolate the endophytic fungi, plant growth-promoting activities (enzymes, IAA, NH_3), and antimicrobial activities. The isolates were characterized by ITS-rDNA, *Penicillium chrysogenum,* and *P. crustosum*, among the foliar endophytes. These endophytes were reported to exhibit plant growth-promoting and antimicrobial activities (Hassan 2017). The maize plant growth response was examined by inoculation with rhizospheric yeasts (*Cryptococcus flavus, Candida railenensis*) and arbuscular mycorrhiza for mineral phosphorus fertilization. The results showed the mineral P fertilization was modulated with interactions between rhizosphere yeasts and AM fungi (Sarabia et al. 2017).

The rice roots from irrigated and upland ecosystems were evaluated for endophytic fungi (ITS) and characterize their role for plant growth and resistance to parasitic nematodes. Seventy-five isolates were identified as *Fusarium* spp. (*F. oxysporum, Gibberella fujikuroi*). The results proved that endophytes were more in irrigated ecosystem than upland (Pili et al. 2016).

The sebacinoid root-associated fungi (*Piriformospora indica, Sebacina vermifera*) were investigated for compatible interaction in *Arabidopsis thaliana* root and shoot growth. The interactions were correlated with secondary metabolites via ultra-performance LC coupled to electrospray ionization quadrupole time-of-flight MS (UPLC/ESI-QTOFMS), and genomic and transcriptomic analysis showed the mutualistic interaction (Lahrmann et al. 2015).

The endophytic fungi are known to release a variety of hydrolytic enzymes that solubilize nutrients from soil and translocate through hyphae to host plant that improve the plant growth. The endophytic fungi *Trametes versicolor* (isolate from

Gallium album) and *Piriformospora indica* (isolate from barley roots) were studied for increase of the grain yield and uptake of phosphorus (P) in wheat (*Triticum aestivum*) grown in soil with P deficiency and rich conditions. The results proved better grain yield, straw yield, and P content by colonization of test endophytes than control and P-rich soil conditions. Among the two isolates, *T. versicolor* showed high grain yield and could be efficient in improving soil fertility in nutrient-deficient soils (Taghinasab et al. 2018).

Endophytes possess potential to increase nutrient availability that enhances the crop plant's growth and were reported to employ them in soil fertility. *Phomopsis liquidambari* endophyte isolate from plant *Bischofia polycarpa* (family: Phyllanthaceae) was studied to enhance the nitrogen (N) uptake and metabolism by regulation of phytohormones in rice (*Oryza sativa*). The results showed that colonization of endophyte increased contents of auxin, cytokinin, and ethylene in test plant with N levels under N-deficient conditions (Li et al. 2018).

The stem and leaf sample segments of *Dendrobium loddigesii* (orchid), a medicinal plant, were investigated for endophytic fungi by inoculating on PDA plates. The isolates were identified based on morphology and molecular analysis for isolates that failed to sporulate with internal transcribed region (ITS) of 5.8S rDNA (universal primers—ITS1 and ITS4). The dominant genera among the total 48 isolates were *Fusarium* and *Pyrenochaeta.* The isolates *Fusarium* and *Pyrenochaeta* were used for plant growth in seedling tray experiment with alder bark-humus medium. The pot experiment with seven different growth media and endophytes was performed to prove the suitable growth media for plant-endophyte symbionts. The results of both the experiments proved that the endophytes exhibit ability to enhance the plant growth influenced by growth media (Chen et al. 2010b).

The cool-season grass (*Lolium multiflorum*) was infected with fungal endophyte *Neotyphodium occultans* in glasshouse experiment. The host-plant biomass, accumulation of nutrients (N and P), and competitive ability were assessed. The test parameters were increased in presence of endophytes and promoted the plant growth (Omacini et al. 2006).

The in vitro culture of *Atractylodes lancea* (Compositae) was used for isolation of fungal endophyte *Gilmaniella* sp. by culture on PDA. The suspension of endophyte mycelial mass was used as elicitor that was investigated for growth promotion of host plantlets. The inoculant and elicitor enhanced photosynthetic rate, carbohydrate levels, chlorophyll content, defense-mediated enzymes (phenylalanine ammonia lyase, polyphenol oxidase, peroxidase, chitinase, β-1,3glucanase), and volatile oil. The results proved endophyte was effective to improve defense and metabolic responses in host plants (Wang et al. 2012b).

The medicinal plants *Spilanthes calva* and *Withania somnifera* were studied for growth and yield by inoculating a plant growth-promoting root endophyte *Piriformospora indica* in nursery. The parameters like shoot and root length, biomass, basal stem, leaf area, overall size, number of inflorescence and flowers, and seed production were enhanced in inoculated than control test plants. The control plants showed high net primary productivity, and the endophytes were suitable for large-scale cultivation of test plants (Rai et al. 2001).

The roots of riparian plants *Murraya koenigii, Equisetum* sp., *Barberis* sp., *Lyonia ovalifolia, Aesculus indica,* and *Eupatorium adenophorum* were used to characterize the fungal endophytes by two methods—direct inoculation in sterile water and malt extract agar (MEA) with streptomycin and penicillin (250 mg/l) or terramycine (250 mg/l). The endophytes isolated were *Acaulopage tetraceros, Alatospora acuminata, A. pulchella, Anguillospora crassa, A. longissima, Campylospora chaetocladia, C. parvula, Clavariopsis aquatica, Cylindrocarpon aquaticum, Heliscus lugdunensis, Lemonniera cornuta, L. pseudofloscula, L. terrestris, Lunulospora curvula, Pestalotiopsis submersus, Tetrachaetum elegans, Tetracladium marchalianum,* and *T. setigerum*. The growth promotion study was carried in two plants—*Solanum melongena* and *Hibiscus esculentus*—by inoculating endophyte isolates *Heliscus lugdunensis, Tetrachaetum elegans,* and *Tetracladium nainitalense* in pot experiments. The results showed significant effect in fresh weight, dry weight, and shoot and root length due to fungal endophytes in test plants (Sati and Arya 2010).

Two grass species—*Poa leptocoma* (marsh blue grass) and *P. reflexa* (nodding blue grass)—were tested for species distribution in presence of fungal endophyte *Epichloë* sp. and endophyte-symbiont relationship. The plant seeds with endophyte symbiosis germinated with greater frequency and high growth rate than endophyte-free test plants (Kazenel et al. 2015).

The perennial ryegrass (*Festuca rubra*) and fungal endophyte *Epichloë festucae* were reported for expression profiles through study-based RNA sequencing for impact of infection in infected and noninfected plants. The host development (trichome and cell wall) and metabolism were triggered in infected plants (Dupont et al. 2015).

The sterile discs of leaves from *Acer truncatum* were inoculated on MEA with benzylpenicillin (50 mg/l) and decomposed sterile leaves on 2% agar plates to isolate fungal endophytes. The endophyte taxa (38 from disc fragment, 9 from leaves, and 35 from twigs) *Alternaria alternata* and *Phomopsis archeri* were dominant among the isolates. The flask incubation method for leaf decomposition and production of extracellular degrading enzymes (amylase, cellulose, laccase, lipase, pectate transeliminase, pectinase, protease, and tyrosinase) were studied with endophytes. The results suggested the role of endophytes in recycling of nutrients in natural ecosystems (Sun et al. 2011).

26.5 Secondary Metabolites

Endophytic fungi in association with medicinal plants are known to release secondary metabolites that gain popularity as natural bioactive compounds (Table 26.3). *Rumex gmelini* (family: Polygonaceae) was evaluated for endophytic fungi that release secondary metabolites. This plant root and rhizome were known for production of secondary metabolites like resveratrol, polydatin, emodin, chrysophanol, chrysophaein, physcion, and musizin. The endophytic fungi were isolated from roots, rhizomes, leaves, and stem and found to produce secondary metabolites.

Table 26.3 Metabolite production

Plant	Plant part	Endophytic fungi isolated	Significance	Reference
Coffea arabica, C. congensis, C. dewevrei, C. liberica	Leaves, stems, roots	*Penicillium brevicompactum, P. brocae, P. cecidicola, P. citrinum, P. coffeae, P. crustosum, P. janthinellum, P. olsonii, P. oxalicum, P. sclerotiorum, P. steckii, P. diversum, P. roseopurpureum*	Ochratoxin A producer (*Penicillium brevicompactum, P. crustosum, P. olsonii, P. oxalicum*)	Vega et al. (2006)
Forsythia suspensa	Leaves, stems, fruits	*Colletotrichum gloeosporioides*	Phillyrin (secondary metabolite in leaves and fruits: antioxidant, anti-inflammatory, anti-hyperlipidemia, and antipyretic)	Zhang et al. (2012b)
Nothapodytes nimmoniana Graham (Icacinaceae)	Bark	*Botryosphaeria parva, Diaporthe conorum, Fusarium* sp., *F. solani, F. sacchari, F. subglutinans, F. oxysporum, F. verticillioides, Galactomyces* sp., *Phomopsis* sp., *Irpex lacteus*	Camptothecin (CPT)—anticancer alkaloid (colon cancer, cervical cancer, and ovarian cancer)	Gurudatt et al. (2010)
Rice and sugarcane	Leaves	Yeasts (*Hannaella sinensis, Cryptococcus flavus, Rhodosporidium paludigenum, Torulaspora globosa*)	Indole-3-acetic acid (IAA)	Nutaratat et al. (2014)
Taxus wallichiana var. mairei	Root, bark	*Trichoderma* sp.	10-Deacetylbaccatin III (10-DABIII)	Li et al. (2015)
Salix sitchensis (Sitka willow) *Populus trichocarpa* (black cottonwood)	Leaves	Diazotrophic endophytes	Enhance plant growth in nutrient-poor sites	Knoth et al. (2013)
Deschampsia flexuosa	Roots	*Phialocephala fortinii, Cryptosporiopsis* sp.	Enhance nitrogen uptake	Zijlstra et al. (2005)

(continued)

Table 26.3 (continued)

Plant	Plant part	Endophytic fungi isolated	Significance	Reference
Hevea brasiliensis, H. guianensis (Euphorbiaceae)	Leaves, sapwood	*Perisporiopsis, Clonostachys, Colletotrichum, Fusarium, Lasiodiplodia, Pestalotiopsis, Phomopsis, Trichoderma,* and *Xylaria*	Growth and metabolism	Chaverri and Gazis (2011)
Curcuma wenyujin		*Gibberella fujikuroi* (*Fusarium moniliforme*)	Secondary metabolites	Yan et al. (2014)
Dendrobium nobile	Leaves, stems, roots	*Guignardia mangiferae, Xylaria* spp., *Colletotrichum, Phomopsis, Fusarium, Clonostachys rosea, Trichoderma chlorosporum*	Physiological functions	Yuan et al. (2009)
Murraya koenigii, Lyonia ovalifolia, Equisetum, Barberis, Aesculus indica, Eupatorium adenophyllum,	Roots	*Alatospora acuminata, A. pulchella, Acaulopage tetraceros, Anguillispora crassa, A. longissima, Campylospora chaetocladia, C. parvula, Clavariopsis aquatica, Cylindrocarpon aquaticum, Heliscus lugdunensis, Lemonniera cornuta, L. pseudofloscula, L. terrestris, Lunulospora curvula, Pestalotiopsis submerses, Tetrachaetum elegans, Tetracladium marchalianum, T. setigerum*		Sati and Belwal (2005)

(continued)

Table 26.3 (continued)

Plant	Plant part	Endophytic fungi isolated	Significance	Reference
Atractylodes lancea (Compositae)	Buds of young stem	*Gilmaniella* sp.	Improves the growth and enhances the volatile oil in plant	Wang et al. (2012b)
Camptotheca acuminata (Nyssaceae)	Seedlings	Nine fungal isolates	Anti-camptothecin resistance	Liu and Reinscheid (2004)
Espeletia spp.	Fruit	*Penicillium glabrum*	Cellulose enzyme production	Cabezas et al. (2012)
Salvia miltiorrhiza Bunge	Leaf	*Alternaria alternata, A. franseriae, Fusarium solani, Phoma pomorum, Phoma* sp., *Eriospermum declinatum, Suaeda salsa, Salsola collina*	20 Medicinal components	Sun et al. (2014a)
Cleistocalyx operculatus (Roxb.) Merr and Perry (Myrtaceae)	Buds	*Ceriporia lacerata*	2′,4′-Dihydroxy-6′-methoxy-3′,5′-dimethylchalcone (DMC)	Wang et al. (2013a)
Arisaema erubescens		*Phoma* sp.	Natural bioactive and novel metabolites	Wang et al. (2012a)
Adiantum capillus-veneris	Leaves	*Chaetomium globosum*	Secondary metabolites to butyryl cholinesterase	Selim et al. (2014)
Moringa oleifera Lam.	Root	*Nigrospora* sp.	Antifungal secondary metabolites	Zhao et al. (2012)
Panax quinquefolius L.	Stems, roots	27 Taxa	Plant growth	Xing et al. (2010)
Gloriosa superba Linn.	Different parts	*Aspergillus* sp.	Bioactive metabolites	Budhiraja et al. (2013)
Solanum melongena, Hibiscus esculentus	Root	*Heliscus lugdunensis, Tetrachaetum elegans, Tetracladium nainitalense*	Promoted plant growth	Sati and Arya (2010)

(continued)

Table 26.3 (continued)

Plant	Plant part	Endophytic fungi isolated	Significance	Reference
Sea grasses, marine algae, forest trees	Leaves/bark (trees)	*Aspergillus terreus, Curvularia* sp., *C. lunata, C. tuberculata, Fusarium* sp., *Nigrospora oryzae, Pestalotiopsis, Sordaria, Phomopsis, Trichoderma, Drechslera papendorphii, Colletotrichum, Pithomyces maydicus*	Antiplasmodial activity	Kaushik et al. (2014)
Nothapodytes foetida		*Neurospora crassa*	Camptothecin (anticancer drug)	Rehman et al. (2008)
Vitis vinifera, V. quinquangularis, Polygonum cuspidatum	Fruits, stems, stem tuber	*Botryosphaeria, Penicillium, Cephalosporium, Aspergillus, Geotrichum, Mucor, Alternaria*	Resveratrol	Shi et al. (2012)
Terminalia arjuna	Leaf	*Chaetomella raphigera*	Taxol (anticancer drug)	Gangadevi and Muthumary (2009)
Panax ginseng	Roots	*Nectria, Aspergillus, Fusarium, Verticillium, Engyodontium, Plectosphaerella, Penicillium, Cladosporium*	Saponin production	Wu et al. (2013a)
Terminalia arjuna	Bark, twig	*Pestalotiopsis, Chaetomium, Myrothecium*	Promote plant growth	Tejesvi et al. (2005)
Lycium barbarum L.	Root	*Paraphoma chrysanthemicola*	Promote plant growth	Zhang et al. (2012a)
Solanum lycopersicum	Seeds	*Leptodontidium orchidicola*	Growth and development	Andrade-Linares et al. (2011)
Atractylodes lancea	Roots, leaves	*Cunninghamella* sp., *Gilmaniella* sp.	Endophytes enhanced the diversity and size of microbes in phyllosphere and rhizosphere	Yang et al. (2013)

(continued)

Table 26.3 (continued)

Plant	Plant part	Endophytic fungi isolated	Significance	Reference
Eremophila longifolia	Leaves	*Aspergillus oryzae, Alternaria* sp., *Preussia minima*	Production of enzymes: amylase, protease	Zaferanloo et al. (2013, 2014)
Annona squamosa	Leaves, twigs, bark, roots	131 Strains	Producers of natural bioactive compounds	Lin et al. (2010)
Spilanthes calva, Withania somnifera			Growth and yield	Rai et al. (2001)
Anoectochilus formosanus (Orchidaceae)	Roots		Promoted growth, secondary metabolites	Zhang et al. (2013)
Digitalis lanata Ehrh.	Stem, leaves	35 Isolates	Production of bioactive glycoside digoxin	Kaul et al. (2013)
Aegle marmelos Correa ex Roxb.	Leaves	*Bartalinia robillardoides*	Production of Taxol (anticancer drug)	Gangadevi and Muthumary (2008b)
Eucommia ulmoides Oliver	Stem	*Sordariomycetes* sp.	Production of chlorogenic acid	Chen et al. (2010a)
Prumnopitys andina (Lleuque) (Podocarpaceae)	Wood	*Penicillium janczewskii*	Secondary metabolites (peniprequinolone and gliovictin)	Schmeda-Hirschmann et al. (2005)
Astragalus mongholicus	Roots	*Fusarium oxysporum, Bionectria ochroleuca*	Production of bioactive metabolites	Ma et al. (2014)
Azadirachta indica A. Juss	Leaves, bark, stem	*Phomopsis oblonga, Cladosporium cladosporioides, Pestalotiopsis* sp., *Trichoderma* sp., *Aspergillus* sp., *Periconia, Stenella, Drechslera*	Bioactive compounds (azadirachtin)	Verma et al. (2007)
Trapa japonica	Root	*Galactomyces geotrichum*	Phytohormone production	Waqas et al. (2014)
Rhododendron fortunei	Root	*Oidiodendron maius, Phialocephala fortinii*	Growth and development	Zhang et al. (2009)

(continued)

Table 26.3 (continued)

Plant	Plant part	Endophytic fungi isolated	Significance	Reference
Drosera rotundifolia	Roots	*Articulospora tetracladia, Alatospora acuminata, Penicillium pinophilum, Trichoderma viride*	Nutrient signaling	Quilliam and Jones (2010)
Putterlickia retrospinosa, P. verrucosa	Roots	*Penicillium avellaneum*	Maytansine (anticancer and cytotoxic compound)	Kusari et al. (2014)
Cymodocea serrulata, Cymodocea spp., *Halodule uninervis, Halodule* spp., *Syringodium* spp., *Enhalus acoroides, Thalassia* spp.	Leaves, rhizome	*Aspergillus* sp., *Paecilomyces* sp., *Penicillium* sp.	Growth and development	Venkatachalam et al. (2015)
Crataeva magna (Lour.) DC. (Capparidaceae)	Bark, twig	*Verticillium, Nigrospora oryzae, Fusarium verticillioides*	Secondary metabolites	Nalini et al. (2005)
Dendrobium spp.		*Acremonium, Alternaria, Ampelomyces, Bionectria, Cladosporium, Colletotrichum, Fusarium, Verticillium, Xylaria*	Natural products with bioactivity	Chen et al. (2011)
Nerium oleander L. (Apocynaceae)	Leaves, stems	*Chaetomium* sp.	Antioxidant activity	Huang et al. (2007)
Ginkgo biloba	Leaves		Flavonoid accumulation	Hao et al. (2010)
Vinca minor		*Aspergillus niveoglaucus, Paecilomyces lilacinus, Trichoderma harzianum*	Growth and production of roots and cell suspensions	Verma et al. (2014a)

(continued)

Table 26.3 (continued)

Plant	Plant part	Endophytic fungi isolated	Significance	Reference
Potentilla fulgens, Osbeckia stellata, O. chinensis, Camellia caduca, Schima khasiana	Root, stem	*Talaromyces flavus, Mortierella hyalina, Paecilomyces variabilis, Penicillium* sp.	Enzymatic activity and as biocontrol agents	Bhagobaty and Joshi (2012)
Scoparia dulcis Linn.	Roots, stem, leaves, fruits	*Penicillium citrinum*	Secondary metabolites	Mathew et al. (2010)
Acer truncatum (Aceraceae)	Leaf	*Alternaria alternata, Ascochytopsis vignae, Coniothyrium olivaceum, Coniothyrium* sp., *Paraconiothyrium brasiliense, Phoma glomerata,* and *Phomopsis archeri*	Production of extracellular enzymes	Sun et al. (2011)
Curcuma amada	Rhizome	*Fusarium oxysporum*	Isolation and characterization of fungal metabolite with antiaging potential (in vivo and in silico)	Tiwari et al. (2014)
Juniperus recurva	Stem, root, leaves	*Fusarium oxysporum*	Anticancer agent—podophyllotoxin	Kour et al. (2008)
Gymnema sylvestre	leaves	*Penicillium oxalicum*	Produce antidiabetic agent—gymnemagenin	Parthasarathy and Sathiyabama (2014)
Solanum nigrum	Leaves	*Fusarium tricinctum, Alternaria alternata*	Characterize endophytes that produce promoters of plant growth and phytohormones	Khan et al. (2015b)
Musa acuminata, Oryza rufipogon, Cinnamomum bejolghota, Equisetum debile		*Muscodor musae, M. oryzae, M. suthepensis, M. equiseti*	Volatile compounds and antimicrobials	Suwannarach et al. (2013)
Huperzia serrata	Leaves, root, stem	*Aspergillus, Podospora, Penicillium, Colletotrichum, Acremonium, Shiraia*	Huperzine A (HupA), cholinesterase inhibitor (ChEI)	Wang et al. (2011), Shu et al. (2014)

(continued)

Table 26.3 (continued)

Plant	Plant part	Endophytic fungi isolated	Significance	Reference
Camptotheca acuminata	Stems	*Botryosphaeria, Fusarium, Xylaria, Diaporthe, Rhizopus, Epicoccum, Preussia*	9-Methoxy camptothecin (9-MCPT)	Ding et al. (2013)
Prestonia trifida	Twig	*Muscodor sutura*	Volatile antibiotics	Kudalkar et al. (2012)
Bulbophyllum neilgherrense, Vanda testacea	Root, stem, leaves	*Aspergillus flavus, A. niger, A. ochraceus, Gliocladium viride, Penicillium* sp.	Explore the endophytic communities in plant tissues	Sudheep and Sridhar (2012)
Pereskia bleo, Murraya koenigii, Oldenlandia diffusa, Cymbopogon citratus	Leaves, stems	*Colletotrichum, Fusarium, Phoma, Penicillium*	Anticancer enzyme-asparaginase activity	Chow and Ting (2015)
Tabernaemontana heyneana	Bark, twig, leaf, fruits, and seed	*Colletotrichum, Curvularia, Fusarium, Phomopsis, Verticillium,* and *Volutella*	Asparaginase activity	Manasa and Nalini (2014)
Rhinacanthus nasutus, Annona squamosa, Goniothalamus amuyon, Catharanthus roseus, Emilia sonchifolia, Artemisia argyi, Ixeris chinensis, Euonymus spraguei, Orthosiphon spiralis, Prunella vulgaris, Ludwigia octovalvis, Plantago asiatica, Cardiospermum halicacabum	Leaves, stem, root	*Phialemonium, Penicillium, Colletotrichum, Fusarium, Rhodotorula*	Cytotoxicity, antiplatelet aggregation, anti-inflammatory activity	Hsieh et al. (2009)
Catharanthus roseus	Leaves	*Fusarium oxysporum*	Anticancer alkaloids	Kumar et al. (2013a, b)

(continued)

Table 26.3 (continued)

Plant	Plant part	Endophytic fungi isolated	Significance	Reference
Taxus mairei, Cephalotaxus fortunei, Torreya grandis	Bark	*Neurospora* sp., *Trichoderma* sp., *Paecilomyces* sp., *Fusarium* sp.	Antitumor, antifungal activities	Huang et al. (2001)
29 Chinese medicinal plants	Leaf, stem, flower, fruit, root	1160 Endophytic fungi	Antioxidant activity	Huang et al. (2007)
3198 Plants (51 orders, 105 families, 232 genera—angiosperms and ferns)	Leaves	2700 Isolates	Bioactivity against human breast cancer cell line (MCF-7), causative agents: malaria, leishmaniasis, and Chagas' disease	Higginbotham et al. (2013)
Miquelia dentata	Fruits	*Fomitopsis, Alternaria, Phomopsis*	Anti-cancer alkaloid—camptothecin (CPT)	Shweta et al. (2012)
Hibiscus sabdariffa	Stem, root, leaves	*Aspergillus niger, Corynespora cassiicola, Colletotrichum acutatum, Glomerella acutata*	Antibacterial activity	Nath and Joshi (2015)
Dysoxylum binectariferum, Amoora rohituka	Flowers, leaves	*Fusarium proliferatum*	Rohitukine, alkaloid-anticancer activity	Kumara et al. (2012, 2014)
Eugenia jambolana	Leaf, petiole, stem	*Aspergillus, Chaetomium*	Antioxidant activity	Yadav et al. (2014)
Moringa peregrina	Bark	*Aspergillus caespitosus, Phoma*	Plant growth-promoting substances	Khan et al. (2014)
Mentha arvensis	Pure culture	*Trichoderma harzianum*	Anti-nematode activity	Pandey et al. (2011)
Nothapodytes foetida, Hypericum mysorense	Leaves	27 Isolates	Free radical scavenging activity	Samaga and Rai (2013)
Panax ginseng	Root	*Penicillium melinii, P. janthinellum*	Cytotoxic metabolites	Zheng et al. (2013)
Dysoxylum binectariferum, Amoora rohituka	Bark, leaves, fruit	*Gibberella fujikuroi, Fusarium solani, Fusarium oxysporum*	Rohitukine, alkaloid-anticancer activity	Kumara et al. (2014)
Apodytes dimidiata	Bark	*Fusarium proliferatum, Fusarium solani*	Camptothecin	Shweta et al. (2010)

(continued)

Table 26.3 (continued)

Plant	Plant part	Endophytic fungi isolated	Significance	Reference
Nothapodytes nimmoniana	Bark	*Diaporthe conorum, Fusarium solani, F. verticillioides, F. subglutinans, F. oxysporum, Irpex lacteus, F. sacchari, Phomopsis* sp., *Botryosphaeria parva, Galactomyces* sp.	Anticancer alkaloid, camptothecin	Gurudatt et al. (2010)
Rosa rugosa, Camellia japonica, Delonix regia, Dianthus caryophyllus, Rosa hybrid	Stems, leaves	*Alternaria, Phomopsis, Cladosporium, Colletotrichum*	Diversity and plant growth-promoting ability	Zhou et al. (2014)
Justicia gendarussa	Leaves	*Colletotrichum gloeosporioides*	Taxol production	Gangadevi and Muthumary (2008a)
Centella asiatica, Murraya koenigii	Leaves, roots, shoots/ branches	*Colletotrichum gloeosporioides, Aspergillus oryzae*	Bioactivity assessment	Nath et al. (2014)
Cardiospermum halicacabum	Leaves	*Pestalotiopsis pauciseta*	Taxol production	Gangadevi et al. (2008)

Among the 300 isolates, secondary metabolite production was found in *Aspergillus* sp. (emodin), *Fusarium* sp. (polydatin), *Ramularia* sp. (chrysophanol), and *Berkleasmium* sp. (diepoxin zeta) (Ding et al. 2018).

The endophytic fungi *Phomopsis castaneae-mollissimae* (identified based on morphological and rDNA sequence analysis) isolate from *Artemisia annua* was evaluated for secondary metabolite production (NMR, MS). The endophyte was found to produce secondary metabolites (curcumol, oleanolic acid, quercetin, luteolin, naringenin) (Qian et al. 2018). The stem and leaf tissues of *Vitis vinifera* (varieties: merlot, wild, pinot, noir, shiraz, muscat) were evaluated for endophytic fungi and screened for resveratrol production. The total endophytic fungi isolated were 53 belonging to genera *Aspergillus, Alternaria, Arcopilus, Botryosphaeria, Fusarium,* and *Lasiodiplodia*. Among the isolates, 29 were positive for metabolite production; *Arcopilus aureus* was capable to produce resveratrol under free fermenting conditions (Dwibedi and Saxena 2018). The endophytic fungus *Diaporthe* sp. PF20 isolate from *Piper nigrum* L. was reported for production of piperine (liquid chromatography tandem mass spectroscopy). The endophyte was subjected for epigenetic treatment to enhance the metabolite production. The results proved enhancement of metabolite in the endophyte (Jasim et al. 2019).

The heartwood stems of *Aquilaria subintegra* were evaluated for endophytic fungi and production of volatile compound (agarwood oil). The extracts from isolates of endophytic fungi (*Colletotrichum, Pestalotiopsis, Fusarium, Russula, Arthrinium, Diaporthe,* and *Cladosporium*) were found to produce volatile compound and antioxidant activity (DPPH). The study proved that the endophytes were source of volatile compound and natural antioxidant (Monggoot et al. 2017). The fungal endophytes *Hypoxylon* sp. (three species) and *Daldinia eschscholzii* were screened for volatile organic compounds (VOCs) with fossil fuels. The results proved that the endophytes were useful for consolidated bioprocessing (CBP) (Wu et al. 2017). The endophytic fungus *Xylariaceae* sp. isolated from stem of *Quercus gilva* Blume was evaluated for α-glucosidase inhibitor. The extracts from fungus were proved as potential source for antidiabetic medicine (Indriangsih and Tachibana 2017). The stems and roots of *Nerium indicum* (family: Apocynaceae) were evaluated for endophytic fungi and screened for vincamine indole alkaloids (TLC, HPLC, LC-MS) and acetylcholinesterase (AchE) inhibitory activity. The results showed 11 isolates among which *Geomyces* sp. was found to produce vincamine and vincamine analogues with AchE inhibitory activity (Na et al. 2016).

26.6 Cytotoxic, Antioxidant, and Antibacterial Activities

Multidrug resistance (MDR) has become common with progress in time as the pathogenic organisms were known to lose the sensitivity against variety of drugs used in therapeutic procedures. The MDR presently serves as emerging threat in therapeutics of infectious diseases. The pathogenesis of infectious (bacterial) diseases was reported due to oxidative stress (reactive species or oxygen radicals). Fungi produce diverse group of secondary metabolites including alkaloids, phenols, steroids, terpenes, and flavonoids that are well-known therapeutics to treat various infections. The endophytic fungi are choice for production of bioactive natural products (Alurappa et al. 2018).

Mitrephora wangii (family: Annonaceae) mature leaves, stems, and flowers were used to isolate endophytic fungi and test antibacterial activity (*Staphylococcus aureus, S. epidermidis, S. agalactiae, Bacillus subtilis, B. cereus, Salmonella typhi, Klebsiella pneumoniae, Pseudomonas aeruginosa,* and *Shigella flexneri*). The isolates of endophytic fungi were from genera *Agrocybe, Aspergillus, Colletotrichum, Nigrospora, Puccinia,* and *Ustilago* were among the 22. Among the isolates, *Aspergillus* sp. was found to produce β-thujaplicin (antibacterial). The extracts from the isolates were proved to be antibacterial (Monggoot et al. 2018).

The fresh flowers of *Melodorum fruticosum* were investigated for the endophytic fungi and their antibacterial activity (*Staphylococcus aureus, Bacillus subtilis, B. cereus, Klebsiella pneumoniae, Pseudomonas aeruginosa, Vibrio cholera, V. parahaemolyticus,* and *Shigella flexneri*) and antioxidant capacity (DPPH). The 52 isolates of endophytic fungi were from genera *Aspergillus, Alternaria, Colletotrichum, Diaporthe, Fusarium, Greeneria,* and *Nigrospora.* The extracts from 43 isolates were antibacterial (abienol, 4-methoxy stilbene, phenethyl cinnamate, 2Z, 6Z-farnesal), and 16 were antioxidants (benzene acetaldehyde, benzyl benzoate, salicylaldehyde, benzoin, and benzyl cinnamate). Among the isolates, two *Diaporthe* spp. were found to be antibacterial and antioxidant (Tanapichatsakul et al. 2018).

The leaves, stems, and roots of six plants (*Equisetum hyemale* Linn., *Gentiana macrophylla* Pall., *Myrrha, Peganum harmala* Linn., *Radix stephaniae tetrandrae,* and *Rheum officinale* Baill.) were evaluated for endophytic fungi and screened for production of adenosine deaminase (ADA) inhibitors and cytotoxicity on HepG2 and SMCC-7721 cells. Among the 54 endophytic fungi, *Aspergillus niger* sp. was the strain found to produce the highest ADA inhibition and cytotoxic activity due to 3-(4-nitrophenyl)-5-phenyl isoxazole (NMR) (Zhang et al. 2018b).

Humboldtia brunonis Wall. leaves and stems were evaluated for isolation of endophytic fungi and potential for antimicrobial and DPPH-free radical scavenging activity. The isolates from leaves were *Aspergillus* sp., *Curvularia clavata, C. pallescens, Debaryomyces hansenii, Guignardia* sp., *Hypoxylon anthochroum, Meyerozyma caribbica,* and *Paecilomyces lilacinus*, and the stem isolates were *Alternaria alternata, Cunninghamella echinulate, Fusarium fusarioides, F. oxysporum,* and *Phanerochaete* sp.; among the isolates, *Lasiodiplodia theobromae* was common in leaf and stem. The extracts from isolate of *C. pallescens* were reported to exhibit both antimicrobial and antioxidant activities (Sheik and Chandrashekar 2018).

The leaves of *Olea europaea* L. were evaluated for fungal endophytes and their extracts for antimicrobial (gram positive: *Bacillus cereus, B. subtilis, Staphylococcus aureus*; gram negative: *E. coli, Pseudomonas aeruginosa*; yeasts: *Candida albicans, C. glabrata, C. parapsilosis*) potential. *Penicillium commune, P. canescens,* and *Alternaria alternata* were endophytes with antimicrobial potential (Malhadas et al. 2017). The fungal endophyte *Epichloë uncinatum* was evaluated for decomposition of *Schedonorus pratensis* plant litter. The litter inoculated with endophyte showed higher decomposition (Gundel et al. 2017).

Entada abyssinica (family: Fabaceae) leaves were used to isolate endophytic fungi and their extracts used to evaluate antibacterial (*Bacillus cereus, Salmonella typhimurium*) and cytotoxic activities (Vero, THP-1, RAW 264.7). Among the isolates of endophytic fungi, *Epicoccum nigrum* was found to produce bioactive compounds, i.e., beauvericin, para-hydroxybenzaldehyde, indole-3-carboxylic acid, and quinizarin. The results of the study showed that the compounds were weak cytotoxic agents, but beauvericin was an active antibacterial agent (Dzoyem et al. 2017). The fresh leaves and stems from *Elaeocarpus sylvestris* were evaluated for endophytic fungi and their antioxidant compounds (DPPH) for radical scavenging activity. Among the isolates, four antioxidant endophytic fungi were identified as *Pestalotiopsis* sp., *Diaporthe* sp., *Meyerozyma* sp., and *Pseudocercospora* sp., from stem and leaves (Prihantini and Tachibana 2017).

Bauhinia guianensis (common name: ladder tortoise) is a well-known plant used as a drug in folklore for treatment of infections and diabetes. This plant is a rich source of glycosides, sterols, triterpenes, lactones, and flavonoids. This plant is explored for isolation of endophytic fungi and their antimicrobial activity (*E. coli, Pseudomonas aeruginosa, B. subtilis, S. aureus, S. typhimurium*). The isolates of endophytes identified were *Colletotrichum* sp., *Aspergillus* sp., *Pestalotiopsis* sp., *Scedosporium* sp., *Paecilomyces* sp., *Xylaria* sp., and *Exserohilum rostratum.* Among the isolates, *E. rostratum* was active to control the test bacteria. The extract of endophyte was characterized by polyketide monocerin through chromatography (silica gel column), nuclear magnetic resonance (1H and 2H NMR), and mass spectrometry (Pinheiro et al. 2017). *Plectophomella* sp., *Physalospora* sp., and *Crataegus monogyna* were endophytic fungus screened for antimicrobial (*Chlorella fusca, Ustilago violacea, Eurotium repens, Fusarium oxysporum, Mycotypha microspora, E. coli,* and *B. megaterium*) and herbicidal activities. The results showed the potential for bioactive compounds that were antibacterial and herbicidal activities (Hussain et al. 2014).

The endophytic fungus *Cladosporium* sp. was isolated from leaves of *Rauwolfia serpentina* (L.) Benth. ex Kurz (family: Apocynaceae) and screened for cytotoxic (human leukemia cell K-562) and antibacterial (*Bacillus megaterium, Staphylococcus aureus, E. coli, Pseudomonas aeruginosa*) activities. The results showed anhydrofusarubin and methyl ether of fusarubin as compounds produced by endophyte that have potential cytotoxic and antibacterial activities (Khan et al. 2016a). The six plants *Tinospora cordifolia* (Willd.), *Piper nigrum* L., *Piper longum* L., *Zingiber officinale, Hedychium coronarium,* and *H. flavescens* were evaluated for endophytic fungi and production of extracellular enzymes (amylase, cellulase, pectinase, asparaginase, laccase) and 112 endophytes of 25 genera: *Acremonium* sp., *Alternaria* sp., *Aspergillus* sp., *Bipolaris* sp., *Chaetomium* sp., *Cladosporium* sp., *Colletotrichum* sp., *Curvularia* sp., *Cylindrocephalum* sp., *Drechslera* sp., *Fusarium* sp., *Lasiodiplodia* sp., *Mucor* sp., *Myrothecium* sp., *Nigrospora* sp., *Paecilomyces* sp., *Penicillium* sp., *Pestalotiopsis* sp., *Phoma* sp., *Phomopsis* sp., *Pithomyces* sp., *Rhizopus* sp., *Sordaria* sp., *Torula* sp., and *Trichoderma* sp. (Uzma et al. 2016).

The soursop leaf (*Annona muricata* L., family: Annonaceae) was evaluated for endophytic fungi that possess anticancer activity (MTT assay) against MCF-7 cells. The fungal extracts from four isolates showed inhibition of cancer cells. The potential isolate was characterized by molecular approach as *Phomopsis* sp. (Minarni et al. 2017). The endophytic fungus *Achaetomium* sp. was isolated from *Euphorbia hirta* L. (family: Euphorbiaceae). This isolate was evaluated for antioxidant, antimicrobial (*Staphylococcus aureus, Pseudomonas aeruginosa, Klebsiella pneumoniae*), hepatoprotective, total phenol, flavonoid, and tannin content in extract from endophyte. The endophyte produced metabolites tested and proved to be novel natural products releasing strain (Anitha and Mythili 2017). The bulbs of *Fritillaria unibracteata* var. *wabuensis* were evaluated for diversity of endophytic fungi, and the extracts were tested for antioxidant capacity (DPPH, ABTS, FRAP), total phenol content, flavonoid, and saponins. The antioxidants reported in the extract were gallic acid, rutin, phlorizin, 2,4-di-tert-butylphenol, and 2,6-di-tert-butyl hydroquinone (TLC-bioautography, HPLC, and GC-MS). The endophytes were potential bioactive antioxidant (phenolics, flavonoids, and saponins) resources (Pan et al. 2017).

26.7 Conclusion

The plant-microbe interactions with positive impact have been cited in literature reviewed. During the course of time, sufficient research is ongoing, and effective studies were carried to prove the role of endophytic fungi for betterment of sustainable agriculture. There are several reports at in vitro level that showed effectiveness of fungal endophytes as tools to replace the synthetic chemical compounds used as fertilizers, pesticides, herbicides, and fungicides. The endophytes were used to ameliorate the abiotic stresses (drought, salinity, and heavy metals) that are common in various parts of the globe.

References

Alurappa R, Chowdappa S, Narayanaswamy R, Sinniah UR, Mohanty SK, Swamy MK (2018) Endophytic fungi and bioactive metabolites production: an update. In: Patra J, Das G, Shin HS (eds) Microbial biotechnology. Springer, Singapore, pp 455–482

Ambrose KV, Tian Z, Wang Y, Smith J, Zylstra G, Huang B, Belanger FC (2015) Functional characterization of salicylate hydroxylase from the fungal endophyte *Epichloe festucae*. Sci Rep 5:10939

Andrade-Linares DR, Grosch R, Restrepo S, Krumbein A, Franken P (2011) Effects of dark septate endophytes on tomato plant performance. Mycorrhiza 21:413–422

Anisha C, Sachidanandan P, Radhakrishnan EK (2018) Endophytic *Paraconiothyrium* sp. from *Zingiber officinale* Rosc. displays broad-spectrum antimicrobial activity by production of danthron. Curr Microbiol 75:343–352

Anitha KPGU, Mythili S (2017) Antioxidant and hepatoprotective potentials of novel endophytic fungus *Achaetomium* sp., from *Euphorbia hirta*. Asian Pac J Trop Med 10(6):588–593

Berestetskiy AO, Gasich EL, Poluektova EV, Nikolaeva EV, Sokornova SV, Khlopunova LB (2014) Biological activity of fungi from the phyllosphere of weeds and wild herbaceous plants. Microbiology 83(5):523–530

Berthelot C, Blaudez D, Beguiristain T, Chalot M, Leyval C (2018) Co-inoculation of *Lolium perenne* with *Funneliformis mosseae* and the dark septate endophyte *Cadophora* sp. in a trace element-polluted soil. Mycorrhiza 28:301–314

Bezerra JDP, Santos MGS, Barbosa RN, Svedese VM, Lima DMM, Fernandes MJS, Gomes BS, Paiva LM, Almeida-Cortez JS, Souza-Motta CM (2013) Fungal endophytes from cactus *Cereus jamacaru* in Brazilian tropical dry forest: a first study. Symbiosis 60:53–63

Bhagat J, Kaur A, Sharma M, Saxena AK, Chadha BS (2012) Molecular and functional characterization of endophytic fungi from traditional medicinal plants. World J Microbiol Biotechnol 28:963–971

Bhagobaty RK, Joshi SR (2012) Enzymatic activity of fungi endophytic on five medicinal plant species of the pristine sacred forests of Meghalaya, India. Biotechnol Bioprocess Eng 17:33–40

Bogner C, Kamdem RST, Sichtermann G, Matthaus C, Holscher D, Popp J, Proksch P, Grundler FMW, Schouten A (2017) Bioactive secondary metabolites with multiple activities from a fungal endophyte. Microb Biotechnol 10(1):175–188

Borrell A, Shi Y, Gan Y, Bainard LD, Germida JJ, Hamel C (2017) Fungal diversity associated with pulses and its influence on the subsequent wheat crop in the Canadian prairies. Plant Soil 414:13–31

Budhiraja A, Nepali K, Sapra S, Gupta S, Kumar S, Dhar KL (2013) Bioactive metabolites from an endophytic fungus of *Aspergillus* species isolated from seeds of *Gloriosa superba* Linn. Med Chem Res 22:323–329

Busby PE, Zimmerman N, Weston DJ, Jawdy SS, Houbraken J, Newcombe G (2013) Leaf endophytes and Populus genotype affect severity of damage from the necrotrophic leaf pathogen, *Drepanopeziza populi*. Ecosphere 4(10):1–12

Cabezas L, Calderon C, Medina LM, Bahamon I, Cardenas M, Bernal AJ, Gonzalez A, Restrepo S (2012) Characterization of cellulases of fungal endophytes isolated from *Espeletia* spp. J Microbiol 50(6):1009–1013

Carvalho CR, Goncalves VN, Pereira CB, Johann S, Galliza IV, Alves TMA, Rabello A, Sobral MEG, Zani CL, Rosa CA, Rosa LH (2012) The diversity, antimicrobial and anticancer activity of endophytic fungi associated with the medicinal plant *Stryphnodendron adstringens* (Mart.) Coville (Fabaceae) from the Brazilian savannah. Symbiosis 57:95–107

Chaverri P, Gazis RO (2011) Linking *ex planta* fungi with their endophytic stages: *Perisporiopsis*, a common leaf litter and soil fungus, is a frequent endophyte of *Hevea* spp. and other plants. Fungal Ecol 4:94–102

Chen J, Hu K-X, Hou X-Q, Guo S-X (2011) Endophytic fungi assemblages from 10 *Dendrobium* medicinal plants (Orchidaceae). World J Microbiol Biotechnol 27:1009–1016

Chen X, Sang X, Li S, Zhang S, Bai L (2010a) Studies on a chlorogenic acid-producing endophytic fungi isolated from *Eucommia ulmoides* Oliver. J Ind Microbiol Biotechnol 37:447–454

Chen XM, Dong HL, Hu KX, Sun ZR, Chen J, Guo SX (2010b) Diversity and antimicrobial and plant-growth-promoting activities of endophytic fungi in *Dendrobium loddigesii* Rolfe. J Plant Growth Regul 29:328–337

Chobba B, Elleuch A, Ayadi I, Khannous L, Namsi A, Cerqueira F, Drira N, Gharsallah N, Vallaeys T (2013) Fungal diversity in adult in adult date palm revealed by culture-dependent and culture-independent approaches. J Zhejiang Univ Sci B 14(12):1084–1099

Chow YY, Rahman S, Ting ASY (2017) Understanding colonization and proliferation potential of endophytes and pathogen *in plants* via plating, polymerase chain reaction, and ergosterol assay. J Adv Res 8:13–21

Chow YY, Ting ASY (2015) Endophytic L-asparaginase-producing fungi from plants associated with anticancer properties. J Adv Res 6:869–876. https://doi.org/10.1016/j.jare.2014.07.005

Coblentz KE, Van Bael SA (2013) Field colonies of leaf-cutting ants select plant materials containing low abundances of endophytic fungi. Ecosphere 4(5):1–10

Cocq KL, Gurr SJ, Hirsch PR, Mauchline TH (2017) Exploitation of endophytes for sustainable agricultural intensification. Mol Plant Pathol 18(3):469–473

Correia AML, Lira SP, Assis MA, Rodrigues A (2018) Fungal endophyte communities in *Begonia* species from the Brazilian Atlantic rainforest. Curr Microbiol 75:441–449

Cui J-L, Guo S-X, Xiao P-G (2011) Antitumor and antimicrobial activities of endophytic fungi from medicinal parts of *Aquilaria sinensis*. J Zhejiang Univ Sci B (Biomed Biotechnol) 12(5):385–392

de Siqueira VM, Conti R, de Araujo JM, Souza-Motta CM (2011) Endophytic fungi from the medicinal plant *Lippia sidoides* Cham. and their antimicrobial activity. Symbiosis 53:89–95

Ding CH, Wang QB, Go S, Wang ZY (2018) The improvement of bioactive secondary metabolites accumulation in *Rumex gmelini* Turcz through co-culture with endophytic fungi. Braz J Microbiol 49:362–369

Ding X, Liu K, Deng B, Chen W, Li W, Liu F (2013) Isolation and characterization of endophytic fungi from *Camptotheca acuminata*. World J Microbiol Biotechnol 29:1831–1838

Dissanayake AJ, Purahong W, Wubet T, Hyde KD, Zhang W, Xu H, Zhang G, Fu C, Liu Q, Li X, Yan J (2018) Direct comparison of culture-dependent and culture-independent molecular approaches reveal the diversity of fungal endophytic communities in stems of grapevine (*Vitis vinifera*). Fungal Divers 90:85–107. https://doi.org/10.1007/s13225-018-0399-3

Dolatabad HK, Javan-Nikkhah M, Shier WT (2017) Evaluation of antifungal, phosphate solubilization, and siderophore and chitinase release activities of endophytic fungi from *Pistacia vera*. Mycol Prog 16:777–790

Dupont PV, Eaton CJ, Wargent JJ, Fechtner S, Solomon P, Schmid J, Day RC, Scott B, Cox MP (2015) Fungal endophyte infection of ryegrass reprograms host metabolism and alters development. New Phytol 208:1227–1240

Dwibedi V, Saxena S (2018) *Arcopilus aureus*, a resveratrol-producing endophyte from *Vitis vinifera*. Appl Biochem Biotechnol 186:476–495. https://doi.org/10.1007/s12010-018-2755-x

Dzoyem JP, Melong R, Tsamo AT, Maffo T, Kapche DGWF, Ngadjui BT, McGaw LJ, Eloff JN (2017) Cytotoxicity, antioxidant and antibacterial activity of four compounds produced by an endophytic fungus *Epicoccum nigrum* associated with *Entada abyssinica*. Braz J Pharmacogn 27:251–253

Estrada C, Rojas EI, Wcislo WT, Van Bael SA (2014) Fungal endophyte effects on leaf chemistry alter the *in vitro* growth rates of leaf-cutting ants fungal mutualist, *Leucocoprinus gongylophorus*. Fungal Biol 8:37–45

Estrada C, Wcislo WT, Van Bael SA (2013) Symbiotic fungi alter plant chemistry that discourages leaf-cutting ants. New Phytol 198(1):241–251

Faville MJ, Briggs L, Cao M, Koulman A, Jahufer MZZ, Koolaard HDE (2015) A QTL analysis of host plant effects on fungal endophyte biomass and alkaloid expression in perennial ryegrass. Mol Breed 35:161

Ferreira MC, de Lourdes Almeida Vieira M, Zani CL, de Almeida Alves TM, Junior PAS, Murta SMF, Romanha AJ, Gil LHVG, de Oliveira Carvalho AG, Zilli JE, Vital MJS, Rosa CA, Rosa LH (2015) Molecular phylogeny, diversity, symbiosis and discover of bioactive compounds of endophytic fungi associated with the medicinal Amazonian plant Carapa guianensis Aublet (Meliaceae). Biochem Syst Ecol 59:36–44

Fierro-Cruz JE, Jimenez P, Coy-Barrera E (2017) Fungal endophytes isolated from *Protium heptaphyllum* and *Trattinnickia rhoifolia* as antagonists of *Fusarium oxysporum*. Rev Argent Microbiol 49(3):255–263

Fuchs B, Krischke M, Mueller MJ, Krauss J (2017) Plant age and seasonal timing determine endophyte growth and alkaloid biosynthesis. Fungal Ecol 29:52–58

Gangadevi V, Murugan M, Muthumary J (2008) Taxol determination from *Pestalotiopsis pauciseta,* a fungal endophyte of a medicinal plant. Chin J Biotechnol 24(8):1433–1438

Gangadevi V, Muthumary J (2008a) Isolation of *Colletotrichum gloeosporioides*, a novel endophytic taxol-producing fungus from the leaves of a medicinal plant, *Justicia gendarussa*. Mycol Balcanica 5:1–4

Gangadevi V, Muthumary J (2008b) Taxol, an anticancer drug produced by an endophytic fungus *Bartalinia robillardoides* Tassi, isolated from a medicinal plant, *Aegle marmelos* Correa ex Roxb. World J Microbiol Biotechnol 24:717–724

Gangadevi V, Muthumary J (2009) A novel endophytic taxol-producing fungus *Chaetomella raphigera* isolated from a medicinal plant, *Terminalia arjuna*. Appl Biochem Biotechnol 158:675–684

Gange AC, Eschen R, Wearn JA, Thawer A, Sutton BC (2012) Differential effects of foliar endophytic fungi on insect herbivores attacking a herbaceous plant. Oecologia 168(4):1023–1031

Gautam AK (2014) Diversity of fungal endophytes in some medicinal plants of Himachal Pradesh, India. Arch Phytopathol Plant Protect 47(5):537–544

Gherbawy YA, Elhariry HM (2014) Endophytic fungi associated with high-altitude Juniperus trees and their antimicrobial activities. Plant Biosyst 150(1):131–140. https://doi.org/10.1080/11263504.2014.984011

Gherbawy YA, Gashgari RM (2014) Molecular characterization of fungal endophytes from *Calotropis procera* plants in Taif region (Saudi Arabia) and their antifungal activities. Plant Biosyst 148(6):1085–1092

Gomes T, Pereira JA, Benhadi J, Lino-Neto T, Baptista P (2018) Endophytic and epiphytic phyllosphere fungal communities are shaped by different environmental factors in a Mediterranean Ecosystem. Microb Ecol 76:668–679. https://doi.org/10.1007/s00248-018-1161-9

Gond SK, Mishra A, Sharma VK, Verma SK, Kumar J, Kharwar RN, Kumar A (2012) Diversity and antimicrobial activity of endophytic fungi isolated from *Nyctanthes arbor-tristis*, a well-known medicinal plant of India. Mycoscience 53:113–121

Gond SK, Verma VC, Kumar A, Kumar V, Kharwar RN (2007) Study of endophytic fungal community from different parts of *Aegle marmelos* Correae (Rutaceae) from Varanasi (India). World J Microbiol Biotechnol 23:1371–1375

Gong B, Liu G, Liao R, Song J, Zhang H (2017) Endophytic fungus *Purpureocillium* sp. A5 protect mangrove plant *Kandelia candel* under copper stress. Braz J Microbiol 48:530–536

Gonzalez-Teuber M, Urzua A, Plaza P, Bascunan-Godoy L (2018) Effects of root endophytic fungi on response of *Chenopodium quinoa* to drought stress. Plant Ecol 219:231–240

Gonzalez-Teuber M, Viloa C, Bascunan-Godoy L (2017) Molecular characterization of endophytic fungi associated with the roots of *Chenopodium quinoa* inhabiting the Atacama Desert, Chile. Genomics Data 11:109–112

Gundel PE, Helander M, Garibaldi LA, Vazquez-de-Aldana BR, Zabalgogeazcoa I, Saikkonen K (2017) Direct and indirect effects of the fungal endophyte *Epichloe uncinatum* on litter decomposition of the host grass, *Schedonorus pratensis*. Plant Ecol 218:1107–1115

Gurudatt PS, Priti V, Shweta S, Ramesha BT, Ravikanth G, Vasudeva R, Amna T, Deepika S, Ganeshaiah KN, Shanker RU, Puri S, Qazi N (2010) Attenuation of camptothecin production and negative relation between hyphal biomass and camptothecin content in endophytic fungal strains isolated from *Nothapodytes nimmoniana* Graham (Icacinaceae). Curr Sci 98(8):1006–1010

Hao G, Du X, Zhao F, Ji H (2010) Fungal endophytes-induced abscisic acid is required for flavonoid accumulation in suspension cells of *Ginkgo biloba*. Biotechnol Lett 32:305–314

Hartley S, Eschen R, Horwood JM, Gange AC, Hill EM (2015) Infection by a foliar endophyte elicits novel arabidopside-based plant defence reactions in its host, *Cirsium arvense*. New Phytol 205(2):816–827

Hassan SED (2017) Plant growth-promoting activities for bacterial and fungal endophytes isolated from medicinal plant of *Teucrium polium* L. J Adv Res 8:687–695

Hazalin NAMN, Ramasamy K, Lim SM, Wahab IA, Cole ALJ, Majeed ABA (2009) Cytotoxic and antibacterial activities of endophytic fungi isolated from plants at the National Park, Pahang, Malaysia. BMC Complement Altern Med 9:46

Helander M, Wali P, Kuuluvainen T, Saikkonen K (2006) Birch leaf endophytes in managed and natural boreal forests. Can J For Res 36:3239–3245

Higginbotham SJ, Arnold AE, Ibanez A, Spadafora C, Coley PD, Kursar TA (2013) Bioactivity of fungal endophytes as a function of endophyte taxonomy and the taxonomy and distribution of their host plants. PLoS One 8(9):e73192. https://doi.org/10.1371/journal.pone.0073192

Ho MY, Chung WC, Huang HC, Chung WH, Chung WH (2012) Identification of endophytic fungi of medicinal herbs of Lauraceae and Rutaceae with antimicrobial property. Taiwania 57(3):229–241

Hodgson S, de Cates C, Hodgson J, Morley NJ, Sutton BC, Gange AC (2014) Vertical transmission of fungal endophytes is widespread in forbs. Ecol Evol 4(8):1199–1208

Hosseini F, Mosaddeghi MR, Dexter AR, Sepehri M (2018) Maize water status and physiological traits as affected by root endophytic fungus *Piriformospora indica* under combined drought and mechanical stresses. Planta 247:1229–1245

Hsieh PW, Hsu LC, Lai CH, Wu CC, Hwang TL, Lin YK, Wu YC (2009) Evaluation of the bioactivities of extracts of endophytes isolated from Taiwanese herbal plants. World J Microbiol Biotechnol 25:1461–1469

Huang W-Y, Cai Y-Z, Hyde KD, Corke H, Sun M (2007) A potential antioxidant resource: endophytic fungi from medicinal plants. Econ Bot 61(1):14–30

Huang Y, Wang J, Li G, Zheng Z, Su W (2001) Antitumor and antifungal activities in endophytic fungi isolated from pharmaceutical plants *Taxus mairei, Cephalotaxus fortunei* and *Torreya grandis*. FEMS Immunol Med Microbiol 31(2):163–167

Hussain H, Kliche-Spory C, Al-Harrasi A, Al-Rawahi A, Abbas G, Green IR, Schulz KK, Shah A (2014) Antimicrobial constituents from three endophytic fungi. Asian Pac J Trop Med 7(1):S224–S227

Ibrahim M, Kaushik N, Sowemimo A, Chhipa H, Koekemoer T, Venter M, Odukoya OA (2017) Antifungal and antiproliferative activities of endophytic fungi isolated from the leaves of *Markhamia tomentosa*. Pharm Biol 55(1):590–595

Ikeda A, Matsuoka S, Masuya H, Mori AS, Hirose D, Osono T (2014) Comparison of the diversity, composition, and host recurrence of xylariaceous endophytes in subtropical, cool temperate, and subboreal regions in Japan. Popul Ecol 56(2):289–300

Indriangsih AW, Tachibana S (2017) α-Glucosidase inhibitor produced by an endophytic fungus, *Xylariaceae* sp. QGS 01 from *Quercus gilva* Blume. Food Sci Human Wellness 6:88–95

Jaber LR, Ownley BH (2018) Can we use Entomopathogenic fungi as endophytes for dual biological control of insect pests and plant pathogens? Biol Control 116:36–45

Jasim B, Sahadevan N, Chithra S, Mathew J, Radhakrishnan EK (2019) Epigenetic modifier based enhancement of piperine production in endophytic *Diaporthe* sp. PF20. Proc Natl Acad Sci India Sect B Biol Sci 89:671–677. https://doi.org/10.1007/s40011-018-0982-0

Jin HQ, Liu HB, Xie YY, Zhang YG, Xu QQ, Mao LJ, Li XJ, Lin FC, Zhang CL (2018) Effect of the dark septate endophytic fungus *Acrocalymma vagum* on heavy metal content in tobacco leaves. Symbiosis 74:89–95

Kazenel MR, Debban CL, Ranelli L, Hendricks WQ, Chung YA, Pendergast TH, Charlton ND, Young CA, Rudgers JA (2015) A mutualistic endophyte alters the niche dimensions of its host plant. AoB Plants 7

Kato S, Fukasawa Y, Seiwa K (2017) Canopy tree species and openness affect foliar endophytic fungal communities of understory seedlings. Ecol Res 32:157–162

Katoch M, Singh G, Sharma S, Gupta N, Sangwan PL, Saxena AK (2014) Cytotoxic and antimicrobial activities of endophytic fungi isolated from *Bacopa monnieri* (L.) Pennell (Scrophulariaceae). BMC Complement Altern Med 14:52

Kaul S, Ahmed M, Zargar K, Sharma P, Dhar MK (2013) Prospecting endophytic fungal assemblage of digitalis lanata Ehrh. (foxglove) as a novel source of digoxin: a cardiac glycoside. 3 Biotech 3:335–340

Kaushik NK, Murali TS, Sahal D, Suryanarayanan TS (2014) A search for antiplasmodial metabolites among fungal endophytes of terrestrial and marine plants of southern India. Acta Parasitol 59(4):745–757

Khan AL, Waqas M, Hussain J, Al-Harrasi A, Al-Rawahi A, Al-Hosni K, Kim MJ, Adnan M, Lee IJ (2014) Endophytes *Aspergillus caespitosus* LK12 and *Phoma* sp. LK13 of *Moringa peregrina* produce gibberellins and improve rice plant growth. J Plant Interact 9(1):731–737

Khan AL, Waqas M, Lee IJ (2015a) Resilience of *Penicillium resedanum* LK6 and exogenous gibberellin in improving *Capsicum annuum* growth under abiotic stresses. J Plant Res 128:259–268

Khan AR, Ullah I, Waqas M, Shahzad R, Hong S-J, Park G-S, Jung BK, Lee I-J, Shin J-H (2015b) Plant growth-promoting potential of endophytic fungi isolated from *Solanum nigrum* leaves. World J Microbiol Biotechnol 31(9):1461–1466

Khan MIH, Sohrab MH, Rony SR, Tareq FS, Hasan CM, Mazid MA (2016a) Cytotoxic and antibacterial naphthoquinones from an endophytic fungus, *Cladosporium* sp. Toxicol Rep 3:861–865

Khan R, Shahzad S, Choudhary MI, Khan SA, Ahmad A (2010) Communities of endophytic fungi in medicinal plant *Withania somnifera*. Pak J Bot 42(2):1281–1287

Khan Z, Rho H, Firrincieli A, Hung SH, Luna V, Masciarelli O, Kim SH, Doty SL (2016b) Growth enhancement and drought tolerance of hybrid poplar upon inoculation with endophyte consortia. Curr Plant Biol 6:38–47

Kharwar RN, Gond SK, Kumar A, Mishra A (2010) A comparative study of endophytic and epiphytic fungal association with leaf of *Eucalyptus citriodora* Hook., and their antimicrobial activity. World J Microbiol Biotechnol 26:1941–1948

Kharwar RN, Maurya AL, Verma VC, Kumar A, Gond SK, Mishra A (2012) Diversity and antimicrobial activity of endophytic fungal community isolated from medicinal plant *Cinnamomum camphora*. Proc Natl Acad Sci India Sect B Biol Sci 82(4):557–565

Kharwar RN, Mishra A, Sharma VK, Gond SK, Verma SK, Kumar J, Singh DK, Goutam J (2014) Diversity and biopotential of endophytic fungal flora isolated from eight medicinal plants of Uttar Pradesh, India. In: Kharwar RN, Upadhyay RS, Dubey K, Raghuwanshi R (eds) Microbial diversity and biotechnology in food security. Springer, New Delhi, pp 23–39

Kharwar RN, Verma SK, Mishra A, Gond SK, Sharma VK, Afreen T, Kumar A (2011) Assessment of diversity, distribution and antibacterial activity of endophytic fungi isolated from a medicinal plant *Adenocalymma alliaceum* Miers. Symbiosis 55:39–46

Kharwar RN, Verma VC, Strobel G, Ezra D (2008) The endophytic fungal complex of *Catharanthus roseus* (L.) G. Don. Curr Sci 95(2):228–233

Knoth JL, Kim SH, Ettl GJ, Doty SL (2013) Effects of cross host species inoculation of nitrogen-fixing endophytes on growth and leaf physiology of maize. GCB Bioenergy 5:408–418

Kour A, Shawl AS, Rehman S, Sultan P, Qazi PH, Suden P, Khajuria RK, Verma V (2008) Isolation and identification of an endophytic strain of *Fusarium oxysporum* producing podophyllotoxin from *Juniperus recurva*. World J Microbiol Biotechnol 24:1115–1121

Kudalkar P, Strobel G, Hassan SRU, Geary B, Sears J (2012) *Muscodor sutura*, a novel endophytic fungus with volatile antibiotic activities. Mycoscience 53:319–325

Kumar A, Abnave PB, Ahmad A (2013b) Cultural, morphological and molecular characterization of vinca alkaloids producing endophytic fungus *Fusarium solani* isolated from *Catharanthus roseus*. Int J Bot Res 3(2):1–12

Kumar A, Patil D, Rajamohanan PR, Ahmad A (2013a) Isolation, purification and characterization of vinblastine and vincristine from endophytic fungus *Fusarium oxysporum* isolated from *Catharanthus roseus*. PLoS One 8(9):e71805. https://doi.org/10.1371/journal.pone.0071805

Kumar S, Kaushik N (2013) Endophytic fungi isolated from oil-seed crop *Jatropha curcas* produces oil and exhibit antifungal activity. PLoS One 8(2):e56202. https://doi.org/10.1371/journal.pone.0056202

Kumara PM, Soujanya KN, Ravikanth G, Vasudeva R, Ganeshaiah KN, Shaanker RU (2014) Rohitukine, a chromone alkaloid and a precursor of flavopiridol, is produced by endophytic fungi isolated from *Dysoxylum binectariferum* Hook.f and *Amoora rohituka* (Roxb). Wight & Arn. Phytomedicine 21:541–546

Kumara PM, Zuehlke S, Priti V, Ramesha BT, Shweta S, Ravikanth G, Vasudeva R, Santhoshkumar TR, Spiteller M, Shaanker RU (2012) *Fusarium proliferatum*, an endophytic fungus from *Dysoxylum binectariferum* Hook.f, produces rohitukine, a chromane alkaloid possessing anti-cancer activity. Antonie Van Leeuwenhoek 101:323–329

Kumaresan V, Suryanarayanan TS (2001) Occurrence and distribution of endophytic fungi in a mangrove community. Mycol Res 105(11):1388–1391

Kusari S, Lamshoft M, Kusari P, Gottfried S, Zuhlke S, Louven K, Hentschel U, Kayser O, Spiteller M (2014) Endophytes are hidden producers of maytansine in *Putterlickia* roots. J Nat Prod 77(12):2577–2584

Lacerda LT, Gusmao LFP, Rodrigues A (2018) Diversity of endophytic fungi in *Eucalyptus microcorys* assessed by complementary isolation methods. Mycol Prog 17:719–727. https://doi.org/10.1007/s11557-018-1385-6

Lahrmann U, Strehmel N, Langen G, Frerigmann LL, Ding Y, Scheel D, Herklotz S, Hillbert M, Zuccaro A (2015) Mutualistic root endophytism is not associated with the reduction of saprotrophic traits and requires a noncompromised plant innate immunity. New Phytol 207(3):841–857

Lakra NS, Koul M, Chandra R, Chandra S (2013) Histological investigations of healthy tissues of *Catharanthus roseus* to localize fungal endophytes. Int J Pharm Sci Rev Res 20(1):205–209

Larriba E, Jaime MDLA, Nislow C, Martin-Nieto J, Lopez-Llorca LV (2015) Endophytic colonization of barley (*Hordeum vulgare*) roots by the nematophagous fungus *Pochonia chlamydosporia* reveals plant growth promotion and a general defense and stress transcriptomic response. J Plant Res 128:665–678

Lata R, Chowdhury S, Gond SK, White JF Jr (2018) Induction of abiotic stress tolerance in plants by endophytic microbes. Lett Appl Microbiol 66:268–276

Li H, Qing C, Zhang Y, Zhao Z (2005a) Screening for endophytic fungi with antitumour and antifungal activities from Chinese medicinal plants. World J Microbiol Biotechnol 21:1515–1519

Li H-Y, Zhao C-A, Liu C-J, Xu X-F (2010) Endophytic fungi diversity of aquatic/riparian plants and their antifungal activity *in vitro*. J Microbiol 48(1):1–6

Li J, Zhao J, Xu L, Li X, Wang J (2008) Endophytic fungi from rhizomes of *Paris polyphylla* var. *yunnanensis*. World J Microbiol Biotechnol 24:733–737

Li P, Luo H, Meng J, Sun W, Wang X, Lu S, Peng Y, Zhou L (2014) Effects of oligosaccharides from endophytic *Fusarium oxysporum* Dzf17 on activities of defense-related enzymes in *Dioscorea zingiberensis* suspension cell and seedling cultures. Electron J Biotechnol 17:156–161

Li X, Zhou J, Xu RS, Meng MY, Yu X, Dai CC (2018) Auxin, cytokinin and ethylene involved in rice N availability improvement caused by endophyte *Phomopsis liquidambari*. J Plant Growth Regul 37:128–143

Li Y, Song YC, Liu JY, Ma YM, Tan RX (2005b) Anti-*Helicobacter pylori* substances from endophytic fungal cultures. World J Microbiol Biotechnol 21:553–558

Li Y, Yang J, Zhou X, Zhao W, Jian Z (2015) Isolation and identification of a 10-Deacetyl Baccatin-III-producing endophyte from *Taxus wallichiana*. Appl Biochem Biotechnol 175:2224–2231

Liang H, Xing Y, Chen J, Zhang D, Guo S, Wang C (2012) Antimicrobial activities of endophytic fungi isolated from *Ophiopogon japonicas* (Liliaceae). BMC Complement Altern Med 12:238

Lin X, Huang YJ, Zheng ZH, Su WJ, Qian XM, Shen YM (2010) Endophytes from the pharmaceutical plant, *Annona squamosa*: isolation, bioactivity, identification and diversity of its polyketide synthase gene. Fungal Divers 41:41–51

Liu T, Greenslade A, Yang S (2017) Levels of rhizome endophytic fungi fluctuate in *Paris polyphylla* var. *yunnanensis* as plants age. Plant Divers 39:60–64

Liu W, Reinscheid UM (2004) Camptothecin-resistant fungal endophytes of *Camptotheca acuminata*. Mycol Prog 3(3):189–192

Lopez DC, Zhu-Salzman K, Ek-Ramos MJ, Sword GA (2014) The entomopathogenic fungal endophytes *Purpureocillium lilacinum* (Formerly *Paecilomyces lilacinus*) and *Beauveria bassiana* negatively affect cotton aphid reproduction under both greenhouse and field conditions. PLoS One 9(8):e103891. https://doi.org/10.1371/journal.pone.0103891

Lugo MA, Reinhart KO, Menoyo E, Crespo EM, Urcelay C (2015) Plant functional traits and phylogenetic relatedness explain variation in associations with root fungal endophytes in an extreme arid environment. Mycorrhiza 25(2):85–95

Lugtenberg BJJ, Caradus JR, Johnson LJ (2016) Fungal endophytes for sustainable crop production. FEMS Microb Ecol 92(12):1–16

Ma W, Liu X, Jiao J, Zhang L, Ren W, Ma L, Kong X, Zhang N, Zhang X (2014) Purification of four strains of endophytic fungi from *Astragalus* and their optimized liquid fermentations. J For Res 25(3):701–706

Macia-Vincente JG, Jansson HB, Abdullah SK, Descals E, Salinas J, Lopez-Llorca LV (2008) Fungal root endophytes from natural vegetation in Mediterranean environments with special reference to *Fusarium* spp. FEMS Microbiol Ecol 64(1):90–105

Macia-Vicente JG, Rosso LC, Ciancio A, Jansson HB, Lopez-Liorca LV (2009) Colonization of barley roots by endophytic *Fusarium equiseti* and *Pochonia chlamydosporia*: effects on plant growth and disease. Ann Appl Biol 155:391–401

Malhadas C, Malheiro R, Pereira JA, de Pinho PG, Baptista P (2017) Antimicrobial activity of endophytic fungi from olive tree leaves. World J Microbiol Biotechnol 33:46

Manasa C, Nalini MS (2014) L-Asparaginase activity of fungal endophytes from *Tabernaemontana heyneana* Wall. (Apocynaceae), endemic to the Western Ghats (India). Int Schol Res Not 2014:925131, 7 p. https://doi.org/10.1155/2014/925131

Mathew AJ, Jayachandran K, Mathew J (2010) Endophytic *Penicillium citrinum* Thom. from *Scoparia dulcis* Linn. Indian J Microbiol 50(1):99–102

Marshall D, Tunali B, Nelson LR (1999) Occurrence of fungal endophytes in species of wild *Triticum*. Crop Sci 39:1507–1512

Mejia LC, Herre EA, Sparks JP, Winter K, Garcia MN, Van Bael SA, Stitt J, Shi Z, Zhang Y, Guiltinan MJ, Maximova SN (2014) Pervasive effects of a dominant foliar endophytic fungus on host genetic and phenotypic expression in a tropical tree. Front Microbiol 5:479

Meli CD, Langer E (2012) Diversity and molecular phylogeny of fungal endophytes associated with *Diospyros crassiflora*. Mycology 3(3):175–187

Miller KI, Qing C, Sze DM-Y, Roufogalis BD, Neilan BA (2012) Culturable endophytes of medicinal plants and the genetic basis for their bioactivity. Microb Ecol 64:431–449

Minarni, Artika IM, Julistiono H, Bermawie N, Riyanti EI, Hasim, Hasan AEZ (2017) Anticancer activity test of ethyl acetate extract of endophytic fungi isolated from soursop leaf (*Annona muricata* L.). Asian Pac J Trop Med 10(6):566–571

Mishra A, Gond SK, Kumar A, Sharma VK, Verma SK, Kharwar RN, Sieber TN (2012) Season and tissue type affect fungal endophyte communities of the Indian medicinal plant *Tinospora cordifolia* more strongly than geographic location. Microb Ecol 64:388–398

Momsia P, Momsia T (2013) Isolation, frequency distribution and diversity of *novel fungal endophytes inhabiting leaves* of *Catharanthus roseus*. Int J Life Sci Biotechnol Pharma Res 2(4):83–86

Monggoot S, Pichaitam T, Tanapichatsakul C, Pripdeevech P (2018) Antibacterial potential of secondary metabolites produced by *Aspergillus* sp., an endophyte of *Mitrephora wangii*. Arch Microbiol 200:951–959. https://doi.org/10.1007/s00203-018-1511-5

Monggoot S, Popluechai S, Gentekaki E, Pripdeevech P (2017) Fungal endophytes: an alternative source for production of volatile compounds from agarwood oil of *Aquilaria subintegra*. Microb Ecol 74:54–61

Mookherjee A, Bera P, Mitra A, Maiti MK (2018) Characterization and synergistic effect of antifungal volatile organic compounds emitted by the *Geotrichum candidum* PF005, an endophytic fungus from the eggplant. Microb Ecol 75:647–661

Musetti R, Grisan S, Polizzotto R, Martini M, Paduano C, Osler R (2011) Interactions between '*Candidatus* Phytoplasma mali' and the apple endophyte *Epicoccum nigrum* in *Catharanthus roseus* plants. J Appl Microbiol 110:746–756

Musetti R, Polizzotto R, Grisan S, Martini M, Borselli S, Carraro L, Osler R (2007) Effects induced by fungal endophytes in *Catharanthus roseus* tissues infected by phytoplasmas. Bull Insectol 60(2):293–294

Na R, Jiajia L, Dongliang Y, Yingzi P, Juan H, Xiong L, Nana Z, Jing Z, Yitian L (2016) Identification of vincamine indole alkaloids producing endophytic fungi isolated from *Nerium indicum*, Apocynaceae. Microbiol Res 192:114–121

Naik BS, Krishnappa M, Krishnamurthy YL (2014) Biodiversity of endophytic fungi from seven herbaceous medicinal plants of Malnad region, Western Ghats, Southern India. J For Res 25(3):707–711

Naik BS, Krishnappa M, Krishnamurthy YL (2015) Endophytic assemblage in *Strychnos nux-vomica* L. and antagonistic activities *in vitro*. Arch Phytopathol Plant Protect 48(1):28–33

Naik BS, Shashikala J, Krishnamurthy YL (2009) Study on the diversity of endophytic communities from rice (*Oryza sativa* L.) and their antagonistic activities *in vitro*. Microbiol Res 164(3):290–296

Nalini MS, Mahesh B, Tejesvi MV, Prakash HS, Kini KR, Shetty HS (2005) Fungal endophytes from the three-leaved caper, *Crataeva magna* (Lour.) DC. (Capparidaceae). Mycopathologia 159:245–249

Nascimento TL, Oki Y, Lima DMM, Almeida-Cortez JS, Fernandes GW, Souza-Motta CM (2015) Biodiversity of endophytic fungi in different leaf ages of *Calotropis procera* and their antimicrobial activity. Fungal Ecol 14:79–86

Nath A, Pathak J, Joshi SR (2014) Bioactivity assessment of endophytic fungfi associated with *Centella asiatica* and *Murraya koengii*. J Appl Biol Biotechnol 2(5):6–11. 10.732JABB.2014.2502

Nath A, Joshi SR (2015) Ultrastructural effect on mastitis pathogens by extract of endophytic fungi associated with ethnoveterinary plant, *Hibiscus sabdariffa* L. J Microsc Ultrastruct 3(1):38–43

Newman JA, Abner ML, Dado RG, Gibson DJ, Brookings A, Parsons AJ (2003) Effects of elevated CO_2, nitrogen and fungal endophyte-infection on tall fescue: growth, photosynthesis, chemical composition and digestibility. Glob Chang Biol 9:425–437

Nutaratat P, Srisuk N, Arunrattiyakorn P, Limtong S (2014) Plant growth-promoting traits of epiphytic and endophytic yeasts isolated from rice and sugarcane leaves in Thailand. Fungal Biol 118(8):683–694

Omacini M, Eggers T, Bonkowski M, Gange AC, Jones TH (2006) Leaf endophytes affect mycorrhizal status and growth of co-infected and neighbouring plants. Funct Ecol 20:226–232

Padhi S, Tayung K (2013) Antimicrobial activity and molecular characterization of an endophytic fungus, *Quambalaria* sp. isolated from *Ipomoea carnea*. Ann Microbiol 63:793–800

Pan F, Liu ZQ, Chen Q, Xu YW, Hou K, Wu W (2016) Endophytic fungus strain 28 isolated from *Houttuynia cordata* possesses wide-spectrum antifungal activity. Braz J Microbiol 47:480–488

Pan F, Su TJ, Cai SM, Wu W (2017) Fungal endophyte-derived *Fritillaria unibracteata* var. *wabuensis*: diversity, antioxidant capacities *in vitro* and relations to phenolic, flavonoid or saponins compounds. Sci Rep 7:42008

Pandey R, Mishra AK, Tiwari S, Singh HN, Kalra A (2011) Enhanced tolerance of *Mentha arvensis* against *Meloidogyne incognita* (Kofoid and White) Chitwood through mutualistic endophytes and PGPRs. J Plant Interact 6(4):247–253

Park YH, Mishra RC, Yoon S, Kim H, Park C, Seo ST, Bae H (2019) Endophytic *Trichoderma citrinoviride* isolated from mountain-cultivated ginseng (*Panax ginseng*) has great potential as a biocontrol agent against ginseng pathogens. J Ginseng Res 43:408–420. https://doi.org/10.1016/j.jgr.2018.03.002

Parsa S, Ortiz V, Gomez-Jimenez MI, Kramer M, Vega FE (2018) Root environment is a key determinant of fungal entomopathogens endophytism following seed treatment in the common bean, *Phaseolus vulgaris*. Biol Control 116:74–81

Parthasarathy R, Sathiyabama M (2014) Gymnemagenin-producing endophytic fungus isolated from a medicinal plant *Gymnema sylvestre* R.Br. Appl Biochem Biotechnol 172:3141–3152

Pecoraro L, Girlanda M, Liu Z-J, Huang L, Perotto S (2015) Molecular analysis of fungi associated with the Mediterranean orchid *Ophrys bertolonii* Mor. Ann Microbiol 65:2001–2007

Petrini O (1991) Fungal endophytes of tree leaves. In: Andrews J, Hirano SS (eds) Microbial ecology of leaves. Springer, New York, pp 179–197

Pili NN, Franca SC, Kyndt T, Makumba BA, Skilton R, Hofte M, Mibey RK, Gheysen G (2016) Analysis of fungal endophytes associated with rice roots from irrigated and upland ecosystems in Kenya. Plant Soil 405:371–380

Pinheiro EAA, Pina JRS, Feitosa AO, Carvalho JM, Borges C, Marinho PSB, Marinho AMR (2017) Bioprospecting of antimicrobial activity of extracts of endophytic fungi from *Bauhinia guianensis*. Rev Argent Microbiol 49(1):3–6

Premalatha K, Gokul S, Kumar A, Mishra P, Mishra P, Ravikumar K, Kalra A (2014) Molecular profiling of fungal assemblages in the healthy and infected roots of *Decalepis arayalpathra* (J. Joseph & V. Chandras) Venter, an endemic and endangered ethnomedicinal plant from Western ghats, India. Ann Microbiol 65(2):785–797

Prihantini AI, Tachibana S (2017) Antioxidant compounds produced by *Pseudocercospora* sp. ESL 02, an endophytic fungus isolated from *Elaeocarpus sylvestris*. Asian Pac J Trop Biomed 7(2):110–115

Qadri M, Johri S, Shah BA, Khajuria A, Sidiq T, Lattoo SK, Abdin MZ, Hassan SR-U (2013) Identification and bioactive potential of endophytic fungi isolated from selected plants of the Western Himalayas. Springerplus 2:8, 1–14

Qadri M, Rajput R, Abdin MZ, Vishwakarma RA, Hassan SR-U (2014) Diversity, molecular phylogeny and bioactive potential with the Himalayan Blue Pine (*Pinus wallichiana*). Microb Ecol 67:877–887

Qian YX, Kang JC, Luo YK, He J, Wang L, Li QR (2018) Secondary metabolites of an endophytic fungus *Phomopsis castaneae-mollissimae*. Chem Nat Compd 54(2):346–347

Qiu SJ, Dong GL, Wei Z, Xiang PW, Fu CD (2008) Diversity and ecological distribution of endophytic fungi associated with medicinal plants. Sci China C Life Sci 51(8):751–759

Quilliam RS, Jones DL (2010) Fungal root endophytes of the carnivorous plant *Drosera rotundifolia*. Mycorrhiza 20:341–348

Quilliam RS, Jones DL (2012) Evidence for host-specificity of culturable fungal root endophytes from the carnivorous plant *Pinguicula vulgaris* (Common Butterwort). Mycol Prog 11:583–585

Rai M, Acharya D, Singh A, Varma A (2001) Positive growth responses of the medicinal plants *Spilanthes calva* and *Withania somnifera* to inoculation by *Piriformospora indica* in a field trial. Mycorrhiza 11:123–128

Raja HA, Kaura A, El-Elimata T, Figueroa M, Kumar R, Deep G, Agarwal R, Faeth SH, Cecha NB, Oberlies NH (2015) Phylogenetic and chemical diversity of fungal endophytes isolated from *Silybum marianum* (L) Gaertn. (milk thistle). Mycology 6(1):8–27

Rakotoniriana EF, Munaut F, Decock C, Randriamampionona D, Andriambololoniaina M, Rakotomalala T, Rakotonirina EJ, Rabemanantsoa C, Cheuk K, Ratsimamanga SU, Mahillon J, El-Jaziri M, Quetin-Leclercq J, Corbisier AM (2008) Endophytic fungi from leaves of *Centella asiatica*: occurrence and potential interactions within leaves. Antonie Van Leeuwenhoek 93:27–36

Rasmann C, Graham JH, Chellemi DO, Datnoff LE, Larsen J (2009) Resilient populations of root fungi occur within five tomato production systems in southeast Florida. Appl Soil Ecol 43:22–31

Rehman S, Shawl AS, Kour A, Andrabi R, Sudan P, Sultan P, Verma V, Qazi GN (2008) An endophytic *Neurospora* sp. from *Nothapodytes foetida* producing camptothecin. Appl Biochem Microbiol 44(2):203–209

Rho H, Hsieh M, Kandel SL, Cantillo J, Doty SL, Kim SH (2018) Do endophytes promote growth of host plants under stress? A meta-analysis on plant stress mitigation by endophytes. Microb Ecol 75(2):407–418

Ribeiro AS, Polonio JC, Costa AT, dos Santos CM, Rhoden SA, Azevedo JL, Pamphile JA (2018) Bioprospection of culturable endophytic fungi associated with ornamental plant *Pachystachys lutea*. Curr Microbiol 75:588–596

Rosa LH, Queiroz SCN, Moraes RM, Wang X, Techen N, Pan Z, Cantrell CL, Wedge DE (2013) *Coniochaeta ligniaria:* antifungal activity of the cryptic endophytic fungus associated with autotrophic tissue cultures of the medicinal plant *Smallanthus sonchifolius* (Asteraceae). Symbiosis 60:133–142

Rosa LH, Tabanca N, Techen N, Wedge DE, Pan Z, Bernier UR, Becnel JJ, Agramonte NM, Walker LA, Moraes RM (2012) Diversity and biological activities of endophytic fungi associated with micropropagated medicinal plant *Echinacea purpurea* (L.) Moench. Am J Plant Sci 3:1105–1114

Russo ML, Pelizza SA, Cabello MN, Stenglein SA, Vianna MF, Scorsetti AC (2016) Endophtyic fungi from selected varieties of soybean (*Glycine max* L. Merr.) and corn (*Zea mays* L.) grown in an agricultural area of Argentina. Rev Argent Microbiol 48(2):154–150

Russo ML, Pelizza SA, Vianna MF, Allegrucci N, Cabello MN, Toledo AV, Mourelos C, Scorsetti AC (2018) Effect of endophytic entomopathogenic fungi on soybean *Glycine max* (L.) Merr. growth and yield. J King Saud Univ Sci. https://doi.org/10.1016/j.jksus.2018.04.008

Samaga PV, Rai VR (2013) Free radical scavenging activity and active metabolite profiling of endophytic fungi from *Nothapodytes foetida* and *Hypericum mysorense*. Int J Chem Anal Sci 4:96–101

Samaga PV, Rai VR, Rai KM (2014) *Bionectria ochroleuca* NOTL33-an endophytic fungus from *Nothapodytes foetida* producing antimicrobial and free radical scavenging metabolites. Ann Microbiol 64:275–285

Santos SN, Ferraris FK, de Souza AO, Henridues MG, Melo IS (2012) Endophytic fungi from *Combretum leprosum* with potential anticancer and antifungal activity. Symbiosis 58:109–117

Santos-Fo FC, Fill TP, Nakamura J, Monteiro MR, Rodrigues-Fo E (2011) Endophytic fungi as a source of biofuel precursors. J Microbiol Biotechnol 21(7):728–733

Sarabia M, Cornejo P, Azcon R, Carreon-Abud Y, Larsen J (2017) Mineral phosphorus fertilization modulates interactions between maize, rhizosphere yeasts and arbuscular mycorrhizal fungi. Rhizosphere 4:89–93

Sati SC, Arya P (2010) Assessment of root endophytic aquatic hyphomycetous fungi on plant growth. Symbiosis 50:143–149

Sati SC, Belwal M (2005) Aquatic hyphomycetes as endophytes of riparian plant roots. Mycologia 97(1):45–49

Schmeda-Hirschmann G, Hormazabal E, Astudillo L, Rodriguez J, Theoduloz C (2005) Secondary metabolites from endophytic fungi isolated from the Chilean gymnosperm *Prumnopitys andina* (Lleuque). World J Microbiol Biotechnol 21:27–32

Schmidt CS, Mrnka L, Frantik T, Lovecka P, Vosatka M (2018) Plant growth promotion of *Miscanthus* x *giganteus* by endophytic bacteria and fungi on non-polluted and polluted soils. World J Microbiol Biotechnol 34:48

Sebastianes FL, Cabedo N, El Aouad N, Valente AM, Lacava PT, Azevedo JL, Pizzirani-Kleiner AA, Cortes D (2012) 3-Hydroxypropionic acid as an antibacterial agent from endophytic fungi *Diaporthe phaseolorum*. Curr Microbiol 65(5):622–632

Selim KA, El-Beih AA, Abdel-Rahman TM, El-Diwany AI (2014) Biological evaluation of endophytic fungus, *Chaetomium globosum* JN711454, as potential candidate for improving drug discovery. Cell Biochem Biophys 68(1):67–82

Sette LD, Passarini MRZ, Delarmelina C, Salati F, Duarte MCT (2006) Molecular characterization and antimicrobial activity of endophytic fungi from coffee plants. World J Microbiol Biotechnol 22:1185–1195

Sharma S, Tiwari KL, Jadhav SK (2018) Diversity of fungal endophytes in *Typha latifolia* (L.) and their lead biosorption activity. Euro-Medit J Environ Integr 3(4):1–8

Sheik S, Chandrashekar KR (2018) Fungal endophytes of an endemic plant *Humboldtia brunonis* wall. of western ghats (India) and their antimicrobial and DPPH radical scavenging potentiality. Orient Pharm Exp Med 18:115–125. https://doi.org/10.1007/s13596-018-0297-y

Shetty KG, Rivadeneira DV, Jayachandran K, Walker DM (2016) Isolation and molecular characterization of the fungal endophytic microbiome from conventionally and organically grown avocado trees in South Florida. Mycol Prog 15:977–986

Shi J, Zeng Q, Liu Y, Pan Z (2012) *Alternaria* sp. MG1, a resveratrol-producing fungus: isolation, identification, and optimal cultivation conditions for resveratrol production. Appl Microbiol Biotechnol 95:369–379

Shu S, Zhao X, Wang W, Zhang G, Cosoveanu A, Ahn Y, Wang M (2014) Identification of a novel endophytic fungus from *Huperzia serrata* which produces huperzine A. World J Microbiol Biotechnol 30:3101–3109

Shweta S, Gurumurthy BR, Ravikanth G, Ramanan US, Shivanna MB (2012) Endophytic fungi from *Miquelia dentata* Bedd., produce the anti-cancer alkaloid, camptothecin. Phytomedicine 20(3-4):337–342

Shweta S, Zuehlke S, Ramesha BT, Priti V, Kumar PM, Ravikanth G, Spiteller M, Vasudeva R, Shaanker RU (2010) Endophytic fungal strains of *Fusarium solani*, from *Apodytes dimidiata* E. Mey. ex Arn (Icacinaceae) produce camptothecin, 10-hydroxycamptothecin and 9-methoxycamptothecin. Phytochemistry 7:117–122

Soltani J, Moghaddam MSH (2015) Fungal endophyte diversity and bioactivity in the Mediterranean Cypress *Cupressus sempervirens*. Curr Microbiol 70(4):580–586

Strobel GA (2003) Endophytes as sources of bioactive products. Microbes Infect 5:535–544

Sudheep NM, Sridhar KR (2012) Non-mycorrhizal fungal endophytes in two orchids of Kaiga forests (Western ghats), India. J For Res 23(3):453–460

Sun J, Xia F, Cui L, Liang J, Wang Z, Wei Y (2014a) Characteristics of foliar endophyte assemblages and host effective components in *Salvia miltiorrhiza* Bunge. Appl Microbiol Biotechnol 98:3143–3155

Sun S-S, Chen X-M, Guo S-S (2014b) Analysis of endophytic fungi in roots of *Santalum album* Linn. and its host plant *Kuhnia rosmarinifolia* Vent. J Zhejiang Univ Sci B (Biomed Biotechnol) 15(2):109–115

Sun X, Guo LD, Hyde KD (2011) Community composition of endophytic fungi in *Acer truncatum* and their role in decomposition. Fungal Divers 47(1):85–95

Sun Y, Wang Q, Lu X, Okane I, Kakishima M (2012) Endophytic fungal community in stems and leaves of plants from desert areas in China. Mycol Prog 11:781–790

Suwannarach N, Kumala J, Bussaban B, Hyde KD, Matsui K, Lumyong S (2013) Molecular and morphological evidence support four new species in the genus *Muscodor* from northern Thailand. Ann Microbiol 63:1341–1351

Taghinasab M, Imani J, Steffens D, Glaeser SP, Kogel KH (2018) The root endophytes *Trametes versicolor Piriformospora indica* increase grain yield and P content in wheat. Plant Soil 426:339–348. https://doi.org/10.1007/s11104-018-3624-7

Tan XM, Chen XM, Wang CL, Jin XH, Cui JL, Chen J, Guo SX, Zhao LF (2012) Isolation and identification of endophytic fungi in roots of nine *Holcoglossum* plants (Orchidaceae) collected from Yunnan, Guangxi, and Hainan provinces of China. Curr Microbiol 64:140–147

Tanapichatsakul C, Monggoot S, Gentekaki E, Pripdeevech P (2018) Antibacterial and antioxidant metabolites of *Diaporthe* spp. isolated from flowers of *Melodorum fruticosum*. Curr Microbiol 75:476–483

Tayung K, Jha DK (2010) Antimicrobial endophytic fungal assemblages inhabiting bark of *Taxus baccata* L. of Indo-Burma mega biodiversity hotspot. Indian J Microbiol 50(1):S74–S81

Tejesvi MV, Mahesh B, Nalini MS, Prakash HS, Kini KR, Subbiah V, Shetty HS (2005) Endophytic fungal assemblages from inner bark and twig of *Terminalia arjuna* W. & A. (Combretaceae). World J Microbiol Biotechnol 21:1535–1540

Tejesvi MV, Tamhankar SA, Kini KR, Rao VS, Prakash HS (2009) Phylogenetic analysis of endophytic *Pestalotiopsis* species from ethnopharmaceutically important medicinal trees. Fungal Divers 38:167–183

Tellenbach C, Grunig CR, Sieber TN (2011) Negative effects on survival and performance of Norway spruce seedlings colonized by dark septate root endophytes are primarily isolated-dependent. Environ Microbiol 13(9):2508–2517

Tellenbach C, Sieber TN (2012) Do colonization by dark septate endophytes and elevated temperature affect pathogenicity of oomycetes? FEMS Microbiol Ecol 82:157–168

Terhonen E, Kerio S, Sun H, Asiegbu FO (2014) Endophytic fungi of Norway spruce roots in boreal pristine mire, drained peatland and mineral soil and their inhibitory effect on *Heterobasidion parviporum*. Fungal Ecol 9:17–26

Tetrad-Jones C, Edwards R (2015) Potential roles for microbial endophytes in herbicide tolerance in plants. Pest Manag Sci 72:203–209

Thiem D, Piernik A, Hrynkiewicz K (2018) Ectomycorrhizal and endophytic fungi associated with *Alnus glutinosa* growing in a saline area of central Poland. Symbiosis 75:17–28

Tian H, Ma YJ, Li WY, Wang JW (2018) Efficient degradation of triclosan by an endophytic fungus *Penicillium oxalicum* B4. Environ Sci Pollut Res 25:8963–8975

Tiwari S, Singh S, Pandey P, Saikia SK, Negi AS, Gupta SK, Pandey R, Banerjee S (2014) Isolation, structure determination, and antiaging effects of 2,3-pentanediol from endophytic fungus of *Curcuma amada* and docking studies. Protoplasma 251:1089–1098

Toju H, Yamamoto S, Sato H, Tanabe AS, Gilbert GS, Kadowaki K (2013) Community composition of root-associated fungi in a *Quercus*-dominated temperate forest: "codominance" of mycorrhizal and root-endophytic fungi. Ecol Evol 3(5):1281–1293

Unterseher M, Schnittler M (2010) Species richness analysis and ITS rDNA phylogeny revealed the majority of cultivable foliar endophytes from beech (*Fagus sylvatica*). Fungal Ecol 3:366–378

Uzma F, Konappa NM, Chowdappa S (2016) Diversity and extracellular enzyme activities of fungal activities isolated from medicinal plants of Western Ghats, Karnataka. Egypt J Basic Appl Sci 3:335–342

Vaz ABM, Da Costa AGFC, Raad LVV, Goes-Neto A (2014) Fungal endophytes associated with three South American Myrtae (*Myrtaceae*) exhibit preferences in the colonization at leaf level. Fungal Biol 118:277–286

Vega FE, Posada F, Peterson SW, Gianfagna TJ, Chaves F (2006) *Penicillium* species endophytic in coffee plants and ochratoxin A production. Mycologia 98(1):31–42

Venkatachalam A, Thirunavukkarasu N, Suryanarayanan TS (2015) Distribution and diversity of endophytes in seagrasses. Fungal Ecol 13:60–65

Verma P, Khan SA, Mathur AK, Shanker K, Kalra A (2014a) Fugal endophytes enhanced the growth and production kinetics of *Vinca minor* hairy roots and cell suspensions grown in bioreactor. Plant Cell Tissue Organ Cult 118:257–268

Verma SK, Gond SK, Mishra A, Sharma VK, Kumar J, Singh DK, Kumar A, Goutam J, Kharwar RN (2014b) Impact of environmental variables on the isolation, diversity and antibacterial activity of endophytic fungal communities from *Madhuca indica* Gmel. at different locations in India. Ann Microbiol 64(2):721–734

Verma VC, Gond SK, Kumar A, Kharwar RN, Boulanger L-A, Strobel GA (2011) Endophytic fungal flora from roots and fruits of an Indian Neem plant *Azadirachta indica* A. Juss., and impact of culture media on their isolation. Indian J Microbiol 51(4):469–476

Verma VC, Gond SK, Kumar A, Kharwar RN, Strobel G (2007) The endophytic mycoflora of bark, leaf and stem tissues of *Azadirachta indica* A. Juss (Neem) from Varanasi (India). Microb Ecol 54:119–125

Vinayarani G, Prakash HS (2018) Fungal endophytes of turmeric (*Curcuma longa* L.) and their biocontrol potential against pathogens *Pythium aphanidermatum* and *Rhizoctonia solani*. World J Microbiol Biotechnol 34:49

Wang FW, Jiao RH, Cheng AB, Tan SH, Song YC (2007) Antimicrobial potentials of endophytic fungi residing in *Quercus variabilis* and brefeldin A obtained from *Cladosporium* sp. World J Microbiol Biotechnol 23:79–83

Wang J, Yao L-Y, Lu Y-H (2013a) *Ceriporia lacerata* DMC1106, a new endophytic fungus: isolation, identification, and optimal medium for 2′,4′-dihydroxy-6′-methoxy-3′,5′-dimethylchalcone production. Biotechnol Bioprocess Eng 18:669–678

Wang L-W, Xu B-G, Wang J-Y, Su Z-Z, Lin F-C, Zhang C-L, Kubicek CP (2012a) Bioactive metabolites from *Phoma* species, an endophytic fungus from the Chinese medicinal plant *Arisaema erubescens*. Appl Microbiol Biotechnol 93:1231–1239

Wang X, Li Y, Song X, Meng Q, Zhu J, Zhao Y, Yu W (2017) Influence of host tree species on isolation and communities of mycorrhizal and endophytic fungi from roots of a tropical epiphytic orchid, *Dendrobium sinense* (Orchidaceae). Mycorrhiza 27:709–718

Wang XM, Yang B, Wang HW, Yang T, Ren CG, Zheng HL, Dai CC (2013b) Consequences of antagonistic interactions between endophytic fungus and bacterium on plant growth and defense responses in *Atractylodes lancea*. J Basic Microbiol 53:1–12

Wang Y, Dai CC, Cao JL, Xu DS (2012b) Comparison of the effects of fungal endophyte *Gilmaniella* sp. and its elicitor on *Atractylodes lancea* plantlets. World J Microbiol Biotechnol 28:575–584

Wang Y, Zeng QG, Zhang ZB, Yan RM, Wang LY, Zhu D (2011) Isolation and characterization of endophytic huperzine A producing fungi from *Huperzia serrata*. J Ind Microbiol Biotechnol 38:1267–1278

Waqas M, Khan AL, Hamayun M, Shahzad R, Kim YH, Choi KS, Lee IJ (2015) Endophytic infection alleviates biotic stress in sunflower through regulation of defence hormones, antioxidants and functional amino acids. Eur J Plant Pathol 141:803–824

Waqas M, Khan AL, Kang S-M, Kim Y-H, Lee I-J (2014) Phytohormone-producing fungal endophytes and hardwood-derived biochar interact to ameliorate heavy metal stress in soybeans. Biol Fertil Soils 50(7):1155–1167

Wazny R, Rozpadek P, Jedrzejczyk RJ, Sliwa M, Stojakowska A, Anielska T, Turnau K (2018) Does co-inoculation of *Lactuca serriola* with endophytic and arbuscular mycorrhizal fungi improve plant growth in a polluted environment? Mycorrhiza 28:235–246

Wezowicz K, Rozpadek P, Turnau K (2017) Interactions of arbuscular and endophytic fungi improve seedling survival and growth in post-mining waste. Mycorrhiza 27:499–511

Wiyakrutta S, Sriubolmas N, Panphut W, Thongon N, Danwiset-kanjana K, Ruangungsi N, Meevootisom V (2004) Endophytic fungi with anti-microbial, anti-cancer and anti-malarial activities isolated from Thai medicinal plants. World J Microbiol Biotechnol 20:265–272

Wu H, Yang H-Y, You X-L, Li Y-H (2013a) Diversity of endophytic fungi from roots of *Panax ginseng* and their saponin yield capacities. Springerplus 2:107. https://doi.org/10.1186/2193-1801-2-107

Wu L, Han T, Li W, Jia M, Rahman K, Qin L (2013b) Geographic and tissue influences on endophytic fungal communities of *Taxus chinensis* var. *mairei* in China. Curr Microbiol 66(1):40–48

Wu W, Davis RW, Tran-Gyamfi MB, Kuo A, LaButti K, Mihaltcheva S, Hundley H, Chovatia M, Lindquist E, Barry K, Grigoriev IV, Henrissat B, Gladden JM (2017) Characterization of four endophytic fungi as potential consolidated bioprocessing hosts for conversion of lignocellulose into advanced biofuels. Appl Microbiol Biotechnol 101:2603–2618

Xia C, Zhang X, Christensen MJ, Nan Z, Li C (2015) *Epichloe* endophyte affects the ability of powdery mildew (*Blumeria graminis*) to colonize drunken horse grass (*Achnatherum inebrians*). Fungal Ecol 16:26–33

Xing X, Guo S (2011) Fungal endophyte communities in four Rhizophoraceae mangrove species on the south coast of China. Ecol Res 26:403–409

Xing X, Guo S, Fu J (2010) Biodiversity and distribution of endophytic fungi associated with *Panax quinquefolius* L., cultivated in a forest reserve. Symbiosis 51:161–166

Xing Y-M, Chen J, Cui J-L, Chen X-M, Guo S-X (2011) Antimicrobial activity and bioactivity of endophytic fungi in *Dendrobium devonianum* and *Dendrobium thyrsiflorum* from Vietnam. Curr Microbiol 62:1218–1224

Yadav M, Yadav A, Yadav JP (2014) *In vitro* antioxidant activity and total phenolic content of endophytic fungi isolated from *Eugenia jambolana* Lam. Asian Pac J Trop Med 7(1):S256–S261

Yan J, Qi N, Wang S, Gadhave K, Yang S (2014) Characterization of secondary metabolites of an endophytic fungus from *Curcuma wenyujin*. Curr Microbiol 69:740–744

Yang T, Du W, Zhou J, Wang X-X, Dai C-C (2013) Effects of the symbiosis between fungal endophytes and *Atractylodes lancea* on rhizosphere and phyllosphere microbial communities. Symbiosis 61:23–36

Yang T, Ma S, Dai SS (2014) Drought degree constrains the beneficial effects of a fungal endophyte on *Atractylodes lancea*. J Appl Microbiol 117(5):1435–1449

Yenn TW, Lee CC, Ibrahim D, Zakaria L (2012) Enhancement of anti-candidal activity of endophytic fungus *Phomopsis* sp. ED2, isolated from *Orthosiphon stamineus* Benth, by incorporation of host plant extract in culture medium. J Microbiol 50(4):581–585

Yuan Y, Feng H, Wang L, Li Z, Shi Y, Zhao L, Feng Z, Zhu H (2017) Potential of endophytic fungi isolated from cotton roots for biological control against Verticillium Wilt disease. PLoS One 12(1):e0170557

Yuan Z-L, Chen Y-C, Yang Y (2009) Diverse non-mycorrhizal fungal endophytes inhabiting an epiphytic, medicinal orchid (*Dendrobium nobile*): estimation and characterization. World J Microbiol Biotechnol 25:295–303

Yuan Z-L, Su Z-Z, Mao L-J, Peng Y-Q, Yang G-M, Lin F-C, Zhang C-L (2011) Distinctive endophytic fungal assemblage in stems of wild rice (*Oryza granulata*) in China with special reference to two species of *Muscodor* (Xylariaceae). J Microbiol 49(1):15–23

Zabalgogeazcoa I, Alvarez A, Herrero N, Vazquez-de-Aldana BR (2018) Production of fumonisins by endophytic strains of *Tolypocladium cylindrosporum* and its relation to fungal virus infection. Mycotoxin Res 34:49–57

Zabalgogeazcoa I, Gundel PE, Helander M, Saikkonen K (2013) Non-systemic fungal endophytes in *Festuca rubra* plants infected by *Epichloe festucae* in subarctic habitats. Fungal Divers 60(1):25–32

Zaferanloo B, Quang TD, Daumoo S, Ghorbani MM, Mahon PJ, Palombo EA (2014) Optimization of protease production by endophytic fungus, *Alternaria alternata,* isolated from an Australian native plant. World J Microbiol Biotechnol 30:1755–1762

Zaferanloo B, Virkar A, Mahon PJ, Palombo EA (2013) Endophytes from an Australian native plant are a promising source of industrially useful enzymes. World J Microbiol Biotechnol 29:335–345

Zhang C, Yin L, Dai S (2009) Diversity of root-associated fungal endophytes in *Rhododendron fortunei* in subtropical forests of China. Mycorrhiza 19(6):417–423

Zhang F-S, Lv Y-L, Zhao Y, Guo S-X (2013) Promoting role of an endophyte on the growth and contents of kinsenosides and flavonoids of *Anoectochilus formosanus* Hayata, a rare and threatened medicinal Orchidaceae plant. J Zhejiang Univ Sci B (Biomed Biotechnol) 14(9):785–792

Zhang H, Ruan C, Bai X, Chen J, Wang H (2018c) Heterocyclic alkaloids as antimicrobial agents of *Aspergillus fumigatus* D endophytic on *Edgeworthia chrysantha*. Chem Nat Compd 54(2):411–414

Zhang H-H, Tang M, Chen H, Wang Y-J (2012a) Effects of a dark-septate endophytic isolate LBF-2 on the medicinal plant *Lycium barbarum* L. J Microbiol 50(1):91–96

Zhang Q, Wei X, Wang J (2012b) Phillyrin produced by *Colletotrichum gloeosporioides,* an endophytic fungus isolated from *Forsythia suspensa*. Fitoterapia 83(8):1500–1505

Zhang Q, Zhang M, Zhou P, Fang Y, Ji Y (2018a) Impact of tree species on barley rhizosphere-associated fungi in an agroforestry ecosystem as revealed by 18S rDNA PCR-DGGE. Agrofor Syst 92:541–554

Zhang X-G, Liu J-W, Tang P, Liu Z-Y, Guo G-J, Sun Q-Y, Yin J-J (2018b) Identification of a new uncompetitive inhibitor of adenosine deaminase from endophyte *Aspergillus niger* sp. Curr Microbiol 75:565–573

Zhao JH, Zhang YL, Wang LW, Wang JY, Zhang CL (2012) Bioactive secondary metabolites from *Nigrospora* sp. LLGLM003, an endophytic fungus of the medicinal plant *Moringa oleifera* Lam. World J Microbiol Biotechnol 28:2107–2112

Zheng CJ, Xu LL, Li YY, Han T, Zhang QY, Ming QL, Rahman K, Qin LP (2013) Cytotoxic metabolites from the cultures of endophytic fungi from *Panax ginseng*. Appl Microbiol Biotechnol 97:7617–7625

Zhou Z, Zhang C, Zhou W, Li W, Chu L, Jinping Yan J, Li H (2014) Diversity and plant growth-promoting ability of endophytic fungi from the five flower plant species collected from Yunnan, Southwest China. J Plant Interact 9(1):585–591

Zijlstra JD, Hof PV, Baar J, Verkley GJM, Summerbell RC, Paradi I, Braakhekke WG, Berendse F (2005) Diversity of symbiotic root endophytes of the *Helotiales* in ericaceous plants and the grass, *Deschampsia flexuosa*. Stud Mycol 53:147–162

Zubek S, Nobis M, Blaszkowski J, Mleczko P, Nowak A (2011) Fungal root endophyte associations of plants endemic to the Pamir Alay mountains of Central Asia. Symbiosis 54:139–149

GC-MS and In Silico Molecular Docking Analysis of Secondary Metabolites Present in Leaf Extract of *Cassia occidentalis* Linn.

27

Amrutha V. Audipudi, Raja Rajeswari Badri, and C. V. S. Bhaskar

Abstract

Solvent extraction of secondary metabolites of *Cassia occidentalis* leaves was investigated. Phytochemical screening of different solvents, petroleum ether (PLCO), ethyl acetate (ELCO), chloroform (CLCO), and aqueous methanol (ALCO) extractions of leaf powder of *C. occidentalis*, was investigated as preliminary experiment. Secondary metabolites of potential solvent extraction were then confirmed using gas chromatography-mass spectrometry (GC-MS). A total of eight compounds were identified with qualitative differences in GC MS analysis. Out of eight constituents, *octadecanoic acid* was identified as one of the major components in chloroform extraction (CLCO). In silico evaluation by molecular docking with alpha amylase revealed that *octadecanoic acid* is a promising antidiabetic compound.

Keywords

Alpha amylase · *Cassia occidentalis* · GC-MS · In silico study · Octadecanoic acid

27.1 Introduction

In conventional therapy, type I diabetes is treated with exogenous insulin and type 2 with oral hypoglycemic agents like sulfonylureas and biguanides (Verma et al. 2010). In traditional practice, medicinal plants are used in many countries to control diabetes mellitus (Alacron-Aguillara et al. 1993). Diabetes mellitus is a chronic

A. V. Audipudi (✉) · R. Rajeswari Badri
Department of Botany and Microbiology, Acharya Nagarjuna University, Guntur, India

C. V. S. Bhaskar
Department of Botany, V.R. College, Nellore, India

S. M. Khasim et al. (eds.), *Medicinal Plants: Biodiversity, Sustainable Utilization and Conservation*, https://doi.org/10.1007/978-981-15-1636-8_27

metabolic disorder resulting from insulin deficiency, characterized by hyperglycemia; altered metabolism of carbohydrates, protein, and lipids; and an increased risk of vascular complication (Barar 2004). Diabetes mellitus has recently been identified by the Indian Council of Medical Research (ICMR) as one of the refractory diseases for which satisfactory treatment is not available in modern allopathic system of medicine and suitable herbal preparations are to be investigated. A large number of plant preparations have been reported to possess antidiabetic activity over the last several decades. Researchers in India have documented the use of over 150 plants in various families with hypoglycemic activity (Patel et al. 2006).

C. occidentalis Linn. (COL) of Caesalpiniaceae family is a common weed scattered from the foothills of Himalayas to West Bengal, South India, Burma, and Sri Lanka (Gupta 2003). The plant is a diffuse (usually annual) undershrub with loosely spreading branches of 60–150 cm long, found throughout India, up to an altitude of 1500 m (The Wealth of India 1998). Different parts of this plant have been reported to possess anti-inflammatory (Kuo et al. 1996), anti-hepatotoxic (Saraf et al. 1994), antibacterial (Samy and Ignacimuthu 2000), and antiplasmodial activities (Tona et al. 2004). They possess purgative, tonic, febrifugal, expectorant, and diuretic properties. The plant is also used to cure sore eyes, hematuria, rheumatism, typhoid, asthma, and disorder of hemoglobin and leprosy. An infusion of the bark is given in diabetes. A wide range of chemical constituents were isolated from *C. occidentalis* including sennoside (Christ et al. 1978), anthraquinone glycoside (Lal and Gupta 1974), fatty oils, flavonoids, glycosides (Purwar et al. 2003), galactomannan, polysaccharides, and tannins (Kudav and Kulkarni 1974). Hence, the present study was designed to evaluate antidiabetic potential of phytochemical compounds of *C. occidentalis* by GSMS analysis and in silico molecular docking of antidiabetic properties of lead molecule.

Cassia occidentalis leaves were collected from the Nellore city, and the samples were authenticated in the Department of Botany, Acharya Nagarjuna University, Guntur, Andhra Pradesh, India. The collected leaves were shade dried and ground to coarse powder. Then, the powdered material was be extracted with petroleum ether, chloroform, and alcohol by Soxhlet extraction method and aqueous methanol by cold maceration method. Petroleum ether (PLCO), ethyl acetate (ELCO), chloroform (CLCO), and aqueous methanol (ALCO) were used separately in the ratio of 80:20 with water. Thereafter, the extracts were concentrated using rotary flash evaporator and named as PLCO, CLCO, ELCO, and ALCO, respectively. Yield percentage of the same was recorded and the results were tabulated.

PLCO, CLCO, ELCO, and ALCO extracts were subjected to standard phytochemical analysis to find the presence of phenols, flavonoids, sterols, tannins (Chitravadivu et al. 2009), terpenoids, cardiac glycosides (Siddiqui and Ali 1997), carbohydrates, amino acids, phlobatannins, and phytosterols (Dipali Somkumar and Vilas Kamble 2013).

The Clarus 680 GC with fused silica column (30 m × 0.25 mm ID × 250 μm DF) packed with Elite-5MS (5% biphenyl and 95% dimethylpolysiloxane) was used in the analysis, and the components were separated using helium as carrier gas at a constant flow of 1 mL/min. The injector temperature was set at 260 °C during the chromatographic run. Extract sample weighing 1 μL was injected into the instrument by regulating the oven temperature as follows: 60 °C (2 min), followed by gradual increase of temperature to 300 °C (at the rate of 10 °C/min), and 300 °C (6 min). The mass detector conditions were set as follows: transfer line temperature, 240 °C; ion source temperature, 240 °C; and ionization mode electron impact, 70 eV; a scan time, 0.2 s; scan interval, 0.1 s; and fragments size from 40 to 600 Da (Chitravadivu et al. 2009). The spectrums of the components were compared with the database of spectrum of known components stored in the GC-MS NIST library (2008).

The National Institute Standard and Technology (NIST) database having more than 62,000 patterns was used for interpretation on mass spectrum GC-MS obtained for the respective plant extract. Spectrum of unknown compounds was compared with spectrum of known compounds stored in the NIST library, and the name, molecular weight, and the structure of the components of the test plant extract were ascertained.

27.2 In Silico Molecular Docking

To study the nature of interactions, binding mode, and selectivity of FTO with secondary metabolites of *C. occidentalis*, docking was carried out with AutoDock 4.2. AutoDockTool was used for creating PDBQT files from traditional PDB files. The sequence of FTO (SwissProt ID, Q9C0B1) was retrieved from SwissProt database. The three-dimensional structure of FTO (PDM ID, 3LFM) was downloaded from PDB database. The domains FTO catalytic domain and C terminal domain (region 35–327 and 329–500) belong to FTO NTD and FTO CTD family. The active sites of FTO were identified using Q-site finder. The drug compound structures were drawn using ACD/ChemSketch and converted into PDB format using Open Babel tool. The 3D structures of FTO were docked with various inhibitors using AutoDock software. The docking results were analyzed using Discovery studio visualize tool. The major chemical constituent present in the *C. occidentalis* was drawn using ChemSketch and optimized. Ligands were prepared in the AutoDock 4.2 for docking studies. The optimized ligands were docked into FTO using "Ligand fit" model in AutoDock 4.2.19. The energy interaction between protein and ligand can be calculated.

Docking can be carried out by Lamarckian genetic algorithm. AutoDock was run several times to get various docked conformations to analyze the predicted docking energy. The binding sites for these molecules were selected based on the ligand-binding pocket of the templates, and the results of docking simulations, such as conformational similarity and binding sites by using AutoDock tools, were analyzed.

27.3 GC-MS Analysis of Secondary Metabolites Present in Leaf Extract of *Cassia occidentalis*

Phytochemical components of PLCO, CLCO, ELCO, and ALCO extracts have been analyzed qualitatively in our lab by using common methods. Plethora of compounds such as flavonoids, glycosides, tannins, terpenoids, steroids, alkaloids, and carbohydrates were found in PLCO, CLCO, ELCO, and ALCO extracts but showed variation in presence of constituents among four different solvent extractions (Table 27.1). Silica gel thin-layer chromatogram revealed a single spot in CLCO (Fig. 27.1) and showed positive tests to alkaloids, anthraquinones, and oils and lipids (Table 27.1) similar to the studies of Lal and Gupta (1974). Hence, CLCO extract was chosen for further analysis of GC-MS and in silico molecular docking (Helaly et al. 2014).

GC-MS chromatogram of the chloroform extract of *C. occidentalis* showed six major peaks at retention time of 16.659, 17.139, 18.480, 20.195, 20.731, and 24.592 min (Fig. 27.2), and ion chromatogram (Fig. 27.3) was obtained from matching all these major peaks with NIST library. The mass of the compounds and fragments recorded were matched with NIST database for identification of probable compounds present in the sample. Twenty five phytochemical compounds were identified in CLCO of *C. occidentalis* by GC-MS. These compounds along with molecular formula and molecular weight are listed (Table 27.1).

Chromatogram of GC MS of CL-CO of third peak with RT:18.480 (Fig. 27.2) revealed that compounds resolved with same molecular formula, and molecular weight showed different distinct peaks and RT values (Table 27.2). The peaks in the chromatogram were integrated and compared with the database of spectrum of known compounds stored in GC-MS library. GC-MS analysis of the extract is given in Table 27.2. Phytochemical analysis revealed the presence of different fatty acids and heterocyclic compounds (Table 27.2).

From the results, it was observed that tridecanoic acid was found to be the major component constituting about 50% of the extract. The second major components in the extract were octadecanoic acid and *N*-hexadecanoic acid constituting about

Table 27.1 Preliminary phytochemical analysis of leaf extracts of *C. occidentalis*

Phytochemicals tested	PLCO	CLCO	ELCO	ALCO
1. Alkaloids	+	+	+	–
2. Saponins	–	–	–	–
3. Flavonoids	–	–	+	+
4. Anthraquinones	+	+	+	–
5. Tannins	–	–	+	+
6. Sterols				
7. Sugars	–	–	+	+
8. Cardiac glycosides	–	–	–	+
9. Lipids and oils	+	+	–	–
10. Coumarins	–	–	+	+
11. Proteins and amino acids	+	–	+	–
12. Terpenoids	+	+	–	–

Fig. 27.1 Thin-layer chromatogram of PLCO, CLCO, ELCO, and ALCO. Line 1, PLCO; line 2, ELCO; line 3, CLCO; and line 4, ALCO

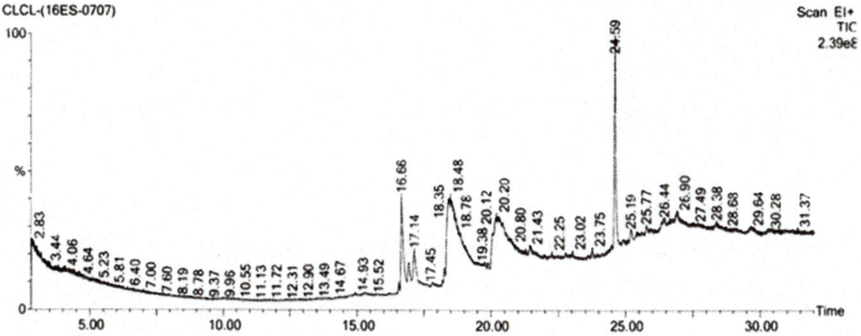

Fig. 27.2 GC-MS chromatogram of CLCO of *C. occidentalis*

37.5% of the extract. Third major components in the extract were tetradecanoic acid and pentadecanoic acid constituting about 25%. Whereas eicosanoic acid, dodecanoic acid, and L-(+)-ascorbic acid 2,6-dihexadecanoate were found to be in less quantity constituting about 12.5% of the extract. The percentage of all constituents present in CLCO is listed in Table 27.3.

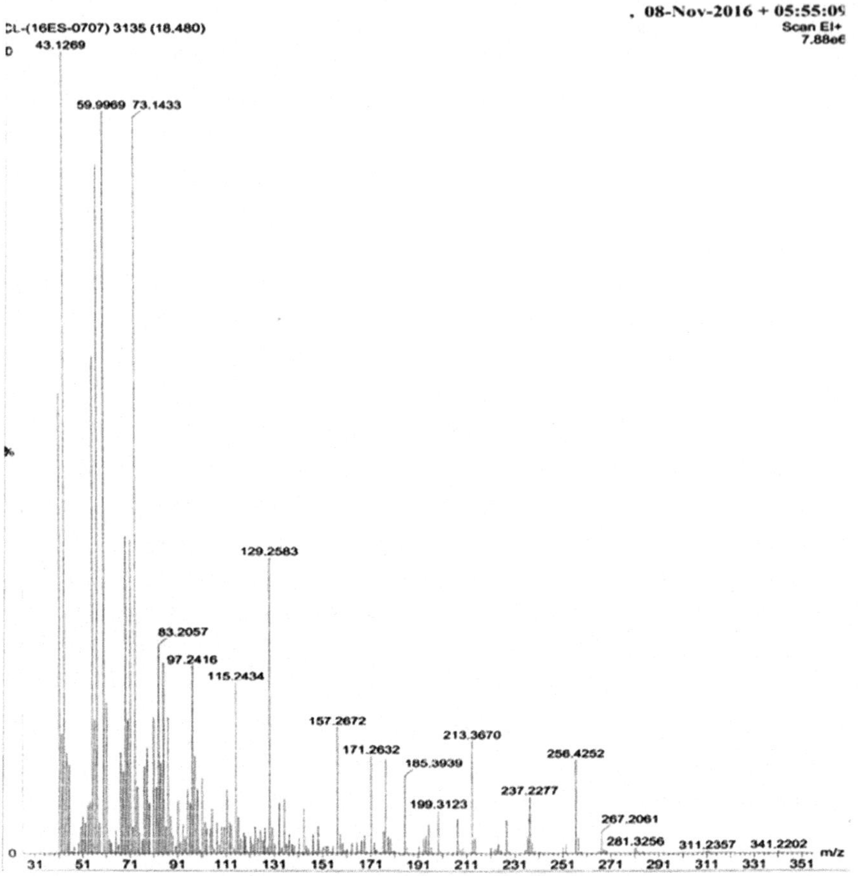

Fig. 27.3 Total ion chromatogram of CLCO of *C. occidentalis*

Table 27.2 Constituents of CLCO resolved in GC-MS chromatogram

#	RT	Scan	Height	Area	Area (%)	Norm (%)
1	16.659	2771	83,243,512	7,292,512.0	9.580	21.72
2	17.139	2867	27,877,410	2,720,471.5	3.574	8.10
3	18.480	3135	69,370,056	33,582,680.0	44.119	100.00
4	20.195	3478	44,131,256	20,000,978.0	26.276	59.56
5	20.731	3585	14,763,632	1,592,544.9	2.092	4.74
6	24.592	4357	181,127,456	10,929,437.0	14.358	32.54

27.3.1 Docking of Antidiabetic Enzymes

Docking study was initiated with octadecanoic acid of first resolved compound in ion chromatogram of CLCO of *C. occidentalis* (Fig. 27.3). Molecular docking of octadecanoic acid with alpha amylase reveals the formation of three different types of interactions: hydrogen bond, Van der Waals interactions, and Pi interactions. The

Table 27.3 Chromatogram of GC MS of CLCO (RT:18.480) compounds resolved with the same molecular formula and molecular weight but with different distinct peaks and RT values

S. no.	Compound name	Molecular formula	Mol. wt.	No of peaks	RT value	% Constituent
1.	Octadecanoic acid	$C_{18}H_{36}O_2$	284	3	917,899,893	37.5
2.	*N*-Hexadecanoic acid	$C_{16}H_{32}O_2$	256	3	943,927,912	37.5
3.	Tridecanoic acid	$C_{13}H_{26}O_2$	214	4	895,891,886,882	50
4.	Tetradecanoic acid	$C_{14}H_{28}O_2$	228	2	893,890	25
5.	Pentadecanoic acid	$C_{15}H_{30}O_2$	242	2	893,890	25
6.	Eicosanoic acid	$C_{20}H_{40}O_2$	312	1	929	12.5
7.	Dodecanoic acid	$C_{12}H_{24}O_2$	200	1	885	12.5
8.	L-(+)-Ascorbic acid 2,6-dihexadecanoate	$C_{38}H_{68}O_8$	652	1	916	12.5

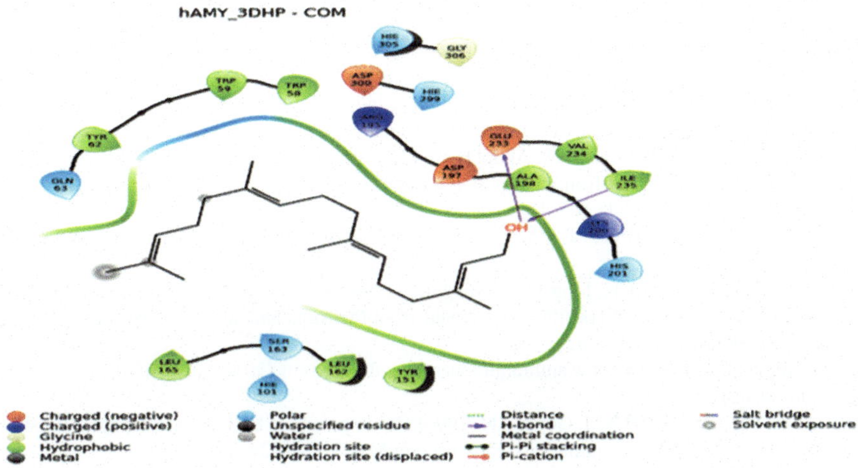

Fig. 27.4 Molecular docking of octadecanoic acid with alpha amylase

aminoacid residue GLU333 and ILE235 hydrogen bond formation, ILU165, LEU162, TYR151, ALA192, VAL239, and ILE235 form Van der Waals interaction and one Pi-Pi stacking between ILU165 and LEU162 (Fig. 27.4). The docking score was −5.052 and glide energy was −37.5 kcal/mol. The docking score and glide is very high with alpha amylase compared with other antioxidant enzymes. The highest docking score and glide energy revealed octadecanoic acid has more stable interactions toward alpha amylase (Helaly et al. 2014)

27.4 Conclusion

GC-MS of CLCO at RT:18.480 resolved into eight compounds with the same molecular formula and molecular weight but with different distinct peaks and RT values. Based on in silico molecular docking analysis, the highest docking score and glide energy revealed octadecanoic acid has more stable interactions toward alpha amylase compared with other antioxidant enzymes and presumed to be promising compound for further investigation.

Acknowledgments Authors are thankful to Ratnam Pharmaceutical College, Nellore, and SVIMS, Tirupati, for GC-MS analysis and molecular docking studies.

References

Alacron-Aguillara FJ, Roman Ramos R, Flores-Saenz JL (1993) Plant medicinales usadas en el control de la diabetes mellitus. Cienc 44:361–363

Barar FS (2004) Essential of pharmacotherapeutics. S. Chand and Company Ltd., New Delhi

Chitravadivu C, Manian S, Kalaichelvi K (2009) Qualitative analysis of selected medicinal plants, Tamil Nadu, India. Middle-East J Sci Res 4(3):144–146

Christ B, Poppinghaus T, Wirtz-Peitz F (1978) Isolation and structural definition of a new sennoside from *Cassia senna* L. Arzneimittelforschung 28:225–231

Dipali Somkumar O, Vilas Kamble A (2013) Phytochemical screening of ethanolic extracts of stem leaves, flower and seed kernel of Mangifera indica L. Int J Pharm Bio Sci 4(2):383–389

Gupta AK (2003) Quality standard of Indian medicinal plants. Indian Council Med Res 1:47–48

Helaly FM, Nashar DEE, Khalaf AI, Abdel Hamid HF, Abd El-Moez SI, Soliman AM (2014) Slow release of 5-fluorouracil from natural polymeric composites as controlled drug delivery. Int J Pharm Sci Rev Res 27(1):105–110

Kudav NA, Kulkarni AB (1974) Chemical investigation of *Cassia occidentalis*. Indian J Chem 12:1042–1044

Kuo SC, Chen SC, La CF, Teng CM, Wang JP (1996) Studies on the anti-inflammatory and antiplatelet activities of constituents isolated from the roots and stem of *Cassia occidentalis* L. Chin Pharm J 48:291–302

Lal J, Gupta PC (1974) Two new anthraquinones from the seeds of *Cassia occidentalis*. Experientia 30:850–851

Patel PM, Patel KN, Patel NM, Goyal RK (2006) Development of HPTLC method for estimation of charantin in herbal formulations. Pharmacogn Mag 8:224–226

Purwar C, Rai R, Srivastava N, Singh J (2003) New flavonoid glycosides from *Cassia occidentalis*. Section B—organic chemistry including medicinal chemistry. Indian J Chem 42:434–436

Samy RP, Ignacimuthu S (2000) Antibacterial activity of some folklore medicinal plants used by tribals in Western Ghats of India. J Ethnopharmacol 69:63–71

Saraf S, Dixit VK, Tripathi SC, Patnaik GK (1994) Antihepatotoxic activity of *Cassia occidentalis*. Pharm Biol 32:178–183

Siddiqui AA, Ali M (1997) Practical pharmaceutical chemistry, 1st edn. CBS Publishers and Distributors, New Delhi, pp 126–131

The Wealth of India (1998) A dictionary of Indian raw material and industrial products. Council of Scientific and Industrial Research, New Delhi

Tona L, Cimanga RK, Mesia K, Musuamba CT, De Bruyne T, Apers S (2004) *In vitro* antiplasmodial activity of extracts and fractions from seven medicinal plants used in the Democratic Republic of Congo. J Ethnopharmacol 93:27–32

Verma L, Singour PK, Chaurasiya PK, Rajak H, Pawar RS, Patil UK (2010) Effect of ethanolic extract of *Cassia occidentalis* Linn. for the management of alloxan-induced diabetic rats. Pharm Res 2(3):132–137

28 Protective Effect of *Mimusops elengi* L. on Renal and Hepatic Markers in STZ-Induced Diabetic Rats

S. K. Jaffar, S. M. Khasim, and M. S. K. Prasad

Abstract

The present study was carried out to investigate the effect of ethanolic leaf extract of *Mimusops elengi* L. on biochemical parameters and enzyme activities in STZ-induced diabetic rats. Leakage of cellular enzymes into the plasma is a clear indication of cell damage. When liver plasma membrane is damaged, a variety of enzymes normally located in the cytosol are released into the bloodstream, and their estimation is a quantitative marker for the extent of damage. The cytoprotective effect of *M. elengi* was evaluated in rats that were rendered diabetic by administration of streptozotocin at a dose of 40 mg/kg b.w. The activities of the marker enzymes were assayed in the serum, liver and kidney. The indicators of renal damage such as urea, uric acid and creatinine were assayed. Treatment of diabetic rats with the ethanolic leaf extract of *M. elengi* (100 mg/kg b.w.) was able to reverse the levels of the marker enzymes in the liver and protect the kidney by reverting back to the normal levels of renal markers (urea, uric acid and creatinine). The present results showed that ethanolic leaf extract of *M. elengi* could alleviate liver and renal damage caused by STZ-induced diabetes.

Keywords

Alanine transaminase · Marker enzymes · *M. elengi* · Renal markers

---

S. K. Jaffar (✉)
Department of Biochemistry, Acharya Nagarjuna University, Guntur, Andhra Pradesh, India

S. M. Khasim
Department of Botany and Microbiology, Acharya Nagarjuna University, Guntur, Andhra Pradesh, India

M. S. K. Prasad
Department of Biochemistry, Kakatiya University, Warangal, Telangana, India

S. M. Khasim et al. (eds.), *Medicinal Plants: Biodiversity, Sustainable Utilization and Conservation*, https://doi.org/10.1007/978-981-15-1636-8_28

28.1 Introduction

Diabetes mellitus is probably the single most important metabolic disease and is widely recognized as one of the leading cause of death and disability. It affects every cell in the body and the body's essential biochemical process, and it is a major public health problem in developed as well as developing countries (Kameswara Rao et al. 2003). Diabetes mellitus (DM) is a group of metabolic disorder characterized by hyperglycaemia, hypercholesterolemia and hypertriglyceridemia resulting from defects in insulin secretion or action or both (Craig et al. 2009).

Streptozotocin (STZ) is a metabolite of the soil organism *Streptomyces achromogenes*. Its structure resembles D-glucose; it is taken up by the glucose transporter (GLUT-2), which is expressed in the pancreas, liver and kidney of rodents (Schnedl et al. 1994). β-cells are necrotized within few hours after STZ administration (Modak et al. 2007). Intracellular action of STZ results in nitric oxide (NO) liberation, alkylation and fragmentation of DNA, leading to cell death. DNA damage causes the poly-ADP-ribosylation which leads to the depletion of cellular NAD+, reduction of ATP and subsequent inhibition of insulin synthesis and secretion. Six hours after injection, hyperglycaemia develops, and blood insulin levels decrease (Szkudelski 2001).

Medicinal plants continue to provide valuable therapeutic agents, in both modern medicine and traditional system. The drawbacks about the efficacy and safety of the oral hypoglycaemic agents have prompted a search for safer and more effective drugs in the treatment of diabetes (Reaven et al. 1983). A multitude of herbs, spices and other plant materials have been described for the treatment of diabetes throughout the world (Kesari et al. 2005). *Mimusops elengi* L. is commonly called Bakul (belongs to Sapotaceae family). It is distributed extensively in the Western Ghats of India. Natural products are known to play an important role in pharmaceutical biology. Plants have been an important source of medicine for thousands of years. Even today, the World Health Organization estimates that up to 80% of people still rely mainly on traditional medicines. In fact, many of the current drugs either mimic naturally occurring molecules or have structures that are fully or in part derived from natural motifs (Baby and Raj 2010). *Mimusops elengi* is considered as a sacred plant and has obtained important place in Hindu religious texts as well as in ancient Sanskrit literature. Its fragrant flowers are celebrated in the Puranas and even placed amongst the flowers of the Hindu paradise. Kalidasa has also mentioned this plant in his classical Sanskrit literature as a symbol of love and beauty (Mitra 1981). It is a small-to-large evergreen tree found in the Deccan Peninsula and Andaman Islands and frequently cultivated in gardens for ornament; it is grown also as an avenue or shade tree throughout the greater parts of India. It has made important contribution to the field of science from ancient times as also to modern research due to its large number of medicinal properties (Nadkarni 1996; Sharma et al. 2000; Manjeshwar et al. 2011). *Mimusops elengi* possessed antianxiety activity (Ganu et al. 2011), antihyperlipidaemic activity (Ghaisas et al. 2008), antiulcer activity (Prakash et al. 2011; Shah et al. 2003), anticonvulsant activity (Ganu et al. 2011), antioxidant activity (Rao et al. 2011), antiurolithiatic activity (Ashok et al. 2010), antiatherosclerotic activity (Satishchandra and Sumithra 2011), cytotoxic activity (Karmakar et al.

2011), antidiabetic activity (Mamatha et al. 2011; Jerline et al. 2009; Jaffar et al. 2011) and diuretic activity (Koti and Ashok 2010).

The present study was undertaken to evaluate the cytoprotective effect of *M. elengi* against the toxicity induced by streptozotocin in rats by using biological markers, since it permits the early identification of the adverse effects and the potentiality of the drug used in the treatment of disease.

Fresh leaves of *M. elengi* L. were collected during the month of May and June from Acharya Nagarjuna University campus, India. The plant was identified by using flora of Madras Presidency (Gamble 1915–1936), and voucher specimen was deposited at Botany Department in Acharya Nagarjuna University (ANU: 1323).

The plant leaves were shade dried at room temperature (32 ± 2 °C), and the dried leaves were ground into fine powder using pulverizer. The powder was stored in deep freezer. To prepare the ethanolic leaf extract, 100 g of dry fine powder was suspended in 400 ml of 95% ethanol for 72 h. The extract was filtered using a muslin cloth and concentrated at 40 ± 5 °C.

Male albino rats of Wistar strain (150–200 g) were used for the study; the animals were maintained in an air-conditioned room under controlled temperature and humidity. They were fed standard rat pellet diet supplied by Hindustan Unilever Ltd., Bangalore, India. Animals were deprived of food for 16 h but allowed free access to water. Ethical clearance for the handling of experimental animals is obtained from the Institutional Animal Ethics Committee (IAEC) constituted for this purpose, and the care of laboratory animals was taken as per the guidelines of the committee for the purpose of control and supervision on experiments on animals (CPCSEA).

28.1.1 Experimental Induction of Diabetes in Animals

The animals were rendered diabetic by a single intraperitoneal injection of STZ (40 mg/kg b.w.) in freshly prepared citrate buffer (0.1 M, pH 4.5) after an overnight fast. STZ-injected animals were given 20% glucose solution for 24 h to prevent initial drug-induced hypoglycaemic mortality. STZ-injected animals exhibited massive glycosuria (determined by Benedict's qualitative test) and hyperglycaemia (by glucose oxidase method), and diabetes in STZ rats was confirmed by measuring the fasting blood glucose levels, 96 h after injection with STZ. The animals with blood glucose more than 220 mg/dl were considered diabetic and used for the experiments.

Blood was collected by retroorbital puncture after 12 h of fasting and 2 h after giving the extract in gum acacia for the estimation of plasma glucose. Autoanalyzer was used in this study to minimize the sample requirement.

28.1.2 Experimental Design

Induction of diabetes was achieved as described above. Rats with the fasting blood glucose levels of above 200 mg/dl were taken for the experiment. In these experiments, rats were divided into five groups of six rats each as follows:

Group 1. Normal rats
Group 2. Normal rats treated with 100 mg *M. elengi* leaf extract/kg b.w.
Group 3. Diabetic control rats
Group 4. Diabetic rats treated with 100 mg *M. elengi* leaf extract/kg b.w.
Group 5. Diabetic rats treated with 0.6 mg glibenclamide/kg b.w.

The animals in groups 2 and 4 were given daily oral dose of 100 mg/kg b.w. of the *M. elengi* leaf extract, while group 1 and 3 rats were given water alone, and group 5 rats were treated with glibenclamide at a dose of 0.6 mg/kg b.w. in the morning time for a period of 30 days.

28.1.3 Biochemical Analysis

After the experimental period, the animals were killed by cervical decapitation. The tissues were excised immediately and immersed in ice-cold physiological saline. The present homogenate was prepared with fresh tissue in 0.01 M tris–HCl buffer (pH 7.4), and these were used for the assays. The plasma was separated and also used for the analysis.

Alanine and aspartate aminotransferases were assayed by the method of Reitman and Frankel (Reitman and Frankel 1957). Alkaline phosphatase (ALP) activity was assayed by the method of King (Kind and King 1954); γ-glutamyl transferase (GGT) was assayed by the method of Rosalki (Rosalki and Rau 1972), and albumin was estimated by the method of Lowry (Lowry et al. 1951).

Urea was estimated by the method of Fawcett (Fawcett and Scott 1960). Uric acid was estimated by the method of Caraway (Caraway 1963), and creatinine was estimated by the method of Tietz (Tietz 1987) using Jaffe's colour reaction (Jaffe 1886).

28.1.4 Statistical Analysis

The values were expressed as Mean ± SD for six rats in each group; statistically significant differences between the groups were calculated using one-way analysis of variance (ANOVA). Value of $p < 0.05$ was considered to be significant.

28.1.5 Effect of Ethanolic Leaf Extract of *M. elengi* on Hepatic Marker Enzymes in Normal and Diabetic Rats

The hepatic marker enzymes such as ALT, AST, ALP and GGT in plasma of diabetic rats are significantly elevated when compared with normal rats (Table 28.1; Fig. 28.1). The oral administration of the ethanolic leaf extract of *M. elengi* (100 mg/kg b.w.) and glibenclamide significantly decreased hepatic marker enzymes when compared with control diabetic rats.

Table 28.1 Effect of long-term treatment with the ethanolic leaf extract of *M. elengi* on the activity of hepatic marker enzymes in the plasma of control and STZ-induced diabetic rats after 30 days

Groups	ALT (IU/l)[a]	AST (IU/l)[a]	ALP (IU/l)[b]	GGT (IU/l)[c]
Control	33.40 ± 1.90[d]	68.88 ± 3.18[d]	89.74 ± 3.20[d]	16.67 ± 1.34[d]
Control + *M. elengi* leaf extract (100 mg/kg b.w.)	32.17 ± 2.10[d]	69.10 ± 2.65[d]	88.88 ± 2.75[d]	16.86 ± 1.17[d]
Diabetic control	70.15 ± 2.35[e]	133.25 ± 6.78[e]	151.29 ± 10.10[e]	27.25 ± 1.40[e]
Diabetic + *M. elengi* leaf extract (100 mg/kg b.w.)	55.20 ± 2.50[f]	105.19 ± 6.50[f]	113.15 ± 8.10[f]	21.80 ± 1.44[f]
Diabetic + glibenclamide (0.6 mg/kg b.w.)	34.10 ± 2.16[d]	71.10 ± 2.75[d]	90.09 ± 5.90[d]	16.77 ± 1.10[d]

Values are SD ± mean of six rats. Values not sharing a common letter differ significantly at $p < 0.05$ (DMRT).
[a]μmol of pyruvate liberated/h
[b]μmol of phenol liberated/min
[c]μmol of *p*-nitroanilide liberated/min
[d]Diabetic rats treated with *Mimusops elengi* leaf extracts
[e]Diabetic control rats
[f]Glibenclamide-treated rats. d and f showed significant values

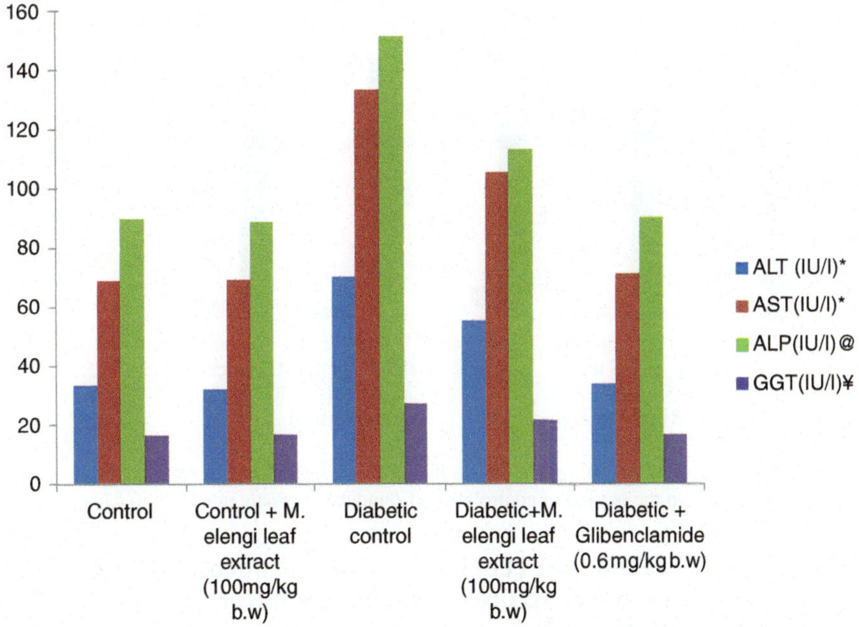

Fig. 28.1 Effect of long-term treatment with the ethanolic leaf extract of *M. elengi* on the activity of hepatic marker enzymes in the plasma of control and STZ-induced diabetic rats after 30 days

Table 28.2 Effect of long-term treatment with the ethanolic leaf extract of *M. elengi* on urea, uric acid and creatinine in the plasma of control and STZ-induced diabetic rats after 30 days

Groups	Urea (mg/dl)	Uric acid (mg/dl)	Creatinine (mg/dl)
Control	24.28 ± 1.77[a]	1.59 ± 0.10[a]	0.95 ± 0.05[a]
Control + *M. elengi* leaf extract (100 mg/kg b.w.)	24.50 ± 1.90[a]	1.61 ± 0.16[a]	0.91 ± 0.06[a]
Diabetic control	40.15 ± 1.75[b]	3.85 ± 0.25[b]	2.12 ± 0.12[b]
Diabetic + *M. elengi* leaf extract (100 mg/kg b.w.)	30.10 ± 1.51[c]	2.20 ± 0.17[c]	1.70 ± 0.15[c]
Diabetic + glibenclamide (0.6 mg/kg b.w.)	29.79 ± 1.49[c]	2.24 ± 0.18[c]	1.68 ± 0.12[c]

Values are SD ± mean of six rats. Values not sharing a common letter differ significantly at $p < 0.05$ (DMRT).

[a]Control

[b]Diabetic control rats

[c]*Mimusops elengi* leaf extract-treated and glibenclamide-treated rats

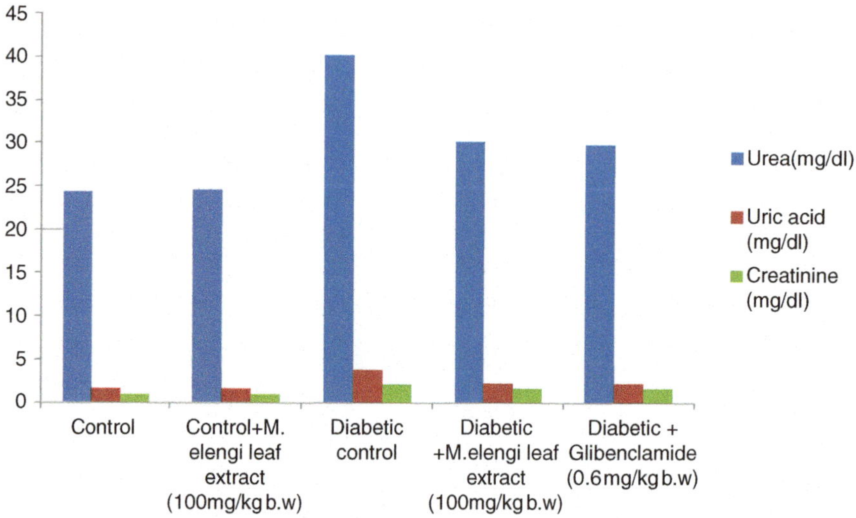

Fig. 28.2 Effect of long-term treatment with the ethanolic leaf extract of *M. elengi* on urea, uric acid and creatinine in the plasma of control and STZ-induced diabetic rats after 30 days

28.1.6 Effect of Ethanolic Leaf Extract of *M. elengi* on Plasma Urea, Uric Acid and Creatinine Levels in Normal and Diabetic Rats

The renal markers such as plasma urea, uric acid and creatinine in diabetic rats significantly increased when compared with normal rats (Table 28.2; Fig. 28.2). The supplementation of ethanolic leaf extract of *M. elengi* (100 mg/kg b.w.) resulted in significant restoration of renal markers to normal levels when compared with control diabetic rats. The standard drug, glibenclamide, also showed similar results.

Table 28.3 Effect of long-term treatment with the ethanolic leaf extract of *M. elengi* on total protein, albumin, globulin and A/G ratio in the plasma of control and STZ-induced diabetic rats after 30 days

	Plasma			
Groups	Total protein (g/dl)	Albumin (g/dl)	Globulin (g/dl)	A/G ratio
Control	8.69 ± 0.57[a]	4.31 ± 0.29[a]	4.18 ± 0.22[a]	1.15 ± 1.10[a]
C + Me (100 mg/kg b.w.)	8.69 ± 0.40[a]	4.22 ± 0.30[a]	4.20 ± 0.18[a]	1.22 ± 1.05[a]
DM	4.92 ± 0.20[b]	1.80 ± 0.11[b]	2.55 ± 0.10[b]	0.75 ± 0.42[b]
DM + Me (100 mg/kg b.w.)	6.88 ± 0.16[c]	3.95 ± 0.14[c]	3.37 ± 0.15[c]	1.01 ± 0.94[c]
DM + glibenclamide (0.6 mg/kg b.w.)	8.03 ± 0.45[d]	4.17 ± 0.20[a]	3.60 ± 0.18[d]	1.11 ± 1.09[a]

Values are SD ± mean of six rats. Values not sharing a common letter differ significantly at $p < 0.05$ (DMRT).
[a]Control
[b]Diabetic control rats
[c]Diabetic control rats treated with *M. elengi* (100 mg/kg b.w.) leaf extract
[d]Glibenaclamide-treated rats

28.1.7 Effect of Ethanolic Leaf Extract of *M. elengi* on Total Proteins, Albumin, Globulin and A/G Ratio Levels in Normal and Diabetic Rats

The total protein, albumin, globulin and A/G ratio in plasma of diabetic rats significantly decreased when compared with normal rats (Table 28.3; Fig. 28.3). Oral administration of the ethanolic leaf extract of *M. elengi* (100 mg/kg b.w.) and glibenclamide could result in significant elevation of total proteins, albumin, globulin and A/G ratio, when compared with control diabetic rats.

28.2 Discussion

Diabetes is a chronic metabolic disorder affecting major population worldwide. A substantial reduction in hyperglycaemia will decrease the risk of developing microvascular diseases and reduce complications (Kim et al. 2006). The treatment of diabetes with medicinal plants proves much safer than synthetic drugs in integral part of many cultures throughout the world. Therefore, the aim of the present study was to investigate the cytoprotective effect of ethanolic leaf extract of *M. elengi* in STZ-induced diabetic rats. STZ-induced hyperglycaemia has been described as a useful experimental model to study the activity of hypoglycaemic agents (Szkudelski 2001; Lenzen 2008). Earlier, we reported, for the first time, the potential antihyperglycaemic activity of ethanolic leaf extract of *M. elengi* in STZ-induced diabetic rats (Jaffar et al. 2011).

The cytotoxicity of various drugs, toxins and xenobiotics could be evaluated using marker enzymes. Leakage of cellular enzymes into the plasma is a clear indication of cell damage. The catalytic activity and their tissue specificity are the best markers of tissue damage. The liver is a versatile organ of the body that regulates the

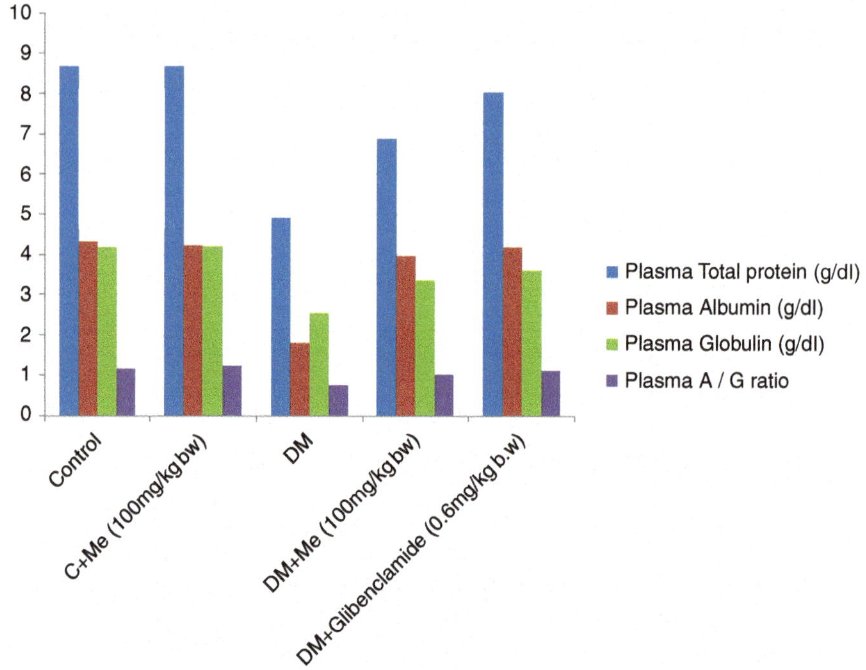

Fig. 28.3 Effect of long-term treatment with the ethanolic leaf extract of *M. elengi* on total protein, albumin, globulin and A/G ratio in the plasma of control and STZ-induced diabetic rats after 30 days

internal chemical environment. Liver injury induced by various hepatotoxins was recognized as a major toxicological problem for years. The liver is an important target of toxicity to xenobiotics, and liver cell membrane is necrotized in diabetic patients (Larcan et al. 1979). A variety of enzymes normally located in the cytosol are released into the bloodstream, and their estimation could reveal the extent of damage. AST and ALT activities in blood are generally accepted as an index of liver damage, and ALT is used as highly liver-specific enzyme (El-Demerdash et al. 2005; Dhanasekaran et al. 2009; Kesari et al. 2007). The change in alanine and aspartate transaminase activities may be due to STZ toxicity. The significant elevation of serum ALT and AST levels in diabetic rats could be related to the excessive accumulation of amino acids in the serum as a result of amino acid mobilization from protein stores (Colev et al. 1994). The increased metabolic pathways such as gluconeogenesis and ketogenesis may be responsible for the increase of these transaminases (Maiti et al. 2004). The reduction in liver enzyme activities (Table 28.1) is mainly due to leakage of these enzymes into the bloodstream as a result of STZ toxicity which leads to liver damage. The decreased levels of ALT and AST after supplementation with ethanolic leaf extract of *M. elengi* indicate the restoration of normal liver function. A possible explanation for the effect of ethanolic leaf extract on the activities of AST and ALT could be due to the restoration of the liver damage caused by STZ.

Diabetic rats showed a significant increase in serum alkaline phosphatase activity which indicates the loss of membrane integrity (Shahraki et al. 2007). Hyperglycaemia may play an important role in the liver functional abnormalities, and association exists between diabetes and liver injury. The increased activity of ALP in group 3 animals and the subsequent decrease upon ethanolic extract administration indicate the membrane-stabilizing property of the ethanolic extract of *M. elengi*. In diabetes, there was a significant increase of GGT, which suggests its role in the pathogenesis of disease (Lee et al. 2003). In the present study, increased activity of GGT observed may be attributed due to the administration of STZ, which is capable of altering the redox balance of the cell by drastic reduction of GSH in the tissues (Arkkila et al. 2001). Supplementation of *M. elengi* could reverse the levels of GGT activity. From these studies, it could be concluded that *M. elengi* ethanolic leaf extract could protect against liver tissue damage in STZ-induced diabetic rats.

The mechanism of hepatoprotective activity of the extract could be attributed to numerous bioactive compounds present in the extract. The preliminary phytochemical screening of the plant revealed the presence of alkaloids, saponins, terpenoids, sterols, flavonoids and steroidal glycosides. The extract also might contain β-sitosterol, a compound known as hepatoprotective agents. These results were in agreement with previous reports (Sarada Devi et al. 2011). The bioactive compounds could improve the function of liver and pancreatic cells; hence, they could restore the activities of liver enzymes (Li et al. 2004).

The kidney removes urea, uric acid and creatinine and ions as metabolic wastes to maintain the optimum chemical composition of body fluids. However, the concentrations of the metabolites increase in blood during renal disease or renal damage associated with uncontrolled diabetes mellitus. The diabetic hyperglycaemia induces elevation of plasma urea and creatinine which are considered to be a significant marker of renal dysfunction (Almdal and Vilstrup 1988). In the present study, there was significant increase of urea, uric acid and creatinine in STZ-induced diabetic rats. This may be due to metabolic disturbances in diabetes reflected in high activities of xanthine oxidase, lipid peroxidation and increased triglycerides and cholesterol (Anwar and Meki 2003). Moreover, protein glycosylation in diabetes may lead to muscle wasting and increased release of purine, the main source of uric acid as well as in activity of xanthine oxidase (Anwar and Meki 2003). The increased elevation and uric acid synthesis can be related to the energetic deficiency in the cells that are unable to utilize glucose in the state of insulin deficiency. The increased renal excretion of uric acid causes impaired glomerular filtration rate (Golembiewska et al. 2005). The metabolic and microvascular complications of diabetes are associated with multiple effects on renal function and creatinine metabolism. Elevated hyperglycaemia, osmotic diuresis and depletion of extracellular fluid volume may cause a decline in glomerular filtration rate and an increase in serum creatinine (Eidi et al. 2006). Treatment with ethanolic leaf extract reversed the alterations in the levels of urea, uric acid and creatinine caused by STZ and increased glycaemia. The results are in agreement with previous reports (Dubey et al. 1994).

Reduction in plasma proteins and albumin levels observed in diabetic rats was in agreement with the results obtained by Bakris (1997) and Tuvemo et al. (1997). The decrease in protein and albumin could be due to microproteinuria and/or due to increased protein catabolism (Mauer et al. 1981). Improvement of plasma total proteins, albumin, globulin and A/G ratio was observed after the oral administration of ethanolic leaf extract of *M. elengi* to experimental diabetic rats. Almdal and Vilstrup (1988) also demonstrated a significant increase in total protein excretion (albuminuria) and glycosuria in impaired renal function. An increased albumin excretion in rats leading to microalbuminuria is widely acknowledged as the earliest index of diabetic nephropathy and as a risk factor for the development of diabetic renal and macrovascular disease (Valmadrid et al. 2000). The decreased albumin excretion seen in *M. elengi*-treated animals proves the renoprotective effect of the extract, which may be due to the downregulation of various autocrine and paracrine factors such as transforming growth factor beta (TGF-β), insulin-like growth factor 1 (IGF-1), type IV collagen and upregulation of rennin (Mitch et al. 1999).

The treatment with ethanolic leaf extract of *M. elengi* could significantly increase the albumin, total protein and globulin and also reduce the elevated serum urea, uric acid and creatinine levels to near normal levels. From the above results, it is evident that ethanolic leaf extract could delay the progression of diabetic nephropathy by improving renal function.

28.3 Conclusion

From the above results, it could be concluded that *M. elengi* leaf ethanolic extract could able to ameliorate the impaired renal function, inhibit the liver damage and induced free radicals associated with STZ diabetes. Further studies on the characterization and active principles responsible for cytoprotective activity of *M. elengi* are highly warranted.

Acknowledgements We thank the Department of Biochemistry, Acharya Nagarjuna University, Guntur, Andhra Pradesh, India, for providing research facilities to carry out this work.

References

Almdal JP, Vilstrup H (1988) Strict insulin therapy normalizes organ nitrogen contents and the capacity of urea nitrogen synthesis in experimental diabetes in rats. Diabetologia 31:114–118

Anwar MM, Meki AR (2003) Oxidative stress in streptozotocin induced diabetic rats; effect of garlic oil and melatonin. Comp Biochem Physiol A Mol Integr Physiol 135:347–539

Arkkila PE, Koskinen PJ, Kantola IM, Ronnemma T, Seppanen E, Viikari JS (2001) Diabetic complications are associated with liver enzyme activities in people with type I diabetes. Diabetes Res Clin Pract 52:113–118

Ashok P, Koti BC, Vishwanathswamy AH (2010) Antiurolithiatic and antioxidant activity of *Mimusops elengi* on ethylene glycol-induced urolithiasis in rats. Ind J Pharmacol 42(6):380–383

Baby J, Raj SJ (2010) Pharmacognostic and phytochemical properties of *Aloe vera* L.—an overview. Int J Pharm Sci Rev Res 4(2):106–110

Bakris GL (1997) Diabetic nephropathy. What you need to know to preserve kidney function. Postgrad Med 93:89

Caraway WT (1963) Uric acid. In: Seligson D (ed) Standard methods of clinical chemistry. Academic Press, New York, pp 239–247

Colev V, Badescu M, Paduraru I, Mandreci I, Bohotin C (1994) The zinc metabolite disorder relation in experimental diabetes mellitus. Rom J Intern Med 32:71–75

Craig ME, Hattersley A, Donghuc KC (2009) Definition and epidemiology and classification of diabetes in children and adolescents. Pediatr Diabetes 10:3–12

Dhanasekaran M, Baskar AA, Ignacimuthu S, Agastian P, Duraipandiyan V (2009) Chemoprotective potential of epoxy clerodane diterpene from *Tinospora cordifolia* against diethylnitrosamine-induced hepatocellular carcinoma. Investig New Drugs 27:347–355

Dubey JJ, Campbell RK, Setter SM, White JR, Rasmussen KA (1994) Diabetic neuropathy—an intensive review. Am J Health Syst Pharm 61:160–176

Eidi A, Eidi M, Esmaelli E (2006) Antidiabetic effect of garlic (*Allium sativum* L.) in normal and streptozotocin-induced diabetic rats. Phytomedicine 13:624–629

El-Demerdash FM, Yousef I, El-Naga NIA (2005) Biochemical study on the hypoglycaemic effects of onion and garlic in alloxan-induced diabetic rats. Food Chem Toxicol 43:57–63

Fawcett JK, Scott JE (1960) Rapid and precise method for the determination of urea. J Clin Pathol 13(2):156–159

Gamble JS (1915–1936) Flora of presidency of Madras, vol 1–3. Adland & Sons Ltd., London

Ganu G, Garud A, Agarwal V, Suralkar U, Jadhav S, Kshirsagar A (2011) Anti-anxiety activity of *Mimusops elengi* barks extract in experimental animals. Res J Pharm Biol Chem Sci 2(3):405

Ghaisas MM, Kadam AH, Kshirsagar BD, Dhote VV, Deshpande AD (2008) Evaluation of antihyperlipidemic activity of *Mimusops elengi* L. in triton WR-1339 induced hyperlipidemia in rats. J Nat Remed 8(2):132–137

Golembiewska E, Ciechanowski K, Safranow K, Kedzierska K, Kabat-Koperska J (2005) Renal handling of uric acid in patients with type I diabetes in relation to glycaemic control. Arch Med Res 36:32–35

Jaffar SK, Khasim SM, Guru Prasad M, Ibrahim MD (2011) Hypoglycaemic activity of ethanolic leaf extract of *Mimusops elengi* L. in streptozotocin induced diabetic rats. The Bioscan 6(4):673–679

Jaffe M (1886) Estimation of creatinine. Z Physiol Chem 10:391–400

Jerline M, Jothi G, Brindha P (2009) Effect of *Mimusops elengi L.* bark extract on alloxan induced hyperglycaemia in albino rats. J Cell Tissue Res 9(3):1985–1988

Kameswara Rao B, Renuka Sudarshan P, Rajasekhar MD, Nagaraju N, Appa Rao C (2003) Antidiabetic activity of *Terminalia palladia* fruit in alloxan induced diabetic rats. J Ethnopharmacol 85:169–172

Karmakar UK, Sultana R, Biswas NN (2011) Antioxidant, analgesic and cytotoxic activities of *Mimusops elengi* L. leaves. Ind J Pharm Sci Res 2(11):2791–2797

Kesari AN, Gupta RK, Watal G (2005) Hypoglycaemic effects of *Murraya koenigii* on normal and alloxan-diabetic rabbits. J Ethnopharmacol 97:245–247

Kesari AN, Kesari S, Singh SK, Gupta RK, Watal G (2007) Studies on the glycaemic and lipidemic effect of *Murraya koenigii* in experimental animals. J Ethnopharmacol 112:305–311

Kim SH, Hyun SH, Choung SY (2006) Antidiabetic effect of cinnamon extract on blood glucose in db/db mice. J Ethnopharmacol 104:119–123

Kind PRN, King EJ (1954) Estimation of plasma phosphatase by determination of hydrolyzed phenol with amino antipyrine. J Clin Pathol 7:330–332

Koti BC, Ashok P (2010) Diuretic activity of extracts of *Mimusops elengi* L. Bark. Int J Green Pharm 4(2):90–92

Larcan A, Lambert H, Laprevote-Heully MC, Delorme N (1979) Light and electron microscopic study of hepatic lesions in the course of hyperlactatemia in diabetic patients. Diabetes Metab 5:103–112

Lee DH, Ha MH, Kim JH (2003) Gamma glutamyltransferase and diabetes: a 4 year follow-up study. Diabetology 46:359–364

Lenzen S (2008) The mechanism of alloxan and streptozotocin-induced diabetes. Diabetologia 51:216–226

Li WL, Zheng HC, Bukuru J, DeKimpe N (2004) Natural medicine used in the traditional chinese medical system for therapy of diabetes mellitus. J Ethnopharmacol 92:1–21

Lowry OH, Rosebrough NJ, Farr AL, Randall RJ (1951) Protein measurement with Folin-phenol reagent. J Biol Chem 193:265–275

Maiti R, Jana D, Das UK, Ghosh D (2004) Antidiabetic effect of aqueous extract of seed of *Tamarindus indica in* streptozotocin-induced diabetic rats. J Ethnopharmacol 92:85–91

Mamatha MK, Thomas T, Yogesh HS, Pramod HJ (2011) Antidiabetic activity of *Mimusops elengi* leaves on alloxan induced diabetic rats. Int J Pharm Sci 3(2):1358–1362

Manjeshwar SB, Ramakrishna JP, Harshith PB, Princy LP, Rekha B (2011) Chemistry and medicinal properties of the Bakul (*Mimusops elengi* L.): a review. Food Res Int 44:1823–1829

Mauer SM, Steffes MW, Brown DM (1981) The kidney in diabetes. Am J Med 70:63

Mitch WE, Bailey JL, Wang X, Jurkovitz C, Newby D, Prince SR (1999) Evaluation of signal activating ubiquitin-proteasome-proteolysis in a model of muscle wasting. Am J Phys 276:C1132–C1138

Mitra R (1981) Bakul—a reputed drugs of ayurveda, its history—uses in Indian medicine. Ind J Hist Sci 16(2):169–180

Modak MA, Datar SP, Bhonde RR, Ghaskadbi SS (2007) Differential susceptibility of chick and mouse islets to streptozotocin and its co-relation with islets antioxidant status. J Comp Physiol B 177:247–257

Nadkarni KM (1996) Indian materia medica, 3rd edn. Popular Prakashan, Mumbai, pp 596–599

Prakash D, Koti BC, Vijay T, Katagi MS (2011) Antiulcer activity of *Mimusops elengi* bark extracts against serotonin induced ulcers in rats. Int Res J Pharm 2(8):173–176

Rao KS, Munjuluri PR, Keshar NK (2011) In vitro antioxidant activity and total phenolic content of *Mimusops elengi* bark. Ind J Pharm Educ Res 45(4):317–323

Reaven E, Wright D, Mondon CE, Solomon R, Ho H, Reaven GM (1983) Effect of age and diet on insulin secretion and insulin action in the rats. Diabetes 32:175–180

Reitman S, Frankel S (1957) Colorimetric method for determination of serum glutamic oxaloacetic acid and glutamic pyruvic transaminases. Am J Clin Pathol 28:56–63

Rosalki SB, Rau D (1972) Serum γ-glutamyl transpeptidase activity in alcoholism. Clin Chem Acta 39:41–47

Sarada Devi KM, Annapoorna S, Ashok kumar K (2011) Hepatic antioxidative potential of ethyl acetate fraction of *Cynodon dactylon* in Balb/c mice. J Med Plant Res 5(6):929–996

Satishchandra A, Sumithra M (2011) Synergistic effect of *Mimusops elengi* and *Moringa oleifera* on high fat diet induced atheroma in rats. Int J Adv Pharm Res 2(6):293–300

Schnedl WL, Ferber S, Johnson JH, Newgard CB (1994) STZ transport and cytotoxicity, specific enhancement in GLUT2-expressing cells. Diabetes 43:1326–1333

Shah PJ, Gandhi MS, Shah MB, Goswami SS, Santani D (2003) Study of *Mimusops elengi* bark in experimental gastric ulcers. J Ethnopharmacol 89(2):305–311

Shahraki MR, Arab MR, Mirimokaddam E, Palan MJ (2007) The effect of *Teucrium polium* (Calpoureh) on liver function, serum lipids and glucose in diabetic male rats. Iran Biomed J 11:65–68

Sharma PC, Yelne MB, Dennis TJ (2000) Database on medicinal plants used in ayurveda, vol I. Central Council for Research in Ayurveda and Siddha, Government of India, New Delhi, pp 65–72

Szkudelski T (2001) The mechanism of alloxan and streptozotocin action in β cells of the rat pancreas. Physiol Res 50:536–546

Tietz NW (1987) Fundamentals of clinical chemistry, 3rd edn. W.B. Saunders Company, Philadelphia, PA, pp 614–620

Tuvemo T, Ewald U, Kobboh M, Proos LA (1997) Serum magnesium and protein concentrations during the first five years of insulin dependent diabetes in children. Acta Paediatr Suppl 418:7–10

Valmadrid CT, Klein R, Moss SE, Klein BE (2000) The risk of cardiovascular disease mortality associated with microalbuminuria and gross proteinuria in person with older-onset diabetes mellitus. Arch Intern Med 160:1093–1100

Extraction and Purification of Gymnemic Acid from *Gymnema sylvestre* R.Br.

29

Srinivasan Kameswaran and Karuppan Perumal

Abstract

Gymnema sylvestre R.Br. is one of the important medicinal plants of India widely used in indigenous medicine in the treatment of diabetes mellitus. The active principle of the drug is complex mixture of gymnemic acids found in the leaves. Presently there is a huge demand for the plant in national and international market. The recent studies have shown that the extract of *Gymnema sylvestre* is useful in controlling blood sugar to treat type II diabetes. It increases the insulin-producing B-cells of the pancreas and significantly reduces the metabolic effects of sugar by preventing the intestine from absorbing the sugar molecules during the process of digestion. The objective of the present investigation was to isolate and characterize the gymnemic acid from *Gymnema sylvestre* leaves with different solvent systems like petroleum ether, benzene, and methanol. The defatted leaves that were extracted under continuous hot extraction in Soxhlet apparatus with 90% methanol gave the maximum yield of gymnemic acid (42%). Gymnemic acid was purified by preparative chromatographic methods, i.e., TLC and HPLC, SDS-PAGE, and NMR.

Keywords

Gymnema sylvestre · Gymnemic acid · Extraction · HPLC · SDS-PAGE · NMR

S. Kameswaran (✉)
Department of Botany, Vikrama Simhapuri University Post Graduation Centre, Kavali, Andhra Pradesh, India

K. Perumal
Murugappa Chettiar Research Centre (MCRC), Taramani, Chennai, Tamil Nadu, India

S. M. Khasim et al. (eds.), *Medicinal Plants: Biodiversity, Sustainable Utilization and Conservation*, https://doi.org/10.1007/978-981-15-1636-8_29

29.1 Introduction

The conservative estimate shows that there are 250,000–500,000 species of plants on Earth (Borris 1996), and a relatively small percentage (1–10%) of these is used as food, but substantial number of plants are used for medicinal purposes (Moerman 1996). Medicinal plants or their parts are primary sources of products for the pharmaceutical industry. It is estimated that even today, two-thirds of the world population rely on plant-derived drugs. The herbal medicine represents probably the first and certainly the oldest system of human health care.

Gymnema sylvestre R.Br. is a valuable medicinal plant belonging to the family *Asclepiadaceae*, native to central and western India and can also be found growing in tropical Africa and Australia (Stocklin 1969). The Hindi name for *G. sylvestre* is "gurmar," meaning "sugar-destroying," which is suggestive of the anti-sweet agent. Many monographs on pharmacognostic studies have emerged as an aid in the pharmacognostic investigations (Kalidass et al. 2009a; Edward 1950; Borris 1996). The process of standardization can be achieved by stepwise pharmacognostic studies. These studies help in identification and authentication of the plant material (Ozarkar 2005).

Recent clinical trials conducted in India have shown that an extract of *Gymnema sylvestre* is useful for controlling blood sugar. Today, *Gymnema sylvestre* has become increasingly popular in the United States as a supportive treatment for diabetes. Use of gymnema was well-known to the Indian people since ancient days ("meshashring") as a source of antidiabetic drugs. In recent years, it became one of the better known names in the world of herbal medicine. It is a rich source of many bioactive compounds such as gymnemic acid (GA-I-X), quercitol, lupeol, stigmasterol, gymnemin, gymnemagenin, gurmarin, etc. which are mainly effective in lowering of blood sugar. Gymnemic acid, the active ingredient of this plant, is extracted from leaves and used widely as anti-diabetes (Shanmugasundaram et al. 1983), anti-sweetener (Kurihara 1992; Liu et al. 1992), and anti-hypercholesterolemia (Bishayee and Chatterjee 1994). It also has stomachic, diuretic, and cough-suppressant property (Kapoor 1990). The plant has been reported to possess antimicrobial (Sative et al. 2003; Dateo and Long 1973) and ethno-veterinary medicinal properties (Kalidass et al. 2009b). In addition, it possesses antimicrobial, hepato-protective, and anti-saccharine activities (Kalidass et al. 2009b; Komalavalli and Rao 2000; Nadkarni and Nadkarni 1976). Hence, because of these properties, *Gymnema sylvestre* is most important for plant prospecting.

The leaves of gurmar are used medicinally for its unique property to directly mask the tongue's ability to taste sweet foods and at the same time suppress glucose absorption from the intestine. This is the reason why it is known as "destroyer of sugar." The fresh leaves when chewed have the remarkable property of paralyzing the sense of taste of sweet substance for some time (Gent et al. 1999; Persaud et al. 1999). The atomic arrangement of gymnemic acid molecules is similar to that of glucose molecules. These molecules fill the receptor locations on the taste buds, thereby preventing its activation by sugar molecules present in the food. This prevents craving for sugar. Similarly, gymnemic acid molecules fill the receptor location in the absorptive external layers of the intestine, thereby preventing the sugar molecule absorption, which results in low blood sugar level (Sahu et al. 1996).

Traditionally it was recommended for stomach problems, constipation, and liver disease, but the recent studies have shown that the extract of *Gymnema sylvestre* is useful in controlling blood sugar to treat type II diabetes (NIDDM). When gymnema leaf extract is administered to a diabetic patient, it stimulates the pancreas to increase release of insulin (Persaud et al. 1999). These compounds have also been found to increase fecal excretion of cholesterol (Nakamura et al. 1999; Masayuki et al. 1997), but further studies to prove clinical significance in treating hypercholesterolemia (high serum cholesterol) are required. However, the present study is the first ever attempt for the isolation, purification, and characterization of gymnemic acid from four ecotypes of *Gymnema sylvestre* with the purpose to obtain its maximum yield using various techniques.

Seventeen ecotypes of *Gymnema sylvestre* R.Br. (Asclepiadaceae) were collected from various parts of the country and maintained in our garden. The plant material was properly identified and confirmed with help of various floras (Gamble 1991; Matthew 1991). All the chemicals and reagents were used analytical grade purchased from Sigma Chemical Co. and Merck.

About 3 kg cleaned leaves from each ecotype were dried under shade, powdered and passed through 40 meshes, and stored in closed vessel for further use. The dried powder material was subjected to Soxhlet extraction with petroleum ether, chloroform, and methanol for continuous hot extraction.

29.2 Extraction of Gymnemic Acid by Hoopers's Method (Hooper 1887)

Step 1: Extraction with petroleum ether

One kg of dry leaf powder was packed into a clean Soxhlet extraction unit. Seven liters of petroleum ether (60–80 °C) was added and extracted for 24–36 h till all the components are soluble in petroleum. Petroleum extract is collected and distilled in a distillation unit. Then a net weight of 250 g of petroleum ether extracts was obtained. Petroleum ether extraction was used for defatting dried leaf power.

Step 2: Extraction with 90% methanol

The plant material is then extracted with 90% methanol. Ninety percent methanol was added, and the extraction was carried out for 24–36 h till the total methanol-soluble extract was obtained. The methanol-soluble extract was distilled and finally 175 g of the thick paste were obtained.

Step 3: Isolation of pure gymnemic acid from methanol extract

One hundred seventy-five g thick paste of methanol-soluble extract was dissolved in 1% aqueous KOH solution on continuously stirring for 45 min to 1 h. The solution is then filtered through filter paper to separate the undissolved particles. Diluted HCl was added slowly under constant stirring, during which the gymnemic acids were precipitated. Precipitated solution was filtered under suction, and precipitate was dried. The pure gymnemic acid was obtained.

29.3 Biochemical Tests to Confirm the Gymnemic Acid

Gymnemic acid gave positive test for phenolics, steroids, and glycoside:

Phenolic test: A pinch of gymnemic acid was taken into a clean test tube and dissolved in 2 ml of methanol. Then a few drops of 1% alcoholic ferric chloride were added.

Steroid test: A pinch of gymnemic acid was added to a solution of 2 ml $CHCl_3$ and 1 ml of acetic anhydride. A few drops of conc. H_2SO_4 were added from the sides of the tubes.

Glycoside test: A pinch of gymnemic acid was taken in a dried test tube and dissolved in 2 ml of methanol. One ml of alpha naphthol alcoholic solution was added from the sides of the test tube.

The identification and separation of the components present in different extracts of *Gymnema sylvestre* were carried out by thin-layer chromatography. The TLC of gymnemic acid was performed using different solvent systems, i.e., chloroform/acetone, chloroform/methanol, toluene/ethyl acetate/diethylamine, and ethyl acetate/petroleum ether. The chromatograms were dried to remove the solvent, cooled, and sprayed with the detecting reagents. The plates were dried at 105 °C for 5 min to enable the full color of the spots to develop.

Known weight of *Gymnema sylvestre* leaf samples was homogenized with 70% ethanol followed by centrifugation at 2000 × *g* for 10 min. Supernatant was collected and incubated at −20 °C for 12 h. Further the content was centrifuged at 10,000 × *g* for 15 min, and the supernatant was discarded. The pellet was dissolved in 100 μl of sample buffer containing 5 ml of Tris-HCl buffer (pH 6.8), 0.5 g of SDS, 5 g of sucrose, 0.25 ml of mercaptoethanol, and 1 ml of 0.5% Bromophenol Blue (W/V). The content was made up to 10 ml with distilled water. About 30 μl of samples were loaded onto the 15% SDS-polyacrylamide gels prepared by the method of Laemmli (1970) and run at 70 mA for 2 h. Protein bands were visualized by silver staining procedure.

The leaf extracts were filtered through Sartorius RC-membrane syringe filter (0.20 μm), and 20 μl was injected. Chromatography was carried out using Shimadzu HPLC (Model SPD-10A UV-VIS Detector) and Supelcosil LC-18 C18 column (25 cm × 4.6 mm, 5 μm) with mobile phase consisting of acetonitrile, water, and acetic acid (50:50:0.1). Flow rate was 1.0 ml/min, back pressures 250 psi, and compounds read at 210 nm in a UV detector. The total run time was 40 min, but it is preferable if the time is extended to 60 min (Shimizu et al. 1997).

Extraction and purification of the active principle for ^{1}H-NMR study: The leaves of *G. sylvestre* was air-dried at 70 °C and stored at –20 °C before use. The fine powder of the biomass weighing about 100 g was immersed in 1 L of water at about 40 °C for 1 h with continuous stirring and filtered with thin tissue paper lined with gauge. The water extraction was repeated three times. The combined filtrate was acidified to pH 2.0 by adding 4 N H_2SO_4. The resulting precipitate was collected by centrifugation at 10,000 rpm for 10 min. The precipitate was then subjected to extraction with ethanol repeatedly to remove hydrophobic materials including gymnemic acids. The remaining ethanol-insoluble fraction was suspended in 300 ml distilled water and neutralized with 2 N NaOH. The deep-brown supernatant obtained by centrifugation was used for NMR studies (Imoto et al. 1991).

Nuclear magnetic resonance (^{1}H-NMR) studies: The use of nuclear magnetic resonance (NMR) measurement was carried out by 400 MHz Fourier transform NMS system (JEOL GSX 400 NB 400 Hz FT-NMR system, RSIC, IIT, Madras). The sample was prepared by dissolving about 0.2 mg of the semi-purified material in 0.4 ml of D_2O and adjusting pH to 7.5 with NaOD. Temperature of the sample was kept at 23 °C by blowing temperature-controlled air into the cavity.

29.4 Gymnemic Acid from *Gymnema sylvestre*

The detailed and systematic pharmacognostic evaluation would give valuable information for the future studies. The work carried out on this plant was mainly on the methods of extraction of gymnemic acid in order to obtain its higher yields, separation, identification, and purification of the gymnemic acid by TLC. The extractions were carried out with different solvent systems like petroleum ether, benzene, and methanol and were extracted under continuous hot extraction in Soxhlet apparatus. Out of all the three solvents tested, the extraction with 90% methanol gave the maximum yield of gymnemic acid. The yields of gymnemic acid from five ecotypes were calculated and presented in Table 29.1.

The result obtained on conducting the phenolic test is a dark blue color was developed which is the positive test indicating the presence of –OH group in the molecule. A pink/red color ring was formed when few drops of conc. H_2SO_4 were added from the sides of the tube containing a pinch of gymnemic acid in a solution of 2 ml $CHCl_3$. This is the positive test for steroid presence in the gymnemic acid. The glycosidic nature of gymnemic acid was a disputed question when it was first isolated. Hooper (1887) isolated it and proved it to be a glycoside. To confirm the glycosidic nature in the present study, a small pinch of gymnemic acid was taken in a dried test tube and dissolved in 2 ml of methanol. One ml of alpha naphthol alcoholic solution was added from the sides of the test tube. A bluish red ring was developed at the junction of the two layers indicating the presence of glycoside.

Table 29.1 Acquisition of four ecotypes of *Gymnema sylvestre* from different parts of Jharkhand and the percentage of gymnemic acid

S. no.	Name of the ecotype	Place of collection	Percentage of gymnemic acid
1	Silent Valley	Kerala	40.8
2	Kolli Hills	Tamil Nadu	38.6
3	Venkatachalam	Nellore	27.9
4	Rapur	Nellore	25.6

Table 29.2 Different R_f values of gymnemic acid

S. no.	Name of the ecotype	R_f value
1	Silent Valley	0.84
2	Kolli Hills	0.81
3	Venkatachalam	0.75
4	Rapur	0.73

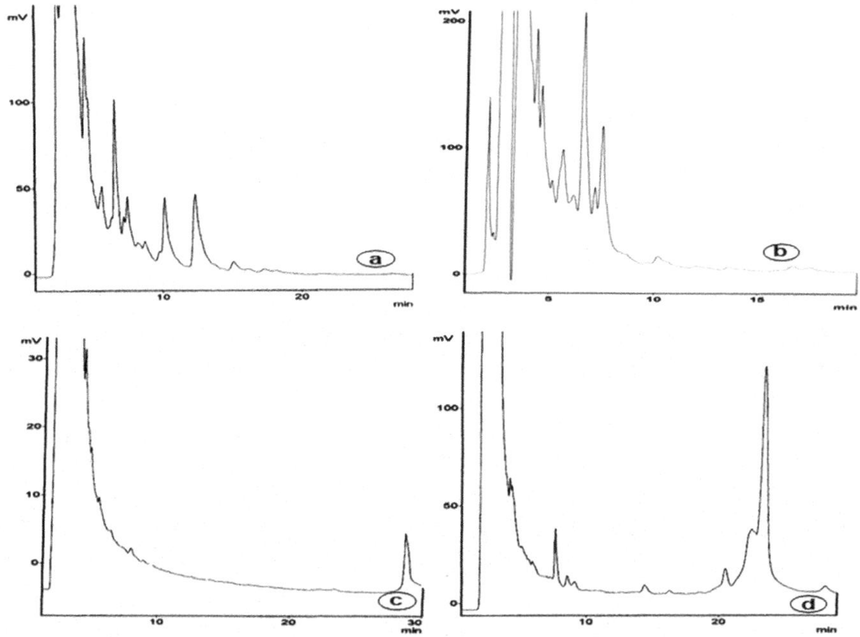

Fig. 29.1 HPLC chromatogram of leaf samples. (**a**) *Gymnema sylvestre* R.Br. (Silent Valley); (**b**) *Gymnema sylvestre* (Kolli Hills); (**c**) *Gymnema sylvestre* (Venkatachalam); (**d**) *Gymnema sylvestre* (Rapur)

Thin-layer chromatography studies were carried with different solvent systems, i.e., chloroform/acetone, chloroform/methanol, toluene/ethyl acetate/diethylamine, and ethyl acetate/petroleum ether. All the samples are gymnemic acid with different R_f values (Table 29.2). The solvent system chloroform/methanol (6:5) gave better results when compared with the other solvent systems. TLC studies revealed that the profiles are similar when compared with the standard gymnemic acid having R_f 0.71.

29.4.1 HPLC Analysis of Active Principles of *Gymnema sylvestre* R.Br.

In this study, the samples of active principles extracted from dried leaves were used for HPLC analysis (Shimizu et al. 1997). HPLC chromatogram of *G. sylvestre* extract samples is presented (Fig. 29.1):

1. *Gymnema sylvestre* R.Br. (Silent Valley) acid precipitate (GSA, concentration of gymnemic acids: 258 mg/g gymnemic acid II or 161.6 mg/g as gymnemagenin) (Fig. 29.1a)
2. *Gymnema sylvestre* (Kolli Hills) leaf extract (Fig. 29.1b)
3. *Gymnema sylvestre* (Venkatachalam) leaf extract (Fig. 29.1c)
4. *Gymnema sylvestre* (Rapur) leaf extract (Fig. 29.1d)

In this study, the overall active components from *G. sylvestre* leaf sample eluted through HPLC were separated into four fractions based on the retention time of standard samples. Each fraction and cumulative value of all four fractions of leaves was taken as percent. Water-soluble compounds in leaf extract sample were found to be higher than other samples.

29.4.2 SDS-PAGE

SDS-PAGE was carried out to study the polypeptide- or proteinaceous-based compound present in the *G. sylvestre*. The samples were taken from the callus grown in different combinations of growth hormone-supplemented media and elicitor-treated media (Fig. 29.2) as follows: Lane-1, standard; Lane-2, Silent Valley sample; Lane-3, Kolli Hills sample; Lane-4, Venkatachalam sample; and Lane-5, Rapur sample.

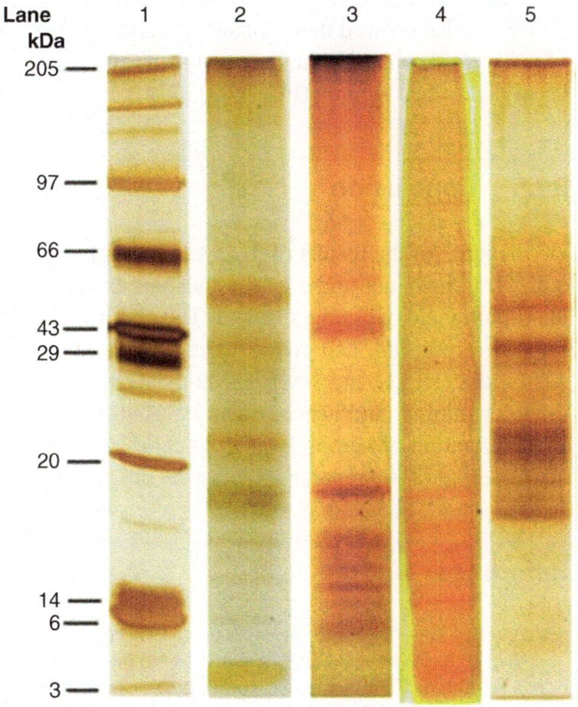

Fig. 29.2 SDS-PAGE protein profiles

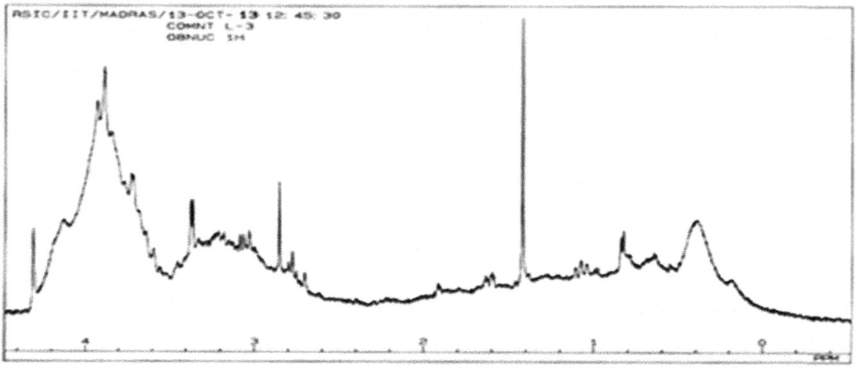

Fig. 29.3 NMR spectrum of *Gymnema sylvestre* R.Br.

The Kolli Hills sample showed more prominent bands approximately at 4, 18, 22, 37, and 56 kDa (Lane-2). Venkatachalam samples showed bands approximately at 5, 8, 12, 14, 16, 18, 45, and 62 kDa (Lane-3). Rapur sample showed bands approximately at 1, 2, 4, 8, 14, 16, and 18 kDa (Lane-4). Silent Valley sample showed bands at 17, 20, 22, 29, and 50 kDa (Lane-5). Lane-1 is standard sample.

29.4.2.1 H-NMR Studies

H-NMR study carried out with callus sample grown in growth hormone standardized medium was compared with leaf sample. Spectrum of the sample taken at 400 MHz also showed the resonance pattern characteristic of compounds consisting of polypeptide chain. From the results, the spectrum of callus sample was very similar to that of leaf sample (Fig. 29.3).

29.5 Conclusion

On the basis of the results of the present study, it was concluded that the extraction with 90% methanol under continuous hot extraction in Soxhlet apparatus gave the maximum yield of gymnemic acid. The gymnemic acid thus obtained can be further identified, purified, and characterized using TLC, HPLC, SDS-PAGE, and NMR techniques. HPLC and NMR methods are found to be accurate, precise, and less time-consuming, and hence, it can be used for analysis of gymnemic acid and for standardization of herbal drugs in general laboratory conditions. These parameters could be useful in preparation of herbal drugs.

References

Bishayee A, Chatterjee M (1994) Hypo-lipidaemic and antiatherosclerotic effects of oral *Gymnema sylvestre* R. Br. Leaf extract in albino rats fed on a high fat diet. Phytother Res 8:118–120

Borris RP (1996) Natural products research: perspectives from a major pharmaceutical company. J Ethnopharmacol 51:29–38

Dateo GP, Long L (1973) Gymnemic acid, the antisaccharine principle of *Gymnema sylvestre*. Studies on isolation and heterogeneity of gymnemic acid A1. J Agric Food Chem 21:899–903
Edward PC (1950) Pharmacognosy, 3rd edn. Lea and Febiger, Philadelphia, PA, p 731
Gamble JS (1991) Flora of the Presidency of Madras, vol I–III, Reprinted edn. Bishen Singh Mahenra Pal Singh, Dehradun
Gent JF, Thomas P, Hettinger TP (1999) Taste confusions following gymnemic acid rinse. Chem Senses 24:393–403
Hooper D (1887) Isolation and antiviral activity of gymnemic acid. Pharm J Trans 17:867–868
Imoto T, Miyasaka A, Ishima R, Akasaka K (1991) A novel peptide isolated from the leaves of *Gymnema sylvestre*. 1. Characterisation and its suppressive effect on the neural responses to sweet taste stimuli in the rat. Comp Biochem Physiol 100:309–314
Kalidass C, Shanmugasundaram R, Mohan VR (2009a) Pharmacognostic studies of *Capparis sepiaria* (L.) R.Br. Pharmacogn J 1(2):121–125
Kalidass C, Muthukumar K, Mohan VR, Manickam VS (2009b) Ethnoveterinary medicinal uses of plants from Agasthyamalai Biosphere Reserve (KMTR), Tirunelveli, Tamil Nadu, India. My Forest 45(1):7–14
Kapoor LD (1990) CRC handbook of ayurvedic medicinal plants. CRC Press, Boca Raton, FL, pp 200–201
Komalavalli N, Rao MV (2000) *In vitro* propagation of *Gymnema sylvestre*—a multipurpose medicinal plant. Plant Cell Tissue Organ Cult 61(2):97–105
Kurihara Y (1992) Characteristics of anti-sweet substances, sweet proteins and sweetness inducing protein. Crit Rev Food Sci Nutr 32:231–252
Laemmli UK (1970) Cleavage of structural proteins during the assembly of the head of bacteriophage T4. Nature 227(5259):680–685
Liu HM, Kiuchi F, Tsuda Y (1992) Isolation and structure elucidation of Gymnemic acids, antisweet principles of *Gymnema sylvestre*. Chem Pharm Bull 40:1366–1375
Masayuki Y, Toshiyuki M, Hisashi M (1997) Structures of new triterpene glycosides, gymnemasides C, D, E and F from the leaves of *Gymnema sylvestre* R Br.: influence of Gymnema glycosides on glucose uptake in rat small intestinal fragments. Chem Pharm Bull 45:2034–2038
Matthew KM (1991) An excursion flora of Central Tamil Nadu, India. Oxford and IBH Publishing Co., New Delhi
Moerman DE (1996) An analysis of the food plants and drug plants of native North America. J Ethnopharmacol 52:1–22
Nadkarni AK, Nadkarni KM (1976) The Indian materia medica. Popular Prakashan, Bombay, pp 430–431
Nakamura Y, Tsumura Y, Tonogai Y, Shibata T (1999) Fecal steroid excretion is increased in rats by oral administration of gymnemic acids contained in *Gymnema sylvestre* leaves. J Nutr 129:1214–1222
Ozarkar KR (2005) Studies on anti-inflammatory effects of two herbs *Cissus quadrangularis* L. and *Valeriana wallichii* DC. Using mouse model. PhD thesis, University of Mumbai, Mumbai
Persaud SJ, Al-Majed H, Raman A, Jones PM (1999) Gymnema sylvestre stimulates insulin release in vitro by increased membrane permeability. J Endocrinol 163:207–212
Sahu NP, Mahato SB, Sarkar SK, Poddar G (1996) Triterpenoid Saponins from Gymnema sylvestre. Phytochemistry 41(1):1181–1185
Sative RK, Abhilash P, Fulzele DD (2003) Antimicrobial activity of *Gymnema sylvestre* leaf extract. Fitotherapia 74:699–701
Shanmugasundaram KR, Panneerselvam C, Samudram P, Shanmugasundaram ER (1983) Enzyme changes and glucose utilization in diabetic rabbits the effect of *Gymnema sylvestre* R Br. J Ethnopharmacol 7:205–234
Shimizu K, Iino A, Nakajima J, Tanaka K, Nkajyo S, Urakawa N, Atsuchi M, Wada T, Ymashita C (1997) Suppression of glucose absorption by some fractions extracted from *Gymnema sylvestre* leaves. J Vet Med Sci 59(4):245–251
Stocklin W (1969) Chemistry and physiological properties of gymnemic acid, the antisaccharine principle of the leaves of *Gymnema sylvestre*. J Agric Food Chem 17:704–708

30 GC-MS Profile of the Unsaponifiable and Saponifiable Matters of *Coldenia procumbens* Linn. Leaves

Gundabolu Usha Rani, Thirupathi Azmeera, Shiva Shanker Kaki, and Bhattiprolu Kesava Rao

Abstract

A lipoid is an oily organic compound soluble in organic solvents which are essential structural components of all living cells. The lipoidal matter, i.e., unsaponifiable and saponifiable fractions of hexane extract of *Coldenia procumbens* Linn. leaves, was characterized by gas chromatography and mass spectrometric (GC-MS) analysis. The unsaponifiable matter was found to contain sterols (61.06%) and hydrocarbons (34.4%), while the saponifiable matter was found to contain about 20 compounds composed of 49.43% saturated fatty acid methyl esters and 50.57% unsaturated fatty acids.

Keywords

Coldenia procumbens Linn. · GC-MS · Steroids · Hydrocarbons

30.1 Introduction

Coldenia procumbens Linn. (Pullaiah and Ali Moulali 1998) grows like an annual herb, and is a common weed in India (Nadkarni and Nadkarni 1955) that belongs to Boraginaceae (Ge-Ling et al. 1995) family, which has around 150 genera and almost 2500 species across the globe. It is found widely in South India on wastelands and common in dry paddy field grounds. The genus has 24 species of plants, and this plant is reported to be widely used in traditional medicines in India, Africa, and Malaysia (The Wealth of India 1950). *Coldenia procumbens* Linn. is the only

G. Usha Rani · B. Kesava Rao (✉)
Department of Chemistry, Acharya Nagarjuna University, Guntur, Andhra Pradesh, India

T. Azmeera · S. S. Kaki
Centre for Lipid Research, CSIR-Indian Institute of Chemical Technology, Hyderabad, Telangana, India

S. M. Khasim et al. (eds.), *Medicinal Plants: Biodiversity, Sustainable Utilization and Conservation*, https://doi.org/10.1007/978-981-15-1636-8_30

species of its genus which has a place both in the Hortus engalensi's and Moon's Catalogue of Ceylon plants (WhiteLaw Anisile 1826).

The plant is known to be efficacious in treating fever, piles, and scorpion stings. In the traditional system of medicine, the plant was reported to be used as anti-inflammatory (Arul et al. 2005), antimicrobial (Beena 2005), analgesic (Senthamarai et al. 2001), antidiabetic (Patel et al. 2007), and CNS depressant (Naga Rani et al. 1991). Fresh leaves of *Coldenia procumbens* Linn. are reported to be powdered and applied to rheumatic swellings, and equal parts of dried powder mixed with seeds of fenugreek is known to cause suppurations of boils (Chopra et al. 1958). Acetone, water, and methanolic extract of dried aerial parts showed weak angiotensin-converting enzyme inhibition in vitro (Schmelzer et al. 2008; Aleemuddin et al. 2011).

The preliminary investigation of this plant has shown the presence of flavonoids, carbohydrates and glycosides, steroids, and alkaloids (Senthamari et al. 2002). The active constituents like coumestan derivative wedelolactone (Beena et al. 2011) and rare cyano glucosides (Niranjan Kumar et al. 2013) were also extracted from this plant. A GC-MS analysis of volatile components of leaves of *Coldenia procumbens* Linn. has shown 20 compounds with 9,12-octadecadienoic acid and hexadecanoic acid as major compounds (Usha Rani and Kesava Rao 2016). Whereas, the hexane extract of leaves of *Coldenia procumbens* Linn. was not explored for the chemical composition of lipoidal matter till now. The present study completely describes the chemical composition of lipoidal matter and their concentrations in hexane extract of *Coldenia procumbens* Linn. for the first time.

30.1.1 Experimental

The aerial parts of the *Coldenia procumbens* Linn. were collected at Nunna near Vijayawada, Andhra Pradesh, India, from moist place of agricultural land. The plant was authenticated by Prof. VS Raju, and voucher specimen was kept in the Department of Botany, Kakatiya University, and Warangal, India, with accession number 1877.

Hexane, ethyl acetate, and methanol were purchased from Avra Synthesis Pvt. Ltd., (Hyderabad), and potassium hydroxide, sodium sulfate, diethyl ether, and sulfuric acid were supplied by Merck Life Science Private Limited (Mumbai). Silica gel G for thin-layer chromatography was supplied by Acme Synthetic Chemicals (Mumbai).

30.1.2 Extraction of Lipoidal Matter

The air-dried and powdered leaves (3.5 kg) of *Coldenia procumbens* Linn. were exhaustively extracted by hexane by Soxhlet extraction technique. The solvent was concentrated under reduced pressure using a rotary evaporator to get the crude extract (76 g).

30.1.3 Saponification of Lipoidal Matter

30.1.3.1 Preparation of Unsaponifiable Matter

Hexane extract weighing 25 g was saponified with 75 ml of 3% alcoholic KOH by refluxing it for 6 h. After cooling, alcohol was completely removed and the residue was diluted with distilled water. Then, the unsaponifiable matter was extracted with diethyl ether and dried over anhydrous sodium sulfate. Evaporation of ether completely afforded the unsaponifiable matter of about 10 g (Johnson and Davenport 1971).

30.1.3.2 Separation of Unsaponifiable Matter

The crude unsaponifiable matter was applied on TLC silica plates and developed with hexane and ethyl acetate in 8:2 (v/v) ratios. The plates were taken out from the chamber and visualized under UV light to identify and mark the bands. Eight bands were identified, marked, and scraped from the plates with a spatula. The bands were dissolved in ethyl acetate and filtered through vacuum to get the pure compounds separately. All the eight bands were concentrated and then analyzed by GC-MS, while the seventh band was fluorescent and phytosterols were separated as white needlelike crystals.

30.1.3.3 Preparation of Saponifiable Matter

The aqueous alkaline layer remained after the removal of unsaponifiable matter was acidified with sulfuric acid (10%) and the liberated fatty acids were extracted with ether, and then washed with water until neutral and dried over anhydrous sodium sulfate followed by evaporation of ether to afford total fatty acids (TFA) residue (6 g) (Vogel 1966). The remaining 9 g of extract was left in water part when fractionated with diethyl ether.

30.1.4 Preparation of Fatty Acid Methyl Esters

The free fatty acids were converted to their methyl esters by refluxing them with 100 ml absolute methanol and 5 ml sulfuric acid for 1 h; the alcohol was distilled, and the residue was dissolved in water and then extracted several times with ether. The combined ethereal extracts were washed with water till free from acidity. The ethereal extract was concentrated, and the residue was dried over anhydrous sodium sulfate to get fatty acid methyl esters (FAME) (Elsaid and Amer 1965). The experimental procedure was shown in Fig. 30.1.

30.1.5 Gas Chromatography-Mass Spectrometric (GC-MS) Analysis

The GC-MS analysis was performed on an Agilent 6890N Gas Chromatograph equipped with an Agilent HP-1 ms capillary column (30 m × 25 mm id; 0.25 μm) and connected to Agilent Mass Spectrometer operating in EI mode (70 eV; *m/z* 50–650; source temperature, 230 °C; quadruple temperature, 150 °C). The column

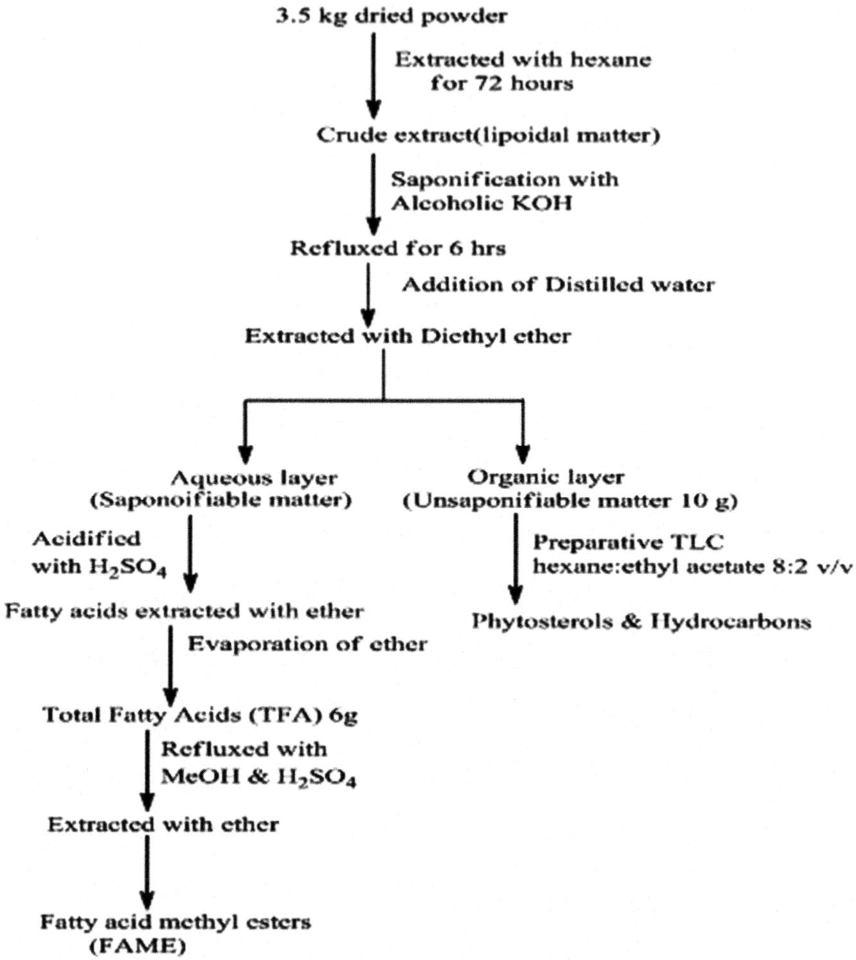

Fig. 30.1 Saponification of hexane crude extract of *Coldenia procumbens* Linn

temperature used was 150 °C for 2 min, increased to 300 °C at 10 °C/min, and maintained for 20 min at 300 °C. The carrier gas was helium at a flow rate of 1.0 ml/min. The inlet temperature was maintained at 280 °C and the split ratio was 50:1. Structural assignments were based on interpretation of mass spectrometric fragmentation and confirmed by comparison of retention times as well as fragmentation pattern of authentic compounds like sterols and alcohols and the spectral data obtained from the Wiley and NIST libraries.

30.2 Results and Discussion

The investigation of lipoidal matter of hexane extract of leaves of *Coldenia procumbens* Linn. was carried out by GC-MS which has shown higher percentage of phytosterols, like β-sitosterol, stigmasterol, and campesterol, unsaturated fatty acids, and saturated fatty acids, respectively.

Hexane extract of leaves of *Coldenia procumbens* Linn. was found to contain unsaponifiable matter and saponifiable matter in 40% and 24%, respectively. The results of GC-MS analysis of unsaponifiable matter of leaves of *Coldenia procumbens* Linn. are shown in Fig. 30.2, and the composition is given in Table 30.1. The analysis revealed the presence of total 19 compounds with identified compounds 94.81% and unidentified compounds with 5.19%. The identified compounds were classified as hydrocarbons (33.45%) and sterols (61.06%). Among all the compounds, β-sitosterol, stigmasterol, gamma tocopherol, and vitamin E were found maximum in unsaponifiable matter of hexane extract. The saponifiable matter of hexane extract of leaves of *Coldenia procumbens* Linn. is shown in Fig. 30.3. It was observed that the saponifiable matter was composed with saturated fatty acids with 49.43% and unsaturated fatty acids with 50.57%. Among the fatty acids identified,

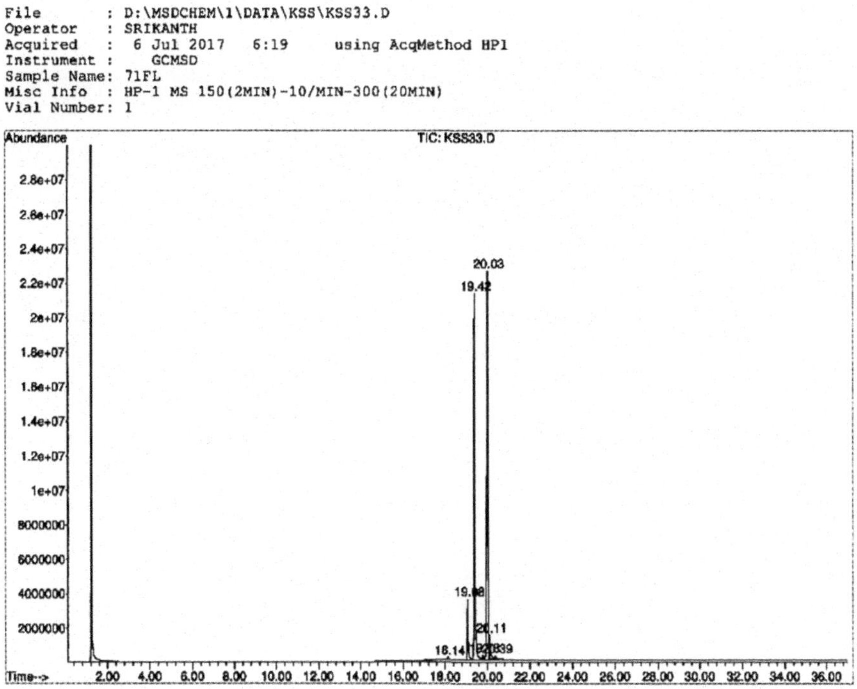

Fig. 30.2 GC chromatogram of phytosterols isolated from unsaponifiable matter of hexane extract of *Coldenia procumbens* Linn

Table 30.1 Chemical composition of unsaponifiable matter of hexane extract of leaves of *Coldenia procumbens* Linn

S. no.	Compounds	Structure	MF	MW	RT	Relative area %
1	9-Octadecene		$C_{18}H_{36}$	252	5.32	0.30
2	Hexadecane		$C_{16}H_{34}$	226	5.42	0.95
3	1-Octadecene		$C_{18}H_{36}$	252	7.40	0.38
4	Octadecane	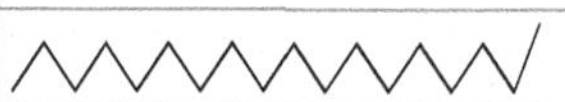	$C_{18}H_{38}$	254	7.50	2.74
5	2-Pentadecanone	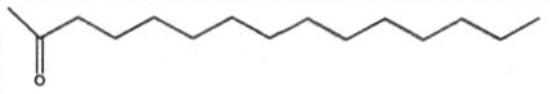	$C_{15}H_{30}O$	226	7.82	12.89
6	2-Heptadecanone	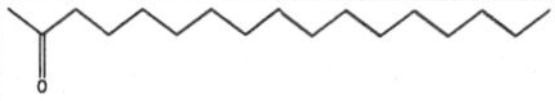	$C_{17}H_{34}O_2$	270	8.34	0.89
7	5,9,13-Pentadecatrien-2-one	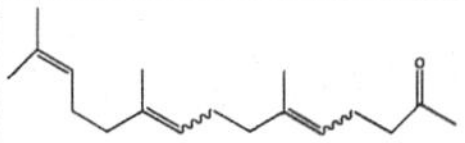	$C_{15}H_{24}O$	220	8.46	3.68
8	3-Eicosene		$C_{20}H_{40}$	280	9.41	0.31
9	13-Octadecenal	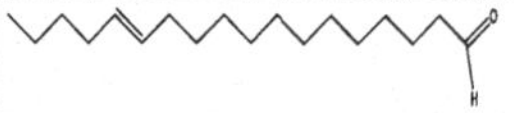	$C_{18}H_{34}O$	266	9.66	0.74
10	2-Nonadecanone	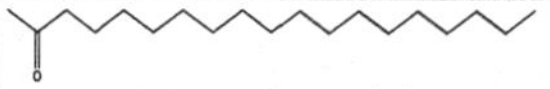	$C_{19}H_{38}O$	282	10.31	3.01
11	9-Octadecyne	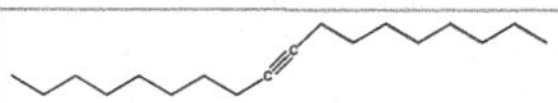	$C_{18}H_{34}$	250	10.48	8.16
12	9-Eicosyne		$C_{20}H_{38}$	278	10.49	3.23
13	Phytol	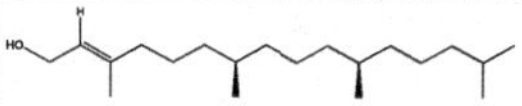	$C_{20}H_{40}O$	296	10.53	6.46
14	β-Tocopherol	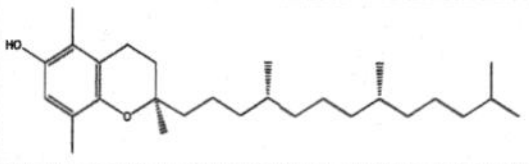	$C_{28}H_{48}O_2$	416	17.60	2.97
15	γ-Tocopherol	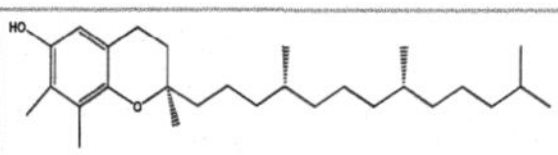	$C_{28}H_{48}O_2$	416	17.61	27.10
16	Vitamin E	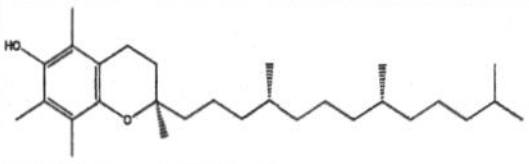	$C_{29}H_{50}O_2$	430	18.21	20.66
17	Campesterol	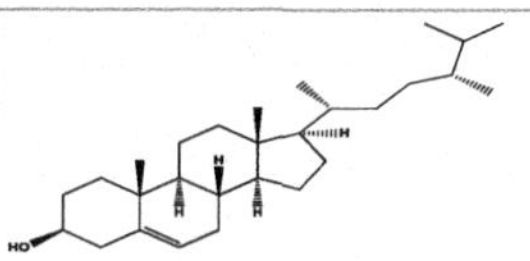	$C_{28}H_{48}O$	400	19.36	11.72
18	Stigmasterol	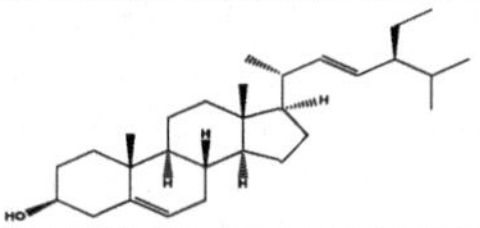	$C_{29}H_{48}O$	412	19.46	65.46

(continued)

Table 30.1 (continued)

S. no.	Compounds	Structure	MF	MW	RT	Relative area %
19	β-Sitosterol	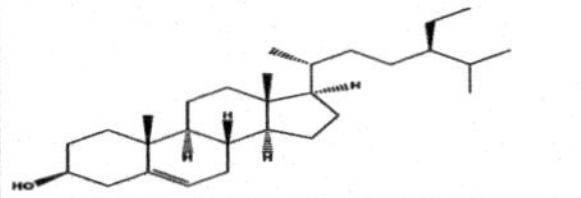	$C_{29}H_{50}O$	414	20.14	95.24
20	Unidentified				23.67	15.48
Total identified hydrocarbons						33.45
Total identified sterols						61.06
Total unidentified compounds						5.48

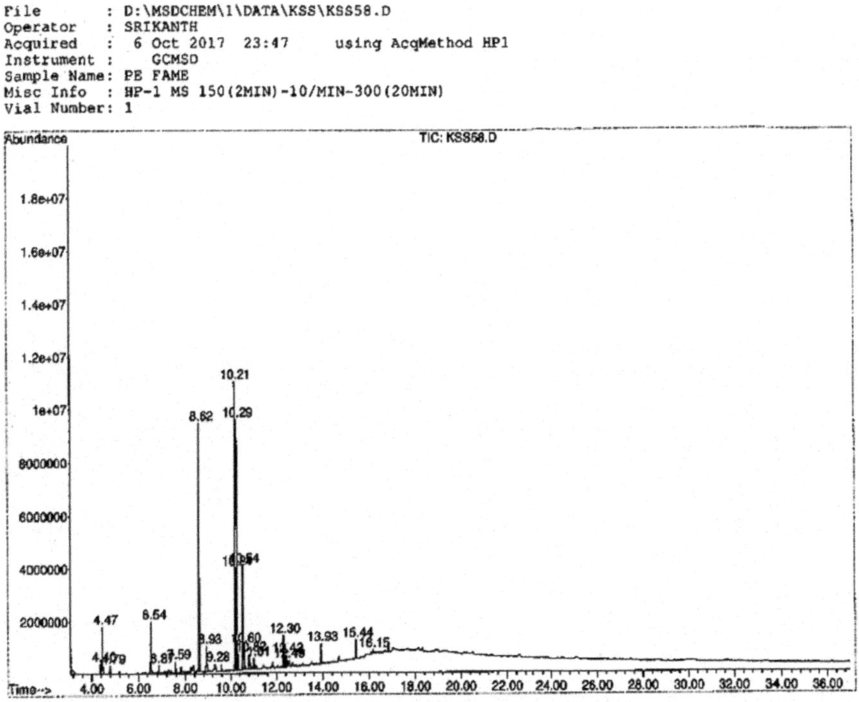

Fig. 30.3 Gas chromatogram of saponifiable matter of hexane extract of leaves of *Coldenia procumbens* Linn

palmitic acid and linoleic acid were found to be maximum with 17.65% and 25.28%, respectively, followed by other fatty acids.

The crude hexane extract of *Coldenia procumbens* Linn. analyzed by GC-MS summarizes the % composition of hydrocarbons with 8.9%, fatty acids with 67.90%, steroids with 13.31%, terpenoids with 8.585%, and fatty alcohols with 1.325% (Table 30.2). The higher concentration of sterols was found to be more in lipoidal matter than in crude hexane extract of *Coldenia procumbens* Linn., whereas the linoleic acid concentration was high in both the crude extract and the lipoidal matter.

Table 30.2 Chemical composition of saponifiable matter of hexane extract of leaves *Coldenia procumbens* Linn

S. no.	Compounds	Structure		Relative area %
1	Dodecanoic acid (lauric acid)	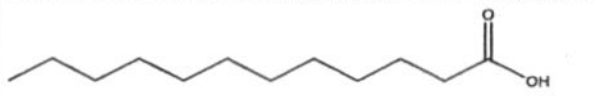	C(12:0)	3.39
2	Myristic acid methyl ester	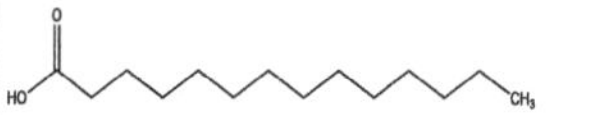	C(14:0)	3.85
3	Pentadecanoic acid methyl ester	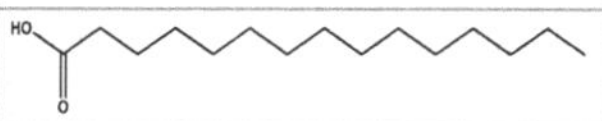	C(15:0)	0.78
4	Hexadecanoic acid (palmitic acid) methyl ester	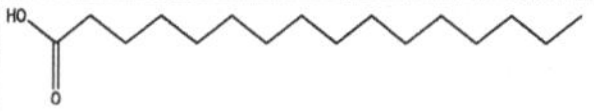	C(16:0)	17.65
5	Hexadecanoic acid ethyl ester	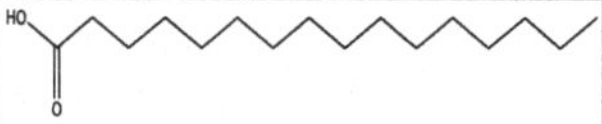	C(16:0)	0.13
6	9,12-Octadecadienoic acid methyl ester	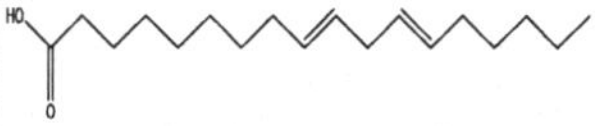	C(18:2)	25.28
7	9-Octadecenoic acid methyl ester	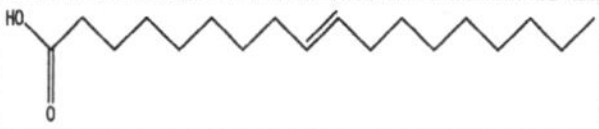	C(18:1)	17.85
8	Octadecanoic acid (stearic acid) methyl ester	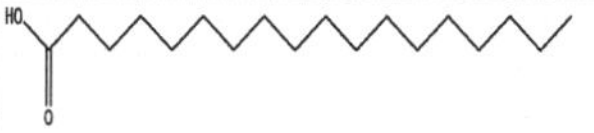	C(18:0)	9.56
9	Linoleic acid ethyl ester	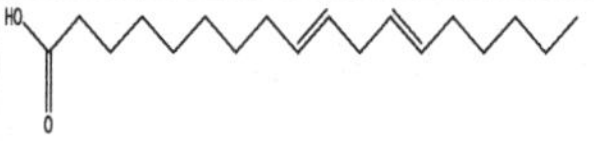	C(18:2)	1.46
10	Eicosanoic acid (arachidic acid) methyl ester	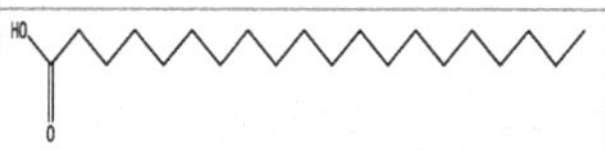	C(20:0)	2.29
11	Docosanoic acid (behenic acid) methyl ester	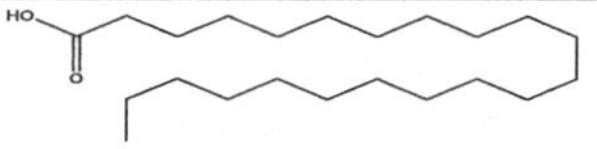	C(23:0)	1.54
12	Tetracosanoic acid (lignoceric acid) methyl ester	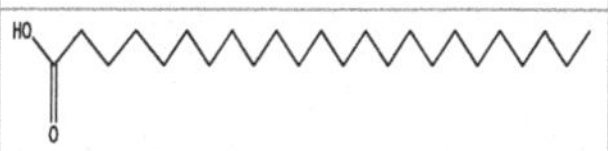	C(18:0)	1.47
Saturated fatty acids				49.43%
Unsaturated fatty acids				50.57%

This proves that the saponification of lipoidal matter will give better results in the separation of bioactive sterols from other compounds like hydrocarbons, fatty acids, and fatty alcohols. Some of the species of Boraginaceae family were also studied for the chemical composition of lipid matter like *Cordia* and *Echium* species, but the qualitative and quantitative determination of lipoidal matter was described in detail for the first time in the species *Coldenia procumbens* Linn.

The GC-MS analysis of *Echium wildpretii* revealed that the esters of hexadecanoic acid, octadecanoic acid, and hydrocarbons like pentacosane and hexacosane were the main compounds (Santana et al. 2012). Forty five phytochemicals were identified in GC-MS studies of *Cordia rothii* roots (Khan et al. 2016). Phytosterols and their derivatives are important products applied in pharmaceutical, food, and cosmetic industry due to their anti-inflammatory, antibacterial, antifungal, anti-ulcerative, and antitumor activities (Ling and Jones 1995).

The palmitic acid is reported to have antibacterial, antifungal, antioxidant, anti-inflammatory, and hypocholesterolemic effects (Agoramoorthy et al. 2007; Elagbar et al. 2016; Abubakar and Majinda 2016). The linoleic acid, i.e., an omega-6 fatty acid, which the body is not able to produce, can be consumed through diet. It is needed for growth and repair and possesses pharmacological activities like antioxidant and anti-inflammatory activities (Henry et al. 2002). Hence, the linoleic acid is responsible for anti-inflammatory activity of this species *Coldenia procumbens* Linn. This species is available as common weed and not edible as the leaves are covered with coarse hairs, i.e., trichomes and produce irritation, provoke sneezing when pulverized (Quisumbing et al. 1951). The pharmacological activities of these bioactive compounds from the lipoidal matter of the hexane extract of the *Coldenia procumbens* Linn. support the medicinal application of the plant leading to develop further biological and pharmacological studies.

Acknowledgments We are very much thankful to all the faculty members and the administration of the Department of Chemistry, Acharya Nagarjuna University. G. Usha Rani is very much thankful to UGC for sanctioning her UGC-BSR Fellowship, and we express our special thanks to Dr. Shiva Shanker Kaki, Senior Scientist-Center for Lipid Research, CSIR-IICT, Hyderabad, and Mr. A. Tirupathi, for getting GC-MS analysis results.

References

Abubakar MN, Majinda RRT (2016) GC-MS analysis and preliminary antimicrobial activity of Albizia adianthifolia and Pterocarpus angolensis. Medicines 3(1):3

Agoramoorthy G, Chandrasekaran M, Venkatesalu V, Hsu MJ (2007) Antibacterial and antifungal activities of fatty acid methyl esters of the blind-your-eye mangrove from India. Braz J Microbiol 38:739

Aleemuddin MA, Karthikeyan M, Rajasekar S (2011) *Coldenia procumbens* Linn.—a phytopharmacological review. Int J Pharm Sci Rev Res 11:133

Arul B, Kothai R, Suresh Kumar K, Christiana AJM (2005) Anti-inflammatory activity of *Coldenia procumbens* Linn. Pak J Pharm Sci 18(3):17–20

Beena P (2005) Anti bacterial activity of Coldenia procumbens. Anc Sci Life 24(3):1–3

Beena P, Purnima S, Kokilavani R (2011) Identification and estimation of Wedelolactone in Coldenia procumbens. Der Pharm Lett 3(4):320–324

Chopra RN, Nayar SL, Chopra IC (1958) Glossary of Indian medicinal plants. Council of Scientific and Industrial Research (CSIR), New Delhi, p 74

Elagbar ZA, Naik RR, Shakya AK, Bardaweel SK (2016) Fatty acids analysis, antioxidant and biological activity of fixed oil of *Annona muricata* L. seeds. J Chem 2016:1–6

Elsaid ME, Amer MM (1965) Oils, fats, waxes and surfactants. In: Anglo Egyptian Book Shop, Cairo, pp 130–132

Ge-Ling Z, Riedi H, Kamelin R (1995) Boraginaceae. Flora China 16:32

Henry GE, Momin RA, Nair MG, Dewitt DL (2002) Antioxidant and cyclooxygenase activities of fatty acids found in food. J Agric Food Chem 50:2231–2234

Johnson AR, Davenport JB (1971) Biochemistry and methodology of lipids. Wiley, New York, p 31

Khan K, Firdous S, Ahmad A, Fayyaz N, Nadir M, Rasheed M, Faizi S (2016) GC-MS profile of antimicrobial and antioxidant fractions from Cordia rothii roots. Pharm Biol 54:2597–2605

Ling WH, Jones PJH (1995) Dietary phytosterols: a review of metabolism, benefits and side effects. Life Sci 57(3):195

Nadkarni KM, Nadkarni AK (1955) Indian materia medica, vol 1, 3rd edn. Popular Book Depot, Bombay, p 371

Naga Rani MA, Vijayasekaran V, Lalitha K (1991) Central nervous system effects of Coldenia procumbens. Ind J Pharm 23:261

Niranjan Kumar A, Satya Srinivas KVN, Kotesh Kumar J, Sastry KP (2013) Rare cyano glucosides from *Coldenia procumbens* Linn. J Chem Pharma Res 5(12):1394–1397

Patel N, Raval S, Goriya H, Jhala M, Joshi B (2007) Evaluation of anti diabetic activity of Coldenia procumbens in alloxan-induced diabetes in rat. J Herb Pharmacother 7(1):13–23

Pullaiah T, Ali Moulali D (1998) Flora of Andhra Pradesh (India), vol 2. Scientific Publishers, Jodhpur, p 611

Santana O, Reina M, Fraga BM, Sanz J, Gonzalez-Coloma A (2012) Antifeedant activity of fatty acid esters and phytosterols from Echium wildpretii. Chem Biodivers 9:567

Schmelzer GH, Gurib-Fakim A, Arroo R, Bosch CH (2008) Plant resources of tropical Africa, Medicinal plants 1, vol 11(1). Backhuys Publishers, Leiden, pp 186–187

Senthamarai R, Kavimani S, Jaykar B, Uvarni M (2001) Analgesic activity of leaf extract of *Coldenia procumbens* Linn. Hamd Med 44(3):20–23

Senthamari R, Uvarani M, Jayakar B (2002) Pharmacognostical studies on leaf of *Coldenia procumbens* Linn. Anc Sci Life 2(1):67–75

The Wealth of India (1950) A dictionary of Indian raw materials and industrial products, vol IA–B. Council of Scientific and Industrial Research (CSIR), New Delhi, p 307

Usha Rani G, Kesava Rao B (2016) GC-MS analysis of volatile components in petroleum ether extracts of *Coldenia procumbens* Linn. Int J Pharm Bio Sci 7(2):241–245

Vogel A (1966) Textbook of practical organic chemistry, 3rd edn. Longman's Green and Co. Ltd., London, pp 133–136

WhiteLaw Ainslie (1826) Materia indica, vol II. Longman, Rees, Orme, Brown, and Green, Paternoster- Row, London, pp 435–436

Quisumbing E (1951) Medicinal plants of the Philippines, technical bulletin 16. Bureau of printing, Manila, p 1234

Isolation of Sterols from the Bark Hexane Extract of *Cordia dichotoma*

31

G. Swapna and Bhattiprolu Kesava Rao

Abstract

The plant *Cordia dichotoma*, which belongs to the family Boraginaceae, is known for its biological activities. The plant is considered as the medicinally valuable tree in the genus *Cordia*. The entire plant parts have many biological activities like diuretic, antibacterial, contraceptive, antidiabetic, and so many other pharmacological properties. This plant's bark powder is used to treat ulcers. The bark has antimicrobial and anti-implantation activities. GC-MS analysis of *Cordia dichotoma* bark shows hydrocarbons, fatty acids, phytosterols, and phthalates. Sterols are known to have hypercholesterolemic activity. The separation of sterols from hydrocarbons using special techniques and column chromatography is described in this paper.

Keywords

Column chromatography · Hydrocarbons · Phthalates and phytosterols

31.1 Introduction

The plant *Cordia dichotoma* belongs to the family Boraginaceae. It is native to India, Myanmar, and Nepal. Different synonyms are *Cordia myxa* Forsk, *Cordia oblique* Wild, *Cordia myxa* Roxb, etc. The scientific importance of this plant is that it is a highly medicinally valuable tree, but there is no scientific validation to prove that all the plant parts are medicinally valuable. It was authenticated by Dr. G.V.S. Murthy, Scientist BSI-Southern Region, Coimbatore, on 12-08-2015, and the plant identification number is given as 177158. The aim of this study was to isolate sterols from the bark of *C. dichotoma* hexane extract by column chromatography.

G. Swapna · B. Kesava Rao (✉)
Department of Chemistry, University College of Sciences, Acharya Nagarjuna University, Guntur, Andhra Pradesh, India

S. M. Khasim et al. (eds.), *Medicinal Plants: Biodiversity, Sustainable Utilization and Conservation*, https://doi.org/10.1007/978-981-15-1636-8_31

2.5 kg of *C. dichotoma* bark was collected from village Pulluru of Krishna district in the month of May 2017, shade dried, and made into coarse powder. The plant material was soaked with hexane for 48 h, and the cold hexane extract was collected. Cold extractions were done five times till the extract became colourless, and subsequently the cold extract was collected. The hexane was removed under vacuum distillation, and the crude sample was submitted to GC-MS.

Hot siphonings were done till the extract became colourless. The hexane extract was subjected for vacuum distillation, and the TLC was monitored for both cold and hot extracts; the TLC of cold and hot extract crude was found to be similar; therefore these two crudes were mixed. The total weight of cold and hot extracts crude was found to be 5 g.

The GC-MS analysis was done in IICT, Hyderabad. GC-MS analysis identifies the different number of compounds in a mixture and was used to analyse compounds in trace levels. Phytosterols are major components followed by fatty acids and hydrocarbons. Among the phytosterols, stigmasterol was the major sterol followed by β-sitosterol. In the fatty acids, palmitic, oleic, and linoleic acids were identified. The hydrocarbon range was observed from C14 to C26. Composition of the bark hexane extract by GC-MS is given in Table 31.1.

The given values are calculated on the basis of GC-MS spectrum of the *C. dichotoma* bark hexane extract. The GC-MS spectral analysis is performed only once; since the analysis is effective, there is no need to record the chromatogram many times. This spectrum gives information about the separated volatile and thermally stable compounds in the test sample. In the GC-MS chromatogram of the *C. dichotoma* bark hexane extract, the phytosterols are the major compounds. It combines gas chromatography and mass spectral analysis; the volatile compounds eluted in the GC undergo mass spectral analysis, which in turn gives the *m*/*z* values of the eluted compounds. So by seeing the GC-MS, one can easily know which molecular weight compounds are eluted at their retention times. As this paper mainly deals with the separation of the phytosterols, we limit this discussion about GC-MS at this point.

As the plant bark hexane extract crude consists of phytosterols, fatty acids, and phthalates, we chose the process of saponification to separate the fatty acids from the other two major components, i.e. from phytosterols and phthalates.

Table 31.1 Composition of bark hexane extract sample by GC-MS

Component	%
Hydrocarbons	15.0
Fatty acids	18.3
Phytosterols	47.5
Phthalates	16.8
Unidentified	2.4

31.1.1 Saponification of *C. dichotoma* Bark Hexane Crude

C. dichotoma bark hexane crude was subjected to the process of saponification; after getting the GC-MS data, the total hexane crude 5 g was dissolved in 50 ml of 20% methanolic KOH and kept at room temperature overnight and refluxed for 4 h at 70 °C. After 4 h the methanol in the reaction mass is distilled; to this crude 500 ml of distilled water is added. The water layer was extracted with ether thrice, each time with 100 ml of ether. The ethereal layers were combined and washed with distilled water till ether layer shows neutral on litmus paper. This is known as unsaponified matter. After this workup, the colour of the unsaponified layer became orangish red. Anhydrous sodium sulphate was added to this unsaponified matter and filtered, which gave positive test for steroids. The unsaponified matter was distilled, and the weight was found to be 0.9 g.

31.1.2 Column Chromatography of Unsaponified Matter

The unsaponified matter was subsequently coated with 100–200 mesh silica gel and loaded into the column, which was prepacked with 100–200 mesh silica gel using hexane as solvent. The fractions are as shown in Table 31.2.

The column 16th fraction was concentrated, added ethanol, and kept at 0 °C for 5 min. The compound was crystallized as white colour crystals, and the white crystals were filtered using cold ethanol and cold hexane. The filtered weight was found to be 76 mg. The code of this compound is B.H.C-16A. The code B.H.C-16A means the white crystals separated from the unsaponified fraction of bark hexane crude's column fraction number 16.

Table 31.2 Column fractions

S. No.	Fraction no.	Solvent	Quantity of solvent (ml)	Colour of the band
1	1 and 2	Hexane	200	No colour
2	3 and 4	Hexane(100 ml):EtOAC(2.5 ml)	200	Light yellow
3	5–11	Hexane(100 ml):EtOAC(5 ml)	400	Light yellow
4	12–15	Hexane(100 ml):EtOAC(10 ml)	100	Light yellow
5	16	Hexane(100 ml):EtOAC(10 ml)	200	Light yellow
6	17–18	Hexane(100 ml):EtOAC(12.5 ml)	200	Light yellow
7	19–23	Hexane(100 ml):EtOAC(15 ml)	600	Pale yellow
8	22–23	Hexane(100 ml):EtOAC(20 ml)	200	Pale yellow
9	24–25	Hexane(100 ml):EtOAC(25 ml)	200	Pale yellow
10	26	Hexane(100 ml):EtOAC(35 ml)	100	Pale yellow
11	27–28	Hexane(100 ml):EtOAC(55 ml)	100	Pale yellow
12	29	Hexane(100 ml):EtOAC(75 ml)	100	Pale yellow
13	30	Hexane(100 ml):EtOAC(80 ml)	50	Colourless
14	31–33	EtOAC (100 ml)	50	Orange
15	34–38	Methanol	500	Colourless

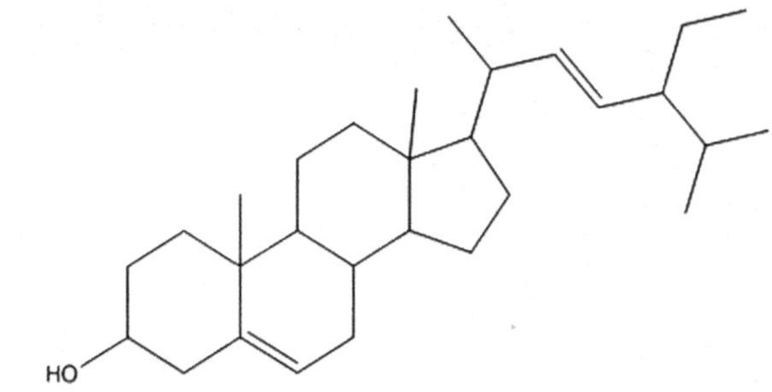

a. Stigmasterol

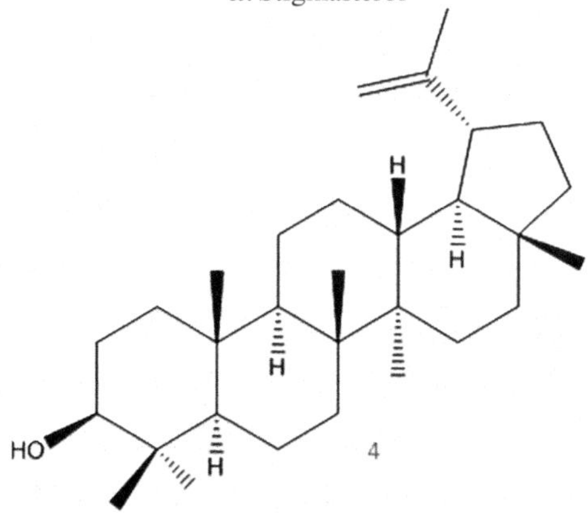

b.Lupeol

Fig. 31.1 (**a**) Stigmasterol and (**b**) lupeol

31.2 Results and Discussions

The B.H.C-16A fraction was given for mass spectral analysis. Mass spectrum analysis showed the presence of stigmasterol and lupeol. Structures of stigmasterol and lupeol are given in Fig. 31.1 and mass spectrum is given Fig. 31.2. The further fractions which were separated in column are under investigation.

The mass spectrum of B.H.C-16A is showing *m/z* values of fragment ions on *x*-axis and the relative abundances of the molecular ions on *y*-axis. The mass chromatogram of B.H.C-16A showed peaks and indicated the presence of β-stigmasterol and lupeol at M+1413.4 and 427.4, respectively. The spectrum was showing loss of water molecule (M-H_2O) peak from stigmasterol by giving peak at 395 *m/z*, (M-CH_3) group at 413 *m/z*. The remaining fractions were separated in column under investigation. Structures of stigmasterol and lupeol are given in Fig. 31.1 and mass spectrum is given Fig. 31.2.

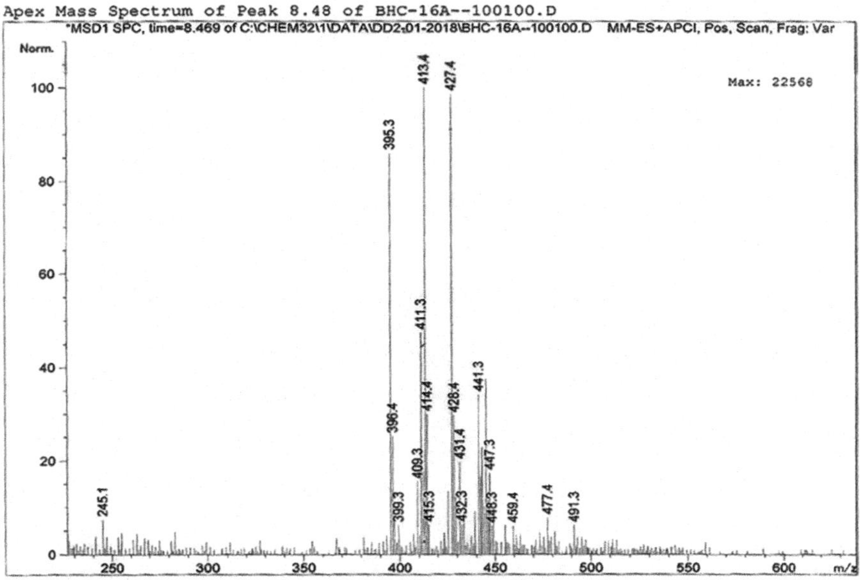

Fig. 31.2 Mass spectrum of 16A

Table 31.3 Biological activities of stigmasterol

Biological activity	Mechanism of action
Cytotoxic activity	It acts against Hep-2 and McCoy cells
Antioxidant activity	It increases the activity of antioxidant enzymes
Anti-hypercholesterolemic activity	Competes with cholesterol for intestinal absorption
Anti-inflammatory activity	Reduces the oedema induced by 12-*O*-tetradecanoyl phorbol acetate

31.2.1 Biological Significance of Stigmasterol

Plant sterols are known to lower the plasma total and LDL cholesterol. Due to this fact, many food products, mostly margarines, have been enriched with phytosterols and used in management of moderate hypercholesterolemia.

The biological activities of stigmasterol are listed in Table 31.3.

Stigmasterol is the precursor for the synthesis of semisynthetic progesterone. It has cytotoxic activity against Hep-2 and McCoy cells (Gomez and Garcia 2001). It shows antioxidant property as it increases the activities of catalase and superoxide dismutase (Panda et al. 2009). It also exhibits anti-hypercholesterolemic activity as it is found to compete with cholesterol for intestinal absorption and lowers the plasma inhibition of cholesterol (Batta et al. 2006). Stigmasterol also reduces the oedema induced by 12-*O*-tetradecanoyl phorbol acetate in rat (Garcia et al. 1999).

Table 31.4 Biological activities of lupeol

Biological activity	Mechanism of action
Anti-inflammatory activity	It decreases the interleukin-4 production by T2 cells
Antitumour activity	It inhibits topoisomerase II
Anti-leukaemia activity	It induces apoptosis in melanomas and leukaemias
Cardioprotective activity	Prevents LDL oxidation

31.2.2 Biological Significance of Lupeol

Lupeol has anti-inflammatory activity, and it decreases the interleukin-4 production by T2 cells (Bani et al. 2006). It also shows the antitumour activity; it is due to its ability to inhibit topoisomerase II. Lupeol can be a lead molecule for many potent antiviral drugs (Debra et al. 1998). Lupeol and its derivatives have anti-leukaemic properties. It induces apoptosis in melanomas and leukaemias (Liu et al. 2004). Lupeol has also been investigated, and it provides 34.4% protection against in vitro low-density lipoprotein (LDL) oxidation. Lupeol and lupeol acetate have also shown hypotensive activity (Saleem et al. 2003). Lupeol is known to have many biological activities; some of them are listed below and shown in Table 31.4.

31.3 Conclusion

The plant bark *C. dichotoma* has unexplored potential of having many biological properties, which is being used by folklore. The mass spectra of these compounds showed stigmasterol and lupeol account for its biological activities scientifically. Further this work supports the anti-oxidant, anti-inflammatory, and cytotoxic activity of *C. dichotoma* plant crude.

Acknowledgements Mrs. G. Swapna is expressing her heartfelt thanks to U.G.C. BSR for providing her financial assistance through BSR-JRF Fellowship, Acharya Nagarjuna University, and all the faculty members of Department of Chemistry for providing the facilities in our research lab.

References

Bani S, Kaul A, Khan B, Ahmad SF, Suri KA (2006) Suppression of T lymphocyte activity by lupeol isolated from Crataeva religiosa. Phytother Res 20:279–287

Batta AK, Xuao G, Honda A, Miyazaki T, Salen G (2006) Stigmasterol reduces plasma cholesterol and inhibit hepatic synthesis and intestinal absorption in rat. J Pharm Sci 55(3):292–299

Debra M, Huang J, Yancey CA, Jhang P, Setzer WN (1998) Lupeol is the cytotoxic principle in the leaf extract of Dendropanax cf. querceti. Planta Med 64(4):370–372

Garcia MD, Saenz MT, Gomez MA, Fernandez MA (1999) Topical anti-inflammatory activity of phytosterols isolated from Eryngium foetidum on chronic and acute inflammatory models. Phytother Res 13(1):78–80

Gomez MA, Garcia MD (2001) Cytostatic activity of Achillea ageratum. Phytother Res 15(9):633–634

Liu W, Ho JC, Cheung FW, Liu BP, Ye WC, Che CT (2004) Apoptotic activity of betulinic acid derivatives on murine melanoma B16 cell line. Eur J Pharmacol 498:71–78

Panda S, Jafri M, kar A, Mehata BK (2009) Thyroid inhibitory, anti-peroxidase and hypoglycemic effects of stigmasterol isolated from Butea monosperma. Fitoterapia 80(2):123–126

Saleem R, Ahmad SI, Ahmed M (2003) Hypotensive activity and toxicology of constituents from Bombax ceiba stem bark. Biol Pharm Bull 26:41–46

Isolation and Characterization of Pharmacologically Active Tannins from Stem Bark of *Syzygium samarangense*

32

P. Asha Bhanu, M. Mohan Krishna Reddy,
N. Sadhana Reddy, and Bhattiprolu Kesava Rao

Abstract

Tannins are high-molecular-weight and water-soluble naturally occurring polyphenols that are widely distributed in many species of plants at higher levels. *Syzygium samarangense* (Blume) Merr. & L.M. Perry is a famous deciduous tree belonging to Myrtaceae family, which is being cultivated in India mainly for its edible fruits. It is also well known by other names, e.g. java apple and water apple. Based on literature survey, we found that *S. samarangense* is having a rich content of tannins. In this study, we have isolated two compounds, i.e. methyl gallate and 3-methoxy ellagic acid, from acetone extract of stem bark of *S. samarangense* by using two different methods; and further, these fractions were individually separated and purified by using RP-HPLC. The structures of these purified compounds were confirmed by using different spectral techniques.

Keywords

Myrtaceae · RP-HPLC · *Syzygium samarangense* · Tannins

P. A. Bhanu · B. Kesava Rao (✉)
Department of Chemistry, Acharya Nagarjuna University, Guntur, Andhra Pradesh, India

M. Mohan Krishna Reddy
Analytical Chemistry and Mass Spectrometry Division, CSIR—Indian Institute of Chemical Technology, Hyderabad, Telangana, India

N. Sadhana Reddy
Natural Products Division, CSIR—Indian Institute of Chemical Technology, Hyderabad, Telangana, India

S. M. Khasim et al. (eds.), *Medicinal Plants: Biodiversity, Sustainable Utilization and Conservation*, https://doi.org/10.1007/978-981-15-1636-8_32

32.1 Introduction

According to Bate-Smith and Swain (1962), vegetable tannins were defined as water-soluble phenolic compounds having a molecular weight ranging 500–3000 Da. These polyphenols contain a large number of hydroxyl or other functional groups and are capable of forming cross-linkages with proteins and other macromolecules (Bate-Smith and Swain 1962). The name tannins came from the ancient Celtic word for oak, a typical source of tannins for leather making. According to Haslam (1989), the term polyphenol for tannins is to emphasize the multiplicity of phenolic groups which is the major characteristic nature of these compounds. He notes that molecular weight as high as 20,000 Da has been reported and that tannins form complexes not only with proteins and alkaloids but also with polysaccharides. The name tannins was given to plant extracts exhibiting astringency without knowing their chemical structure. The features distinguishing tannins from plant polyphenols of the other types are binding to proteins, basic compounds, pigments, large molecular compounds, and metallic ions and also antioxidant activities. These features of tannins lead to qualitative and quantitative differences between tannins and other polyphenols. High tannin concentrations are found in nearly all parts of the plant, i.e. bark, wood, leaves, fruits, roots, and seed. Tannins are responsible for the astringent taste of the many fruits and vegetables with a number of OH groups (Haslam 1989). If the number of OH increases, astringency is increased. Increased tannin production can be associated with sickness of the plant (Khanbabaee and van Ree 2001). Therefore, it is assumed that the biological role of tannins in plants is related to protection against infection, insects, or animal herbivore (Porter 1989).

32.2 Astringency

Astringency plays an important role in the sensory experience of many foods and beverages ranging from wine to nuts. Astringency can be described as a dry, puckering, mouth feel sensation that can be elicited by eating unripe fruits that contain mainly tannins which bind to the salivary proteins causing them to precipitate or aggregate and leading to a rough or dry sensation in the mouth. Astringency is a sensory characteristic of food and beverages rich in polyphenols mainly resulting from the formation of complexes between polyphenols and salivary proteins causing a reduction of the lubricating properties of saliva (Bajec and Pickering 2008). Astringency is attributed to the presence of polyphenols mainly catechins such as epigallocatechin gallate. Saliva contains diverse types of proteins such as α-amylase, lactoferrin, mucins, and some small peptides (Tabak 1995).

32.2.1 Classification of Tannins

Due to the enormous structural diversity of tannins, a systematic classification system based on specific structural characteristics and chemical properties would

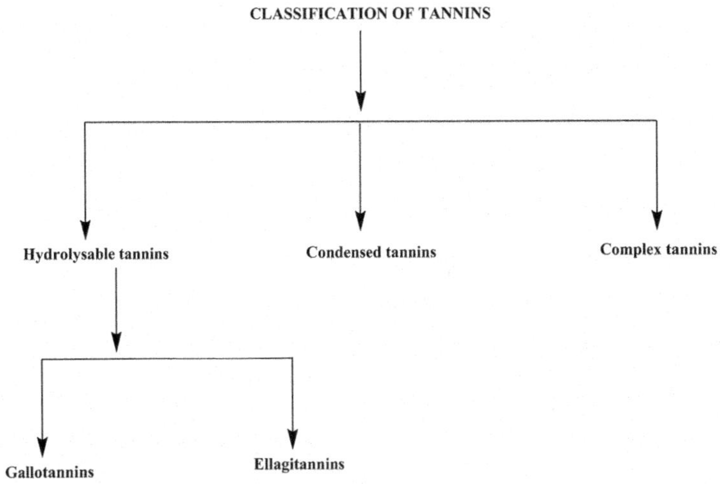

Fig. 32.1 Classification of tannins

provide a convenient framework for further study. There are three major classes of tannins (Khanbabaee and van Ree 2001), namely, (1) hydrolysable tannins, (2) condensed tannins, and (3) complex tannins (Fig. 32.1).

32.2.1.1 Hydrolysable Tannins

The hydrolysable tannins are hydrolysed by weak acids or weak bases to produce carbohydrate and phenolic acids. At the centre of a hydrolysable tannin molecule, there is a carbohydrate (usually D-glucose). The hydroxyl groups of the carbohydrate are partially or totally esterified with phenolic groups such as gallic acid (in gallotannins) or ellagic acid (in ellagitannins). The hydrolysable tannins on dry distillation yield pyrogallol and on reaction with $FeCl_3$ produce bluish-black precipitate. These are further subdivided into gallotannins and ellagitannins on the basis of their product of hydrolysis. Pentagalloylglucose (PGG) (Fig. 32.2) is a basic unit of

Fig. 32.2 Pentagalloylglucose (PGG)

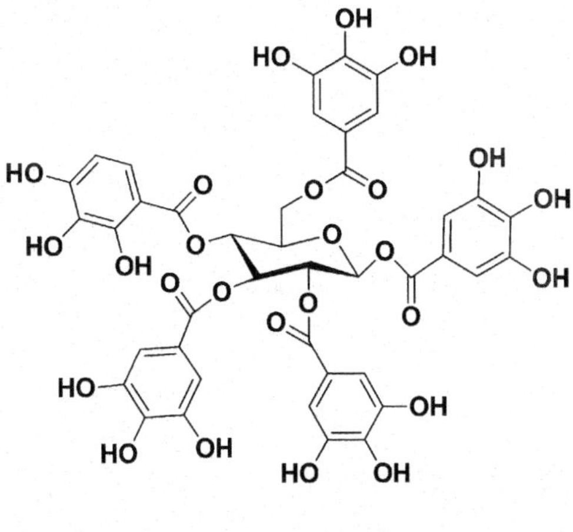

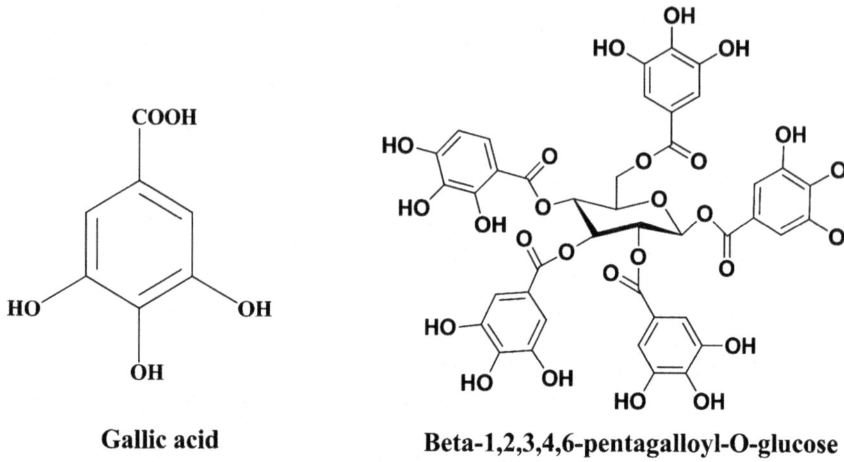

Fig. 32.3 Gallic acid, beta-1,2,3,4,6-pentagalloyl-*o*-glucopyranose

the hydrolysable tannins from which other molecules are derived. The hydrolysable tannins are comprised of phenolic acids such as gallic acid or hexahydroxydiphenic acid (HHDP).

Gallotannins: The gallotannins are the simplest hydrolysable tannins formed, when a polyphenol monomer, i.e. gallic acid, esterifies and binds with hydroxyl group of polyol carbohydrate, i.e. glucose. On hydrolysis, gallotannins yield gallic acids and glucose molecule. The prototypical gallotannin is pentagalloylglucose (β-1,2,3,4,6-pentagalloyl-*o*-glucopyranose) (Fig. 32.3) which has five identical ester linkages that involve aliphatic hydroxyl groups of core sugar. Pentagalloylglucose has many isomers, and molecular weights of all isomers are the same, but chemical properties such as sensitivity to hydrolysis, chromatographic behaviour, and ability to precipitate protein are structure dependent (Arbenz and Avérous 2015).

Ellagitannins: Ellagitannins are formed from gallotannins by intermolecular carbon-carbon coupling between two galloyl units and forms a hexahydroxydiphenic acid (HHDP) and spontaneously lactonizes to ellagic acid (Fig. 32.4) (Landete 2011).

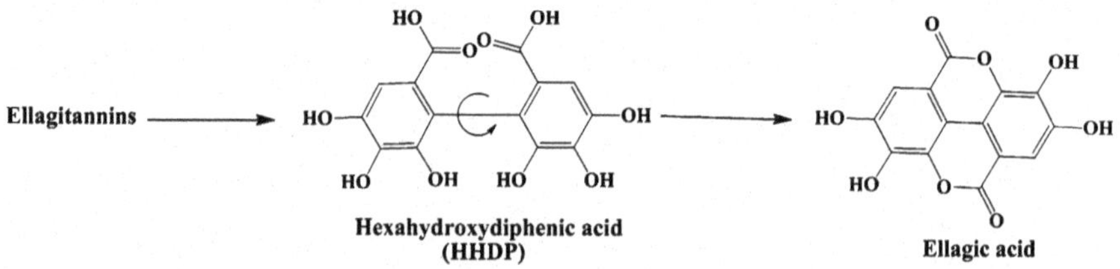

Fig. 32.4 Mechanism of formation of ellagic acid

32.2.1.2 Condensed Tannins

The condensed tannins are oligomeric and polymeric proanthocyanidins consisting of coupled flavan-3-ol (catechin) units with molecular weight ranging between 1000 and 3000 Da. Condensed tannins are resistant to hydrolysis, which do not contain any sugar molecule and are the derivatives of flavones like catechin, flavan-3-ol, flavan-3,4-diol, etc. Condensed tannins on treatment with acids or enzymes converted to a red water-insoluble compound called phlobaphene which imparts typical brownish colour to many of the plant materials, mainly bark. In condensed tannins there is a coupling of C-4 of the first unit linked with C-8 (or C-6) of the second unit, which may have a different substitution pattern (Schofield et al. 2001).

32.2.1.3 Complex Tannins

These are formed by the combination of hydrolysable and condensed tannins. The formation of complex tannins occurs by the union of C–C bond between C1 of glucose unit of ellagitannins and C8 or C6 of flavones-3-ol (Okuda 2005). They are present in dicotyledon plants, mainly in the subclasses Rosidae and Dilleniidae (Okuda 2005).

32.2.2 Medicinal Properties of Tannins

The therapeutic or medicinal activity of tannins is due to the astringent properties. Astringent herbs draw together or constrict body tissues and are effective in stopping the flow of blood or other secretions. They are tightening, healing and drying, reducing irritation and inflammation and creating a barrier against infection that is helpful in wounds and burns. Tannins not only heal burns and stop bleeding, but they also stop infection while they continue to heal the wound internally. The ability of tannins to form a protective layer over the exposed tissue keeps the wound from being infected even more. Tannins can also be effective in protecting the kidneys. Tannins have been used for immediate relief of sore throats, diarrhoea, dysentery, haemorrhaging, fatigue, and skin ulcers. Tannins can cause regression of tumours that are already present in tissue, but if used excessively over time, they can cause tumours in healthy tissue (Pankaj Jain et al. 2013). These were also reported to have anti-HIV (Lin et al. 2004), antibacterial (Akiyama et al. 2001), and antimicrobial activity (Funatogawa et al. 2004).

Syzygium samarangense (syn. *Eugenia javanica*) is a plant species in the Myrtaceae, native to an area that includes the Malay Peninsula and the Andaman Nicobar Islands but introduced in prehistoric times in a wider area and is now under cultivation in different states of India for their edible fruits (Morton 1987). It is also called as wax apple or java apple and in Telugu is well known as gulaabijaamichettu. The plant has been reported to have analgesic and anti-inflammatory (Raga et al. 2011), immunopharmacological (Kuo et al. 2004), antidiarrhoeal (Ghayur et al. 2006), and antioxidant activities (Soubir 2007). The previous chemical work reveals the presence of triterpenoids and chalcones (Srivastava et al. 1995), tannins (Nonaka et al. 1992), flavonol glycosides (Nair et al. 1999), and multi-metals (Kesava Rao et al. 2016).

Fig. 32.5 (**a**) Flowers and leaves and (**b**) stem bark of *Syzygium samarangense*

The fully mature stem bark (Fig. 32.5) of plant material was collected by us from Tirumala Hills in Tirupati, Chittoor District, Andhra Pradesh. The plant was authenticated by Dr. G.V.S. Murthy on April 28, 2015, and voucher specimen is kept in Madras Herbarium (MH) in Botanical Survey of India (BSI), Coimbatore, India. He confirmed the plant as *Syzygium samarangense* (Blume) Merr. & L.M. Perry with accession number 177157. The stem bark was dried at room temperature and made into fine powder with an electrical blender.

32.2.3 Extraction and Isolation

Stem bark powder (1 kg) was sequentially extracted with hexane, dichloromethane, ethyl acetate, acetone, and methanol in an aspirator. The excess of solvent was distilled off. After the distillation, the acetone extract gives a dark red colour liquid. The dark red colour liquid obtained from acetone extract gave a positive result for tannins, i.e. it has given a bluish-black colour precipitate with $FeCl_3$. It is an indication for the presence of hydrolysable tannins. Based on the literature, tannins are very difficult to separate in normal chromatographic methods. We have selected two methods, i.e. (1) alkaline hydrolysis, and so, first it was subjected to liquid-liquid extraction to reduce the complex nature of the tannins. Acetone crude partitioned with ethyl acetate and forms a brown precipitate (CAE-BP) along with gold colour liquid. The ethyl acetate extracts, (2) i.e. gold colour liquids were combined together and concentrated by using rotary vacuum evaporator to yield dark reddish-brown powder. This powder was dissolved in 300 ml of 70% aq. acetone and filtrated by using Whatman filter paper (42). The reddish-brown water extract on acetone evaporation was divided into two equal parts, i.e. 75 ml each. The first 75 ml reddish-brown water filtrate was partitioned with n-butanol; no layers were separated. To the second 75 ml reddish-brown water, 1.5 g of sodium bisulphite was added and kept overnight. Sodium bisulphite was added for the extraction of high purity tannins. Sediments were removed by centrifugation at 10,000 rpm for 10 min as greenish-yellow undissolved solid. The remaining reddish-brown liquid was subjected to acid hydrolysis by using 2 N HCl placed in oil bath for 8 h and neutralized with 30%

aqueous sodium carbonate solution. After the neutralization, 3 g of sodium bisulphate was added and kept overnight. Sediments were collected by centrifugation at 10,000 rpm for 10 min as greenish-yellow undissolved solid. The remaining dark reddish-brown liquid was partitioned with ethyl acetate in 1:3 ratios, i.e. to 1 ml of extract, 3 ml of ethyl acetate was added. Ethyl acetate layers were combined concentrated and used for further chromatographic analysis, i.e. thin layer chromatography (TLC) and paper chromatography. The dark reddish-brown liquid was kept for TLC using benzene and acetone as mobile phase in the ratio of 8:2, respectively. Reddish-brown streaky spots were obtained. After that the TLC plate was sprayed with $FeCl_3$, and the reddish-brown colour spots turned to blue colour which indicates the presence of gallic acid. The dark reddish-brown liquid was subjected to two-dimensional paper TLC by using butanol/acetic acid/water (3:1:1) for first direction and 6% acetic acid for second direction. Totally four streaky bands were formed. The dark reddish-brown liquid was again subjected to TLC by using benzene and acetone using 8:2 ratios. Brown colour streaky spots were formed, and the spots were sprayed by using ammonia vapours; the brown colour spots turned to blue colour which indicates the presence of gallic acid. Basing on all these identifications, the dark reddish-brown liquid is having the flavan-3,4-diols, i.e. leucoanthocyanins. The reddish-brown colour liquid was concentrated and partitioned with chloroform and forms a brown precipitate (methanol soluble) along with reddish-purple colour liquid (RCL-CAE-BP), and these were further separated and purified by using HPLC and preparative HPLC (Fig. 32.6).

32.2.4 Instrument Conditions

Column: 150 × 90 mm × Select Phenyl Hexyl; Flow rate: 14 ml/min; Wave length: 254 nm; Mobile phase: (a) Water and 0.1% Formic acid, (b) Acetonitrile.

In preparative HPLC, totally three fractions were collected and distilled, and these were subjected to different spectral analysis, i.e. ^{1}H-NMR and MASS Spectroscopy.

32.2.5 Alkaline Hydrolysis

To the 4 g of extract, 9 ml of distilled water and 4.2 ml of 40% NaOH were added to obtain the pH 12–13, and it was kept for reflux in an oil bath for 8 h. It was neutralized with 62% H_2SO_4 until it reached pH 6.8–7. The reddish-brown liquid was partitioned with ethyl acetate in 1:3 ratio, i.e. to 1 ml of extract, 3 ml of ethyl acetate was added. Ethyl acetate layers were combined and concentrated by using rotary vacuum evaporator which formed a red colour liquid. The red colour liquid was concentrated and partitioned with chloroform until it formed a brown precipitate (CAE-BP-ETOAC-BP) (methanol soluble) along with yellow colour liquid (CAE-BP-ETOAC-YCL), and these were further separated and purified by using HPLC and preparative HPLC (Fig. 32.7). In preparative HPLC totally three fractions were collected from CAE-BP-ETOAC-YCL and distilled, and these were submitted for ^{1}H-NMR and mass spectral analysis.

INDIAN INSTITUTE OF CHEMICAL TECHNOLOGY
HABSIGUDA, HYDERABAD-500607
HPLC ANALYSIS REPORT

F:\swamy AMK\ANU\14061774.lcd

Acquired by	: Admin
Sample Name	: RCL-CAE-BP
Vail #	: 34
Injection Volume	: 10 uL
Data File Name	: 14061774.lcd
Method File Name	: mgn.lcm
Data Acquired	: 22-07-2017 12:24:35

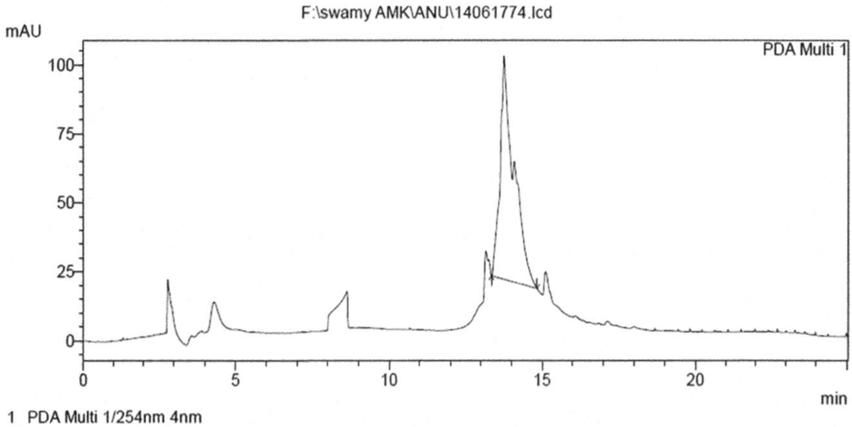

PeakTable

Chl 254nm 4nm

Peak#	Ret. Time	Area	Height	Area %	Height %
1	13.738	2534684	80788	100.000	100.000
Total		2534684	80788	100.000	100.000

Fig. 32.6 HPLC chromatogram of reddish-purple colour liquid (RCL-CAE-BP) isolated from acetone extract of stem bark of *Syzygium samarangense*

32.3 Result and Discussion

RCL-CAE-BP-3: The proton NMR showed the presence of a methoxy group by a singlet δ 4.08. The aromatic proton signal appeared as a singlet at 7.53 with 1 integration indicating that the one aromatic proton and also second aromatic proton appeared as a singlet at 7.49 with 1 integration, indicating that the one aromatic proton was present. Based on these spectral data, the compounds have two aromatic rings along with methoxy group. The mass spectroscopy data showed an ion peak at *m/z* 316 (M+1) which indicate a molecular weight of 3-methoxy ellagic acid. Basing on these studies, RCL-CAE-BP-3 was conformed as 3-methoxy ellagic acid (Figs. 32.8, 32.9, and 32.10).

INDIAN INSTITUTE OF CHEMICAL TECHNOLOGY
HABSIGUDA, HYDERABAD-500607
HPLC ANALYSIS REPORT

F:\swamy AMK\ANU\14061771.lcd

Acquired by : Admin
Sample Name : CAE-BP-EtoAc-YOL
Vail # : 31
Injection Volume : 10 uL
Data File Name : 14061771.lcd
Method File Name : mgn.lcm
Data Acquired : 21-07-2017 19:53:04

F:\swamy AMK\ANU\14061771.lcd

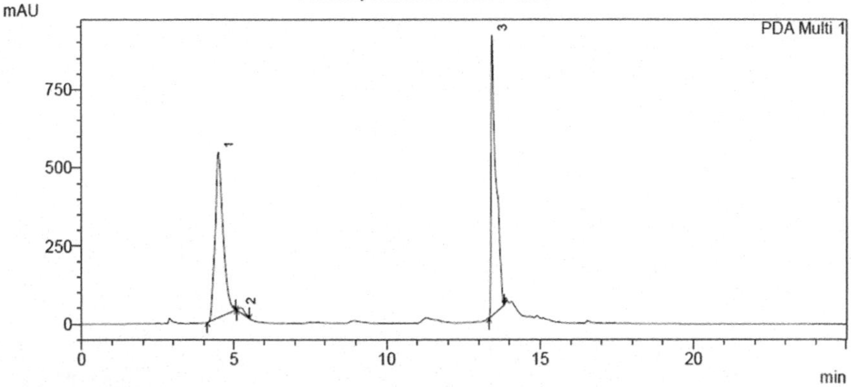

1 PDA Multi 1/280nm 4nm

PeakTable

Chl 280nm 4nm

Peak#	Ret. Time	Area	Height	Area %	Height %
1	4.480	9651280	529135	49.034	36.674
2	5.211	228698	15880	1.162	1.101
3	13.438	9802885	897803	49.804	62.226
Total		19682863	1442818	100.000	100.000

Fig. 32.7 HPLC chromatogram of brown precipitate (CAE-BP-ETOAC-YCL) isolated from acetone extract of stem bark of *Syzygium samarangense*

Fig. 32.8 3-Methoxy ellagic acid

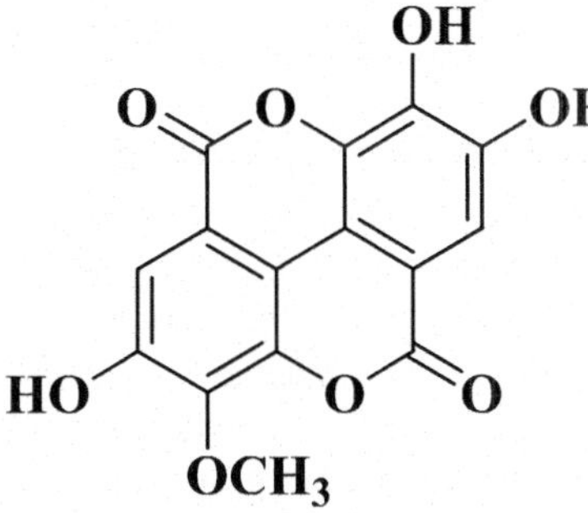

3-methoxy ellagic acid

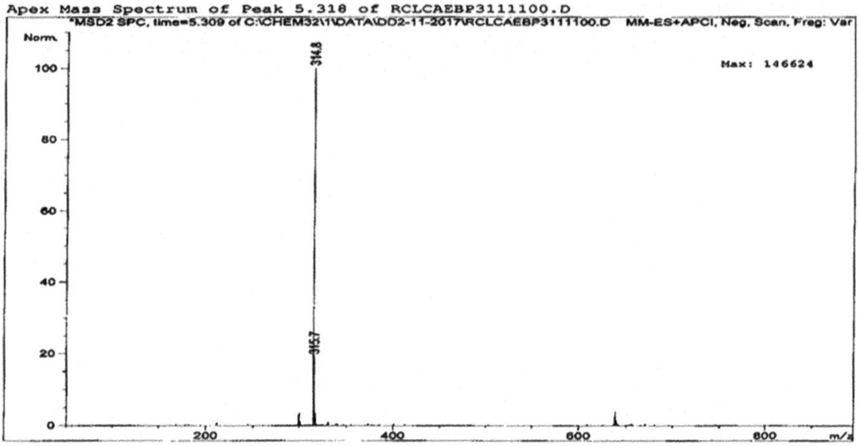

Fig. 32.9 Mass spectrum of RCL-CAE-BP-3

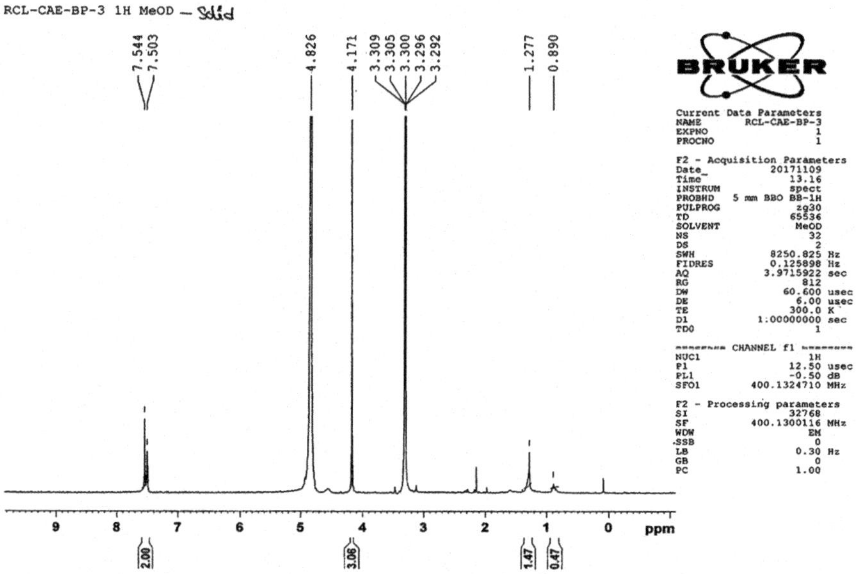

Fig. 32.10 ^{1}H-NMR chromatograms of RCL-CAE-BP-3

CAE-BP-ETOAC-YCL-3: The proton NMR of CAE-BP-ETOAC-YCL-3 shows the presence of a methoxy group by a singlet δ 3.80. The aromatic proton signal appeared as a singlet at δ 7.1 with two integrations, indicating that the two aromatic protons were present within the molecule, i.e. H-2 and H-6 which are symmetrical to each other. The mass spectroscopy data showed an ion peak at *m*/*z* 185 (M+1) which indicate a molecular weight of methyl gallate. The base peak at *m*/*z* 153

Fig. 32.11 Methyl gallate

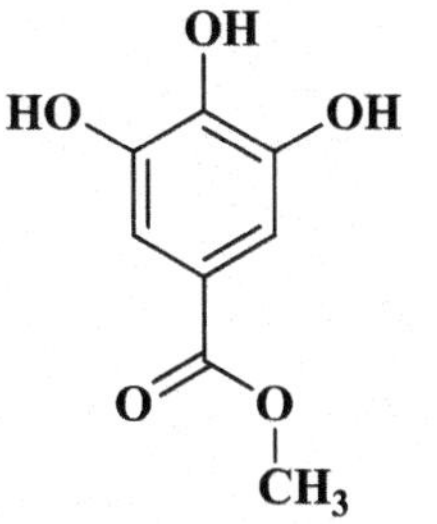

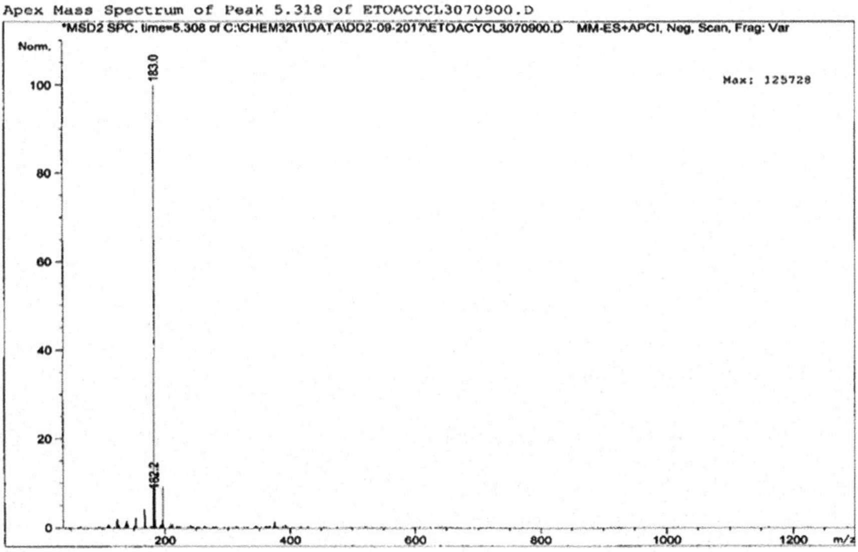

Fig. 32.12 Mass spectrum of CAE-BP-ETOAC-YCL-3

indicated the release of methoxy group (OCH_3) from the compound. Basing on these studies, CAE-BP-ETOAC-YCL-3 was conformed as methyl gallate (Figs. 32.11, 32.12, and 32.13).

32.4 Conclusion

The tannin compounds, methyl gallate and 3-methoxy ellagic acid, which were isolated from the stem bark of *Syzygium samarangense* by using two different methods are having various bioactivities. The activity studies of these two compounds are under process.

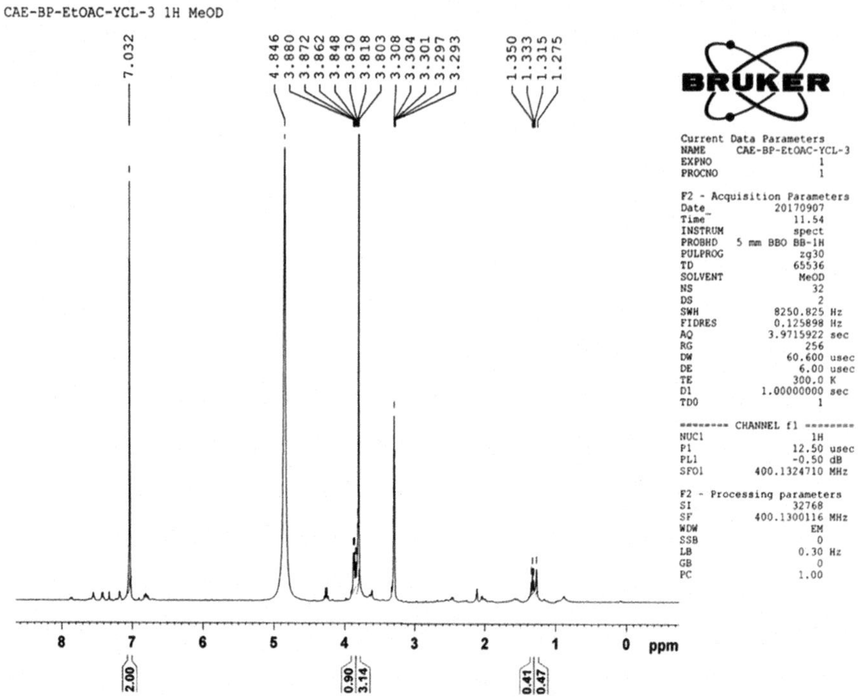

Fig. 32.13 ^{1}H-NMR chromatograms of CAE-BP-ETOAC-YCL-3

Acknowledgements We are very much thankful to all the faculty members of the Department of Chemistry and to the administration of Acharya Nagarjuna University, for their constant encouragement. P. Asha Bhanu is very much thankful to University Grants Commission for sanctioning her Rajiv Gandhi National Fellowship for the year 2015–2017.

References

Akiyama H, Kazuyasu F, Yamasaki O, Oono T, Iwatsuki K (2001) Antibacterial action of several tannins against Staphylococcus aureus. J Antimicrob Chemother 48:487–491

Arbenz A, Avérous L (2015) Chemical modification of tannins to elaborate aromatic biobased macromolecular architectures. Green Chem 17:2626–2646

Bajec MR, Pickering GJ (2008) Astringency: mechanisms and perception. Crit Rev Food Sci Nutr 48:858–875

Bate-Smith EC, Swain T (1962) In: Mason HS, Florkin M (eds) Comparative biochemistry, vol 3. Academic Press, New York, p 764

Funatogawa K, Hayashi S, Shimomura H, Yoshida T, Hatano T, Ito H (2004) Antibacterial activity of hydrolysable tannins derived from medicinal plants against Helicobacter pylori. Microbiol Immunol 48(4):251–261

Ghayur MN, Gilani AH, Khan A, Amor EC, Villaseñor IM, Choudhary MI (2006) Presence of calcium antagonist activity explains the use of Syzygium samarangense in diarrhea. Phytother Res 20(1):49–52

Haslam E (1989) Plant polyphenols-vegetable tannins revisited-chemistry and pharmacology of natural products. Cambridge University Press, Cambridge, p 165

Jain P, Jain S, Pareek A, Sharma S (2013) A comprehensive study on the natural plant phenols: perception to current scenario. Bull Pharm Res 3(2):90–106

Kesava Rao B, Asha Bhanu P, Usha Rani G (2016) Atomic absorption spectral analysis & pharmacological properties of multi-metals isolated from the stem bark of syzygium samarangense (myrtaceae family). Int J Rec Sci Res 7(4):10633–10635

Khanbabaee K, van Ree T (2001) Tannins: classification and definition. Nat Prod Rep 18:641–649

Kuo YC, Yang LM, Lin LC (2004) Isolation and immunomodulatory effect of flavonoids from Syzygium samarangense. Planta Med 70(12):1237–1239

Landete JM (2011) Ellagitannins, ellagic acid and their derived metabolites a review about source, metabolism, functions and health. Food Res Int 44(5):1150–1160

Lin LU, Shu-wen L, Shi-bo J, Shu-guang W (2004) Tannin inhibits HIV-1 entry by targeting gp 41. Acta Pharm Sin 25(2):213–218

Morton J (1987) Fruits of warm climates. Julia F. Morton Publication, Miami, pp 381–382

Nair AGRS, Krishnan S, Ravikrishna C, Madhusudanan KP (1999) New and rare flavonol glycosides from leaves of Syzygium samarangense. Fitoterapia 70:148–151

Nonaka G-I, Aiko Y, Aritake K, Nishioka I (1992) Tannins and related compounds: CXIX. Samarangenins A and B, novel proanthocyanidins with doubly bonded structures from Syzygium samarangense and Syzygium aqueum. Chem Pharm Bull 40(10):2671–2673

Okuda T (2005) Systematics and health effects of chemically distinct tannins in medicinal plants. Phytochemistry 66:2012–2031

Porter LJ (1989) In: Dey PM, Harborne JB (eds) Methods in plant biochemistry—plant phenolics series, vol 1. Academic Press, London, p 389

Raga DD, Cheng CL, Lee KC, Olaziman WZ, De Guzman VJ, Shen CC, Franco FC Jr, Ragasa CY (2011) Bioactivities of triterpenes and a sterol from Syzygium samarangense. Z Nat C 66(5–6):235–244

Schofield P, Mbugua DM, Pell AN (2001) Analysis of condensed tannins a review. Anim Feed Sci Technol 91:21–40

Soubir T (2007) Antioxidant activities of some local bangladeshi fruits (Artocarpus heterophyllus, Annona squamosa, Terminalia bellirica, Syzygium samarangense, Averrhoa carambola and Olea europa). Sheng Wu Gong Cheng XueBao 23(2):257–261

Srivastava R, Shaw AK, Kulshreshtha DK (1995) Triterpenoids and chalcone from Syzygium samarangense. Phytochemistry 38(3):687–689

Tabak LA (1995) In defense of the oral cavity: structure, biosynthesis & function of salivary mucins. Annu Rev Physiol 57(1):547–564

Rare Actinobacteria *Nocardiopsis lucentensis* VLK-104 Isolated from Mangrove Ecosystem of Krishna District, Andhra Pradesh

33

Krishna Naragani

Abstract

The aim of the study was to isolate, identify, and analyze the phylogenetic relationship of the secondary metabolite-producing actinobacterial strains isolated from the mangrove ecosystem of Krishna district, Andhra Pradesh. The potent bioactive metabolite-producing strain was isolated and designated as VLK-104. The identification of the strain was carried out by employing the micromorphological, cultural, physiological, and biochemical methods. The antimicrobial efficacy of the strain was evaluated by using four solvents such as chloroform, ethyl acetate, methanol, and acetone. Among the solvents used, ethyl acetate extract exhibited maximum antimicrobial activity, whereas the other solvent extracts showed moderate to minimum activity against the Gram-positive and Gram-negative bacteria and fungi. Phylogenetic analysis of 16S rRNA gene sequence showed that the strain VLK-104 forms a distinct clade within the *Nocardiopsis* 16S rRNA gene tree and is closely related to *Nocardiopsis lucentensis*.

Keywords

Mangrove ecosystem · *Nocardiopsis lucentensis* · Phylogenetic analysis · Antimicrobial activity

K. Naragani (✉)
Department of Botany and Microbiology, Acharya Nagarjuna University, Guntur, Andhra Pradesh, India

S. M. Khasim et al. (eds.), *Medicinal Plants: Biodiversity, Sustainable Utilization and Conservation*, https://doi.org/10.1007/978-981-15-1636-8_33

33.1 Introduction

Mangrove ecosystem has diverse groups of microorganisms. Among the microorganisms, Actinobacteria are the microbial poulation with a broad variety of secondary metabolites with various biological activities. It has rich biological diversity due to tremendous conditions like high moisture, salinity, pH, and temperature (Amrita et al. 2012). It is essential that new groups of microbes from unexplored habitats are pursued as sources of novel antibiotics and other bioactive compounds (Goodfellow and Fiedler 2010). The importance of cultivation of these microorganisms is necessary for a viable opportunity to bio-discovery (Joint et al. 2010). The phylum Actinobacteria stands for the most prominent group of microorganisms for the production of bioactive compounds, especially antibiotics and antitumor agents (Stach et al. 2003). Close to 40% of all microbial bioactive compounds derived from Actinobacteria, where approximately 80% of them are produced by the genus *Streptomyces* (Bérdy 2012). In fact, two of the four new classes of antibiotics discovered in recent years have been derived from actinobacterial strains (Hardesty and Juang 2011). In view of that, the mangrove environment is a good source for actinobacterial diversity and secondary metabolites. Hence, I have switched over to extreme environments to identify rare actinobacteria with promising bioactive compounds from mangrove sediments of Krishna district, Andhra Pradesh, India.

The soil samples were collected from the coastal region of Krishna district of Andhra Pradesh. The collected soil samples were air-dried at room temperature and transported to the laboratory in sterile bags. The air-dried mangrove soil samples were pretreated with calcium carbonate to selectively isolate rare actinobacteria. The identification of the strain was carried out by employing the micromorphological, cultural, physiological, and biochemical methods. The antimicrobial efficacy of the strain was evaluated by using four solvents such as chloroform, ethyl acetate, methanol, and acetone. After incubation for a week at 30 °C, distinct strains were selected for subculturing to maintain pure culture on agar slants.

33.2 Polyphasic Characterization of the Actinobacterial Strains

The potent actinobacterial strain was identified by using cultural, morphological, biochemical, and physiological characteristics together with genomic (16S rRNA gene sequencing) analysis. Morphological characteristics of the strain were observed by employing slide culture as well as scanning electron microscopy (SEM) as per the procedure described by Bozzola and Russell. Biochemical and physiological characters of the strain such as starch hydrolysis, urea hydrolysis, gelatin hydrolysis, acid production, temperature tolerance test, sodium chloride tolerance test, citrate utilization test, indole test, methyl red and Voges-Proskauer test, hydrogen sulfide test, and melanin tests were tested according to various standard methods (Holding and Collee 1971). Cultural characteristics were studied by culturing the strains on different International *Streptomyces* Project (ISP) media including ISP-1 (tryptone-yeast extract agar), ISP-2 (yeast extract-malt extract-dextrose agar), ISP-3 (oatmeal agar), ISP-4 (inorganic salts-starch agar), ISP-5 (glycerol-asparagine salts agar), ISP-6 (starch-casein agar), and ISP-7 (tyrosine agar) as well as on maltose-tryptone agar, Czapek-Dox agar, and nutrient agar (Shirling and Gottlieb 1966). In addition, the sensitivity of the strain to different antibiotics was determined by paper disc method (Williams 1989). Molecular identification of the strain based on 16S rRNA sequencing analysis was carried out.

33.3 Screening of Potent Actinobacterial Strains for Bioactive Metabolites

Pure culture of the actinobacterial strain was tested for secondary metabolites by the method of agar well diffusion (Cappuccino and Sherman 2004). The pure culture of the strain was transferred aseptically into the seed medium. Fermentation was carried out at 30 °C for 1 week under agitation at 120 rpm. At every 24 h interval, the flasks were harvested. Solvent extracts were evaporated to dry in water bath, and residues obtained were used to determine antimicrobial assay by employing seeded plate techniques. The inoculated plates were examined for zones of inhibition after incubation period. Diameter of the inhibition zone against the test microorganisms was taken as criteria for determining the antimicrobial potential of the actinobacterial strains.

33.4 Test Microorganisms

The test bacteria and fungi used for tetsing antimicrobial activity such as *Bacillus megaterium* (ATCC-10231), *B. subtilis* (ATCC 6633), *Escherichia coli* (ATCC-15597), and *Staphylococcus aureus* (ATCC-6538) and fungi *Candida albicans* (NCIM 2187). The inoculated plates were examined after 24–48 h of incubation at 37 °C for bacteria and 48–72 h at 28 °C for fungi.

Sample was fixed in 5.5% glutaraldehyde in 8.1 M phosphate buffer (pH 7.2) for 24 h at 40 °C and postfixed in 2% aqueous osmium tetraoxide for 4h, dehydrated in series of graded alcohols, and dried to critical point drying with CPD unit. The processed samples were mounted over the stubs with double-sided carbon conductivity tape, and a thin layer of gold coat over the sample was done by using an automated sputter coater (Model—JOEL JFC-5600) for 3 min and scanned under scanning electron microscope (SEM) (Model— JOEL-JSM 1600) at required magnifications as per the standard procedures (Bozzola and Russell 1999) at RUSKA Lab's College of Veterinary Science, SVVU, Rajendranagar, Hyderabad, India.

33.5 Molecular Identification

The molecular identification of the potential strain was carried out by using 18S rRNA gene sequencing analysis. The gene sequence of the strain was aligned using MEGA against the gene library available for potential actinobacteria strains in the NCBI. The phylogenetic tree was constructed using the maximum parsimony method. The closely related homologous strains were identified, retrieved, and compared to the sequence of the isolated strain using BLAST available with the MEGA 6.0 version (Tamura et al. 2013).

The 16S rRNA gene sequence of the strain has been deposited in the National Center for Biotechnology Information (NCBI).

33.6 *Nocardiopsis lucentensis* VLK-104 from Mangrove Ecosystem

33.6.1 Sample Collection

The mangrove sediment samples were collected at bimonthly intervals from mangrove ecosystems of the different places of the coastal region of Krishna district located along the east coast of Andhra Pradesh, India. Samples were collected from 6–10 cm depth and transported to the laboratory in sterile bags and air-dried at room temperature.

Table 33.1 Cultural characteristics of the strain VLK-104

S. no.	Medium	Growth	Aerial mycelium	Substrate mycelium	Pigmentation
1	Tryptone yeast-extract agar	Moderate	White to grayish	Nil	Nil
2	Yeast extract-malt extract-dextrose agar	Good	White to grayish	Brown	Nil
3	Oatmeal agar	Poor	Nil	Brown	Nil
4	Inorganic salts-starch agar	Good	White to grayish	Brown	Nil
5	Glycerol asparagine agar	Moderate	White to grayish	Brown	Nil
6	Starch-casein agar	Good	White to grayish	Brown	Nil
7	Tyrosine agar	Good	White to grayish	Nil	Nil
8	Maltose-tryptone agar	Moderate	White to grayish	Brown	Nil
9	Nutrient agar	Moderate	White to grayish	Nil	Nil
10	Czapek-Dox agar	Moderate	White to grayish	Brown	Nil

33.6.2 Isolation of Actinobacteria

During our search for potent actinobacterial strains from the coastal region of Krishna district, Andhra Pradesh, a total of 520 strains were isolated by using different pretreatment techniques and the selective media. Among them, one of the potent strains was designated as VLK-108, and it was identified by cultural, morphological, physiological, and biochemical characteristics.

33.6.2.1 Cultural Morphological, Physiological, and Biochemical Characteristics

The cultural characteristics of the strain VLK-108 grown on ten culture media are presented in Table 33.1. The strain exhibited good growth on ISP-2, ISP-4, and ISP-6 media out of ten culture media tested. The growth was moderate on ISP-1, ISP-5, ISP-7, nutrient agar, maltose-tryptone agar, and Czapek-Dox agar media, while it was poor on ISP-9. The color of aerial mycelium was white to ash when cultured on different media, while the substrate mycelium was brown. However, melanin pigmentation was not found on ISP-7.

Morphological Characteristics of the Strain VLK-104

The strain VLK-108 exhibited typical morphological characteristics of the genus *Nocardiopsis* spp. Micromorphology of the strain was examined by cover slip method and SEM analysis (Fig. 33.1). The strain showed zigzag hyphae and spore with a smooth surface. The culture was grown on ISP-2 medium supplemented with 3% NaCl for 6 days.

Physiological and Biochemical Characteristics of the Strain VLK-104

Physiological and biochemical characteristics of the strain VLK-104 are recorded in Table 33.2. Kampfer et al. (1991) suggested the physiological tests as indispensable tools for classification and identification of actinobacteria. The temperature range for growth was 25–40 °C with the optimum at 30 °C. NaCl tolerance is also serves as an important characteristic for species identification. VLK-104 was sensitive to

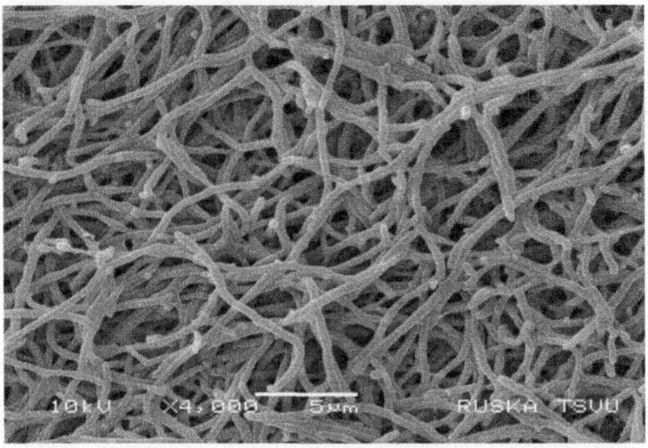

Fig. 33.1 Scanning electron micrograph of strain VLK-104

Table 33.2 Morphological, physiological, and biochemical characteristics of strain VLK-104

Characteristic	Response
Morphological characteristics	
Color of aerial mycelium	White to grayish
Color of substrate mycelium	Brown
Physiological characteristics	
Gram's reaction	+
Production of melanin pigment	–
Optimum temperature for growth	40 °C
NaCl tolerance	9%
Biochemical characteristics	
Catalase production	+
Urease production	+
Hydrogen sulfide production	–
Nitrate reduction	–
Starch hydrolysis	+
Gelatin liquefaction	–
Methyl red test	–
Voges-Proskauer test	–
Indole production	–
Citrate utilization	+

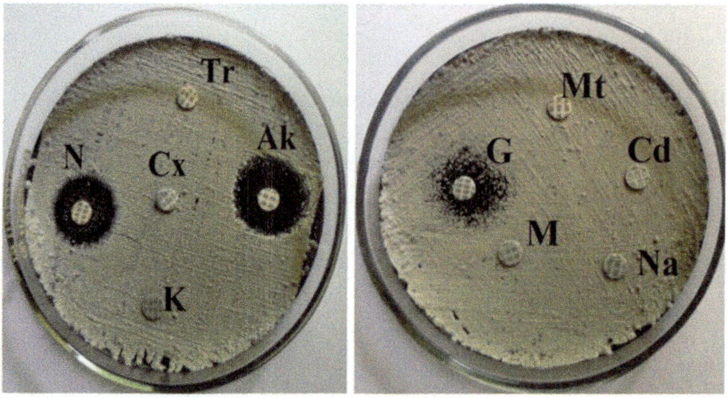

Fig. 33.2 Antibiotic susceptibility of the strain VLK-108. *N* neomycin, *Tr* trimethoprim, *Cx* cloxacillin, *Ak* amikacin, *K* kanamycin, *G* gentamycin, *Mt* metronidazole, *Cd* clindamycin, *M* methicillin, *Na* nalidixic acid

amikacin, gentamicin, and neomycin but resistant to kanamycin, trimethoprim, metronidazole, methicillin, nalidixic acid, clindamycin, and cloxacillin (Fig. 33.2; Table 33.3).

The strain VLK-104 has the ability to hydrolyze starch and exhibited positive response to citrate utilization and catalase production but negative for indole, methyl red, Voges-Proskauer, and nitrate reduction tests and also hydrogen sulfide

Table 33.3 Antibiotic susceptibility testing of VLK-104

S. no.	Name of the antibiotic (μg/disc)	Response
1	Gentamycin (10)	S
2	Kanamycin (30)	R
3	Trimethoprim (5)	R
4	Metronidazole (5)	R
5	Neomycin (30)	S
6	Amikacin (30)	S
7	Methicillin (5)	R
8	Nalidixic acid (30)	R
9	Cloxacillin (1)	R
10	Clindamycin (2)	R

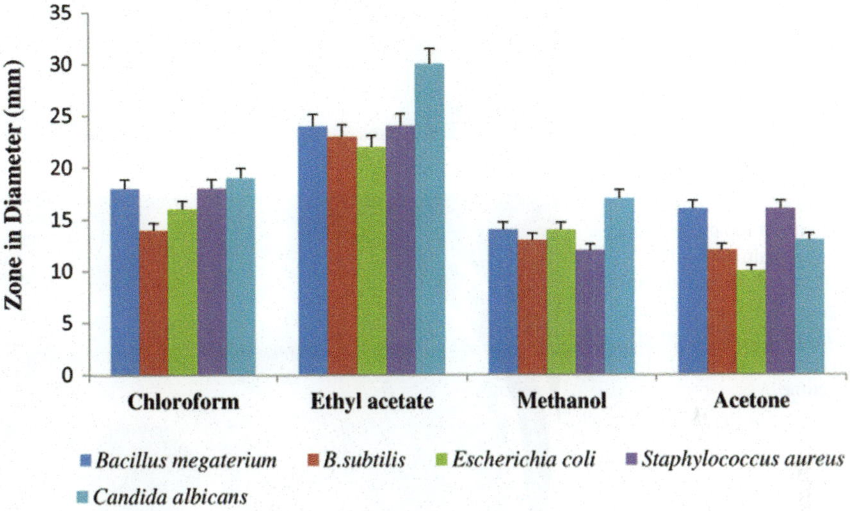

Fig. 33.3 Antimicrobial activity of the strain 104 by using different solvent extracts

production. The utilization of starch revealed the ability of the strain to produce extracellular amylase and protease respectively. Positive reaction with catalase revealed its potential to withstand the stress generated by reactive oxygen species.

33.6.3 Antimicrobial Activity

The antimicrobial efficacy of the strain was evaluated by using four solvents such as chloroform, ethyl acetate, methanol, and acetone (Fig. 33.3). Among the solvents used, ethyl acetate extract exhibited maximum antimicrobial activity, whereas the other solvent extracts showed moderate to minimum activity against the test microorganisms (Figs. 33.4 and 33.5). The ethyl acetate extract is highly effective against *Candida albicans*, *Staphylococcus aureus*, *Bacillus megaterium*, *Bacillus subtilis*, and *Escherichia coli*.

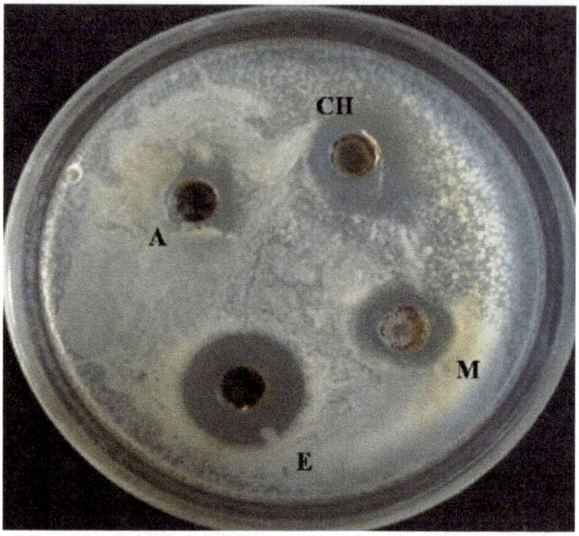

Fig. 33.4 Antibacterial activity of the crude extract of the strain 104 with different solvent extracts against *E. coli*. *CH* chloroform, *E* ethyl acetate, *M* methanol, *A* acetone

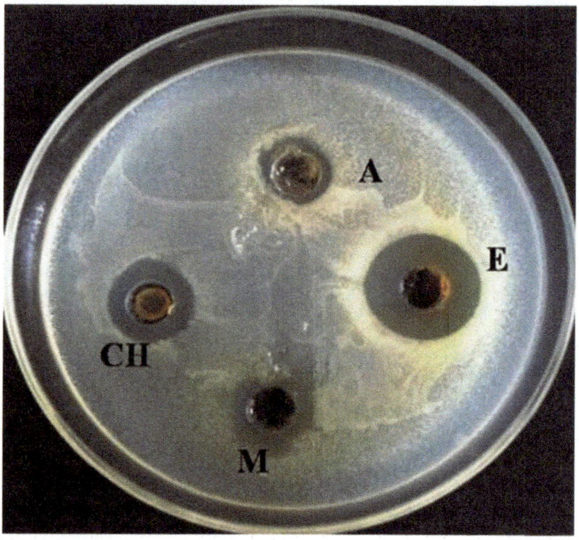

Fig. 33.5 Antifungal activity of the crude extract of the strain 104 with different solvent extracts against *Candida albicans*. *CH* chloroform, *E* ethyl acetate, *M* methanol, *A* acetone

33.6.4 Analysis of the 16S rRNA Gene Sequence of the Strain VLK-104

The 16S rRNA sequence data supported the assignment of this strain VLK-104 to the genus *Nocardiopsis* and species *lucentensis* (Fig. 33.6). The partial 16S rRNA sequence of the strain VLK-104 was submitted to the GenBank database under an accession number **KF317772**. The partial sequence was aligned and compared with all the 16S rRNA gene sequence available in the GenBank database by using the multi-sequence advanced BLAST comparison tool.

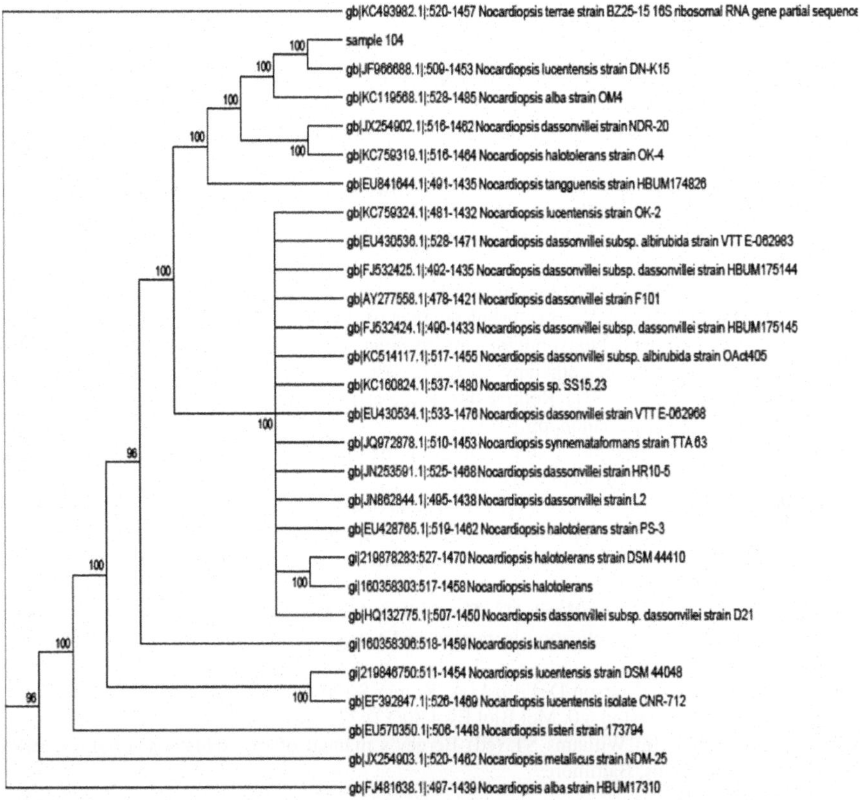

Fig. 33.6 Maximum parsimony tree based on partial 16S rRNA gene sequence showing relationship between isolate (Sample 108) VLK-104 and related members of the genus *Nocardiopsis*. The numbers at the nodes indicate the level of bootstrap support based on maximum parsimony analysis of 1000 resampled datasets; only values above 50% are given

33.7 Conclusion

The present study was aimed at the isolation of novel actinobacterium *Nocardiopsis lucentensis* VLK-104 having potent antimicrobial properties from mangrove ecosystem of Krishna district, Andhra Pradesh, and its identification based on cultural, physiological, and biochemical characteristics. Further study on optimization, purification, and chemical characterization of bioactive compounds of the strain is in progress.

Acknowledgment The author was thankful to SERB (Science and Engineering Research Board), New Delhi, for providing the financial support for the present study.

References

Amrita K, Nitin J, Subathra Devi C (2012) Novel bioactive compounds from mangrove derived actinomycetes. Int Res J Pharm 3:25–29

Bérdy J (2012) Thoughts and facts about antibiotics: where we are now and where we are heading. J Antibiot 65:1–11

Bozzola JJ, Russell LD (1999) Electron microscopy principles and techniques for biologists, 2nd edn. Jones & Bartlett, Sudbury

Cappuccino JG, Sherman N (2004) Microbiology, a laboratory manual. Pearson Education, New Delhi, pp 282–283

Goodfellow M, Fiedler HP (2010) A guide to successful bioprospecting: informed by actinobacterial systematics. Antonie Van Leeuwenhoek 98:119–142

Hardesty J, Juang P (2011) Fidaxomicin: a macrocyclic antibiotic for the treatment of Clostridium difficile infection. Pharmacotherapy 31(9):877–886

Holding AJ, Collee JG (1971) Routine biochemical tests. In: Methods in microbiology, vol 6A. Academic Press, London, pp 1–131

Joint I, Mühling M, Querellou J (2010) Culturing marine bacteria—an essential prerequisite for biodiscovery. Microb Biotechnol 3:564–575

Kampfer P, Kroppenstedt RM, Dott W (1991) A numerical classification of the genera *Streptomyces* and *Streptoverticillium* using miniaturized physiological tests. J Gen Microbiol 137:1831–1891

Shirling EB, Gottlieb D (1966) Methods for characterization of *Streptomyces* sp. Int J Syst Bacteriol 16:313–340

Stach JEM, Maldonado LA, Ward AC, Goodfellow M, Bull AT (2003) New primers for the class Actinobacteria: application to marine and terrestrial environments. Environ Microbiol 5:828–841

Tamura K, Stecher G, Peterson D, Filipski A, Kumar S (2013) MEGA6: molecular evolutionary genetics analysis version 6.0. Mol Biol Evol 30(12):2725–2729

Williams ST (1989) In: Williams ST (ed) Bergey's manual of systematic bacteriology, vol 4. Williams & Wilkins, Baltimore

Aegle marmelos (Rutaceae): Evaluation of Root Phytochemical Constituents for Antimicrobial Activity

34

Aniel Kumar Owk and Mutyala Naidu Lagudu

Abstract

The phytochemical investigation of the roots of *Aegle marmelos* revealed the presence of secondary metabolites like alkaloids, anthraquinones, cardiac glycosides, flavonoids, glycosides, phenols, saponins, steroids, tannins, terpenoids, amino acids, and carbohydrates. The antimicrobial activity and minimum inhibition concentration values were determined for these phytochemical constituents as crude extracts using the agar well diffusion and twofold serial dilution methods, respectively. The results indicated that *Bacillus subtilis* was the most susceptible bacterium with high inhibition zones for the methanol and chloroform extracts of 31 and 26 mm, respectively. The minimum inhibitory concentration values indicated that extracts possess good phytochemicals with significant minimum inhibitory concentration value against *Bacillus subtilis, Klebsiella pneumoniae*, and *Staphylococcus aureus* at 31.2 μg/ml concentration. The extracts showed marked antimicrobial activity against both bacteria and fungi. Among the bacterial strains, Gram-positive bacteria were more susceptible than the Gram-negative. All the tested 12 microorganisms showed dose-dependent susceptibility towards the phytochemicals present in the solvent as well as aqueous crude extracts. The present study suggests that *Aegle marmelos* roots possess potent antimicrobial activity and can be a good source for the development of new antibiotics.

Keywords

Aegle marmelos · Solvent extracts · Phytochemicals · Antimicrobial activity

A. K. Owk
Department of Botany, Andhra University, Visakhapatnam, Andhra Pradesh, India

M. N. Lagudu (✉)
Department of Botany, Adikavi Nannaya University, Rajamahendravaram, Andhra Pradesh, India

S. M. Khasim et al. (eds.), *Medicinal Plants: Biodiversity, Sustainable Utilization and Conservation*, https://doi.org/10.1007/978-981-15-1636-8_34

34.1 Introduction

Medicinal plants have been identified and used throughout human history. They have the ability to synthesize a wide variety of chemical compounds that are used to perform important biological functions and to depend against attack from predators such as insects, fungi, and bacteria. The World Health Organization (WHO) estimates 80% of the population of Asian and African countries presently use herbal medicine for some aspects of primary health care. Studies in United States and Europe have shown that their use is less common in clinical settings, but has become increasingly more common in recent years as scientific evidence about the effectiveness of herbal medicine has become more widely available. The annual global export value of pharmaceutical plants in 2011 accounted for over US $2.2 billion. India is rich in medicinal plant diversity because agroclimatic, ecological, and edaphic conditions are found. In India, from ancient times, different parts of plants have been used to cure specific ailments. The increased interest in medicinal plant cures is because, primarily plants as medicines are safe, less rigorous, and more affordable than synthetic drugs.

Aegle marmelos L. (Rutaceae) is commonly known as bael (Hindi) and golden apple (English) (Fig. 34.1a). It is a subtropical plant and grows up to an altitude of 1200 m from sea level. It is widely distributed throughout India and is known from prehistoric time. The plant is generally grown as a temple garden plant and its leaves are used to pray to Lord Shiva. It is an important medicinal plant with several

Fig. 34.1 (**a**) *Aegle marmelos* L. (**b**) Inhibition zones against *E. coli*. (**c**) Inhibition zones against *S. aureus*. (**d**) Inhibition zones against *S. pneumoniae*

ethnomedicinal applications in traditional and folk medicine system. Traditionally, it is used in the treatment of diarrhea and dysentery. Leaves of this plant are used to cause infertility/abortion (Dhankhar et al. 2011). Leaves, fruits, stem, and roots of the tree at all stages of maturity are used as ethnomedicinal properties against various human diseases such as astringent, laxative, digestive, ophthalmia, hearing loss, inflammation, diabetes, diarrhea, detoxification, skin ailments, asthma, cold, fever, jaundice, chronic diarrhea, dysentery, ulcers, swellings, antioxidant, anticancer, antipyretic effects (Badam et al. 2002; Rani and Khullar 2004; Baliga et al. 2011; Gautam et al. 2013; Shaili et al. 2015), and antimicrobial activity (Poonkothai and Saravanan 2008; Kothari et al. 2011; Pandey and Mishra 2011; Behera et al. 2014; Nayaka and Londonkar 2015; Surender et al. 2015; Karumaran et al. 2016; Ganapathy and Karpagam 2016). Based on the background knowledge of antimicrobial activity of *A. marmelos* and there are no reports on the antimicrobial activity particularly in roots and considering the urgent need to develop alternative therapeutic options. Therefore, the present study is focused to assess the in vitro antimicrobial activity and minimum inhibitory concentration (MIC) of *A. marmelos* roots against selected microorganisms followed by preliminary phytochemicals in the root extracts responsible for antimicrobial activity.

34.2 Collection and Extraction of Roots

The roots of *A. marmelos* were collected from Kambalaakonda forest, Visakhapatnam district, Andhra Pradesh, India. The specimen was authenticated by Prof. Vatsavaya S. Raju, Plant Systematics Lab, Kakatiya University, Warangal, and voucher specimen (L. Mutyala Naidu–3625) was deposited in the Herbarium of Botany Department (BDH), Andhra University, Visakhapatnam, India. The collected roots were made into small pieces and dried in the shadow until it gets dried completely. Then it was powdered in the mixture grinder and stored in airtight bottles. The shade dried and powdered root material (10 g) of *A. marmelos* was extracted with hexane, chloroform, followed by methanol by using sequential extraction method (Aniel Kumar et al. 2014). Thereafter it was filtered by rotary evaporator at 40 °C and to give the crude dried extract. Simultaneously, the aqueous extract of the roots was prepared by adding boiled water to the powder in a beaker on water bath, with occasional stirring for 4 h. The aqueous extract was then filtered and centrifuged at 5000 rpm for 15 min. The supernatant was collected and evaporated to dryness to give the crude dried extract. The extracts were dissolved in DMSO to get the known concentrations of 25, 50, and 100 mg/ml.

34.3 Antimicrobial Activity

The tested microbial strains used in the study were procured from Microbial Type Culture and Collection (MTCC), Chandigarh, India. The bacterial and fungal strains were *Bacillus subtilis* MTCC B2274, *Enterococcus faecalis* MTCC B3159,

Escherichia coli MTCC B1560, *Klebsiella pneumoniae* MTCC B4030, *Micrococcus luteus* MTCC B1538, *Pseudomonas aeruginosa* MTCC B2297, *Proteus vulgaris* MTCC B7299, *Staphylococcus aureus* MTCC B3160, *Streptococcus pneumoniae* MTCC B2672, *Aspergillus niger* MTCC F4325, *Candida albicans* MTCC F7315, and *Saccharomyces cerevisiae* MTCC F2567. The bacterial strains were grown in the nutrient broth and maintained on nutrient agar slants at 4 °C while fungal strains were grown in Sabouraud broth and maintained on Sabouraud agar slants (*C. albicans* and *S. cerevisiae*) and potato dextrose agar slants (*A. niger*) at 4 °C.

The antimicrobial activity of hexane, chloroform, methanol, and aqueous extracts of roots of *A. marmelos* were determined by agar well diffusion method (Aniel Kumar et al. 2015) and agar disc diffusion method for standard antibiotics. The lyophilized culture was subcultured and concentration of working stock culture was assessed as 10^{-6} CFU/ml. For susceptibility test, 100 μl of inoculum was mixed with 6 ml of sterilized nutrient agar and poured immediately into the sterile petri dishes. The petri dishes were left to solidify for 10 min. A sterilized 6 mm metal borer was used to make wells in the center of the divided areas. About 50 μl of each extract was then pipette into the wells. The petri dishes were incubated at 28 °C for 24 h. The experiment was done three times to minimize error. After incubation period the antimicrobial activity was evaluated by measuring the inhibition zones by using an antibiotic zone reader scale (HiAntibiotic Zonescale-c).

For the fungal strains, the same method as for bacteria was adopted of nutrient agar, Sabouraud agar was used. The inoculated petri dishes were incubated at 25 °C for 2 days for the *C. albicans* and *S. cerevisiae* and 3 days for *A. niger*. About 500 μg of nystatin was dissolved in 1 ml of sterile deionized water. About 10 μl of 0.5 mg/ml of nystatin (equivalent to 5 μg dose) was pipette into the wells for comparison with fungal inhibition zones. The bacterial inhibition zones were compared with tetracycline disc (5 μg/disc) of multidrug disc (Axiom Laboratories Ltd., India). About 50 μl of DMSO was pipette into each well for bacteria and fungi as a negative control.

The extracts that exhibited inhibition zones were subjected to MIC assay by using twofold serial dilution (Kumar et al. 2010). A quantity of 0.6 g of each extract was dissolved in 300 ml sterile nutrient broth which yields initial concentration of 2000 μg/ml. Subsequently, twofold serial dilution was made from the stock to obtain 1000, 500, 250, 125, 62.5, and 31.2 μg/ml concentrations. One ml of standardized inoculums of each test organism was introduced into each extract nutrient broth mixture and then incubated at 37 °C. The lowest concentration inhibiting growth was regarded as the MIC of the extracts. For the fungi, the inoculated medium was incubated at 25 °C for 2 (*C. albicans* and *S. cerevisiae*) to 3 (*A. niger*) days.

34.4 Statistical Analysis

Each experimental data from triplicates of standard error was subjected to one-way ANOVA using Minitab version 15. A significant level of $p < 0.001$ was used for all statistical analyses.

34.5 Phytochemical Constituents from *Aegle marmelos* and Their Antimicrobial Activity

The antimicrobial activity of the different concentrations of hexane, chloroform, methanol, and aqueous extracts of roots of *A. marmelos* revealed significant antimicrobial activity against tested nine bacterial and three fungal strains (Table 34.1). Methanol extract exhibited the highest, while aqueous and chloroform showed moderate, and least antimicrobial activity for hexane extracts of *A. marmelos* roots (Figs. 34.1b–d and 34.2a–c). The antimicrobial activity was increased with increasing concentration of extracts. The results of the present study were significant at level of $p > 0.05$.

The zone of inhibitions indicated that *Bacillus subtilis* was the most susceptible bacterium with high inhibition zones for the methanol and chloroform extracts of 31 and 26 mm, respectively. Hexane extract showed high zone of inhibition against *Staphylococcus aureus* whereas aqueous extract against *Streptococcus pneumoniae*. The aqueous extract did not shown inhibition zones against *Klebsiella pneumoniae, Pseudomonas aeruginosa*, and *Aspergillus niger*. The fungal strain *Aspergillus niger* is also resistant to hexane and chloroform extracts. Methanol and chloroform extracts exhibited zone of inhibition values more or similar to standard antibiotics (Fig. 34.2d), whereas DMSO, a negative, control did not shown growth inhibition of bacteria and fungi.

From the MIC values (Table 34.2), it was observed that *B. subtilis, K. pneumonia*, and *S. aureus* exhibited the least MIC value for methanol extract, while chloroform extract showed the least MIC value against *B. subtilis*, and hexane extract against *S. aureus* and aqueous extract against *C. albicans*. The preliminary phytochemical analysis revealed the presence of alkaloids, amino acids, anthraquinones, carbohydrates, cardiac glycosides, flavonoids, glycosides, phenols, saponins, steroids, tannins, and terpenoids in the hexane, chloroform, methanol, and aqueous extracts of *A. marmelos* roots (Table 34.3).

Preliminary phytochemical analysis was noticed that in some cases chemical constituents fail to answer due to trace amount or other reason. The extract obtained from successive solvents is subjected to phytochemical tests to reveal the presence of different phytochemicals, especially the primary and secondary metabolites present in the extract. The phytochemicals present in the extracts are responsible for the antimicrobial activity by inhibiting the growth of microorganisms. Different solvents have been reported to have the capacity to extract different phytochemical constituents depending on their solubility or polarity and property of the solvent.

A. marmelos root extracts exhibited varying degree of broad-spectrum antimicrobial activity against tested microorganisms. Plants have a wide range of compounds, viz. alkaloids, steroids, saponins, tannins, phenols, flavonoids, quinines, terpenes, terpenoids, glycosides, carbohydrates, amino acids, etc. which are responsible for bioactivity. Antimicrobial activity of aqueous, hexane, and chloroform extracts showed variable antimicrobial activity and that could be attributed to the presence of phenols and sterols as such activities with these compounds are reported (Chiambara et al. 2003; Shimada 2006). The antimicrobial activity of methanol extract may be due to

Table 34.1 Antimicrobial activity of *Aegle marmelos* roots

	Zone of inhibition (mm)[a]													
	Hexane extract			Chloroform extract			Methanol extract			Aqueous extract				
Organisms	25	50	100	25	50	100	25	50	100	25	50	100	S	D
B. subtilis	10 ± 0.20	12 ± 0.50	15 ± 0.90	22 ± 0.50	24 ± 0.50	26 ± 0.44	26 ± 0.90	29 ± 0.64	31 ± 0.28	16 ± 0.50	18 ± 0.50	20 ± 0.50	18^T	–
E. coli	–	10 ± 0.28	13 ± 0.28	11 ± 0.10	12 ± 0.60	18 ± 0.16	15 ± 0.50	20 ± 0.90	24 ± 0.45	11 ± 0.52	12 ± 0.52	14 ± 0.28	22^T	–
K. pneumoniae	–	–	12 ± 0.90	12 ± 0.76	16 ± 0.50	20 ± 0.44	16 ± 0.90	21 ± 0.90	24 ± 0.50	–	–	–	24^T	–
P. aeruginosa	11 ± 0.45	13 ± 0.44	15 ± 0.19	16 ± 0.19	18 ± 0.90	20 ± 0.16	15 ± 0.28	17 ± 0.19	19 ± 0.45	10 ± 0.90	12 ± 0.90	14 ± 0.44	25^T	–
P. vulgaris	–	10 ± 0.08	12 ± 0.28	12 ± 0.76	14 ± 0.19	16 ± 0.50	13 ± 0.52	17 ± 0.76	24 ± 0.52	–	–	–	22^T	–
S. aureus	10 ± 0.20	12 ± 0.50	16 ± 0.20	12 ± 0.20	15 ± 0.20	18 ± 0.90	14 ± 0.45	21 ± 0.50	25 ± 0.50	10 ± 0.28	12 ± 0.19	14 ± 0.50	24^T	–
S. pneumoniae	–	–	–	11 ± 0.22	13 ± 0.22	19 ± 0.22	14 ± 0.90	19 ± 0.90	24 ± 0.22	12 ± 0.20	16 ± 0.20	21 ± 0.20	22^T	–
E. faecalis	–	–	10 ± 0.40	10 ± 0.50	11 ± 0.22	13 ± 0.90	12 ± 0.64	15 ± 0.28	16 ± 0.90	10 ± 0.44	11 ± 0.52	13 ± 0.50	22^T	–
M. luteus	–	–	–	10 ± 0.20	12 ± 0.20	14 ± 0.20	12 ± 0.28	14 ± 0.76	16 ± 0.28	10 ± 0.20	12 ± 0.20	14 ± 0.20	22^T	–
A. niger	–	–	–	–	–	–	12 ± 0.20	16 ± 0.20	22 ± 0.20	–	–	–	18^N	–
C. albicans	10 ± 0.16	12 ± 0.10	16 ± 0.46	12 ± 0.90	13 ± 0.44	19 ± 0.44	17 ± 0.64	21 ± 0.76	25 ± 0.45	18 ± 0.44	20 ± 0.45	21 ± 0.44	23^N	–
S. cerevisiae	–	10 ± 0.60	12 ± 0.16	10 ± 0.20	10 ± 0.16	12 ± 0.22	15 ± 0.19	18 ± 0.22	21 ± 0.45	20 ± 0.22	22 ± 0.45	24 ± 0.19	20^N	–

p Value is <0.05 extremely significant when compared with standard
S standard antibiotics, *T* tetracycline, *N* nystatin, *D* DMSO, – no activity
[a]Each value is the mean of three replicates with standard deviation

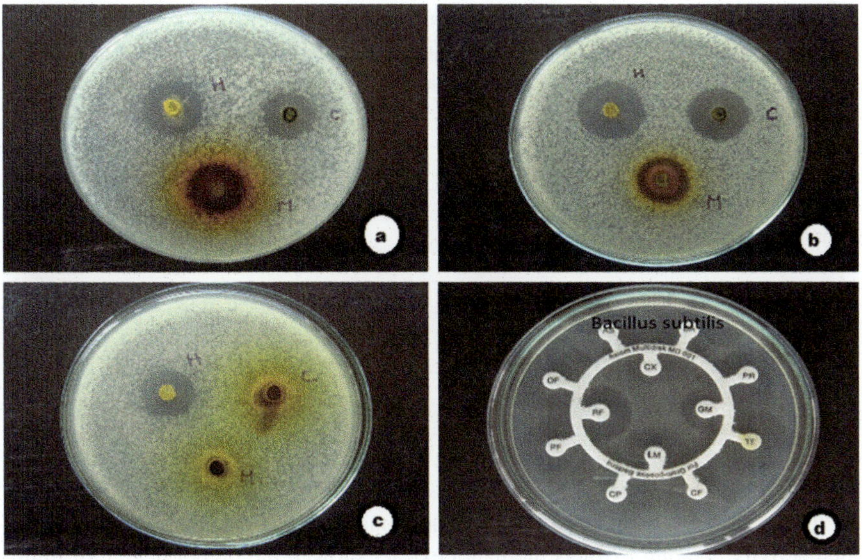

Fig. 34.2 (**a**) Inhibition zones against *K. pneumonia*, (**b**) inhibition zones against *B. subtilis*, (**c**) inhibition zones against *C. albicans*, (**d**) inhibition zones by multidrug

Table 34.2 MIC values of different extracts of *A. marmelos* roots

Microorganisms	Hexane extract	Chloroform extract	Methanol extract	Aqueous extract
B. subtilis	500	31.2	31.2	1000
E. coli	>1000	>1000	62.5	>1000
K. pneumoniae	>1000	125	31.2	250
P. aeruginosa	1000	62.5	250	1000
P. vulgaris	1000	1000	500	>1000
S. aureus	62.5	62.5	31.2	>1000
S. pneumoniae	>1000	1000	125	>1000
E. faecalis	1000	1000	250	1000
M. luteus	1000	>1000	500	>1000
A. niger	>1000	>1000	1000	>1000
C. albicans	125	1000	125	125
S. cerevisiae	>1000	>1000	500	>1000

the presence of tannins, triterpenoids, and flavonoids. Tannins have been known to form irreversible complexes with proline-rich protein resulting in the inhibition of cell wall synthesis (Mamtha et al. 2004). Triterpenoids are known to weaken the membranous tissue, which results in dissolving cell wall of microorganism (Hemandez et al. 2000). Flavonoids have the ability to bind with extracellular and soluble proteins and complexes with microbial cell walls. Antimicrobial activity of steroids is specifically associated with membrane lipids and cause leakage from liposomes (Epand et al. 2007). The antifungal activity may be contributed due to the presence of coumarins and this is supported by earlier work (Ojala et al. 2000) showing antifungal activity of herbal plants containing coumarins.

Table 34.3 Phytochemical constituents of *A. marmelos* roots

Phytochemical constituents	Hexane extract	Chloroform extract	Methanol extract	Aqueous extract
Alkaloids	+	+	+	+
Amino acids	+	+	+	+
Anthraquinones	–	–	+	–
Carbohydrates	–	+	+	+
Cardiac glycosides	–	+	+	–
Flavonoids	–	–	+	+
Glycosides	–	+	+	–
Phenols	–	–	+	+
Saponins	–	+	+	+
Steroids	+	+	+	–
Tannins	–	–	+	+
Terpenoids	+	+	+	–

The results revealed that solvent and aqueous extracts possessed good antimicrobial activity. Among extracts methanol extract showed the higher degree of inhibition zones against tested microorganisms. This is in accordance with the previous study which reported that methanol is the most effective solvent for plant than hexane and chloroform (Johnson et al. 2010). All the root extracts have worked in a dose-dependent manner; as the concentration of the extract increased, the activity also increased. This is due to susceptibility of the microorganisms towards concentration of the extracts, after which the extract damages that microbe which is not tolerable for it (Ordonez et al. 2006).

It is clear from the present study that the Gram-positive bacteria have found to be the most sensitive than Gram-negative bacteria. Generally Gram-positive bacteria are more susceptible to commercial antibiotics, crude extracts, and isolated compounds from natural sources, which may be related to cell wall structure. According to Tortora et al. (2001) the cell wall of Gram-negative bacteria acts as a barrier to a number of substances, including antibiotics.

34.6 Conclusions

From the results obtained, the present study highlights the significant antimicrobial activity of methanol extracts of roots of *A. marmelos* against selected bacterial and fungal strains. The study serves as a preliminary scientific validation of *A. marmelos* methanol extract as an important source for the development of therapeutic antimicrobial compounds against infectious pathogenic microbial strains. However the exact structure of chemical components and mode of action of phytochemicals are currently not clear. Our next line of further investigations is fractionation, purification, and characterization of the bioactive components in the methanol extracts of roots of *A. marmelos*.

References

Aniel Kumar O, Krishna Rao M, Mutyala Naidu L (2014) Evaluation of antimicrobial activity of chemical constituents of *Achyranthes aspera* L. roots against human pathogens. Indian J Nat Prod Resour 5(3):278–281

Aniel Kumar O, Krishna Rao M, Mutyala Naidu L (2015) Evaluation of antimicrobial activity of root extracts of *Abutilon indicum*. Not Sci Biol 7(2):160–163

Badam L, Bedekar SS, Sonaware KB, Joshi SP (2002) *In vitro* antiviral activity of Bael (*Aegle marmelos* Corr) upon human coxsackie viruses. J Commun Dis 34:88

Baliga MS, Bhat HP, Joseph N, Fazal F (2011) Phytochemistry and medicinal uses of the bael fruit (*Aegle marmelos* Correa), a concise review. Food Res Int 44:1768–1775

Behera P, Raj V, Basavaraju R (2014) Phytochemical and antimicrobial activity of fruit pulp of *Aegle marmelos*. J Chem Pharm Res 6(8):319–326

Chiambara KNM, Vanitha M, Swamy M, Ravishankar GA (2003) Antioxidant and antimicrobial activity of *Cissus quadrangularis* L. J Med Food 6:99–106

Dhankhar S, Ruhil S, Balhara M, Dhankhar S, Chhillar AK (2011) *Aegle marmelos* (Linn.) Correa: a potential source of phytomedicine. J Med Plants Res 5(9):1497–1507

Epand RF, Savage PB, Epand RM (2007) Bacterial lipid composition and the antimicrobial efficacy of cationic steroid compounds (Ceragenins). Biochem Biophys Acta 1768(10):2500–2509

Ganapathy S, Karpagam S (2016) Phytochemical and *in vitro* antibacterial activity of leaf and fruit extracts of *Aegle marmelos* against Methicillin resistant *Staphylococcus aureus* (MRSA). Mint J Pharm Med Sci 5(1):19–21

Gautam MK, Ghatule RR, Singh A, Purohit V, Gangwar M, Kumar M (2013) Healing effects of *Aegle marmelos* (L.) Correa fruit extract on experimental colitis. Indian J Exp Biol 51:157–164

Hemandez NE, Tereschuk ML, Abdala LR (2000) Antimicrobial activity of flavonoids in medicinal plants from Tafi del Valle (*Tucuman Argentina*). J Ethnopharmacol 73:317–322

Johnson M, Wesely EG, Zafir HMI, Selvan N (2010) *In vitro* phytochemical and antibacterial efficacy of *Balisopermum montanum* Muell. Asian Pac J Trop Med 3:894–897

Karumaran S, Nethaji S, Rajakumar R (2016) Antimicrobial and antioxidant activity of leaf extracts of *Aegle marmelos*. Adv Appl Sci Res 7(3):205–208

Kothari S, Mishra V, Bharat S, Tonpay S (2011) Antimicrobial activity and phytochemical screening of serial extracts from leaves of *Aegle marmelos* (Linn.). Acta Pol Pharm Drug Res 68(5):687–692

Kumar OA, Naidu LM, Rao KG (2010) *In vitro* antibacterial activity in the extracts of *Andrographis paniculata* Burm. F. Int J PharmTech Res 2(2):1383–1385

Mamtha B, Kavitha K, Srinivasan KK, Shivannda PG (2004) An *in vitro* study of *Centella asiatica* (Indian pennywort) on enteric pathogens. Indian J Pharmacol 36(1):41–47

Nayaka HB, Londonkar RL (2015) Phytochemical screening and potential *in vitro* antibacterial activity of *Aegle marmelos* leaf extract against Gram negative pathogenic bacteria. Int J Pharm Clin Res 7(1):64–67

Ojala T, Remes S, Haansuu P, Vuorella H, Hiltunen R (2000) Antimicrobial activity of some coumarin containing herbal plants growing in Finland. J Ethnopharmacol 73(1–2):299–305

Ordonez RM, Ordonez AA, Sayago JE, Moreno MIN, Isla MI (2006) Antimicrobial activity of glycosidase inhibitory protein isolated from *Cyphomandra betacea* Sendt. fruit. Peptides 27:1187–1191

Pandey A, Mishra R (2011) Antibacterial properties of *Aegle marmelos* leaves, fruits and peels against various pathogens. JPBMS 13(13):1–6

Poonkothai M, Saravanan M (2008) Antibacterial activity of *Aegle marmelos* against leaf, bark and fruit extracts. Anc Sci Life 27:15–18

Rani P, Khullar N (2004) Antimicrobial evaluation of some medicinal plants for their antienteric potential against multidrug resistant *Salmonella typhi*. Phytother Res 18:670–675

Shaili Y, Neeti S, Jyothi S (2015) Curative aspects of *Aegle marmelos* (Bael) in drug bioavailability. World J Pharm Pharm Sci 4(3):621–633

Shimada T (2006) Salivary proteins as a defense against dietary tannins. J Chem Ecol 32:1149–1163

Surender Y, Dahiya K, Ganif S, Gulia S (2015) Antibacterial activity of *Aegle marmelos* (L) Correa. Int J Pharm Sci 7(3):462–464

Tortora GJ, Funke BR, Case CL (2001) Microbiology: an introduction. Benjamin Cummings, San Francisco, CA, p 88

Qualitative and Quantitative Phytochemical Studies in Different Parts of *Sesamum indicum* L.

35

Mukta Nagpurkar and Neeta M. Patil

Abstract

Sesamum indicum L. is an oil crop mentioned in early Hindu legends and shows seeds of different colours. Sesame seeds, oil and other parts of plant are ethnomedicinally significant and have numerous uses from preparations of hair oils to treatments of various ailments in human and cattle. Qualitative and quantitative phytochemical investigations provide the comprehensive account of different types and classes of metabolites and their quantities present in different parts of plant. In preliminary phytochemical investigation in seed, leaf, shoot and root of certified variety of white sesame G1 alkaloids were detected. Quantification of alkaloids present in different parts of G1 variety was performed by non-spectrophotometric (weighing) method and spectrophotometric method. According to both methods, shoots of G1 variety exhibited highest amount of alkaloids. Amount quantified per gram of tissue by weighing method was found to be the lowest in roots whereas the same quantified per millilitre of extract by spectrophotometric method was observed to be the lowest in seeds. Alkaloids present in this plant were further analysed using thin layer chromatography and advanced techniques like liquid chromatography mass spectrophotometry. The former showed the presence of caffeine or compounds of same molecular mass whereas the latter indicated compounds present in proportionately highest amount in all four tissues. The study proposed use of herbage of an agriculture crop as a source of valuable compounds.

Keywords

Sesame · Alkaloids · Metabolites · TLC · LCMS

M. Nagpurkar (✉)
Department of Botany, Post Graduate Research Centre,, Modern College of Arts, Science and Commerce, Pune, Maharashtra, India

N. M. Patil
Department of Botany, Modern College of Arts, Science and Commerce, Pune, Maharashtra, India

S. M. Khasim et al. (eds.), *Medicinal Plants: Biodiversity, Sustainable Utilization and Conservation*, https://doi.org/10.1007/978-981-15-1636-8_35

35.1 Introduction

Sesamum indicum L. (family—Pedaliaceae) seeds contain 40–60% of oil, making it an economically important oil crop (Hassan 2013). The semi-drying oil contains palmitic and stearic acids (Asghar and Majeed 2013). It is one of the oldest and world's most important oil seed crop, cultivated in tropics and temperate zones throughout the world. India is one of the largest commercial producers of sesame along with China and Mexico (Morris 2002). Sesame ranks ninth among the top 13 oilseed crops which make up 90% of the world, in production of edible oil (Saha et al. 2014).

This drought tolerant crop is an ancient spice and one of the first recorded plants used for its seeds. It has been used for thousands of years and is still an oil seed of worldwide significance (Sani et al. 2014). The primary marketable products of sesame are the whole seeds, seed oil and meal (Anilakumar et al. 2010).

Sesame seed oil is a natural salad oil, requiring little or no winterization, can be used without refining as it is stable and the most resistant to rancidity owing to endogenous phenolic antioxidants such as lignans and tocopherols (γ-tocopherol) preventing autoxidation, which also show protective effects against neurodegenerative diseases and prevent ageing (Jannat et al. 2013; Sani et al. 2013, 2014). In European countries it is used as a substitute for olive oil (Anilakumar et al. 2010).

Ethno-medicinally sesame oil is used in many formulations to treat skin, teeth, bone and lung problems (Patil et al. 2008; Anilakumar et al. 2010). It is an important part of Ayurvedic, Chinese and Tibetian traditional medicinal systems (Reshma et al. 2013).

Sesame seeds showed the presence of metabolites like flavonoids, phenolic acids, alkaloids, tannins, saponins, steroids, terpenoids and minerals like calcium, iron, magnesium, manganese, copper, zinc and phosphorus. The compounds like sesamin, sesaminol, gamma tocopherol, cephalin and lecithin are also found in sesame, involved in many pharmacological activities like antioxidant, antibacterial, cardio tonic, antidiabetic, hypocholesterolemic, antitumor, antiulcer, anti-inflammatory, anti-allergic, analgesic and diuretic properties. Seeds and oil have desirable physiological effects on blood pressure and serum lipid lowering potential as proven in experimental animals and humans. Sesame oil is mildly laxative, emollient and demulcent. The seeds and fresh leaves are used as a poultice. It is naturally antibacterial for common skin pathogens such as Staphylococcus and Streptococcus as well as common skin fungi such as athlete's foot fungus (Anilakumar et al. 2010; Kaliyamoorthy et al. 2015). The lignans such as sesamin, episesamin, sesaminol and sesamolin are major constituents of sesame oil and all have chemically methylene dioxyphenyl group (Gokbulut 2010).

The tannins show astringent, antiviral and antibacterial activities useful in wound and burns healing whereas saponins have anticancer and immunomodulatory properties. As mentioned in traditional medicines sesame seeds cure anaemia, protect digestive and respiratory systems, bones, oral, and eye health, relieve from rheumatoid arthritis and protect from radiation damage and is useful in the treatment of anxiety and as a skin detoxifier (Kaliyamoorthy et al. 2015).

The oil has wide medical and pharmaceutical applications such as oleaginous vehicle for drugs and solvent for intramuscular injections. Sesame oil is used in products like perfumes, cosmetics, hair oils and soaps (Anilakumar et al. 2010).

In Pedaliaceae family alkaloids were found to be present in 20 genera or species out of 80 and from one genus or species alkaloids were isolated and total 3 alkaloids were isolated (Cordell et al. 2001).

The present phytochemical investigation includes qualitative and quantitative comparison of secondary metabolites present in different parts—seed, leaf, shoot and root—of G1 variety of white sesame.

Certified G1 variety of white sesame was procured from Mahabeej (Ahmednagar), Government of Maharashtra. Seeds were sown in pots and full grown plants after 2–2.5 months were harvested. All plant parts were separated and shade dried. Plant material was stored in dry and cool place in labelled bags. For all analyses shade dried material was powdered in mortar-pestle or in an electrical grinder and extracted in different solvents. All solvents used were of analytical grade.

35.2 Qualitative Phytochemical Analysis

The preliminary phytochemical examination in 4 plant parts, seed, leaf, shoot and root, was carried out. In qualitative analysis 24 divergent classes of metabolites were tested using 44 different tests (Thamaraiselvi and Jayanthi 2012; Mamta and Jyoti 2012; Somkuwar and Kamble 2013; Okoh-Esene et al. 2012; Savithramma et al. 2011; Kumar et al. 2013). Out of those 7 were the primary metabolites (reducing sugars, carbohydrates, starch, mucilage, amino acids, proteins, fats/fixed oils) and 17 were secondary metabolites (volatile oils, alkaloids, tannins, steroids, flavonoids, saponins, anthraquinones, phlobatannins, anthocyanins, coumarins, terpenoids, cardiac glycosides, emodins, glycosides, phenols, leucoanthocyanins, resins). For this analysis extracts were prepared using solvents methanol, distilled water and chloroform.

10 g shade dried tissue (seeds/leaves/shoots/roots separately) were finely powdered and soaked in 50 ml of methanol, chloroform and distilled water. After 24 h, all the extracts were filtered using Buckner's funnel through Whatman filter paper no. 1 (70 mm). Methanol and chloroform residues were taken to dryness and then the residues were dissolved in 2 N HCl. All extracts were stored in labelled bottles at 0–4 °C. Qualitative phytochemical tests were performed as per standard methods.

35.3 Quantitative Phytochemical Analysis

35.3.1 By Non-spectrophotometric (Weighing) Method

2 g of dried and powdered tissue (seeds/leaves/shoots/roots separately) was extracted in 10 ml of extracting solvent, i.e. 10% acetic acid in ethanol. Extracts were put on shaker for 24 h at 110–150 rpm and 28–32 °C temperature. After 24 h extracts were

filtered through Whatman filter paper no. 1 (70 mm) and a pinch of sodium sulphite was added to each extract and filtered again. Extracts were concentrated on hot water bath (at 70 °C) to reduce volume to quarter to original volume. Remaining extracts were transferred to pre-weighed micro vials (2 ml). Concentrated NH_4OH was added drop-wise to extracts to precipitate the alkaloids till white and cloudy precipitate was visible (500–700 μl). Extracts were put on a short spin for 2–2.5 min to collect precipitate at bottom. All micro vials were kept standing overnight to settle down precipitate. Upper layer was discarded and pellet or residues were dried overnight. All eppendorfs were weighed and weight of residues in each eppendorf was calculated. The residues in each eppendorf were dissolved in ethanol (0.5–1.0 ml) and were stored in fridge at 0–4 °C in labelled and sealed glass sample bottles (Gracelin et al. 2013; Biradar and Rachetti 2013; Sani et al. 2013).

35.3.2 By Spectrophotometric Method (John et al. 2014; Shamsa et al. 2008; Tabasum et al. 2016; Sadasivam 1996)

For calibration curve 1 mg/ml stock of caffeine (anhydrous pure, Merck) was prepared in distilled water. Bromocresol green (LobaChemie) (5 ml), phosphate buffer (pH 4.5) (5 ml) and chloroform (2 ml) were added, respectively, to 8 different aliquots of different concentrations of caffeine (0.0, 0.5, 1.0, 1.5, 2.0, 2.5, 3.0, 3.5 ml) and absorbance of lower organic or chloroform layer was recorded at 470 nm after thorough mixing of all added components by vortex.

For preparation of samples 2 g of tissue was soaked in 30 ml of methanol and kept on shaker at 120 rpm and 30 °C temperature for 48 h. Extracts were filtered through Whatman filter paper (70 mm discs) using Bückner's funnel and dried in Petri dishes for 24 h. After drying residues were dissolved in 2 N HCl and filtered again in the same way as before. Final methanol residues were stored in labelled bottles at 0–4 °C.

To test samples chloroform (10 ml) was added to 1 ml of extract (methanol residues) and shaken vigorously. Bromocresol green (5 ml) and phosphate buffer (pH 7.0) (5 ml) were added to the lower organic or chloroform layer and shaken vigorously. Chloroform (2 ml) was added and mixed thoroughly. Absorbance of lower organic chloroform layer was recorded at 470 nm using UV/Visible spectrophotometer (Systronics PC based double beam spectrophotometer 2202).

35.4 Thin Layer Chromatography (TLC) (Harborne 1998)

For thin layer chromatography the same extracts of quantitative analysis by non-spectrophotometric method were used. Methanol, liquor ammonia, ethyl acetate and hexane were used for the preparation of mobile system. TLC Silica Gel plates (MERK TLC Silica Gel 60 F_{254}, 20 × 20 cm Aluminium plates) was stationary phase and iodine chamber was used for detection of spots. Different solvent systems used were E—ethyl acetate, A.3—0.5% liq. NH_3, E 20—methanol:ethyl acetate:: 20:80, EH 50—ethyl acetate:hexane:: 50:50 and E20H—ethyl acetate:hexane:: 20:80.

Stock of standard caffeine 0.1 μg/ml ethanol was prepared. The silica gel plates of appropriate size were cut and samples were loaded. The plate was put in chromatography chamber saturated with selected solvent system and was developed within appropriate time. The plate was removed from chamber dried and put in iodine chamber for detection of spots. The brown spots were observed. Total number of separated spots was counted to find possible number of compounds.

35.5 Liquid Chromatography Mass Spectrophotometry (LCMS)

The same extracts used for TLC were subjected to further analysis using LCMS technique. This analysis was carried out at Venture Centre, 100, NCL Innovation Park, Dr. Homi Bhabha Road, Pashan, Pune-411008, Maharashtra, India. Web: www.venturecenter.co.in/cams. (Material and methodology used for this technique is attached separately in email in PDF format.)

35.6 Phytochemicals from *Sesamum indicum*

In seeds of G1 variety, *primary metabolites* like proteins and fixed oils were detected in all four extracts, methanolic and aqueous extracts and methanol and chloroform residues, whereas carbohydrates were found to be present in methanolic and aqueous extracts. Amongst *secondary metabolites* tannins were detected in chloroform residues and flavonoids were present in methanolic and aqueous extracts and methanol residues. In methanolic extract resins were detected and in aqueous extracts along with resins phenols were found. Volatile oils were found to be present in all four extracts. Alkaloids were detected in aqueous extract and chloroform residues when Wagner's and Mayer's reagents were used while with Dragendorff's reagent alkaloids were found in both methanolic and aqueous extracts. Using Hager's reagent alkaloids were noticed in all four extracts (Tables 35.1 and 35.2).

When oil extracts from the seeds of some selected indigenous cultivars from Nigeria were explored, flavonoids, alkaloids, steroids and terpenoids were detected (Warra et al. 2012). The seeds of sesame variety from Tamil Nadu (India) revealed the presence of steroids, terpenoids and quinone indicating its medicinal importance and ethno uses (Kaliyamoorthy et al. 2015).

All the four extracts of leaf tissue showed the presence of volatile oils and coumarins. Tannins, steroids, phenols and glycosides were detected in methanolic extract. Saponins and mucilage were found in aqueous extract. In methanol residues anthraquinones, cardiac glycosides and emodins were observed. Fixed oils and terpenoids were present in both methanolic and aqueous extracts and latter one was also detected in methanol residues. Leucoanthocyanins and flavonoids were observed in aqueous extract and methanol residues whereas flavonoids were also

Table 35.1 A brief composition of seed, leaf, shoot and root of G1 variety

Plant parts used	G1 variety							
	Methanol extract		Water (aqueous) extract		Methanol residue		Chloroform residue	
	Primary	Secondary	Primary	Secondary	Primary	Secondary	Primary	Secondary
Seed	Carbohydrates, proteins, fixed oils	Volatile oils, alkaloids, flavonoids, resins	Carbohydrates, proteins, fixed oils	Volatile oils, alkaloids, flavonoids, resins, phenols	Proteins, fixed oils	Volatile oils, alkaloids, flavonoids	Proteins, fixed oils	Volatile oils, tannins
Leaf	Reducing sugars, fixed oils	Volatile oils, alkaloids, tannins, steroids, coumarins, terpenoids, glycosides, phenols	Fixed oils, mucilage, proteins	Volatile oils, alkaloids, flavonoids, saponins, coumarins, terpenoids, leucoanthocyanins	Reducing sugars, carbohydrates	Volatile oils, alkaloids, flavonoids, coumarins, terpenoids, leucoanthocyanins, cardiac glycosides, emodins, anthraquinones	Proteins, carbohydrates	Volatile oils, alkaloids, flavonoids, coumarins
Shoot	Reducing sugars, carbohydrates, proteins, fixed oils	Volatile oils, alkaloids, tannins, flavonoids, phlobatannins, anthocyanins, coumarins, cardiac glycosides	Reducing sugars, carbohydrates, mucilage	Volatile oils, alkaloids, flavonoids, anthocyanins, coumarins, phenols	Carbohydrates, proteins, fixed oils	Alkaloids, flavonoids, phlobatannins, anthocyanins, coumarins, cardiac glycosides	Carbohydrates, fixed oils	Alkaloids, flavonoids
Root	Reducing sugars, fixed oils	Volatile oils, tannins, steroids, flavonoids, coumarins, alkaloids	–	Volatile oils, steroids, flavonoids, alkaloids	–	Volatile oils, steroids, flavonoids, alkaloids	–	Volatile oils, steroids, flavonoids, alkaloids

present in chloroform residues. Primary metabolites like reducing sugars were present in methanolic extract and methanol residues, carbohydrates were found in both methanol and chloroform residues and proteins were observed in aqueous extract and chloroform residues. The presence of *alkaloids* was noticed in all four extracts with iodine solution and only in methanolic extract. Alkaloids were detected in methanolic extract, methanol and chloroform residues while testing with Hager's reagent whereas with Dragendorff's reagent positive results were obtained in methanol and chloroform residues (Tables 35.1 and 35.2).

The analysis on phytochemical content of ethanolic leaf extract of *Sesamum indicum* L. from Akwa Ibom State, Nigeria, revealed the presence of saponins, flavonoids, tannins, terpenes and cardiac glycosides and absence of alkaloids and anthraquinones using same standard methods and regents as those used in present investigation (Okon and Umoh 2013).

Carbohydrates were found to be present in all four extracts of shoots while reducing sugars were detected in methanolic and aqueous extracts. In aqueous extracts mucilage were noticed. Methanolic extract and methanol residues showed the presence of proteins and fixed oils and the latter one were also observed in chloroform residues. All four extracts showed the presence of flavonoids and aqueous and methanolic extract exhibited traces of volatile oils, anthocyanins and coumarins. Besides the presence of phlobatannins in methanol residues, tannins and phlobatannins were observed to be present in methanolic extract. All four extracts exhibited the presence of alkaloids. Using Hager's reagent alkaloids were noticed in all four extracts of shoots of G1 variety. Tests with Mayer's reagent exhibited the presence of alkaloids in methanolic and aqueous extracts and those with Dragendorff's reagent showed the presence of alkaloids in chloroform residues. Using Wagner's reagent and iodine solution alkaloids were detected in aqueous extract and methanol residues and in methanolic extract, methanol and chloroform residues, respectively (Tables 35.1 and 35.2).

In roots of G1 variety volatile oils, flavonoids and steroids appeared in all four extracts. Secondary metabolites such as coumarins and tannins as well as primary metabolites like reducing sugars and fixed oils were detected in methanolic extract. In roots all four extracts exhibited alkaloids using Hager's reagent. In methanol and chloroform residues, iodine solution showed positive results, and in aqueous extract and methanol and chloroform residues Wagner's reagent revealed the presence of alkaloids, whereas in methanolic and aqueous extracts and methanol residues Mayer's reagent indicated the presence of alkaloids (Tables 35.1 and 35.2).

The seed cake from a Nigerian variety of *Sesamum indicum* L. was analysed for phytochemical properties. The qualitative phytochemical investigation of the seed cake revealed the presence of flavonoids, alkaloids, saponins, phenol, carbohydrate and steroids suggesting its usefulness in drug manufacture (Ajayi and Adeshina 2014).

As alkaloids were detected in all four extracts from all four tissues, this class of metabolites was further selected for quantitative analysis. In the present investigation during thin layer chromatography different solvent systems were standardized using caffeine as a chemical standard. Caffeine was found to be present in all four samples, i.e. seed, leaf, shoot and root (Fig. 35.1). The possible functional

Table 35.2 Detection of alkaloids in different parts of sesame G1 variety using different reagents

Plant parts used	G1 Alkaloids			
Extracts	ME	WE	MR	CR
Seed	D, H	W, M, D, H	H	W, M, H
Leaf	I, H	I	I, H, D	I, H, D
Shoot	I, M, H	W, H, M	W, I, H	I, H, D
Root	M, H	W, M, H	W, I, M, H	W, I, H

ME methanol extract, *WE* aqueous extract, *MR* methanol residue, *CR* chloroform residue, *W* Wagner's reagent, *I* iodine solution, *M* Mayer's reagent, *H* Hager's reagent, *D* Dragendorff's reagent

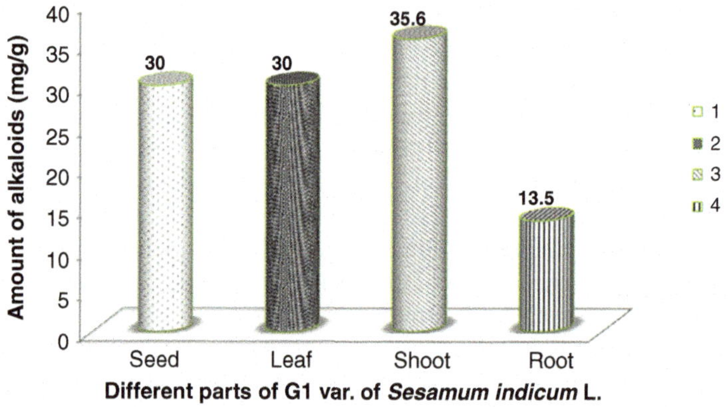

Fig. 35.1 Variations in the amount of alkaloids in different plant parts quantified by non-spectrophotometric method

molecules for therapeutic uses the crude aqueous and methanolic extracts derived from sesame seeds of local variety of Bangladesh were screened using thin layer chromatography and high performance liquid chromatography. The crude aqueous extracts showed the possibilities to present caffeine and cetirizine or its derivatives like molecules. On the other hand, the crude methanolic extract may contain loratadine or its derivatives like molecules (Saha et al. 2014).

The alkaloids present in different parts such as seed, leaf, shoot and root of same variety was quantified by non-spectrophotometric method and amount of alkaloids was represented in terms of mg/g of tissue. In G1 variety the highest amount of alkaloids, i.e. 35.6 ± 0.01799 mg/g were present in shoots whereas in seeds and leaves same amount of alkaloids, i.e. 30.0 ± 0.01124 and 30.0 ± 0.04994 mg/g were indicated, respectively. Roots of this variety showed the lowest amount of alkaloids, i.e. 13.5 ± 0.00163 mg/g (Table 35.3; Fig. 35.2).

Quantification of alkaloids from seed, leaf, shoot and root was accomplished by spectrophotometric method. The calibration curve was plotted using caffeine as a chemical standard and an equation was obtained with *biologically significant* value of R^2. The amount of alkaloids was represented in terms of mg/ml of extract. The highest amount of alkaloids, i.e. 1.6416 ± 0.007 mg/ml was present in shoots. Next

Table 35.3 Quantitative analysis of alkaloids present in different parts of G1 variety by weighing method

Plant parts used	G1 Alkaloid content (g/g)
Seed	0.030 ± 0.01124
Leaf	0.03 ± 0.04994
Shoot	**0.0356 ± 0.01799**
Root	0.0135 ± 0.00163

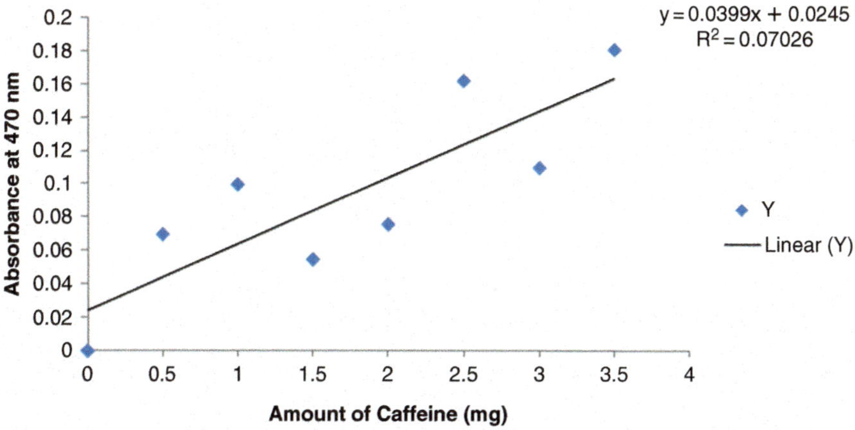

Fig. 35.2 Calibration curve for quantitative analysis by spectrophotometric method

to it in roots and leaves 1.4912 ± 0.008 and 1.2907 ± 0.0384 mg/ml were present, respectively. Seeds of this variety indicated the lowest amount of alkaloids, i.e. 0.66416 ± 0.02497 mg/ml (Table 35.4; Figs. 35.3 and 35.4).

Formula used

$$\text{Amount of alkaloids}\left(g/g\right) = \text{weight of residue} / \text{weight of sample or tissue}$$

Both the methods of quantification expressed the amount of alkaloids present in different terms. Within different parts of G1 variety, the highest amount of alkaloids was observed in shoots by both these methods. Amount quantified per gram of tissue by weighing method was found to be the lowest in roots whereas the same quantified per millilitre of extract by spectrophotometric method was observed to be the lowest in seeds.

Seeds, leaves, roots and whole plant of sesame from Abia State, South Africa, expressed alkaloids in variable amounts such as roots showed the lowest amount 1.96%, in seeds 6.28% and in whole plant 7.72% was found while the highest amount 10.04% was observed in leaves (Mbaebie et al. 2010).

Table 35.4 Quantitative analysis of alkaloids present in different parts of G1 variety by spectrophotometric method

Plant parts used	G1 Alkaloid content (mg/ml)
Seed	0.66416 ± 0.029
Leaf	1.2907± 0.038
Shoot	**1.6416± 0.007**
Root	1.4912± 0.008

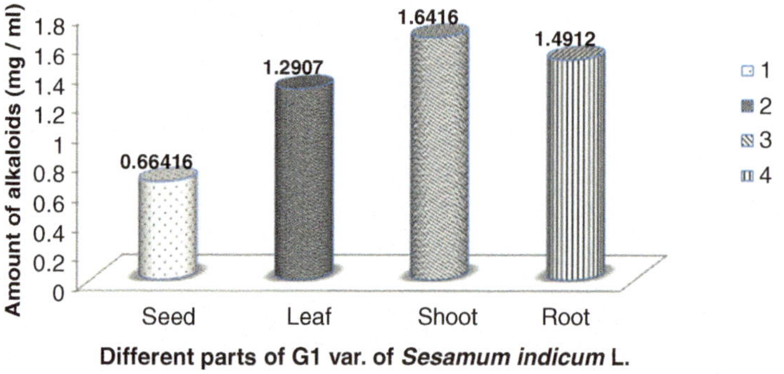

Fig. 35.3 Variations in the amount of alkaloids in different plant parts quantified by spectrophotometric method

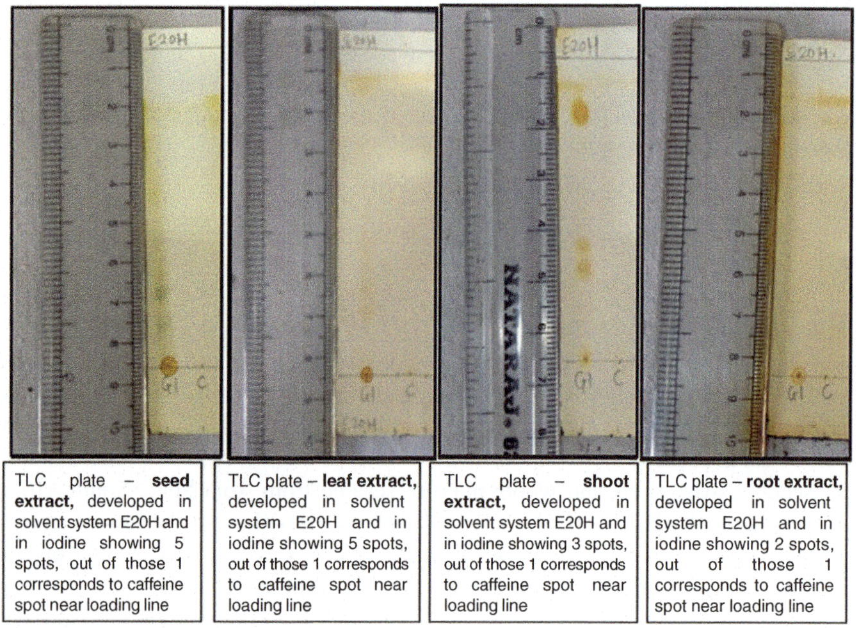

Fig. 35.4 TLC plates showing the presence of caffeine

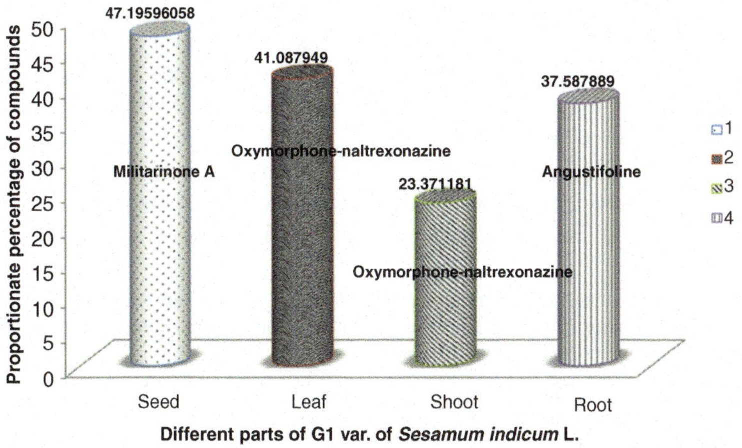

Fig. 35.5 Compounds detected proportionately using LCMS technique

The phytochemical constituents in white sesame seed oil from Kebbi State, Nigeria, extracted using Soxhlet apparatus, were investigated using the same method as of this study. It revealed the presence of alkaloids, saponins, flavonoids, tannins, steroids, terpenoids, anthraquinone and phenols. The amount of alkaloids quantified was 132.80 ± 0.15 mg/g which is four times more than the amount found in seeds and leaves of G1 variety (30.0 mg/g) and much higher than three times of amount present in shoots (35.6 mg/g) while almost 10 times higher than the quantity observed in roots of G1 variety (Sani et al. 2013). The fresh beniseeds from Kogi State of Nigeria showed 11.13% alkaloid content (Momoh et al. 2012).

Using LCMS technique different alkaloids present in different parts of sesame were detected and compounds present in proportionately highest amount were found. Seeds of G1 variety showed the presence of militarinone A in the highest (47.19%) amount whereas in roots angustifoline (37.58%) was present. In case of both leaves (41.08%) and shoots (23.37%) oxymorphone-naltrexonazine is the compound present in the highest amount but in leaves its amount is appreciably higher than in shoots (Fig. 35.5).

For the first time the identification of volatile compounds was performed in Indian genotypes of *Sesamum indicum* DT-46 (Uttar Pradesh), RT-46 (Rajasthan), RT-346 (Rajasthan), RT-127 (Rajasthan), T-12 (Uttar Pradesh), T-13(Uttar Pradesh) and TMV3 (Tamil Nadu). The oil extracted from seeds by Soxhlet method, was analysed by gas chromatography-mass spectrometry (GC-MS). A total of 84 different components were identified and out of which 12 are fatty acid components (Tyagi and Sharma 2014).

The different plant parts varied in composition of their primary and secondary metabolites. Quantification provides the scaling up for further separation and isolation of compounds through columns using chromatography technique. Techniques like thin layer chromatography and liquid chromatography–mass spectrophotometry can be used to explore diversity of metabolites.

Preliminary phytochemical screening of different plant extracts is very useful for determination of the active constituents in different solvents and their yields. Plants synthesize primary metabolites for their normal growth and development. Many primary metabolites can play the role of a precursors or pharmacologically active metabolites (Sharma and Sarin 2012)

The knowledge of chemical constituents in plants will be valuable for the synthesis of complex chemical substances and it is necessary to study pharmacological activities and medicinal importance of plants (Savithramma et al. 2011).

Acknowledgements I am thankful to Dr. R.S. Zunjarrao, Head and Principal, Department of Botany, Modern College of Arts, Science and Commerce, Shivajinagar, Pune-411005, Maharashtra, India, for all the support to carry out this study.

References

Ajayi IA, Adeshina AI (2014) Chemical analysis and toxicological assessment of *Sesamum indicum* seed cake on albino rats. IOSR-JESTFT 8(7):60–66

Anilakumar KR, Pal A, Khanum F, Bawa AS (2010) Nutritional, medicinal and industrial uses of sesame (*Sesamum indicum* L.) seeds—an overview. Agric Conspec Sci 75(4):159–168

Asghar A, Majeed MN (2013) Chemical characterization and fatty acid profile of different sesame verities in Pakistan. Am J Sci Ind Res 2013:540–545

Biradar SR, Rachetti BD (2013) Extraction of some secondary metabolites & thin layer chromatography from different parts of *Centella asiatica* L. (URB). Am J Life Sci 1(6):243–247

Cordell GA, Quinn-Beattie ML, Farnsworth NR (2001) The potential of alkaloids in drug discovery. Phytother Res 15(3):183–205

Gokbulut C (2010) Sesame oil: potential interaction with P450 isozymes. J Pharmacol Toxicol 5(8):469–472

Gracelin DH, Britto A, Kumar BJ (2013) Qualitative and quantitative analysis of phytochemicals in five *Pteris* species. Int J Pharm Pharm Sci 5:105–107

Harborne JB (1998) Phytochemical methods a guide to modern techniques of plant analysis, 3rd edn. Springer Science & Business Media, New Delhi, pp 207–214

Hassan MAM (2013) Studies on Egyptian sesame seeds (*Sesamum indicum* L) and its products. 2. Effect of roasting conditions on peroxide value, free acidity, iodine value and antioxidant activity of sesame seeds (*Sesamum indicum* L.). World J Dairy Food Sci 8(1):11–17

Jannat B, Oveisi MR, Sadeghi N, Hajimahmoodi M, Behzad M, Nahavandi B, Tehrani S, Sadeghi F, Oveisi M (2013) Effect of roasting process on total phenolic compounds and γ-tocopherol contents of Iranian sesame seeds (*Sesamum indicum*). Iran J Pharm Res 12(4):751

John BI, Sulaiman CT, George S, Reddy VR (2014) Spectrophotometric estimation of total alkaloids in selected *Justicia* species. Int J Pharm Pharm Sci 6(5):647–648

Kaliyamoorthy J, Satheeskannan TM, Thamizhiniyan P, Vijayarengan P (2015) Evaluation of phytochemicals analysis, medicinal properties and nutritional value of sesame seeds (*Sesamum indicum.* L). Int J Mod Biol Med 6(2):129–135

Kumar MS, Selvakumar S, Rao MR, Anbuselvi S (2013) Preliminary phytochemical analysis of *Dodonaea viscosa* leaves. Asian J Plant Sci Res 3:43–46

Mamta S, Jyoti S (2012) Phytochemical screening of *Acorus calamus* and *Lantana camara*. Int Res J Pharm 3(5):324–326

Mbaebie B, Omosun G, Uti A, Oyedemi S (2010) Chemical composition of *Sesamum indicum* L. (Sesame) grown in Southeastern Nigeria and the physicochemical properties of the seed oil. Seed Sci Biotechnol 4(1):69–72

Momoh AO, Adebolu TT, Ogundare AO (2012) The immunostimulatory potential of fermented beniseeds (*Sesamum Indicum* Linn) liqour on albino rats. Sci J Microbiol 2012:sjmb-277, 5 p. https://doi.org/10.7237/sjmb/277

Morris JB (2002) Food, industrial, nutraceutical, and pharmaceutical uses of sesame genetic resources. Trends New Crops New Uses 2002:153–156

Okoh-Esene RU, Okogun JI, Okwute SK, Thomas SA (2012) Preliminary phytochemical and mineral analysis of the root of *Hippocratea welwitschii*. Arch Appl Sci Res 4:315–322

Okon JE, Umoh NS (2013) Influence of *Sesamum indicum* L. ethanolic leaf extract on haematological paramaters on albino rats in akwaibom state, Nigeria. Indian J Fundam Appl Life Sci 3(1):106–110

Patil GG, Mali PY, Bhadane VV (2008) Folk remedies used against respiratory disorders in Jalgaon district, Maharashtra. Nat Prod Rad 7(4):354–358

Reshma MV, Namitha LK, Sundaresan A, Ravi Kiran C (2013) Total phenol content, antioxidant activities and α-glucosidase inhibition of sesame cake extracts. J Food Biochem 37(6):723–731

Sadasivam S (1996) Biochemical methods, 3rd edn. New Age International, New Delhi, pp 256–257

Saha RK, Dinar MA, Nabila KA, Roy P (2014) HPLC analysis and cell surface receptor binding activities of the crude aqueous and methanolic extract of *Sesamum indicum*. Asian Pac J Trop Biomed 1(4):S 516–S 520

Sani I, Sule FA, Warra AA, Bello F, Fakai IM, Abdulhamid A (2013) Phytochemicals and mineral elements composition of white *Sesamum indicum* L. seed oil. Int J Trad Nat Med 2(2):118–130

Sani I, Okpalaoka CC, Bello F, Warra AA, Abdulhamid A (2014) Flavonoid content and antioxidant potential of white and brown sesame seed oils. EJBPS 1(3):56–63

Savithramma N, Rao ML, Suhrulatha D (2011) Screening of medicinal plants for secondary metabolites. Middle-East J Sci Res 8(3):579–584

Shamsa F, Monsef H, Ghamooshi R, Verdian-rizi M (2008) Spectrophotometric determination of total alkaloids in some Iranian medicinal plants. Thai J Pharm Sci 32:17–20

Sharma P, Sarin R (2012) In vivo and in vitro phytochemical evaluation and quantification of primary metabolites from *Sesamum indicum*. Int J Res Pharm Biomed Sci 3(3):1164–1166

Somkuwar DO, Kamble VA (2013) Phytochemical screening of ethanolic extracts of stem, leaves, flower and seed kernel of *Mangifera indica* L. Int J Pharm Bio Sci 4(2):383–389

Tabasum S, Khare S, Jain K (2016) Spectrophotometric quantification of total phenolic, flavonoid, and alkaloid contents of *Abrus precatorius* L. seeds. Asian J Pharm Clin Res 9(2):371–374

Thamaraiselvi LP, Jayanthi P (2012) Preliminary studies on phytochemicals and antimicrobial activity of solvent extracts of *Eichhornia crassipes* (Mart.) Solms. Asian J Plant Sci Res 2(2):115–122

Tyagi R, Sharma V (2014) A comparison of volatile compounds in different genotypes of *Sesamum indicum* L. by GC-MS. Int J Pharm Sci Res 5(1):249

Warra A, Umar R, Atiku F, Nasiru A, Gafar M (2012) Physical and phytochemical characteristics of seed oils from selected cultivars grown in northern Nigeria. Res Rev J Agric Allied Sci 1:4–7

36 Phytochemical Investigation and Comparative Evaluation of Various Market Samples of *Triphala* Powder from India with References to Their Free Scavenging and Anti-diabetic Activity: An In Vitro Approach

P. Pavani, P. Rohini, S. M. Khasim, and P. Bhagyasree

Abstract

During the past few decades, there has been exponential growth in the field of herbal medicines. Most of the traditional system of medicine is effective but they lack standardization. So there is a need to develop standardization techniques. Standardization of herbal formulation is essential in order to assess the quality, purity, safety and efficacy of the drug based on their concentration of their principles. This paper reports on standardization of *Triphala* churna. The present research study deals with the comparative study of four samples of marketed *Triphala* churna from Guntur, Vijayawada local made and also Dabur and Baidyanath brands. The comparative study of 4 samples of organoleptic characteristics and physical properties such as moisture content, ash value, extractive values and crude fibre content has been carried out. Phytochemical screening tests were also carried out to ascertain the quality, purity and safety of these herbal formulations. The in vitro studies on percentage of α-amylase inhibition activity as well as scavenged hydrogen peroxide activity results were tabulated by concentration and absorbance. Thus, the *Triphala* powder possesses very good potential activity for anti-diabetic and antioxidant activities. The present study indicates that Dabur brand (*Triphala* powder) would be the best formulation for anti-diabetic and antioxidant activity. It would not be surprising therefore to use the plant samples to cure certain types of illness in humans and animals.

P. Pavani · S. M. Khasim (✉)
Department of Botany and Microbiology, University College of Sciences, Acharya Nagarjuna University, Guntur, India

P. Rohini · P. Bhagyasree
Department of Pharmaceutics, University College of Pharmaceutical Sciences, Acharya Nagarjuna University, Guntur, India

S. M. Khasim et al. (eds.), *Medicinal Plants: Biodiversity, Sustainable Utilization and Conservation*, https://doi.org/10.1007/978-981-15-1636-8_36

This obtained information will be helpful as a primary platform for further phytochemical and pharmacological studies.

Keywords
Triphala churna · α-Amylase inhibition · Scavenged hydrogen peroxide · Ayurveda

36.1 Introduction

Most of the traditional system of medicine is very effective but they lack standardization. So there is a need to develop standardization techniques. Standardization of herbal formulation is essential in order to assess the quality, purity, safety and efficacy of the drug. Nature always stands as a golden mark to exemplify the outstanding phenomenon of symbiosis. Today about 80% of people in developing countries still relay on traditional medicine based largely on the different plants species for their primary health care. About 500 plants with medicinal uses are mentioned in ancient literature and 800 plants have been used in indigenous system of medicine (Chopra et al. 1956). The various indigenous systems such as Ayurveda, Siddha and Unani have been using several plant species to treat different ailments (Sane 2002).

Triphala is an Ayurvedic herbal formula consisting of a mixture of equal proportions of three ingredients, e.g. Amalaki (*Emblica officinalis*), Bibhitaki *(Terminalia bellirica*) and Haritaki (*Terminalia chebula*) (Jeganathan and Kannan 2008). *Triphala* has been reported to possess anti-ageing properties and improves the mental faculties (The Ayurvedic Formulary of India 2003). *Triphala* is reported to have immunomodulatory (The Wealth of India 2005), anticancer (Srikumar et al. 2005), antimicrobial (Sandhya et al. 2006), wound healing (Sandhya and Mishra 2006), hypolipidemic (Biradar et al. 2008), anti-inflammatory (Kumar et al. 2008), chondroprotective (Saravanan et al. 2007a), radioprotective (The Ayurvedic Formulary of India 2003), anti-diabetic (Rasool and Sabina 2007) and antioxidant (Sumantran et al. 2007) properties. *Triphala* contains many biological compounds. It mainly consists of tannins, gallic acid, chebulagic acid and chebulinic acid. In Ayurvedic terms, *Triphala*, used in moderation, is said to have a beneficial effect on all three doshas—vata, pitta and kapha. The synergy of the three "fruits" (Amalaki—*Phyllanthus emblica*, Bibhitaki—*Terminalia bellirica* and Haritaki—*T. chebula*) produces the tonifying, detoxifying, mildly antiseptic, rejuvenative and laxative effects of this classic formulation.

36.2 Triphala

Amalaki is known by the botanical name *Emblica officinalis* and also known in Sanskrit as Dhatri (The nurse), which is a reference to its incredible healing properties. It has its beneficial role in cancer, diabetes, liver treatment, heart trouble, ulcer, anaemia and various other diseases. Similarly, it has application as immunomodulatory, antipyretic, analgesic, cytoprotective, antitussive and gastroprotective agent.

T. chebula belongs to the family Combretaceae. The fruit of the tree has been used as traditional medicine for household remedy against various human ailments. It has been extensively used in Ayurveda, Unani and Homoeopathic medicine and has become a cynosure of modern medicine. It has also shown anti-mutagenic/anti-carcinogenic activity, antioxidant activity, adaptogenic and anti-anaphylactic activities and immunomodulatory activity and is also effective in hypolipidemia/hypercholesterolemia, improving gastro-intestinal motility with antispasmodic activity, diabetes, retinopathy, and wound healing (Chattopadhyay and Bhattacharyya 2007).

T. bellirica (Combretaceae), commonly known as "bellericmyrobalan" (locally as "bahera"), is a large deciduous tree. Its fruit is used in folk medicine to treat asthma, cancer, colic, diarrhoea, dysuria, headache, hypertension, inflammations and pain. *T. bellirica* possesses antioxidant, antispasmodic, bronchodilatory, hypercholesterolemic, antibacterial, cardioprotective, hepatoprotective, hypoglycemic and hypotensive properties (Khan and Gilani 2010).

36.2.1 Sample Collection and Preparation of Crude Extract

Triphala powder consisting of *Terminalia chebula*, *T. bellirica* and *Phyllanthus emblica* (*Triphala*) was bought from the local markets in Guntur and Vijayawada. Besides this, standard brands, such as Dabur and Baidyanath, were also employed in this study. The four samples of *Triphala*, each 1 kg of powder, were soaked in ethanol for 7 days with intermittent shaking and the solvent was filtered with Whatman filter paper. The filtrate was evaporated under vacuum drier and the brown mass residue obtained was stored at −4 °C for further use.

Sample 1—Guntur local; Sample 2—Vijayawada local; Sample 3—Dabur brand; Sample 4—Baidyanath brand.

UV-Visible spectrophotometer (Teccomp UV-2301 double beam), Ultrasonicator (Sonicator 1.3L), Analytical balance (Denver electronic analytical balance SI-234) and Rota evaporator have been used in this study.

36.2.2 Development of Standardization Parameters for *Triphala* churna

36.2.2.1 Study of Organoleptic Characters

The polyhedral formulation is studied for organoleptic characters like colour, odour and taste using the sensory organs of our body.

36.2.2.2 Determination of Physicochemical Parameters (Indian Pharmacopoeia 1996)

Determination of Total ash: About 2–3 g of sample was accurately weighed in a tarred silica dish at a temperature not exceeding 450 °C until it was free from carbon. Then it was cooled and weighed. The percentage of total ash was calculated with reference to the air-dried drug.

Determination of acid-insoluble ash: The total ash obtained was boiled for 5 min with 25 mL of dilute hydrochloric acid; the insoluble matter obtained was collected on an ash-less filter paper, washed with hot water and ignited to constant weight. The percentage of acid-insoluble ash was calculated with reference to the air-dried drug.

Determination of Water-soluble ash: The ash obtained in the determination of total ash was boiled for 5 min with 25 mL of water. The insoluble matter was collected on an ash-less filter paper and washed with hot water. The insoluble ash was transferred into a tarred silica crucible and ignited for 15 min at a temperature not exceeding 450 °C. The weight of the insoluble matter was subtracted from the weight of the total ash. The difference in weight was considered as the water-soluble ash and it was calculated with reference to the air-dried drug.

Determination of loss and drying: 10 g of the sample (without preliminary drying) was weighed and placed in a tarred evaporating dish. It was dried at 105 °C for 5 h and at 1 h interval until difference between two successive weighing corresponded to not more than 0.25%.

Determination of Water-soluble extractive: 5 g of the test sample was weighed and macerated with 100 mL of chloroform water in a closed flask for 24 h, shaking frequently for 6 h and allowing standing for 18 h. It was filtered rapidly, taking precautions against the loss of solvent. 25 mL of the filtrate was taken and evaporated to dryness in a tarred flat-bottomed shallow dish at 105 °C and was weighed. The percentage of water-soluble extractive was calculated with reference to the air-dried sample.

Determination of alcohol-soluble extractive: Procedure for water-soluble extractive was followed for the determination of alcohol-soluble extractive but 90% ethanol was used instead of chloroform water.

Determination of crude fibre content: Mix about 2 g of the powdered drug in no. 60 with 50 mL of 10% nitric acid. Bring to boil and maintain at the boiling point for 30 s. Dilute with water and strain through a fine filter cloth held over the mouth of filter funnel. Transfer the washed residue to beaker and boil further 30 s with 50 mL of a 2.5% solution of sodium hydroxide. Collect and wash residue as before, mount and examine.

36.2.2.3 Determination of Physical Characteristics (USP-31 2008)

Bulk density: It is the ratio of given mass of powder and its bulk volume. It is determined by transferring an accurately weighed amount of powder sample to the graduated cylinder with the aid of a funnel. The initial volume was noted. The ratio of weight of the volume it occupied was calculated.

$$\text{Bulk density} = w / V_0 \ \text{g} / \text{mL}$$

where w = mass of the powder and V_0 = untapped volume.

Tapped density: It is measured by transferring a known quantity (25 g) of powder into a graduated cylinder and tapping it for a specific number of times. The initial volume was noted. The graduated cylinder was tapped continuously for a period of 10–15 min. The density can be determined as the ratio of mass of the powder to the tapped volume.

$$\text{Tapped volume} = w / V_{\mathrm{f}} \text{ g} / \text{mL}$$

where w = mass of the powder, V_f = tapped volume.

Compressibility index/Carr's index: It is the propensity of the powder to be compressed. Based on the apparent bulk density and tapped density the percentage compressibility of the powder can be determined using the following formula.

$$\text{Compressibility / Carr's index} = \left[\left(V_0 - V_{\mathrm{f}} \right) / V_0 \right] \times 100$$

Hausner's ratio: It indicates the flow properties of the powder. The ratio of tapped density to the bulk density of the powder is called Hausner's ratio.

$$\text{Hausner's ratio} = \text{Tapped density / bulk density}$$

Angle of repose: The internal angle between the surface of the pile of powder and the horizontal surface is known as the angle of repose. The powder is passed through funnel fixed to a burette at *s* height of 4 cm. A graph paper is placed below the funnel on the table. The height and the radius of the pile were measured. Angle of repose of the powder was calculated using the formula:

$$\text{Angle of repose} = \tan - 1(h / r)$$

where h = height of the pile; r = radius of the pile.

36.2.3 Qualitative Phytochemical Screening (Saravanan et al. 2007b)

The prepared test extracts were analysed for the presence of alkaloids, glycosides, saponins, fixed oils, phytosterols, phenols, flavonoids, gum, mucilages, etc. The presence of phytochemicals extracted in different solvents was confirmed by standard protocols.

Test for carbohydrates (Test for tannins)

To the extracts, a few drops of 10% ferric chloride solution were added. Appearance of a green or blue colour indicates the presence of tannins.

Test for terpenoids (Salkowski test)

Five mL of the extracts were mixed with 2 mL of chloroform and 3 mL of conc. H_2SO_4 solution. A reddish brown colour at the interphase indicates the presence of terpenoids.

Test for alkaloids (Mayer's test)

Extracts were treated with Mayer's reagent (potassium mercuric chloride). Formation of a yellow coloured precipitate indicates the presence of alkaloids.

Test for proteins (Xanthoprotein test)

To the leaf extracts 20% NaOH solution were added and the formation of an orange colour confirms the presence of proteins which is characteristic for ammonia formation.

Test for cardiac glycosides (Keller-Killani test)

Five mL of test extracts were treated with 2 mL of glacial acetic acid containing 2–3 drops of ferric chloride solution and 1 mL of conc. H_2SO_4 solution. Appearance of a green ring initially which first turns violet and then to brown at the interphase indicates the presence of cardiac glycosides.

Test for saponins (Foam test)

Two mL of the extracts were diluted with 20 mL of distilled water, shaken vigorously and was observed for a stable persistent froth.

Test for phenolic compounds (Ferric chloride test)

Two mL of diluted extracts were treated with dil. $FeCl_3$ solution. Appearance of a violet colour indicates the presence of phenol-like compounds.

Test for amino acids (Ninhydrin test)

Two drops of ninhydrin solution (10 mg of ninhydrin in 200 mL of acetone) were added to 2 mL of aqueous filtrates. A characteristic purple colour indicates the presence of amino acids.

36.2.4 In Vitro Studies by Spectrophotometric Method

36.2.4.1 α-Amylase Inhibition Activity

Extracts (250 μL) and 0.02 mol/L sodium phosphate buffer (pH 6.9 with 0.006 mol/L NaCl) containing porcine pancreatic α-amylase enzyme (0.5 mg/mL) was incubated at 25 °C for 10 min. 250 μL of 1% starch solution in 0.02 M sodium phosphate buffer (pH 6.9 with 0.006 mol/L NaCl) was added to the reaction mixture after incubation. Subsequently, the reaction mixture was incubated at 25 °C for 10 min, followed by the addition of 2 mL of dinitrosalicylic acid (DNS). Finally the reaction was stopped by incubation of boiling water bath for 5 min and cooled to room temperature. The reaction mixture was diluted with 10 mL distilled water and the absorbance was measured at 540 nm. The mixture of all other reagents and the enzyme except the test sample was used as a control and the results of α-amylase inhibition activity were expressed in terms of inhibition percentage (Sharma et al. 2014).

$$\%\ \text{Inhibition} = \frac{\text{Enzyme activity of control} - \text{Enzyme activity of extract}}{\text{Enzyme activity of control}} \times 100$$

36.2.4.2 Percentage of Scavenged Hydrogen Peroxide

A solution of hydrogen peroxide (20 mM) was prepared in phosphate buffer saline (PBS, pH 7.4). Different concentrations of the extract (20–100 μg/mL) in ethanol (1 mL) were added to 2 mL of hydrogen peroxide solution in PBS. After 10 min the absorbance was measured at 230 nm against a blank solution that contained hydrogen peroxide solution without the extract.

The percentage of H_2O_2 scavenging of the plant extract was calculated as follows:

$$\% \text{ Scavenged hydrogen peroxide} = \frac{\text{Absorbance control} - \text{Absorbance sample}}{\text{Absorbance control}} \times 100$$

36.3 Triphala: Free Scavenging and Anti-diabetic Activity

The following *Triphala* powder samples have been taken to study their free scavenging and anti-diabetic properties.

Sample 1—Guntur local; Sample 2—Vijayawada local; Sample 3—Dabur brand; Sample 4—Baidyanath brand.

36.3.1 Determination of Organoleptic Characters

Organoleptic characters were determined by the sensory organs of our body of different samples through the sensations. Organoleptic organs of our body, i.e. *nose* stands for smell of the given powder, *eyes* for vision of the powder colour and *tongue* with taste buds for the taste of the powder. The characteristic determination of the four samples was observed as below (Table 36.1).

36.3.2 Determination of Physicochemical Parameters

The sample powders of *Triphala* churna were evaluated in order to establish its quality control parameters and also be controlled in order to ensure the stability of product. The results are shown in Table 36.2.

36.3.3 Evaluation of Physical Characteristics

Results of physical evaluation are shown in Table 36.3. The bulk density indicates the size for a best choice for its packaging. The angle of repose shows the excellent flow ability from the package, thereby it can be easily taken out from the packaging/container. The Carr's index and Hausner's ratio indicate the possible flow ability which could be employed as a useful parameter.

Table 36.1 Determination of organoleptic characters

Characteristics	Sample 1	Sample 2	Sample 3	Sample 4
Colour	Light yellow	Yellowish	Yellowish	Greenish
Odour	Characteristics	Characteristics	Characteristics	Characteristics
Taste	Very bitter	Astringent	Astringent	Astringent

Table 36.2 Determination of physiochemical parameters

Characteristics	Sample 1	Sample 2	Sample 3	Sample 4
Total ash	5.21	6.48	6.56	5.74
Acid-insoluble ash	2.5	3.1	3.8	2.9
Water-soluble ash	2.25	3.75	4.31	3.18
Moisture content/loss on drying	0.89	1.2	1.6	1.1
Water-soluble extractive	2.28	2.76	3.35	3.14
Alcohol-soluble extractive	1.29	1.59	2.14	1.8
Crude fibre content	3.85	4.09	4.18	4.01

Table 36.3 Determination of physical characteristics

Characteristics	Sample 1	Sample 2	Sample 3	Sample 4
Bulk density	0.479	0.613	0.535	0.564
Tapped density	0.627	0.781	0.895	0.653
Angle of repose	34.69	35.47	39.16	38.69
Compressibility	23.73	28.73	30.67	27.59
Hausner's ratio	1.24	1.28	1.46	1.31

36.3.4 Qualitative Phytochemical Screening Tests

Results of phytochemical analysis are shown in Table 36.4. Ethanolic extract from the samples showed the presence of carbohydrates, tannins, terpenoids and phenolic compounds while the alkaloids, proteins, cardiac glycosides, saponins and amino acids were found to be absent.

36.3.5 In Vitro Studies by Spectrophotometric Method

36.3.5.1 Comparative Study Among Various Samples for Their Various Activities

Comparative studies have been made among different samples for their α-amylase inhibition and scavenging properties (Figs. 36.1 and 36.2).

(a) *Percentage of amylase inhibition activity*: The present study indicates that better performance of α-amylase inhibitory activity showed by the Dabur brand, followed by Baidyanath and others (Table 36.5, Fig. 36.1).
(b) *Percentage of scavenged hydrogen peroxide activity*: Comparative study showed that Dabur sample had performed better with respect to radical scavenging (Table 36.6, Fig. 36.2).

In the past few years, interest in the search of new natural antioxidants has grown because of many diseases caused by reactive oxygen species (ROS) production and oxidative stress. The use of synthetic antioxidants generally leads to problems of toxicity. Free radical is a molecule with an unpaired electron and is involved in bacterial and

Table 36.4 Qualitative phytochemical screening tests

Tests	Sample 1	Sample 2	Sample 3	Sample 4
Tannins	+	+	+	+
Terpenoids	+	+	+	+
Alkaloids	–	–	–	–
Proteins	–	–	–	–
Cardiac glycosides	–	–	–	–
Saponins	–	–	–	–
Phenolic compounds	+	+	+	+
Amino acids	–	–	–	–

(+) present, (–) absent

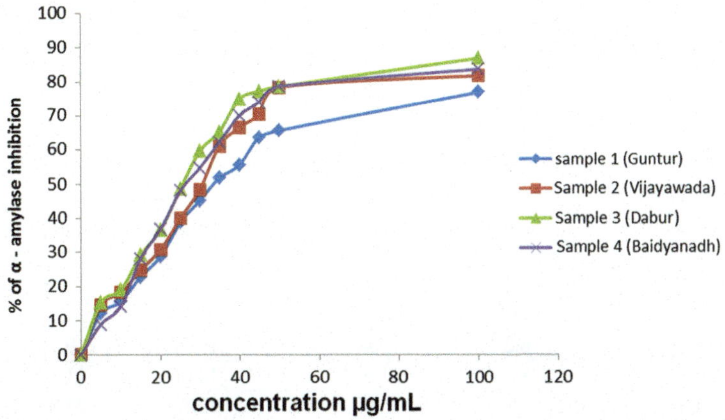

Fig. 36.1 Graphical representation of comparative study of different samples of % α-amylase inhibition activity

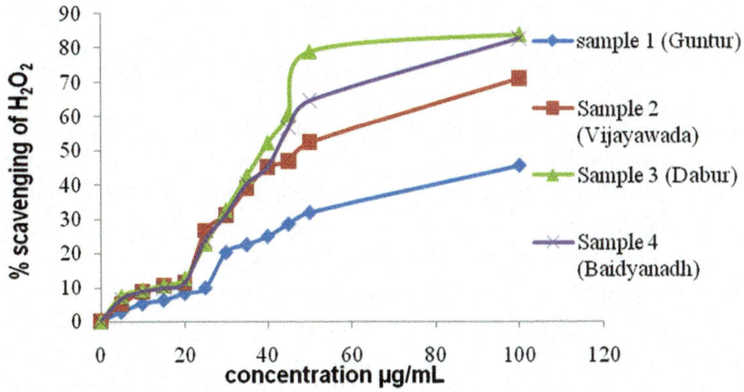

Fig. 36.2 Graphical representation of comparative study of different samples of % scavenged hydrogen peroxide activity

Table 36.5 Comparative study of different samples showing the percentage of amylase inhibition activity

	α-Amylase inhibition activity				
S. no.	Concentration (μg/mL)	Sample 1 (Guntur)	Sample 2 (Vijayawada)	Sample 3 (Dabur)	Sample 4 (Baidyanath)
1	5	12.02	14.59	15.29	8.87
2	10	15.52	18.2	18.9	14.12
3	15	22.75	24.74	29.05	28
4	20	28.7	30.8	36.64	36.87
5	25	38.74	40.02	48.54	47.958
6	30	45.04	48.31	59.51	54.61
7	35	51.69	61.03	65.23	62.19
8	40	55.42	66.51	74.8	69.89
9	45	63.59	70.36	77.13	73.98
10	50	65.58	78.3	78.53	78.53
11	100	76.78	81.68	86.81	83.55

Table 36.6 Comparative study of different samples showing the percentage of scavenged hydrogen peroxide activity

S. no.	Concentration (μg/mL)	Sample 1 (Guntur)	Sample 2 (Vijayawada)	Sample 3 (Dabur)	Sample 4 (Baidyanath)
1	5	2.8	5.2	7.4	6.49
2	10	5.2	8.7	9.17	8.71
3	15	6.2	10.46	10.57	9.76
4	20	8.2	11.39	12.56	11.39
5	25	9.8	26.6	22.8	23.88
6	30	20.2	31.1	32.5	31.11
7	35	22.5	39.04	42.4	40.21
8	40	24.9	45.11	52.1	45.11
9	45	28.5	46.9	60.5	56.55
10	50	31.8	52.3	78.7	64.48
11	100	45.4	71.02	83.7	82.6

parasitic infections, lung damage, inflammation, reperfusion injury, cardiovascular disorders, atherosclerosis, ageing and neoplastic diseases (Thomas and Kalyanaraman 1997). They are also involved in autoimmune disorders like rheumatoid arthritis (Halliwell and Gutteridge 1989; Beckman and Ames 1998). Therefore, research for the determination of the natural antioxidants is need of the hour.

In living systems, free radicals are constantly generated and they can cause extensive damage to tissues and biomolecules leading to various disease conditions, especially degenerative diseases and extensive lysis (Oyaizu 1986). Many synthetic drugs protect against oxidative damage but they have adverse side effects. An alternative solution to the problem is to consume natural antioxidants from food supplements and traditional medicines (Halliwell and Gutteridge 1989; Yazdanparast and Ardestani 2008). Recently, many natural antioxidants have been isolated from different plants materials (Yazdanparast and Ardestani 2008; Packer and Ong 1997).

Each individual samples of *Triphala* churna is subjected to determination of moisture content, tapped bulk density and angle of repose and their values are shown in Table 36.2. Angle of repose values ranged from different samples of ingredient; this indicates the ingredient possess fair to possible flow properties inherently. *Triphala* churna is prepared by mixing equal quantities of herbal powders of Amla, Bibhitaki and Haritaki. Total ash values, water-soluble ash, acid-insoluble ash, water-soluble extract and alcohol-soluble extract of *Triphala* churna are determined and their values are mentioned in Table 36.3.

Triphala also contains other bioactive compounds such as flavonoids (e.g. quercetin and luteolin), saponins, anthraquinones, amino acids, fatty acids and various carbohydrates (Belapurkar et al. 2014). In addition, *Triphala*-derived polyphenols such as chebulinic acid are also transformed by the human gut microbiota into bioactive metabolites, which have demonstrated potential in vitro to prevent oxidative damage (Olennikov et al. 2015).

36.4 Conclusion

Triphala is a polyherbal formulation made up of fruit of *Emblica officinalis*, *Terminalia chebula* and *Terminalia bellirica*. It is rich in tannins, terpenoids and phenolic compounds. The study shows the linear progression of the α-amylase inhibition and scavenged hydrogen peroxide activities in all the four samples (local samples of Guntur and Vijayawada, Dabur and Baidyanath brands). Among these four samples Dabur brand performs well, followed by Vijayawada local sample and Baidyanath sample. This is only a preliminary in vitro study. Further elaborate studies including in vivo are needed to assess the various therapeutic properties of market samples.

References

Anonymous (1996) Indian Pharmacopoeia. Govt. of India, Ministry of Health, Controller of Publication, New Delhi

Anonymous (2008) The United State Pharmacopeia (USP 31), The National Formulary (NF 26), Asian edition, by authority of The United States Pharmacopeial Convention, vol I, 188,189,231,639,640

Beckman KB, Ames BN (1998) The free radical theory of aging matures. Physiol Rev 78(2):547–581

Belapurkar P, Goyal P, Tiwari-Barua P (2014) Immuno modulatory effects of *triphala* and its individual constituents: a review. Indian J Pharm Sci 76(1):467–475

Biradar YS, Jagatap S, Khandelwal KR, Singhania SS (2008) Exploring of antimicrobial activity of *Triphala* Mashi—an ayurvedic formulation. Evid Based Complement Alternat Med 5(1):107–113

Chattopadhyay RR, Bhattacharyya SK (2007) *Terminalia chebula*: an update. Pharmacogn Rev 1(6):151–156

Chopra RN, Nayar SL, Chopra IC (1956) Glossary of Indian medicinal plants. Council of Scientific and Industrial Research, New Delhi, p 17

Halliwell B, Gutteridge JMC (1989) Free radicals in biology and medicine, vol 3. Oxford University Press, New York, pp 617–783

Jeganathan NS, Kannan K (2008) HPTLC method for the estimation of ellagicand gallic acid in *Triphala* churanum formulations. Res J Phytochem 2(1):1–9

Khan A, Gilani AH (2010) Antisecretory and analgesic activities of *Terminalia bellerica*. Afr J Biotechnol 9(1):717–719

Kumar MS, Kirubanandan S, Sripriya R, Sehgal PK (2008) *Triphala* promotes healing of infected full-thickness dermal wound. J Surg Res 144(2):94–101

Olennikov DN, Kashchenko NI, Chirikova NK (2015) *In vitro* bio accessibility, human gut microbiota metabolites and hepatoprotective potential of chebulic ellagitannins: a case of Padma Hepaten formulation. Nutrients 7(1):8456–8477

Oyaizu M (1986) Studies on products of browning reaction: antioxidative activities of products of browning reaction prepared from glucosamine. Jpn J Nutr 44(4):307–315

Packer L, Ong AS (1997) Biological oxidants and antioxidants: molecular mechanisms and health effects. AOCS Press, Champaign, IL, pp 419–422

Rasool M, Sabina EP (2007) Antiinflammatory effect of the Indian ayurvedic herbal *Triphala* on adjuvant-induced arthritis in mice. Phytother Res 21(2):889–894

Sandhya T, Mishra KP (2006) Cytotoxic response of breast cancer cell lines, MCF 7 and T 47 D to *triphala* and its modification by antioxidants. Cancer Lett 238(1):304–313

Sandhya T, Lathika KM, Pandey BN, Mishra KP (2006) Potential of traditional ayurvedic formulation, *Triphala*, as a novel anticancer drug. Cancer Lett 231(3):206–214

Sane RT (2002) Standardization, quality control and GMP for herbal drug. Indian Drugs 39(3):184–190

Saravanan S, Srikumar R, Manikandan S (2007a) Hypolipidemic effect of *Triphala* in experimentally induced hypercholesteremic. Yakugaku Zasshi 27(9):385–388

Saravanan S, Srikumar R, Manikandan S, Jeya Parthasarathy N, Sheela Devi R (2007b) Hypolipidemic effect of *Triphala* in experimentally induced hypercholesteremic rats. Yakugaku Zasshi 127(2):385–388

Sharma S, Gupta M, Bhadhuria R (2014) Mycobiota of commercially available *Triphala* powder: a well known dietary supplement of indian system of medicine. J Mycol 1(1):836036

Srikumar R, Jeya Parthasarathy N, Sheela Devi R (2005) Immunomodulatory activity of *triphala* on neutrophil functions. Biol Pharm Bull 28(1):1398–1403

Sumantran VN, Kulkarni AA, Harsulkar A, Wele A, Koppikar SJ, Chandwaskar R, Gaire V, Dalvi M, Wagh UV (2007) Hyaluronidase and collagenase inhibitory activities of the herbal formulation *Triphala* guggulu. J Biosci 32(1):755–761

The Ayurvedic Formulary of India (2003) Part I and 2nd revised English edition. Ministry of Health and Family Welfare, Dept. of Indian System of Medicine & Homoeopathy, p 110

The Wealth of India (2005) A dictionary of Indian raw materials and industrial products Ist supplement series, vol 3(3), pp 77–79, CSIR, New Delhi

Thomas CE, Kalyanaraman B (1997) Oxygen radicals and the disease process, vol 4(3). Hardwood Academic Publishers, Amsterdam, pp 46–51

Yazdanparast R, Ardestani A (2008) Nosturtium officinale reduces oxidative stress and enhances antioxidant capacity in hypercholesterolemic rats. Chem Biol Interact 172:176–184

In Vitro Anticancer Activity of *Canthium parviflorum* Lam. Extracts Against Cancer Cell Lines

37

Sirigiri Chandra Kala and Kandru Ammani

Abstract

In this study, the methanol extracts of *Canthium parviflorum* leaf, leaf-derived callus, and root extracts were evaluated for the anticancer activity against the Caco2 human epithelial colorectal adenocarcinoma cell lines and liver cancer (HepG2) cell lines were used to screen the in vitro anticancer screening with the *C. parviflorum* extracts. The concentration ranges from 5 to 100 μg of *C. parviflorum* crude extracts were screened against Caco2 and HepG2 cells. The antiproliferative activity of *C. parviflorum* was performed by 3-[4,5-Di methyl thiazol-2-yl]-2,5-diphenyltetrazolium Bromide (MTT) assay. The cytotoxicity was highly significant in the Caco2 cell lines using the methanolic leaf callus extracts; comparatively leaf callus extracts showed potent cytotoxicity than the leaf extracts of the *C. parviflorum*; finally the leaf, leaf callus extracts, and root extracts showed maximum cytotoxicity against Caco2 and HepG2 cancer cell lines.

Keywords

Caco2 cancer cell line · HepG2 cancer cell line · *Canthium parviflorum* · Leaf extracts · Leaf callus extracts · Root extracts · Cytotoxicity

37.1 Introduction

Natural substances have afforded a rich repository of remedies with diverse chemical structures and bioactivities against several health disorders including cancer. Natural products and their derivatives represent more than 50% of all the drugs in

S. Chandra Kala · K. Ammani (✉)
Department of Botany and Microbiology, Acharya Nagarjuna University, Guntur, Andhra Pradesh, India

S. M. Khasim et al. (eds.), *Medicinal Plants: Biodiversity, Sustainable Utilization and Conservation*, https://doi.org/10.1007/978-981-15-1636-8_37

clinical use of the world higher plants contribute not less than 25% of the total. Almost 60% of drugs approved for cancer treatment are of natural origin. More than 3000 plants are known to possess anticancer properties and are subsequently used as potent anticancer drugs (Jain and Kumar 2012). Medicinal plants play a vital role for the development of new drugs (Kala et al. 2015). During 1950–1970 approximately 100 plant-based drugs were introduced in the USA including deserpidine, rescinnamine, vinblastine, and vincristine which are derived from higher plants. From 1971 to 1990, new drugs such as etoposide, teniposide, nabilone, z-guggulsterone, artemisinin, and ginkgolides appeared all over the world. The alkaloid Serpentine isolated from the roots of Indian plants *Rauwolfia serpentina* and *Catharanthus roseus* is used for the treatment of Hodgkin's and non-Hodgkin's lymphomas, choriocarcinoma, leukaemia in children, and testicular and neck cancer. Vincristine is recommended for acute lymphocytic leukaemia in childhood and advanced stages of Hodgkin's, lymphosarcoma, and lung, cervical, and breast cancer (Farnsworth and Bingel 1977).

Canthium parviflorum Lam. (syn: *Plectoria parviflora*) of Rubiaceae is commonly called as Balusu in Telugu. The leaves and roots are astringent, sweet, thermogenic, constipating, and tonic (Sirigiri et al. 2014). They are astringent and effective against cough and indigestion (Wealth of India 1992). Traditionally the roots and leaves are used to cure vitiated conditions of Kapha in fever and constipation (Kirtikar and Basu 2001). Leaves and roots of this plant are used as astringent, diuretic, febrifuge, anthelmintic, and anti-diarrhoea and for leucorrhoea (Warrier et al. 1994), and it is used for the treatment of diabetes among major tribal groups in South Tamil Nadu (Ayyanar et al. 2008). Tribes of Orissa state in India use fruits of this plant to treat headache.

The *C. parviflorum* roots are astringent, sweet, thermogenic, diuretic, febrifuge, constipating, anthelmintic, and tonic (Sirigiri et al. 2012). They are used in vitiated conditions of diarrhoea, strangury, fever, leucorrhoea, intestinal worms, and general debility (Sirigiri and Mallikarjuna 2017). Though the various bioactive phytochemicals and diverse medicinal activities have been attributed to this plant (Sirigiri et al. 2017), no *In-house* studies have been carried out to shed light on the role of this plant in cancers. In light of the above, the current study was undertaken to investigate *C. parviflorum* crude extracts against the liver cancer cell line (HEPG-2) and Caco2 human epithelial colorectal adenocarcinoma cell line.

Caco-2 cells, a cell line derived from colorectal adenocarcinoma, are recommended by FDA to evaluate the permeation of drugs (Awortwe et al. 2014). These cells have the ability to grow in a monolayer, developing all the enzymatic machinery similar to the cells present in the small intestine (Engle et al. 1998).

Hepatocellular carcinoma (HCC) is the most widespread type of cancer in the world. The induction of HCC is preceded by the occurrence of hepatocellular damage via reactive oxygen species (ROS) and the generation of chronic inflammation related to hepatocarcinogenesis. The liver cell line HEPG2 is the perpetual cell line which was derived from the liver tissue of a 15-year-old Caucasian American male with a well-differentiated hepatocellular carcinoma. This cell line was screened to evaluate the cytotoxicity effect in normal cell line.

37.2 Callus Culture

Fresh, young leaf material was collected and washed thoroughly under running tap water to remove dust particles. Leaf explants were excised aseptically and cultured on MS (Murashige and Skoog) medium (1962) supplemented with BA (benzyl adenine) (0.2 mg/l) and NAA (naphthalene acetic acid) (0.5 and 1.0 mg/l).

37.2.1 Extraction from Callus Cultures

A 6–8-week-old callus derived from the leaf cuttings were collected, and 25 g of wild leaf explants were dried in an oven at 40 ± 1 °C for 5 h. 25 g of leaf callus powder were extracted with 150 ml of solvent methanol for 24 h by using Soxhlet apparatus. 100 mg/ml were prepared by redissolving the extracted powder in the same solvent which was used in the extraction. This callus extract was used for anticancer analysis.

37.2.1.1 MTT Assay

The chemical for MTT assay, DMEM (Dulbecco's Modified Eagle's Medium), MTT [3-(4,5-dimethylthiazol-2-yl)-2,5-diphenyl tetrazolium bromide], trypsin, and EDTA phosphate-buffered saline (PBS) were purchased from Sigma Chemicals Co. (St. Louis, MO), and fetal bovine serum (FBS) was purchased from Gibco.

37.2.1.2 Maintenance of Cell Line

The Caco2 human epithelial colorectal adenocarcinoma cell line and Hepg2 human liver cancer cell line were purchased from NCCS, Pune, and the cells were maintained in DMEM supplemented with 10% FBS and antibiotics (penicillin/streptomycin) (0.5 mL^{-1}), in atmosphere of 5% CO_2/95% air at 37 °C.

37.2.1.3 Preparation of Test Compound

The test compounds were weighed separately and dissolved in DMSO and the final concentration was made to 1 mg/ml and the cells were treated with series of concentrations from 10 to 100 µg/ml.

37.2.1.4 Caco2 and HEPG2 Cell Viability by MTT Assay

Cell viability was evaluated by the MTT assay with three independent experiments with six concentrations of compounds in triplicates. Human epithelial colorectal adenocarcinoma cells and Hepg2 cells were trypsinized, and the trypan blue assay was performed to know viable cells in cell suspension. Cells were counted by haemocytometer and seeded at density of 5.0×10^3 cells/well in 100 µl media in 96 well plate culture medium and incubated overnight at 37 °C. After incubation, the old media was removed and fresh media (100 µl) was added with different concentrations of test compound in representative wells in 96 plate. After 48 h, the drug solution was discarded and the fresh media with MTT solution (0.5 mg/mL) were added to each well and plates were incubated at 37 °C for 3 h. After incubation precipitates

Table 37.1 Cytotoxicity effect of *C. parviflorum* extracts on the Caco2 and HepG2 cell line

Concentration (µg/ml)	Leaf extract				Leaf callus extract				Root extract			
	Hepg2		Caco2		Hepg2		Caco2		Hepg2		Caco2	
	Absorbance at 570 nm	% Cytotoxicity	Absorbance at 570 nm	% Cytotoxicity	Absorbance at 570 nm	% Cytotoxicity	Absorbance at 570 nm	%_Cytotoxicity	Absorbance at 570 nm	% Cytotoxicity	Absorbance at 570 nm	% Cytotoxicity
5	0.177 ± 0.012	5.18 ± 0.43	0.782 ± 0.022	4.56 ± 0.43	0.148 ± 0.012	19.97 ± 0.23	0.324 ± 0.008	0	0.202 ± 0.070	0	1.416 ± 0.022	0
10	0.179 ± 0.022	3.57 ± 0.23	0.775 ± 0.012	17.46 ± 0.42	0.150 ± 0.072	22.82 ± 0.23	0.355 ± 0.012	0	0.208 ± 0.140	0	1.411 ± 0.014	6.48 ± 0.42
25	0.204 ± 0.041	0	0.792 ± 0.032	7.59 ± 0.26	0.151 ± 0.062	9.28 ± 0.23	0.336 ± 0.011	19.08 ± 0.42	0.182 ± 0.102	1.79 ± 0.32	1.514 ± 0.009	8.38 ± 0.33
50	0.177 ± 0.009	4.28 ± 0.29	1.184 ± 0.012	29.09 ± 0.23	0.144 ± 0.022	16.23 ± 0.44	0.182 ± 0.088	78.59 ± 0.23	0.186 ± 0.104	1.08 ± 0.43	1.321 ± 0.014	0
75	0.150 ± 0.014	**18.01** ± 0.09	1.185 ± 0.024	**53.35** ± 0.32	0.143 ± 0.024	22.11 ± 0.46	0.185 ± 0.028	**88.86** ± 0.34	0.128 ± 0.12	30.85 ± 0.53	1.356 ± 0.026	15.34 ± 0.13
100	0.175 ± 0.012	6.25 ± 0.22	1.195 ± 0.070	28.76 ± 0.23	0.146 ± 0.009	**24.25** ± 0.22	0.196 ± 0.012	80.33 ± 0.43	0.118 ± 0.022	**41.56** ± 0.43	1.399 ± 0.032	**20.62** ± 0.43

are formed as a result of the reduction of the MTT salt to chromophore, formazan crystals by the cells with metabolically active mitochondria. The optical density of solubilized crystals in DMSO was measured at 570 nm on a microplate reader. The percentage of growth inhibition was calculated using the following formula, and concentration of test drug needed to inhibit cell growth by 50% values is generated from the dose-response curves for each cell line using with origin software.

$$\%\text{Inhibition} = \frac{100(\text{Control} - \text{Treatment})}{\text{Control}}$$

37.2.2 Anticancer Activity of *Canthium parviflorum*

The present study revealed anticancer potentials of in vitro leaf and leaf derived callus and root methanolic extract of *C. parviflorum* tested against Caco2 human epithelial colorectal adenocarcinoma cell line and liver cancer (HepG2) cell lines by MTT assay and DNA fragmentation analysis. The different concentrations of IC50 values recorded potential anticancer property. The induction of cytotoxicity (alteration in morphologic characteristics) and DNA fragmentation were recorded in a concentration-dependent manner; the increased concentration of extracts induced a higher rate of cellular damages (MTT and DNA fragmentation assay).

The leaf methanolic extracts of *C. parviflorum* have minimum cytotoxicity with 53.36 against the tested Caco2 cancer cells (Table 37.1, Figs. 37.1, 37.2 and 37.3). The leaf callus extracts of *C. parviflorum* have higher inhibitory effect with IC50 value of 88.86 against the tested Caco2 cancer cells at a lower concentration (about 75 μg/ml) followed by other sample extracts (Table 37.2). The inhibitory concentration IC50 of the extracts illustrated the significant growth inhibition (Figs. 37.4, 37.5, 37.6) at 75 μg/ml concentration. The leaf callus extracts showed potent anticancer activity with IC50 value of 48.43 μg/ml than the leaf extracts on the Caco2

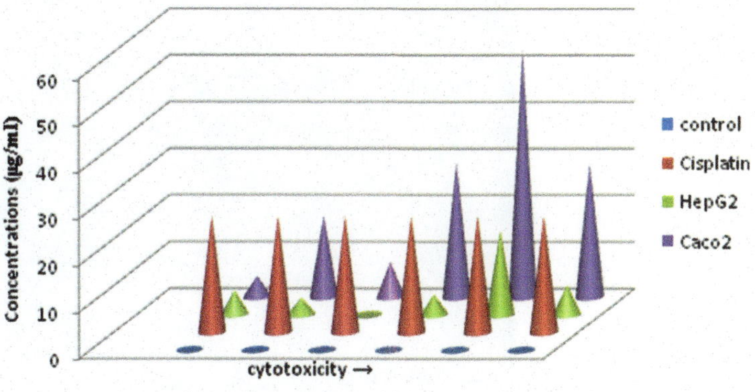

Fig. 37.1 Cytotoxicity effect of *C. parviflorum* leaf extract on the Caoc2 and HepG2 cell line

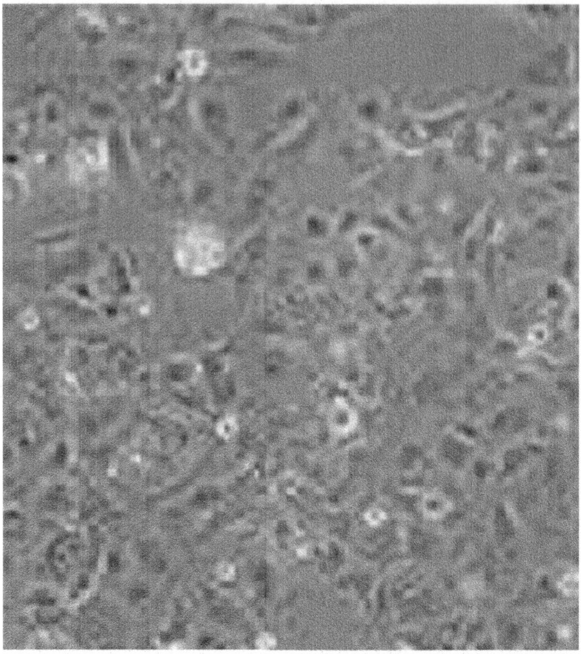

Fig. 37.2 Caco2 control cell line

Fig. 37.3 Caco2 cells treated with methanolic leaf extract 100 μg/ml of *C. parviflorum*

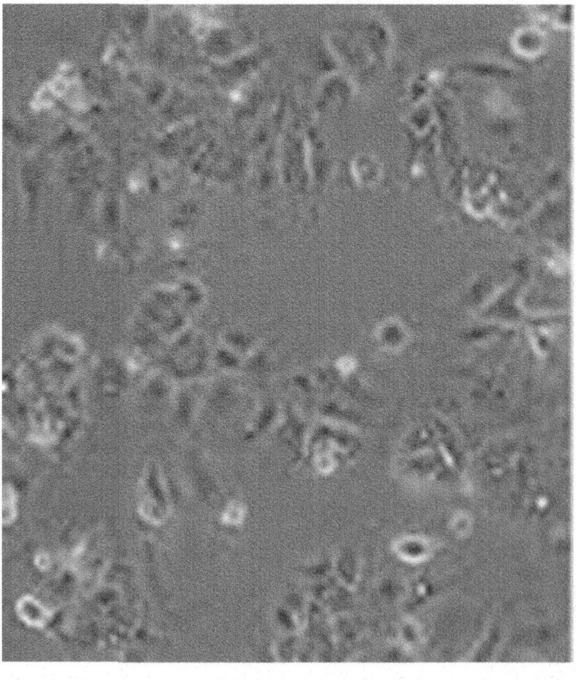

Table 37.2 Cytotoxicity effect of *C. parviflorum* extracts on the Caoc2 and HepG2 cell line

S. no.	Sample	IC50 (μg/ml) By using Caco2	IC50 (μg/ml) By using HepG2
1	CpL	120.48 ± 1.6	534.2 ± 1.2
2	Cp.Lc	**48.43** ± 1.4	498.6 ± 1.2
3	CpR	330.58 ± 1.2	**77.77** ± 1.4
4	Cisplatin	25 ± 1.2	25 ± 1.2

CpL Canthium parviflorum leaf extracts, *Cp.Lc Canthium parviflorum* leaf callus extracts, *CpR Canthium parviflorum* root extracts

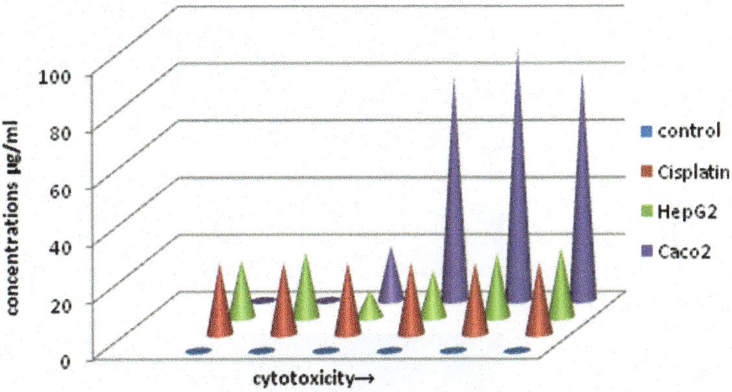

Fig. 37.4 Cytotoxicity effect of *C. parviflorum* leaf callus extract on the Caoc2 and HepG2 cell line

cell lines (Table 37.2, Figs. 37.4–37.6). This is the first report on anticancer properties of leaf callus extracts of *C. parviflorum species*. The lower concentration of the extracts demonstrates good cell viability against the Caco2 cell monolayers. The callus extracts exhibited gradual increase in antiproliferative activity in a dose-dependent manner.

The methanolic root extracts revealed significant cytotoxicity with IC50 value of 77.77 μg/ml on the HepG2 liver cancer cell lines (Table 37.2, Figs. 37.7, 37.8 and 37.9); comparatively the root extracts showed high cytotoxicity than the leaf and leaf callus extracts.

Cancer is becoming one of the major public health burdens on both developed and developing countries. Several synthetic agents are used to reverse, suppress, or prevent the disease, but they have their own side effects. Thousands of herbal and traditional compounds are being screened worldwide to confirm their use as anticancerous drugs (Garg and Shrivastava 2013). Some medicinal plants have shown significant inhibition of cancer cell proliferation. Bachrach (2012) reported anticancer properties of *Withania somnifera, Crocus sativus, and Vitex agnus-castus.*

In recent years it has been used clinically to treat cancers and demonstrated an ability to greatly inhibit tumour development and growth with limited side effects (Niu and He 1992). In the present study, the leaf callus extracts exhibited good

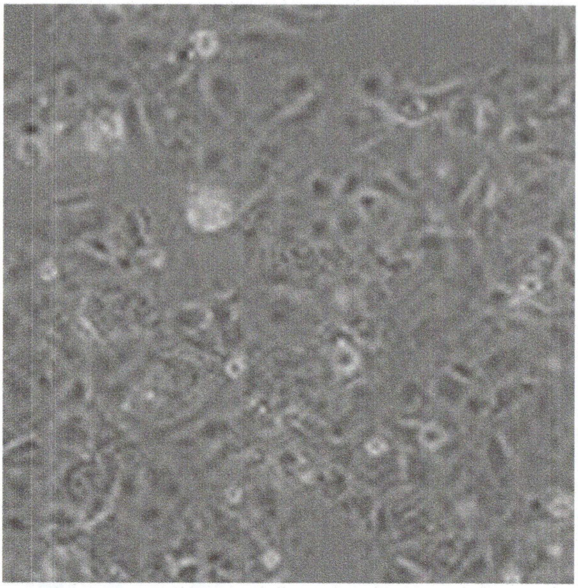

Fig. 37.5 Caco2 control cell line

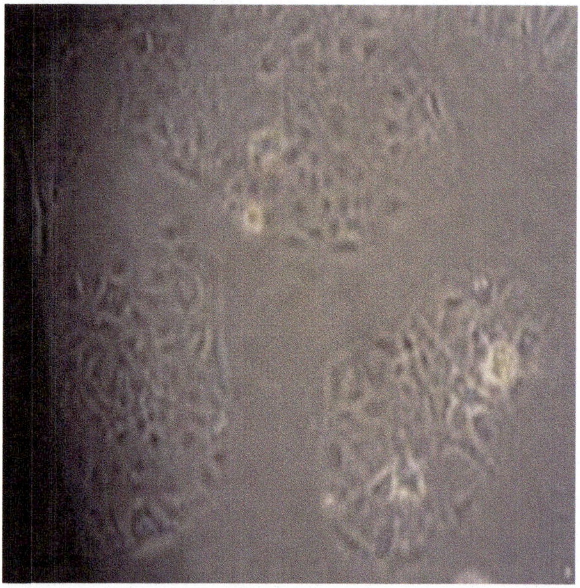

Fig. 37.6 Caco2 cells treated with methanolic leaf callus extract 75 µg/ml of *C. parviflorum*

cytotoxic activity on the Caco2 human adenoma cancer cell lines; this is the first report. The cancer cell line studies reported the in vitro anticancer activity of leaf ethanolic extract of *Canthium parviflorum* on DLA and Hela cell lines (Purushoth Prabhu et al. 2011). Pasumarthi et al. (2011) reported *C. parviflorum* leaves with methanol extracts exhibits flavonoids showed a potent cytotoxic activity due to presence of flavonoids may be one of the reasons to show the cytotoxicity.

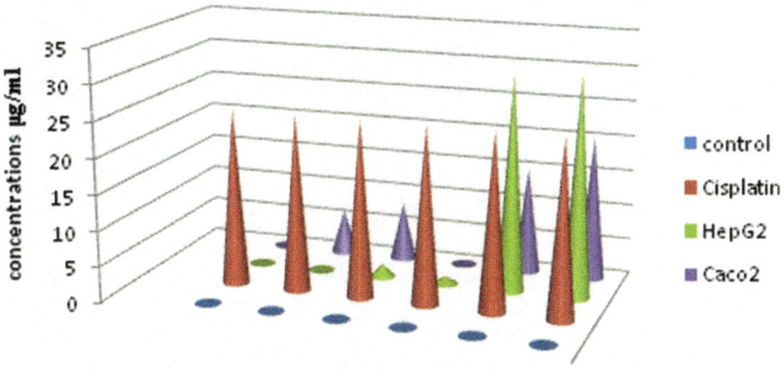

Fig. 37.7 Cytotoxicity effect of *C. parviflorum* root extract on the Caoc2 and HepG2 cell line

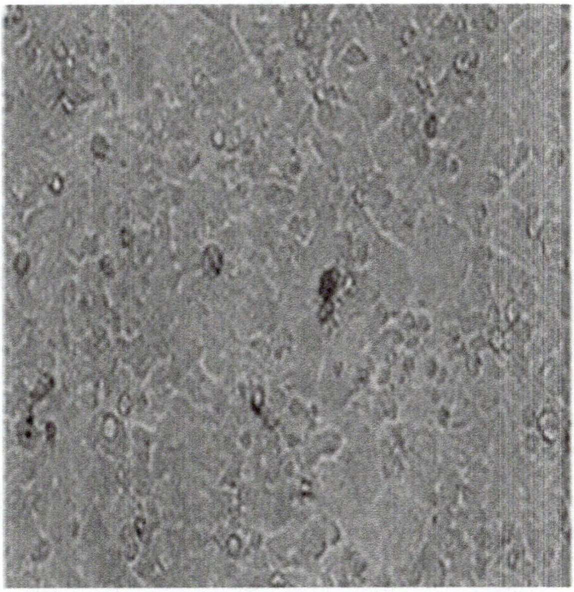

Fig. 37.8 HepG2 control cell line

The ethanolic extracts of leaves revealed having flavonoid content (Karthik et al. 2013), and most important flavonoids are present in the leaf extracts (Kumar et al. 2008). Sirigiri and Ammani (2017) reported in the GC-MS analysis the methanolic leaf callus extracts of *C. parviflorum* contained anticancer compounds.

The methanolic extracts of roots have shown superior anticancer activity comparatively to the leaf and leaf-derived callus extracts on the HepG2 human liver cancer cell lines. The acetone root extract of *Canthium parviflorum* showed maximum anticancer activity both in in vitro and in vivo studies carried out in male Swiss albino mice (Milton and Jose 2014). Krishna et al. (2014) reported the roots with ethyl acetate and crude methanol extracts of *Canthium parviflorum* showed

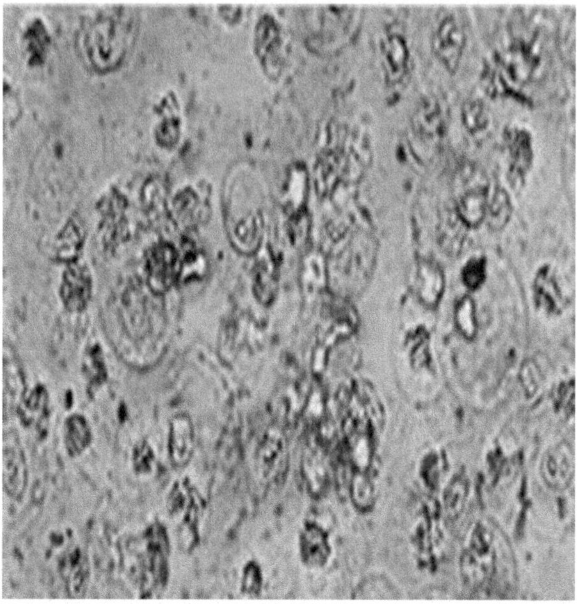

Fig. 37.9 HepG2 cells treated with methanolic root extract 100 μg/ml of *C. parviflorum*

flavonoids. Based on the qualitative phytochemical analysis, the root material of *Canthium parviflorum* methanolic extracts revealed flavonoids (Shabi Ruskin et al. 2014) and it is one of the reason to show the cytotoxicity. *C. parviflorum* leaves revealed D-mannitol compound, one of the anticancer agent (Beena et al. 2008). Henceforth, our results also ensure the anticancer potential of leaf, root, and leaf callus methanolic extracts of *C. parviflorum* on Caco2 and Hepg2 human cancer. Similarly, cytotoxic activity of *Broussonetia papyrifera* on MCF-7, HELA, and HepG2 cell lines was reported by Naveen Kumar et al. (2014).

Finally the present study suggests that the methanolic extract of *C. parviflorum* leaf and leaf-derived callus and root extracts comprised novel anticancer compounds which will be potent therapeutic candidate for liver and human epithelial colorectal adenocarcinoma cell lines.

37.3 Conclusion

Anticancer activity of *Canthium parviflorum* may be due to flavonoid present in the leaf, leaf-derived callus, and root extracts. Further studies are also in process to evaluate the most potent fraction of the plant and to isolate the constituents of the fractions.

Acknowledgement The first author, Dr. Sirigiri Chandra Kala, is thankful to the University Grants Commission, New Delhi, for providing financial supports with Post-Doctoral Fellowship.

References

Awortwe C, Fasinu PS, Rosenkranz B (2014) Application of Caco-2 cell line in herb-drug interaction studies: current approaches and challenges. J Pharm Pharm Sci 17(1):19

Ayyanar M, Sankarasivaraman K, Ignacimuthu S (2008) Traditional herbal medicines used for the treatment of diabetes among two major tribal groups in south Tamil Nadu, India. Ethnobot Leaflets 12:276–280

Bachrach Z (2012) Contribution of selected medicinal plants for cancer prevention and therapy. Acta Fac Med Naissensis 29:117–123

Beena J, Leopold J, Shafi PM (2008) Phytochemical studies on the thorns and leaves of *Canthium parviflorum* Lam. Asian J Chem 20:5372–5376

Engle MJ, Goetz GS, Alpers DH (1998) Caco-2 cells express a combination of colonocyte and enterocyte phenotypes. J Cell Physiol 174:362–369

Farnsworth NR, Bingel AS (1977) Problems and prospects of discovery new drugs from higher plants by pharmacological screening. In: Wagner H, Wolf P (eds) New natural products and drugs with pharmacological, biological and therapeutical activity. Springer, Berlin, pp 1–22

Garg HK, Shrivastava A (2013) Application of nanotechnology and biocomputation for treatment of cancer. Pharma Innov 2:63–66

Jain D, Kumar S (2012) Snake venom: a potent anticancer agent. Asian Pac J Cancer Prev 13:4855

Karthik K, Bharath Rathna Kumar P, Venu Priya R, Sunil Kumar K, Ranjith S, Rathore B (2013) Evaluation of anti-inflammatory activity of *Canthium parviflorum* by *in vitro* method. Indian J Res Pharm Biotechnol 1:729–730

Kirtikar KR, Basu BD (2001) Indian medicinal plants, vol 1. Lalit Mohan Basu, Allahabad, pp 35–45

Krishna B, Sanjeeva Kumar A, Praneetha R, Rathnakar E, Venkata Ramana Murthy B, Pavan Kumar G (2014) Phytochemical screening and anthelminthic activity of *Canthium parviflorum* roots. World J Pharm Pharm Sci 3:566–570

Kumar S, Shanmugam T, Palvannan S, Bharathi Kumar VM (2008) Evaluation of antioxidant properties of *Canthium parviflorum* Lam. Leaves. Indian J Nat Prod Resour 7:122–126

Milton JD, Jose MA (2014) *In vitro* and *in vivo* anticancer evaluation of *Canthium parviflorum* root extracts. Int J Adv Pharm Biol Sci 4:17–24

Murashige T, Skoog F (1962) A revised medium for rapid growth and bioassays with tobacco tissue cultures. Plant Physiol 150:473–497

Naveen Kumar N, Ramakrishnaiah H, Krishna V, Radhika M (2014) Cytotoxic activity of *Broussonetia papyrifera* (L.) Vent on MCF-7, HELA and HEPG2 cell lines. Int J Pharm Pharm Sci 6:339–342

Niu CQ, He LY (1992) HPTLC separation and fluorodensitometric determination of isoquinoline alkaloids in *Chelidonine Majus* L. Yao Xue Xue Bao 27:69–73

Pasumarthi S, Chimata MK, Chetty CS, Challa S (2011) Screening of phytochemical compounds in selected medicinal plants of Deccan Plateau and their viability effects on Caco-2 cells. J Med Plants Res 5:6955–6962

Purushoth Prabhu T, Panneerselvam P, Selvakumari S, Sivaraman D (2011) *In vitro* and *in vivo* anticancer activity of an ethanolic extract of *Canthium parviflorum* Lam. on DLA and Hela cell lines. Int J Drug Dev Res 3:280–285

Shabi Ruskin R, Vasantha Kumari B, Chitarasu T (2014) Qualitative phytochemical and ftir analysis of root extracts of *Canthium parviflorum* Lam. J Chem Pharm Sci 2:122–127

Sirigiri CK, Ammani K (2017) GC-MS analysis of biologically active compounds in *Canthium parviflorum* Lam leaf and callus extracts. Int J Chemtech Res 10:1039–1058

Sirigiri CK, Mallikarjuna K (2017) High frequency direct regeneration of *Canthium parviflorum* Lam. using mature nodal explants. Plant Cell Biotechnol Mol Biol 18:473–480

Sirigiri CK, Mallikarjuna K, Aruna P (2012) Qualitative phytochemical analysis of seed and leaf callus extracts of *Canthium parviflorum* Lam. Guntur District, Andhra Pradesh. Int J Pharm Bio Sci 3:177–182

Sirigiri CK, Mallikarjuna K, Siva Reddy C (2014) In vitro pharmacological investigations of using leaf callus extracts of *Canthium parviflorum* Lam. Br Biotechnol J 4:325–338

Sirigiri CK, Mallikarjuna K, Aruna P (2015) Antimicrobial screening of callus extracts of *Canthium parviflorum* Lam. Int J Med Pharm Sci 4:1–10

Sirigiri CK, Ratnakar Reddi KVN, Akila CS, Chandra ORP, Mallikarjuna K (2017) Assessment of genetic diversity of *Canthium parviflorum* Lam by RAPD and ISSR markers. Ann Plant Sci 6(11):1775–1783

Warrier PK, Nambiar VPK, Raman Kutty C (1994) Indian medicinal plants, vol I. Orient Longman, Madras, p 366

Wealth of India (1992) A dictionary of Indian raw materials and industrials products—raw materials, vol 3. Revised Series, Ca-Ci. Publications & Information Directorate, CSIR, New Delhi, p 210

Bioactive Metabolites from *Streptomyces nanhaiensis* VSM-1: Polyphasic Taxonomy, Optimization, and Evaluation of Antimicrobial Metabolites by GC-MS Analysis

38

Ushakiranmayi Managamuri, Muvva Vijayalakshmi, Sudhakar Poda, V. S. Rama Krishna Ganduri, and R. Satish Babu

Abstract

Actinomycetes strain VSM-1 isolated from the marine sediment samples of north coastal Bay of Bengal was identified as *Streptomyces nanhaiensis* by conventional and molecular approaches. *S. nanhaiensis* recorded significant broad-spectrum activity against all human and plant pathogenic microorganisms tested. An attempt has been made to evaluate the influence of medium components on the antimicrobial metabolite production of VSM-1 and to optimize an effective fermentation medium to enhance its bioactive metabolite yield. Bioactive metabolite production by VSM-1 was initially optimized using one-factor-at-a-time (OFAT) method where the yield of metabolites was recorded to generate inhibition zones of 30, 29, 28, 27, and 29 mm against test microorganisms, i.e., *Shigella flexneri*, *Serratia marcescens*, *Proteus vulgaris*, *Pseudomonas*

U. Managamuri · M. Vijayalakshmi (✉)
Department of Botany and Microbiology, Acharya Nagarjuna University, Guntur, Andhra Pradesh, India

S. Poda
Department of Biotechnology, Acharya Nagarjuna University, Guntur, Andhra Pradesh, India

V. S. Rama Krishna Ganduri
Department of Biotechnology, Acharya Nagarjuna University, Guntur, Andhra Pradesh, India

Department of Biotechnology, K L University, Guntur, Andhra Pradesh, India

R. Satish Babu
Department of Biotechnology, National Institute of Technology, Warangal, Telangana, India

S. M. Khasim et al. (eds.), *Medicinal Plants: Biodiversity, Sustainable Utilization and Conservation*, https://doi.org/10.1007/978-981-15-1636-8_38

aeruginosa, and medically important dermatophyte *Candida albicans*. The optimal values obtained from OFAT were selected, and the experimental model was designed using response surface methodology. Central composite design (CCD) was employed to study the influence of the variables on the production of bioactive metabolites by VSM-1 and their effect on the responses (test microorganisms). The statistical analysis showed that the variables which have a significant effect ($P < 0.001$) on the metabolite production at both the interactive and individual levels were measured in terms of antimicrobial activity. Maximum yield of metabolites was recorded to generate increased zones of inhibition, i.e., 34.2, 32, 31.6, 32.3, and 33.6 mm, against test microorganisms *Shigella flexneri*, *Serratia marcescens*, *Proteus vulgaris*, *Pseudomonas aeruginosa*, and *Candida albicans*, respectively. Unstructured kinetic models were used to fit the results obtained from experiments, and kinetic parameters were also estimated. Further, chemometric profile of the ethyl acetate extract of the strain was performed by gas chromatography and mass spectroscopy (GC-MS). Interpretation on mass spectrum of GC-MS was carried out using NIST Mass Spectral Data Base, with NIST MS search v.2.0. The results of the present study revealed the presence of various active principles with a diverse range of positive pharmacological actions, and hence this strain could be a possible source of novel bioactive compounds.

Keywords
Actinomycetes · *Streptomyces nanhaiensis* · Optimization · Modeling · Bioactive metabolites · GC-MS

38.1 Introduction

Natural products and their derivatives have been deep rooted as precious source of therapeutic agents. The natural product structures have the traits of excessive chemical diversity, biochemical specificity, and adequate amount of chiral centers, aggravated steric complexity, and exclusive molecular properties that make them favorable lead structures for drug discovery and differentiate them from synthetic and combinatorial compounds. Natural products have been deemed to be a natural combinatorial chemical library which could provide unrivaled compounds with chemical diversity and proper pharmacological chemical profiles (Bredholdt et al. 2007). The abundance and diversity of natural products having biological activity ends in the opportunity for the invention of drugs and the drugs derived from the natural assets play a widespread position in the prevention and treatment of human illnesses (Koehn and Carter 2005). Natural products continue to be the prominent source in view to the urgent need and demand for the new drugs, especially antibiotics and anticancer drugs. Hence, natural products have been witnessed as a highly compelled source to develop promising new drugs (Appendino and Banfi 2011).

The potential sources of the commercially essential bioactive compounds are the actinomycetes which have been historically wealthy in generating the vital bioactive compounds which provide a direct evidence to be a reliable supply of novel compounds (Parthasarathi et al. 2010). The list of novel actinobacteria and products derived from poorly explored areas of the world stresses the importance of investigating new habitats. The probability of locating novel bioactive compounds relies on the number of strains screened and their degree of diversity, uniqueness, and capability to produce secondary metabolites (Donadio et al. 2002). In this context awareness on rare resources which includes marine actinomycetes has captivated unique attention. This reality is due to their potency to produce biologically active secondary metabolites with many of them as potent antibiotics and lead compounds that cannot be produced and discovered by terrestrial microbes (Xiong et al. 2013).

Marine ecosystems harbor many unique forms of actinomycetes which appear to be widely distributed through the ocean and found in intertidal zones, sea water, sponges, and ocean sediments (Ramesh and Mathivanan 2009; Sun et al. 2010). Therefore, there is a need to bio-prospect untapped geographical assets and explore new strains of actinomycetes to maximize the discovery of novel bioactive metabolites (Leal et al. 2012). The terrestrial soils have been the predominant and extensively exploited source, and investigations on marine actinobacteria are few and inconclusive, despite the fact that they are the critical assets for new bioactive compounds (Newman and Cragg 2007). In recent years, there has been a growing awareness of the potential value of deep sea sediments as sources of actinobacteria that produce useful bioactive metabolic products. Literature on the isolation of actinobacteria from marine sediments suggested that these sources may be treasured for the isolation of novel actinobacteria with the potential to yield useful bioactive products (Jensen et al. 2007).

The search for filamentous actinobacterium offering incredibly precious bioactive compounds has been remarkably successful, and about two thirds of the naturally occurring antibiotics have been purified from actinomycetes species. Among the filamentous actinomycetes species that accounts for a major fraction of the bioactive metabolites, *Streptomyces* is by far the most prolific genus that the pharmacy industry has isolated and screened millions of strains over several decades. It has been advised that only a fraction of the bioactive compound producers of *Streptomyces* strains have been discovered and figuring out the undiscovered portion will require substantial effort (Solanki et al. 2008).

Optimization of the process variables can be computed by response surface methodology (RSM) effectively, if the mathematical formulations of RSM model are close to physics (being modeled) significantly (Wang et al. 2017). The drafted article of RSM by Box and Hunter (1957) sketched the principles for the design of the model. RSM designs the experiments using the mathematical and statistical techniques; it also builds models, evaluates the effects of parameters of the system, and engineers the optimum conditions for the target responses (Fatma et al. 2013). RSM is an up-to-date approach for constructing the model approximation that optimizes a response (output variable result) which is influenced by many independent process variables. The traditional methods-based optimization follows the optimization by one-factor-at-a-time (OFAT), but optimization using RSM for any production has an edge by reduction in the number of the experimental runs, time reduction for the execution of the experiments, and analyzing the influence of the variable interactions against the response (Wang et al. 2017). In addition RSM evaluates the conjunction between the variables and responses and defines the effect of variables against responses either as a single variable or via combination in process (Zafari et al. 2013).

Fermentation process provides a controlled environment to carry out complex reactions within the cell. Careful understanding of its metabolic stoichiometry in general is supported by mathematical expression with good approximations. These developed kinetic models pave the way for better elucidation of bioactive metabolite synthesis in terms of their yield and production rates. Simulation of the chosen models allows the bioengineers to estimate the kinetic parameters (or constants), which in turn validates the kinetic models. Several researchers have used unstructured models to explain the biological systems for effective synthesis of microbial products (Rama Krishna et al. 2016). In this study, estimation of kinetic parameters in bioactive metabolite production using *Streptomyces nanhaiensis* VSM-1 (KU507594) was attempted.

Research toward the exploration of marine actinobacteria has not progressed much in India. As a result, an attempt has been made to isolate potent indigenous actinobacterium strain VSM-1 with precise bioactive capability that was isolated from marine sediment samples of Bay of Bengal (Visakhapatnam) of Andhra Pradesh. Attempts have been additionally made to optimize the cultural and environmental conditions of the potent isolate that facilitate improved production of biologically active compounds. The present study additionally describes the extraction and biological evaluation of the antimicrobial compounds (by GC-MS analysis) produced by the strain.

38.2 Materials and Methods

38.2.1 Sampling and Isolation

The marine sediment samples were collected at different depths of the Bay of Bengal of coastal Andhra Pradesh, India. The collected sediment samples were transported to the laboratory in sterile bags and air-dried at room temperature for further isolation studies. The air-dried sediment sample was pretreated with calcium carbonate (10:1 w/w) and incubated at 37 °C for 4 days (Alferova et al. 1989). The treated sediment samples were suspended in sterile distilled water (1 g in 100 mL), homogenized by vortexing and 0.1 mL of serially diluted sample (10^{-4} dilution) that was spread over the surface of Bennet's agar (0.1% yeast extract, 0.1% beef extract, 0.2% casein enzymic hydrolysate, 1% dextrose, and 2% agar) (pH 8) supplemented with nalidixic acid (50 μg/mL) and secnidazole (50 μg/mL). After incubation for 3 weeks at 30 °C, distinct strains were selected and maintained by subculturing on yeast extract malt extract dextrose agar (ISP-2) slants.

38.2.2 Identification

Characterization of the isolate was carried out as described by the International *Streptomyces* Project by microscopy, morphological, cultural, physiological, and biochemical characteristics. The cultural characteristics of the strain were studied on different media such as tryptone yeast extract agar (ISP-1), YMD agar (ISP-2), oatmeal agar (ISP-3), starch inorganic salts agar (ISP-4), glycerol asparagine agar (ISP-5), peptone yeast extract iron agar (ISP-6), tyrosine agar (ISP-7), starch-casein agar, Czapek-Dox agar, and nutrient agar media (Shirling and Gottlieb 1966). The micromorphology of the strain was examined under scanning electron microscopy (model JOEL-JSM 5600) (Bozzola and Russell 1999). The utilization of carbon sources by the strain was carried out in minimal medium containing different carbon sources at 1% concentration according to the method described by Isik et al. (1999). Biochemical tests which include IMVIC, H_2S production (Cowan 1974), nitrate reduction (Gordon 1966), catalase (Jones 1949), and starch and gelatin hydrolysis (Waksman 1961) were also evaluated. Physiological characterization such as the effect of pH (5–9), temperature (20–60 °C), and salinity tolerance were analyzed. In addition the sensitivity of the strain to different antibiotics was determined by paper disc method (Williams 1989). Molecular identification of the strain was carried out according to the procedure of Nilsson and Strom (2002).

38.2.3 Optimization

38.2.3.1 Incubation Period

The growth pattern and bioactive metabolite production of the strain were studied at regular intervals up to 10 days. Two-week-old culture of the strain was cultivated in seed medium (ISP-2 broth) comprising of yeast extract (0.4%), malt extract (1%), dextrose (0.4%), and $CaCO_3$ (0.2%) at room temperature for 48 h. Seed culture at a rate of 10% was inoculated into Bennet's broth (production medium) consisting of 0.1% yeast extract, 0.1% beef extract, 0.2% casein enzymic hydrolysate, and 1% dextrose with pH 7.3. The fermentation process was carried out for 10 days under agitation at 120 rpm. At every 24 h interval, the flasks were harvested, and the biomass was separated from the culture filtrate. Biomass was determined in terms of total cell dry weight. Antimicrobial metabolite production was determined in terms of their antimicrobial spectrum (Narayana and Vijayalakshmi 2008). The culture filtrates were extracted with ethyl acetate and evaporated to dryness in a water bath at 80 °C. The solvent extracts were concentrated, and 50 μL of crude extract was tested for antimicrobial activity by employing agar well diffusion method against test organisms like *Shigella flexneri* (MTCC 1457), *Serratia marcescens* (MTCC 118), *Proteus vulgaris* (ATCC 6380), *Pseudomonas aeruginosa* (ATCC 9027), and *Candida albicans* (ATCC 10231).

38.2.3.2 Culture Conditions for the Optimum Production of Bioactive Metabolites

Bioactive metabolite production of the strain was optimized by using different parameters including pH, temperature, salt concentration, carbon and nitrogen sources, and minerals.

pH and Temperature

To determine the influence of initial pH on growth and bioactive metabolite production, *Streptomyces nanhaiensis* VSM-1 was cultured in the medium with different initial pH, ranging from 4 to 10, and at specific starting temperatures, from 20 to 50 °C. The biomass and bioactive metabolite production were envisioned to determine optimal pH and temperature conditions which were used in this study.

NaCl Concentration

The impact of salinity on growth and bioactive metabolite production by *Streptomyces nanhaiensis* VSM-1 was recorded by cultivating the strain in the fermentation medium amended with different concentrations of NaCl (0.1–9%) at optimum pH and temperature for 9 days. The salt concentration in which the strain exhibits optimum levels of bioactive metabolites was fixed for further studies.

Carbon and Nitrogen Sources

To determine the impact of carbon sources on biomass and bioactive metabolite production, different carbon sources like maltose, lactose, fructose, sucrose, glucose, starch, mannitol, arabinose, raffinose, and rhamnose each at a concentration of 1% were added separately into the production medium, maintaining all other conditions at optimum levels. The impact of varying concentrations of the best carbon source (0.5–5%) on the bioactive metabolite production was evaluated. Similarly, the influence of various nitrogen sources on antimicrobial metabolite production was evaluated by amending different nitrogen sources like soya peptone, Bio peptone, asparagine, meat extract, yeast extract, tryptone, peptone, casein, beef extract, and malt extract each at a concentration of 0.5% that were individually supplemented into the production medium containing an optimum amount of the superior carbon source. The growth and bioactive metabolite production were determined after 9 days of incubation at optimum pH, temperature, and salt concentration. Further, the impact of varying concentrations of optimized nitrogen source (0.1–2%) was studied to standardize the maximum antimicrobial metabolite production.

Minerals

Impact of minerals on the production of biomass and bioactive metabolites was studied by supplementing different minerals like $MgSO_4$, $FeSO_4$, $ZnSO_4$, K_2HPO_4, and KH_2PO_4 each at a concentration of 0.05% (w/v) to the optimized medium.

Evaluation of Antimicrobial Activity

The bioactive metabolites of the strain produced under optimized conditions were tested in opposition to various strains of microorganisms, viz., *Shigella flexneri*

(MTCC 1457), *Serratia marcescens* (MTCC 118), *Proteus vulgaris* (ATCC 6380), *Pseudomonas aeruginosa* (ATCC 9027), and medically important dermatophyte *Candida albicans* (ATCC 10231). The test microorganisms were procured from ATCC and MTCC and preserved at 4 °C.

Experimental Design by RSM

Experimental design applied for optimization of process variables (incubation time in days, pH, temperature, concentrations of mannitol and peptone) is executed using RSM. RSM is a combination of mathematical and statistical methods applied commonly for optimization of industrial, agricultural, and chemical reaction experimental designs (Sathish and Vivekanandan 2016). RSM is applied to classify the relationship among the process variables and the response (zone of inhibition against the pathogenic microorganisms by the bioactive metabolite produced by VSM-1), as well as optimize the best condition of the variables to predict model and to probe the best values of the responses (Güray et al. 2008). Central composite design (CCD) of RSM is an ideal constructive experimentation as it allows the reasonable amount of data to test lack of fit when sufficient number of experimental values exists. The five significant independent variables selected for the present study include incubation time in days, pH, temperature, and concentrations of mannitol and peptone, where *A*, *B*, *C*, *D*, and *E* are the values of the respective variable. The experimental ranges of the actual and the coded values of the process variables are shown in Table 38.1.

The independent variables experimental ranges were selected as low (*−1*), middle (*0*), and high (*+1*) based on some preliminary experimental studies (Myers et al. 2009). The experiments were designed and carried out as per central composite design (CCD) for 5 independent variables, of which a total 50 ($=2^5 + 2 \times 5 + 8$) experiments consists of 32 factorial, 10 axial, and 8 center points, based on the following equation:

$$N = 2^n + 2n + n_c \quad (38.1)$$

where N is the total number of experiments, n the number of independent variables, and n_c the number of central points (replicates). ANOVA (analysis of variance) is applied for prediction of equation, and the information obtained from the experiments was fitted with the second-order polynomial equation (Wang et al. 2016):

Table 38.1 Experimental range and level coded of independent variables

		Actual levels of coded factors		
Factors	Symbols	*−1* (low)	*0* (middle)	*+1* (high)
Time of incubation (days)	*A*	8	9	10
pH	*B*	7	8	9
Temperature (°C)	*C*	25	30	35
Concentration of mannitol (%w/v)	*D*	0.5	1	1.5
Concentration of peptone (%w/v)	*E*	0.5	1	1.5

$$Y = \beta_0 + \sum_{i=1}^{n} \beta_i X_i + \left(\sum_{i=1}^{n} \beta_{ii} X_i \right)^2 + \sum_{i=1}^{n-1} \sum_{j=i+1}^{n} \beta_{ij} X_i X_j \tag{38.2}$$

where Y is the predicted response, β_0 is the intercept coefficient, β_i is the linear coefficient, β_{ij} is the interaction coefficients, β_{ii} is the quadratic coefficients, and X_i and X_j are the coded values of the five additive variables under study.

Model Adequacy Verification

Analysis of variance (ANOVA) is crucial and useful for evaluating the fitted model because the fitted mathematical model sometimes is not sufficient in explaining the experimental region of the independent variable (Marcos et al. 2008). In order to exercise the statistical discrepancy between at least three or more independent group means and their associated source of variations, Table 38.7 presents the results of ANOVA for the main effects, the quadratic terms, and the obtained data interactions. The sum of squares is the sum of the squares of the deviation from the mean. Mean square is an estimate of population variance that is obtained by dividing the associated sum of squares by the degrees of freedom. The mean square is compared with the residual mean square by the F value. The probability P value (Prob > F) is the probability of obtaining the F value or what we observed in the sample that determine the trueness of the null hypothesis (no significant difference of factor effects). Determining the coefficient determination (R^2) is another common way to validate how well is the model's goodness of fit. The R^2 is measured on the scale of 1; if the R^2 is close to 1, it means that the predicted values are in close agreement with the original experimental values. The 3D graphs were obtained using the Design-Expert software. The interactive effect between the process variables and their effect against the responses was plotted by 3D surface figure. Finally by solving the polynomial equation derived and the designed quadratic model and the grid search in RSM 3D plots, the optimal conditions were obtained.

Unstructured Mathematical Modeling

Basic mathematical and unstructured kinetic models which quantitatively describe the substrate utilization and growth-associated production formation kinetics in a batch system were taken from Ushakiranmayi et al. (2017). Models of logistic and Luedeking-Piret were used to simulate the cell growth and bioactive metabolite productions of *Streptomyces nanhaiensis* VSM-1 (KU507594). The data acquired from the logistic (L) model were used to calculate the maximum specific cell growth rate (μ_{max}), d^{-1}; the specific production rate of bioactive metabolite, d^{-1}; and the maximum biomass concentration (X_m), g/L. Constants of growth-associated substrate consumption, γ, and non-growth-associated substrate consumption, η, were calculated using the logistic incorporated modified Luedeking-Piret (LIMLP) model. Similarly, growth-associated product formation constant, α, and non-growth-associated product formation constant, β, were determined from stationary phase data (where $\frac{dX}{dt} = 0$), using the logistic incorporated Luedeking-Piret (LILP) model.

Statistical Analysis

Results on the growth and the production of antimicrobial metabolites by *S. nanhaiensis* VSM-1 exposed to different cultural conditions are statistically analyzed with two-way analysis of variance (ANOVA).

Fermentation

A loopful of properly sporulated culture of *S. nanhaiensis* VSM-1 was inoculated into 100 mL of seed medium (YMD broth) and incubated on a rotary shaker at 30 °C for 48 h. Seed culture at the rate of 10% (v/v) was transferred to the optimized production medium consisting of 1% (w/v) mannitol, 1% (w/v) peptone, 0.05% (w/v) K_2HPO_4, and 3% (w/v) NaCl at pH 8. The inoculated flasks were incubated on rotary shaker at 30 °C and incubated for 9 days. The fermented broth collected at the end of 9 days was centrifuged, and the culture filtrate obtained was extracted with ethyl acetate. The solvent extracts were pooled and concentrated in vacuum to dryness, and the resultant crude extract was subjected to gas chromatography and mass spectroscopy.

Identification of Antimicrobial Metabolites by GC-MS

Identification of the chemical compounds present in the crude extract (CE) was carried out by GC-MS. Analysis was conducted on Agilent GC-MS machine (GC: 5890 series II; MSD 5972). The fused silica HP-5 capillary column (30 m × 0.25 mm, ID, film thickness of 0.25 μm) was directly coupled to the MS. The carrier gas was helium with a flow rate of 1.2 mL/min. Oven temperature was programmed (50 °C for 1 min and then 50–280 °C at a rate of 5 °C/min) and, subsequently, held isothermally for 20 min. The temperature of the injector port was maintained at 250 °C and that of detector at 280 °C (Boussada et al. 2008). The peaks of the obtained compounds in the gas chromatography were subjected to mass spectral evaluation. The spectra have been analyzed from the library data, NIST MS search (version 2.0) (included with NIST'02 mass spectral library, Agilent p/n G1033 A).

38.3 Results and Discussion

Taxonomic position of the strain VSM-1 was described on the basis of conventional and molecular approaches. Cultural characteristics of the strain were studied by growing the isolate on ten selective media (seven ISP media and three non-ISP media), and the results are tabulated in Table 38.2. The strain exhibited good growth on ISP-2, ISP-3, ISP-4, ISP-5, Czapek-Dox agar medium, and nutrient agar medium. The strain is aerobic actinomycetes that form grayish white aerial mycelium, while the substrate mycelium varied from pale yellow to white. No pigment was observed on any of the tested medium including melanin pigmentation on ISP-7. The biochemical and physiological characteristics of the isolate are summarized in Table 38.3. The strain is gram positive, and it has shown positive results for indole, Voges-Proskauer, citrate utilization, starch, casein hydrolysis, nitrate reduction, and catalase test but negative for gelatin hydrolysis, H_2S production, and methyl red and

Table 38.2 Cultural characteristics of the strain on various ISP and non-ISP media

S. no.	Medium	Growth	Aerial mycelium	Substrate mycelium	Pigmentation
1	Tryptone yeast extract agar (ISP-1)	Moderate	Gray	White	Nil
2	Yeast extract malt extract dextrose agar (ISP-2)	Good	Grayish white	Yellow	Nil
3	Oatmeal agar (ISP-3)	Good	White	Light gray	Nil
4	Inorganic salts starch agar (ISP-4)	Good	Gray	Yellow	Nil
5	Glycerol asparagine agar (ISP-5)	Good	Gray	Pale yellow	Nil
6	Peptone yeast extract iron agar (ISP-6)	Poor	–	–	Nil
7	Tyrosine agar (ISP-7)	Moderate	Creamy white	Yellow	Nil
8	Starch casein agar	Moderate	Grayish white	White	Nil
9	Nutrient agar	Good	White	Pale yellow	Nil
10	Czapek-Dox agar	Good	Grayish white	Yellow	–

coagulase tests. Growth of the strain occurred in the pH range of 6–10 with optimum growth at pH 7.5. The temperature range for growth was 20–45 ° C with the optimum growth at 30 °C. Tolerance of the strain to NaCl concentration additionally serves as an important character for species identification. VSM-1 exhibited salt tolerance up to 8% with optimum growth at 3% NaCl; therefore, the strain may be positioned in intermediate salt tolerance group. The salt concentration has a profound effect on the production of antibiotic from microorganisms due to its effect on the osmotic pressure to the medium (Tresner et al. 1968). The utilization of diverse carbon sources by the strain indicated its wide pattern of carbon assimilation ability (Table 38.4). The strain utilized a wide range of carbon sources such as fructose, galactose, glucose, maltose, sucrose, and xylose but did not utilize arabinose, inositol, lactose, rhamnose, and raffinose. Carbohydrate utilization plays a distinguished role in the taxonomic characterization of actinomycetes strains (Pridham and Gottlieb 1948). The strain was sensitive to the majority of the antibiotics tested but showed resistance to streptomycin, gentamycin, and ampicillin (Table 38.5).

The spore chain ornamentation and spore morphology of the isolate were observed by scanning electron microscopy. The arrangement of spores was in spiral fashion, and the surface of the spore was rough (Fig. 38.1), and hence it can be placed in the spira group of the family *Streptomycetaceae* and the genus *Streptomyces* (Pridham et al. 1958). These results were confirmed by the identification of the culture by 16S rDNA sequencing. Gene sequence of 16S rRNA of VSM-1 was blasted against nucleotide database of the NCBI. The library search reported matching strains, and the sequences had been aligned with the set of published sequence on the basis of the conserved primary sequence and additionally with the aid of

Table 38.3 Morphological, physiological, and biochemical characteristics of the strain VSM-1

S. no.		Response
	Morphological characters	
1	Cell shape	Mycelial
2	Sporophore morphology	Spiral
3	Color of aerial mycelium (ISP-2)	Grayish white
4	Color of substrate mycelium (ISP-2)	Yellow
	Physiological characters	
5	Gram reaction	+
6	Acid-fast reaction	–
7	Production of melanin pigment	–
8	Range of temperature for growth	20–45 °C
9	Optimum temperature for growth	30 °C
10	Range of pH for growth	6–10
11	Optimum pH for growth	7.5
12	NaCl tolerance	8%
	Biochemical characters	
13	Catalase production	+
14	Nitrate reduction	+
15	Hydrogen sulfide production	–
16	Starch hydrolysis	+
17	Gelatin liquefaction	–
18	Methyl red test	–
19	Voges-Proskauer test	+
20	Indole production	+
21	Citrate utilization	+
22	Casein hydrolysis	+
23	Coagulase	–

+ positive; – negative

Table 38.4 Utilization of the carbon sources by the strain VSM-1

S. no.	Carbon source	Response
1	Fructose	+
2	Galactose	+
3	Glucose	+
4	Maltose	+
5	Sucrose	+
6	Xylose	+
7	Rhamnose	–
8	Raffinose	–
9	Inositol	–
10	Lactose	–
11	Arabinose	–

+ positive; – negative

Table 38.5 Growth in the presence of antibiotics

S. no.	Carbon source	Response
1	Streptomycin	R
2	Gentamycin	R
3	Ampicillin	R
4	Tetracycline	S
5	Vancomycin	S
6	Novobiocin	S
7	Amoxicillin	S
8	Ciprofloxacin	S
9	Chloramphenicol	S
10	Neomycin	S
11	Erythromycin	S

S sensitive, *R* resistant

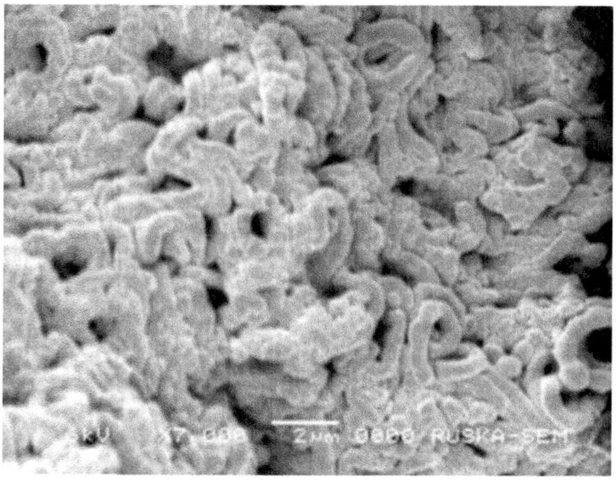

Fig. 38.1 Scanning electron micrograph of the marine actinobacterium VSM-1 grown on ISP-2 medium

nucleotide blast similarity search evaluation. The 16S rDNA gene sequence of the isolate VSM-1 showed a close relation with *Streptomyces nanhaiensis* VSM-1 (Fig. 38.2). The rDNA sequence was deposited in the GenBank database of NCBI with an accession number KU507594.

38.3.1 Media Optimization

38.3.1.1 Incubation Period

The growth pattern of *S. nanhaiensis* VSM-1 was studied on Bennet's broth. The stationary phase of the strain extended from 168 to 216 h of incubation, after which it declined (Fig. 38.3a). The secondary metabolites obtained from 9-day-old culture exhibited high antimicrobial activity against the test microorganisms. The incubation period for the production of bioactive metabolites appears to differ among *Streptomyces* strains. Saurav and Kannabiran (2010) stated that the maximum

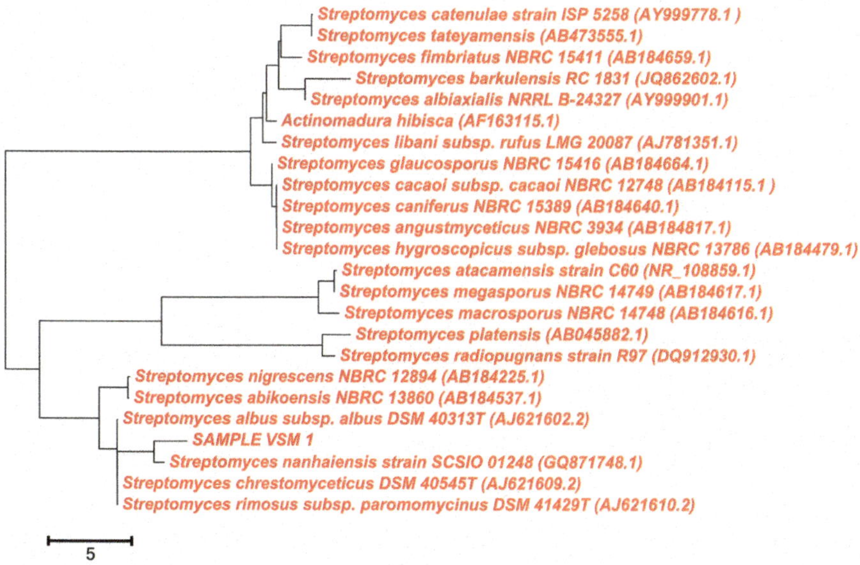

Fig. 38.2 Maximum parsimony tree based on partial 16S rRNA gene sequence showing relationship between *Streptomyces* isolate VSM-1 and related members of the genus *Streptomyces*

incubation duration required for optimum growth and antibiotic yield by the isolate *Streptomyces* VITSVK9 sp. was 9 days which was in complete accordance with the earlier report, whereas metabolites accumulated from 10-day-old culture of *S. crystallinus* AZ-A151 producing hygromycin B exhibited good antimicrobial activity (Atta et al. 2012).

38.3.1.2 pH and Temperature

The bioactive metabolite production was found to be influenced by pH of the medium. The maximum growth as well as bioactive metabolite production of the strain was obtained at pH 8 (Fig. 38.3b). This result suggests the inclusion of this strain in the alkaliphilic actinomycetes group. Medium maintained at pH 8 was reported to support enhanced antimicrobial metabolite production by *Streptomyces tanashiensis* strain A2D (Singh et al. 2009). The results are also comparable with some *Streptomyces* spp. reported to produce antibiotics against bacteria, fungi, and yeast at alkaline pH (Basilio et al. 2003). The impact of temperature on biomass and bioactive metabolite production of the strain was studied. Highest growth in addition to antimicrobial metabolite production was obtained at 30 °C (Fig. 38.3c). In terms of its optimum temperature for growth, the organism appeared to be mesophilic in nature. This is in agreement with earlier reports for several of *Streptomyces* spp. (Ellaiah et al. 2004; Mustafa 2011; Atta et al. 2011).

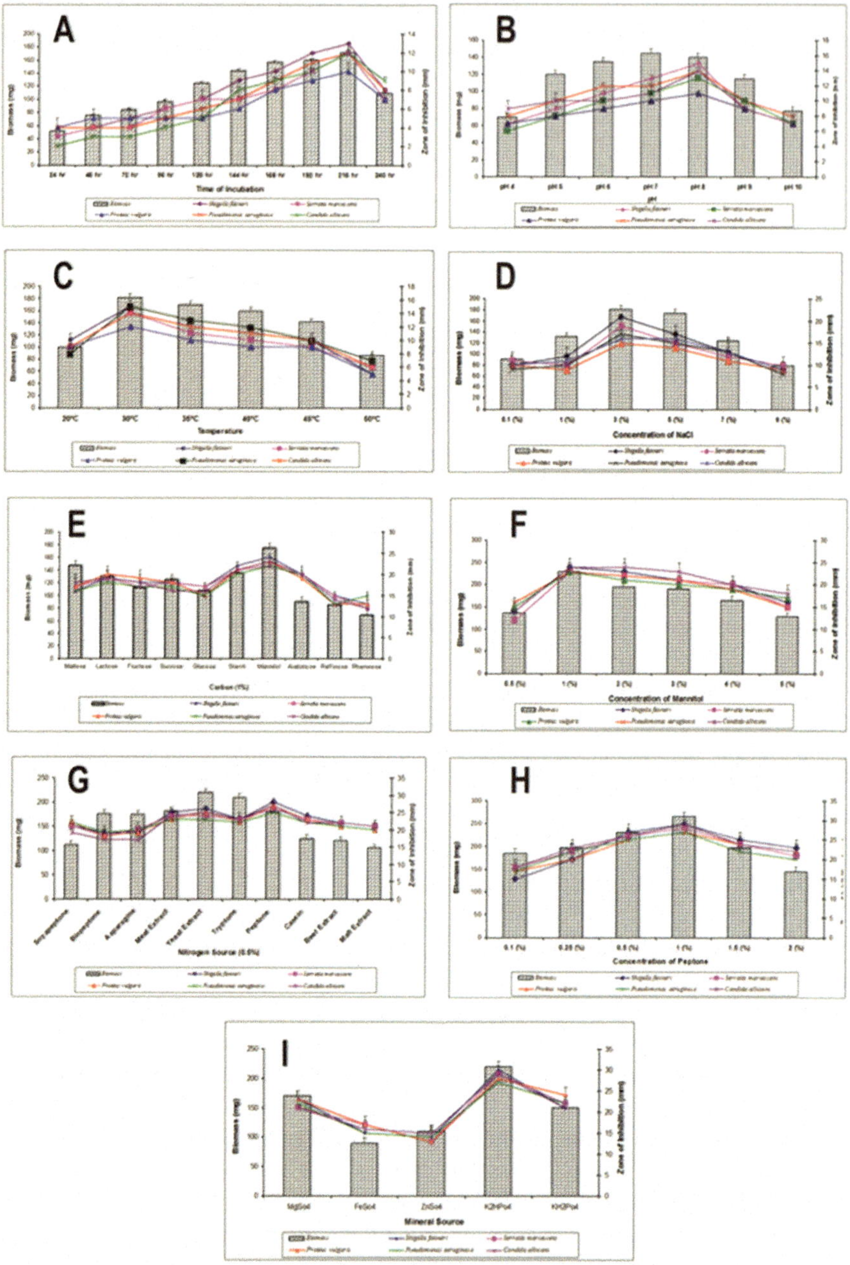

Fig. 38.3 (**a**) Growth pattern and antimicrobial activity of *S. nanhaiensis* VSM-1. (**b**) Effect of pH on growth and bioactive metabolite yield of *S. nanhaiensis* VSM-1. (**c**) Effect of temperature on growth and bioactive metabolite yield of *S. nanhaiensis* VSM-1. (**d**) Effect of NaCl on growth and bioactive metabolite yield of *S. nanhaiensis* VSM-1. (**e**) Effect of different carbon sources on growth and bioactive metabolite yield of *S. nanhaiensis* VSM-1. (**f**) Effect of different concentrations of

38.3.1.3 NaCl Concentration

Optimum salt requirement for antimicrobial metabolite production was examined in liquid basal medium supplemented with different salt concentrations ranging from 0.1% to 9%. 3% NaCl was found to be optimum for maximum growth as well as antimicrobial compound production of the strain VSM-1 (Fig. 38.3d). Further increase in salt concentration reduced the antimicrobial agent biosynthesis. The requirement of NaCl for the production of bioactive metabolites seems to be different among actinomycetes strains. Singh et al. (2009) recorded that antibiotic production by *S. tanashiensis* was maximum at 2% NaCl in the medium. Culture medium with 5% NaCl was reported to support high antimicrobial metabolite production by *Streptomyces* VITSVK9 (Saurav and Kannabiran 2010).

38.3.1.4 Carbon and Nitrogen Sources

Impact of several carbon sources on biomass and bioactive metabolite yield was shown in Fig. 38.3e. The strain was able to grow in all the tested carbon sources. The strain produced highest biomass as well as antimicrobial compound in mannitol-supplemented medium. Addition of other carbon sources to the medium separately also favored growth, but the metabolite production was less when compared with mannitol. As mannitol is the preferred carbon source for bioactive metabolite production by the strain, different concentrations of mannitol (0.5–5%) were examined to determine its optimal concentration on metabolite production. Medium containing 1% mannitol supported high levels of biomass and bioactive metabolite production by the strain (Fig. 38.3f). Mannitol at the concentration of 1.5% as the best carbon source for high rates of antibiotic production by *Streptomyces* spp. 201 was reported by Thakur et al. (2009).

Of all the examined nitrogen sources, yeast extract and tryptone have been favored as good nitrogen sources for biomass production, whereas peptone was found to be the excellent nitrogen source for antimicrobial agent production, of the strain VSM-1 (Fig. 38.3g). Influence of different concentrations of peptone on the production of bioactive metabolites is represented in Fig. 38.3h. Peptone at the concentration of 1% exhibited optimal production of biomass and bioactive metabolite production. This is in conformity with the findings of Chattopadhyay and Sen (1997) and Han et al. (2004) for which peptone was proved to be the best nitrogen source for antibiotic production by *Streptomyces rochei* G 164 and *Streptomyces scabiei* PK-A41, respectively.

In the present study, the optimal culture conditions for the production of bioactive metabolites by strain VSM-1 were reported. The yield of metabolites by the strain was high in production medium supplemented with 1% mannitol and 1% peptone with initial pH 8 at temperature 30 °C for 9 days of incubation.

Fig. 38.3 (continued) mannitol on growth and bioactive metabolite yield of *S. nanhaiensis* VSM-1. (**g**) Effect of different nitrogen sources on growth and bioactive metabolite yield of *S. nanhaiensis* VSM-1. (**h**) Effect of different concentrations of peptone on growth and bioactive metabolite yield of *S. nanhaiensis* VSM-1. (**i**) Effect of different minerals on growth and bioactive metabolite yield of *S. nanhaiensis* VSM-1 (Data on cell growth and bioactive metabolite yield were statistically analyzed by two-way ANOVA and found to be significant at 1%)

38.3.1.5 Minerals

The impact of minerals on growth and bioactive metabolite production of the strain is represented in Fig. 38.3i. K_2HPO_4 at the concentration of 0.05% enhanced the production of biomass and bioactive metabolites. In contrast the metabolite production was low with other minerals tested. Similar results were obtained with *Streptomyces fradiae* in neomycin production as mentioned by Majumdar and Majumdar (1965).

38.3.2 RSM Modeling and Optimization of Bioactive Metabolite Production by *Streptomyces nanhaiensis* VSM-1

Response surface methodology has been used for model designing and to optimize the independent process variables in various fields such as food industry, dye stuff industry, biotechnology, and also wastewater treatment (Hong et al. 2017). CCD design for the production of the bioactive metabolite produced by *Streptomyces nanhaiensis* VSM-1 and its effect against the five responses (zone of inhibition measured in mm) is given in Appendix Table 38.6 along with the experimental and predicted responses. The experimental results were verified using statistical analysis and modified quadratic models. The regression model equation for the bioactive metabolite production was expressed as follows (Payam et al. 2017).

$$
\begin{aligned}
Y_1 = & -73.00000 + 10.40000A + 12.40000B + 0.70000C + 3.60000D + 4.40000E \\
& + 0.000000AB + 0.000000AC + 0.000000AD + 0.000000AE + 0.000000BC \\
& + 0.000000BD + 0.000000BE + 0.000000CD + 0.000000CE + 0.000000DE \\
& - 0.60000A^2 - 0.80000B^2 - 0.012000C^2 - 1.60000D^2 - 2.40000E^2
\end{aligned}
$$

$$
\begin{aligned}
Y_2 = & -61.00000 + 12.30000A + 7.90000B + 0.40000C + 3.60000D + 4.40000E \\
& + 0.000000AB + 0.000000AC + 0.000000AD + 0.000000AE + 0.000000BC \\
& + 0.000000BD + 0.000000BE + 0.000000CD + 0.000000CE + 0.000000DE \\
& - 0.70000A^2 - 0.50000B^2 - 8.00000E-003C^2 - 1.60000D^2 - 2.40000E^2
\end{aligned}
$$

$$
\begin{aligned}
Y_3 = & -27.40000 + 7.00000A + 4.70000B + 0.44000C + 4.20000D + 3.80000E \\
& + 0.000000AB + 0.000000AC + 0.000000AD + 0.000000AE + 0.000000BC \\
& + 0.000000BD + 0.000000BE + 0.000000CD + 0.000000CE + 0.000000DE \\
& - 0.40000A^2 - 0.30000B^2 - 8.00000E-003C^2 - 2.00000D^2 - 2.00000E^2
\end{aligned}
$$

$$Y_4 = -70.80000 + 8.70000A + 9.40000B + 1.74000C + 0.60000D + 5.00000E + 0.000000AB + 0.000000AC + 0.000000AD + 0.000000AE + 0.000000BC + 0.000000BD + 0.000000BE + 0.000000CD + 0.000000CE + 0.000000DE - 0.50000A^2 - 0.60000B^2 - 0.028000C^2 - 0.40000D^2 - 2.80000E^2$$

$$Y_5 = -78.60000 + 8.90000A + 12.60000B + 1.22000C + 3.00000D + 6.00000E + 0.000000AB + 0.000000AC + 0.000000AD + 0.000000AE + 0.000000BC + 0.000000BD + 0.000000BE + 0.000000CD + 0.000000CE + 0.000000DE - 0.50000A^2 - 0.80000B^2 - 0.020000C^2 - 1.20000D^2 - 3.20000E^2$$

where Y_1 is *Shigella flexneri*, Y_2 *Serratia marcescens*, Y_3 *Proteus vulgaris*, Y_4 *Pseudomonas aeruginosa*, and Y_5 *Candida albicans*. *A* is the incubation time (days), *B* pH, and *C* temperature, and *D* and *E* are the concentrations of mannitol and peptone, respectively.

Predicted values calculated for Eq. (38.2) are in good agreement with the experimental values (Appendix Table 38.6). Hence, the designed quadratic model is well suited for the present experimental setup. Model adequacy was analyzed using ANOVA that determines the significance and the goodness of fit of the model (Almeida et al. 2017). As shown from Table 38.7, the *F* value of all the five responses *S. flexneri*, *S. marcescens*, *P. vulgaris*, *P. aeruginosa*, and *C. albicans* simply that the model is highly significant that there is only 0.01 chance that this large value (Model *F* value) could occur due to noise (Almeida et al. 2017). Based on the *F* values, the optimum values of all the five variables for the bioactive metabolite production was found to be incubation time (9 days), pH (8), temperature (30°C), and concentrations of mannitol and peptone (1% and 1%) and their effect against the five responses (effect of the bioactive metabolite produced by *Streptomyces nanhaiensis* VSM-1 and its effect is measured as zone of inhibition in mm) *S. flexneri* (34.2 mm), *S. marcescens* (32 mm), *P. vulgaris* (31.6), *P. aeruginosa* (32.3), and *C. albicans* (33.6).

The model terms are said to be significant if the probability values Prob > *F* less than 0.05 and values exceeding greater than 0.1 indicate that model terms are not significant. The *P* values of all the five variables (Table 38.8) are significant since the *P* value is 0.0001 (Ka et al. 2017). The lack-of-fit value of all the five responses is not significant relative to the pure error. Non-significant lack of fit is desired as it

Table 38.7 Analysis of variance (ANOVA) to test the adequacy of the model

	Response				
Statistics	*S. flexneri*	*S. marcescens*	*P. vulgaris*	*P. aeruginosa*	*C. albicans*
R^2	0.9978	0.9472	0.9409	0.9768	0.9675
Adj R^2	0.9879	0.9383	0.9234	0.9269	0.9129
Predicted R^2	0.9564	0.9577	0.8678	0.9567	0.9416
Adequate Precession	20.988	26.004	18.354	12.158	18.184
CV %	1.57	1.87	2.00	2.31	2.21

Table 38.8 Sequential model fitting for all the responses (in terms of inhibition zone from the bioactive metabolite produced by *Streptomyces nanhaiensis* VSM-1)

Model parameter	*Shigella flexneri*	*Serratia marcescens*	*Proteus vulgaris*	*Pseudomonas aeruginosa*	*Candida albicans*
Sequential model sum of squares—quadratic vs 2FI (suggested)					
Sum of squares	67.96	53.75	33.64	63.23	78.42
d.f.[a]	5	5	5	5	5
Mean square	13.59	10.75	6.73	12.65	15.68
F-value	6.366E+007	6.366E+007	6.366E+007	6.366E+007	6.366E+007
P-value (Prob > *F*)	<0.0001	<0.0001	<0.0001	<0.0001	<0.0001
Lack-of-fit tests—quadratic (suggested)					
Sum of squares	1.667	4.12	5.62	16.74	7.89
d.f.[a]	22	22	22	22	22
Mean square	0.16	0.19	0.16	0.89	0.76
F-value	–	–	–	–	–
P-value (Prob > *F*)	–	–	–	–	–
Model summary statistics—quadratic (suggested)					
Std. dev.	0.52	0.35	0.46	0.16	0.39

[a]Degrees of freedom

is mandatory to make the goodness of the model fit (Zhang et al. 2017). The lack-of-fit value of the five responses is given in Table 38.8. The predicted response values estimated by the designed model are shown by the determination of the coefficient (R^2). The difference between the R^2 and the adjusted R^2 should be between 0 and 0.2 to determine the adequacy of the model. The R^2 and the adjusted R^2 are in reasonable agreement (Table 38.7) that indicate that the model is significant. Adequate precision determines the signal-to-noise ratio. A ratio greater than 4 is desirable. The adequate precision of the five responses is given in Table 38.7.

38.3.3 3D Plots

The 3D response plots for the bioactive metabolite production by *Streptomyces nanhaiensis* VSM-1 and its effect against the five responses obtained from the above model were given in Figs. 38.4, 38.5, 38.6, 38.7, and 38.8. The 3D plots represent the main and the interactive effects of the variables against the response. Two of the variables are tested while keeping the third variable at zero level. Mutual interaction between each of the two factors all showed elliptical shapes indicating that these factors interacted with each other and produced a combined effect for the production of the bioactive metabolites and its effect against the five responses (Ying et al. 2014).

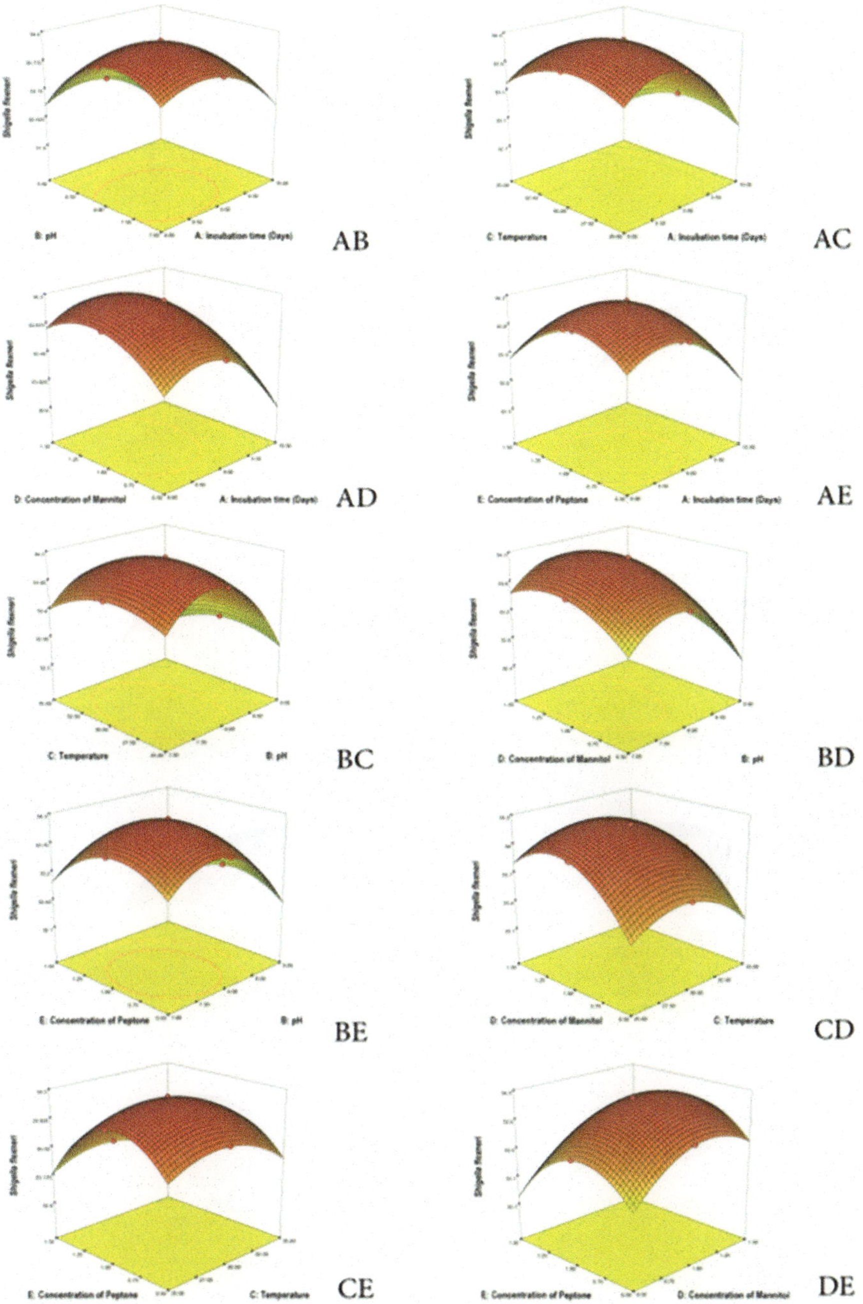

Fig. 38.4 Response surface 3D plots showing the interactive effects of selective variables on zone of inhibition (mm) of the bioactive compound production by *Streptomyces nanhaiensis* VSM-1 against response *Shigella flexneri*

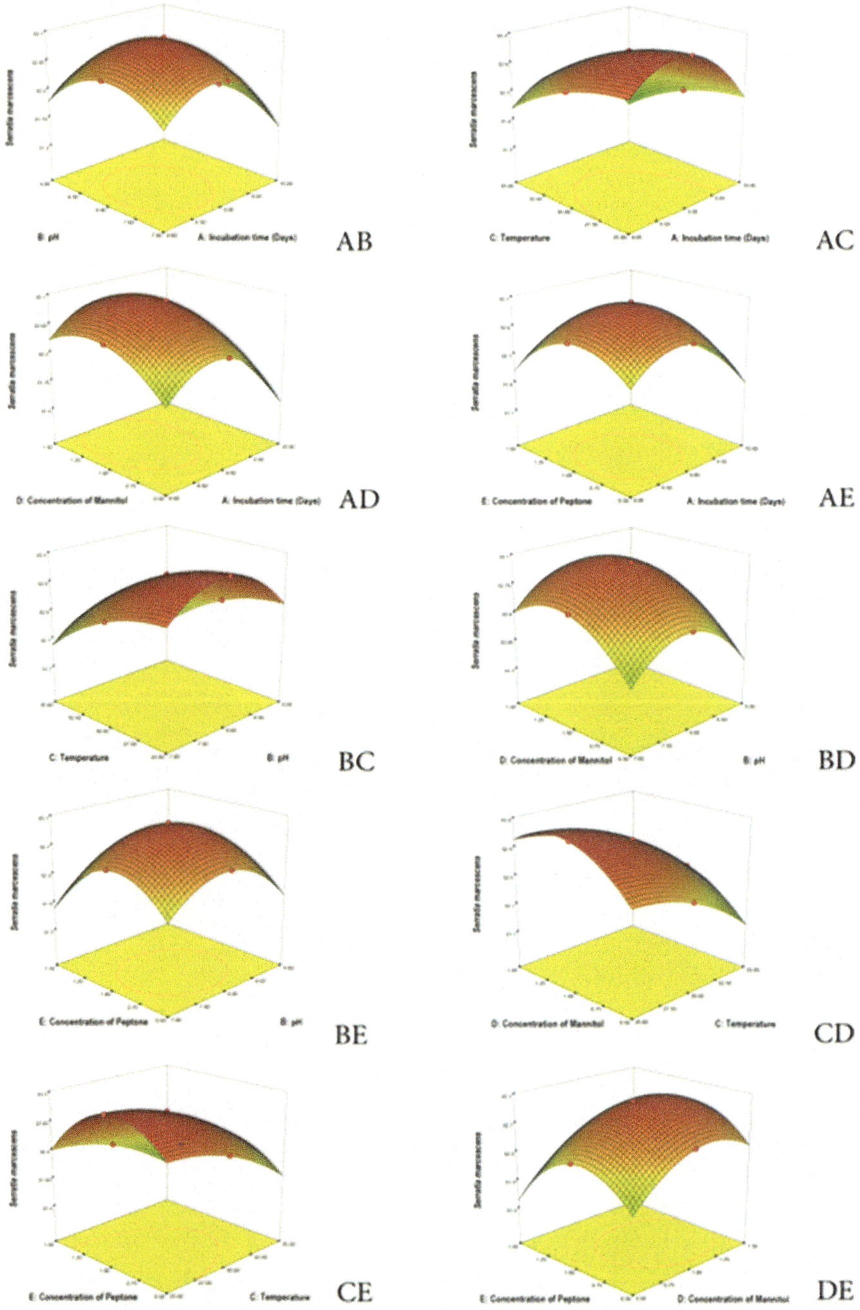

Fig. 38.5 Response surface 3D plots showing the interactive effects of selective variables on zone of inhibition (mm) of the bioactive compound production by *Streptomyces nanhaiensis* VSM-1 against response *Serratia marcescens*

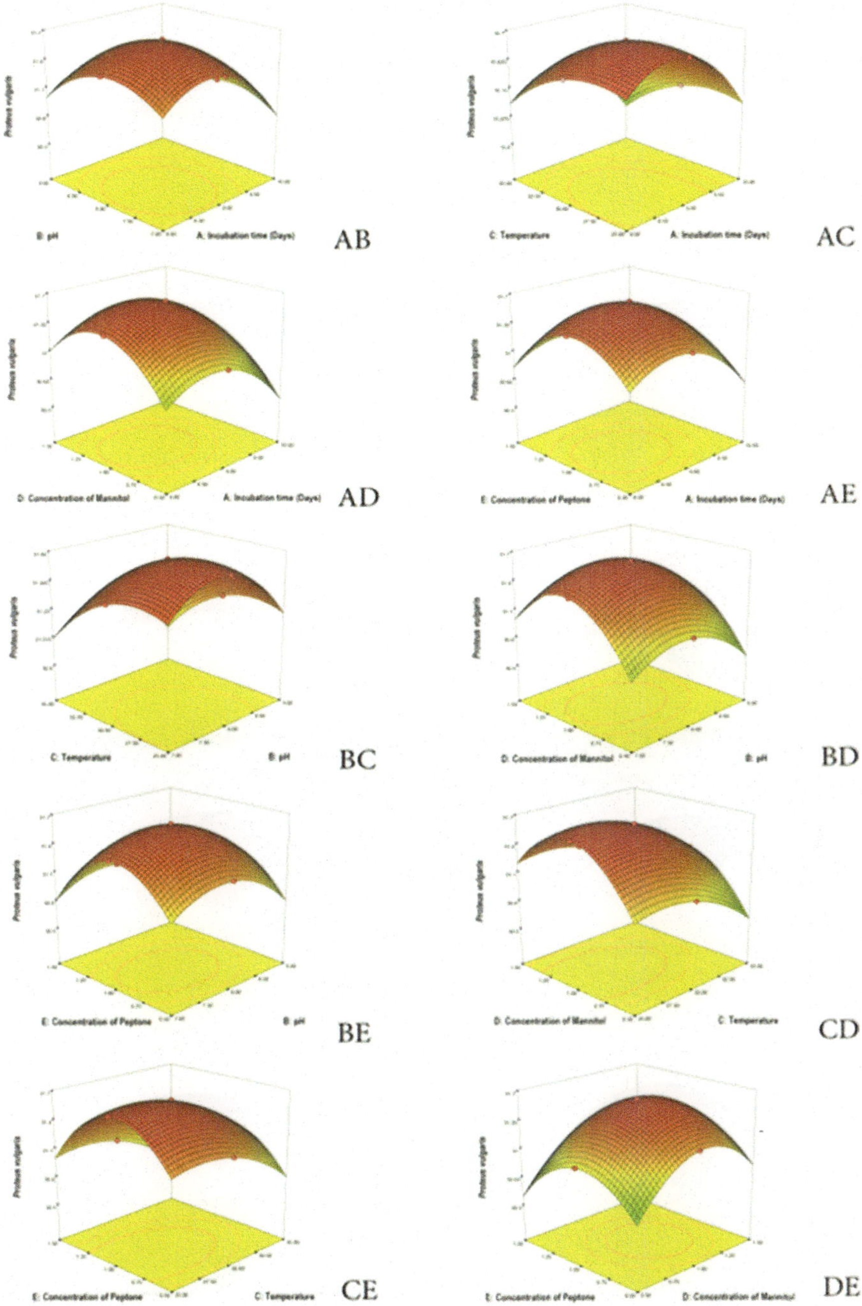

Fig. 38.6 Response surface 3D plots showing the interactive effects of selective variables on zone of inhibition (mm) of the bioactive compound production by *Streptomyces nanhaiensis* VSM-1 against response *Proteus vulgaris*

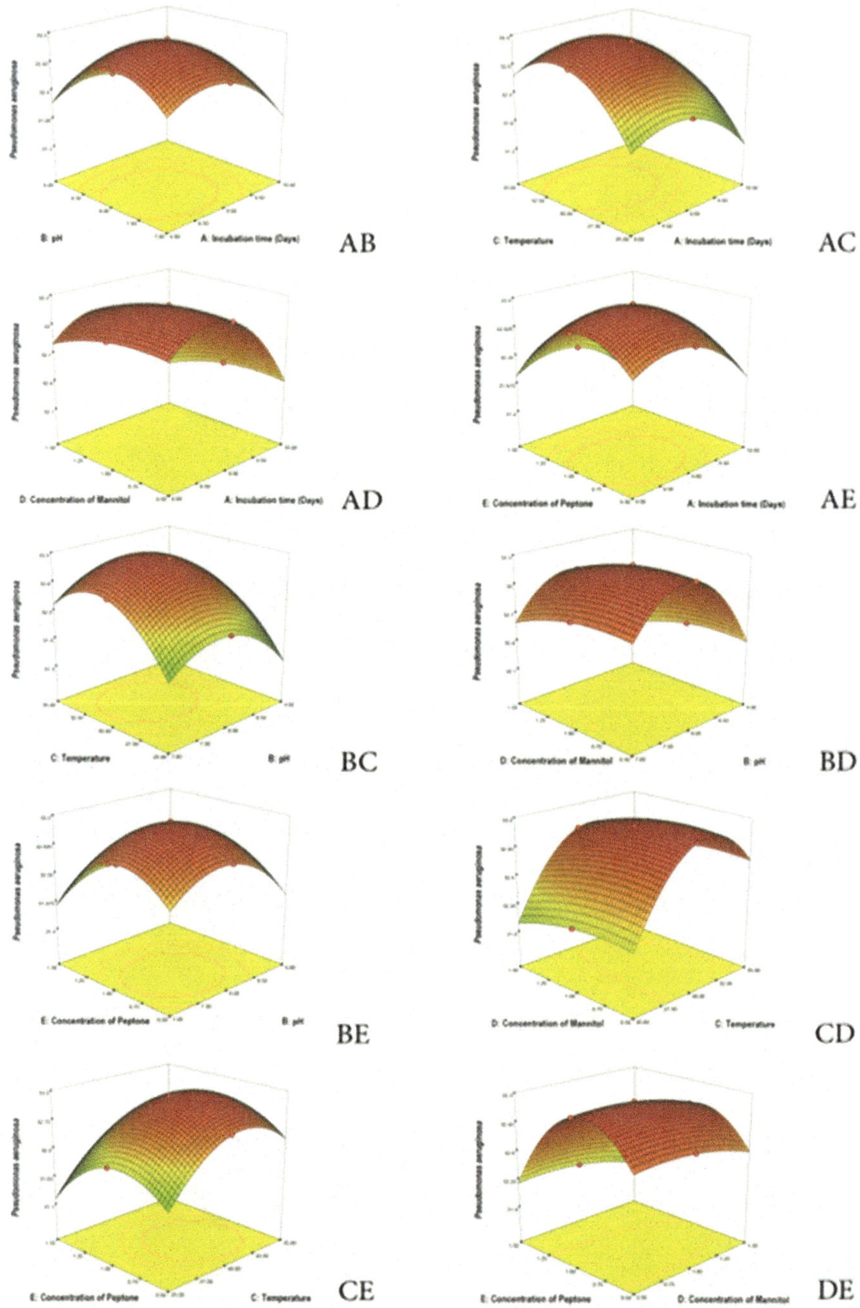

Fig. 38.7 Response surface 3D plots showing the interactive effects of selective variables on zone of inhibition (mm) of the bioactive compound production by *Streptomyces nanhaiensis* VSM-1 against response *Pseudomonas aeruginosa*

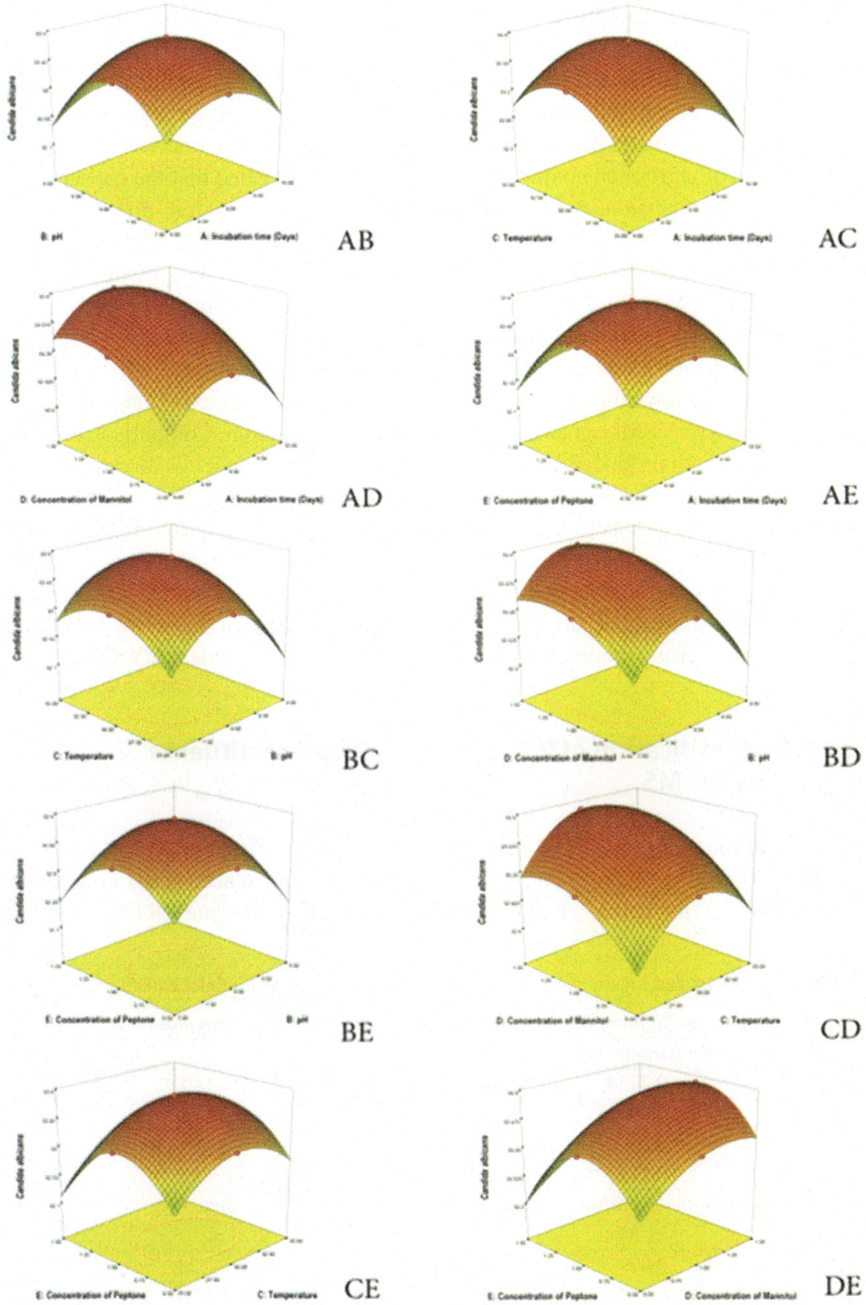

Fig. 38.8 Response surface 3D plots showing the interactive effects of selective variables on zone of inhibition (mm) of the bioactive compound production by *Streptomyces nanhaiensis* VSM-1 against response *Candida albicans*

38.3.4 Kinetic Modeling

In this study, L, LILP, and LIMLP models were fitted for experimental data using non-linear regression of least-square method with the help of Microsoft Excel Solver 2010. Estimated kinetic parameters from model fitting were tabulated in Table 38.9. High values of regression coefficient (R^2) above 0.95 indicated that the chosen models have good precision. A high value (2.68–3.86 fold) of α than β (LILP model parameters) indicates that the bioactive metabolite produced from *Streptomyces nanhaiensis* VSM-1 (KU507594) is more growth associated than non-growth associated in shake flask. LIMLP model parameters γ and η are also having very good agreement with experimental data. Table 38.10 compares the values of zones of inhibition from agar diffusion studies of experimental versus model predicted.

Figure 38.9 shows the comparison of model predicted and experimental profiles of *S. nanhaiensis* VSM-1 growth-limiting substrate utilization. Comparison of zones of inhibition from produced bioactive metabolite on media inoculated with *Shigella flexneri*, *Serratia marcescens*, *Proteus vulgaris*, *Pseudomonas aeruginosa*, and *Candida albicans* strains over the time was shown in Fig. 38.10. Both Figs. 38.9 and 38.10 show very good fit of data for model predicted and experimental results. Thus, the selected unstructured models provided a better approximation of kinetic profiles, and these models can be used as kinetic approximations for bioactive metabolite production (in terms of zones of inhibition studies) by *S. nanhaiensis* VSM-1.

38.3.5 Identification of Potential Bioactive Constituents by GC-MS

GC-MS chromatogram of the ethyl acetate extracts of *Streptomyces nanhaiensis* VSM-1 showed a total of 30 peaks, indicating the presence of 30 compounds (Fig. 38.11). The active principles with their retention time (RT), molecular

Table 38.9 Estimated kinetic parameters using L, LILP, and LIMLP model equations

Kinetic parameters	*S. flexneri*	*S. marcescens*	*P. vulgaris*	*P. aeruginosa*	*C. albicans*
Logistic (L) model parameters					
μ_{max} (d^{-1})	0.7134				
R^2	0.9952				
X_0 (g/L)	0.03				
X_{max} (g/L)	0.173				
Logistic incorporated modified Luedeking-Piret (LIMLP) model parameters					
γ (g·S/g·X)	21.34				
R^2	0.979				
η (g·S/(g·X·d))	0.231				
Logistic incorporated Luedeking-Piret (LILP) model parameters					
α (mm/g·X)	66.9	59.006	52.317	46.59	53.365
R^2	0.969	0.9532	0.9575	0.9416	0.9627
β (mm/(g·X·d))	17.341				

Table 38.10 Comparison of zones of inhibition (mm) from shake flask experiments and from model

Maximum zone of inhibition (mm)	*S. flexneri*	*S. marcescens*	*P. vulgaris*	*P. aeruginosa*	*C. albicans*
Experimental	30	29	28	27	29
Model (fitted)	29.14	28.01	27.07	26.26	27.22

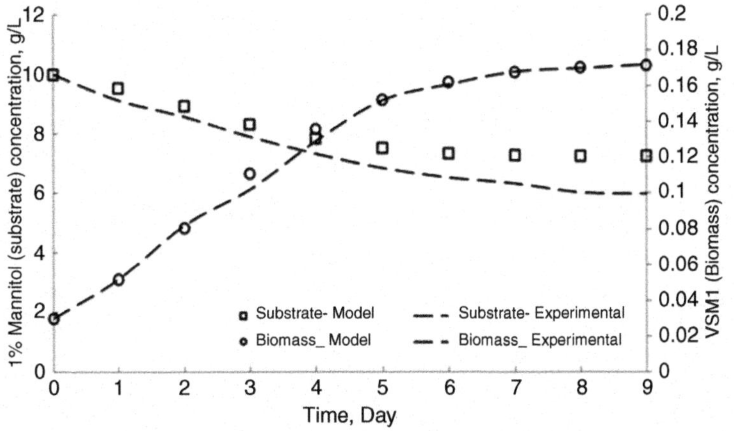

Fig. 38.9 Experimental and model predicted kinetics of biomass growth and substrate utilization. ● model biomass concentration (g/L), ■ model substrate concentration (g/L), ---- experimental predicted values (in each case)

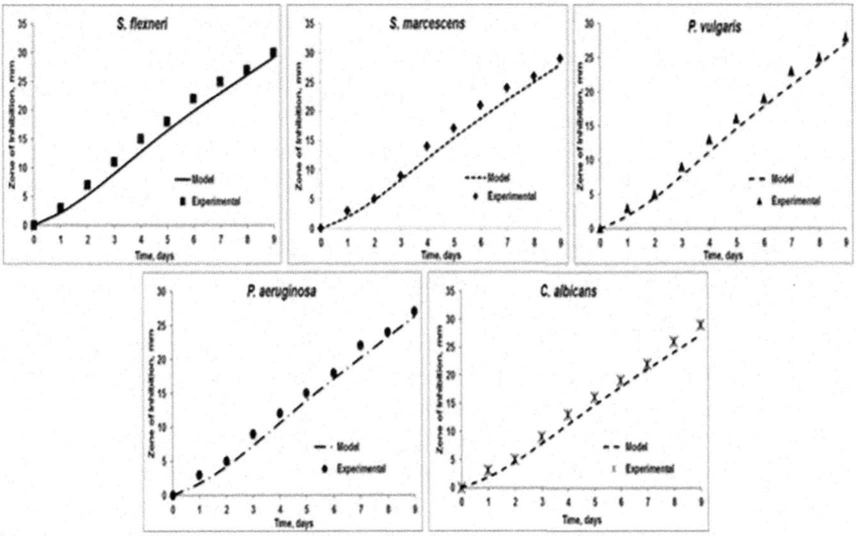

Fig. 38.10 Zones of inhibition (in mm) compared from model and experimental kinetics

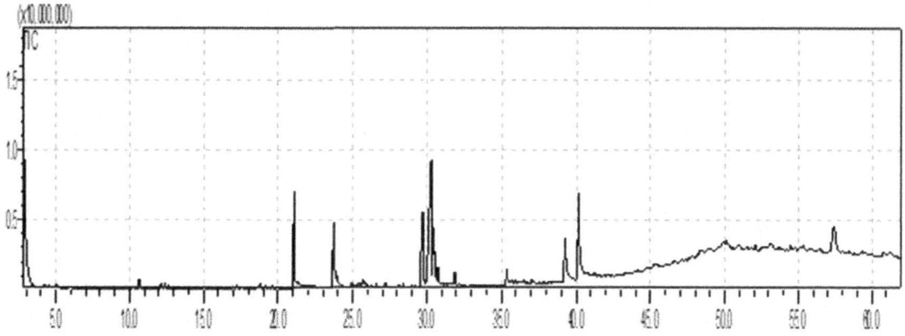

Fig. 38.11 Gas chromatography-mass spectrometry analysis of crude ethyl acetate extract of VSM-1

formula, molecular weight, and concentration (peak area) are reported in Table 38.11. The peak area is directly proportional to the quantity of the compounds present in the ethyl acetate fraction (EAF). When compared with NIST database, the nearest compound hits for those peaks were found. According to the available library data, NIST MS search (ver. 2) (included with NIST'02 mass spectral library, Agilent p/n G1033 A), the compounds were identified as undecane (5.065); 2-propanol, 1-(2-ethoxypropoxy)- (10.643); ethanone, 1-(2,6-dihydroxy-4-methoxyphenyl) (12.141); phenol, 2,4-bis(1,1-dimethylethyl)- (18.673); 1-dodecanol, 3,7,11-trimethyl- (18.826); cis-11,14-eicosadienoic acid, methyl ester (21.056); pentadecanoic acid (23.689); 2-propenoic acid, 2-methyl-, dodecyl ester (23.885); 1-octadecene, (*E*)- (25.691); methyl 2-hydroxydodecanoate (25.919); *N*-[4-cyclooctylaminobutyl]aziridine (27.214); hexadecanoic acid methyl ester (29.715); dibutyl phthalate (30.289); benzene acetic acid, 2-phenyl ethyl ester (30.481); 2,6-dibutyl 1–4-methyl piperidine (30.747); butyric acid, pentadecyl ester (31.891); 1-nonadecene (35.358); octadecane, 6-methyl (35.953); 9-eicosene, (*E*)- (39.252); *n*-tricosane (40.178); heptacosane, 1-chloro- (49.731); sulfurous acid, hexyl pentadecyl ester (49.948); erythro-9,10-dibromopentacosane (50.27); 2-methyl octacosane (50.922); oxalic acid, 2-ethylhexyl octadecyl ester (52.091); docosane, 11-decyl- (53.21); triacontane (53.717); tetratriacontane (54.475); 10-methyldotriacontane (55.362); and 13-*n*-undecylpentacosane (60.827). NCBI PubChem BioAssay Database (https://pubchem.ncbi.nlm.nih.gov) and Dr. Duke's Phytochemical and Ethnobotanical Databases (http://www.ars-gringov/duke) revealed that among the 30 peaks, 14 compounds indicate the presence of active principles which have been previously reported for their antimicrobial, antioxidant, anti-inflammatory, and cytotoxic properties (Table 38.12).

GC-MS analysis of the strain VSM-1 indicated the presence of at least 14 bioactive compounds which makes this particular strain very potential for further study. Among the active principles, phenol, 2,4-bis(1,1-dimethylethyl)-; dibutyl phthalate; 9-eicosene, (*E*)-; and tetratriacontane have both antibacterial and antifungal properties (Salini et al. 2014; Rangel-Sánchez et al. 2014; Roy et al. 2006; Hsouna et al. 2011; Mahmood et al. 2009), while 1-octadecene, (*E*)-, and *n*-tricosane act only on

Table 38.11 Active principles identified in the ethyl acetate extract of the strain VSM-1 by GC-MS

S. no.	Name of the compound	Molecular formula	Molecular weight	RT	Area	SI
1	Undecane	$C_{11}H_{24}$	156	5.065	1,323,965	99
2	2-Propanol,1-(2-ethoxypropoxy)-	$C_8H_{18}O_3$	162	10.643	1,325,519	99
3	Ethanone, 1-(2,6-dihydroxy-4-methoxy phenyl)	$C_9H_{10}O_4$	182	12.141	276,428	99
4	Phenol,2,4-bis(1,1-dimethylethyl)-	$C_{14}H_{22}O$	206	18.673	510,939	96
5	1-Dodecanol, 3,7,11-trimethyl-	$C_{15}H_{32}$	228	18.826	152,064	99
6	Cis-11,14-eicosadienoic acid, methyl ester	$C_{15}H_{26}O$	222	21.056	13,475,957	99
7	Pentadecanoic acid	$C_{15}H_{30}O_2$	242	23.689	4,918,727	99
8	2-Propenoic acid, 2-methyl-, dodecyl ester	$C_{16}H_{30}O_2$	254	23.885	798,012	85
9	1-Octadecene, (*E*)-	$C_{18}H_{36}$	252	25.691	323,589	98
10	Methyl 2-hydroxydodecanoate	$C_{13}H_{26}\ O_3$	230	25.919	159,296	99
11	*N*-[4-Cyclooctylaminobutyl] aziridine	$C_{14}H_{28}N_2$	224	27.214	421,137	99
12	Hexadecanoic acid, methyl ester	$C_{17}H_{34}O_2$	270	29.715	13,766,600	99
13	Dibutyl phthalate	$C_{16}H_{22}O_4$	278	30.289	25,918,364	99
14	Benzene acetic acid, 2-phenyl ethyl ester	$C_{16}H_{16}O_2$	240	30.481	6,978,986	97
15	2,6-Dibutyl 1–4-methyl piperidine	$C_{14}H_{29}N$	211	30.747	587,102	97
16	Butyric acid, pentadecyl ester	$C_{19}H_{38}O_2$	298	31.891	580,287	99
17	1-Nonadecene	$C_{19}H_{38}$	266	35.358	680,266	99
18	Octadecane, 6-methyl	$C_{19}H_{40}$	268	35.953	147,112	99
19	9-Eicosene, (*E*)-	$C_{20}H_{40}$	280	39.252	6,679,597	99
20	*n*-Tricosane	$C_{23}H_{48}$	324	40.178	10,438,891	99
21	Heptacosane, 1-chloro-	$C_{27}H_{55}Cl$	415	49.731	28,937	89
22	Sulfurous acid, hexyl pentadecyl ester	$C_{21}H_{44}O_3S$	376	49.948	973,353	98
23	Erythro-9,10-dibromopentacosane	$C_{25}H_{50}Br_2$	510	50.27	97,360	83
24	2-Methyl octacosane	$C_{29}H_{60}$	408	50.922	401,213	94
25	Oxalic acid, 2-ethylhexyl octadecyl ester	$C_{28}H_{54}O_4$	454	52.091	469,936	99
26	Docosane, 11-decyl-	$C_{32}H_{66}$	450	53.21	339,245	92
27	Triacontane	$C_{30}H_{62}$	422	53.717	288,915	98
28	Tetratriacontane	$C_{34}H_{70}$	478	54.475	579,382	99
29	10-Methyldotriacontane	$C_{33}H_{68}$	464	55.362	441,700	95
30	13-*n*-Undecylpentacosane	$C_{36}H_{74}$	506	60.827	407,251	99

Table 38.12 Identification of bioactive compounds in the ethyl acetate extract of the strain VSM-1 by GC-MS

S. no.	Name of the compound	Biological activity
1	Phenol, 2,4-bis(1,1-dimethylethyl)-	Antimicrobial; antioxidant
2	Cis-11,14-eicosadienoic acid, methyl ester	Anti-inflammatory; antioxidant
3	Pentadecanoic acid	Antioxidant
4	1-Octadecene, (*E*)-	Antibacterial; antioxidant; anticancer
5	Hexadecanoic acid, methyl ester	Antioxidant, anti-inflammatory, decrease blood cholesterol
6	Dibutyl phthalate	Antimicrobial
7	Benzene acetic acid, 2-phenyl ethyl ester	Active plant hormone
8	Butyric acid, pentadecyl ester	Analgesics
9	1-Nonadecene	Antifungal activity
10	9-Eicosene, (*E*)-	Antimicrobial, cytotoxic
11	*n*-Tricosane	Antibacterial
12	Heptacosane, 1-chloro-	Antioxidant
13	2-Methyl Octacosane	Antifungal
14	Triacontane	Antibacterial, anti-diabetic, and anti-tumor activities
15	Tetratriacontane	Antibacterial and antifungal

Source: NCBI PubChem BioAssay Database; Dr. Duke's Phytochemical and Ethnobotanical Databases

bacteria (Mishra and Sree 2007; Mihailovi et al. 2011). The compounds like 2-methyl octacosane and 1-nonadecene were previously reported for their antifungal properties (Yu et al. 2009). Several studies have attributed the antioxidant effect to phenol, 2,4-bis(1,1-dimethylethyl)-; cis-11,14-eicosadienoic acid, methyl ester; pentadecanoic acid; heptacosane, 1-chloro-; 1-octadecene, (*E*)-; and hexadecanoic acid, methyl ester (Ajayi et al. 2011; Mishra and Sree 2007; Hema et al. 2011). Cis-11,14-eicosadienoic acid, methyl ester, and hexadecanoic acid, methyl ester, are also reported for their anti-inflammatory active principles (Hema et al. 2011), while triacontane was reported for its antibacterial, anti-diabetic, and anti-tumor properties (Khare 2004; Mammen et al. 2010).

38.4 Conclusion

In the pursuit of fruitful bioactive metabolites, the study was designed to investigate the unexplored marine sediment samples of the Bay of Bengal of coastal Andhra Pradesh, India, for novel actinomycetes and their antimicrobial properties. An actinobacterium isolate VSM-1 having broad-spectrum activity was isolated from the marine sediment sample using $CaCO_3$ pretreatment technique on Bennet's agar media. The isolate was identified as *Streptomyces nanhaiensis* based on the polyphasic taxonomy. Bioactive metabolite production was initially optimized using one-factor-at-a-time (OFAT) method to select the variables. CCD of RSM was employed for further optimization that enables the evaluation of the effects of the individual factors, and their interactive influence on the responses and metabolite yield was enhanced which was represented in the form of zones of inhibition. The experimental values are in good agreement with the predicted values. Very good agreement between experimental results and model fitted data also confirmed the use of unstructured kinetic models could be better approximations for further enhancement in productivities. Lastly, the existence of various bioactive principles identified through GC-MS which is ascertained to have broad spectrum of pharmacological properties also supported that the strain VSM-1 was the most active strain. However, further studies will need to be undertaken to isolate and characterize the individual bioactive chemotypes will definitely give promising results.

Acknowledgments The first author (UKM) is grateful to the University Grants Commission (UGC), New Delhi, Government of India, for providing financial assistance in the form of fellowship to carry out this work.

Appendix

Table 38.6 Central composite factor experimental design along with experimental and predicted values for the production of bioactive metabolite (Zone of inhibition against the responses) by Streptomyces nanhaiensis VSM-1

Run	*A*-Time	*B*-pH	C-Temperature	D-Mannitol	E-Peptone	*S.flexneri*		*S.marcescens*		*P.vulgaris*		*P.aeruginos*		*C. albicans*	
						Actual	RSM	Actual	RSM	Actual	RSM	Actual	RSM	Actual	RSM
1	9.00	8.00	25.00	1.00	1.00	32.40	32.40	31.40	31.40	30.20	30.20	31.20	31.20	31.00	31.00
2	10.00	9.00	35.00	0.50	0.50	31.60	31.60	30.80	30.80	29.80	29.80	30.60	30.60	30.80	30.80
3	8.00	9.00	35.00	1.50	1.50	31.60	31.60	31.20	31.20	30.00	30.00	30.80	30.80	30.60	30.60
4	10.00	7.00	35.00	1.50	0.50	30.80	30.80	30.60	30.60	29.60	29.60	30.20	30.20	30.40	30.40
5	10.00	9.00	35.00	1.50	0.50	32.20	32.20	30.60	30.60	29.80	29.80	31.80	31.80	31.20	31.20
6	9.00	8.00	30.00	1.00	0.50	31.40	31.40	30.00	30.00	29.40	29.40	31.20	31.20	31.00	31.00
7	8.00	8.00	30.00	1.00	1.00	31.40	31.40	30.40	30.40	29.60	29.60	31.40	31.40	30.80	30.80
8	8.00	9.00	35.00	1.50	0.50	30.60	30.60	29.80	29.80	29.20	29.20	30.80	30.80	30.60	30.60
9	9.00	8.00	30.00	1.00	1.00	32.80	32.80	31.80	31.80	30.40	30.40	31.00	31.00	31.60	31.60
10	10.00	7.00	25.00	0.50	0.50	32.00	32.00	31.20	31.20	30.00	30.00	30.40	30.40	31.40	31.40
11	10.00	7.00	35.00	1.50	1.50	32.00	32.00	31.60	31.60	30.20	30.20	30.60	30.60	31.20	31.20
12	8.00	7.00	35.00	1.50	1.50	31.20	31.20	31.00	31.00	29.80	29.80	30.00	30.00	31.00	31.00
13	9.00	8.00	30.00	0.50	1.00	32.60	32.60	31.00	31.00	30.00	30.00	31.60	31.60	31.80	31.80
14	9.00	8.00	30.00	1.00	1.00	31.80	31.80	30.40	30.40	29.60	29.60	31.00	31.00	31.60	31.60
15	9.00	9.00	30.00	1.00	1.00	31.80	31.80	30.80	30.80	29.80	29.80	31.20	31.20	31.40	31.40
16	10.00	7.00	35.00	0.50	1.50	31.00	31.00	30.20	30.20	29.40	29.40	30.60	30.60	31.20	31.20
17	10.00	7.00	25.00	0.50	1.50	32.00	32.00	31.00	31.00	30.00	30.00	30.60	30.60	30.60	30.60
18	8.00	9.00	35.00	0.50	0.50	31.20	31.20	30.40	30.40	29.60	29.60	30.00	30.00	30.40	30.40
19	9.00	8.00	35.00	1.00	1.00	31.20	31.20	30.80	30.80	29.80	29.80	30.20	30.20	30.20	30.20
20	10.00	9.00	35.00	1.50	1.50	30.40	30.40	30.20	30.20	29.40	29.40	29.60	29.60	30.00	30.00
21	9.00	7.00	30.00	1.00	1.00	31.80	31.80	30.20	30.20	29.60	29.60	31.20	31.20	30.80	30.80
22	8.00	7.00	35.00	1.50	0.50	31.00	31.00	29.60	29.60	29.20	29.20	30.60	30.60	30.60	30.60

23	8.00	7.00	25.00	1.50	1.50	31.00	31.00	30.00	30.00	29.40	29.40	30.80	30.80	30.40	30.40
24	10.00	9.00	25.00	1.50	1.50	30.20	30.20	29.40	29.40	29.00	29.00	30.20	30.20	30.20	30.20
25	8.00	9.00	25.00	1.50	1.50	32.40	32.40	31.40	31.40	30.20	30.20	30.40	30.40	31.20	31.20
26	10.00	9.00	35.00	0.50	1.50	31.60	31.60	30.80	30.80	29.80	29.80	29.80	29.80	31.00	31.00
27	8.00	7.00	25.00	1.50	0.50	31.60	31.60	31.20	31.20	30.00	30.00	30.00	30.00	30.80	30.80
28	8.00	7.00	25.00	0.50	0.50	30.80	30.80	30.60	30.60	29.60	29.60	29.40	29.40	30.60	30.60
29	10.00	7.00	35.00	0.50	0.50	32.20	32.20	30.60	30.60	29.80	29.80	31.00	31.00	31.40	31.40
30	8.00	9.00	35.00	0.50	1.50	31.40	31.40	30.00	30.00	29.40	29.40	30.40	30.40	31.20	31.20
31	10.00	9.00	25.00	0.50	0.50	31.40	31.40	30.40	30.40	29.60	29.60	30.60	30.60	31.00	31.00
32	8.00	9.00	25.00	0.50	1.50	30.60	30.60	29.80	29.80	29.20	29.20	30.00	30.00	30.80	30.80
33	10.00	7.00	25.00	1.50	0.50	34.00	34.00	32.60	32.60	31.40	31.40	33.00	33.00	33.40	33.40
34	10.00	9.00	25.00	0.50	1.50	33.20	33.20	32.00	32.00	31.00	31.00	32.40	32.40	33.20	33.20
35	10.00	8.00	30.00	1.00	1.00	33.80	33.80	32.60	32.60	31.40	31.40	32.80	32.80	33.20	33.20
36	10.00	7.00	25.00	1.50	1.50	33.00	33.00	32.40	32.40	31.20	31.20	32.40	32.40	32.80	32.80
37	9.00	8.00	30.00	1.50	1.00	34.00	34.00	33.20	33.20	31.60	31.60	32.20	32.20	33.20	33.20
38	8.00	7.00	25.00	0.50	1.50	33.80	33.80	32.40	32.40	31.20	31.20	32.80	32.80	33.40	33.40
39	9.00	8.00	30.00	1.00	1.50	33.60	33.60	32.40	32.40	31.00	31.00	33.20	33.20	33.20	33.20
40	8.00	9.00	25.00	1.50	0.50	34.00	34.00	32.80	32.80	31.20	31.20	33.00	33.00	33.80	33.80
41	9.00	8.00	30.00	1.00	1.00	33.80	33.80	32.60	32.60	31.20	31.20	32.80	32.80	33.20	33.20
42	8.00	7.00	35.00	0.50	0.50	33.40	33.40	32.20	32.20	31.00	31.00	32.20	32.20	32.80	32.80
43	9.00	8.00	30.00	1.00	1.00	34.20	34.20	33.00	33.00	31.60	31.60	33.20	33.20	33.80	33.80
44	8.00	7.00	35.00	0.50	1.50	34.20	34.20	33.00	33.00	31.60	31.60	33.20	33.20	33.80	33.80
45	9.00	8.00	30.00	1.00	1.00	34.20	34.20	33.00	33.00	31.60	31.60	33.20	33.20	33.80	33.80
46	10.00	9.00	25.00	1.50	0.50	34.20	34.20	33.00	33.00	31.60	31.60	33.20	33.20	33.80	33.80
47	9.00	8.00	30.00	1.00	1.00	34.20	34.20	33.00	33.00	31.60	31.60	33.20	33.20	33.80	33.80
48	9.00	8.00	30.00	1.00	1.00	34.20	34.20	33.00	33.00	31.60	31.60	33.20	33.20	33.80	33.80
49	8.00	9.00	25.00	0.50	0.50	34.20	34.20	33.00	33.00	31.60	31.60	33.20	33.20	33.80	33.80
50	9.00	8.00	30.00	1.00	1.00	34.20	34.20	33.00	33.00	31.60	31.60	33.20	33.20	33.80	33.80

References

Ajayi GO, Olagunju JA, Ademuyiwa O (2011) Gas chromatography-mass spectrometry analysis and phytochemical screening of ethanolic root extract of *Plumbago zeylanica,* Linn. Med Plant Res 5:1756–1761

Alferova IV, Terekhova LP, Prauzer KH (1989) Selective medium with nalidixic acid for isolating antibiotic producing actinomycetes. Antibiot Khimioter 34:344–348

Almeida DG, Soares da Silva R d CF, Luna JM, Rufino RD, Santos VA, Sarubbo LA (2017) Response surface methodology for optimizing the production of biosurfactant by *Candida tropicalis* on industrial waste substrates. Front Microbiol 8:157

Appendino G, Banfi L (2011) Molecular diversity and natural products. Mol Divers 15:291–292

Atta HM, Bahobail AS, El-Sehrawi MH (2011) Studies on isolation, classification and phylogenetic characterization of antifungal substances produced by *Streptomycesalbidoflavus*-143. New York Sci J 4:40–53

Atta HM, Afifi MM, Elshanawany AA (2012) Production of Hygromycin-B antibiotic from *Streptomyces crystallinus.* AZ-A151: II. Parameters controlling of antibiotic production. Acad Arena 4:37–52

Basilio A, Gonzalez I, Vicente MF (2003) Patterns of antimicrobial activities from soil actinomycetes isolated under different conditions of pH and salinity. J Appl Microbiol 95:814–823

Boussada O, Ammar A, Saidana D (2008) Chemical composition and antimicrobial activity of volatile components from capitula and aerial parts of *Rhaponticum acaule* DC growing wild in Tunisia. Microbiol Res 163:87–95

Box GEP, Hunter JS (1957) Multi-factor experimental designs for exploring response surfaces. Ann Math Stat 28:1957) 195–1957) 241

Bozzola JJ, Russell LD (1999) Electron microscopy principles and techniques for biologists. Jones and Bartlett Publishers, Sudbury

Bredholdt H, Galatenko OA, Engelhardt K (2007) Rare actinomycete bacteria from the shallow water sediments of the Trondheim fjord, Norway: isolation, diversity, biological activity. Environ Microbiol 9:2756–2764

Chattopadhyay D, Sen SK (1997) Optimization of cultural conditions for antifungal antibiotic accumulated by *Streptomyces rochei* G 164. Hindustan Antibiot Bull 39:64–71

Cowan ST (1974) Cowan and Steel's manual for the identification of medical bacteria. Cambridge University Press, Cambridge

Donadio S, Sosio M, Lancini G (2002) Impact of the first *Streptomyces* genome sequence on the discovery and production of bioactive substances. Appl Microbiol Biotechnol 60:377–380

Ellaiah P, Srinivasulu B, Adinarayana K (2004) Optimization studies on neomycin production by a mutant strain of *Streptomyces marinensis* in solid state fermentation. Process Biochem 39:529–534

Fatma GSB, Isil IG, Osman O (2013) Analysis of the polymer composite bipolar plate properties on the performance of PEMFC (polymer electrolyte membrane fuel cells) by RSM (response surface methodology). Energy 55:1067–1075

Gordon RE (1966) Some criteria for the recognition of *Nocardia madura* (Vincent) Blanchord. J Gen Microbiol 45:355–364

Güray G, Altunay P, Abdurrahman T (2008) Electrochemical treatment of deproteinated whey waste water and optimization of treatment conditions with response surface methodology. J Hazard Mater 157:69–78

Han WC, Lee JY, Park DH (2004) Isolation and antifungal and antioomycete activity of *Streptomyces scabiei* strain PK-A41, the causal agent of common scab disease. Plant Pathol J 20:115–126

Hema R, Kumaravel S, Alagusundaram K (2011) GC/MS determination of bioactive components of *Murraya koenigii*. J Am Sci 7:80–83

Hong L, Yi-min Z, Jing H (2017) Optimization of vanadium (IV) extraction from stone coal leaching solution by emulsion liquid membrane using response surface methodology. Chem Eng Res Des 123:111–119

Hsouna AB, Trigui M, Mansour RB (2011) Chemical composition, cytotoxicity effect and antimicrobial activity of *Ceratonia siliqua* essential oil with preservative effects against *Listeria* inoculated in minced beef meat. Int J Food Microbiol 148:66–72

Isik K, Chun J, Hah YC, Goodfellow M (1999) *Nocardia salmonicida* nom. rev., a fish pathogen. Int J Syst Bacteriol 49:833–837

Jensen PR, Williams PG, Oh DC (2007) Species specific secondary metabolite production in marine actinomycetes of the genus *Salinispora*. Appl Environ Microbiol 73:1146–1152

Jones K (1949) Fresh isolates of actinomycetes in which the presence of sporogenous aerial mycelia is a fluctuating characteristics. J Bacteriol 57:141–145

Ka HW, George Q, Kong M (2017) Optimisation of Pueraria isoflavonoids by response surface methodology using ultrasonic-assisted extraction. Food Chem 231:231–237

Khare CP (2004) Indian herbal remedies: rational western therapy, ayurvedic and other traditional usage, botany. Springer, Berlin

Koehn FE, Carter GT (2005) The evolving role of natural products in drug discovery. Nat Rev Drug Discov 4:206–220

Leal MC, Puga J, Serodio J (2012) Trends in the discovery of new marine natural products from invertebrates over the last two decades—where and what are we bioprospecting? PLoS One 7:1–15

Mahmood A, Ahmed R, Kosar S (2009) Phytochemical screening and biological activities of the oil components of *Prunus domestica* Linn. J Saudi Chem Soc 13:273–277

Majumdar MK, Majumdar SK (1965) Effect of minerals on neomycin production by *Streptomyces fradiae*. Appl Microbiol 13:190–193

Mammen D, Daniel M, Sane RT (2010) Seasonal and geographical variations in chemical constituents of *Leptadenia reticulata*. Int J Pharm Sci Rev Res 4:111–116

Marcos AB, Ricardo ES, Eliane PO (2008) Response surface methodology (RSM) as a tool for optimization in analytical chemistry. Talanta 76:965–977

Mihailovi V, Vukovic N, Niciforovic N (2011) Studies on the antimicrobial activity and chemical composition of the essential oils and alcoholic extracts of *Gentiana asclepiadea* L. J Med Plant Res 5:1164–1174

Mishra PM, Sree A (2007) Antibacterial activity and GC-MS analysis of the extract of leaves of *Finlaysonia obovata* (a Mangrove Plant). Asian J Plant Sci 6:168–172

Mustafa O (2011) Effects of some environmental conditions on biomass and antimicrobial metabolite production by *Streptomyces* spp. KGG32. Int J Agric Biol 3:317–324

Myers RH, Montgomery DC, Anderson-Cook CM (2009) Response surface methodology: process and product optimization using designed experiments. Wiley, Hoboken, NJ

Narayana KJP, Vijayalakshmi M (2008) Optimization of antimicrobial metabolites production by *Streptomyces albidoflavus*. Res J Pharmacol 2:4–7

Newman DJ, Cragg GM (2007) Natural products as sources of new drugs over the last 25 years. J Nat Prod 70:461–477

Nilsson WB, Strom MS (2002) Detection and identification of bacterial pathogens of fish in kidney tissue using terminal restriction fragment length polymorphism (T-RFLP) analysis of 16S rRNA genes. Dis Aquat Org 48:175–185

Parthasarathi S, Kim CJ, Kim PK, Sathya S, Manikandan M, Manikandan T, Balakrishnan K (2010) Taxonomic characterization and UV/VIS analysis of antagonistic marine actinomycete isolated from South West Coast of South Korea. Int J Med Res 1:99–105

Payam R, Shohreh M, Saeid V (2017) Modeling and optimization of gelatin-chitosan microcarriers preparation for soft tissue engineering: using response surface methodology. Mater Sci Eng C 75:545–553

Pridham TG, Gottlieb D (1948) The utilization of carbon compounds by some Actinomycetales as an aid for species determination. J Bacteriol 56:107–114

Pridham TG, Hesseltine CW, Benedict RG (1958) A guide for the classification of *Streptomyces* according to selected groups. Appl Microbiol 6:52–79

Rama Krishna VSG, Ushakiranmayi M, Vijayalakshmi M, Sudhakar P (2016) Model-based kinetic parameters estimation in batch Pullulan fermentation using Jaggery as substrate. J Chem Pharm Res 8:217–224

Ramesh S, Mathivanan N (2009) Screening of marine actinomycetes isolated from the Bay of Bengal, India for antimicrobial activity and industrial enzymes. World J Microbiol Biotechnol 25:2103–2111

Rangel-Sánchez G, Castro-Mercado E, García-Pineda E (2014) Avocado roots treated with salicylic acid produce phenol-2,4-bis (1,1-dimethylethyl), a compound with antifungal activity. J Plant Physiol 17:189–198

Roy RRN, Laskar S, Sen SK (2006) Dibutyl phthalate, the bioactive compound produced by *Streptomyces albidoflavus* 321.2. Microbiol Res 161:121–126

Salini TS, Divakaran D, Shabanamol S (2014) Antimicrobial and Immuno modulatory potential of endophytic fungi *Fusarium solani* isolated from *Withania somnifera*. World J Pharm Res 3:879–890

Sathish S, Vivekanandan S (2016) Parametric optimization for floating drum anaerobic biodigester using response surface methodology and artificial neural network. Alexandria Eng J 55:3297–3307

Saurav K, Kannabiran K (2010) Diversity and optimization of process parameters for the growth of *Streptomyces* VITSVK 9 spp. isolated from Bay of Bengal, India. J Nat Environ Sci 1:56–65

Shirling EB, Gottlieb D (1966) Methods for characterization of *Streptomyces* species. Int J Syst Bacteriol 16:313–340

Singh LS, Majumdar S, Bora TC (2009) Optimization of process parameters for growth and bioactive metabolite production by a salt tolerant and alkaliphilic actinomycete, *Streptomyces tanashiensis* strain A2D. J Mycol Med 19:223–225

Solanki R, Khanna M, Cal R (2008) Bioactive compounds from marine actinomycetes. Curr Sci 48:410–431

Sun W, Dai S, Jiang S (2010) Culture dependent and culture independent diversity of actinobacteria associated with the marine sponge *Hymeniacidon perleve* from the South China Sea. Antonie Van Leeuwenhoek 98:65–75

Thakur D, Bora TC, Bordoloi GN (2009) Influence of nutrition and culturing conditions for optimum growth and antimicrobial metabolite production by *Streptomyces* spp. 201. J Mycol Med 19:161–167

Tresner HD, Hayes JA, Backus EJ (1968) Differential tolerance of *Streptomyces* to sodium chloride as a taxonomic aid. Appl Microbiol 16:1134–1136

Ushakiranmayi M, Vijayalakshmi M, Sudhakar P, Rama Krishna Ganduri VS, Satish Babu R (2017) Optimization of culture conditions by response surface methodology and unstructured kinetic modeling for L-Asparaginase production by *Pseudonocardia endophytica* VUK-10. J Appl Pharm Sci 7:42–50

Waksman SA (1961) The actinomycetes, classification, identification and descriptions of genera and species. The Williams and Wilkins Co., Baltimore, MD

Wang G, Deng Y, Xu X (2016) Optimization of air jet impingement drying of okara using response surface methodology. Food Control 59:743–749

Wang Y, Qin X, Huang S, Feng J (2017) Structural-borne acoustics analysis and multi-objective optimization by using panel acoustic participation and response surface methodology. Appl Acoust 116:139–151

Williams ST (1989) In: Williams ST (ed) Bergey's manual of systematic bacteriology, vol 4. Williams & Wilkins, Baltimore, MD

Xiong ZQ, Wang JF, Hao YY, Wang Y (2013) Recent advances in the discovery and development of marine microbial natural products. Mar Drugs 11:700–717

Ying Z, Chengyou C, Mei P (2014) Diversity of nitrogen-fixing, ammonia-oxidizing, and denitrifying bacteria in biological soil crusts of a revegetation area in Horqin Sandy Land, Northeast China. Ecol Eng 71:71–79

Yu H, Zhao X, Yue P (2009) Chemical communication in mammal population: urinary olfactory chemosignals in lactating female root voles (Microtus oeconomus pallas). Polish J Ecol 58:153–165

Zafari A, Kianmehr MH, Abdolahzadeh R (2013) Modeling the effect of extrusion parameters on density of biomass pellet using artificial neural network. Int J Recycl Org Waste Agric 2(1):9

Zhang B, Han X, Pengju G, ShuqiFang JB, Zhang B, Han X, Gu P, Fang S, Bai J (2017) Response surface methodology approach for optimization of ciprofloxacin adsorption using activated carbon derived from the residue of desilicated rice husk. J Mol Liq 238:316–325

In Vitro Cultured Cells as an Option for Enhancing the Production of Bioactive Compounds: Some Selected Case Studies

39

D. Madhavi, G. Jawahar, P. Komaraiah, G. Rajasheker, K. Sujatha, P. Sita Kumari, and P. B. Kavi Kishor

Abstract

Many of the drugs that are sold in the market today are mostly the synthetic modifications of the naturally available plant products. Many have been resorting to natural sources of remedies since they have less side effects. Since a large number of secondary plant products have commercial importance, there is a renewed interest in the possibility of altering the production of bioactive compounds using in vitro techniques. Plants produce low amounts of secondary plant products or bioactive compounds, and hence in vitro grown callus, suspensions, and morphogenic cultures are being exploited for enhanced production. Detection and increased production of secondary metabolites have been reported in diverse medicinal plants. The present review is aimed to cover recent advances for the production of some important plant pharmaceuticals from selected plant species.

Keywords

Callus · Elicitation · Immobilization · Medicinal plants · Permeabilization · Secondary plant products

---

D. Madhavi · G. Jawahar · G. Rajasheker · P. B. Kavi Kishor (✉)
Department of Genetics, Osmania University, Hyderabad, India

P. Komaraiah
Department of Oncologic Sciences, USA Mitchell Cancer Institute, University of South Alabama, Mobile, AL, USA

K. Sujatha · P. Sita Kumari
Department of Chemistry, K.T.R. Women's College, Gudivada, India

S. M. Khasim et al. (eds.), *Medicinal Plants: Biodiversity, Sustainable Utilization and Conservation*, https://doi.org/10.1007/978-981-15-1636-8_39

39.1 Introduction

Plants are a rich source of not only food and fodder but also many secondary plant products. Growing interest in natural product research was prompted by their great utilization as antibiotics, drugs, dyes, essential oils, fibers, flavoring agents, perfumes, and polymers. Natural products can be divided into terpenoids, alkaloids, and the phenolic compounds. While terpenoids (nearly 25,000) are derived from the five-carbon precursor isopentenyl diphosphate (IPP), alkaloids (roughly 12,000) contain one or more nitrogen atoms and are derived from amino acids. On the other hand, phenolic compounds (8000 or more) are synthesized either from shikimic acid pathway or the mevalonate/acetate pathway (Rodney et al. 2000). Medicinal plants are the exclusive source of life-saving drugs for majority of the rural populations in India as well as in many countries. Secondary plant products play a vital role in the adaptation of plants to their environment and in combat mechanism (Rao and Ravishankar 2002). But there is a threat for the natural habitat of the plants due to ruthless exploitation by humans and degradation of the forest land at an alarming rate. Cultivation of medicinal plants is increasingly becoming difficult due to competition for land use, decreasing water resources, and increasing labor costs. Hence, utilization of plant cell or organ cultures for the production of natural or bioactive compounds has gained increasing attention (Canter et al. 2005). Plant tissue culture technology provides continuous, controlled production and reliable source of myriad plant pharmaceuticals on demand. The current yields and productivity of the cultured cells for bioactive compounds do not meet the commercial goals, and hence recent advances/strategies in triggering the secondary plant products need to be critically examined. While the biosynthetic pathways are not completely known in majority of the cases, stability of the cultured cells for production of the desired compounds is another bottleneck. Desired metabolite accumulation is limited by the lack of availability of specific precursors and lack of our understanding about the stable production of compounds over a period of time. For triggering the biosynthesis and accumulation of secondary products using biotic and abiotic elicitors, immobilization and permeabilization are being widely practiced now. However, immobilization of cells in suitable bioreactors is still a major challenge. Permeabilization of cells with natural or artificial agents and accumulation of bioactive compound into the surrounding medium are being carried out currently, but in situ adsorption of the compounds using amberlite or XAD-7 is crucial.

39.2 Advances Made in the Production of Secondary Metabolites Through In Vitro Cultures

Rao and Ravishankar (2002) and Vijaya et al. (2010) have reviewed the production of secondary plant products and found that they are being produced at a rate similar or superior to that of intact plants by including diverse elicitors into the medium and by immobilization of cells. Further, optimization of the cultural conditions and utilization of morphogenic (roots, hairy roots, and shooty teratomas or shoot cultures)

cultures are aiding the technology. Selection of strain and its improvement is cumbersome and laborious, but has been reported and found vital for enhanced production of bioactive compounds (Komaraiah et al. 2003a). Though unorganized callus or suspensions are commonly used, in certain cases, production requires differentiated structures (Davioud et al. 1989). This is mostly because only specialized plant tissues or glands are able to produce metabolites of interest. Interesting examples of such production systems are *Panax ginseng* and *Hypericum perforatum*. While ginseng is accumulated in roots, hypericins and hyperforins are accumulated in foliar glands, but not in undifferentiated callus or suspensions (Davioud et al. 1989; Smetanska 2008). Transgenic hairy roots and shooty teratomas revolutionized the in vitro production of secondary metabolites because they are easily maintained in culture and are biosynthetically stable (Giri and Narasu 2000). Not many compounds were commercialized since the inception of plant tissue culture technology. The major examples of commercialization are shikonin, a reddish-purple pigment with antiseptic properties isolated from cell cultures of *Lithospermum erythrorhizon* (Boraginaceae) (Fujita 1988), and berberine manufactured by the Mitsui Petrochemical Industries, Japan. Berberine is produced from cells of *Coptis japonica* (Sato and Yamada 1984) and is used in the treatment of stomach disorders and for removing kidney stones. But, commercialization of secondary plant products from cultured cells is picking up worldwide. The development of a strategy to exploit the ability of cultured plant cells to produce useful compounds must now be an essential prerequisite to further progress in extending the spectrum of these compounds (Buitelaar and Tramper 1992; Lipsky 1992). To achieve higher amounts of secondary metabolites in culture, properties of the plant cells, cell line selection, and manipulation of the physical and chemical environment (choice of culture system, nutrient limitation, plant growth regulators, elicitation, permeabilization, precursor feeding, light, pH of the medium, etc.) must be studied and the associated mechanisms be understood.

39.3 Elicitation of Cultured Cells

We report here elicitation of cell or hairy root cultures initiated from *Plumbago rosea*, *Tinospora cordifolia*, *Morinda citrifolia*, *Withania somnifera*, and *Gloriosa superba*.

39.3.1 *Plumbago rosea*

Plumbago rosea (synonymous *P. indica*) produces plumbagin (Fig. 39.1a), a naphthoquinone which is commercially important for its broad range of pharmacological activities, such as insecticidal (Kubo et al. 1983), antifertility (Bhargava 1984), anticancer (Parimala and Sachdanandam 1993), and antimicrobial (Didry et al. 1994). Roots of *Plumbago* species are the main source of plumbagin production, and its synthesis from some species of Droseraceae, Ebenaceae, and Euphorbiaceae was

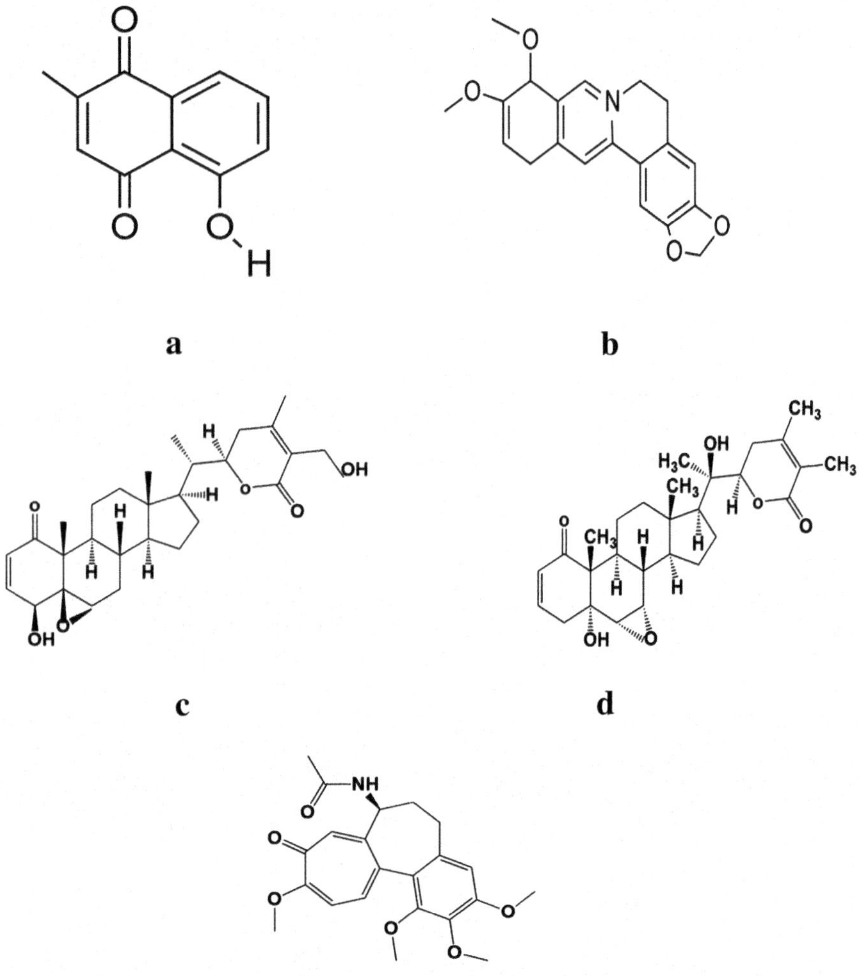

Fig. 39.1 (a) Chemical structures of plumbagin, (b) berberine, (c) withaferin A, (d) withanolide, and (e) colchicine

reported earlier (Thomson 1971). Chemical structures of different bioactive compounds are shown in Fig. 39.1. Production of plumbagin from plant cell cultures is of great importance because intact plants produce this compound in small quantities after 2–6 years of growth period. Callus and suspension cultures derived from leaf explants of *Plumbago rosea* were established, and plumbagin was isolated from them and confirmed by ^{1}H NMR and electron-ionization mass spectroscopy. Maximum content of plumbagin was obtained in the stationary phase of growth (4.3 mg/g dry cell wt.). Media pH, phytohormones, and carbon sources were optimized for biomass and plumbagin accumulation. Cell aggregates, measuring 500 μm in diameter, produced 8.2 g dry cell weight/l, but larger aggregates (above

500 μm) favored plumbagin accumulation with a yield of 4.5 mg/g dry cell weight. Cell cultures were treated with the elicitors prepared from the fungi (*Aspergillus niger* and *Rhizopus oryzae*), bacteria (*Bacillus subtilis* and *Pseudomonas aeruginosa*), yeast extract, and chitosan to induce and enhance the synthesis of plumbagin (Komaraiah et al. 2001, 2002, 2003b). Elicitation of plumbagin production in chitosan-treated cells was 6.71-fold higher compared to control suspension cultures. The treatment of cells with *A. niger*, *R. oryzae*, and yeast elicitors resulted in two- to threefold more plumbagin over control cells. Bacterial elicitors did not show much (<2-fold) influence on plumbagin accumulation. Chitosan at 150 mg/l dose level enhanced the permeability of plumbagin from the cell to the exterior and also elicited plumbagin synthesis significantly. Likewise, Jin et al. (1994) increased the anthraquinone production in *Rubia akane* cell cultures using chitosan, a biotic elicitor. Both biotic and abiotic elicitors enhanced the growth and saponin biosynthesis in the hairy roots of *P. ginseng* (Jeong and Park 2006). Similarly, Staniszewska et al. (2003) improved the production of secondary metabolites in *Ammi majus* by using elicitors such as benzo(1,2,3)-thiadiazole-7-carbothionic acid *S*-methyl ester and also suspensions of bacteria (*Enterobacter sakazakii*).

39.3.2 *Tinospora cordifolia*

Tinospora cordifolia Miers belongs to the family Menispermaceae and produces many alkaloids. The plant is used for the treatment of jaundice, rheumatism, and urinary diseases (Hanuman et al. 1988). Berberine is widely used as a tonic and for stomach disorders in the Far East (Misawa 1977). Berberine is an antibacterial isoquinoline alkaloid (Fig. 39.1b) which was earlier isolated from *Coptis japonica* (Sato and Yamada 1984), *Thalictrum minus* (Nakagawa et al. 1988), and *Berberis wilsoniae* (Breuling et al. 1985). Chopra *et al.* (1982) have ascribed the medicinal properties of *T. cordifolia* and for the presence of berberine. But, *in vitro* production of berberine from *T. cordifolia* was first reported by Padhya (1986). The plant produces berberine, an isoquinoline alkaloid, together with its related analogs protoberberine and palmatine. These compounds were detected in cell suspension cultures derived from leaf explants of *T. cordifolia*. Maximum production of biomass (9.9 g dry cell weight/l) was noticed on day 36, while berberine was the highest (5.5 mg/g dry weight) at day 24 (Rao et al. 2008). Media pH, plant growth regulators, and carbon sources were found to play a vital role not only for biomass but also for the accumulation of berberine (Rao et al. 2008) in Linsmaier and Skoog's medium (Linsmaier and Skoog 1965). Out of the eight cell lines selected, only one line accumulated high berberine (13.9 mg/g dry weight) in 4 weeks (Table 39.1). While cell

Table 39.1 *Variation in berberine content indifferent selected cell lines of T. cordifolia*

Cell line	Berberine content (mg/g dry wt.)
Tc-1Tc-2Tc-3 Tc-4 Tc-5Tc-6 Tc-7Tc-8	0.8 (±0.2)1.6 (±0.3)2.9 (±0.4)4.3 (±0.4)13.9 (±1.1)5.4 (±0.6)3.2 (±0.5)2.7 (±0.3)

Figures in parenthesis represent standard error ($n = 12$)

aggregates of 500 μm in diameter promoted production of biomass (9.6 g dry cell weight/l), larger cell aggregates (above 500 μm) favored berberine accumulation with 3.8 mg/g dry weight of tissue (Rao et al. 2008). The yield of berberine in cell suspensions of *Tinospora* was 5–14-fold higher than that of intact plants.

39.3.3 *Morinda citrifolia*

M. citrifolia or noni (Rubiaceae) is a medicinal plant called Indian mulberry in India, ba ji tian in China, nono in Tahiti, and noni in Hawaii (Abbott and Shimazu 1985; Morton 1992) and extensively used in folk medicine by Polynesians for over 2000 years and reported to have anticancer activity (Wang and Su 2001). *M. citrifolia* has anti-inflammatory, antihistamine, anti-fungal, antibiotic, antiviral, anticancer, and hypotensive properties (Habrah and Eveleth 1996; Hirazumi 1997). More than 160 chemicals were identified in noni, but the major components are terpene compounds, anthraquinones, morindone, morindin, asperuloside, aucubin, caproic acid, caprylic acid, damnacanthal, and scopoletin (Heinicke 1985). The production and metabolism of anthraquinones in *M. citrifolia* and related species were studied (Stalman et al. 2003; Han et al. 2001). Anthraquinones are normally stored mainly in the roots in the form of aglycones and also as glycosides. Accumulation of anthraquinones in suspension cultures was reported earlier (Kieran et al. 1995; Bassetti et al. 1996). Pagnussat et al. (2002) showed that a variety of elicitors alter the cell metabolism and metabolite productivity. Weber (2002), on the other hand, reported that methyl jasmonate and polyunsaturated fatty acids (PUFAs) are effective in triggering defense-related signals and secondary metabolite production capacity. We reported enhanced accumulation of anthraquinones in *M. citrifolia* by treating the suspension cultures with elicitors, by ultrasonication, and by controlled feeding of the carbon source in the growth medium (Komaraiah et al. 2005). Elicitations were carried out using linoleic acid, α-linolenic acid, arachidonic acid (PUFAs), methyl jasmonate, salicylate, and nitric oxide at different concentrations. Accumulation of anthraquinones ranged from 5 to 12 mg/g dry weight of cells, which was two- to threefold of what was attained in control suspension cultures. Short pulses of ultrasonication (16 s) enhanced the anthraquinone accumulation by 2.5-fold. But a synergistic effect was noticed when elicitation was applied alongside the addition of sucrose. The content was enhanced to 16.74 mg/g dry weight of tissue (fourfold increase). Further, minor fluorescence was noticed in suspensions undergoing lag phase (1–3 days), while higher fluorescence was recorded from cells in the stationary phase of growth (10–14 days after culture). Likewise, spherical cells emitted less fluorescence in comparison with elongated cells as visualized under the confocal microscope. The fluorescence was assumed to be a result of autofluorescent properties of suspension cells containing anthraquinone molecules (Komaraiah et al. 2005). This points out that cell line selection is important for producing cultures with high amounts of secondary plant product accumulation.

39.3.4 *Withania somnifera*

W. somnifera belongs to the family Solanaceae and is an important perennial medicinal plant (Singh and Kumar 1998). This plant produces several bioactive compounds mainly withanolides that have commercial importance. Madhavi et al. (2012) reported carbohydrate and elicitor enhanced withanolide (withaferin A and withanolide A) accumulation in hairy root cultures of *W. somnifera*. Differential accumulation of either withaferin A (Fig. 39.1c) or withanolide A (Fig. 39.1d) was also reported with the manipulation of quantity and quality of carbohydrates in the Murashige and Skoog's (Murashige and Skoog 1962) medium (Madhavi et al. 2012). This suggests that elicitors and carbohydrates play a profound role in enhancing the with anolide accumulation.

39.3.5 *Gloriosa superba*

G. superba is a climbing herb belonging to the family Liliaceae. It is native to the African and Asian continents, but is an ornamental plant worldwide. It is an important medicinal herb, and the underground tubers are a rich source of colchicine (Fig. 39.1e) (phenylethyl isoquinoline derived), colchicoside, and gloriosine. The plant is used against gout, kidney problems, cancer, and leprosy (Mahendran et al. 2017). The alkaloid is accumulated not only in tubers but also in leaves and seeds. Kavina et al. (2011) reported 0.019 mg/g dry weight of colchicine in seeds and 0.0056 mg/g dry tissue in tubers. Hairy root cultures were initiated and used by Bai and Agastian (2013) for in vitro accumulation of colchicine. Ghosh et al. (2002) studied the effect of precursors on the synthesis and accumulation of colchicine in vitro. They noticed 0.19% increase in colchicine content by feeding precursors to the cultures. Jawahar et al. (2018) recently reported ~12 mg of colchicine per g dry weight of tissue by incorporating diverse elicitors like rubidium chloride and sodium chloride. Addition of phenylalanine, tryptophan, and tyrosine improved the accumulation of colchicine. But a combination of elicitors displayed a synergistic effect on the accumulation (Jawahar et al. 2018). These studies indicate that optimum concentrations and combinations of elicitors help to enhance the accumulation of bioactive compounds significantly in suspensions or in morphogenic cultures.

39.4 Conclusions

The above studies clearly indicate that elicitation, immobilization, and permeabilization of cells and optimum concentration of carbohydrate and precursor feeding certainly improve the accumulation of secondary plant products in culture. However, stable accumulation of bioactive compounds over a long period of time in culture is necessary for subsequent commercial exploitation of such important bioactive compounds.

Acknowledgments The authors like to thank the UGC, New Delhi, CSIR New Delhi, for financial assistance. M.D. is thankful to the CSIR for providing fellowship. J.G. is grateful to the UGC, New Delhi, for providing SRF fellowship. G.R. is thankful to the DST, New Delhi, for providing fellowship. PBK is grateful to the CSIR, New Delhi, for providing CSIR-Emeritus fellowship.

References

Abbott IA, Shimazu C (1985) The geographic origin of the plants most commonly used for medicine by Hawaiians. J Ethanopharmacol 14:213–222

Bai ALG, Agastian P (2013) *Agrobacterium rhizogenes* mediated hairy root induction for increased colchicine content in *Gloriosa superba* L. J Acad Ind Res 2:68–73

Bassetti L, Pijnenburg J, Tramper J (1996) Silicone-stimulated anthraquinone production and release by *Morinda citrifolia* in a two-liquid phase system. Biotechnol Lett 18:377–382

Bhargava SK (1984) Effects of plumbagin on reproductive function of male dog. Indian J Exp Biol 22:153–156

Breuling M, Alfermann AW, Reinhard E (1985) Cultivation of cell cultures of *Berberis wilsoniae* in airlift bioreactors. Plant Cell Rep 4:220–223

Buitelaar RM, Tramper J (1992) Strategies to improve the production of secondary metabolites with plant cell cultures: a literature review. J Biotechnol 23:111–141

Canter PH, Thomas H, Ernst E (2005) Bringing medicinal plants into cultivation: opportunities and challenges for biotechnology. Trends Biotechnol 23:180–185

Chopra R, Chopra C, Handa KL, Kapur LD (1982) Indigenous drugs of India. B.K. Dhur of Academic Publishers, Calcutta, pp 426–428

Davioud E, Kan C, Hamon J, Tempe J, Husson HP (1989) Production of indole alkaloids by *in vitro* root cultures from *Catharanthus trichophyllus*. Phytochemistry 28:2675–2680

Didry N, Dubrevil L, Pinkas M (1994) Activity of anthroquinonic and naphthoquinonic compounds on oral bacteria. Pharmazie 49:681–683

Fujita Y (1988) Shikonin: production by plant (*Lithospermum erythrorhizon*) cell cultures. In: Medicinal and aromatic plants. Springer, Berlin, pp 225–236

Ghosh B, Mukherjee M, Jha TB, Jha S (2002) Enhanced colchicine production in root cultures of *Gloriosa superba* by direct and indirect precursors of the biosynthetic pathway. Biotechnol Lett 24:231–234

Giri A, Narasu ML (2000) Transgenic hairy roots: recent trends and applications. Biotechnol Adv 18:1–22

Habrah FL, Eveleth MB (1996) Evaluation of the effectiveness of ancient Hawaiian medicine. Hawaii Med J 25(3):223

Han YS, van der Heijden R, Verpoote R (2001) Biosynthesis of anthraquinones in cell cultures of the Rubiaceae. Plant Cell Tissue Org Cult 67:201–220

Hanuman JB, Bhatt RK, Sabata BK (1988) A clerodane furano-diterpene from *Tinospora cordifolia*. J Nat Prod 51:197–201

Heinicke RM (1985) The pharmacologically active ingredient of Noni. Bulletin of the National Tropical Botanical Garden

Hirazumi AY (1997) Antitumor studies of a traditional Hawaiian medical plant, *Morinda citrifolia* (Noni), *in vitro* and *in vivo*. PhD thesis, University of Hawaii

Jawahar G, Punita DL, Rajasheker G, Manoharachary C, Venkatachalam P, Kavi Kishor PB (2018) Feeding elicitors and precursors enhance colchicine accumulation in morphogenic cultures of *Gloriosa superba* L. Plant Cell Tissue Organ Cult 135:235. https://doi.org/10.1007/s11240-018-1459-9

Jeong GA, Park DH (2006) Enhanced secondary metabolite biosynthesis by elicitation in transformed plant root system: effect of abiotic elicitors. Appl Biochem Biotechnol 129:436–446

Jin JH, Shin JH, Kim JH, Chung IS, Lee HJ (1994) Effect of chitosan elicitation and media components on the production of anthraquinone colorants in madder (*Rubia akane* Nakai) cell culture. Biotechnol Bioprocess Eng 4:300–304

Kavina J, Gopi JR, Pannerselvam R (2011) Quantification of colchicine in seed and tuber samples of *Gloriosa superba* by high performance liquid chromatography method. J Appl Pharm Sci 1:116–119

Kieran PM, O'Donnell HJ, Malone DM, MacLoughlin PF (1995) Fluid shear effects on suspension cultures of *Morinda citrifolia*. Biotechnol Bioeng 45:415–425

Komaraiah P, Kavi Kishor PB, Ramakrishna SV (2001) Production of plumbagin from cell cultures of *Plumbago rosea* L. Biotechnol Lett 23:1269–1272

Komaraiah P, Naga Amrutha R, Kavi Kishor PB, Ramakrishna SV (2002) Elicitor enhanced production of plumbagin in suspension cultures of *Plumbago rosea* L. Enzyme Microb Technol 31:634–639

Komaraiah P, Jogeswar G, Naga Amrutha R, Sri Laxmi P, Lavanya B, Rama Krishna SV, Kavi Kishor PB (2003a) Influence of hormones and selection of stable cell lines of *Plumbago rosea* for accumulation of plumbagin. Plant Biotechnol 5:171–175

Komaraiah P, Ramakrishna SV, Reddanna P, Kavi Kishor PB (2003b) Production of plumbagin in immobilized cells of *Plumbago rosea* by elicitation and *in situ* adsorption. J Biotechnol 101:181–187

Komaraiah P, Kavi Kishor PB, Carlson M, Magnusson KE, Mandenius CF (2005) Enhancement of anthraquinone accumulation in *Morinda citrifolia* suspension cultures. Plant Sci 168:1337–1344

Kubo I, Uchida M, Klocke JA (1983) An insect ecdysis inhibitor from the African medicinal plant, *Plumbago capensis* (Plumbaginaceae). Agric Biol Chem 47:911–913

Linsmaier EM, Skoog F (1965) Organic growth factor requirements of tobacco tissue cultures. Physiol Plant 18:100–127

Lipsky AK (1992) Problems of optimisation of plant cell culture processes. J Biotechnol 26:83–97

Madhavi D, Abhayankar G, Reddy VD, Kavi Kishor PB (2012) Carbohydrate and elicitor enhanced withanolide (withaferin A and withanolide A) accumulation in hairy root cultures of Withania somnifera (L.). Ind J Exp Biol 50:484–490

Mahendran D, Kavi Kishor PB, Natesan G, Venkatachalam P (2017) Phycomolecules coated silver nanoparticles and seaweed extracts induced high frequency somatic embryogenesis and plant regeneration from *Gloriosa superba* L. J Appl Phycol 30:1425–1436

Misawa M (1977) In: Barz W, Reinhard E, Zenk MH (eds) Plant tissue culture and its biotechnological applications, Proceedings of the first international congress on medicinal plant research, University of Munich. Springer, Berlin, pp 17–26

Morton J (1992) The ocean-going noni, or Indian mulberry (*Morinda citrifolia*, Rubiaceae) and some of its "colorful" relatives. Econ Bot 46:241–256

Murashige T, Skoog F (1962) A revised medium for rapid growth and bioassays with tobacco tissue culture. Physiol Plant 15:473–497

Nakagawa K, Konagai A, Fukui H, Tabata M (1988) Release and crystallization of berberine in the liquid medium of *Thalictrum minus* cell suspension culture. Plant Cell Rep 7:254–257

Padhya MA (1986) Biosynthesis of isoquinoline alkaloid berberine in tissue cultures of *Tinospora cordifolia*. Ind Drugs 24:47–48

Pagnussat GC, Simontacchi M, Puntarulo S, Lamattina L (2002) Nitric oxide is required for root organogenesis. Plant Physiol 129:954–956

Parimala R, Sachdanandam P (1993) Effect of plumbagin on some glucose metabolizing enzymes studied in rats in experimental hepatoma. Mol Cell Biochem 125:59–63

Rao SR, Ravishankar GA (2002) Plant cell cultures: chemical factories of secondary metabolites. Biotechnol Adv 20:101–153

Rao BR, Vijay Kumar D, Naga Amrutha R, Jalaja N, Vaidyanath K, Maruthi Rao A, Rao S, Rathnagiri P, Kavi Kishor PB (2008) Effect of growth regulators, carbon source and cell aggregate size on berberine production from cell cultures of *Tinospora cordifolia* Miers. Curr Trends Biotechnol Pharm 3:269–276

Rodney C, Toni M, Kutchan N, Lewis G (2000) Natural products. In: Buchanan B, Gruissem W, Jones R (eds) Biochemistry and molecular biology of plants. American Society of Plant Physiologists, Rockville, MD, pp 1253–1348

Sato F, Yamada Y (1984) High berberine producing cultures of *Coptis japonica* cells. Phytochemistry 23:281–285

Singh S, Kumar S (1998) *Withania somnifera*. The Indian ginseng ashwagandha. Central Institute of Medicinal and Aromatic Plants, Lucknow, p 293

Smetanska I (2008) Production of secondary metabolites using plant cell cultures. Adv Biochem Eng Biotechnol 111:187–228

Stalman M, Koskamp A, Luderer R, Vernooy JHJ, Wind JC, Wullems GJ, Croes AF (2003) Regulation of anthraquinone biosynthesis in cell cultures of *Morinda citrifolia*. J Plant Physiol 160:607–614

Staniszewska I, Krolicka A, Mali E, Ojkowska E, Szafranek J (2003) Elicitation of secondary metabolites in *in vitro* cultures of *Ammi majus* L. Enzyme Microbial Technol 33:565–568

Thomson RH (1971) Naturally occurring naphthoquinones, 2nd edn. Academic Press, London

Vijaya SN, Udayasri PV, Aswani KY, Ravi BB, Phani KY, Vijay VM (2010) Advancements in the production of secondary metabolites. J Nat Prod 3:112–123

Wang MY, Su C (2001) Cancer preventive effect of *Morinda citrifolia* (Noni). Ann N Y Acad Sci 952:161–168

Weber H (2002) Fatty acid-derived signals in plants. Trends Plant Sci 7:217–224

Antidiabetic Studies of the Leaf Extract of *Enicostemma littorale* (Blume) Using Wistar Rats

40

A. Kantha Sri

Abstract

The leaves of *Enicostemma littorale* are used in traditional folk medicine to treat diabetes mellitus, rheumatism, abdominal ulcers, anti-bacterial, and antifungal. The plants were collected from Talakona forest near Tirupati to evaluate antidiabetic by ethyl acetate active fraction (EAAF) (200 mg) using animal models. Leaf materials were shade dried and coarsely powdered in order to extract the ethyl acetate active fraction, and were observed for treatment with ethyl acetate active fraction (EAAF 200 mg/kg b.wt.) which significantly normalized the blood glucose levels in STA-diabetic rats. Light microscopic studies also revealed the protective effect of EAAF on pancreatic β-cells and the results were comparable with the reference drug glibenclamide. The hematological parameters like erythrocyte, leukocyte, and hemoglobin were thoroughly studied.

Keywords

Enicostemma littorale · Antidiabetic · Wistar rats

40.1 Introduction

Medicinal plants are the source of the medicine, and herbal medicines have been practiced over the generations. Modern medicine does however make use of many plant-derived compounds as the basis for evidence-tested pharmaceutical drugs. In the last few decades, eco-friendly, bio-friendly, cost-effective, and relatively safe plant-based medicines have gained importance with the increased research in the field of traditional medicine (Sen et al. 2011). Every culture on earth, through written or oral tradition, has been using a variety of healing plants for their therapeutic

A. Kantha Sri (✉)
Department of Botany, Mahatma Gandhi College, Guntur, Andhra Pradesh, India

S. M. Khasim et al. (eds.), *Medicinal Plants: Biodiversity, Sustainable Utilization and Conservation*, https://doi.org/10.1007/978-981-15-1636-8_40

properties. The World Health Organization (WHO) states that 4 billion people, 80% of the world's population, presently use plant-based medicine for some aspect of their primary health care. India is the largest producer of medicinal herbs and is called the botanical garden of the world (Modak et al. 2007).

The World Health Organization (WHO) has enumerated 21,000 plants which are used for medicinal purposes around the world. India comprises 2500 species, out of which 150 species are used commercially. In traditional medicine, diabetes mellitus is treated with diet, physical exercise, and medicinal plants, more than 1200 plants are used around the world in the control of diabetes mellitus, and approximately 30% of the traditionally used antidiabetic plants were pharmacologically and chemically studied (Alarcon-Aguilar et al. 2002).

Studies conducted in India in the last decade have highlighted that the prevalence of diabetes is high and that it is increasing rapidly in the urban population (Ramachandran et al. 2002). In traditional medicine, diabetes mellitus is treated with diet, physical exercise, and medicinal plants. More than 1200 plants are in use around the world in the control of diabetes mellitus, and approximately 30% of the traditionally used antidiabetic plants were pharmacologically and chemically studied (Alarcon-Aguilar et al. 2002). Diabetes patients of type I take daily doses of insulin. Type II (non-insulin-dependent) diabetes mellitus is defined by failing to produce enough quantity of, or appropriately use, insulin. This type constitutes the epidemic form of the disease (Lei et al. 2004). The increased free radical activity is suggested to play an important role in lipid peroxidation and protein oxidation of cellular structures resulting in cell injury and is involved in the pathogenesis of vascular disease which is the main cause of morbidity and mortality in both type I and type II diabetes (Ramakrishna and Jailkhani 2008).

Enicostemma littorale Blume (chhota chirayata) has been selected for antidiabetic studies. It is a well-known medicinal plant in Indian ayurvedic medicine used for the present study. *E. littorale* is also called as Nagajihva which belongs to the family Gentianaceae.

Table 40.1 Preparation of *E. littorale* crude extracts

Solvent used	Solvent recovery	Extract obtained
Hexane—20 L	15 L	10 g
Hexane—20 L	18 L	6 g
Hexane—20 L	16 L	4 g
Total Txt—20 g		
Ethyl acetate—20 L	15 L	15 g
Ethyl acetate—18 L	18 L	10 g
Ethyl acetate—15 L	15 L	5 g
Total Ext—30 g		
Methanol—20 L	18 L	50 g
Methanol—15 L	16 L	20 g
Methanol—18 L	15 L	5 g
Total Ext—75 g		

Enicostemma littorale (Blume) was collected, shade dried, coarsely powdered, and used for extraction; 6.5 kg of dry powder was successively extracted with hexane (20 L) by percolation method for 3 days, and the extract is filtered. The residue was re-percolated again for four times with fresh solvent (20 L). The extracts were filtered, and the solvent was removed by rotary evaporator at 35 °C under reduced pressure. The remaining plant residue was successively extracted with ethyl acetate (20 L) and methanol (20 L) sequentially as mentioned above. The obtained hexane, ethyl acetate, and methanol extracts were concentrated under reduced pressure at a temperature not exceeding 45 °C to yield a dry hexane extract (20 g), ethyl acetate (30 g), and methanol extract (75 g) (Table 40.1). The concentrated extracts of aerial parts of *E. littorale* were subjected to following preliminary phytochemical studies.

40.1.1 Thin Layer Chromatography

Thin layer chromatography (TLC) is the common method used to detect the phytoconstituents (Marston et al. 1997). TLC is the solid-liquid technique in which the two phases are a stationary phase (solid) and a mobile phase (liquid). Solids most commonly used in TLC are silica gel ($SiO_2{\cdot}H_2O$) and alumina ($Al_2O_3{\cdot}H_2O$).

Plates were also sprayed with 5% sulfuric acid in ethanol and 0.36% sulfuric acid in methanol and heated for 3 min at 100 °C to allow for the development of color changes (Carr and Rojers 1986).

40.1.2 Column Chromatography

Column chromatography is the most widely used technique to isolate the components of complex mixture (preparative chromatography). It can also be used to determine the quantity of different compounds present in the mixture (analytical chromatography). Various stationary phases were used to separate the phytoconstituents according to either the polarity or size of the compound.

The active crude extract (g) was adsorbed on silica gel (Acmae's 60–120 mesh) and chromatographed on a silica gel (Acmae's 100–200 mesh) column; initially it was eluted with continuous suitable system and gradually with increasing the polarity of mixture of solvents. Eluted fractions were evaluated using TLC and similar TLC patterns were pooled into major fractions.

40.1.3 Experimental Animals

Adult male rats of Wistar strain weighing approximately 110–180 g were obtained from Mahaveera Enterprises, a licensed breeding center at Hyderabad, India. All the animals were kept and maintained under laboratory conditions of temperature (22 °C ± 2), humidity (45 ± 5%), and 12 h day:12 h night cycle and fed with commercial pelleted rat chow (Hindustan Unilever Ltd., Bangalore, India) and had free access to water.

Table 40.2 Effect of *E. littorale* ethyl acetate active fraction on hemoglobin (Hb), glycosylated hemoglobin (HbA1C), plasma insulin, and hepatic glycogen levels in control and streptozotocin (STZ)-induced diabetic rats

	Group I	Group II	Group III	Group IV	Group V
Hb (mg/dL)	14.31 ± 2.67	13.21 ± 3.70[a]	6.93 ± 1.178	15.3±1[a]	15.51 ± 2.28[a]
HbAlc (mg/g of Hb)	0.76 ± 0.38	0.67 ± 0.32[a]	3.77 ± 0.60	1.12 ± 0.91[a]	1.05 ± 0.91[a]
Plasma insulin (IU/L)	15.35 ± 2.60	15.05 ± 3.63[a]	6.5 ± 1.38	11.53±1[a]	12.38 ± 2.40[a]
Hepatic glycogen (g/100 g wet tissue)	4.98 ± 2.09	5.46 ± 1.92[a]	2.76 ± 1.02	3.15 ± 1.6[b]	3.78 ± 1.49[a,b]

Each value is mean ± SD for six rats in each group
–: No significance
[a]$p < 0.05$ by comparison with normal rat
[b]$p < 0.05$ by comparison with streptozotocin-induced diabetic rats

For all the antidiabetic experiments, five groups of animals, each group consisting of six animals, were chosen. The dosage of the plant extract has been fixed (Table 40.2).

40.2 Effect of *E. littorale* Ethyl Acetate Active Fractions on Fasting Plasma Glucose and Plasma Insulin Levels in STZ-Induced Diabetic Rats

Diabetes mellitus was induced by single intraperitoneal injection of freshly prepared streptozotocin (STZ) (30 mg/kg b.wt.) in 0.1 M citrate buffer (pH—4.5) in a volume of 1 mL/kg b.wt. Rats were supplied with 5% glucose solution for 48 h after STZ injection in order to prevent hypoglycemia. The control animals were treated with citrate buffer (pH—4.5). Diabetes was developed and stabilized in these STZ-treated rats over a period of 7 days. The control animals were treated with citrate buffer (pH—4.5). After 7 days of STZ administration, plasma glucose levels of each rat were determined. Rats with fasting plasma glucose (FPG) range of 280–350 mg/dL were considered diabetic and included in the study. Blood was collected by sinocular puncture (Table 40.3).

40.3 Assessment of Total Hemoglobin, Glycosylated Hemoglobin, Plasma Insulin, and Hepatic Glycogen Levels

In the control group of animals in Group I and Group II, the hemoglobin level (14.31 and 13.21 mg/dL) was higher than the STZ-induced diabetic group of animals in Group III (6.93 mg/dL). As in the case of 200 mg/kg/b.wt. EAAF-treated group of diabetic animals, the hemoglobin level (15.3 mg/dL) was like that of the

Table 40.3 Effect of *E. littorale* ethyl acetate active fraction on fasting blood glucose in normal and STZ-induced diabetic rats

	Group I	Group II	Group III	Group IV	Group V
Day 1	80 ± 6.05	83.6 ± 3.90[a]	403.5 ± 5.79	442.5 ± 6.99[b]	438.5 ± 4.71[b]
Day 15	81 ± 6.13	82.16 ± 2.91[a]	400.3 ± 7.3	322.5 ± 6.9[b]	268.16 ± 7.42
Day 30	81.5 ± 5.79	80.5 ± 6.8[a]	379.8 ± 5.39	147.3 ± 7.11[a]	131.5 ± 7.95[a]
Day 45	81.83 ± 8.27	82.6 ± 4.81[a]	356 ± 7.23	128.6 ± 4.71[a]	121.5 ± 6.02[a]

Each value is mean ± SD for six rats in each group
Group I: normal rats treated with vehicle alone. Group II: normal rats + *Ethyl* active fraction (100 mg/kg b.wt.). Group III: normal rats + *Ethyl* active fraction (200 mg/kg b.wt.). Group IV: STZ-induced diabetic rats treated with vehicle alone. Group V: STZ-induced diabetic rats + *Ethyl* active fraction (100 mg/kg b.wt.)
–: No significance
[a]$p < 0.05$ by comparison with normal rats
[b]$p < 0.05$ by comparison with streptozotocin-induced diabetic rats

control group of animals. Similarly in the glibenclamide-treated group of diabetic rats in Group V (15.51 mg/dL), the hemoglobin level was like that of the control group of animals.

In the control group of animals in Group I and Group II, the glycosylated hemoglobin level (0.76 mg/g of Hb and 0.67 mg/g of Hb) was higher than the STZ-induced diabetic group of animals in Group III (3.77 mg/g of Hb). As in the case of 200 mg/kg/b.wt. EAAF-treated group of diabetic animals, the glycosylated hemoglobin level (1.12 mg/g of Hb) was like that of the control group of animals. Similarly in the glibenclamide-treated group of diabetic rats in Group V (1.05 mg/g of Hb), the glycosylated hemoglobin level was like that of the control group of animals.

In the control group of animals in Group I and Group II, the plasma insulin level (15.35 and 15.05 IU/L) was higher than the STZ-induced diabetic group of animals in Group III (6.5 IU/L). As in the case of 200 mg/kg/b.wt. EAAF-treated group of diabetic animals, the plasma insulin level (11.53 IU/L) was like that of the control group of animals. Similarly in the glibenclamide-treated group of diabetic rats in Group V (12.38 IU/L), the plasma insulin level was like that of the control group of animals.

In the control group of animals in Group I and Group II, the hepatic glycogen level (4.9 g/100 g wet tissue and 5.46 g/100 g wet tissue) was higher than the STZ-induced diabetic group of animals in Group III (2.76 g/100 g wet tissue). As in the case of 200 mg/kg/b.wt. EAAF-treated group of diabetic animals, the plasma insulin level (3.15 g/100 g wet tissue) was like that of the control group of animals. Similarly in the glibenclamide-treated group of diabetic rats in Group V (3.78 g/100 g wet tissue), the plasma insulin level was like that of the control group of animals. Thus, in the untreated diabetic rats and in diabetic rats, glycosylated hemoglobin levels were significantly elevated. Glycosylated hemoglobin, plasma insulin, total hemoglobin, and hepatic glycogen were decreased in diabetic rats as compared with normal rats. Oral administration of EAAF 200 mg/kg/b.wt. for 45 days significantly ($p < 0.05$) maintained all the parameters in near-normal status in diabetic rats.

40.4 Histopathological Observation in Antidiabetic Study of the Experimental Rat Pancreas

The acute and sub-acute toxicity study of EAAF of *E. littorale* has been designed and carried out using male Wistar rats. The histopathological evidences have been recorded for five groups of animals. As in the case of diseased untreated diabetic control group of rats, presence of pancreatic acini and small atrophic islets can be seen. The architecture of pancreatic islets in control groups of animals was normal (Fig. 40.1a–e).

40.4.1 Determination of Effective Dose of *E. littorale* Ethyl Acetate Active Fraction and Fasting Glucose Test

Based on the result obtained from dose determination study, treatment with 200 mg/kg/b.wt. of EAAF for 15 days significantly decreased the blood glucose levels in STZ experimental rats. The blood glucose-lowering effect may be due to the insulin secretory potential of the *E. littorale.* In this instance, 200 mg/kg/b.wt. of ethyl acetate active fraction was fixed as the effective dose.

Glycolysis is a process wherein glucose is degraded, and gluconeogenesis is a pathway wherein glucose gets synthesized from lactate, amino acids, and glycerol (Mohammed et al. 2007). Reduction in the plasma glucose level was accompanied with decrease in the level of glycosylated hemoglobin and glucose-6-phosphatase activity in the liver was observed after 30 days of study and suggesting its use as hypoglycemic agent (Vijayvargia et al. 2000).

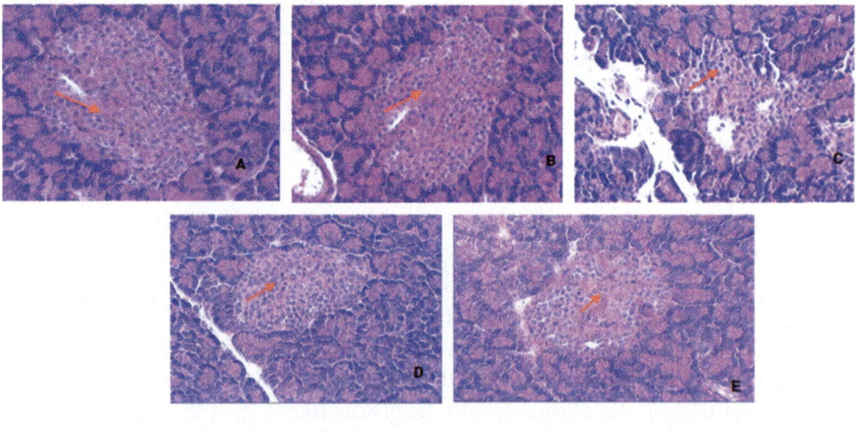

Fig. 40.1 Histopathological observation of normal and experimental rat pancreas (H&E, 400×). (**a**) Normal architecture of pancreatic islets. (**b**) Normal + EAAF 200 mg/kg b.wt.: normal architecture of pancreatic islets. (**c**) Diabetic control: presence of pancreatic acini and small atrophic islet cells. (**d**) Diabetic + EAAF 200 mg/kg: mild expansion pancreatic islets prominent hyperplastic of islets. (**e**) Diabetic + glibenclamide 600 μg/kg b.wt. absence of dilation and prominent hyperplastic of islets

References

Alarcon-Aguilar FJ, Flores Saenz FL, Aguirre Garcia F, Roman Ramos R (2002) Investigation on the hypoglycemic effects of extracts of four Mexican medicinal plants about diabetes. World Health Organization, Geneva. Retrieved 4 Apr 2014

Carr I, Rojers CB (1986) Chemosystematics studies of the genus Combretum (Combretaceae). I. A convenient method of identifying species of this genus by a comparison of the polar constituents extracted from leaf material. S Afr J Bot 53:173–176

Lei L, Zhaohong Y, Masaharu S, Itaru K (2004) Activin A and betacellulin effect on regeneration of pancreatic β-cells in neonatal streptozotocin-treated rats. Diabetes 53:608–615

Marston A, Maillard M, Hostettmann K (1997) GIT Laboratory Journal 1, pp 36–39

Modak M, Dixi P, Londhe J, Ghaskadbi S, Devasagayam TPA (2007) Indian herbs and herbal drugs use for the treatment of diabetes. J Clin Biochem Nutr 40:163–117

Mohammed A, Tanko Y, Okasha MA, Magaji RA, Yaro AH (2007) Effects of aqueous leaves extract of *Ocimum gratissimum* on blood glucose levels of streptozocin-induced diabetic wistar rats. Afr J Biotechnol 6(18):2087–2090

Ramachandran HD, Narasimhamurthy K, Raina PL (2002) Effect of oxidative stress on serum and antioxidant enzymes in liver and kidney of rats and their modulation through dietary factors. Indian J Exp Biol 40:1010–1015

Ramakrishna V, Jailkhani R (2008) Oxidative stress in non-insulin-dependent diabetes mellitus (NIDDM) patients. Acta Diabetol 45:41–46

Sen S, Chakraborty R, De B (2011) Challenges and opportunities in the advancement of herbal medicine: India's position and role in a global context. J Herb Med 1(3–4):67–75

Vijayvargia R, Kumar M, Gupta S (2000) Hypoglycemic effect of aqueous extract of *Enicostemma littorale* Blume on alloxan induced diabetes mellitus in rats. Indian J Exp Biol 38:781–784

Part IV

Biotechnology

41 Molecular and Cytogenetical Approaches for Genetic Diversity Analysis of Wild and Cultivated Medicinal Plant Species from North-East India with Focus on Genus *Curcuma*

Rama Rao Satyawada, Daniel Regie Wahlang, and Judith Mary Lamo

Abstract

The north-eastern region of India is a well-known major hot spot of the world, with a rich reservoir of diversity of plants with medicinal importance. There is an ever-increasing demand of medicinally important plant species resulting in their overexploitation and degradation and loss of genetic diversity. The understanding of genetic variability and diversity is a pre-requisite for sustainable utilization of such scant genetic resources. There is an urgent need to characterize the vast amount of plant genetic resources, which are crucial for developing strategies for their effective conservation, management and sustainable utilization. Cytogenetical data as well as molecular marker-based approaches and a combination of plastid and nuclear genes are considered to assess the genetic diversity at both intra- and interspecific levels and therefore identify medicinally important species and varieties. The present study on genus *Curcuma* encompassing 15 species based on genomic DNA-based molecular markers shows a significant level of molecular variability and genetic diversity among the turmeric collections. A comparison of the degree of polymorphism and efficacy of the different molecular markers showed a high degree of polymorphism in *C. longa* ranging from 93.32% (DAMD) to 95.79% (ISSR). Cytogenetical analysis of various species has confirmed the somatic chromosome number as $2n = 42$, 63 and 105 indicating the occurrences of polyploidy based on $x = 21$. Male meiotic analysis has furthered confirmed the allopolyploid nature of various species but also supported our view of $x = 21$ as basic number of the genus *Curcuma*.

R. R. Satyawada (✉) · D. R. Wahlang
Plant Biotechnology Laboratory, Department of Biotechnology and Bioinformatics, North-Eastern Hill University, Shillong, India

J. M. Lamo
Department of Biotechnology, St. Anthony's College, Shillong, India

S. M. Khasim et al. (eds.), *Medicinal Plants: Biodiversity, Sustainable Utilization and Conservation*, https://doi.org/10.1007/978-981-15-1636-8_41

Keywords
Curcuma · Genetic diversity · North-east India · Molecular approach

41.1 Introduction

The genus *Curcuma* L. belonging to the tribe Zingibereae of the family Zingiberaceae (Kress et al. 2002) is a rhizomatous perennial herb and consists of about 120 species (Sasikumar 2005; Škorničková et al. 2008; Záveská et al. 2012). The generic name finds its origin from 'Kurkum', an Arabic word (yellow), due to the prominent yellow colour of the rhizome (Salvi et al. 2000; Shirgurkar et al. 2001). It is widely distributed in the tropical and subtropical regions of South and Southeast Asia, while a few *Curcuma* species have been reported from China, Australia and the South Pacific islands (Joseph et al. 1999; Škorničková et al. 2007; Záveská et al. 2012).

Curcuma L., a diverse polyploid complex, contains many taxa with multifaceted uses, and quite a few species are used as spice, dyes, medicines, cosmetics, ornamentals and as a source for starch (Sasikumar 2005; Velayudhan et al. 2012). One of the best known examples is turmeric (*C. longa*) which is well known all over the world as a condiment and as a colouring agent in many cuisines. It is locally known as 'shynrai/shyrmit' in Khasi, 'haldi' in Hindi and 'pasupa' in Telugu. Turmeric is used for medicinal purposes. Besides its economical and medicinal uses, turmeric is also used in religious and cultural rites in India and certain Southeast Asian countries (Sasikumar 2005; Velayudhan et al. 2012). The members of the genus are important resources with great potential and utility. Therefore, knowledge of genetic variability in *Curcuma* species is a prerequisite for genetic improvement, sustainable utilization and effective conservation (Islam et al. 2007; Rout and Mohapatra 2006).

Characterization and assessment of the genetic variation and/or diversity using cytogenetical as well as molecular markers is crucial for effective breeding and commercialization, as well as their sustainable utilization and conservation of plant genetic resources. Cytogenetical studies such as chromosome numbers, karyomorphology, chromosomes behaviour and association during meiosis and heterochromatin banding pattern are reliable tools for elucidating the genome architecture and to some extent its organization in related plant taxa. They provide important keys for solving taxonomic riddles, phylogenetic relationships (Joseph et al. 1999; Samaddar et al. 2015) as well as for understanding delineation of taxa (Yashikane and Naohiro 1991).

Earlier cytogenetical observations on *Curcuma* indicate the existence of polyploidy (amphidiploidy) (Raghavan and Venkatasubban 1943; Venkatasubban 1946; Ramachandran 1961, 1969; Nambiar 1979) with significant variations in chromosome numbers and ploidy level within and among the species of *Curcuma*. The lowest chromosome number being reported was $2n = 20$ for *C. parviflora* (Sirisawad et al. 2003), while the highest chromosome number was $2n = 105$ for *C. raktakanta* (Škorničková et al. 2007). Detailed meiotic analysis in the genus *Curcuma* is sparse, and restricted to only a few species (Ramachandran 1961, 1969; Nambiar 1979; Puangpairote et al. 2016).

Understanding the distribution and extent of plant genetic diversity is imperative for crop selection and improvement and will aid in devising effective management programmes for sustainable utilization and conservation (Rao and Hodgkin 2001). Genetic variations between and among individuals are often defined by differences in gross morphological characters, biochemical characteristics and DNA sequence. DNA-based molecular markers, unlike morphological and biochemical markers, are not influenced by environmental conditions, thus offering a more reliable estimation of genetic diversity. DNA-based molecular markers such as directed amplification of minisatellite region DNA (DAMD), inter-simple sequence repeat (ISSR), intron-targeted intron-exon splice junction (IT-ISJ) and sequence-related amplified polymorphism (SRAP) targets specific loci of the DNA are of great value in assessing polymorphism and genetic variability of both wild and cultivated species (Lewandowska 2001; Han et al. 2008; Fu et al. 2010; Amirmoradi et al. 2012; Rajkumari et al. 2013; Li et al. 2014; Singh et al. 2014).

In the present study, comprehensive and integrated approaches involving cytogenetical analysis and DNA-based molecular markers have been adopted to assess the genetic diversity of wild and cultivated species of *Curcuma.* Multidisciplinary approaches rather than reliance through on a single technique might provide strategic framework to evaluate the useful traits in *Curcuma* species and transfer these to other cultivars which are economically important.

41.2 Plant Materials and Study Area

Curcuma germplasm was obtained from Indian Institute of Spices Research (IISR), Kozhikode, and by personal exploration trips. The plants were grown and maintained in polyhouse conditions at Plant Biotechnology Laboratory, Department of Biotechnology and Bioinformatics, North-Eastern Hill University, Shillong. The details of germplasm along with accession numbers are given in Tables 41.1 and 41.2 and Fig. 41.1.

Table 41.1 *Curcuma* germplasm obtained from IISR, Kozhikode, along with their specimen voucher number

Sl. no.	Species	Voucher numbers	Sl. no.	Species	Voucher numbers
1	*C. aeruginosa* Roxb.	1147	9	*C. longa* L.	Pratibha
2	*C. amada* Roxb.	521	10	*C. mangga* Valeton and Zijp	1049
3	*C. aromatica* Salisb.	711	11	*C. montana* Roxb.	649
4	*C. caesia* Roxb.	700	12	*C. raktakanta* Mangaly and M. Sabu	1137
5	*C. comosa* Roxb.	644			
6	*C. haritha* Mangaly and M. Sabu	1136	13	*C. sylvatica* Valeton	526
7	*C. latifolia* Roscoe	638	14	*C. zanthorrhiza* Roxb.	1108
8	*C. leucorrhiza* Roxb.	1169	15	*C. zedoaria* (Christm.) Roscoe	760

Table 41.2 *Curcuma longa* collections used in the present study

Sl. no.	Place of collection	Collection number	District	Sl. no.	Place of collection	Collection number	District
1	Umpling	CuL-1	East Khasi Hills (EKH)	21	Bhoirymbong	CuL-21	RiBhoi (RB)
2	Rynjah	CuL-2		22	Mynsain	CuL-22	
3	Pohkseh	CuL-3		23	Nonglum	CuL-23	
4	Nongpiur	CuL-4		24	Umden Umroi	CuL-24	
5	Polo	CuL-5		25	Umiam	CuL-25	
6	Polo	CuL-6		26	Umran	CuL-26	
7	Mawtawar	CuL-7		27	Nongpoh	CuL-27	
8	NEHU Campus	CuL-8		28	Umsning	CuL-28	
9	Malki	CuL-9		29	Nongsder	CuL-29	
10	Wahkhen	CuL-10		30	Umden	CuL-30	
11	Sutnga	CuL-11	Jaintia Hills (JH)	31	Nongstoin	CuL-31	West Khasi Hills (WKH)
12	Mawkyndeng	CuL-12		32	Nongstoin	CuL-32	
13	Mawkyndeng	CuL-13		33	Nongstoin	CuL-33	
14	Amlarem	CuL-14		34	Nongkhlaw	CuL-34	
15	Mawrynkneng	CuL-15		35	Mawkyrwat	CuL-35	
16	Ummulong	CuL-16		36	Mairang	CuL-36	
17	Mynkrem	CuL-17		37	Myriaw	CuL-37	
18	Nongjngi	CuL-18		38	Nonglwai	CuL-38	
19	Kdohkule	CuL-19		39	Manai	CuL-39	
20	Puriang	CuL-20		40	Nongktieh Manai	CuL-40	

The details of methodology followed for chromosome counts in root tip cells and analysis of chromosome association, recombination frequency and anaphase distribution of chromosome(s) are detailed elsewhere (Lamo 2016).

Genomic DNA extraction, PCR amplification and genetic diversity analysis methods were also detailed elsewhere (Lamo 2016).

Total curcuminoid content of *C. longa* rhizomes was determined by solvent extraction and spectrophotometric method (ASTA Method 1968). The percentage of total curcuminoid content in turmeric was calculated using the formula (Sasikumar et al. 2004):

$$\text{Percentage of curcuminoid} = \frac{\text{Absorbance at 425 nm} \times 125}{\text{Cell length}\left(\text{cm}\right) \times A \times \text{sample weight}}$$

where A = absorptivity of the standard curcumin at 425 nm.

41.3 Genetic Diversity Analysis of Wild and Cultivated Curcuma: Molecular and Cytogenetical Approach

The knowledge about cytology and chromosome behaviour of a species provides information about its genetic architecture which is imperative for genetic improvement as well as initiating hybridization programmes (Normann et al. 1989).

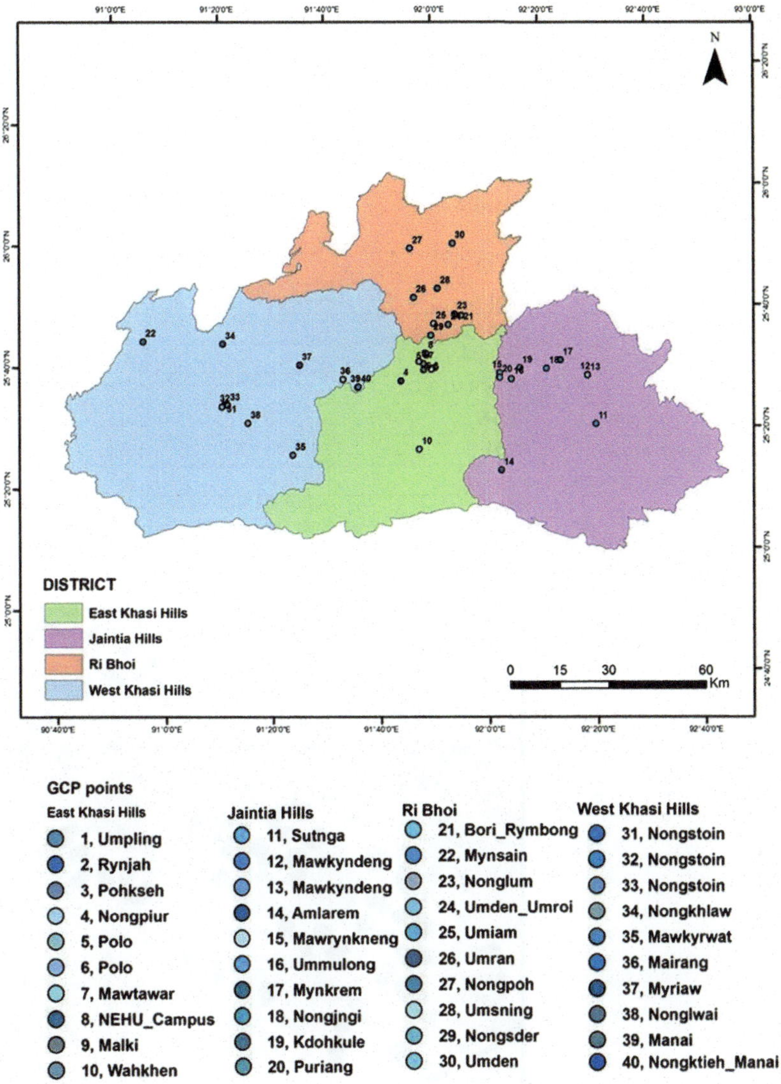

Fig. 41.1 Map showing collection sites of *C. longa* from different geographical locations of Meghalaya

Chromosome counts in 15 *Curcuma* species revealed the presence of $2n = 42$, 63 and 105 somatic chromosome numbers (Lamo and Rao 2014, 2017a), which is summarized in Table 41.3. The somatic chromosome number of *C. amada* was previously reported to be $2n = 40$ and 42; however, only $2n = 42$ has been recorded in the present study and therefore confirmed the same number (Fig. 41.2). These observations are in line with those reported by Bhadra and Bandyopadhyaya (2015). Several unrelated chromosome numbers in *C. aromatica* have been reported (Ramachandran 1961, 1969; Nambiar et al. 1982; Chen and Chen 1984;

Table 41.3 Somatic chromosome number in *Curcuma* species

Sl. no.	Species	Voucher numbers/accession numbers	2*n*	Groups
1	*C. amada*	521	42	I
2	*C. aromatica*	711	42	
3	*C. comosa*	644	42	
4	*C. haritha*	1136	42	
5	*C. mangga*	1049	42	
6	*C. montana*	649	42	
7	*C. aeruginosa*	1144	63	II
8	*C. caesia*	751	63	
9	*C. latifolia*	638	63	
10	*C. longa*	Pratibha	63	
		NEHU-12037		
11	*C. leucorrhiza*	1169	63	
12	*C. sylvatica*	526	63	
13	*C. zanthorrhiza*	1108	63	
14	*C. zedoaria*	760	63	
15	*C. raktakanta*	1137	105	III

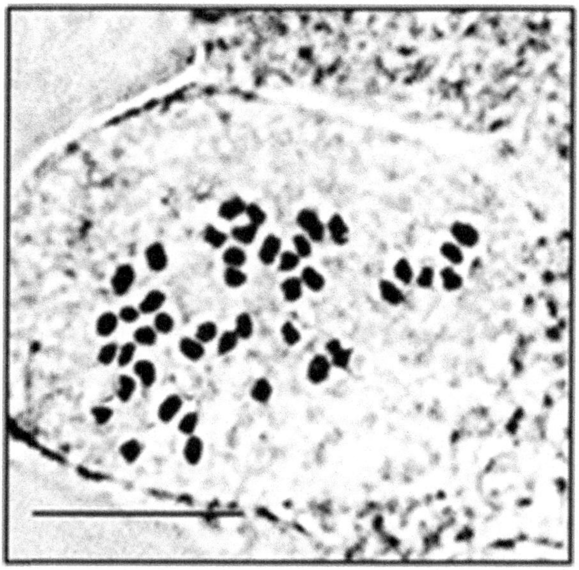

Fig. 41.2 Mitotic complements of *C. amada* ($2n = 42$). Scale bar = 10 μm

Paisookasantivatana and Thepsen 2001). On the other hand, we have recorded only the expected chromosome number of $2n = 42$, both in *C. aromatica* and *C. comosa* which are in full agreement with the earlier reports of Raghavan and Venkatasubban (1943) and Joseph et al. (1999). Similarly, we confirm the somatic chromosome number of $2n = 42$ for other species, viz. *C. haritha*, *C. mangga* and *C. montana* which are in line with the published reports of Prana (1978), Joseph et al. (1999) and Škorničková et al. (2007). On the other hand, $2n = 63$ was also observed in *C. aeruginosa*, *C. caesia*, *C. latifolia*, *C. longa*, *C. leucorrhiza*, *C. zanthorrhiza* and *C.*

zedoaria in our present study, which derived support previous findings of Ramachandran (1961), Apavatjrut et al. (1996), Joseph et al. (1999) and Škorničková et al. (2007). Contrary to this, several unrelated chromosome numbers such as $2n = 84$ in *C. aeruginosa* (Islam 2004), $2n = 22$ in *C. caesia* (Das et al. 1999), $2n = 62$ and 84 in *C. latifolia* (Soontornchainaksaeng and Jenjittikul 2010) and $2n = 42$ in *C. zedoaria* (Paisookasantivatana and Thepsen 2001) were reported by various workers; thus, our observations stand distinctly at variance with these findings. It is interesting to note that *C. longa* with its large number of varieties/cultivars is known to exhibit great diversity of chromosome number at intraspecific level. The reports by different workers for *C. longa* indicate the existence of $2n = 22$, 48, 62, 64 and 93 as observed by Sharma and Bhattacharya (1959), Sato (1960) and Škorničková et al. (2007). In a classical but convincing report, Nair and Sasikumar (2009) observed $2n = 61$, 63, 72, 77, 78, 80, 84 and 86 in different accessions of *C. longa* maintained at the germplasm repository of the Indian Institute of Spices Research, Kozhikode. The somatic chromosome number of $2n = 105$ of *C. raktakanta* observed by the present authors is in agreement with the report of Škorničková et al. (2007), but is at varaiance with the observation ($2n = 63$) of Joseph et al. (1999).

Based on somatic chromosome number, the species can possibly be divided into three groups, viz. group I, group II and group III with $2n = 42$, $2n = 63$ and $2n = 105$, respectively. Such grouping might give some useful keys for elucidating species diversification and evolution of *Curcuma*. However, unless we have authentic and unambiguous chromosome counts of all the 120 *Curcuma* species, it is promiscuous to approve the above grouping of *Curcuma* species.

Meiosis, a highly specialized and conserved mechanism, can offer some insight into the evolutionary origin of species (Lenormand et al. 2016). Meiotic analysis was carried out in eight *Curcuma* species belonging to the group I and group II with varying degree of chromosome association(s) (Lamo and Rao 2017b). Group I species revealed that bivalents outnumbered any other chromosome associations with near-normal meiosis. However, group II showed the prevalence of trivalents, besides a few bivalent, quadrivalent and occasional univalent associations (Table 41.4; Fig. 41.3a–d).Comprehensive and combined investigation of both meiosis and mitosis strongly supports that *Curcuma* is an allopolyploid complex which is evident from the low frequency of multivalent associations and in view of the fact that chromosome associations at the first meiotic division are the usual source of information concerning the type of polyploidy in a given plant (Swaminathan 1953; Lamo and Rao 2017a, b). Moreover, we speculate that the speciation in *Curcuma*

Table 41.4 Relative percentages of chromosome associations during male meiosis in eight *Curcuma* species (modified from Lamo and Rao 2017b)

	C. amada	*C. comosa*	*C. haritha*	*C. mangga*	*C. Montana*	*C. caesia*	*C. longa*	*C. sylvatica*
Quadrivalents	7.57	3.13	2.11	2.89	5.14	5.30	4.52	2.05
Trivalents	–	–	–	–	–	31.32	35.86	31.14
Bivalents	92.03	83.98	64.12	94.02	77.23	53.09	44.31	61.99
Univalents	0.40	12.89	33.77	3.89	17.63	10.29	15.31	4.82

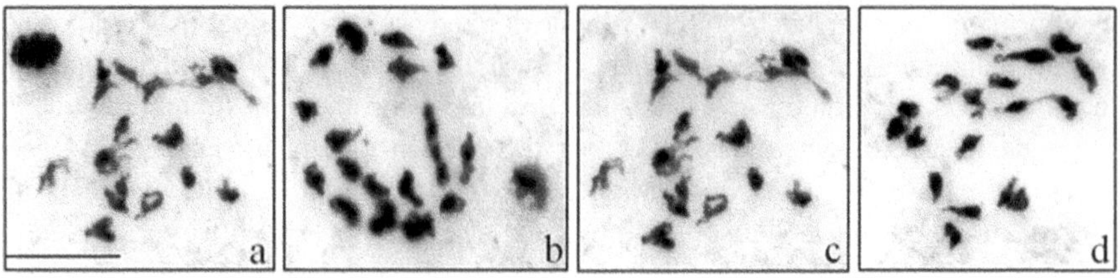

Fig. 41.3 (**a**–**d**) Male meiosis in *C. amada*. (**a**, **b**) Diplotene, (**c**) diakinesis, (**d**) metaphase I. Scale bar = 10 μm

might have been influenced by inter-specific crosses and propose $x = 21$ as the basic number which is in accordance with the observations of Ramachandran (1961, 1969) and Nambiar (1979).

C. longa is great in demand for therapeutic and pharmaceutical purposes due to the prevalance of high value curcumin in its rhizome (Thaikert and Paisooksantivatana 2009). Curcumin in its pure form can be isolated (90–95%) from dried turmeric rhizomes by solvent extraction method (Madhavamenon and Paulose 2014). However, the solvent extract consists of a mixture of curcuminoids, viz. curcumin, demethoxycurcumin and bisdemethoxycurcumin (Ashraf et al. 2012). In the present study, the total curcuminoid content was evaluated in 40 collections of *C. longa* from Meghalaya and was expressed as the mean ± SD and was taken in three replicates. Total curcuminoid content ranged from 2.2 ± 0.25 (CuL-1) to 11.7 ± 0.10 (CuL-11) (Table 41.5). The collections from Jaintia Hills (JH) showed the highest total curcuminoid content with an average of 6.8%, and the lowest (3.32%) was recorded in the collections from East Khasi Hills (EKH). Such variations in total curcumin/curcuminoid content have also been reported by previous workers. Lynrah et al. (1998) reported a high variation of 0.08–8.58% total curcumin content in 25 genotypes of *Curcuma*. Niar et al. (2010) reported that 'ISSR Kedaram' a high-yielding variety released in India has a high curcumin (5.67%), while Kulkarni et al. (2012) reported a much higher curcumin yield (12.39%) in turmeric samples from Satara district in India. The variation in total curcumin content might be attributed to their eco-geographical distribution and derives support from those of Lynrah et al. (1998), Sasikumar (2005), Pothitirat and Gritsanapan (2006), Singh et al. (2013) and Anandaraj et al. (2014), conclusively reported a direct relationship between eco-geographical and curcuminoid content. Other factors affecting the total curcuminoid content might be age of the plant, time of harvesting, etc. It is well documented that plant maturity has significant impact on chemical constituents of turmeric rhizomes of *Curcuma* (Chavalittumrong and Jirawattanapong 1992; Cooray et al. 1988; Remadevi et al. 2007; Li et al. 2011).

Characterization and assessment of the genetic variation and/or diversity using DNA-based molecular markers is important for effective breeding, commercialization as well as sustainable utilization and conservation of turmeric genetic resources. Therefore, four maker systems, viz. DAMD, IT-ISJ, ISSR and SRAP approaches (Table 41.6), were employed for estimating the genetic diversity in *C. longa* collections from Meghalaya. These marker systems being cost-effective with high reproducibility and reliability are highly useful for evaluating genetic

Table 41.5 Total curcuminoid content in *C. longa* collections

Collection name	Location[a]	Curcuminoid content (%)	Average	Collection name	Location	Curcuminoid content (%)	Average
CuL-1	EKH	2.2 ± 0.25	**3.32 ± 1.11**	CuL-21	RB	6 ± 0.36	**5.46 ± 1.38**
CuL-2		2.3 ± 0.25		CuL-22		6.2 ± 0.15	
CuL-3		6 ± 0		CuL-23		5.2 ± 0.1	
CuL-4		3 ± 0		CuL-24		6.3 ± 0.23	
CuL-5		3.4 ± 0.32		CuL-25		3.6 ± 0.25	
CuL-6		3.4 ± 0.35		CuL-26		6.5 ± 0.31	
CuL-7		2.4 ± 0.49		CuL-27		6.6 ± 0.36	
CuL-8		3.4 ± 0.49		CuL-28		4.6 ± 0.63	
CuL-9		2.5 ± 0		CuL-29		6.8 ± 0.32	
CuL-10		3.7 ± 0.1		CuL-30		2.8 ± 0.29	
CuL-11	JH	11.7 ± 0.1	**6.8 ± 1.82**	CuL-31	WKH	4 ± 0	**4.76 ± 0.7**
CuL-12		6.8 ± 4.37		CuL-32		5.1 ± 0.27	
CuL-13		6.3 ± 0.45		CuL-33		4.1 ± 0.15	
CuL-14		7.2 ± 0.25		CuL-34		5.9 ± 0.17	
CuL-15		5.5 ± 0.1		CuL-35		4.2 ± 0.29	
CuL-16		6.2 ± 0.77		CuL-36		5.9 ± 0.12	
CuL-17		6.2 ± 0.17		CuL-37		4.9 ± 0.29	
CuL-18		6.6 ± 0.06		CuL-38		4.2 ± 0.12	
CuL-19		5.2 ± 0.15		CuL-39		4.6 ± 0.35	
CuL-20		6.3 ± 0.25		CuL-40		4.7 ± 0.26	

[a] *EKH* East Khasi Hills, *JH* Jaintia Hills, *RB* Ri-Bhoi, *WKH* West Khasi Hills

Table 41.6 Sequence information for DAMD, IT-ISJ, ISSR and SRAP markers

Markers	Primer name	Primer sequence 5′–3′	Markers	Primer name	Primer sequence 5′–3′
DAMD	D1	ATCCAAGG CCGAGACAACC	IT-ISJ	B1	GAGCCCAGAACGACGCGGG
	D4	AGGACTCGATAACAGGCTCC		E2	GGAATTCCACGTCCA
	D6	ATGTGTGCGATCAGTTGCTG		H1	ACGCGTCAGGTAAGTACA
	D9	GATGTGTTCTTGGAGCC GT		H2	ACGCGTCAGGTAAGTACC
	D10	GGACAAGAAGAGGATGTGGA		H6	ACGCGTCAGGTAAGTAGC
	D14	CCTCCTCCCTCCT		H8	ACGCGTCAGGTAAGTAGT
	D18	ACAGGGGTGTGGGG		H11	GAAGCTTCACCTGCAACG
	D19	CCCGTGGGGCCGCCG		H16	GAAGCTTCACCTGCAAGT
	HBV	GGTGTAGAGAGGGGT		R1	TCGTGGCTGACTCTACTG
	HBY	GCCTTTCCCGAG		R2	TGCGTGTGTGTGTGTGTGCAGGT
ISSR*	807	AGAGAGAGAGAGAGAGT	SRAP	1F/1R	TGAGTCCAAACCGGATA/GACTGCGTACGAATTAAT
	808	AGAGAGAGAGAGAGAGC		1F/5R	TGAGTCCAAACCGGATA/GACTGCGTACGAATTAAC
	809	AGAGAGAGAGAGAGAGG		1F/8R	TGAGTCCAAACCGGATA/GACTGCGTACGAATTCCG
	810	GAGAGAGAGAGAGAGAT		1F/10R	TGAGTCCAAACCGGATA/GACTGCGTACGAATTTCA
	811	GAGAGAGAGAGAGAGAC		2F/2R	TGAGTCCAAACCGGAGC/GACTGCGTACGAATTGAC
	812	GAGAGAGAGAGAGAGAA		2F/6R	TGAGTCCAAACCGGAGC/GACTGCGTACGAATTGCA
	816	CACACACACACACACAT		3F/3R	TGAGTCCAAACCGGAAT/GACTGCGTACGAATTTGC
	821	GTGTGTGTGTGTGTGTT		3F/6R	TGAGTCCAAACCGGAAT/GACTGCGTACGAATTGCA
	826	ACACACACACACACACC		4F/1R	TGAGTCCAAACCGGACC/GACTGCGTACGAATTAAT
	834	AGAGAGAGAGAGAGAGYT		5F/5R	TGAGTCCAAACCGGAAG/GACTGCGTACGAATTAAC
	835	AGAGAGAGAGAGAGAGYC		6F/6R	TGAGTCCAAACCGGAGG/GACTGCGTACGAATTGCA
	841	GAGAGAGAGAGAGAGAYC		6F/7R	TGAGTCCAAACCGGAGG/GACTGCGTACGAATTC
	842	GAGAGAGAGAGAGAGAYG		7F/2R	TGAGTCCAAACCGGAGG/GACTGCGTACGAATTGAC
	847	CACACACACACACACARC		7F/7R	TGAGTCCAAACCGGAGG/GACTGCGTACGAATTC
	857	ACACACACACACACACYG		8F/7R	TGAGTCCAAACCGGA/GACTGCGTACGAATTC
	861	ACCACCACCACCACCACC		10F/3R	TGAGTCCAAACCGGTGT/ACTGCGTACGAATTTGC
	862	AGCAGCAGCAGCAGCAGC		10F/4R	TGAGTCCAAACCGGTGT/GACTGCGTACGAATTTGA
	864	ATGATGATGATGATGATG		10F/6R	TGAGTCCAAACCGGTGT/GACTGCGTACGAATTGCA
	873	GACAGACAGACAGACA		10F/7R	TGAGTCCAAACCGGTGT/GACTGCGTACGAATTC
	900	ACTTCCCCACAGGTTAACACA		10F/10R	TGAGTCCAAACCGGTGT/GACTGCGTACGAATTTCA

*In case of ISSR primers: R = A/G; Y = C/T

diversity at both intra- and interspecific levels among wild and cultivated taxa. Though vegetatively propagated species like of *Curcuma* and other members of the zingibers, are expected and should theoretically are expected to give a low genetic variation (Singh et al. 2013). On the contrary, our study revealed high degree of polymorphism (>93.32%) (Table 41.7), reflecting a significant level of molecular variability and genetic diversity among turmeric collections from Meghalaya. Similar findings were also reported by many workers using various molecular markers, viz. RAPD, ISSR, DAMD and SSR (Nayak et al. 2006; Jatoi et al. 2010; Jan et al. 2011; Singh et al. 2012 Verma et al. 2015). Possibly, this can be due to the selection and cultivation practices adopted by the local populace resulting in diverse planting material, capable of adaptive response to a wide range of eco-geographical conditions (Hangelbroek et al. 2002).

UPGMA clustering pattern of DAMD, IT-ISJ, ISSR and SRAP data set revealed similar clustering pattern with three main clusters (Cluster I, II and III) (Fig. 41.4a–d). This might be due to geographically isolated population which tend to accumulate genetic differences as they adapt to different environmental conditions (Sarwat et al. 2008). Though *C. longa* collections from West Khasi Hills always clustered together in all the analyses, there was no clear distinction of clustering pattern in collections from other regions. Similar observations were also reported in *Trigonella* (Dangi et al. 2004) and *Curcuma* (Singh et al. 2012; Verma et al. 2015).

Discriminating power of a given primer can be considered as a reliable indicator of the efficiency of that primer and/or marker in diversity analysis. To test the efficiency and utility of the molecular approaches used in the study, mean PIC, mean Rp and mean MI were compared for their discriminatory power (Table 41.7).

Table 41.7 Summary of the various genetic diversity indices

Markers	Percentage of polymorphic bands	PIC	Resolving power (Rp)	Marker Index (MI)
DAMD	95.4	0.29	5.20	4.96
IT-ISJ	93.32	0.30	3.75	2.15
ISSR	95.79	0.26	4.59	4.59
SRAP	95.62	0.24	4.34	4.01

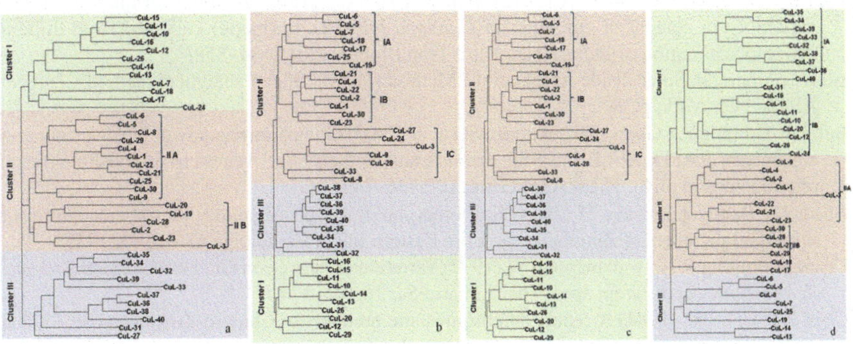

Fig. 41.4 (a–d) UPGMA clustering pattern. (a) IT-ISJ, (b) DAMD, (c) ISSR and (d) SRAP

The PIC value is generally used to measure the overall utility of a marker system in determination of polymorphism across the taxa (Powell et al. 1996). PIC value ranged from 0.24 (SRAP) to 0.30 (IT-ISJ). Comparative analysis of all the molecular approaches revealed hypervariability of DAMD and ISSR. Such phenomenon exhibited by DAMD and ISSR markers reflects the extent of genome coverage (Tautz and Schlötterer 1994). The genetic profiles of both DAMD and ISSR can provide information about the diversity in repetitive DNA sequences whereby DAMD targets the core minisatellite regions while ISSR targets the microsatellites regions (Al-Turki and Basahi 2015). Such regions are hypervariable, and the most probable reason for their occurrence might be due to 'replication slippage' which is thought to occur more frequently than single-nucleotide mutations and insertion/deletion events (Belaj et al. 2003). From our study, we proposed that DAMD and SRAP markers are highly informative and could be used in assessing genetic diversity of *C. longa* accession and/or varieties.

Thus, the present study provides an account on genetic diversity and variability at inter- and intraspecific level of both cultivated and wild *Curcuma* species. Such information, though not directly, offers scope for genetic improvement among the cultivars for increasing genetic diversity in the cultivated turmeric pool as well as their wild relatives.

References

Al-Turki TA, Basahi MA (2015) Assessment of ISSR based molecular genetic diversity of Hassawi rice in Saudi Arabia. Saudi J Biol Sci 22:591–599

Amirmoradi B, Talebi R, Kamari E (2012) Comparison of genetic variation and differentiation among annual *Cicer* species using start codon targeted (SCoT) polymorphism, DAMD-PCR, and ISSR markers. Plant Syst Evol 298:1679. https://doi.org/10.1007/s00606-012-0669-6

Anandaraj M, Prasath D, Kandiannan K et al (2014) Genotype × environment interactions for yield and curcumin in turmeric (*Curcuma longa* L.) multienvironment trials in India. Ind Crop Prod 53:358–364

Apavatjrut P, Sirisawad T, Sirirugsa P, Voraurai P, Suwanthada C (1996) Studies on chromosome number of seventeen Thai *Curcuma* species. In: Proceedings of second national conference on flower and ornamental plant, vol 2, pp 86–99

Ashraf K, Mujeeb M, Ahmad A (2012) Validated HPTLC analysis method for quantification of variability in content of curcumin in *Curcuma longa* L. (turmeric) collected from different geographical region of India. Asian Pac J Trop Biomed 2(2):S584–S588

ASTA (American Spice Trade Association) Method (1968) Official analytical methods, 2nd edn. ASTA, New York, p 38

Belaj A, Dominguez-García MC, Atienza SG et al (2003) Comparative study of the discriminating capacity of RAPD, AFLP and SSR markers and of their effectiveness in establishing genetic relationships in olive. Theor Appl Genet 107:736–744

Bhadra S, Bandyopadhyaya M (2015) Karyomorphological investigations on some economically important members of Zingiberaceae from Eastern India. Caryologia 68(3):184–192

Chavalittumrong P, Jirawattanapong W (1992) Variation of active constituents of *Curcuma domestica* rhizomes at different ages. Thai J Pharm Sci 16:165–174

Chen ZY, Chen SJ (1984) A report on chromosome numbers of Chinese Zingiberaceae. Guihaia 4:13–18

Cooray NF, Jansz ER, Ranatunga J, Wimalasena S (1988) Effect of maturity on some chemical constituents of turmeric (*Curcuma longa* L.). J Natl Sci Council Sri Lanka 16:39–51
Dangi RS, Lagu MD, Choudhary LB, Ranjekar PK, Gupta VS (2004) Assessment of genetic diversity in *Trigonella foenum-graecum* and *Trigonella caerulea* using ISSR and RAPD markers. BMC Plant Biol 4:13. https://doi.org/10.1186/1471-2229-4-13
Das AB, Rai S, Das P (1999) Karyotype analysis and cytophotometric estimation of nuclear DNA content in some members of the Zingiberaceae. Cytobios 97:23–33
Fu LZ, Zhang HY, Wu XW (2010) Evaluation of genetic diversity in *Lentinula edodes* strains using RAPD, ISSR and SRAP markers. World J Microbiol Biotechnol 26:709–716
Han XY, Wang LS, Shu QY (2008) Molecular characterization of tree peony germplasm using sequence-related amplified polymorphism markers. Biochem Genet 46:162–179
Hangelbroek HH, Ouborg NJ, Santamaria L, Schwenk I (2002) Clonal diversity and structure within a population of the pondweed *Potamogeton pectinatus* foraged by Bewick's swans. Mol Ecol 11:2137–2150
Islam MA (2004) Genetic diversity of the genus *Curcuma* in Bangladesh and further biotechnological approaches for *in vitro* regeneration and long-term conservation of *C. longa* germplasm. PhD thesis, University of Hannover
Islam MA, Meister A, Schubert V, Kloppstech K, Esch E (2007) Genetic diversity and cytogenetic analyses in *Curcuma zedoaria* (Christm.) Roscoe from Bangladesh. Genet Resour Crop Evol 54:149–156
Jan UH, Rabbani MA, Shinwari ZK (2011) Assessment of genetic diversity of indigenous turmeric (*Curcuma longa* L.) germplasm from Pakistan using RAPD markers. Med. Plant Res 5(5):823–830
Jatoi SA, Kikuchi A, Yi SS, Naing KW (2010) Use of SSR markers as RAPD markers for genetic diversity analysis in Zingiberaceae. Breed Sci 56:107–111
Joseph R, Joseph T, Joseph J (1999) Karyomorphological studies in the genus *Curcuma* Linn. Cytologia 33:313–317
Kress WJ, Price LM, Williams KJ (2002) The phylogeny and a new classification of the gingers (Zingiberaceae): evidence from molecular data. Am J Bot 89:1682–1696
Kulkarni SJ, Maske KN, Budre MP, Mahajan RP (2012) Extraction and purification of curcuminoids from turmeric (*Curcuma longa* L.). Int J Pharmacol Pharm Technol 1:81–84
Lamo J.M (2016) Assessment of genetic diversity and species interrelationship in the genus *Curcuma* L. PhD thesis, North-Eastern Hill University
Lamo JM, Rao SR (2014) Chromosome counts in two species of *Curcuma* Linnaeus (Zingiberaceae) from North-East India. Pleione 8(2):435–438
Lamo JM, Rao SR (2017a) Chromosome counts in wild and cultivated species of *Curcuma* Linn. Cytologia 82(2):141–144
Lamo JM, Rao SR (2017b) Meiotic behaviour and its implication on species inter-relationship in the genus *Curcuma* (Linnaeus, 1753) (Zingiberaceae). Comp Cytogenet 11(4):691–702
Lenormand T, Engelstädter J, Johnston SE, Wijnker E, Haag CR (2016) Evolutionary mysteries in meiosis. Phil Trans R Soc B 371:20160001. https://doi.org/10.1101/050831
Lewandowska K (2001) Estimation genetic similarity of *Lolium perenne* L. and *L. multiflorum* Lam, on the basis of molecular markers. PhD thesis, University of Warmia and Mazury
Li S, Yuan W, Deng G, Wang P, Yang P, Aggarwal BB (2011) Chemical composition and product quality control of turmeric (*Curcuma longa* L.). Pharm Crops 2:28–54
Li M, Zhoa Z, Miao X (2014) Genetic diversity and relationships of apricot cultivars in North China revealed by ISSR and SRAP markers. Sci Hort 173:20–28
Lynrah PG, Barua PK, Chakrabarty BK (1998) Pattern of genetic variability in a collection of turmeric (*Curcuma* spp.) genotypes. Indian J Genet 58(2):201–207
Madhavamenon KI, Paulose MB (2014) Curcuminoid composition with enhanced bioavailability and a process for its preparation. Patent Application Publication US2014/0010903 A1
Nair RR, Sasikumar B (2009) Chromosome number variation among germplasm collections and seedling progenies in turmeric, *Curcuma longa* L. Cytologia 74:153–157

Niar RR, Shiva KN, Anchu S, Zachariah TJ (2010) Characterization of open pollinated seedling progeniesof turmeric (*Curcuma longa* L.) based on chromosome number, plant morphology, rhizome yield and rhizome quality. Cytologia 75:43–449

Nambiar MC (1979) Morphological and cytological investigations in the genus *Curcuma* L. PhD thesis, University of Bombay

Nambiar MC, Pillai PKT, Sharma YN (1982) Seedling propagation in turmeric (*Curcuma aromatic* Salisb.). J Plant Crops 10:81–85

Nayak S, Naik PK, Acharya LK, Pattnaik AK (2006) Detection and evaluation of genetic variation in 17 promising cultivars of turmeric (*Curcuma longa* L.) using 4c nuclear DNA content and RAPD markers. Cytologia 71(1):49–55

Normann GA, Quarin CL, Burson BL (1989) Cytogenetics and reproductive behavior of different chromosome races in six *Paspalum* species. J Herd 80:24–28

Paisookasantivatana Y, Thepsen O (2001) Phenetic relationships of some Thai *Curcuma* species (Zingiberaceae) based on morphological, palynological and cytological evidences. Thai J Agric Sci 34:47–57

Pothitirat W, Gritsanapan W (2006) Variation of bioactive components in *Curcuma longa* in Thailand. Curr Sci 91(10):1397–1400

Powell W, Morgante M, Andre C, Hanafey M, Vogel J, Tingey S (1996) The comparison of RFLP, RAPD, AFLP and SSR (microsatellite) markers for germplasm analysis. Mol Breed 2:225–238

Prana MS (1978) Studies on some Indonesian *Curcuma* species. PhD thesis, University of Birmingham

Puangpairote T, Maknoi C, Jenjittikul T, Anamthawat-Jónsson K, Soontornchainaksaeng P (2016) Natural triploidy in phyto-estrogen producing species and cultivars from Thailand. Euphytica 208(1):47–61

Raghavan TS, Venkatasubban KR (1943) Cytological studies in family Zingiberaceae with special reference to chromosome number and cytotaxonomy. Proc Indian Acad Sci B 17:118–132

Rajkumari K, Sharma SK, Rao SR (2013) Assaying polymorphism by intron targeted intron-exon specific junction (ISJ) DNA marker for genetic diversity diagnostics of the *Cucumis* L. taxa. Nucleus 56(1):15–21

Ramachandran K (1961) Chromosome numbers in the genus *Curcuma* Linn. Curr Sci 30:194–196

Ramachandran K (1969) Chromosome numbers in Zingiberaceae. Cytologia 34:213–221

Rao VR, Hodgkin T (2001) Genetic diversity and conservation and utilization of plant genetic resources. Plant Cell Tissue Organ Cult 68:1–19

Remadevi R, Surendran E, Kimura T (2007) Turmeric in traditional medicine. In: Ravindran PN, Nirmal Babu K, Sivaraman K (eds) Turmeric: the genus *Curcuma*. CRC Press, Boca Raton, FL, pp 409–436

Rout GR, Mohapatra A (2006) Use of molecular markers in ornamental plants: a critical reappraisal. Eur J Hort Sci 71(2):53–68

Salvi ND, Geoge L, Eapen S (2000) Direct regeneration of shoots from immature inflorescence cultures of turmeric. Plant Cell Tissue Organ Cult 62:235–238

Samaddar T, Than MMM, Jha TB, Jha S (2015) Cytogenetic and DNA fingerprinting analysis in three species of *Swertia* from Eastern Himalaya. Carylogia 68(3):207–216

Sarwat M, Das S, Srivastava PS (2008) Analysis of genetic diversity through AFLP, SAMPL, ISSR and RAPD markers in *Tribulus terrestris*, a medicinal herb. Plant Cell Rep 27:519–528

Sasikumar B (2005) Genetic resources of *Curcuma*: diversity, characterization and utilization. Plant Genet Res 3:230–251

Sasikumar B, Syamkumar S, Remya R, John Zachariah T (2004) PCR based detection of adulteration in the market samples of turmeric powder. Food Biotechnol 18:299. https://doi.org/10.1081/FBT-200035022

Sato D (1960) The karyotype analysis in Zingiberales with special reference to the protokaryotype and stable karyotype. Sci Papers Coll Gen Educ Univ Tokyo 10:225–243

Sharma AK, Bhattacharya NK (1959) Cytology of several members of Zingiberaceae and a study of the inconstancy of their chromosome complements. Cellule 59:299–346

Shirgurkar MV, John CK, Nadgauda RS (2001) Factors affecting in vitro microrhizome production in turmeric. Plant Cell Tissue Organ Cult 64:5–11

Singh S, Panda MK, Nayak S (2012) Evaluation of genetic diversity in turmeric (*C. longa*) using RAPD and ISSR markers. Ind Crop Prod 37:284–291

Singh S, Panda MK, Nayak S (2013) Evaluation of genetic diversity in turmeric (*Curcuma longa* L.) using RAPD and ISSR markers. Ind Crop Prod 37:284–291

Singh N, Bajpai R, Mahar KS, Tiwari V, Upreti DK, Rana TS (2014) ISSR and DAMD markers revealed high genetic variability within *Flavoparmelia caperata* in Western Himalaya (India). Physiol Mol Biol Plants 20(4):501–508

Sirisawad T, Sirirugsa P, Suwanthada C, Apavtjrut P (2003) Investigation of chromosome numbers in 20 taxa of *Curcuma*. In: Chantaranothai P, Larsen K, Sirirugsa P, Simpson D (eds) Proceedings of the third symposium on the family Zingiberaceae. Applied Taxonomic Research Centre, Khon Kaen University, Khon Kaen, pp 54–62

Škorničková J, Šída O, Jarolimova V (2007) Chromosome numbers and genome size variation in Indian species of *Curcuma* (Zingiberaceae). Ann Bot 100:505–526

Škorničková J, Šída O, Sabu M, Marhold K (2008) Taxonomic and nomenclatural puzzles in Indian *Curcuma*: the identity and nomenclatural history of *C. zedoaria* (Christm.) Roscoe and *C. zerumbet* Roxb. (Zingiberaceae). Taxon 57(3):949–962

Soontornchainaksaeng P, Jenjittikul T (2010) Chromosome number variation of phytoestrogen-producing *Curcuma* (Zingiberaceae) from Thailand. J Nat Med 64:370–377

Swaminathan MS (1953) Nature of polyploidy in some 48-chromosome species of the genus *Solanum*, section Tuberarium. Genetics 39(1):59–76

Tautz D, Schlötterer C (1994) Simple sequences. Curr Opin Genet Dev 4:834–837

Thaikert R, Paisooksantivatana Y (2009) Variation of total curcuminoids content, antioxidant activity and genetic diversity in turmeric (*Curcuma longa* L.) collections. Kasetsart J (Nat Sci) 43:507–518

Velayudhan KC, Dikshit N, Abdul Nizar M (2012) Ethnobotany of turmeric (*Curcuma longa* L.). Indian J Trad Knowl 11(4):607–614

Venkatasubban KR (1946) A preliminary survey of chromosome numbers in Scitamineae of Bentham & Hooker. Proc Indian Acad Sci B 23:281–300

Verma S, Singh S, Sharma S (2015) Assessment of genetic diversity in indigenous turmeric (*Curcuma longa*) germplasm from India using molecular markers. Physiol Mol Biol Plants 21(2):233–242

Yashikane I, Naohiro N (1991) Karyotypes of *Fragaria nubicola* and *F. daltoniana*. Cytologia 56:453–457

Záveská E, Fér T, Šída O, Krak K, Marhold K, Škorničková J (2012) Phylogeny of *Curcuma* (Zingiberaceae) based on plastid and nuclear sequences: proposal of the new subgenus Ecomata. Taxon 61(4):747–763

Mutagenic Effect of Chemicals on Certain Biochemical Parameters in Two Cultivars of Sunflower (*Helianthus annuus* L.)

42

T. Padmavathi, M. Uma Devi, and B. Prathibha Devi

Abstract

The effect of three chemicals, viz., dES (di-Ethyl Sulphate); MH (Maleic Hydrazide) and NN (1-Nitroso-2,naphthol) in different concentrations on two varieties of sunflower (*Helianthus annuus* L.): EC-68415 (Armavirskij) and Morden were studied for three generations, and further extensive biochemical studies conducted on these cultivars including certain mutants that were isolated during the course of the research work are discussed in the present paper. All the seed type mutants and the seeds from the malformed heads and bisected heads of EC-68415 and Morden, respectively, were subjected to intensive biochemical studies by standard analytical procedures. A remarkable difference in leaf protein was seen between the control and the mutant types. The leaf enzyme studies included both the qualitative and quantitative analysis of peroxidase and catalase isoenzymes. The enzyme activity was very low in the round seed type mutants of both EC-68415 and Morden, while a high enzyme activity was seen in the small seed type mutants of the two cultivars. The zymogram studies of peroxidase and catalase revealed a distinct variation in the banding pattern among the mutants and the controls of EC-68415 and Morden. The seed protein content and seed protein profile were determined by standard biochemical procedures. High contents of protein was seen in the malformed head, flat seed and small seed type mutants of EC-68415, and also in the bisected head, striped hull type and small seed type mutants of Morden. The electrophorogram studies of seed protein revealed fast mobility protein fractions in the controls and slow pattern in the

T. Padmavathi (✉)
Department of Botany, UCST, Adikavi Nannaya University, Rajamahendravaram, Andhra Pradesh, India

M. Uma Devi
Department of Botany, UCSS, Osmania University, Hyderabad, Telangana, India

B. Prathibha Devi
Department of Botany, UCS, Osmania University, Hyderabad, Telangana, India

S. M. Khasim et al. (eds.), *Medicinal Plants: Biodiversity, Sustainable Utilization and Conservation*, https://doi.org/10.1007/978-981-15-1636-8_42

mutants. This shows that the mutants are more advanced genetically than the controls. The oil content in the mutants varied quite remarkably when compared to the respective controls. The bold seed type and round seed type mutants observed in the two cultivars possessed higher oil content, while the malformed head type, bisected head type and small seed type mutants of EC-68415 and Morden, respectively, had very low quantities of oil. Fatty acid profile revealed high percentage of linoleic acid in the round seed type followed by bold seed type mutants of the two cultivars. The malformed head type, flat seed type, bisected head type, striped hull type and small seed type mutants of EC-68415 and Morden, respectively, yielded oil with high oleic acid content.

Keywords
Helianthus annuus · Seed type mutants · Biochemical studies · Leaf proteins and enzymes · Seed protein · Oil content and Oil profile

42.1 Introduction

When genetic variability is narrowed using traditional breeding methods for a long period, induced mutations are one of the most important approaches for broadening the genetic variations to circumvent the bottleneck conditions (Toker et al. 2007). The number of mutant varieties of all species officially released and recorded in FAO/IAEA mutant varieties Database is over 2300. Gamma rays were the most frequently used technique to alter genes. The ionizing radiations are believed to act in a mechanical way by producing a high or low proportion of chromosomal alterations and low or high proportion of point mutations—depending on whether they are densely or sparsely ionizing types, respectively. Though chemicals form potential mutagens, the number of mutants released after their treatment is much less comparatively. The chemicals produced both chromosomal alterations and point mutations due to the action of chemical forces. Fishbein et al. (1970) classified a number of chemicals into different groups of mutagens based on their mode of action.

The cultivated sunflower (*Helianthus annuus* L.) is one of the 67 species of the genus Helianthus and is a dicotyledonous plant and member of the Compositae (Asteraceae) family, having a typical composite flower. Sunflower seed oils are high in saturated fatty acids (lauric acid C12:0, maristic C14:0, palmitic C16:0 and stearic C18:0), monounsaturated fatty acids (oleic acid C18:1, n-9) and polyunsaturated fatty acids (linoleic acid C18:2, n-6, and α-linolenic C18:3, n-3) (Heiser 1976).

It is an important oil-yielding crop in temperate countries like USSR, Bulgaria, Romania, Canada and the USA. Sunflower was introduced in India in the year 1969, from Russia, as an edible oil seed crop. It ranked fourth in the world among sources of vegetable oils, soybean, groundnut and cotton-seed oil being the other three leading sources. The commercial sunflower of the present day is believed to have originated in Peru/Mexico. In India, Karnataka is the largest producer of sunflower seed with a production of 0.17 M tonnes making it to 69.15% of the All India production (Directorate of Economics and Statistics 2015–16). Sunflower is a potential crop to replace low yielding and uneconomical crops like rabi cotton, kharif groundnut and bajra under rainfed conditions. It is also grown as a mixed crop with groundnut, ragi and blackgram. It forms a good substitute crop for groundnut crop when there is a delay in sowing of groundnut due to delay in monsoon. Unlike groundnut, sunflower has low seed rate and higher multiplication ratio. Hence within a short period, a large area can be covered with improved seeds. The crop can make an important contribution to the income of dryland farmers, thereby reducing the edible oil shortage in the country.

Sunflower oil has an ideal combination of saturated and polyunsaturated fatty acids (PUFA). The latter in fact lower the low–density lipoproteins (LDLs) in the blood cholesterol, thereby improving the rate of the heart. Corn and Safflower oils also have a similar nature but sunflower oil scores over them because it has a better balance of polyunsaturated, monounsaturated and saturated fatty acids. Sunflower oil which helps reduce the level of cholesterol in blood is rich in linoleic acid (of PUFA group). Soybean on the other hand, though rich in polyunsaturated fatty acids, has linolenic acids, which causes oxidation of the oil, resulting in poor-quality oil.

An effort was made to study the effect of three chemicals, viz., dES (di-Ethyl Sulphate); MH (Maleic Hydrazide) and NN (1-Nitroso-2,naphthol), in different concentrations on two varieties of sunflower: EC-68415 (Armavirskij) and Morden. These studies comprised micro- and macro-mutational levels for three generations—M_1, M_2 and M_3. Further, extensive biochemical studies were conducted on the two cultivars and certain mutants that were isolated during the course of the research work, which are being discussed in the present paper.

42.2 Materials and Methods

The two cultivars of sunflower, EC-68415 and Morden, chosen as experimental material was procured from Sunflower Breeding Centre, Rajendranagar, Hyderabad, AP, India. The three chemicals, dES, MH and NN, were manufactured by Loba-Chemie Indo Austranal Co., Bombay.

About 250 seeds were used per treatment comprising four concentrations per mutagen: 0.2%, 0.4%, 0.6% and 0.8%. Seeds of uniform size (moisture content 10%) were selected and soaked for 6–8 h in distilled water prior to chemical mutagen treatment. The water-soaked seeds were then wrapped in a filter paper to remove excess water. These seeds were then soaked in different concentrations of each chemical mutagen for at least 3 h at room temperature (25 ± 2 °C). Intermittent shaking was necessary for uniform diffusions. These mutagen-treated seeds were initially washed thoroughly in continuous flow of tap water for half an hour, and later in distilled water to remove the excess chemical.

Necessary controls were maintained by soaking equal number of seeds in distilled water for 9 h. 150 seeds per each concentration and control were sown in the field in randomized block design (RBD), in three replications each of 50 seeds. They were sown with a spacing of 38 cm between the rows and 23–30 cm between the plants. Each inflorescence was hand-pollinated and bagged to avoid outcrossing. The M_1 generation plants were harvested on a single plant basis. The M_3 was thoroughly screened for the trend in the quantitative characters and were subsequently analysed statistically. However, the viable mutant plants were sown separately on single mutant basis to further field test and characterize their quantitative traits. These were then subjected to different biochemical analyses.

The biochemical studies comprised estimation of leaf protein and estimation and analysis of leaf enzyme activity (isoperoxidases and catalase isoenzymes), seed protein content and seed protein profile (Electrophoresis), seed oil content and its fatty acid profile.

42.2.1 Determination of Leaf Protein

The protein content of different mutants along with respective parents was assayed by Lowry et al. (1951) method at 750 nm reading of extinction. Freshly collected leaves of controls and mutants were ground in a mortar and pestle with 0.5 ml of trichloro-acetic acid to precipitate the proteins. The contents were centrifuged at 3000 rpm for 15 min and the precipitate was dissolved in 5 ml of sodium hydroxide. This was taken as the protein extract. The following solutions were prepared

- Alkaline sodium carbonate solution in 2% of sodium carbonate in 0.1 N sodium hydroxide.
- Copper sulphate in sodium potassium tartarate solution (0.5% copper sulphate in 0.1% sodium potassium tartarate). This is freshly prepared by mixing stock solution.
- Alkaline solution freshly prepared each time by mixing 30 ml of the first and 1 ml of the second solutions.
- Folin–Ciocalteu reagent, the commercial reagents diluted with equal volume of water.
- Standard protein, Bovine albumin solution 0.2 mg ml.

For analysis of the leaf protein to 1 ml of the dissolved protein extract, 5 ml of the alkaline solution was added and mixed thoroughly and allowed to stand at room temperature for 10 min. 0.5 ml of Folin's reagent was added and mixed rapidly. Change in colour was read in spectrophotometer at 750 nm after 30 min. The protein value was expressed as mg/g fresh weight.

42.2.2 Quantitative and Qualitative Estimation of Enzyme Activity in the 10-Day-Old Seedlings

42.2.2.1 Estimation of Peroxidase Activity

Peroxidase activity was assayed with the modified procedure of Kar and Mishra (1976). Assay mixture contained 0.1 M phosphate buffer, 0.01 M pyrogallol, 0.005 M H_2O_2 and 1 ml of enzyme extract. After incubation for 5 min, the reaction was stopped with 2.5 N H_2SO_4 and estimated by measuring the absorbance at 420 nm in LKB ultra spectrophotometer. The enzyme activity was expressed in

$$\text{Absorbancy units/mg} = \frac{OD\ 460/\text{min}}{11.3 \times \text{mg enzyme/ml reaction mixture}}$$

42.2.2.2 Isolation of Peroxidases

Ten-day-old seedlings of the mutants selected along with their respective controls were washed with water, cut into small pieces and homogenized with 0.1 M K_2HPO_4 in a Waring blender. The resulting homogenate was then squeezed through eight layers of cheesecloth. The filtrate was then brought to 35% saturation with $(NH_4)_2SO_4$. After standing for 18 h, the solution was centrifuged at 10,000 rpm for 15 min. The pellet was dissolved in 5 mM Tris HCl buffer (pH 7) and then dialysed against 5 mM Tris HCl (pH 8) containing 0.1 M KCl for 48 h. The dialysate was centrifuged at 10,000 rpm for 15 min and the supernatant was lyophilized and designated as crude peroxidase fraction. The crude peroxidase was separated into different isoperoxidases on a slab gel, by following Davis' (1964) PAGE electrophoretic method. The gels, however, were stained with a solution consisting of benzidine, NH_4Cl (30%), 0.2% H_2O_2 (50:10:20). The gels were incubated for 20 min at room temperature till the bands became distinct, then destained in a solution of 7% acetic acid and fixed in 2% acetic acid. The R_f values of each isoperoxidase was calculated by

$$R_f = \frac{\text{Distance travelled } by \text{ the band}}{\text{Distance travelled } by \text{ tracking dye}} \times 100$$

42.2.2.3 Determination of Catalase Activity

Catalase activity was assayed by the method of Barber (1980). The reaction mixture consisted of 1 ml of enzyme extract, 2 ml of H_2O_2 and 3 ml of phosphate buffer (pH 7). Reaction was stopped with 0.1 N H_2SO_4 and titrated with 0.01 M $KMNO_4$. Catalase activity was expressed as mM H_2O_2 oxidized/min/g fresh weight.

42.2.2.4 Isolation of Catalase Isoenzymes

The young seedlings of the mutants under study and their controls were homogenized in an extraction medium containing 0.05 M Tris HCl buffer (pH 7.5), 0.02% ß-mercapto-ethanol and 0.05% EDTA. The extract was centrifuged at 10,000 rpm for 15 min. The clear supernatant was taken for the qualitative study by the PAGE electrophoretic method. A modified Hale's method was followed to visualize catalase (Scandalios 1974). The gels were first soaked for a few seconds in 0.5% H_2O_2 solution, washed twice in running water, then immersed for 2 min in 0.1% KI solution acidified with glacial acetic acid. The peroxidase releases the iodine which stains the acrylamide gel polymerized with starch as blue areas except where the catalase activity has been destroyed. The R_f values for the catalase isoenzymes were calculated by using the formula

$$R_f = \frac{\text{Distance travelled } by \text{ the band}}{\text{Distance travelled } by \text{ tracking dye}} \times 100$$

42.2.3 Quantitative and Qualitative Estimation of Seed Protein

42.2.3.1 Determination of Seed Protein

The protein content of certain mutants along with respective controls was estimated by using the Lowry's method. This method was based on the principle that different proteins contain different amounts of aromatic residue which react with Folin's reagent, giving a blue colour. Optical densities were recorded at 540 nm on a Spectronic 20 spectrophotometer.

42.2.3.2 Protein Profile Analysis by Electrophoresis

The water-soluble seed proteins of mutants and controls were analysed by the polyacrylamide gel electrophoresis (PAGE) (Davis 1964). The dry seeds were homogenized by grinding with 50 mM Tris buffer, pH 7.4 and 250 mM sucrose. The homogenate was centrifuged for 15 min at 0 °C at 10,000 rpm. The clear supernatant served as the source of soluble proteins for SDS–PAGE. Aliquots of the extract were mixed with an appropriate volume of diluent consisting 0.05 M Tris, pH 7, 20% glycerol, 2% SDS, 5% 2-mercapto ethanol and 0.25 mg/ml Coomassie Brilliant Blue to obtain a final protein concentration of 1 μg/ml. The mixer was heated at 100 °C for 3 min before loading 75 mg protein in each track on 12% homogeneous discontinuous SDS polyacrylamide slab gel system according to the method of Laemmli (1970).

42.2.3.3 Preparation of Gel

The discontinuous SDS polyacrylamide gel was prepared as follows

(a) Running gel (12%)

Acrylamide + Bis (29.2% + 0.8%) 30%	32 ml
Tris HCl buffer 1.875 M pH 8.8	16 ml
Distilled water	27.2 ml
Ammonium per sulphate (1%)	4 ml
SDS (Sodium Dodecyl Sulphate) 10%	0.8 ml
TEMED 30 μl	80.0 ml

(b) Stacking gel (4%)

Acrylamide + Bis (10% + 2.5%) 12.5%	7.5 ml
Tris HCl buffer 1.25 M pH 8.8	3.0 ml
Distilled water	17.7 ml
Ammonium per sulphate (1%)	1.5 ml
SDS (Sodium Dodecyl Sulphate) 10%	0.3 ml
TEMED 8 μl	30.0 ml

(c) Running buffer

Tris buffer	0.025 M pH 8.3
Glycine	0.129 M
SDS	0.1%

Dilute 10× before use.

Electrophoresis was performed at constant voltage for 650–700 V h in a Pharmacia GE 2/4 apparatus. After electrophoresis gels were fixed and stained in 0.2% Coomassie Brilliant Blue R-250 in methanol, acetic acid and water (30:10:60) and destained in the same solution without the dye. The apparent molecular weights of the polypeptides were determined from the calibration curve prepared using the Pharmacia molecular weight calibration kit. The R_f value was expressed as

$$R_f = \frac{\text{Distance travelled } by \text{ the band from the top of the gel}}{\text{Distance travelled } by \text{ the dye end}}$$

42.2.4 Quantitative and Qualitative Estimation of Seed Oil

42.2.4.1 Determination of Oil Content (Nuclear Magnetic Resonance)

Among the various methods available for the determination of oil content, NMR method is the only non-destructive type where the total oil content is estimated in intact seeds. Nuclear magnetic resonance (NMR) is a form of radio frequency spectroscopy. It provides accurate counts of hydrogen nuclei in liquid oils even in surrounding matrix of starch, protein, etc. (Alexander et al. 1967). Seeds of different seed type mutants along with their parents were weighed to about 15 g and oven dried at 40–50 °C prior to the estimation of oil by NMR analyser. The NMR analysis was conducted at the Directorate of Oil Seeds Research, ICAR, Rajendranagar, Hyderabad.

Two grams of the seed material was oven dried at 65 °C for 4 h and ground in a glass mortar to a fine powder. The powder is put in the thimble of the Soxhlet extraction apparatus and enough petroleum ether was added to cover the thimble of the Soxhlet extraction apparatus to flow down into the flask. The apparatus was fixed up and heated uniformly at 40–45 °C for about 4 h allowing at least 20 siphonings. Finally ether was evaporated completely leaving the oil in the flask. The thimble was removed with petroleum ether. The flask with the oil was oven dried for about 12 h. The weight was then recorded. Percentage of oil was calculated as

$$\%\ \text{Oil} = \frac{W_o}{W_s} \times 100$$

where W_o = weight of oil extracted and W_s = weight of seeds taken.

42.2.4.2 Determination of Fatty Acids by GLC

1.3 g of oil was taken in a 15 × 45 bial, 0.25 ml of 0.4 N sodium methylate and anhydrous methanol was added to the vial and closed with a cap. The vial was immersed in a water bath at 65 °C to the depth of 1.5 cm for 30 s. The vial was shaken constantly. When the contents of the vial became homogenous the conversion of methyl esters was said to be complete. Heating was continued for about 90 s, then cooled down to room temperature. 0.6 g each of silica gel and anhydrous calcium chloride was added to the vial. The contents were thoroughly mixed. Then 0.3 ml of carbon disulphide was added, the vial closed and shaken for 2 min. The suspended material was allowed to settle down. Clear carbon disulphide layer was withdrawn and evaporated along with methyl esters to be isolated. Thus, the esterification was complete. Then 1.5 ml of the sample was injected to the gas liquid chromatograph. The fatty acids were separated depending on the degree of saturation and the chain length. The boiling point increases with the length of the chain. Short-chain fatty acids get volatilized fast and occur early in the recorder. Also the fatty acids having greater unsaturation is held more tightly by the liquid coated on a solid support (saturation phase) and is released slowly than a less unsaturated fatty acid of the same chain length. When the volatilization of the fatty acid is completed these are carried away by the gas or mobile phase to the flame and are burnt and ionized. The flame is supplied with hydrogen gas and air. The ionization generates a current. It is amplified and fed to the recorder. The information is printed by a recorder on a chart in the form of peaks. The peaks of the unknown sample are compared with the known standards. The area under each peak is calculated by the planimeter. It can also be calculated by the following method.

(a) Area of peak is equal to height of the peak multiplied by width of the peak at half the height.
(b) Sum up the area under different peaks.
(c) Percentage of fatty acids equal to 100 multiplied by area of the peak and divided by total area of peaks.

The sample was injected to the GLC unit with the following conditions

Column—¼ OD 8′ length stainless steel
Packing = 20% DEGS on chromosorb P
Temperature
- (a) Column = 200 °C
- (b) Detector = 300 °C
- (c) Injection part = 300 °C

FID detector chart speed 60 cm/h, gas flow rate = 80 ml N_2/min. The GLC of the fatty acids was conducted at the Andhra Pradesh State Food Laboratories, Nacharam, Hyderabad.

42.3 Results

Quite a few plants with varying morphological characters were observed in M_2 generation. Each mutant plant was found to be affected differently, viz., plant height, stem type, branching pattern, shape of the head, seed shape, size, structure and colour. Such plants, recognized as mutants, were harvested and sown separately as M_3 generation to test their true breeding nature. Altogether, six morphological and four seed type mutants were isolated in each of the plant cultivars.

42.3.1 Biochemical Studies

The seeds of the malformed and bisected head mutants, because of the peculiar nature, along with the four seed type mutants were selected for the different biochemical analyses. These included the estimation of protein, enzyme activity and analysis in the leaves, seed protein content and protein profile analysis (Electrophoretic), seed oil content and its fatty acid profile. Thus, altogether five mutant types from each variety were selected.

42.3.2 Estimation of Leaf Protein

The leaf protein content in the different mutants, along with the controls, was estimated by using Lowry's method (1951). A significant variation in the protein content of the mutants and the parents was observed (Tables 42.1 and 42.2). The values for protein were expressed as protein mg/g of fresh weight.

Bold seed type mutants and round seed type mutants in EC-68415 (Table 42.1) recorded the lowest content of protein, viz., 2.112 and 2.621 mg/g, respectively, as

Table 42.1 Levels of peroxidase and catalase activity and leaf protein content in certain mutants of sunflower (EC-68415)

		Peroxidase	Catalase	Protein
Type of mutant	Treatment	SPA units/fresh mg protein	mM H_2O_2 oxidized/min/weight g fresh weight	mg/g
Control	–	1.434	2.44	4.915
Malformed head mutant	0.8%NN	1.154	2.21	5.673
Bold seed type mutant	0.4%NN	1.181	2.28	2.112
Round seed type mutant	0.6%NN	0.717	1.23	2.621
Flat seed type mutant	0.4%dES	1.434	2.39	5.668
Small seed type mutant	0.2%NN	1.481	2.91	5.823

Table 42.2 Levels of peroxidase and catalase activity and leaf protein content in certain mutants of sunflower (Morden)

Type of mutant	Treatment	Peroxidase SPA units/mg protein	Catalase mM H_2O_2 oxidized/min/weight g fresh weight	Protein mg/g
Control	–	1.241	2.18	4.723
Malformed head mutant	0.8%NN	1.142	2.01	5.312
Bold seed type mutant	0.4%NN	1.204	2.21	2.401
Round seed type mutant	0.6%NN	0.752	1.25	2.230
Flat seed type mutant	0.4%dES	1.281	2.23	5.401
Small seed type mutant	0.2%NN	1.284	2.38	5.723

against the control (4.915). The highest amount of protein was observed in the small seed type mutants. The protein content in the other two mutants was higher than the control.

Similarly, in Morden (Table 42.2) the round and the bold seed type mutants contained very low quantity of protein as compared to the control of 4.723 64 mg/g fresh weight. The highest value was recorded in the small seed type mutant.

42.3.3 Quantitative and Qualitative Estimation of Enzyme Activity in the 10-Day-Old Seedlings

The enzyme activity, specifically for the isoenzymes like peroxidases and catalases, was determined by the methods proposed by Kar and Mishra (1976) and Barber (1980), respectively. Leaves from the 10-day-old seedlings were selected for the quantitative estimation of the enzymes in the five mutants and their controls of the M_3 generation. The peroxidase activity was expressed as specific activity units (SPA units) and the catalase activity was expressed as mM H_2O_2 oxidized/min/g fresh weight. Tables 42.1 and 42.2 show significant variation for the enzyme activity between the mutants and the parents.

42.3.3.1 Estimation of Peroxidase Activity

The peroxidase activity in EC-68415 (Table 42.1) control was 1.434 SPA units/mg protein. The lowest and highest amounts were recorded in the round seed and small seed type mutants, respectively. The peroxidase activity in Morden (Table 42.2) control was 1.241 SPA units. The lowest and the highest values were recorded in the round and small seed type mutants, respectively, in both the varieties.

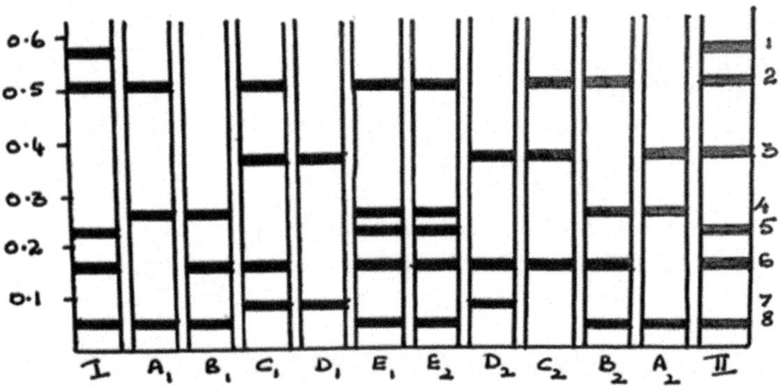

Fig. 42.1 Idiogram of peroxidase isoenzymes—controls and mutants of EC-68415 and Morden, where I = EC-68415 control, A_1 = malformed head mutant, B_1 = flat seed type mutant, C_1 = bold seed type mutant, D_1 = round seed type mutant, E_1 = small seed type mutant, A_1–E_1 = mutants of EC-68415, II = Morden control, A_2 = bisected head mutant, B_2 = striped hull mutant, C_2 = bold seed type mutant, D_2 = round seed type mutant, E_2 = small seed type mutant and A_2–E_2 = mutants of Morden

42.3.3.2 Analysis of Peroxidase Isoenzymes by PAGE

The zymogram (Fig. 42.1) of the peroxidase enzymes studied in the 10-day-old seedlings exhibited a total number of 8 bands with R_f values ranging between 0.04 and 0.57. A minimum of 2 bands were traced in the round seed type mutant of the EC-68415. A maximum number of 6 bands were observed in the control seedlings of the variety Morden. Band 1 was found only in the controls of the two cultivars. Band 2 was found to be absent in the flat seed and round seed type mutants of EC-68415 and in the bisected head and the round seed type mutants of the variety Morden. Band 3 was observed to be present only in bold and round seed type mutants of EC-68415 and Morden, respectively, and also in the flat seed type mutant and the control seedlings of Morden. Band 4 was found in malformed head mutant, flat seed and small seed type mutants of EC-68415. In Morden, band 4 was observed in bisected head type mutant, striped hull and small seed type mutants. Band 5 was seen in both the control and small seed type mutants of the two cultivars. Band 6 was present in both the control seedlings and almost all the mutants except the malformed head type mutant and round seed type mutant of EC-68415 and the bisected head type mutant of Morden. Band 7 was present in just three mutants, two of EC-68415 and one mutant of Morden. Band 8 was absent only in bold and round seed type mutants of both EC-68415 and Morden.

42.3.3.3 Determination of Catalase Activity

The catalase activity recorded was 2.44 and 2.18 mM H_2O_2 oxidized/min/g fresh weight for EC-68415 and Morden, respectively (Tables 42.1 and 42.2). The lowest amount was observed in the round seed type mutants of both the cultivars and the highest value recorded in the small seed type mutants of the two varieties.

42.3.3.4 Isolation of Catalase Isoenzymes

The zymogram (Fig. 42.2) for catalase isoenzymes recorded a total of 5 bands. These bands were present within an R_f ranging between 0.16 and 0.48. A maximum of 4 bands (1, 3, 4 and 5) was observed in the controls of both EC-68415 and Morden. A minimum of 2 bands was observed in 3 mutant types (flat, bold and round seed type mutants) of EC-68415 and 2 mutant types (striped hull and round seed type mutants) of Morden. Each of the other mutant types of the two cultivars was observed to exhibit three bands at different R_f levels.

42.3.4 Quantitative and Qualitative Estimation of Seed Protein

42.3.4.1 Determination of Seed Protein Content

The seeds of each of the five mutants of the two varieties were subjected to protein content estimation in comparison with the respective controls by the Lowry's method. The seed protein estimates for both the cultivars (EC-68415 and Morden) and their mutants are presented in Table 42.3 and expressed as mg/ml of homogenate. The protein content for each mutant type varied significantly from their respective controls.

The control values for protein in EC-68415 and Morden were 12.86 and 13.82 mg/ml, respectively. The lowest protein content was observed in the round seed type mutants of the two varieties. The maximum values, on the other hand, were observed in the malformed head mutant of EC-68415 and bisected head mutant of Morden. The bold seed type mutants isolated in both the cultivars were found to contain lower seed protein as compared to their controls. The other two seed type mutants of each cultivar were observed to record higher values than their control seeds.

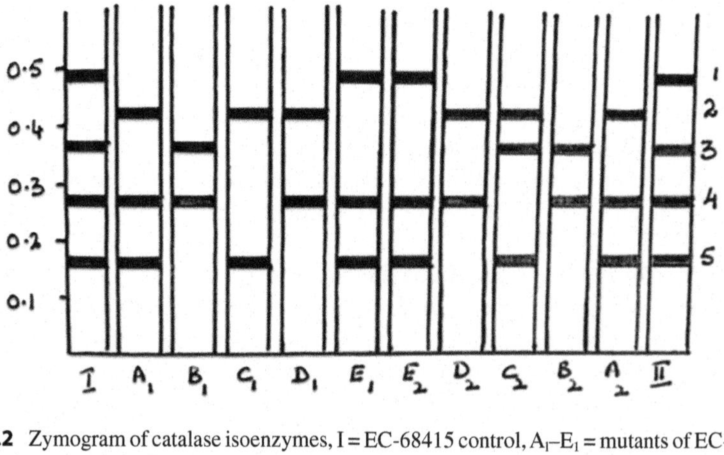

Fig. 42.2 Zymogram of catalase isoenzymes, I = EC-68415 control, A_1–E_1 = mutants of EC-68415, II = Morden control, A_2–E_2 = mutants of Morden

42.3.4.2 Analysis of Seed Protein Profile

The protein profile pattern on polyacrylamide gel was scanned for different mutant types and their controls (Fig. 42.3). Electrophorogram was prepared on the basis of protein mobility and density expressed in R_f values. The R_f values were calculated for each band as

$$R_f = \frac{\text{Distance travelled } by \text{ the band the top of the running gel}}{\text{Distance travelled } by \text{ tracking dye}}$$

Table 42.3 Seed protein content in certain mutants of EC-68415 and Morden

Type of mutant	Protein content mg/ml of homogenate
Control (EC-68415)	12.86
Malformed head mutant	13.97
Bold seed type mutant	11.20
Round seed type mutant	9.97
Flat seed type mutant	13.06
Small seed type mutant	13.32
Control (Morden)	13.82
Bisected head mutant	15.50
Bold seed type mutant	12.70
Round seed type mutant	10.35
Striped hull type mutant	13.97
Small seed type mutant	14.61

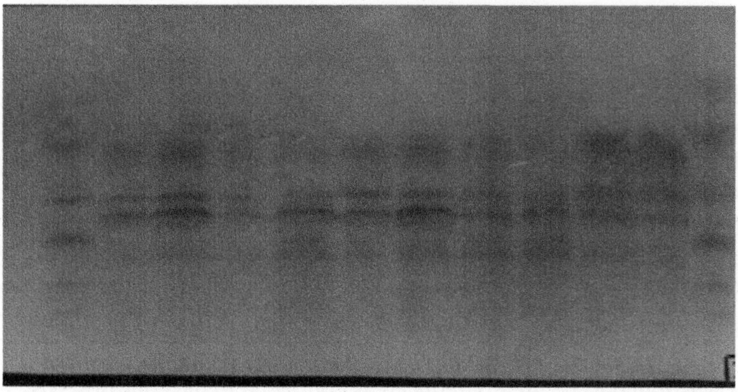

Fig. 42.3 Seed protein profile pattern on PAGE Gel—control and mutants of EC-68415 and Morden

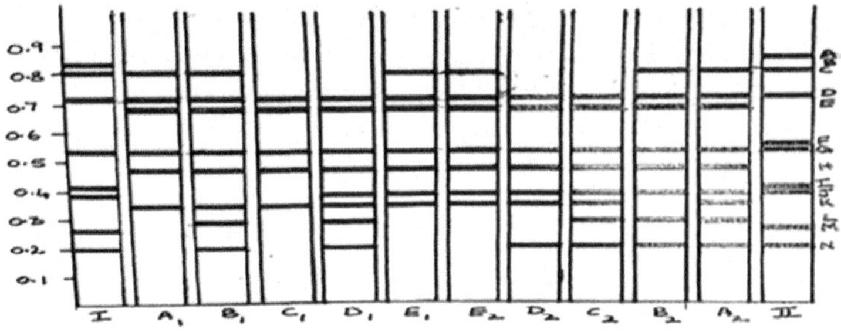

Fig. 42.4 Electrophorogram of seed protein—control and mutants of EC-68415 and Morden, I = EC-68415 control, A_1–E_1 = mutants of EC-68415, A_1 = malformed head mutant, B_1 = flat seed type mutant, C_1 = bold seed type mutant, D_1 = round seed type mutant, E_1 = small seed type mutant, II = Morden control, A_2–E_2 = mutants of Morden, A_2 = bisected head mutant, B_2 = striped hull mutant, C_2 = bold seed type mutant, D_2 = round seed type mutant, E_2 = small seed type mutant

The electrophoretic seed protein profiles of the representative mutants and controls have been outlined in the form of electrophorogram (Fig. 42.4). Altogether 14 protein bands have been identified and the R_f values ranged from 0.42 to 0.85 indicating that the bands were present in all the regions of mobility, i.e. from slow to fast migrating bands (0.01–0.2 slow; 0.21–0.5 intermediate; 0.51–0.8 fast and 0.81–1.0 very fast mobility).

Figure 42.4 indicated that the total bands for any mutant type or variety were limited to 5–9 bands. The total bands present in variety Morden were 9 and those in variety EC-68415 were 8. The lowest number of 5 bands was found in the bold seed type mutant of EC-68415. Two mutant types, viz., the bisected head and striped hull type mutants of Morden were observed to have 9 bands of similar mobility. The bands of D and G type were observed to be present in all the 12 samples studied. However, the bands of E, H and K types were observed to be present only in the mutant types of the two varieties.

42.3.5 Quantitative and Qualitative Estimation of Oil

42.3.5.1 Determination of Seed Oil

The estimation of the oil in the different mutant seeds and their control seeds was done by using the NMR analyser. The results for the different mutants of both EC-68415 and Morden are presented in Table 42.4. Quite a variation in the oil content of the mutants was observed in both the cultivar types.

The oil content in the control seeds of EC-68415 and Morden was 41.95% and 38.81%, respectively. An increase of 0.88% and 1.17% of oil was recorded in the bold seed type and round seed type mutants, respectively, of EC-68415. The oil content in the malformed head mutant, flat seed type mutant and small seed type mutant was found to be greatly reduced as compared to the control seeds. The bold seed type and the round seed type mutants of Morden were also observed to exhibit an increase in the oil content by 1.11% and 1.69%, respectively. The bisected head

Table 42.4 Percentage oil content in certain seed type mutants in EC-68415 and Morden

Type of mutant	Oil content (%)	Difference from control (%)
Control (EC-68415)	41.95	–
Malformed head mutant	23.20	−18.75
Bold seed type mutant	42.83	0.88
Round seed type mutant	43.12	1.17
Flat seed type mutant	29.18	−12.77
Small seed type mutant	24.35	−17.60
Control (Morden)	38.81	–
Bisected head mutant	20.00	−18.81
Bold seed type mutant	39.92	1.11
Round seed type mutant	40.50	1.69
Striped hull type mutant	31.82	−6.99
Small seed type mutant	22.52	−16.29

Table 42.5 Fatty acid profile of certain mutant seeds of sunflower (EC-68415 and Morden)

Type of mutant	16:0 Palmitic	18:1 Oleic	18:2 Linoleic	Others
Control (EC-68415)	7.4	28.0	63.5	1.1
Malformed head mutant	4.8	54.2	39.3	1.7
Bold seed type mutant	4.8	24.6	69.6	1.0
Round seed type mutant	4.8	18.8	75.4	1.0
Flat seed type mutant	4.8	57.3	36.1	1.8
Small seed type mutant	3.9	55.1	39.2	1.8
Control (Morden)	7.3	30.9	60.2	1.6
Bisected head mutant	4.5	53.3	40.3	1.9
Bold seed type mutant	4.8	26.9	67.2	1.1
Round seed type mutant	4.8	19.5	74.7	1.0
Striped hull type mutant	4.6	51.4	42.2	1.8
Small seed type mutant	4.1	56.4	37.7	1.8

mutant, striped hull type mutant and the small seed type mutants recorded a very low percentage of oil.

42.3.5.2 Analysis of Fatty Acid Profile

The oil extracted from the different mutant type seeds, of the two cultivars, was analysed for the different fatty acid composition to determine the quality of oil in the individual sample and the results are presented in Table 42.5. The different fatty acids analysed in the cultivars and mutants of sunflower were palmitic, oleic and linoleic acids. Traces of myristic, stearic, arachidic and lignoceric acids were also recorded which were grouped together as 'others' in Table 42.5.

The percentage of each fatty acid recorded in the control seeds of EC-68415 were 7.4% palmitic, 28.0% oleic and 63.5% linoleic acids. The total percentage of other fatty acids was 1.1%. The percentage of palmitic acid in the five mutant types of EC-68415 was observed to be very low as compared to the control. The percentage of oleic acid was observed to be higher than control in all the mutants except in round seed type mutants. The highest quantity of oleic acid was observed in the flat seed type mutant. The percentage of linoleic acid in the mutants in general was lower than the control except in the round seed and bold seed type mutants. The lowest value was observed in the flat seed type mutant. The cumulative percentage of the other fatty acids seemed to be higher than control in malformed head mutant, flat seed type mutant and small seed type mutant, whereas in bold and round seed type mutants it was lower than the control value.

The fatty acid composition in Morden was 7.3%, 30.9%, 60.2% and 1.6% of palmitic, oleic, linoleic and the others in that order. A fall in the percentage of palmitic acid in the mutants of Morden too was observed, the lowest value being 4.1% in the small seed type mutant. The oleic acid, on the other hand, was higher in all the mutants except in the round seed type and bold seed type mutants. The highest percentage of oleic acid was recorded in the small seed type mutant. The linoleic acid was observed to be lower than control in all the mutants except the round and bold seed type mutants. The lowest value was recorded in the small seed type mutant. The bold and the round seed type mutants were the only two which showed lower percentages of the other fatty acids than the control.

From the observations recorded in Table 42.5, it was noted that when there was an increase in percentage of oleic acid, in any mutant, a corresponding decrease was observed in the percentage of linoleic acid or vice versa.

42.4 Discussion

Mutants with altered seed morphology have been isolated in the course of the present investigation. Such seed type mutants have been reported earlier by a few workers, Rai (1957), Chary Narasimha (1983) and Zareen (1991).

42.4.1 Biochemical Studies

Plant breeders are of the opinion that for the production of improved varieties, improvement of quality, especially protein and oil, is necessary apart from improvement of yield. However, several environmental factors seem to influence the qualitative characters. For example, irrigation, nitrogen and phosphorus (Chaniara et al. 1989; Shinde et al. 1990), seeding dates (Patil et al. 1989), or climatic conditions (Ishida et al. 1989; Lajara et al. 1990) in sunflower are some of the factors.

42.4.2 Estimation of Leaf Protein

The protein content in the present study was observed to increase in relation to increase in peroxidase activity in certain mutants. Results to this effect were reported by Narahari and Bhatia (1975), George (1982), Khadeer (1989) and Zareen (1991).

42.4.3 Estimation and Analysis of Isoenzymes

The remarkable specificity of the isoenzyme pattern at the developmental and differentiational stage that is characteristic of each tissue implies the significant physiological role for their different molecular forms (Markert and Muller 1959; Scandalios 1974). Isoenzymes provide a natural built-in marker system for biochemical, developmental and genetic studies. An isoenzyme is thought to perform a particular physiological action at a particular phase of development, although the exact physiological role of peroxidases in plants is still obscure, due to the multiplicity of its functions. The activity of peroxidase was often correlated with growth, development and hormonal activity (Jermyn and Thomas 1954; Galston and McCune 1967; Shannon 1968). Peroxidase activity is often related to the biosynthesis of lignin (Harkin and Obst 1973), respiration (Nicholls 1965), inactivation of proteins (Sizer 1953), degradation of anthocyanin (Grommeck and Markakis 1964) and metabolism of hormones (Birecka and Galstone 1970). The enhanced peroxidase activity observed in some of the mutants, in the present study, might be a secondary consequence since the primary effect is on the gene (Nelson and Burr 1973). Quantitative differences in the isoenzymes, in the present study, were found to be more pronounced than the qualitative differences (McCune 1961; Schertz et al. 1971). High peroxidase content is a symptom of an overall increase in dissimilative metabolism of the plant. Several investigations have shown that peroxidase activity in dwarf plants was in general higher than the peroxidase activity in the tall plants (Overbeek 1935; Kamerbeek 1956; McCune and Galstone 1959). The differential activity of the isoenzymes in the present study revealed the differential function of the genes during development and differentiation (Shannon 1968). Peroxidases in plants were known to be regulated by the growth hormone indole acetic acid, by repression of certain isoenzyme forms.

Morphological characters have several disadvantages when used as markers in plant genetic studies or breeding schemes. Isoenzymes have been widely used to screen the variants present in plant populations produced through breeding and to select desirable genotypes from them. The findings in several mutation studies of altered morphological characters such as growth and differentiation were supposed to be due to interaction between peroxidase and plant hormones resulting in abnormal peroxidase activity (Gupta and Stebbins 1969; Balasimha et al. 1978).

42.4.4 Determination of Seed Protein Content and Profile

The protein content, in the present study, varied in the different mutant types. Genetic manipulation for the improvement of protein content and composition was suggested by Gottschalk and Muller (1970) while mutation breeding for the improvement of protein qualities was suggested by Narahari and Bhatia (1975). Nelson (1968) opined that though mutants with high protein quality could be induced, the yield ability of those mutants would be low. Dumanovic et al. (1970) reported chemical mutagens like EI and EMS as more effective in inducing variability in protein than the physical mutagens.

Although many techniques have been used in attempts to characterize the protein components present in the seeds, polyacrylamide gel electrophsoresis (PAGE) has become the method of choice in recent years. It has often demonstrated superior capability for component resolution (Savoy 1977). Though both the proteins and enzymes are equally important considerations in their application to biochemical taxonomy, the former has an upper hand over the latter (McDaniel 1970).

A unique pattern of protein fraction is probably consequence of specific gene arrangement, structure and activity in different mutants. Duke and Glassman (1968), while working on Drosophila, proposed that electrophoretic mobility of isozymes tended to be reduced as the species became advanced. Although the present study does not confine to any structurally or functionally definite protein bands, the gross fraction suggests that the mutants differ considerably in their mobility pattern. Both the control varieties possess more fast mobility protein fractions, thus strengthening the view that they are of wild nature, while slow mobility pattern, observed in some of the mutants, indicates that they are more advanced or derived ones.

42.4.5 Oil Estimation and Fatty Acid Profile

According to Robbelen (1980), the major requirement of economic success in case of oil seeds is the quantity and quality of the edible oil. Oil quantity can be increased in two ways: (1) by decreasing the proportion of the hull or seed coat and (2) by increasing the oil content of the embryo. Rubis (1967), Urie and Zimmer (1970) and Urie (1981) observed an increase in oil content in the seeds with reduced hull. As expected in the present study, the oil content was observed to be greatly reduced in the seeds of malformed and bisected head type mutants of EC-68415 and Morden, respectively. In the seeds of these two head type mutants, the seed coat was observed to be quite thick enveloping a tiny embryo. However, the round seed type mutants of both the varieties possessed thin seed coats and were observed to contain higher oil content than the control. This fact supports the observation of Downey et al. (1975), who reported higher oil content in flax seeds with thinner seed coats. Bold seed type mutants observed in the two cultivars recorded comparatively higher oil content than the controls even though they were observed to possess slightly thick seed coat. This was possibly because of the mutagenic induction of increase in the oil content of the embryo.

The round and bold seed type mutants in both the varieties were observed to contain higher amounts of linoleic acid and lower amounts of oleic acid when compared to respective controls. However, high amount of oleic acid and low amount of linoleic acid was observed in flat seed type and small seed type mutants in case of EC-68415 and Morden, respectively.

Fatty acid composition is the main determinant of oil quality. Final ideal sunflower oil fatty acid composition depends on its end use (Rondanini et al. 2003). High oleic sunflower hybrids have been developed through conventional breeding with oils that approach up to 89% of oleic acid percentage (Dorrel and Vick 1997) as compared to traditional hybrids containing less than 50% of oleic acid (Sadras and Villalobos 1996). High oleic hybrids have been released to the seed market since 1996 to an industry demand of seed oil with high stability and nutritional value.

Evidences from different crop plants indicate that oleic acid is the precursor of linoleic acid (Canvin 1965; Dutton and Mounts 1966; McMohan and Stumpf 1964; McMahon and Stumpf 1966). Oleic acid was suggested as a reservoir fatty acid as it responds to the changes in stearic and linoleic acid contents. Therefore, it was interesting to note that the mutants exhibiting increased linoleic acid had relatively less oleic and palmitic acid than their parents. A combination of 'St' and 'Ol' alleles affect the proportion of oleic and linoleic acids (Ladd and Knowles 1970).

A new sunflower high oleic mutation, identified as the mutant 29,066 (NM1), has been recently described by Leon et al. (2013). It involves in a gene insertion of 4872 bp at nucleotide position 201 (Zambelli et al. 2015). The consequence of the insertion was the generation of a premature stop codon producing a truncated oleate desaturase protein (ODS), accounting for the accumulation of oleic acid in grain oil at the expense of linoleic acid. Due to the nature of the mutation, a low response to MNT could be expected in genotypes carrying this mutation. On the other hand, in varieties carrying the mutation Pervenets, the high oleic acid percentage is due to a duplication affecting the 3′ end of oleoyl-phosphatidyl choline desaturase-1 (FAD2-1) gene that produces a transcription silencing and therefore an increase of oleic acid percentage when compared to traditional hybrids (Lacombe et al. 2009).

Constanza Alberio et al. (2016) studied the effect of temperature on fatty acid composition of the new high oleic mutant compared to traditional and Pervenet genotypes and found it was unaffected by the temperature variations at grain filling with 91.3–92.5% oleic acid content.

Thus, the results in the present investigation of various induced mutants and parent varieties prove the fact that a genotypic difference exists for both the protein and oil contents and their composition. It is also obvious that variability for various biochemical aspects can be induced through different chemical mutagenic treatments. The present results support the view that mutagenesis is one of the best methods for simultaneous improvement of seed yield and oil content would be through mutagenesis. New cultivars can further be developed from the high oil and oleic content seed mutants isolated in the present study.

References

Alberio C, Izquierdo NG, Galella T, Zuil S, Reid R, Zambelli A, Aguirrezabal LAN (2016) A new sunflower high oleic mutant confers stable oil grain fatty acid composition across environments. Eur J Agron 73:25–33

Alexander DE, Silvela SL, Collins FI, Rodger SRC (1967) Analysis of oil contents of maize by wild line NMR. J Am Oil Chem Soc 44:555–558

Balasimha D, Ram L, Tewari MN (1978) Plant. Biochem J 5:77–82

Barber JM (1980) Catalase and peroxidase in primary bean leaves during development and senescence. Z Planzen Physiol 97(2):135–144

Birecka H, Galstone AW (1970) Peroxidase ontogeny in a dwarf pea stem as affected by gibberellin and decapitation. J Expt Bot 21:735–745

Canvin DT (1965) The effect of temperature on oil content and fatty acid composition of the oils from several oil seeds crops. Can J Bot 43(1):63–69

Chaniara NJ, Patel JC, Malavia DD, Baldha NM (1989) Effect of irrigation, nitrogen and phosphorus on the productivity of sunflower. Indian J Agron 34(4):399–401

Chary Narasimha S (1983) Mutagenetic studies in pigeonpea (Cajanus Cajan L. Millsp). PhD thesis, Osmania University, Hyderabad

Davis BJ (1964) Disc electrophoresis - ii method and application to human serum proteins. Ann N Y Acad Sci 21(2):404–427

Directorate of Economics and Statistics (2015–16) DAC & FW: Table 4.20 (b): Sunflower—area, production and yield during 2014–15 and 2015–16 in major producing states along with coverage under irrigation

Dorrel DG, Vick BA (1997) Properties and processing of oilseed sunflower. In: Schneiter AA (ed) Sunflower technology and production. ASA, CSSAY, SSSA, Madison, WI, pp 709–746

Downey RK, Stringam GR, McGregor DI, Stefanson SR (1975) Breeding rape seed and mustard crops. In: Harapicked JT (ed) Oilseeds and pulse crops in western Canada. W.C.F. Ltd., Calgary, pp 157–187

Duke EJ, Glassman E (1968) Evaluation of xanthine dehydrogenase in Drosophila. Genetica 58:101–112

Dumanovic J, Ehrenberg L, Denic M (1970) Induced variation of protein content and composition in hexaploid wheat in improving plant protein by nuclear techniques. Proc Sci STI/PUB/258. IAEA

Dutton HJ, Mounts TL (1966) Desaturation of fatty acids in seeds of higher plants. J Lipid Res 7:221–225

Fishbein L, Flamm WG, Falk HL (1970) Chemical mutagens. Academic Press, New York

Galston AW, McCune DC (1967) Plant growth regulation. Iowa State College Press, Ames, IA, p 611

George A (1982) Studies on mutagenisation in two varieties of Brassica. PhD thesis, Osmnia University, Hyderabad

Gottschalk W, Muller H (1970) Monogenic alteration of seed protein content and protein pattern in X-ray induced pisum mutants from improving plant protein by nuclear techniques. IAEA, Vienna, pp 201–215

Grommeck R, Markakis P (1964) The effect of peroxidase on anthocyanin pigments. J Food Sci 29:53

Gupta V, Stebbins GL (1969) Perioxidase activity in hooded and awned barley at successive stages of development. Biochem Genet 3:15–24

Harkin JM, Obst JR (1973) Lignification in trees: indication of exclusive peroxidase participation. Science 180:296–298

Heiser CB Jr (1976) Sunflowers: *Helianthus* (Compositae-Heliantheae). In: Simmonds NW (ed) Evolution of crop plants. Longman, London

Ishida K, Ujihira Y, Hiramatsu T (1989) The introduction and cultivation of sunflower for oil production in the warm region. Agric Hort 64(1):57–62

Jermyn MA, Thomas R (1954) Multiple components in worse-radish peroxidase. Biochem J 56:631

Kamerbeek CA (1956) Peroxidase content of dwarf and giant types of plants. Acta Bot Neerl 5:257–260

Kar M, Mishra D (1976) Catalase, peroxidase and polyphenal-oxidase activities during rice leaf senescence. Plant Physiol 5:315–319

Khadeer MA (1989) Mutational studies in the improvement of safflower (Carthamus tinctorius L.). PhD thesis, Osmania University, Hyderabad

Lacombe S, Souyris I, Berville AI (2009) An insertion of oleate desaturase homologous sequence silences via siRNA the functional gene leading to high oleic acid content in sunflower seed oil. Mol Gen Genomics 281:43–45

Ladd SL, Knowles PF (1970) Inheritance of stearic acid in the seed oil of safflower (Carthamus tinctorius L.). Crop Sci 10(5):525

Laemmli UK (1970) Cleavage of structural proteins during the assembly of the head of bacteriophage T4. Nature 227(5259):680–685

Lajara FR, Diaz U, Quidiello RD (1990) Definite influence of location and climatic conditions on the fatty acid composition of sunflower seed oil. J Am Oil Chem Soc 67(10):618–623

Leon AI, Zambelli AD, Reid RJ, Morata MM, Kaspar M (2013) Nucleotide sequences mutated by insertion that encode a truncated oleate desaturase protein, proteins, methods and uses. Patent WO 2013/004281A1, 10 Jan 2013

Lowry OH, Rosebrough NJ, Farr AL, Randall RJ (1951) Protein measurements with Folin-Phenol reagent. J Biol Chem 193:265–275

Markert CL, Muller F (1959) Multiple forms of enzymes: tissue ontogenic and species specific patterns. Proc Natl Acad Sci U S A 45:753–763

McCune DC (1961) Multiple peroxidases in corn. Ann N Y Acad Sci 94:723–729

McCune DC, Galstone AW (1959) Inverse effect of gibberellin on peroxidase activity and growth in dwarf strains of peas and corn. Plant Physiol 34:416–418

McDaniel RG (1970) Eletrophoretic characterization of proteins in Hordeum. J Hered 61:243–247

McMahon V, Stumpf PK (1966) Fat metabolism in higher plants XXVI. biosynthesis of fatty acids in tissues of developing seeds and germinating seedlings of safflower (L.). Plant Physiol 41(1):148–156

McMohan V, Stumpf PK (1964) Synthesis of linoleic acid by particulate system from safflower seeds. Biochem Biophys Acta 84:359–361

Narahari P, Bhatia CR (1975) Induced mutations for increasing protein content in wheat and rice. Breeding for seed protein improvement using nuclear techniques. IAEA. P1.570/3, pp 23–29

Nelson AO (1968) The waxy locus in maize. The location of controlling element alleles. Genetica 60:537

Nelson OE, Burr B (1973) Biochemical genetics of higher plants. Ann Rev Plant Physiol 24:493

Nicholls P (1965) Oxidation and peroxidation. J Gen Physiol 49:131

Overbeek J (1935) The growth hormone and the dwarf type in corn. Proc Natl Acad Sci U S A 21:292–299

Patil SD, Pol PS, Shinde SH, Umrani NK (1989) Seed yield and quality of some cultivars of summer sunflower at different seeding dates. Indian J Agron 34(4):430–431

Rai UK (1957) X-ray induced "Appressed pod mutant" in Brassica juncea. Sci Cult 24:45–47

Robbelen G (1980) Proceedings of advisory group meeting. FAO/IAEA, Vienna, 1982, pp 103–117

Rondanini D, Savin R, Hall A (2003) Dynamics of fruit growth and oil quality of sunflower (*Helianthus annuus* L.) exposed to brief intervals of high temperature during grain filling. Field Crop Res 83:79–90

Rubis DD (1967) Genetics of safflower seed characters related to utilization. In: Safflower utilization conference. USDA Agricultural Research Service Publ. 74-93, Albany, CA, pp 23–28

Sadras V, Villalobos F (1996) Physiological characteristics related to yield improvement in sunflower (*Helianthus annuus* L.). In: Slafer G (ed) Genetic improvement of field crops. Marcel Dekker, New York, pp 287–320

Savoy CF (1977) Electrophorogram pattern similarities of seed proteins from 21 different soybean (Glycine max) varieties. Can J Bot 55:2245–2250

Scandalios JG (1974) Isoenzymes in development and differentiation. Annu Rev Plant Physiol 25:225–258

Schertz KF, Sumpter NA, Sakissian IV, Hart GE (1971) Peroxidase regulation by 3-dwarf height locus in sorghum. J Hered 62:235–238

Shannon LM (1968) Plant isoenzymes. Plant Physiol 19:287–310

Shinde SH, Pawan AD, Pol PS (1990) Growth yield and quality parameters of summer sunflower cultivars as influenced by irrigation levels. J Maharashtra Agric Univ 15(1):98–100

Sizer IW (1953) In: Nord FF (ed) Advances in enzymology. Interscience Publishers, New York

Toker C, Yadav SS, Solanki LS (2007) In: Yadav SS et al (eds) Lentil: an ancient crop for modern times, pp 209–224

Urie AL (1981) Abstract: First international safflower conference, Davis, CA

Urie AL, Zimmer DE (1970) Yield reduction in safflower hybrids caused by female selfs. Crop Sci 10:419–422

Zambelli A, Leon A, Garces R (2015) Mutagenesis in sunflower. In: Marinez Force E, Dunford NT, Salas JJ (eds) Sunflower oilseed. Chemistry, production, processing and utilization, AOCS monograph series on oilseeds. AOCS Press, Urbana, IL, p 7

Zareen F (1991) Studies on experimental mutagenisis in Brassica campestris L. Var sorson Praim. PhD thesis, Osmania University, Hyderabad

Mutagenic Effectiveness and Efficiency of Gamma Rays in Musk Okra (*Abelmoschus moschatus* L.)

43

S. Suneetha, B. Tanuja Priya, Syed Sadarunnisa, M. Reddi Sekhar, and P. Latha

Abstract

Musk mallow (*Abelmoschus moschatus* L.) is a medicinal and aromatic herb also known as musk ambrette, and cultivated for its seeds, which have a characteristic musk-like odour. The seeds are the source of ambrette, an aromatic oil used in perfumery. The seeds are valued medicinally for their diuretic, demulcent and stomachic properties. Gamma radiation (GR) is the most widely used physical mutagen in crop improvement. Seeds of musk okra (*Abelmoschus moschatus*) were exposed to different gamma rays doses, viz. 100, 200, 300, 400, 500 and 600 Gy using ^{60}Co as gamma source; untreated seeds were used as control. Plant morphological variations were recorded in field conditions, i.e. plant height, number of branches, stem girth, stem colour, leaf and petiole length, width and colour and yield parameters. Highest number of branches (15.61), stem girth (25.23 mm) and plant height (105.58 cm) were recorded with seeds irradiated at 100 Gy. Seeds on exposure to gamma irradiation at 300 Gy produced plants with dark red stem and green leaf with purple coloured petiole Reduction in leaf petiole size (0.367 cm) was evidenced in plants evolved from seeds treated with 100 Gy. Irradiation had significant effect on flowering when treated with 100 Gy, which recorded minimum number of days (61.67 days) for 50% flowering, while the maximum number of days (70 days) was recorded in seeds irradiated with 600 Gy. Flowers with seven and six petals were observed in

S. Suneetha (✉) · B. Tanuja Priya · S. Sadarunnisa
Department of Plantation, Spices, Medicinal and Aromatic Crops, College of Horticulture (Dr. Y.S.R. Horticultural University), Anantharajupeta, Andhra Pradesh, India

M. Reddi Sekhar
Department of Genetics and Plant Breeding, SV Agriculture College, Tirupathi, Andhra Pradesh, India

P. Latha
Department of Crop Physiology, RARS, Tirupathi, Andhra Pradesh, India

S. M. Khasim et al. (eds.), *Medicinal Plants: Biodiversity, Sustainable Utilization and Conservation*, https://doi.org/10.1007/978-981-15-1636-8_43

100 Gy, while the remaining treatments (0, 300, 400, 500, 600 Gy) recorded six and five petals. Maximum pod yield (480.12) and seed yield (499.06) were recorded in 100 Gy treatment.

Keywords
Gamma rays · Morphological characters · Ambrette

43.1 Introduction

The use of gamma radiation to induce mutation is a method that has been applied in plant breeding to increase genetic variations. Gamma radiation can be useful for the alteration of physiological characters (Kiong et al., 2008). It is a powerful and effective tool in the hands of plant breeders especially for autogamous crops having narrow genetic base. Ambrette (*Abelmoschus moschatus* L.) is a widely grown medicinal and aromatic herb of family Malvaceae and native to India. The seed of this plant yields a volatile oil known as musk seed oil or ambrette seed oil. Medicinally, muskdana is used as a stimulant in the treatment of several diseases, which is in high demand at national, international and pharmaceutical industries level due to its antioxidant, antimicrobial, anti-proliferative, anti-cancerous, anti-depression, antiseptic, antispasmodic and antiemetic properties documented by Verma et al. (2011). To increase the production of this crop, there is a need to have a better understanding of its genetic background. Further, mutation paves a way to widen the genetic base and enables the plant breeder to select a mutant with high yields. Hence, the present investigation was undertaken to study the mutagenic effectiveness and efficiency with different dose treatments on morphological variations.

43.2 Material and Methods

Healthy seeds of musk okra were selected and treated with different doses of gammaravs viz., 100, 200, 300, 400, 500 and 600 Grays and untreated seeds were maintained as control. The treated seeds were observed for various growth and yield parameters. Irradiation treatment was performed at Indian Institute of Horticultural Research, Bengaluru, Karnataka, India. The samples were irradiated in Gamma Chamber 5000 at a dose rate of 9 kGy/h Co6o as a source of gamma rays.

43.2.1 Plant Height

As the growth stages advanced, the increase in plant height was observed (Table 43.1, Fig. 43.1). Significantly superior height was recorded in the treatment 100 Gy (T_1)

Table 43.1 Influence of gamma irradiation on plant height, number of branches per plant and stem girth of muskdana at 30 DAS in field condition

Treatments	Plant height (cm)	Number of branches per plant	Stem girth (mm)
T_1 (100 Gy)	7.45	3.80	4.90
T_2 (200 Gy)	6.78	3.11	5.60
T_3 (300 Gy)	6.12	3.44	5.05
T_4 (400 Gy)	6.34	2.85	4.54
T_5 (500 Gy)	6.32	3.71	5.70
T_6 (600 Gy)	6.43	2.39	4.57
T_7 (Control)	6.60	3.65	4.27
SEM±	0.11	0.30	0.24
C.D. ($P = 0.05$)	0.34	0.91	0.74

DAS days after sowing

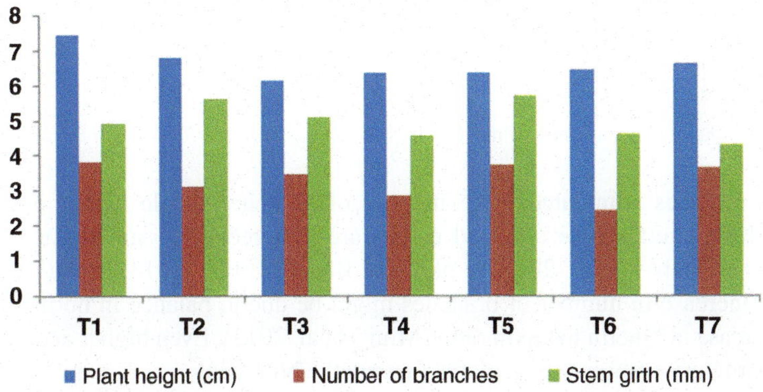

Fig. 43.1 Influence of gamma irradiation on plant height, number of branches per plant and stem girth of muskdana at 30 DAS in field condition

(7.45 cm), while the lowest plant height was recorded in 300 Gy (T_3) (6.12 cm) at 30 days after sowing. These findings are in conformity with the findings of Ashish et al. (2011) in musk okra, Bharathi et al. (2013) in *Withania somnifera*, Rahman (1999) and Al-Ansary et al. (2016) in okra.

It is also explained that plant height depends upon the nature and extent of chromosome damage. Furthermore, inactivation of auxins and a decrease in auxin content with increased irradiation dose also cause plant height reduction (Viveka et al. 2009).

43.2.2 Number of Branches per Plant

Gamma irradiation had significant influence on number of branches per plant (Fig. 43.2). More number of branches were recorded in 100 Gy (T_1) treatment (3.80), while less number was recorded in 600 Gy (T_6) treatment (2.39).

Light red stem

(100Gy, 200Gy, 400Gy, 500Gy, 600Gy and control)

Dark red stem (300Gy)

Fig. 43.2 Qualitative character of muskdana

Lower doses stimulated more number of branches, while higher doses suppressed the laterals. The obtained results are in agreement with Al-Ansary et al. (2016) and Dubey et al. (2007) in okra and Bharathi et al. (2013) in *Withania somnifera.* Increase in number of branches might be due to balance in hormonal level and increase in kinetin to auxin ratio (Minisi et al. 2013). Even higher accumulation of assimilatory compounds in plants favours the branching.

43.2.3 Stem Girth (mm)

Maximum stem girth was recorded in 500 Gy T_5 (5.70 mm) followed by 200 Gy T_2 (5.60 mm) and 300 Gy T_3 (5.05 mm) which were on par with one other. Minimum stem girth was recorded in control T_7 (4.27 mm). The results obtained abide with the results of Priyanka and Animesh (2012) in ambrette. This can be hypothesized as morphogenic changes that occur in crop plants as a protective mechanism against high levels of gamma rays.

43.2.4 Stem Colour

Dark red coloured stem was observed in 300 Gy treatment. The treatments 100, 200, 400, 500, 600 Gy and control had shown light red coloured stem. In general, muskdana possesses red pigmentation in stem due to accumulation of anthocyanin. The obtained results were similar to the results laid out by Sekyere et al. (2011) in okra.

43.2.5 Leaf Characters

43.2.5.1 Leaf Colour

Leaf colour in all gamma-treated plants exhibited broad-spectrum shades of green colour irrespective of doses. T_3 (300 Gy) treatment showed dark green coloured leaves when compared to other treatments. Similar leaf colour variation was visualized by Yadav (2016) in *Canscora decurrens*, and it was concluded that leaf colour variation may be due to adverse effect of radiation on chlorophyll synthesis process.

43.2.5.2 Leaf Length and Width (cm)

Maximum leaf length was recorded in T_5 (500 Gy) (8.71 cm) followed by T_6 (600 Gy) (8.47 cm) which was statistically on par with each other, while minimum leaf length was recorded in T_4 (400 Gy) (6.39 cm). Significantly superior leaf width was recorded in 300 Gy (10.45 cm) and minimum leaf width in 500 Gy (8.14 cm) (Table 43.2, Fig. 43.3).The above results corroborated with the findings of Priyanka and Animesh (2012) in muskdana and Maryam and Kasimu (2016) in okra.

43.2.5.3 Petiole Length and Width (cm)

Increased petiole length was recorded in 500 Gy (5.48 cm), and minimum petiole length was observed in 400 Gy (4.29 cm). Maximum petiole width was recorded in 500 Gy treatment (0.297 cm) followed by 100 Gy treatment (0.271 cm) and 200 Gy treatment (0.267 cm), which were on par with one another. Minimum petiole width was recorded at higher dosage of gamma rays, i.e. 600 Gy (0.239 cm) treatment.

This was in line with the reports of Priyanka and Animesh (2012) in muskdana. It is predicted that increase in time of exposure to gamma rays will increase the petiole length (Sharafia et al. 2013).

Therefore, leaf characters, viz. leaf colour, length and width and petiole length and width, showed significant variation with varied intensities of gamma rays. Further, it was also observed that long petiole was associated with large leaf size. The results are in line with the findings of Priyanka and Animesh (2012) in

Table 43.2 Effect of gamma irradiation on leaf and petiole length and width in muskdana at 30 DAS

Treatments	Leaf length (cm)	Leaf width (cm)	Petiole length (cm)	Petiole width (cm)
T_1 (100 Gy)	5.69	8.54	4.77	0.271
T_2 (200 Gy)	6.84	9.04	4.67	0.267
T_3 (300 Gy)	7.70	10.45	4.65	0.252
T_4 (400 Gy)	6.39	7.90	4.29	0.249
T_5 (500 Gy)	8.71	8.14	5.48	0.297
T_6 (600 Gy)	8.47	8.76	4.85	0.239
T_7 (Control)	7.17	8.89	4.72	0.259
SEM±	0.11	0.10	0.10	0.10
C.D. ($P = 0.05$)	0.34	0.31	0.31	0.30

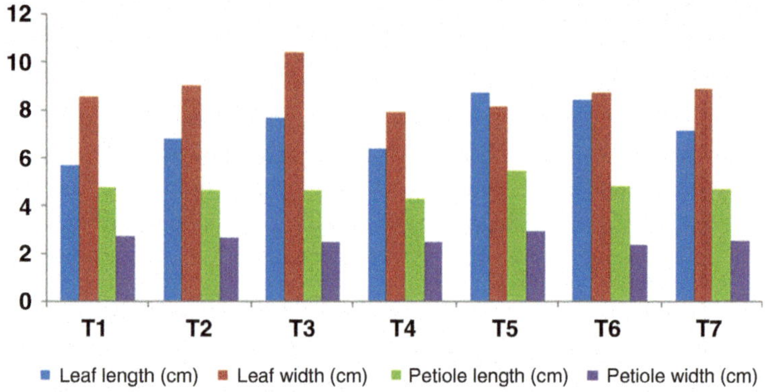

Fig. 43.3 Effect of gamma irradiation on leaf and petiole length and width in muskdana at 30 DAS

muskdana. Different doses of gamma rays produced variable effect in the leaf morphology, and promising variation was observed in 500 Gy. However in contrary, very high doses showed great reduction in leaf size (Yadav 2016). Extreme abnormalities in leaf characters suggested strong interaction and impact of mutagen on the leaf-associated genes.

43.2.5.4 Petiole Colour

Purple coloured petiole was observed in the irradiated doses, viz. 300, 400, 500 and 600 Gy, while 200, 100 Gy and control exhibited green coloured petiole. Plants exposed to lower doses of gamma rays were devoid of pigmentation. It is evident that there is no biosynthesis of anthocyanin due to inactivation of associated enzymes at lower doses (Sekyere et al. 2011).

43.2.6 Flower Characters

43.2.6.1 Number of Petals

Flower mutation varied with the dose or concentration of the mutagens. At flowering stage, the number of petals, viz. 7, 6 and 5, was observed in various treatments. Flowers with seven and six petals were recorded in 100 and 200 Gy, while six and five petals were recorded in the rest of the treatments and control. Concomitant results were presented by Borkar and More (2010) in *Phaseolus vulgaris* (Table 43.3, Fig. 43.4). The floral changes might be attributed to genetic alterations which originated due to loss of gene functions (Sridevi and Mullainathan 2012).

43.2.6.2 Days Taken for 50% Flowering

The results showed that longer duration for 50% of flowering was observed in 600 Gy (70 days), whereas flowering was found advanced in 100 Gy treatment

Table 43.3 Effect of gamma irradiation on number of petals and days taken for 50% flowering in muskdana

Treatments	Number of petals	Days taken for 50% flowering
T_1 (100 Gy)	7 and 6	61.67
T_2 (200 Gy)	7 and 6	63.00
T_3 (300 Gy)	6 and 5	64.00
T_4 (400 Gy)	6 and 5	67.00
T_5 (500 Gy)	6 and 5	67.67
T_6 (600 Gy)	6 and 5	70.00
T_7 (Control)	6 and 5	63.33
SEM±	–	0.51
C.D. ($P = 0.05$)	–	1.57

Fig. 43.4 Effect of gamma irradiation on number of petals in muskdana

(61.67 days). The results are in agreement with those reported by Rahman (1999), Elangovan and Pavadai (2015), Jagajanantham et al. (2012) in okra and Bharathi et al. (2013) in *Withania somnifera*.

Early flowering is thought to be triggered due to stress induction and variation created at genetic level (Maryam and Kasimu 2016).

43.2.6.3 Pod Yield (Number of Pods/Plot)

The maximum number of pods was recorded in 100 Gy treatment (480.12), while the minimum number of pods was recorded in 600 Gy (136.22) (Table 43.4, Fig. 43.5). Decline in the fruit number due to increasing levels of gamma ray exposure was reported earlier by Pushparajan et al. (2014).

43.2.6.4 Seed Yield per Plot (g)

Increased seed yield with irradiation was observed when compared to non-irradiated mutants, and the seed yield per plot ranged from 591.06 to 196.92 g. Among the different treatments, 100 Gy T_1 (499.06 g) recorded maximum seed yield compared to control, while T_6 (600 Gy) treatment recorded minimum seed yield per plot (196.92 g).

Table 43.4 Effect of gamma irradiation on pod yield (number of pods/plot) and seed yield per plot (g) in muskdana

Treatments	Pod yield (number of pods/plot)	Seed yield per plot (g)
T_1 (100 Gy)	480.12	499.06
T_2 (200 Gy)	265.45	229.00
T_3 (300 Gy)	263.48	212.51
T_4 (400 Gy)	225.49	226.18
T_5 (500 Gy)	261.49	202.97
T_6 (600 Gy)	136.22	196.92
T_7 (Control)	419.09	391.42
SEM±	14.38	32.07
C.D. ($P = 0.05$)	44.32	98.80

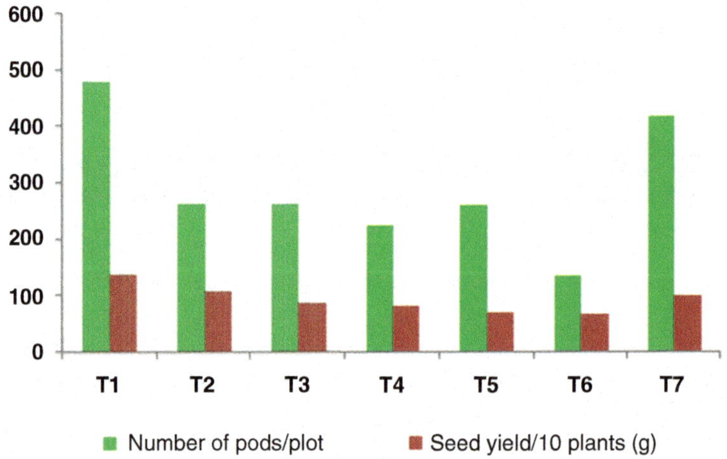

Fig. 43.5 Effect of gamma irradiation on pod yield (number of pods/plot) and seed yield per plot (g) in muskdana

43.3 Conclusion

It could be concluded that the plants with higher yield were observed under low doses of gamma rays. The useful mutant isolated through the present study needs to be tested further on a wider scale to establish any changes in chromosome or allele frequency and also to assess its performance in later generations.

References

Al-Ansary AMF, Hamied NRA, Ottai MES, Mergawi RAE (2016) Gamma irradiation effect on some morphological and chemical characters of Sudani and Masri Roselle varieties. Int J Chem Technol Res 9(3):83–96

Ashish R, Warghat N, Rampure H, Wagh P (2011) Effect of sodium azide and gamma rays treatments on percentage germination, survival, morphological variation and chlorophyll mutation in musk okra (*Abelmoschus moschatus* L.). Int J Pharm Pharm Sci 3(5):483–486

Bharathi T, Gnanamurthy S, Dhanavel D, Murugan S, Ariraman M (2013) Induced physical mutagenesis on seed germination, lethal dosage and morphological mutants of Ashwagandha (*Withania somnifera* (L.) Dunal). Int J Adv Res 1(5):136–141

Borkar AT, More AD (2010) Induced flower colour mutations in *Phaseolus vulgaris* Linn through physical and chemical mutagens. Adv Biores 1(1):22–28

Dubey AK, Yadav JR, Singh B (2007) Studies on induced mutations by gamma irradiation in okra (*Abelmoschus esculentus.* Monch.). Progress Agric 7(1 and 2):46–48

Elangovan R, Pavadai P (2015) Effect of gamma rays on germination, morphological and yield characters of bhendi (*Abelmoschus esculentus* L. Moench). Hort Biotechnol Res 1:35–38

Jagajanantham N, Dhanavel D, Pavadai P, Chidambaram AA (2012) Growth and yield parameters using gamma rays in bhendi (*Abelmoschus esculentus* (L.) Moench) var. arka anamika. Int J Res Plant Sci 2(4):56–58

Kiong ALP, Lai GA, Hussein S, Harun AR (2008) Physiological responses of *Orthosiphon stamineus* plantles to gamma irradiation. American-Eurasian. J Sustain Agric 2(2):135–149

Maryam AH, Kasimu AA (2016) Effect of combined doses of gamma ray and sodium azide (mutagenic agents) on the morphological traits of some varieties of okra (*Abelmoschus esculentus*). Afr J Agric 11(32):2968–2973

Minisi FA, El-mahrouk ME, El-Din RM, Nasr FN (2013) Effects of gamma radiation on germination, growth characteristics and morphological variations of *Moluccella laevis* L. Am Eurasian J Agric Environ Sci 13(5):696–704

Priyanka DK, Animesh DK (2012) Induced mutagenesis in *Abelmoschus moschatus* (L.) Medik. Int Res J Pharm 3(5):432–435

Pushparajan G, Surendran S, Harinarayanan MK (2014) Effect of gamma rays on yield attributing characters of okra (*Abelmoschus esculentus* (L.) Moench). Int J Adv Res 2(5):535–540

Rahman EMEA (1999) The effect of gamma irradiation on growth and yield of okra (*Abelmoschus esculentus* L. moench). MSc thesis, Sudan University of Science and Technology, pp 1–103

Sekyere DO, Akromah R, Nyamah EY, Brenya E, Yeboah S (2011) Characterization of okra (*Abelmoschus spp.* L.) germplasm based on morphological characters in Ghana. J Plant Breed Crop Sci 3(13):367–378

Sharafia Y, Rezaei A, Ghanbari A, Naji AM (2013) Gamma irradiation affected phenotypic characteristics of almond (*Prunus amygdalus* L.). Int J Agron Plant Prod 4(1):25–29

Sridevi A, Mullainathan L (2012) The use of ethyl methane sulfonate to study the flower development in *Capsicum annuum* L. mutants. Bot Res Int 5(1):4–9

Verma A, Singh MK, Kumar B (2011) Phytochemical evaluation and characterization of hypoglycemic activity of various extracts of *Abelmoschus esculentus* Linn. fruit. Int J Pharm Pharm Sci 3(4):60–65

Viveka NS, Banerji BK, Dwivedi AK, Verma AK (2009) Effect of gamma irradiation on African marigold (*Tagetes erecta* L.) cv. Pusa Narangi Gainda. J Hort Sci 4(1):36–40

Yadav V (2016) Effect of gamma radiation on various growth parameters and biomass of *Canscora decurrens* Dalz. Int J Herb Med 4(5):109–115

Detection of Genetic Variation in *Biophytum sensitivum* Linn. by RAPD and ISSR Markers

44

Sirigiri Chandra Kala and Kokkanti Mallikarjuna

Abstract

Biophytum sensitivum is an important medicinal plant widely used in traditional systems of medicine with propagation limitations. In the present work, we are reporting the genetic diversity analysis of naturally occurring and in vitro grown plants by RAPD and ISSR markers. The optimal in vitro multiplication of shoots was observed on MS medium supplemented with BA (2 mg/l) and NAA (0.5 mg/l) using nodal and leaf explants and on MS medium with BA (2 mg/l) and NAA (0.1 mg/l) using leaf explants. DNA-based marker (RAPD and ISSR) analysis of plants present in five different geographical areas indicated that considerable genetic variations are present in naturally occurring plants. It is also indicated that tissue culture plants and their wild relatives show genetic similarity by grouping into one clad. The variations observed in naturally occurring plants could be due to the impact of local environmental factors and accumulation of mutations in the course of evolution. This is the first report on genetic diversity analysis.

Keywords

Biophytum sensitivum · Genetic diversity · RAPD · ISSR · In vitro

44.1 Introduction

Biophytum sensitivum common name is life plant, little tree plant, and sensitive plant in English, attapatti, chumi, jala puspa in Telugu, Lakshmana, Lajalu in Hindi. The work on *B. sensitivum* has reported various pharmacological activities

S. Chandra Kala · K. Mallikarjuna (✉)
Department of Botany and Microbiology, Acharya Nagarjuna University, Guntur, Andhra Pradesh, India

S. M. Khasim et al. (eds.), *Medicinal Plants: Biodiversity, Sustainable Utilization and Conservation*, https://doi.org/10.1007/978-981-15-1636-8_44

including protective effect on radiation-induced damage in mice (Guruvayoorappan and Kuttan 2008g), immunomodulatory activity (Guruvayoorappan and Kuttan 2008d), inhibition of tumor-specific angiogenesis (Guruvayoorappan and Kuttan 2008f), tumor cell invasion and metastasis (Guruvayoorappan and Kuttan 2008a), immunomodulatory activity (Guruvayoorappan and Kuttan 2008b), apoptosis of melanoma cells (Guruvayoorappan and Kuttan 2008c), antioxidant potential (Guruvayoorappan et al. 2006), inhibition of metastatic tumor (Guruvayoorappan and Kuttan 2008e), and anti-inflammatory activity (Jachak et al. 1994). The biological activity of the plant shows hypoglycemic (Puri and Baral 1998), immunomodulatory (Guruvayoorappan and Kuttan 2007b), chemoprotective (Guruvayoorappan and Kuttan 2007a), and hypocholesterolemic (Puri 2003) effects. Powdered dry leaves of the plant are a known traditional remedy for the treatment of madhumeha (Puri and Baral 1998). It possesses a wide spectrum of medicinal properties for curing diseases like asthma, phthisis and inflammatory diseases (Mitra and Ambasta 1988; Kirtikar and Basu 1984; Pullaiah 2002). Diabetic drink formulation from this plant is effective in treating diabetes.

Global concern about the loss of valuable genetic resources has stimulated many new programs for the conservation of plant genetic resources. Within past decade, several conservation strategies were developed. Wild life conservation is based mainly on in situ conservation. Ex situ conservation involves preservation and protection of samples of living organisms outside their natural habitat, in the form of whole plants, seed, pollen, vegetative propagates, and tissue or cell cultures. Ex situ techniques are generally used to complement in situ methods but in some cases are the only possible techniques to conserve certain species (Ramsay et al. 2000). Among ex situ conservation methods, the most common are cultivation in botanical gardens, seed storage, and in vitro cultivation. Among the various ex situ conservation methods, seed storage seems to be one of the most convenient for long-term conservation. This involves desiccation and storage at low temperatures. However, there are a large number of threatened species; hence conventional seed storage strategies are not suitable (Paunescu 2009). The loss of plant genetic resources has made necessary the development of new ex situ conservation methods. Advances in biotechnology provide new methods for plant germplasm conservation and evaluation. Biotechnological tools like in vitro culture, cryopreservation, and molecular markers offer a valuable alternative to plant diversity studies, management of genetic resources, and ultimately conservation.

Human activity-mediated unfriendly harvesting techniques, loss of proper growth habits, and uncontrolled exploitation of traditional medicinal herbs from natural ecosystems are causing genetic breach or even causing extinction of many medicinal herbs. There is an urgent need for scientific explorations and identification of natural genetic diversity of important medicinal herbs. Molecular markers play an important role and have been widely used as indicators to study genetic diversity (Joy and Maridass 2008). When compared to morphological markers, isozyme-based biochemical markers and DNA-based molecular markers avoid many of the complications of environmental effects acting upon characters by looking directly at variation controlled by gene or by looking at genetic material itself.

Molecular markers represent a powerful and a potentially rapid method for characterizing genetic diversity for in situ and ex situ conservation. Molecular markers are useful in the construction of genetic maps in breeding programs. In vitro plants are susceptible to changes due to culture stress. The usefulness of AFLP, ISSR, RFLP, RAPD, and cytological analysis in detection of variation in micropropagated plants has been amply demonstrated. Genetic fidelity of micropropagated plants has immense practical utility and commercial implication (Smita et al. 2009; Saha et al. 2012).

To commercially propagate a medicinal plant tissue, it is important to know the genetic stability of propagates produced. A major problem associated with in vitro culture is the occurrence of somaclonal variations among tissue culture plants. This variation is often heritable and is therefore unwanted in clonal propagation. Thus detection of this genetic variation is important to avoid the process of becoming disastrous (Yuan et al. 2009). Among the several markers, RAPD and ISSR have been mostly favored because of their sensitivity, simplicity, and cost-effectiveness. Both RAPD and ISSR markers have been successfully applied to detect genetic similarities or differences in tissue-cultured plants compared to its wild progenitors (Thomas et al. 2006). The use of two types of markers, which amplify different regions of the genome with multiple amplicons, allows better analysis of genetic stability/variation of tissue culture-generated plantlets (Yuan et al. 2009).

44.2 Experimental Design and Data Collection

44.2.1 Genomic DNA Isolation

Genomic DNA was isolated from leaf tissue by modified procedure of CTAB method as described by Murray and Thompson (1980). Total genomic DNA was isolated from tender leaves of the eight diverse lines that were collected from different geographical regions and two tissue culture-generated plants (Tables 44.1 and 44.2) which are maintained in botanical garden in Acharya Nagarjuna University, Guntur. After RNA*se* treatment, DNA solution was purified with the standard phenol:chloroform method. DNA solution was mixed gently with phenol:chloroform (1:1) and centrifuged at 5000 rpm for 10 min at room temperature. The aqueous phase was separated and mixed with an equal volume of chloroform, mixed gently and centrifuged at 5000 rpm for 10 min at room temperature.

The aqueous phase was separated and mixed with two volumes of absolute ethanol and incubated at −20 °C for 20 min. The DNA was spooled out with a glass hook, washed with 70% aqueous ethanol. The DNA pellet was air-dried for 20 min and dissolved in an appropriate volume of sterile TE buffer (10 mM Tris–HCl, pH 8.0, 1 mM EDTA, pH 8.0) and stored at 4 °C. The concentration of DNA was determined spectrophotometrically, and the quality of DNA was checked through

Table 44.1 Tested samples of *Biophytum sensitivum* plants for genetic diversity analysis

Line	Geographical location
1	A.N.U. Campus Guntur, mother plant for explants
2	Directly regenerated plant
3	Indirectly regenerated plant
4	Wild plant of Machilipatnam, Krishna Dt.
5	Wild plant of Tirupati, Chittoor Dt.
6	Wild plant of Karimnagar Dt.
7	Wild plant of Araku, Vizag Dt.
8	Wild plant of Y.V. University, Kadapa Dt.

Table 44.2 RAPD primers, sequence, and annealing temperatures used for diversity analysis of *Biophytum sensitivum*

S. no.	RAPD	Sequence 5′–3′	Annealing temp (°C)	Remarks
1.	OPA9	GGGTAACGCC	35	Amplified
2.	OPB5	TGCGCCCTTC	35	Amplified
3.	OPB10	CTGCTGGGAC	35	Amplified
4.	OPC2	GTGAGGCGTC	35	Amplified
5.	OPC5	GATGACCGCC	35	Amplified
6.	OPD2	GGACCCAACC	35	Amplified
7.	OPF7	CCGATATCCC	35	Amplified
8.	OPA12	TCGGCGATAG	35	Not amplified
9.	OPB6	TGCTCTGCCC	35	Not amplified
10.	OPC3	GGGGGTCTTT	35	Not amplified
11.	OPD6	ACCTGAACGG	35	Not amplified
12.	OPG19	GTCAGGGCAA	35	Not amplified
13.	OPO10	TCAGAGCGCC	35	Not amplified
14.	OPX7	GAGCGAGGCT	35	Not amplified
15.	OPV11	CTCGACAGAG	35	Not amplified

agarose gel electrophoresis. DNA is quantified using double-beam UV-visible spectrophotometer at 260 and 280 nm. The DNA samples were diluted to 10 ng/μl and further used for PCR amplification of RAPD and ISSR primers.

44.2.2 RAPD Studies

Primers obtained from MWG Biotech, Bangalore, India (Table 44.3), were used to amplify DNA of eight varieties of plants with two tissue-regenerated plants along with six other diverse lines (which include parental line from Guntur). PCR reaction was carried out in 20 μl reaction made of 40 ng of DNA as template, 1×PCR buffer, 250 μm of each dNTP, 5 pmol of primer, and 1 unit of Taq DNA polymerase. PCR

Table 44.3 ISSR primers, sequence, and annealing temperatures used for diversity analysis of *Biophytum sensitivum*

S. no.	ISSR	Sequence 5′–3′	Annealing temp (°C)	Remarks
1.	873	GAC AGA CAG ACA GAC A	50	Amplified
2.	864	ATG ATG ATG ATG ATG ATG	50	Amplified
3.	806	TAT ATA TAT ATA TAT AG	45	Not amplified
4.	811	GAG AGA GAG AGA GAG AC	45	Not amplified
5.	814	CTC TCT CTC TCT CTC TA	50	Not amplified
6.	815	CTC TCT CTC TCT CTC TG	45	Not amplified
7.	818	CAC ACA CAC ACA CAC AG	45	Not amplified
8.	822	TCT CTC TCT CTC TCT CA	45	Not amplified

reaction was set with all the above mixture program with 95 °C of initial denaturation for 5 min, followed by 40 cycles of 94 °C of denaturation for 1 min, 35 °C of annealing temperature for 1 min, 72 °C of extension for 1 min, and finally final extension at 72 °C for 10 min. The amplified product is resolved in 1.5% agarose gel stained with ethidium bromide in 0.5×TBE buffer with 100 bp Marker at one end of the gel. The RAPD bands were scored on the basis of the presence or absence of band.

44.2.3 ISSR Studies

ISSR primers that were procured from MWG Biotech, Bangalore, India (Table 44.4), were used to amplify DNA of eight varieties of *Biophytum sensitivum* selected in the study. PCR reaction was carried out in 20 μl reaction made of 40 ng of DNA as template, 1×PCR buffer, 250 μm of each dNTP, 5 pmol of primer, and 1 unit of Taq DNA pol. The PCR reaction was set with all the above cocktail mixture. The program was set with 95 °C for 5 min initial denaturation of template DNA, followed by 40 cycles of 94 °C of denaturation for 1 min, 45–50 °C of annealing temperature for 1 min, 72 °C of extension for 1 min, and final extension at 72 °C for 10 min. The amplified product is resolved in 1.5% agarose gel stained with ethidium bromide in 0.5×TBE buffer with 100 bp Marker at one end of the gel. The ISSR bands were scored on the basis of the presence or absence of band.

44.2.4 Scoring the Data and Analysis

Reproducible bands were scored in all the samples for each of the primer separately. Each amplification was considered as a separate marker. The presence of the amplified product in each variety was recorded as 1 and absence as 0, respectively. All the numerical and taxonomical analysis was conducted using the *ntsys-pc* software version 2.0, and cluster analysis of the complete data is done. Similarly matching dice coefficient values for pairwise comparison between accessions were calculated, and a dice coefficient matrix was constructed using the *simqual* subroutine. This matrix

Table 44.4 Distance matrix values based on RAPD and ISSR data (similarity table) between *Biophytum sensitivum* plants collected from different areas

	Guntur wild mother plant	Directly regenerated plant	Indirectly regenerated plant	Krishna Dt.	Chittoor Dt.	Karimnagar Dt.	Visakhapatnam Dt.	Khammam Dt.
Guntur wild mother plant	1.00							
Directly regenerated plant	0.79	1.00						
Indirectly regenerated plant	0.79	0.82	1.00					
Krishna Dt.	0.63	0.63	0.60	1.00				
Chittoor Dt.	0.82	0.82	0.82	0.57	1.00			
Karimnagar Dt.	0.74	0.76	0.68	0.57	0.82	1.00		
Visakhapatnam Dt.	0.66	0.69	0.72	0.68	0.69	0.63	1.00	
Khammam Dt.	0.81	0.63	0.69	0.62	0.69	0.63	0.62	1.00

was subjected to unweighed pairwise group method using arithmetic average analysis (*upgma*) to generate a dendrogram using *sahn* subroutine and tree plot of ntsys-pc. The similarity indices were calculated across all possible pairwise comparisons of individuals within and among the population, following the method of Nei and Li (1979). The formula was as follows: *si=2nxy/(nx+ny).*

44.3 Genetic Variation in *Biophytum sensitivum* Linn.

44.3.1 RAPD Analysis

Fifteen primers were initially screened and tissue-cultured their mother plants and wild-type plants collected from six areas for their ability to amplify polymorphic DNA. Out of them, seven primers OPA9, OPB5, OPB10, OPC2, OPC5, OPD2, and OPF7 (Table 44.2 and Figs. 44.1, 44.2, 44.3, 44.4, 44.5, 44.6, and 44.7) showed reproducible and distinct polymorphic amplified products. A total of 197 bands were scored of which 60 (30.4%) were polymorphic. The seven selected primers produced comparatively the maximum number of high-intensity bands with minimal smearing, good technical resolution, and sufficient variation among different cultivars. The highest number of bands was generated by primer OPF7 (Fig. 44.4), whereas the least number of bands was produced by primer OPB5 (Table 44.2; Fig. 44.1). In addition to that, the primers OPA9 (Table 44.2, Fig. 44.5), OPB10 (Table 44.2, Fig. 44.6), OPC2 (Table 44.2, Fig. 44.2), OPC-5 (Table 44.2, Fig. 44.7), OPD2 (Table 44.2, Fig. 44.3), and OPF7 (Table 44.2, Fig. 44.4) generated the maximum number of polymorphic bands. OPD2 primer gave a high PIC value of 0.29.

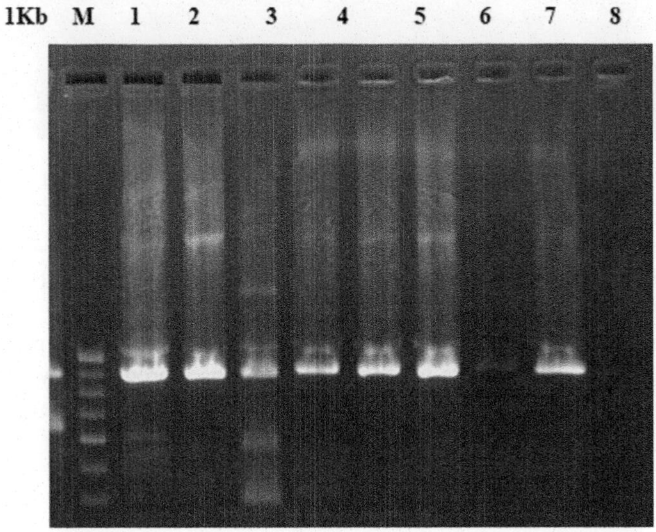

Fig. 44.1 RAPD profile of *Biophytum sensitivum* using OPB5 primer (Table 44.2)

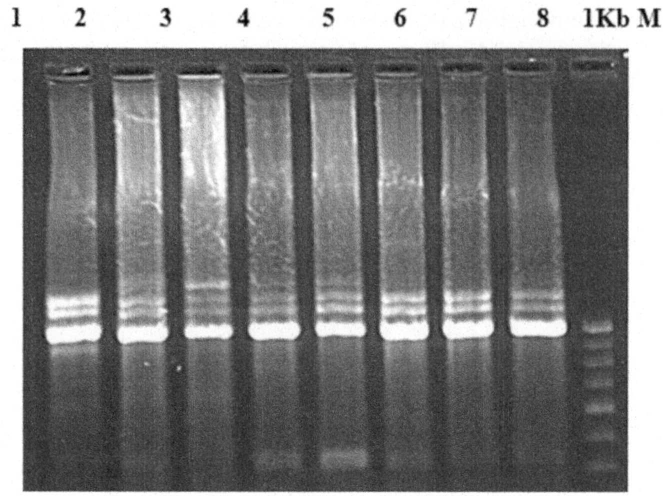

Fig. 44.2 RAPD profile of *Biophytum sensitivum* using OPC2 primer (Table 44.2)

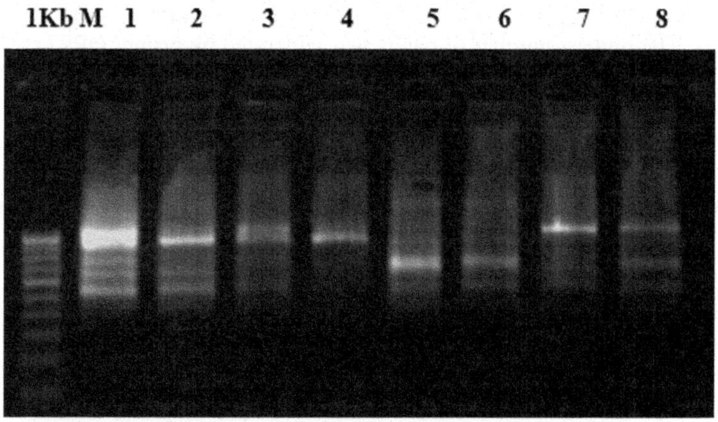

Fig. 44.3 RAPD profile of *Biophytum sensitivum* using OPD2 primer (Table 44.2)

Maximum polymorphic bands were observed in OPF7 primer (Fig. 44.4). While the primer OPB5 (Fig. 44.1) generated the least polymorphic bands. The number of amplified polymorphic bands ranged from 12 to 13. The directly and indirectly regenerated plants are showing genetic similarity with their mother plants in terms of their strong uniform bands in all RAPD gels. The reproducibility of the RAPD banding pattern was confirmed by three replicated reactions with the same primer.

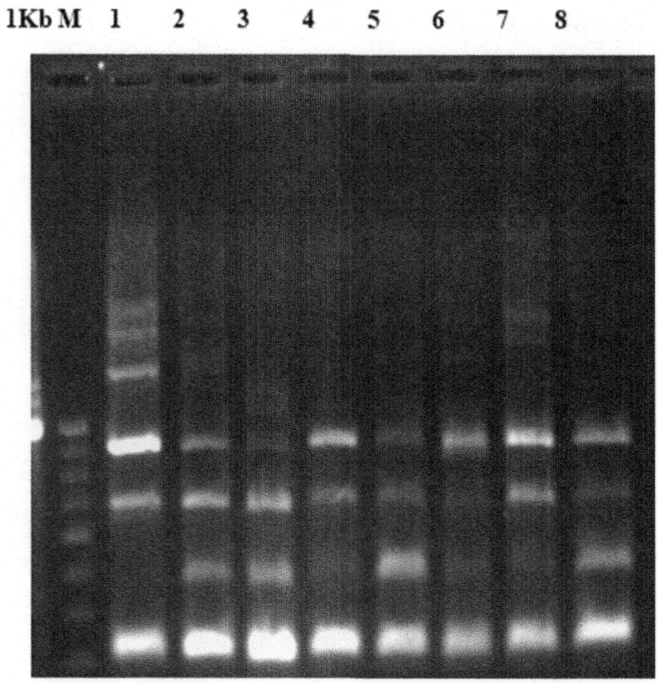

Fig. 44.4 RAPD profile of *Biophytum sensitivum* using OPF7 primer (Table 44.2)

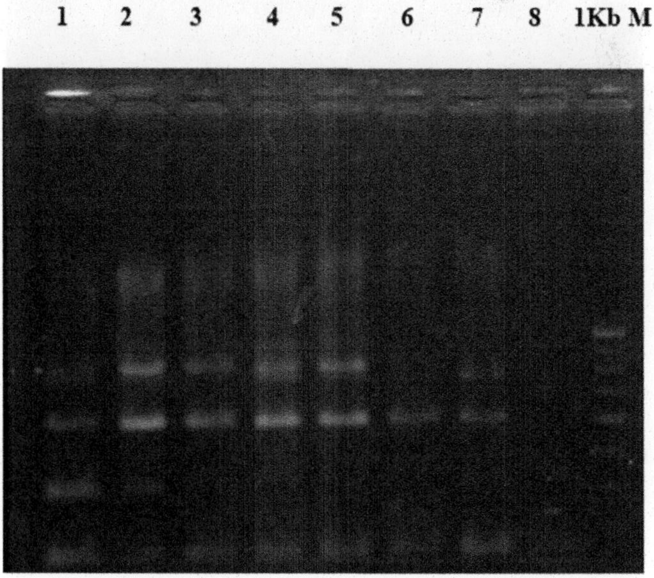

Fig. 44.5 RAPD profile of *Biophytum sensitivum* used OPA9 (c) primer (Table 44.2)

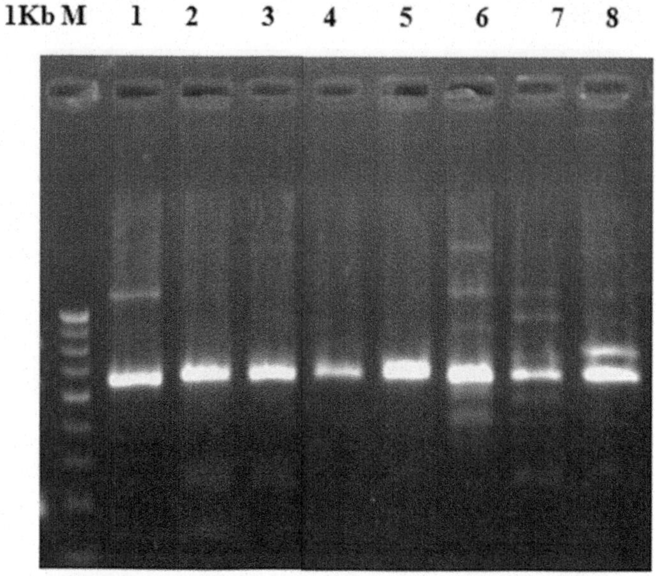

Fig. 44.6 RAPD profile of *Biophytum sensitivum* using OPB10 (b) primer (Table 44.2)

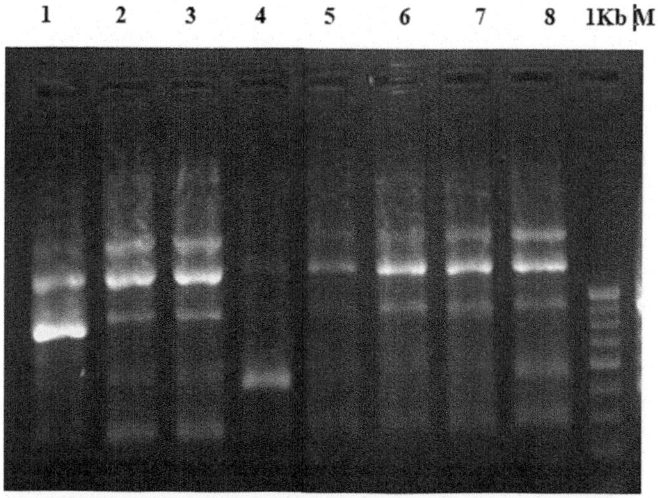

Fig. 44.7 RAPD profile of *Biophytum sensitivum* using OPC5 (a) primer (Table 44.2)

44.3.2 ISSR Analysis

The eight ISSR primers (Table 44.3) were initially screened on tissue-cultured, their parent type plant and plants collected from six areas for their ability to amplify polymorphic DNA. From out of them, two primers ISSR-873 and ISSR-864 showed reproducible and distinct polymorphic amplified products. A total of 51 bands were scored of which 18 (35.2%) were polymorphic. The two selected primers produced comparatively the maximum number of high-intensity bands with minimal smearing, good technical resolution, and sufficient variation among different cultivars. The highest numbers of bands were generated by primer ISSR-864 (Table 44.3, Fig. 44.8), whereas the least number of bands was produced by primer ISSR-873 (Table 44.3, Fig. 44.9). *B. sensitivum* plants developed through tissue culture (either directly or indirectly) and their mother plants are lacking genetic diversity measured as scorable prominent bands in ISSR gels (Lane 1, 2, 3 of Figs. 44.8 and 44.9). ISSR-864 gave a PIC value of 0.36. The number of amplified polymorphic bands remained the same (9) in both gels. The reproducibility of the ISSR banding pattern was confirmed by three replicated reactions with the same primer.

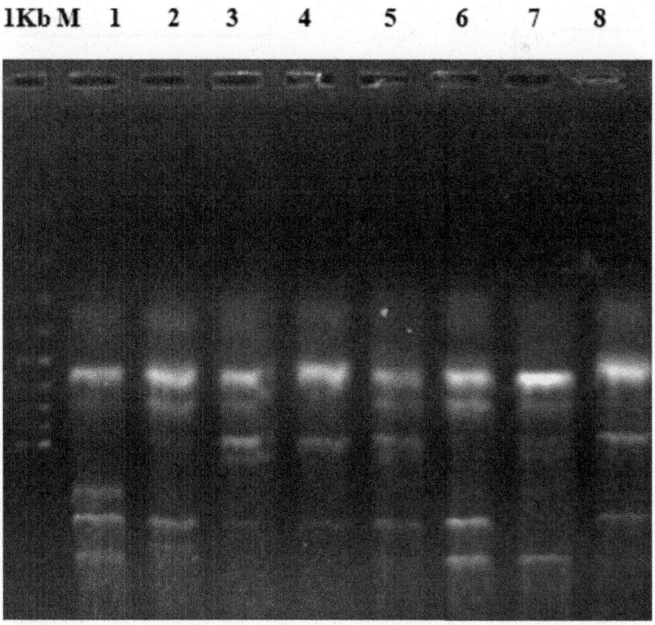

Fig. 44.8 ISSR profile of *Biophytum sensitivum* using 864 primer (Table 44.3)

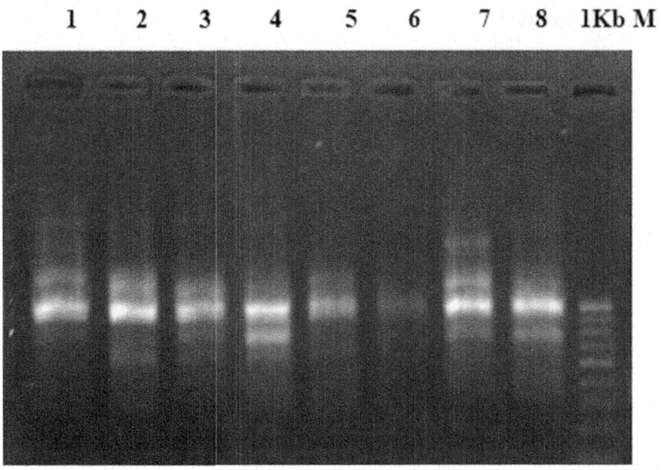

Fig. 44.9 ISSR profile of *Biophytum sensitivum* using 873 primer (Table 44.3)

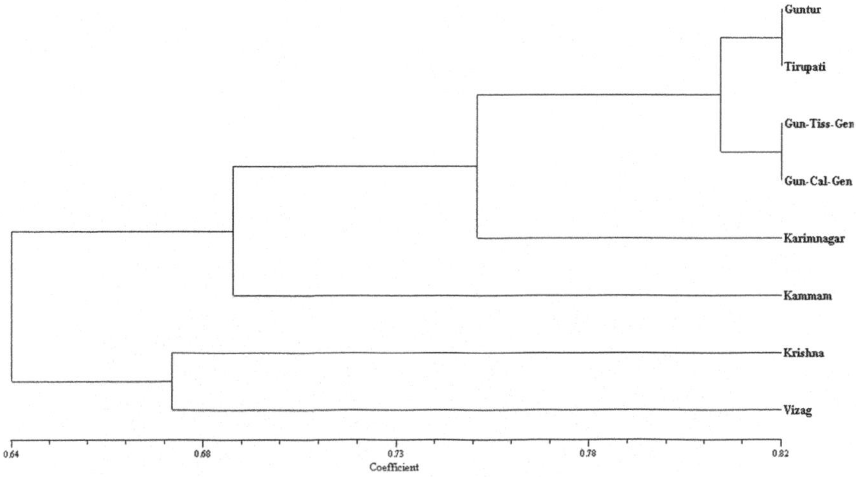

Fig. 44.10 Dendrogram derived from UPGMA cluster analysis using dice coefficient of RAPD- and ISSR-based markers in *Biophytum sensitivum*

44.3.3 Dendrogram of *B. sensitivum*

A dendrogram (Fig. 44.10) derived from UPGMA cluster analysis using coefficient of RAPD and ISSR data in *B. sensitivum* plants indicates the grouping of different types of *B. sensitivum* plants into two main clusters. Cluster one is formed by *B. sensitivum* plants of Visakhapatnam and Krishna districts, indicating that genetically they are different from other types of plants and the reason for this could be due to different climatic conditions or the evolutionary selection pressures that might have acted on mutation rates of DNAs of Visakhapatnam and Krishna district

Table 44.5 Summary of RAPD and ISSR amplified products from eight mixed samples of *Biophytum sensitivum*

Description	RAPD	ISSR
Total bands scored	54	14
No. of monomorphic bands	19	6
No. of polymorphic bands	35	8
Percentage of polymorphic	65	57
No. of primers used	15	8
Average polymorphism per primers	2.06	4.4

plants. The second main cluster is again divided into two sub-clusters, Khammam forming one sub-cluster and wild-type plants of Karimnagar, tissue-cultured plants and wild-type plants of Guntur, and Tirupati forming another sub-cluster. Interesting observation in this case is that wild-type parental plants and tissue-cultured plants are in one cluster, but tissue-cultured plants are distinctly present as one subcluster indicating that indirectly and directly regenerated plants are genetically similar and less genetic variation is there when compared to their parental plant. This data indicates that some extent of genetic variation is present among tissue-cultured plants and their wild relatives.

The same can be evidenced from (Similarity Table 44.4) distance matrix values of RAPD/ISSR data. The segregation of wild-type plants belonging into different areas into different subclusters indicates the level and extent of genetic variation (Fig. 44.10, Tables 44.4 and 44.5) present in them. Among them, the interesting observation is the prevalence of more genetic diversity in the *B. sensitivum* wild plants of Visakhapatnam and Krishna districts. Many reasons could be responsible for this. The usefulness of RAPD and ISSR markers and prevalence of genetic diversity in many plant populations have been demonstrated (Kesari et al. 2010; Yuan et al. 2009; Smita et al. 2009). Among RAPD and ISSR markers, ISSR marker detected more genetic variation (Table 44.5). This study has provided a longer number of reliable and reproducible finger printing profiles for sustainable management, linkage mapping, and genetic improvement of plants.

Tissue culture is a mutagenic technique that causes cytogenetic, genetic, and epigenetic variation (Jain and Lim 2001). There are many factors like length of culture periods, genotypes, and nature of explants that influence the stability of the tissue-cultured plants. RAPD analysis was used to find out the genetic relationship between the mother plant and the variants. The random amplified polymorphic DNA (RAPD) analysis is the simplest and quickest method for genetic stability assessment of in vitro grown plants. Among the different molecular markers, RAPD is user-friendly oligomer which has the potential to amplify multiple amplicons covering whole genome among different genotypes. RAPD technique developed by Williams et al. (1990) is reliable, faster, and easier for exploiting genetic polymorphism within and among species of populations. RAPD markers have been widely used for the identification of genetic relationship among populations and between plant populations (Ba et al. 2004). It is important to estimate the actual molecular genetic diversity of the existing plants to identify whether the lack of genetic variability might be the major constraining factor.

Apart from RAPD markers, several other DNA-based markers have effectively employed to study/monitor genetic stability of in vitro regenerated plants. RAPDs have been mostly used in variety identification, genetic relationship, and diversity studies in many crops species, such as rice (Yu and Nguyen 1994), sorghum (Vierling et al. 1994), soybean (Mienie et al. 2005), and *Jatropha* (Ganesh-Ram et al. 2008). RAPD markers were exploited in genome mapping and gene tagging studies in plants like faba bean (Torres et al. 1993), lettuce (Paran et al. 1992, Kesseli et al. 1994), spring barley (Thomas et al. 1995), tomato (Martin et al. 1991), rice (Virk et al. 1995a, b), and *Arabidopsis* (Reiter et al. 1992).

Several authors have used many molecular markers to monitor the plant genetic stability in in vitro regenerated plants in many species (Yadav et al. 2010) like *Withania somnifera* (Sinha et al. 2010) and *Capparis decidua* (Tyagi et al., 2010). The plant regeneration in in vitro and re-introduction into natural habitat is one strategy for conservation of important plant species (Rout et al. 2000). The genetic diversity analysis in coffee species has been reported (Mishra et al. 2011). RAPD and AFLP have limitations like low reproducibility in RAPD, high cost of AFLP, and the knowledge of flanking sequences to develop species-specific primers (Pradeep Reddy et al. 2002). ISSR-PCR is a technique that overcomes most of these limitations. Many people have used RAPD and ISSR markers to screen for genetic diversity in intra-populations and inter-populations (Kesari et al. 2010; Yuan et al. 2009).

Inter-simple sequence repeat analysis involves polymerase chain reaction (PCR)-derived amplification of regions between adjacent, inversely oriented microsatellites using a single simple sequence repeat (SSR) containing primer (Zietkiewicz et al. 1994). This technique can be applied to any species that contains sufficient number and distribution of SSR motifs. Another advantage of ISSR markers is that genomic sequence data of an organism is not required (Gupta et al. 1994). The primer used in ISSR analysis is based on any of the SSR motifs (di-, tri-, tetra-, or penta-nucleotides) found at microsatellite loci, giving a wide array of possible amplification products, and can be anchored to genomic sequences flanking either side of the targeted simple sequence repeats. The ISSR technique amplifies a large number of DNA fragments per reaction, representing multiple loci across the genome. The ISSR method has proven its usefulness, especially in the Gramineae family, for analysis of near-isogenic lines (Akagi et al. 1996) and varieties (Parsons et al., 1997) of rice, inbred lines of corn (Kantety et al. 1995), populations of finger millet (Salimath et al. 1995), and accessions of sorghum (Yang et al. 1996).

44.4 Conclusion

We, for the first time report the genetic diversity analysis of in vitro regenerated plant and their wild parent and compared with different district plants of *B. sensitivum.* Diversity analysis of various plant populations living in different districts revealed that vast genetic diversity exists in different populations of *B. sensitivum. B. sensitivum* plant populations revealed that the plants of Krishna and Visakhapatnam district are present as separate cluster; this indicates the genetic diversity and evolutionary distance of this plant from other plant populations. Genetic diversity analysis by DNA-based markers revealed that tissue-cultured plants are genetically similar to their wild plants with some variation.

Acknowledgments The first author Dr. Sirigiri Chandra Kala is thankful to University Grants Commission, New Delhi, for providing financial support with **Rajiv Gandhi National Fellowship**. We are very much thankful to Dr. A. Chandra Sekhar, Dr. K.V.N. Ratnakar Reddi, Dr. Puli Chandra Obul Reddy, and Yogi Vemana University, Kadapa, for their help in the present work.

References

Akagi H, Yokozeki Y, Inagaki A, Fujimura T (1996) Microsatellite DNA markers for rice chromosomes. Theor Appl Genet 94:61–67

Ba FS, Pasquet RS, Gepts P (2004) Genetic diversity in cowpea [Vigna unguiculata (L.) Walp] revealed by RAPD markers. Genet Resour Crop Evol 51:539–550

Ganesh-Ram S, Parthiban KT, Senthil-Kumar R, Thiruvengadam V, Paramathma M (2008) Genetic diversity among *Jatropha* species as revealed by RAPD markers. Genet Resour Crop Evol 55:803–809

Gupta M, Chyi YS, Romero-Severson J, Owen JL (1994) Amplification of DNA markers from evolutionarily diverse genomes using single primers of simple-sequence repeats. Theor Appl Genet 89:998–1006

Guruvayoorappan C, Kuttan G (2007a) Effect of *Biophytum sensitivum* on cell-mediated immune response in mice. Immunopharmacol Immunotoxicol 29:337–350

Guruvayoorappan C, Kuttan G (2007b) Immunomodulatory and antitumor activity of *Biophytum sensitivum* extract. Asian Pac J Cancer Prev 8:27–32

Guruvayoorappan C, Kuttan G (2008a) Effect of *Biophytum sensitivum* (L). DC inhibits tumor cell invasion and metastasis through a mechanism involving regulation of MMPs, prolyl hydroxylase, lysyl oxidase, nm23, ERK-1, ERK-2, STAT-1, and proinflammatory cytokine gene expression in metastatic lung tissue. Integr Cancer Ther 7:42–50

Guruvayoorappan C, Kuttan G (2008b) Extract of *Biophytum sensitivum.* Alters the cytokine profile and inhibits iNOS and COX-2 expression in LPS/Con a stimulated macrophages. Drug Chem Toxicol 31:175–188

Guruvayoorappan C, Kuttan G (2008c) Amentoflavone stimulates apoptosis in B16F-10 melanoma cells by regulating bcl-2, p.53, as well as caspase-3 genes and regulates the nitric oxide as well as proinflammatory cytokine production in B16F-10 melanoma cells, tumour associated macrophages and peritoneal macrophages. J Exp Ther Oncol 7:207–218

Guruvayoorappan C, Kuttan G (2008d) Anti-metastatic effect of *Biophytum sensitivum* is exerted through its cytokine and Immunomodulatory activity and its regulatory effect on the activation and nuclear translocation of transcription factors in B16F-10 melanoma cells. J Exp Ther Oncol 7:49–63

Guruvayoorappan C, Kuttan G (2008e) Amentoflavone inhibits experimental tumor metastasis through a regulatory mechanism involving MMP-2, MMP-9, Prolyl hydroxylase, lysyl oxidase, VEGF, ERK-1, ERK-2, STAT-1, nm23 and cytokines in lung tissues of C57BL/6 mice. Immunopharmacol Immunotoxicol 30:711–727

Guruvayoorappan C, Kuttan G (2008f) Inhibition of tumor specific angiogenesis by amentoflavone. Biochemistry (Mosc.) 73:209–218

Guruvayoorappan C, Kuttan G (2008g) Protective effect of *Biophytum sensitivum* (L.) DC on radiation induced damage in mice. Immunopharmacol Immunotoxicol 30:815–835

Guruvayoorappan C, Afira AH, Kuttan G (2006) Antioxidant potential of *Biophytum sensitivum* extract *in vitro* and *in vivo.* J Basic Clin Physiol Pharmacol 17:255–267

Jachak SM, Bucar F, Karting J (1994) Anti-inflammatory activity of extracts of *Biophytum sensitivum* in carrageenan induced paw edema. Phytother Res 13:73–74

Jain SK, Lim G (2001) Pyridoxine and pyrodoxamine inhibits superoxide radicals and prevents lipid peroxidation, protein glycation, and (Na^+ and K^+) –ATPase activity reduction in high glucose treated human erythrocytes. Free Radic Biol Med 30:232–237

Joy P, Maridass M (2008) Inter species relationship of *Cinnamomum* species using RAPD marker analysis. Ethanobot Leaflets 12:476–480

Kantety RV, Zeng X, Bennetzen JL, Zehr BE (1995) Assessment of genetic diversity in dent and popcorn (*Zea mays* L.) inbred lines using inter-simple sequence repeat (ISSR) amplification. Mol Breed 1(4):365–373

Kesari V, Sathyanarayana VM, Parida A, Rangan L (2010) Molecular marker-based characterization in candidate plus trees of *Pongamia pinnata*, a potential biodiesel legume. AoB Plants 2010:plq017

Kesseli RV, Paran I, Michelmore RW (1994) Analysis of a detailed genetic linkage map of Lactuca sativa (Lettuce) constructed from RFLP and RAPD markers. Genetics 136:1435–1446

Kirtikar KR, Basu BD (1984) B.S.M.P. Singh, Dehradun. Indian Med Plants I:440–441

Martin GB, Williams JGK, Tanksley SD (1991) Rapid identification of markers linked to Pseudomonas resistance gene in tomato by using random primers and near isogenic lines. Proc Natl Acad Sci U S A 88:2336–2340

Mienie C, Liebenberg M, Pretorius Z, Miklas P (2005) SCAR markers linked to the common bean rust resistance gene Ur-13. Theor Appl Genet 111:972–979

Mitra AP, Ambasta SP (1988) The wealth of India (raw materials), vol 6. Council of Scientific and Industrial Research, Publication and Information Directorate, New Delhi, pp 151–152

Mishra MK, Nishani S, Jayarama (2011) Molecular identification and genetic relationships among Coffee species (Coffea L.) inferred from ISSR and SRAP marker analyses. Arch Biol Sci Belg 63(3):667–679

Murray MG, Thompson WF (1980) Rapid isolation of high molecular weight plant DNA. Nuclei Acids Res 8:4321–4325

Nei M, Li WH (1979) Mathematical model for studying genetic variation in terms of restriction endonucleases. Proc Natl Acad Sci U S A 76(10):5269–5273

Paran I, Keseli RV, Westphal L, Michelmore RW (1992) Recent amplification of triose phosphate isomerase related sequences in lettuce. Genome 35:627–635

Parsons BJ, Newbury HJ, Jackson MT, Ford-Lloyd BV (1997) Contrasting genetic diversity relationships are revealed in rice (*Oryza sativa* L.) using different marker types. Mol Breed 3:115–125

Paunescu A (2009) Biotechnology for endangered plant conservation. Rom Soc Biol Sci 14:4095–4103

Pradeep Reddy M, Sarla N, Siddiq EA (2002) Inter simple sequence repeat (ISSR) polymorphism and its application in plant breeding. Euphytica 128:9–17

Pullaiah T (2002) Medicinal plants in India, vol 1. Regency Publications, New Delhi, pp 96–97

Puri D (2003) Hypocholestrolemic effect of *Biophytum sensitivum* leaf water extract. Pharm Biol 41(4):253–258

Puri D, Baral N (1998) Hypoglycemic effect of *Biophytum sensitivum* in alloxan diabetic rabbits. Indian J Physiol Pharmacol 42:401–406

Ramsay MM, Jackson AD, Porley RD (2000) A pilot study for ex situ conservation of UK bryophytes. In: BGCI (ed) Euro Gard—EBGC, Canary Islands, Spain. EBGC, Las Palmas de Gran Canaria, pp 52–57

Reiter RS, Williams JGK, Feldman KA, Rafalski JA, Tingey SV, Scolnik PA (1992) Global and local genome mapping in Arabidopsis thaliana by using recombinant inbred lines and random amplified polymorphic DNAs. Proc Natl Acad Sci U S A 89:1477–1481

Rout GR, Samantary S, Das P (2000) *In vitro* manipulation and propagation of medicinal plants. Biotechnol Adv 18:91–20

Saha DP, Chowdhury S, Monda M, Sinha K (2012) Biosorption of direct Red 28 (Congo Red) from aqueous solutions by egg shells: batch and column studies. Sep Sci Technol 47:112–123

Salimath SS, de Oliveira AC, Godwin ID, Bennetzen JL (1995) Assessment of genome origins and genetic diversity in the genus Eleusine with DNA markers. Genome 38:757–763

Sinha A, Jain R, Kachhwaha S, Kothari SL (2010) Optimization of the level of micronutrient copper in the culture medium improves shoot bud regeneration in Indian Ginseng [*Withania somnifera* (L.) Dunal]. Natl Acad Sci Lett (India) 33:11–16

Smita B, Bhosale JLJ, Kad TD, Nadal AB (2009) Genetic diversity assessment in intra an inter populations of *Xylocarpus granatum* koen. A critically endangered and narrowly distributed species of Maharastra. Curr Sci 97(5)

Thomas CM, Vos P, Zabeau M, Jones PA, Norcott KA, Chadwick BP, Jones JDG (1995) Identification of amplified restriction fragment polymorphism (AFLP) markers tightly linked to the tomato CF-9 gene for resistance to Cladosporium fulvum. Plant J 8:785–794

Thomas JB, Vijayan D, Joshi SD, Lopez SJ, Kumar RR (2006) Genetic integrity of somaclonal variants in tea (*Camellia sinensis* (L.) O Kuntze) as revealed by inter simple sequence repeats. J Biotechnol 123:149–154

Torres AM, Weeden NF, Martin A (1993) Linkage among isozyme, RFLP and RAPD markers in Vicia faba. Theor Appl Genet 85:937–945

Tyagi P, Khanduja S, Kothari SL (2010) *In vitro* culture of *Capparis decidua* and assessment of clonal fidelity of the regenerated plants. Biol Plant 54:126–130

Vierling RA, Xiang Z, Joshi CP, Gilbert ML, Nguyen HT (1994) Genetic diversity among elite Sorghum lines revealed by restriction fragment length polymorphisms and random amplified polymorphic DNAs. Theor Appl Genet 87:816–820

Virk PS, Ford-Lloyd BV, Jackson MT, Newbury HJ (1995a) Use of RAPD for the study of diversity within plant germplasm collections. Heredity 74:170–179

Virk PS, Ford-Lloyd BV, Newbury HJ, Jackson MT (1995b) The identification of duplicate accessions within rice germplasm collection using RAPD analysis. Theor Appl Genet 90:1049–1055

Williams JGK, Kubelik AR, Livak KJ, Rafalski JA, Tingey SV (1990) DNA polymorphisms amplified by arbitrary primers are useful as genetic markers. Nucleic Acids Res 18(22):6531–6535

Yang W, de Oliveira AC, Goodwin I, Schertz K, Bennetzenm JL (1996) Comparison of DNA marker technologies in characterizing plant genome diversity: variability in Chinese sorghums. Crop Sci 36:1669–1676

Yadav SK, Kachhawaha S, Kothari SL (2010) Comparison of *in vitro* regeneration efficiency of leaf explants in response to different cytokinins and assessment of genetic uniformity of regenerated plants of *Solanum surattense* Burm. f. Int J Biotechnol 1:97–103

Yu LX, Nguyen NT (1994) Genetic variation detected with RAPD markers among upland and lowland rice cultivars (*Oryza sativa* L.). Theor Appl Genet 87:668–672

Yuan XF, Dai ZH, Wang XD, Zhao B (2009) Assessment of genetic stability in tissue-cultured products and seedlings of *Saussurea involucrata* by RAPD and ISSR markers. Biotechnol Lett 31:1279–1287

Zietkiewicz E, Rafalski A, Labuda D (1994) Genomic fingerprinting by simple sequence repeat (SSR)-anchored polymerase chain reaction amplification. Genomics 20:176–183

Development of Standard Protocols for In Vitro Regeneration of Some Selected Banana Cultivars (*Musa* spp.) from India

45

Saifuldeen Ahmed Hasan, S. M. Khasim, and J. Ramudu

Abstract

Banana is a very popular fruit due to its high nutritive value. It helps in reducing the risk of heart diseases and is also recommended for patients suffering from high blood pressure, arthritis, ulcers, and gastrointestinal and kidney disorders. Micropropagation of selected banana cultivars such as Grand Naine (G9), Monthan, and Red Banana on commercial scale using economical cytokinins (BAP) and an effective auxin (IAA) has been taken up in this study in order to supply them to farmers on an affordable price. Consistent shoot proliferation of commercial standard was best in 5 and 3 mg/L BAP for Grand Naine where the production schedule could be formatted accurately with quality shoots. However, for Monthan 10 and 2 mg/L BAP and for Red Banana 10 and 3 mg/L BAP along with 0.2 mg/L IAA for each, respectively, appeared to be more suitable for obtaining productive shoots for a viable commercial production scheduling.

Keywords

Banana cultivars · *Musa* spp. · In vitro regeneration · Auxins and cytokinins

---

S. Ahmed Hasan · S. M. Khasim (✉)
Department of Botany and Microbiology, Acharya Nagarjuna University, Guntur, Andhra Pradesh, India

J. Ramudu
Department of Botany, VSR Government Degree and Post-Graduate College, Movva, Andhra Pradesh, India

S. M. Khasim et al. (eds.), *Medicinal Plants: Biodiversity, Sustainable Utilization and Conservation*, https://doi.org/10.1007/978-981-15-1636-8_45

45.1 Introduction

Banana is the most popular fresh fruit all over the world, and its name comes from the Arabic word "banan," which means finger. Almost all banana cultivars are derived from *Musa acuminata* and *Musa balbisiana*. India is now the second largest producer of fruits and vegetables in the world, and banana stands in second place for exports (Anonymous 2016). Banana is grown in more than 150 countries, producing 105 million tonnes of fruit per year. The global production of banana is around 102,028.17 thousand tons of which India contributes 29.19%. Banana (*Musa* spp.) is an important fruit crop in India. Main banana-growing states in India are Tamil Nadu, Maharashtra, Gujarat, Andhra Pradesh, and Karnataka.

Banana is a very popular fruit due to its low price and high nutritive value. Its high vitamin B6 content helps fight infection and is essential for the synthesis of "heme," the iron-containing pigment of hemoglobin. The fruit is also rich in carbohydrates and potassium and a great source of fiber too. It is also a good source of phosphorus, calcium, and magnesium. The fruit is easy to digest and free from fat and cholesterol. It helps in reducing the risk of heart diseases when consumed regularly and is recommended for patients suffering from high blood pressure, arthritis, ulcer, gastroenteritis, and kidney disorders.

At present, India is the largest producer of banana in the world with about 30% of total global production. However, the export market share is a meager of 1%. With increased productivity/unit area, the export capabilities can be improved. This is possible by substituting the conventional suckers with virus-indexed and tissue-cultured plants and adopting the scientific methods of cultivation. In Andhra Pradesh, major banana-producing districts are East Godavari, West Godavari, Anantapura, Cuddapah, and Vizianagaram. Most popular varieties cultivated are Dwarf Cavendish, Robusta, Rasthali, Amritpant, Thella Chakkarakeli, Karpoora Poovan, Chakrakeli, Monthan, and Yenagu Bontha. The farmers are cultivating local cultivars which are low yielding, and the productivity of banana is quite low, i.e., 35 tonnes per hectare as against 65 tonnes per hectare in Maharashtra (Karuna and Rao 2016). Micropropagation of selected banana cultivars, such as Grand Naine (G9), Monthan, and Red Banana, on commercial scale using economical cytokinin (BAP) and an effective auxin (IAA) has been taken up in this study in order to supply them to farmers on affordable prices.

45.2 Media Preparation and Sterilization

MS basal medium (Murashige and Skoog 1962) was used for the culture of selected banana cultivars. The basal medium with different concentrations of BAP with a fixed concentration of IAA at 0.2 mg/L was used for optimizing an in vitro production protocol for the different selected banana cultivars. The media compositions consisting of different concentrations of BAP along with fixed concentration of IAA as the auxin at 0.2 mg/L in MS basal medium (Table 45.1) used in the present study

Table 45.1 Different growth regulators used in MS basal medium for optimizing in vitro protocol for selected banana cultivars

Media code	Growth regulators used	Concentration (mg/L)
M1	BAP + IAA	10 + 0.2
M2	BAP + IAA	5 + 0.2
M3	BAP + IAA	3 + 0.2
M4	BAP + IAA	2 + 0.2
M5	BAP + IAA	1 + 0.2

were designated as M1, M2, M3, M4, and M5. One liter of each of the media was dispensed in to 25 glass jars, and all the media was prepared at one stretch.

The final volume of medium was made up to 1 L. Each of the glass jar was dispensed with 40 mL of this medium and capped tightly. The jars were sealed with autoclavable polypropylene wrap and autoclaved for 45 min at a temperature of 121 °C and a pressure of 15 psi in a horizontal autoclave. The process of media preparation was repeated again every 2 or 3 weeks for subsequent culture transfers until the final data recording was completed. The glass jars were incubated at room temperature for a week in a separate clean room to check the sterility of the medium and then used for explant/culture inoculation purpose.

45.3 Micropropagation of Selected Banana Cultivars

45.3.1 Initiation of Explants

All the explants inoculated in the initiation medium started responding within 15 days under the given conditions. The outer leaf sheaths of the explants changed from off-white to green and started to unfold externally. At the base of explants of Grand Naine and Red Banana, blackening was noticed due to phenolic exudation. However, there was no blackening observed in Monthan cultivar, clearly indicating that phenolic exudation was found to be controlled and is not a serious problem (Tables 45.2, 45.3, and 45.4). The shoot tip explants of different banana cultivars exhibited varied response in terms of both encouraging growth signs. The categories of general growth response at this stage were assessed with its ability to unfold the leaf sheath and turn into green color. It was excellent (+++) when the response was in 2–3 weeks of incubation, very good (++) when the response was in 3–4 weeks of incubation, good (+) when the response was in 4 weeks of incubation, and poor (–) when there was no response at all.

In vitro propagation of banana preferably uses sword sucker as explant source. Despite pretreatment in antibiotic solution and surface sterilization in mercuric chloride, some bacterial contamination was observed in all the three banana cultivars. None of the explants had any fungal contamination. This study demonstrates reduction in microbial contamination, saving up to 85% of the explants in aseptic culture establishment supporting the subsequent micropropagation of the banana suckers.

Table 45.2 Response of Grand Naine to different initiation media

Media code	No. of explants initiated	No. of explants survived free of contamination	% Survival	General growth response	Distinct remarks
M1	5	4	80	++	Slight phenolic exudation
M2	5	5	100	+++	Slight phenolic exudation
M3	5	4	80	++	Slight phenolic exudation
M4	5	4	80	++	Slight phenolic exudation
M5	5	5	100	+++	Slight phenolic exudation
Total/ ave.	25	22	88		

+++ Excellent; ++ very good; + good; – poor

Table 45.3 Response of Monthan to different initiation media

Media code	No. of explants initiated	No. of explants survived free of contamination	% Survival	General growth response	Distinct remarks
M1	5	5	100	+++	No phenolic exudation
M2	5	4	80	++	No phenolic exudation
M3	5	4	80	++	No phenolic exudation
M4	5	5	100	+++	No phenolic exudation
M5	5	3	60	+	No phenolic exudation
Total/ ave.	25	21	84		

+++ Excellent; ++ very good; + good; – poor

In the present study, the overall bacterial contamination in Grand Naine was only about 12%, while that in Monthan was 16%, and in Red Banana, it was 20% based on survival rate. It is worth noting that the suckers of Grand Naine and Monthan were robust, and many outer leaf sheaths were eliminated at every stage of pretreatment and surface sterilization. However, the suckers of Red Banana when received from the field were smaller, and only a few outer leaf sheaths were removed during

Table 45.4 Response of Red Banana to different initiation media

Media code	No. of explants initiated	No. of explants survived free of contamination	% Survival	General growth response	Distinct remarks
M1	5	4	80	++	Slight phenolic exudation
M2	5	3	60	+	Slight phenolic exudation
M3	5	4	80	++	Slight phenolic exudation
M4	5	4	80	++	Slight phenolic exudation
M5	5	5	100	+++	Slight phenolic exudation
Total/ ave.	25	20	80		

+++ Excellent; ++ very good; + good; – poor

the entire process. The excessive bacterial contamination in Red Banana (20%) can be attributed to this aspect, and it cautions on choosing well-grown robust suckers for improved culture establishment.

45.3.2 In Vitro Multiplication of Initiated Explants in Different Media

During the first 4 weeks in culture, the external leaf sheaths of each explant, which were initially white, later turned green. Some elongation of the explant could be observed in all the three banana cultivars. Browning or blackening at the base of the explants occurred but only externally, which could be removed during the process of subculturing. Most of the axenic explants established in culture formed fresh multiple shoots by 4–8 weeks of longitudinally splitting and re-inoculating into the respective medium. However, each of the explant in each of the cultivar showed a vast variation in their ability to proliferate in different media.

45.3.2.1 The Multiplication Ratio

The establishment of substantial multiplication of the explant approximately took about 120–150 days depending on the cultivar. Eventually, until that period of 120–150 days, the individual explant/cluster was periodically transferred to fresh medium every 3 or 4 weeks in order to provide fresh nutrients and also prevent the ill effects of phenolics and tissue blackening. The base of the explants in the multiplication stage expanded horizontally forming a compact base from which cluster of shoots usually 6–15 in number appeared from the basal mass. Each proliferating cluster is subdivided into smaller clusters of minimum three microshoots in each culture cycle, and the number of such smaller clusters thus produced is known as multiplication ratio throughout the present study. Usually each cluster consisting of three or four small proliferating shoot initials is the requirement as per the commercial standard.

45.3.2.2 Grand Naine

All the explants in M1 to M5 series of media showed considerably good response (Fig. 45.1) in 4 weeks of culturing. However, mixed type of multiplication and shoot elongation was seen in different combinations (Table 45.5). Good multiplication of proliferating clusters was obtained in M2 and M3 medium by three successive transfers in the respective medium. Poor multiplication ratio was seen in M1 medium. The shoots had poor quality often with yellow leaves.

The clumps were neither multiplied vigorously nor elongated sufficiently. Even in M5, horizontal proliferations of shoots were not observed. Instead, all the shoot initials exhibited shoot elongation, and sturdy plantlets were obtained.

Fig. 45.1 Response of Grand Naine banana cultivar in M1 to M5 multiplication media after three subcultures

Fig. 45.2 Response of Monthan in M1 to M5 multiplication media after three subcultures

45.3.2.3 Monthan

The explants of Monthan in M1 to M5 also showed an impressive proliferating pattern (Fig. 45.2) with good multiplication ratio in 4 weeks of culturing (Table 45.6). However, the cultures did not appear to be remarkable as obtained for that of Grand Naine in M1 to M5 series of media. Elongation of shoots and leaf expansion were hardly found in these series of media in 4 weeks of culturing. Most of the cultures were very compact with the emerging shoot tips covered with a blackened scale. None of the clump was more than 2 cm in height.

45.3.2.4 Red Banana

All the explants in M1 to M5 series of media showed considerably good response (Fig. 45.3) in 4 weeks of culturing (Table 45.7). Good multiplication of proliferating clusters was obtained in M1 and M2 medium by three successive transfers in the respective medium. In M3 medium, a mix of shoot elongation and proliferation was found. Somehow, the quality of cultures was not good in M4. Instead, all the shoot initials exhibited shoot elongation and produced sturdy plantlets in M5. In all stages of growth, the red shade was found on the stems once the cultures were incubated in 12-h light.

Table 45.5 Multi-ratio of banana cultivar Grand Naine in different initiation media in three successive growth cycles, each cycle of 4 weeks

Media code	I multi-cycle	II multi-cycle	III multi-cycle	Ave. no. of elongated shoots/jar	Culture quality
M1	2.2	2.0	1.5	0	+
M2	3.0	2.8	2.5	0	+++
M3	2.5	2.2	2.0	3	+++
M4	2.0	1.8	1.5	3	++
M5	3.5	3.5	3.5	5	+++

+++ Excellent; ++ very good; + good; – poor

Table 45.6 Multi-ratio of banana cultivar Monthan in different initiation media in three successive growth cycles each cycle of 4 weeks

Media code	I multi-cycle	II multi-cycle	III multi-cycle	Ave. no. of elongated shoots/jar	Culture quality
M1	3.7	3.8	3.6	0	+++
M2	3.9	3.9	3.9	0	+++
M3	2.5	2.4	2.3	2	++
M4	2.5	2.5	2.3	3	++
M5	1.8	1.7	1.5	4	+

+++ Excellent; ++ very good; + good; – poor

Fig. 45.3 Response of Red Banana in M1 to M5 multiplication media after three subcultures

45.3.3 Response of Banana Shoots in Medium Containing Activated Charcoal

Shooting and rooting stage is a pre-final in vitro growth stage required for banana cultivars. The microshoots produced during multiplication stage will generally carry the effect of cytokinin and hence will not possess sturdy stems although they may or may not have the roots. It is very important to have sturdy stems and sufficient roots prior to transferring them to rooting medium. In order to obtain good quality plantlets, the multiplying cultures will pass through a shooting and rooting stage where individual microshoots are allowed to elongate and mature in the absence of cytokinins. This was achieved by transferring separate single plants extracted from small clusters of 2–3 slightly elongated shoots into MS basal medium with 1% activated charcoal (AC) without any plant growth regulators. The elongated shoots of banana in AC medium further elongated and grew up into sturdy plantlets by 3–4 weeks with an average shoot length of about 6 cm (Table 45.8).

45.3.4 A Model for In Vitro Micropropagation Protocol for Banana

It can be clearly understood from the flow charts (Figs. 45.4, 45.5, and 45.6) that M1 and/or M2 was good for multiplication and bulking of the proliferating clusters over a period of several cycles at least up to six multiplication cycles. The shoot elongation was best in M5 medium for all the three cultivars. However, for Monthan and Red Banana, it was essential to culture the shooting clusters for two successive cycles. The shoots could be completely rooted ready to be sent to greenhouse for acclimatization after 3–4 weeks of culture in MS basal medium containing 1% activated charcoal. The entire process of initiation up to getting the rooted plants in vitro for each of the cultivar carried out in the present study is shown as flow charts (Figs. 45.4–45.6).

Table 45.7 Multi-ratio of banana cultivar Red Banana in different initiation media in three successive growth cycles each cycle of 4 weeks

Media code	I multi-cycle	II multi-cycle	III multi-cycle	Ave. no. of elongated shoots/jar	Culture quality
M1	3.2	3.5	3.4	0	+++
M2	2.8	2.7	2.5	0	++
M3	1.8	1.9	2.0	3	++
M4	1.5	1.3	1.2	3	+
M5	1.5	1.3	1.2	5	++

+++ Excellent; ++ very good; + good; – poor

Table 45.8 Effect of charcoal medium on general growth parameters and quality of shoots on different banana cultivars after 4 weeks of growth

Banana cultivars	Leaf color	Growth vigor	No. of leaves/plant	Shoot length (cm)	Stem girth (mm)	Root quality
Grand Naine	Dark green	+++	5	5–6	6–8	A lot of black elongated roots
Monthan	Pale green	++	3	4–5	4–5	Black roots
Red Banana	Reddish green	+++	4	8	6–8	A lot of white roots, with root hairs

+++ Excellent; ++ very good; + good; – poor

Consistent shoot proliferation of commercial standard was best in 5 and 3 mg/L BAP for Grand Naine where the production scheduling could be formatted accurately, with quality shoots. However, for Monthan 10 and 2 mg/L BAP and for Red Banana 10 and 3 mg/L BAP along with 0.2 mg/L IAA for each, respectively, appeared to be more suitable for obtaining productive shoots for a viable commercial production scheduling.

For all the three cultivars investigated, shoot elongation was best in 1 mg/L BAP wherein the shoot initials elongated and the leaves expanded with profuse rooting. However, for Monthan and Red Banana, one cycle of 4 weeks in lesser BAP was not enough for shoot elongation. A second cycle of another 4 weeks was essential to obtain elongated shoots which could then be transferred to medium containing activated charcoal for development of complete plants.

The ex-agar plants of all three cultivars of banana performed very well in poly-tunnel during the process of primary acclimatization. Later plants were transformed to primary nursery. The net pot plants started to grow very rapidly after shifting them from poly-tunnel to the greenhouse under higher light intensity. Then, the plants are transferred to natural soil conditions.

Plant cells growing in vitro are considered to be under some degree of stress and may be predisposed to direct infection, even by microbes, which are not normally pathogenic to them. Thus, microbial contaminants are the major challenges in

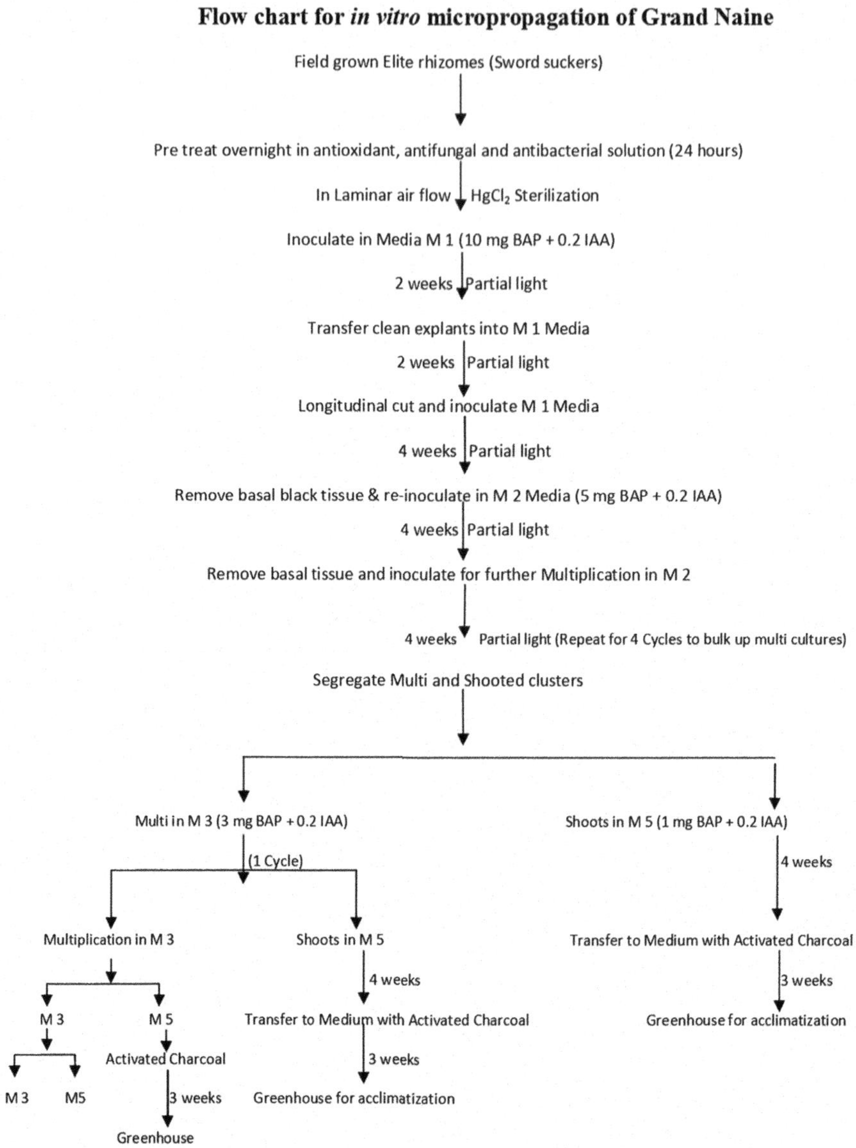

Fig. 45.4 Flow chart for in vitro micropropagation of Grand Naine

plant's in vitro propagation during the different stages of culture processes (Helaly et al. 2014), especially during the initiation stage; to minimize the microbial contamination and to promote healthy growth, explants were treated with 70% absolute alcohol for 6 min, 0.1% mercuric chloride for 10 min and 0.2% for 10 min, 1% sodium hypochlorite for 15 min, 0.1% cefotaxime for 5 min, and 0.05% gentamycin for 5 min (Shashikumar et al. 2015).

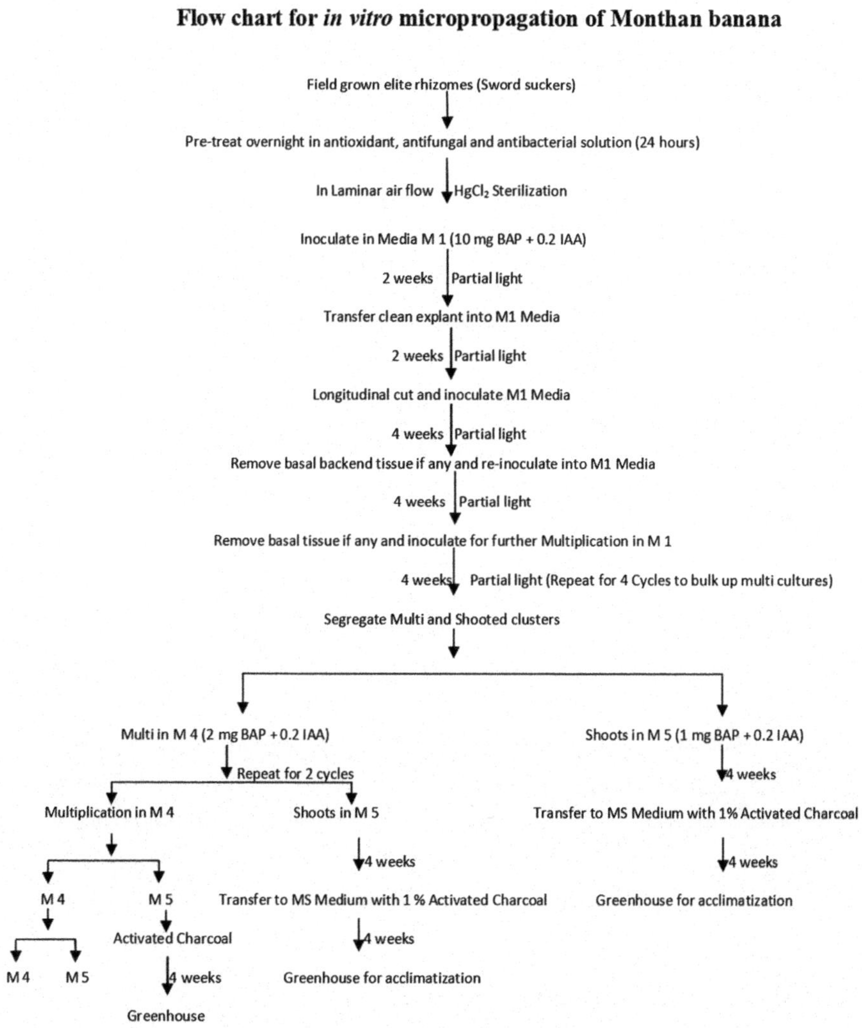

Fig. 45.5 Flow chart for in vitro micropropagation of Monthan banana

45.3.5 Importance of Bud Splitting Technique

In the present study, the axenic explants were longitudinally split to promote rapid multiplication. In this study, bud splitting technique is applied; a remarkable multiplication of shoots is achieved in about 120 days in Grand Naine cultivar and about 150 days in Monthan and Red Banana. The rate of multiplication was more than threefold in all the three cultivars in the present study. The concept of injury to shoot tip buds to promote axillary proliferation by breaking the apical dominance was reported by Woolley and Wareing (1972). Bud splitting is not a common practice in banana tissue culture. But splitting of the shoot tip in banana was reported to

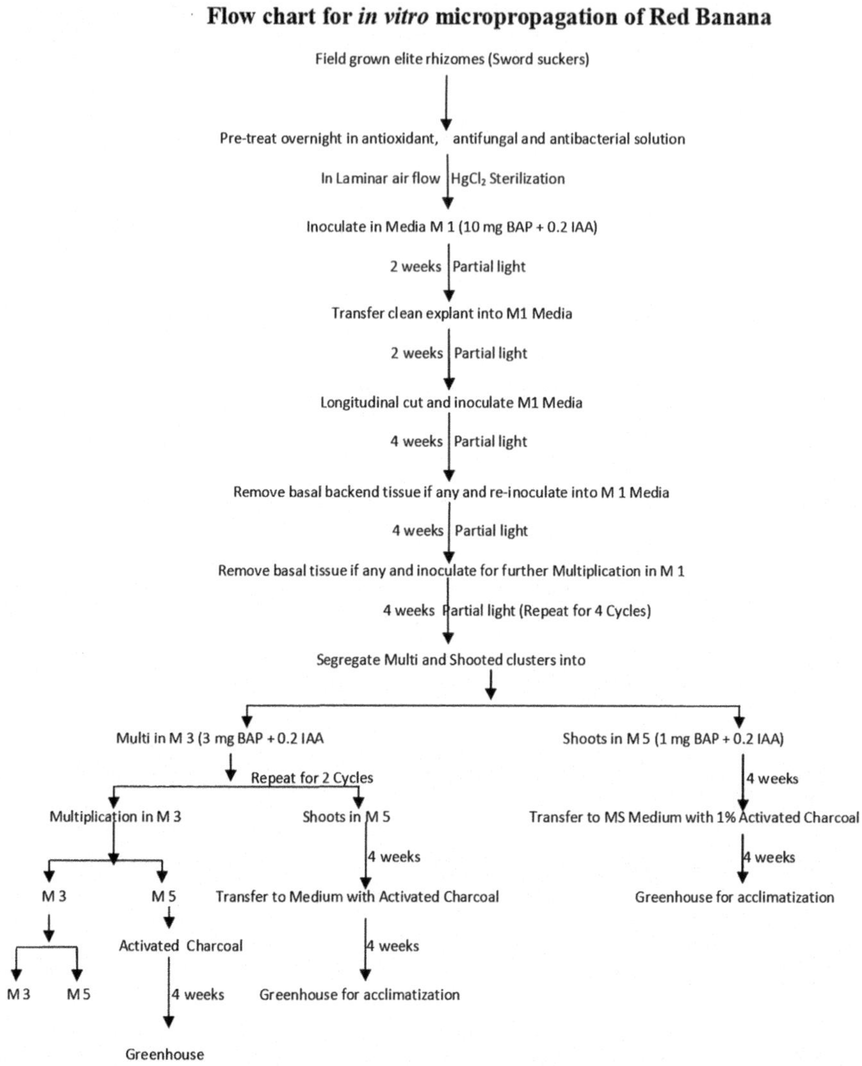

Fig. 45.6 Flow chart for in vitro micropropagation of Red Banana

enormously increase the growth rate and shoot proliferation (Swamy et al. 1983). In the application of bud splitting technique, the meristematic tissues in juvenile state are very competent that are found in young suckers. Generally longitudinal bud splitting promotes the rapid multiplication (Hussein 2012). These auxiliary meristems are commonly the source of formation of buds in nature especially when the apical shoots are damaged or injured (Burrows 1989). In another study, Mateille and Foncelle (1988) noticed that longitudinal cuts of buds induced a threefold increase in multiplication in *Musa* cv. Poyo from the Ivory Coast. The multiplication rate was also found to depend on the origin of suckers, whereby lateral buds doubled in size within 3 weeks, while the apical buds reached three times their size. In general, different cultivars show variation in the degree of their shoot bud proliferation. This is because the multiple budding appears to be linked to genome configuration of a given cultivar (Khatri et al. 1997) and apical dominance which is under control of plant growth regulators. Thus, appropriate concentration of plant growth regulators and bud splitting technique in effort to break dormancy and induce multiple buds development are necessary to ensure production of large number of shoots in vitro, which is a commercial requirement in banana cultivars.

45.3.6 Role of Activated Charcoal (AC) and Antioxidants in Tissue Culture Media

In the present study, phenolic exudation in the initiated explants was found in Grand Naine and Red Banana cultivars in the first few cycles. It was interesting to note that no phenolic exudation was found in Monthan cultivar. Gradually, the phenolic exudation was found to be controlled even in Grand Naine and Red Banana too. Generally, banana and plantain explants are susceptible to tissue blackening with excessive exudation of phenolics, which hinders the normal growth of tissue in vitro. A wide variety of chemicals and antioxidants have been used to prevent the phenolic exudation (Safwat et al. 2015). Antioxidant, i.e., ascorbic acid, has been successfully used by Strosse et al. (2004) and Abdelwahd et al. (2008) to inhibit phenolics and to reduce the oxidative blackening in various plant species.

The use of activated charcoal in micropropagation was first reported by Fridborg et al. (1978), and it was suggested that charcoal was responsible for absorption and desorption which controlled the release of nutrients in the production of synthetic seeds (Ganapathi et al. 1992). Activated charcoal has a very fine network of pores with large inner surface area on which many substances can be adsorbed (Pan and Van-Staden 1998). Safwat et al. (2015) reported in their studies the effect of some antioxidants on blackening and growth of in vitro culture of banana. They were of opinion that the cultivated charcoal could be acted as antioxidant and used to minimize the blackening process of in vitro derived plantlets.

Activated charcoal has some effect on the morphogenesis by the irreversible adsorption of inhibitory compounds in the culture medium, thus substantially decreasing the toxic metabolites, phenolic exudation, and accumulation of brown exudates (Thomas 2008). In the present study, soaking the explants in pretreatment solution consists of citric acid and ascorbic acid at 0.1 g/L for 24 h, and removal of outer leaf sheaths at every stage coupled with frequent washes at every stage during process of initiation has effectively controlled harmful effects of phenolics. Antioxidant growth regulators are considered as one of the most important factors in the development of a standard tissue culture protocol (Dayarani and Dhanarajan 2013), and the present study endorses previous reports. In this present study, it could be seen that all plants in activated charcoal media had sturdy with glossy leaves, which can be attributed to the adsorption capability of charcoal eliminating the remnant effects of cytokinins and auxins.

45.3.7 Role of IAA (Auxin) and BAP (Cytokinin) in Tissue Culture Media

The multiplication rate of adventitious buds under the influence of cytokinin is one of the determining factors deciding on the efficiency of the micropropagation system. It is evident from the present study that the multiplication of the explants inoculated in MS basal medium containing BAP and IAA resulted in axillary shoot proliferation and shoot elongation. Axillary shoot proliferation and shoot elongation required for commercial purpose were successfully achieved for banana cultivars in this investigation. In the methodology adopted here, elongation of the terminal shoot is suppressed by excising the shoot tip, thereby promoting the proliferation of the axillary shoots. The suppression of terminal shoot allows multiplication of a large number of microshoots. The process of axillary proliferation in banana is further augmented by the use of BAP and IAA.

The multiplication ratio of banana cultivars is strongly supported by BAP and IAA (Dagnew et al. 2012; Ahmed et al. 2014; Reddy et al. 2014; Sahoo et al. 2015; Karule et al. 2016) as also obtained in the present study substantiating that the combination of BAP and IAA is an influential cytokinin-auxin combination for in vitro micropropagation of banana. Adenine-based cytokinins are used in several *Musa* spp. for in vitro propagation (Gubbuk and Pekmezci 2004). N6-benzylaminopurine (BAP) is the most commonly preferred cytokinin (Vuylsteke 1989).

Increased levels of cytokinin inhibit apical dominance and promote lateral shoot proliferation. This principle holds good for banana cultivars wherein higher multiratio was obtained at 5 or 10 mg/L of BAP along with 0.2 mg/L of IAA and lesser multiplication ratio with elongation of shoots was obtained at 1 mg/L BAP. However, the multiplying cultures of Grand Naine were not good in medium with 10 mg/L BAP over a period of 3–4 multiplication cycles. Yellowing of leaves occurred after three to four subcultures. Higher concentrations of BAP and kinetin beyond optimum levels were also reported to cause necrosis and reduction in shoot formation during in vitro multiplication of Nendran (Rabbani et al. 1996). Reddy et al. (2014) also stressed the importance of BAP at low concentration about 2 mg/L that had given the best induction of Grand Naine plantlets.

After the initial shoot multiplication in M1 as booster multiplication medium for a few cycles, the normal multiplication for bulking was carried out in M2 for Grand Naine, while M1 continued to be used for Monthan and Red Banana. However, an intermediate shoot elongation medium was adopted for each of the cultivar, i.e., M3 for Grand Naine and Red Banana (both belong to AAA genome) and M4 for Monthan in which a combination of dwarf multiplying clumps and also clumps with elongated shoots was obtained. The dwarf clumps were excised on top and re-inoculated for further multiplication, while the elongated clumps were transferred to M5 without cutting the top for further elongation. Shoot elongation was observed in the lowest concentration tried in M5 medium (1 mg/L BAP + 0.2 mg/L IAA). These results clearly establish the fact that optimal growth and morphogenesis of tissues may vary among the cultivars of same genus based on their nutritional and hormonal requirements.

Sahoo et al. (2015) reported that MS medium supplemented with 2 mg/L BAP and 1 mg/L IBA was found to be ideal for the early shoot elongation after 30 days of inoculation in Grand Naine cultivar. Further, Lalrinsanga et al. (2013) revealed highest multiple shoot induction in MS medium fortified with 5 mg/L BAP with 2.17 shoots while MS with 1 mg/L NAA + 0.2 mg/L BAP with longest regenerated shoots after 45 days of incubation.

Micropropagation of banana through initiation of shoot tip explants has been reported by a number of researchers in MS basal medium supplemented with 5–20 mg/L BAP (Priyono 2001; Noor-Aziah and Khalid 2002; Venkatachalam et al. 2006). Some researchers have reported that a combination of BAP and an auxin has enhanced proliferation and shoot length during the tissue culture of banana (Ngomuo et al. 2014). TDZ, a phenyl urea-based cytokinin, is frequently used along with BAP and IAA (Gubbuk and Pekmezci 2004).

In the previous study with banana cultivar Robusta (Saifuldeen 2015), it was demonstrated that BAP along with an auxin is most suited for commercial in vitro micropropagation. In the present study, among the various BAP concentrations (1–10 mg/L) along with a fixed concentration of IAA (0.2 mg/L), for the tested cultivars, it was found that maximum shoot multiplication was obtained in 5 mg/L BAP for Grand Naine. However, maximum shoot multiplication for Monthan and Red Banana was obtained in 10 mg/L BAP.

45.3.8 Acclimatization of In Vitro Derived Plants

Micropropagated plants are cultured within in vitro environments, with a high relative humidity and low light intensity, taking nutrients and energy from the culture medium. These plants do not possess protective mechanisms against desiccation. Impaired stomatal function (Marín et al. 1988) and reduced epicuticular waxes (Gaspar and Coumans 1987; Sutter 1988) have been noted in these plants. In addition, their photosynthetic competence is reported to be reduced (Preece and Sutter 1991). During acclimatization, physiological and structural changes allow micropropagated plants to adapt to the new environment conditions, mainly to low relative humidity and high light intensity. As a result, plants become autotrophic and develop as normal plants.

Pospisilova et al. (1999) have opined that abnormalities occurring in morphology, anatomy, and physiology of in vitro derived plantlets could be repaired after transfer to ex vitro conditions and further stated that many plant species need gradual changes in environmental conditions to avoid desiccation losses and photoinhibition. By manipulating the in vitro environment, leaves that have greater tolerance to water stress and are photosynthetically competent could be developed as part of the acclimatization process in preparing plantlets for transferring out of culture (Seelye et al. 2003).

45.4 Conclusions

The present work has contributed to the development of suitable set of media for different stages of in vitro production protocol for Grand Naine, Monthan, and Red Banana cultivars. It is concluded from the results that each of the cultivars had specificity in the concentration with plant growth regulator requirement at different stages of growth for obtaining optimum results.

Further, under ideal conditions as per the in vitro production flow chart, each established explant is expected to generate about 10,000 plants in 12 months' time with 6–8 multiplication cycles. The entire laboratory process starting from selection of elite sword suckers followed by initiation to obtaining the ex-agar rooted plants has to be carefully planned and judiciously executed. It is essential to give extreme importance to dissections of multiplying cultures to obtain productive cultures clumps and the final product.

References

Abdelwahd R, Hakam N, Labhilili M, Udupa M (2008) Use of an adsorbent and antioxidants to reduce the effects of leached phenolics in *in vitro* plantlet regeneration of Faba bean. Afr J Biotechnol 7(8):997–1002

Ahmed S, Sharma A, Singh AK, Wali VK, Kumari P (2014) *In vitro* multiplication of banana (*Musa* sp.) cv. Grand Naine. Afr J Biotechnol 13(27):2696–2703

Anonymous (2016) Banana—common varieties of banana. http://www.itfnet.org/v1/2016/03/banana-common-varieties/

Burrows GE (1989) Developmental anatomy of axillary meristems of Araucaria cunninghamii released from apical dominance following shoot apex decapitation in vitro and in vivo. Univ Chicago Press J 150(4):369–377

Dagnew A, Shibru S, Debebe A, Lemma A, Dessalegn L, Berhanu B, Beyene W, Sierra M (2012) Micropropagation of Banana varieties (*Musa* spp.) using shoot-tip culture. Ethiop J Agric Sci 22:14–25

Dayarani M, Dhanarajan MS (2013) Control of excessive browning during *in vitro* regeneration of *Musa laterita*. Int J Pharm Bio Sci 4(3):471–476

Fridborg G, Pedersen M, Landstrom L-E, Eriksson T (1978) The effect of activated charcoal on tissue cultures: adsorption of metabolites inhibiting morphogenesis. Physiol Plant 43(2):104–106

Ganapathi TR, Suprasanna P, Bapat VA, Rao PS (1992) Propagation of banana through encapsulated shoot tips. Plant Cell Rep 11:571–575

Gaspar T, Coumans N (1987) Root formation. In: Bonga JM, Durzan DJ (eds) Cell and tissue culture in forestry, High-tech and micropropagation 1, vol 2. Springer, Berlin, pp 202–217

Gubbuk H, Pekmezci M (2004) *In vitro* propagation of some new banana types (*Musa* spp.). Turk J Agric For 28:355–361

Helaly MN, El-Metwally MA, El-Hoseiny H, Omar SA, El-Sheerys NI (2014) Effect of nanoparticles on biological contamination of *in vitro* cultures and organogenic regeneration of banana. Aust J Crop Sci 8(4):612–624

Hussein N (2012) Effects of nutrient media constituents on growth and development of banana (*Musa* spp.) shoot tips cultured in vitro. Afr J Biotechnol 11(37)

Karule P, Dalvi V, Kadu A, Chaudhari R, Subramaniam VR, Patil AB (2016) A commercial micropropagation protocol for Virupakshi (AAB) banana via apical meristem. Afr J Biotechnol 15(11):401–407

Karuna Y, Rao K (2016) Studies on phenological characters of different banana cultivars (*Musa*) in Visakhapatnam, Andhra Pradesh. Int J Sci Res (IJSR) 5(5):1689–1693

Khatri A, Khan IA, Siddiqui SH, Ahmad M, Siddiqui K (1997) *In vitro* culture of indigenous and exotic banana clones for maximizing multiplication. Pak J Bot 29:143–150

Lalrinsanga R, Vanlaldiki H, Meitei WI (2013) *In vitro* shoot tip culture of banana cultivar Meitei hei. The Bioscan 8(3):839–844

Marín JA, Gella R, Herrero M (1988) Stomatal structure and functioning as a response to environmental changes in acclimatized micropropagated *Prunus cerasus* L. Ann Bot 62:663–670

Mateille T, Foncelle B (1988) Micropropagation of *Musa* AAA cv. Poyo in the Ivory Coast. Trop Agric (Trinidad) 65:325–328

Murashige T, Skoog F (1962) A revised medium for rapid growth & bioassays with tobacco tissue cultures. Physiol Plant 15:473–497

Ngomuo M, Mneney E, Ndakdemi PA (2014) The *in vitro* propagation techniques for producing banana using shoot tip cultures. Am J Plant Sci 5:1614–1622

Noor-Aziah J, Khalid N (2002) Comparative analysis of regenerants between single and naked meristem (scalps) of *Musa Acuminata* Var. Berangan. Asia Pacific J Mol Biol Biotechnol 10:127–131

Pan MJ, Van-Staden J (1998) The use of charcoal in *in vitro* culture—a review. Plant Growth Regul 26:155–163

Pospisilova J, Ticha I, Kadlecek P, Haisel D, Plzakova S (1999) Acclimatization of micropropagated plants to *ex vitro* conditions. Biol Plant 42(4):48–497

Preece JE, Sutter EG (1991) Acclimatization of micropropagated plants to the greenhouse and field. In: Debergh PC, Zimmerman RH (eds) Micropropagation: technology and application. Kluwer Academic Publishers, Dordrecht, pp 71–93

Priyono (2001) Micropropagation of Banana (*Musa paradisiaca*) through cormlet initiation by *in vitro* culture of apical meristem slices. Jurnal Ilmu Dasar 2(1):41–48

Rabbani MG, Ali MH, Mondal MF (1996) Effect of BAP and IBA on micropropagation of some banana cultivars. Bangladesh Hort 25(1 & 2):47–52

Reddy DDR, Suvarna D, Muralidhra Rao D (2014) Effects of 6-benzyl amino purine (6-BAP) on *in vitro* shoot multiplication of Grand Naine (*Musa* sp.). Int J Adv Biotechnol Res 5:36–42

Safwat G, Abdul rahman F, Sharbasy S (2015) The effect of some antioxidants on blackening and growth of *in vitro* culture of Banana (*Musa* spp.cv. Grand Naine). Egypt J Genet Cytol 44:47–59

Sahoo C, Beura S, Rout S, Beura R (2015) High frequency *in vitro* cloning of banana (*Musa acuminata*) cv. Grande Naine. Int J Agric Environ Biotechnol 8(4):943–950

Saifuldeen AH (2015) Evaluation of different cytokinins for *in vitro* multiplication of Banana var. Robusta. Dissertation submitted for the degree of MSc in Botany from Acharya Nagarjuna University and the work carried out at Labland Biotech Private Limited, Mysore

Seelye JS, Burge GK, ED Morgan R (2003) Acclimatizing tissue culture plants: reducing the shock. Proc Int Plant Propag Soc 53:85–90

Shashikumar R, Krishna V, Ramu V (2015) High frequency plant regeneration of *Musa paradisiaca* cv. Karibale Monthan. Int J Appl Sci Biotechnol (IJASBT) 3(2):202–209

Strosse H, Houwe I, Panis B, Jain S, Swennen R (2004) Banana cell and tissue culture—review. In: Banana improvement: cellular, molecular biology, and induced mutations. Proceedings of a meeting, Leuven, 2001, pp 1–12

Sutter E (1988) Stomatal & cuticular water loss from apple, cherry and sweet gum plants after removal from *in vitro* culture. J Am Soc Hort Sci 113:234–238

Swamy RD, Rao N, Chacko EK (1983) Tissue culture propagation of banana. Sci Hort 18:247–252

Thomas T (2008) The role of activated charcoal in plant tissue culture. Biotechnol Adv 26:618–631

Venkatachalam L, Thimmaraju R, Sreedhar RV, Bhagyalakshmi N (2006) Direct shoot and cormlet regeneration from leaf explants of "silk" banana (AAB). In Vitro Cell Dev Biol Plant 42(3):262–269

Vuylsteke DR (1989) Shoot-tip culture for the propagation, conservation and exchange of "Your growing guide to better farming guide"— Manual. Agrilink Series QAL9807

Woolley DJ, Wareing PF (1972) The interaction between growth promoters in apical dominance. New Phytol 71:781–793

In Vitro Method of High-Frequency Plant Regeneration Through Internodal Callus of *Ruta graveolens* L.

46

Malik Aabid Hussain and Varsha Nitin Nathar

Abstract

Ruta graveolens L. is of considerable interest worldwide because of its medicinal properties. In vitro culture is a useful tool for both multiplication and study of important secondary metabolites. It was intended to develop an effective in vitro indirect propagation protocol for *Ruta graveolens* L. from internodal explants. The explants excised from 10-week-old plants were placed on Murashige and Skoog (MS) medium supplemented with various concentrations and combinations of auxins, indole-3-butyric acid (IBA), 2,4-dichlorophenoxy acetic acid (2,4-D), and naphthalene acetic acid (NAA), and cytokinins (benzyl amino purine (BAP) and kinetin) singly and in combinations for callus induction. The callus response was maximum (90.00 ± 1.66) in 2,4-D (1.5 mg/l) + NAA (1.5 mg/l). The friable callus obtained was subcultured on MS full-strength medium fortified with different concentrations of plant growth regulators. The highest percent response (81.22 ± 0.57) per culture for shoot bud regeneration was noted for the concentration of BAP (1.5 mg/l) with IBA (1.0 mg/l). The same concentration effectively increased the number of shoots per culture. Different concentrations of indole-3-butyric acid (IBA) and indole acetic acid (IAA) were used in half-strength MS medium for in vitro rooting of regenerated shoots. The maximum percentage of rooting (86.10 ± 0.50) and the highest number of root formations (8.11 ± 0.19) per shoot were observed on the medium containing 0.50 mg/l of IBA. Plantlets with well-developed root and shoot systems were successfully acclimatized (85%) and established in earthen pots.

Keywords

Callus culture · Organogenesis · Regeneration · *Ruta graveolens* L.

M. A. Hussain · V. N. Nathar (✉)
Plant Tissue Culture Laboratory, Department of Botany, Sant Gadge Baba Amravati University, Amravati, Maharashtra, India

S. M. Khasim et al. (eds.), *Medicinal Plants: Biodiversity, Sustainable Utilization and Conservation*, https://doi.org/10.1007/978-981-15-1636-8_46

46.1 Introduction

Ruta graveolens L. of family Rutaceae (order Sapindales) commonly known as rue (Hindi) and sitab (Marathi) is an herbaceous perennial, originally native to the Mediterranean region. It is an ornamental aromatic culinary of medicinal importance cultivated in gardens. The uses of rue in cuisine and in medicine are well described (Ghirardini et al. 2007). It prefers rocky well-drained soils, and it resists dry weather. The phytochemical profile of rue is quite complex. The plant has, in fact, been known in folk medicine since ancient times and is currently used for the treatment of such varied disorders such as pain, rheumatism, eye problems, and dermatitis. It contains more than 120 compounds of different classes of natural products such as acridone alkaloids, coumarins, essential oils, flavonoids, and fluoroquinolones (Feo et al. 2002). Many of these compounds are physiologically active and therefore of immense pharmaceutical interest. Extracts of rue have been used as antidotes for some snake and scorpion venoms (Sallal and Alkofahi 1996) and to treat many infections and inflammation (Kataki et al. 2014). Rue extracts have been shown to have potent anti-cancer activity, exhibited through strong anti-proliferative and anti-survival effects on cancer cells (Fadlalla et al. 2011). The alkaloid skimmianine has been demonstrated to have anti-inflammatory activity (Ratheesh et al. 2013).

Recently there has been an increased interest in in vitro culture techniques that offer a multiple tool for mass multiplication and germplasm preservation of aromatic and medicinal plants (Faisal and Anis 2003). In vitro cultures of *Ruta graveolens* are a potential source of secondary metabolites (furanocoumarins) of significant medical interest. Massot et al. (2000) reported that micropropagated shoots were richer in furanocoumarins than any other plant material.

Protocols for in vitro plant production via direct and indirect morphogenesis have many potential applications to medicinal species such as *Ruta graveolens*, e.g., true-to-type large-scale propagation of superior genotypes and genotypic improvement via mutagenesis and/or genetic engineering, respectively. Direct shoot bud induction in nodal segments of *Ruta graveolens* through axillary shoot multiplication has been reported earlier (Faisal et al. 2005). The present communication deals the development of a rapid regeneration system from leaf callus (indirect organogenesis) and the subsequent transplantation of the plantlets to natural environmental conditions.

The plant material was collected from Village Kat Amla situated at 337 m above sea level and Melghat forest located at 21°26′45″N 77°11′50″E in northern part of Amravati district of Maharashtra state in India. They were maintained in Botanical Garden, Department of Botany, Sant Gadge Baba Amravati, University, Amravati, under regular watering.

The internodal explants from field-grown plants were sterilized with running tap water followed by soaking in 2% Tween 20 solution for 7 min. After washing, explants were surface sterilized with (0.1%) mercuric chloride for 3 min which was followed by dipping of explants in 70% ethyl alcohol for 2 min. Then explants were washed with distilled water three times in order to remove the traces of surface sterilants. The sterilized explants were inoculated on full-strength Murashige and

Skoog's basal medium (1962) supplemented with different concentrations and combinations of growth regulators such as 2,4-D, NAA, 2,4-D + NAA, NAA + BAP, and BAP + IBA containing 3% (w/v) sucrose and 0.8% (w/v) agar. The pH of the medium was adjusted to 5.8, and it was autoclaved at 121 °C under 15 psi for 20 min. Cultures were incubated at 25 ± 2 °C under 16-h photoperiod and 55 ± 5% relative humidity for callus induction.

The yellowish compact calli formed after 5 weeks of culture were subcultured for shoot differentiation on MS media supplemented with 30 g/l sucrose and 8 g/l agar containing different concentrations/combinations of BAP, kinetin, NAA, and IBA to induce multiple shoots. The cultures were incubated at 16-h light and 8-h dark cycle at 25 ± 2 °C.

In vitro regenerated adventitious shoots (30–50 mm in length) were separated from cultures, were washed thoroughly with sterile distilled water several times, and were transferred to half-strength solid sterile MS medium supplemented with different auxin concentrations alone and in combination with BAP for in vitro rooting.

46.2 Plants Regeneration Through Internodal Callus of *Ruta graveolens* L.

46.2.1 Callus Induction

Different concentrations of plant growth regulators had a significant effect on callus regeneration. Internodal explants of *Ruta graveolens* were cultured on MS media fortified with 2,4-D, 2,4-D + NAA, NAA + BAP, and BAP + IBA in different concentrations to test the efficiency of sprouting of various explants. It was observed that after 9–12 days of inoculation, the explants showed callus initiation in all the concentrations, but the callogenic response was different in various hormonal combinations/concentrations. The maximum callus induction response 90.00 ± 1.67 was achieved in 2,4-D + NAA (1.5 + 1.5) mg/l followed by 87.89 ± 0.77 at 2,4-D 2.0 mg/l, and the lowest callogenic response 42.55 ± 1.39 was observed in 2,4-D 0.5 mg/l. The callus induction response varied from 42.55 ± 1.39 to 90.00 ± 1.67 (Table 46.1, Fig. 46.1a, b).

46.2.2 Shoot Induction

The yellowish compact calli formed were subcultured for shoot differentiation on MS media supplemented with 30 g/l sucrose and 8 g/l agar containing different concentrations/combinations of BAP, kinetin, NAA, and IBA to induce multiple shoots. Multiple shoot initiation started from the callus after 2 weeks when subcultured onto fresh MS medium containing different cytokinins like BAP and kinetin and auxins like NAA and IBA. Among the various combinations of auxins and cytokinins, BAP + IBA (1.5 + 1.0 mg/l) showed maximum response, i.e., 78.78 ± 1.35 for multiple shoot induction. The same concentration effectively

Table 46.1 Effect of plant growth regulators on callus induction from internodal explants

Plant growth regulator (PGR)	Concentration (mg/l)	Mean response percent (%)
Control	–	–
2,4-D	0.5	42.55 ± 1.39
	1.0	63.33 ± 1.67
	1.5	71.66 ± 1.67
	2.0	87.89 ± 0.77
	2.5	68.33 ± 1.67
	3.0	67.55 ± 1.17
NAA	0.5	55.00 ± 1.67
	1.0	63.33 ± 1.67
	1.5	67.77 ± 1.57
	2.0	83.44 ± 1.02
	2.5	63.33 ± 1.67
	3.0	55.00 ± 1.67
2,4-D + NAA	1.0 + 1.0	68.33 ± 1.67
	1.5 + 1.5	90.00 ± 1.67
	1.5 + 1.0	83.44 ± 1.02
NAA + BAP	1.5 + 2.5	75.55 ± 0.96
	2.0 + 2.0	71.66 ± 1.67
	2.5 + 1.5	75.55 ± 0.96
	2.5 + 2.0	71.66 ± 1.67
BAP + IBA	1.0 + 0.5	71.66 ± 1.67
	1.0 + 2.0	83.44 ± 1.02
	2.0 + 1.0	64.67 ± 1.53

Values are mean ± SD of three replicates

increased the number of shoots per culture. The highest mean number of shoots, i.e., 22.55 ± 1.07, and highest mean shoot length (mm), i.e., 30.22 ± 1.84, were observed at same concentration, while the lowest shoot induction response 35.11 ± 1.35, lowest mean number of shoots 6.89 ± 1.26, and lowest shoot length 9.66 ± 1.34 were observed at BAP (0.5 mg/l) (Table 46.2 and Fig. 46.1c, d).

46.2.3 Rooting and Acclimatization

Rooting of the in vitro regenerated shoots (32–79 mm in length) did not occur on full-strength MS medium though the media were supplied with rooting hormones like IAA and IBA, so they were separated from cultures, were washed thoroughly with sterile distilled water several times, and were transferred to half-strength solid sterile MS medium supplemented with different auxin concentrations as well as a cytokinin (BAP) in combination with auxins for in vitro rooting. Rooting was obtained in 3 weeks after transferring the shoots to rooting media containing hormones like IAA, IBA, and BAP. The maximum percentage of root induction 85.22 ± 1.71 was achieved at IBA 0.5 mg/l which was followed by 82.33 ± 1.20 at 1.0 mg/l of IAA, and the highest number of root formations (8.22 ± 1.07) per shoot was observed on the medium containing 0.5 mg/l of IBA followed by 7.44 ± 1.26 at

Fig. 46.1 In vitro studies on *Ruta graveolens* L. (**a**, **b**) Callogenic response from internodal explants at 2,4-D + NAA (1.5 + 1.5) mg/l; (**c**, **d**) multiple shoot formation from callus at BAP (0.5) mg/l and BAP + IBA (1.5 + 1.0) mg/l, respectively; (**e**) multiple root formation at IBA (0.5) mg/l; (**f**) acclimatized plant

IAA 1.0 mg/l. The maximum mean root length 36.00 ± 1.45 per shoot was achieved at 0.5 mg/l IBA followed by 35.22 ± 1.50 at IBA 1.0 mg/l. After 0.1mg/l, though IAA also induces 82% rooting, the mean root number and mean root length were lower than IBA (Table 46.3 and Fig. 46.1e).

The plantlets with well-developed root and shoot systems were first maintained in a programmable environmental chamber for 1 month and then isolated from the culture vessels and washed with sterile double distilled water to remove adhering medium. The hardened plants were then transferred to the earthen pots containing sterilized mixture of soil, sand, and vermicompost and were kept in the greenhouse for acclimatization. The 80% of plantlets were successfully acclimatized (Fig. 46.1f). No variation was observed among the acclimatized plants with respect to morphological characteristics.

Table 46.2 Effect of plant growth regulators on indirect shoot regeneration from internodal callus

Plant growth regulator (PGR)	Concentration (mg/l)	Mean response percent (%)	Mean shoot number	Mean shoot length (mm)
Control	–	–	–	–
BAP	0.5	35.11 ± 1.35	6.89 ± 1.26	9.66 ± 1.34
	1.0	39.22 ± 1.71	9.48 ± 0.74	9.89 ± 1.17
	1.5	56.67 ± 1.53	16.11 ± 1.39	14.11 ± 1.83
	2.0	47.88 ± 1.35	12.77 ± 1.50	12.11 ± 1.35
KIN	0.5	39.99 ± 1.15	8.88 ± 1.57	12.77 ± 1.02
	1.0	46.44 ± 1.17	10.22 ± 1.17	15.66 ± 1.86
	1.5	60.22 ± 1.65	14.89 ± 1.26	17.89 ± 1.50
	2.0	55.44 ± 1.26	11.22 ± 1.17	17.33 ± 1.46
BAP + IBA	1.0 + 1.0	61.88 ± 1.35	15.55 ± 1.26	17.66 ± 1.53
	1.5 + 1.0	78.78 ± 1.35	22.55 ± 1.07	30.22 ± 1.84
	2.0 + 1.0	75.55 ± 1.02	18.00 ± 1.21	28.89 ± 1.64
	2.5 + 1.0	71.33 ± 1.20	15.77 ± 1.17	26.44 ± 1.39
BAP + KIN	1.5 + 1.0	61.66 ± 1.15	18.55 ± 1.07	17.44 ± 1.58
	2.0 + 1.0	68.00 ± 1.34	17.77 ± 1.71	26.44 ± 1.54
	2.5 + 1.0	75.00 ± 1.34	16.00 ± 1.15	27.33 ± 1.86
	1.5 + 1.5	69.44 ± 1.17	15.22 ± 1.07	19.33 ± 1.53

Values are mean ± SD of three replicates

Table 46.3 Effect of plant growth regulators on in vitro root induction of indirectly regenerated shoots of *Ruta graveolens* L.

Plant growth regulator (PGR)	Concentration (mg/l)	Mean response percent (%)	Mean root number	Mean root length (mm)
Control	–	–	–	–
IAA	0.25	72.66 ± 1.21	4.33 ± 1.15	16.67 ± 1.15
	0.50	80.89 ± 1.83	5.11 ± 1.02	26.44 ± 1.26
	1.0	82.33 ± 1.20	7.44 ± 1.26	29.22 ± 1.17
	1.5	75.78 ± 1.35	4.78 ± 1.35	21.55 ± 1.65
IBA	0.25	74.66 ± 1.15	6.55 ± 1.35	31.33 ± 1.53
	0.50	85.22 ± 1.71	8.22 ± 1.07	36.00 ± 1.45
	1.0	77.55 ± 1.39	7.44 ± 1.39	35.22 ± 1.50
	1.5	69.33 ± 1.86	4.89 ± 0.96	24.22 ± 1.34
IBA + BAP	1.0 + 0.5	18.33 ± 1.20	2.22 ± 1.02	9.88 ± 1.35
	1.5 + 0.5	16.33 ± 1.15	2.33 ± 1.00	7.00 ± 1.45
	2.0 + 0.5	15.99 ± 1.53	1.51 ± 0.46	7.33 ± 1.15
IAA + BAP	1.0 + 0.5	14.33 ± 1.20	2.22 ± 1.26	7.44 ± 1.17
	1.5 + 0.5	15.33 ± 1.53	3.99 ± 1.15	10.11 ± 1.5
	2.0 + 0.5	15.33 ± 1.15	3.89 ± 1.02	11.66 ± 1.21

46.2.4 Micropropagation Studies in *R. graveolens* L.

The percentage of cultures showing callus formation was less at lower concentrations of plant growth regulators. However, an increase was observed in the frequency of callus initiation response with increased concentration of plant growth regulators up to a certain level, while a decline in the rate of callus growth was observed when the PGR concentrations were used above 2.5 mg/l. The maximum organogenic callus induction frequency (70.6 ± 2.33%) from leaf explants was observed on Murashige and Skoog (MS) medium supplemented with 10 μM 2,4,5-trichlorophenoxyacetic acid (2,4,5-T) (Ahmad et al. 2010). Also the maximum number of callus induction was achieved from leaf explants on Murashige and Skoog (MS) medium enriched with 0.5 mg/l IAA and 1.0 mg/l of 2,4-D which yielded morphogenic compact yellowish green calli. Shabana et al. (2002) have reported the highest survival values in callus formation from stem and root cultures on MS media supplemented with 0.5 mg/l 2,4-D + 0.5 mg/l kinetin. Callus induction was conducted from leaf discs on MS (Murashige and Skoog) media supplemented with 4.0 mg/l NAA (Al-Ajloun et al. 2015).

Indirect regeneration of plants was influenced by different combinations and concentrations of plant growth regulators. Several studies have clearly revealed the requirement of high auxin:low cytokinin ratio for callus induction and low auxin:high cytokinin ratio for shoot induction from the callus. Ahmad et al. (2010) reported the highest percentage of callus response and the maximum number of shoots per unit callus from leaf explants of *Ruta graveolens* in the media containing higher concentrations of NAA with lower concentrations of BA which is contrasting to the current studies as the explants used are different in both experiments. Faisal et al. (2005) have obtained the highest shoot regeneration frequency (98.5%) and the highest number of shoots per explant (40 ± 2.8) on MS medium supplemented with 10 μM BA and 2.5 μM NAA. Rapid multiplication of *R. graveolens* was achieved by culturing microshoot on MS medium containing TDZ, 2-iP, zeatin, or GA3 at various concentrations. Zeatin at 0.5 mg/l resulted in the highest shoot number and length (Al-Ajloun et al. 2015). Among the various combinations of auxins and cytokinins, BAP + IBA (1.5 + 1.0 mg/l) proved superior to all other combinations for multiple shoot bud induction as well as for shoot number and shoot length.

Half-strength MS media were supplied with IBA and IAA alone as well as in combination with BAP. In every combination, the concentration of BAP was kept constant, while the concentration of IAA and IBA was increased at each level. The response for root induction was very low when used in combination as compared to when used alone. Plantlets significantly developed lengthy roots and root induction strengthened with time. IBA was found better in inducing the roots. Cultures without any plant growth regulator (control) don't show any sign of rooting. Among the various concentrations and combinations of PGRs used, IBA 0.5 mg/l proved best for root induction, and the same concentration effectively increased the number of roots per culture. In vitro shoots were rooted best on half-strength MS medium containing 0.5 μM IBA (Faisal et al. 2005).

46.2.5 Acclimatization of Plantlets

The in vitro regenerated plantlets with sufficient roots were isolated from the culture tubes and washed with sterile double distilled water to remove adhering medium and transferred to the plastic cups containing sterile perlite. The plantlets were initially maintained in a programmable environmental chamber for 5 weeks. The temperature and humidity were adjusted to 20 ± 2 °C and 50%, respectively. The temperature of the programmable environmental chamber was raised by 1 °C every 5 days. After about 4 weeks, the plants were transferred to bigger earthen pots containing normal garden soil in greenhouse for further growth and development and were maintained under natural conditions of day length, temperature, and humidity. The acclimatized plants were irrigated regularly.

The present study developed a rapid and efficient indirect plant regeneration method for *Ruta graveolens* L. The literature reveals different regeneration systems for in vitro propagation of the plant, but mass regeneration from internodal callus has not been reported yet. The protocol developed is very simple and can be reproduced. The in vitro regeneration protocol will be helpful for the conservation of this medicinally important plant and at the same time can be utilized for the extraction and isolation of several bioactive components.

References

Ahmad N, Faisal M, Anis M, Aref IM (2010) *In vitro* callus induction and plant regeneration from leaf explants of *Ruta graveolens* L. S Afr J Bot 76:597–600

Al-Ajloun ZI, Abbas S, Shatnawi M, Al-Makhadmeh I (2015) In vitro propagation, callus induction, and evaluation of active compounds on Ruta graveolens. J Food Agric Environ 13(2):101–106

Fadlalla K, Watson A, Yehualaeshet T, Turner T, Samuel T (2011) Ruta *graveolens* extract induces DNA damage pathways and blocks akt activation to inhibit cancer cell proliferation and survival. Anticancer Res 31:233–241

Faisal M, Anis M (2003) Rapid mass propagation of *Tylophora indica* Merrill via leaf callus culture. Plant Cell Tissue Org Cult 75:125–129

Faisal M, Ahmad N, Anis M (2005) *In vitro* regeneration and mass propagation of *Ruta graveolens* L. - multipurpose shrub. HortScience 40(5):1478–1480

Feo V, Simone F, Senatore F (2002) Eight peak index of mass spectra, potential allelochemicals from the essential oil of *Ruta graveolens*. Phytochemistry 61:573–578

Ghirardini MP, Carli M, del Vecchio N, Rovati A, Cova O, Valigi F, Agnetti G, Gurudeeban S, Satyavani K, Ramanathan T, Balasubramanian T (2007) Effect of antioxidant and anti-aggregating properties of micro-propagated plantlets of *Ruta graveolens*. Afr J Biotechnol 11(6):1497–1504

Kataki MS, Kakoti BB, Bhuyan B, Rajkumari A, Rajak P (2014) Garden rue inhibits the arachidonic acid pathway, scavenges free radicals, and elevates FRAP: role in inflammation. Chin J Nat Med 12:172–179

Massot B, Milesi S, Gontier E, Bourgaud F, Guckert A (2000) Optimized culture conditions for the production of furanocoumarins by micropropagated shoots of Ruta graveolens. Plant Cell Tissue Organ Cult 62:11–19

Ratheesh M, Sindhu G, Helen A (2013) Anti-inflammatory effect of quinoline alkaloid skimmianine isolated from *Ruta graveolens* L. Inflamm Res 62:367–376

Sallal AJ, Alkofahi A (1996) Inhibition of the hemolytic activities of snake and scorpion venoms *in vitro* with plant extracts. Biomed Lett 53:211–215

Shabana MM, El-Alfy TS, El-Tantawy ME, Ibrahim AI, Ibrahim GF (2002) Tissue culture and evaluation of some active constituents of *Ruta graveolens* L. Arab J Biotechnol 5(1):45–56

Conservation of an Endangered Medicinal Forest Tree Species, *Oroxylum indicum* L. Kurz, Through In Vitro Culture: A Review

47

T. Samatha and N. Rama Swamy

Abstract

The *Oroxylum indicum* L. Kurz (Bignoniaceae) is an important medicinal tree rich in secondary metabolites and is employed in various traditional medicinal systems from folk to modern medicine. All the plant parts (root, stem bark, seed, leaf and flower) of the species are known to possess bioactive compounds, and the plant is overexploited by pharmaceutical industries. In view of the increased anthropogenic activities, destruction of natural habitat, these trees are being rapidly eroded from natural ecosystem. Thus, the tree is becoming endangered, vulnerable, will be facing extinction in the near future and is being pushed into the red data book. Hence, there is an urgent need for the development of conservation methodologies through in vitro culture techniques by employing embryo culture, meristem culture, and somatic embryogenesis, etc., for the large-scale production of this valuable medicinal tree. In the present review, we have made an attempt to compile and critically analyze the various reports on micropropagation protocols for the conservation of this endangered valuable medicinal forest tree.

Keywords

Oroxylum indicum · Conservation · Endangered · In vitro culture techniques

47.1 Introduction

Trees are an integral part of human life and a vital component of biodiversity. Forest trees are in particular renewable sources of food, fodder, fuel wood, timber and other valuable non-timber products. Due to the rapid growth of population and the

T. Samatha · N. Rama Swamy (✉)
Department of Botany, Kakatiya University, Warangal, Telangana, India

S. M. Khasim et al. (eds.), *Medicinal Plants: Biodiversity, Sustainable Utilization and Conservation*, https://doi.org/10.1007/978-981-15-1636-8_47

human desire to progress, there has been a tremendous reduction in forest cover from the earth's surface. To maintain and sustain forest vegetation, conventional approaches have been exploited in the past for propagation and improvement. However, such efforts are confronted with several inherent bottlenecks. In order to overcome these barriers, biotechnological interventions for in vitro regeneration, mass propagation and gene transfer methods in forest tree species have been successfully practiced during the last decade.

Plants have always been a major component of traditional system of healing in the developing countries, which have also been an integral part of their history and culture. Medicinal plants offer alternative remedies with tremendous opportunities. Many traditional healing herbs and plant parts have been shown to have medicinal value, especially in the rural areas, and these can be used to prevent and cure several human diseases. Even today, the majority of the world population depends on herbal healthcare practice.

Modern science and technology have an essential role to play in the process. An integrated approach for the cultivation, conservation and preservation of important plant species through in vitro culture technology/plant tissue culture; research on the rationale and methodology of ayurvedic medical practice; isolation of active compounds and their development into new therapeutics; standardization and validation of known herbal medicines and other related aspects need to be focused upon.

Plant biotechnology/plant tissue culture technology offers an important option for effective rapid multiplication, improvement and conservation of an endangered, medicinal and economically important forest tree species. Especially in the case of forest woody/tree species, conventional methods of selection and breeding are very slow, and such programs are difficult because of the prolonged life cycle of forest trees, and it requires a large space and a long time for screening of the elite ones. The potential of plant biotechnology includes: (a) enhanced production of natural products, (b) rapid clonal multiplication of selected genotypes, (c) production of disease-free plants, (d) germplasm preservation and (e) gene manipulation.

We have attempted to multiply and conserve an endangered and medicinally important forest tree species of *Oroxylum indicum* L. Kurz through in vitro culture technology.

The *O. indicum* (Indian trumpet tree) belongs to the family Bignoniaceae. It is a small- to medium-sized deciduous tree measuring up to 12 m in height; with a light greyish brown, soft, spongy bark, which has corky lenticels; large pinnate, bipinnate or tripinnate ovate or elliptic leaves; lurid purple, fleshy, foetid flowers and large, flat, sword-shaped capsules full of many flat and papery thin seeds, which are round with broad papery silvery wings. The tree is a night bloomer, and flowers are adapted to natural pollination by bats (Fujita and Tuttle 1991; Samatha 2013).

The *O. indicum* is an important herbal medicine in many Asian countries and is used in folk medicine to cure various diseases. Each part of this plant possesses medicinal value. The root bark is a well-known tonic and astringent useful in fever, diarrhoea, dysentry, bronchitis, intestinal worms, leucoderma, asthma, inflammation, anal troubles, etc. It is diaphoretic and is used in rheumatism and also for the treatment of tuberculosis. The seed extracts exhibit antimicrobial, analgesic, antitussive and anti-inflammatory properties. Tender fruits and seeds are refreshing and stomachic and used as an expectorant, purgative and bitter tonic. The decoction of root bark is effective in curing nasopharyngeal cancer. Leaves are emollient and contain anthraquinone and aloe-emodin. The fruits are used in treating bronchitis, leucoderma and helminthosis. In addition to its medicinal importance, the young shoots and unripe fruits are eaten as vegetables. The wood of the tree is used to make match boxes. The stem bark and fruits of the tree are used as mordant and yield colour dye (Jain et al. 2003).

Natural regeneration of *O. indicum* is slow due to the presence of high percentage of aborted (70%) seeds; low fruit set is due to poor pollination efficiency (Chirepterophily), only 30% of the seeds were found to be viable and also germination fails frequently due to fungal infection of seeds (Samatha et al. 2016a, b). In addition, the destruction is caused by harvesting roots, stem and also whole plant as they serve as a major source of drug in various medicinal systems. Futher, low seed viability and low rate of germination restricted the natural populations of this tree. These are some of the reasons for the species at the verge of extinction and has become an endangered. Due to its demand in pharmaceutical industry, it has been pushed into the list of *Indian red data book*. The existence of *O. indicum* in natural population is highly threatened and has been categorized as vulnerable by the government of India. The literature survey showed that the species needs conservation and mass-scale multiplication.

Hence, we made an attempt for conservation and mass-scale multiplication of this medicinal plant by using different types of plant biotechnological approaches.

47.2 In Vitro Seed Germination

For in vitro seed germination, different types of media such as MS, B_5 and WPM with varying concentrations of sucrose were used to enhance the percentage of seed germination and better seedling formation (Samatha et al. 2016a, b). The decoated seeds of *O. indicum* were inoculated on different types of media (MS, B_5 and WPM) containing various levels of sucrose concentrations. Absolute percentage of germination was recorded on all types of MS media irrespective of the concentration of sucrose. Among the various types of media used, seed germination was effective (100%) on ½ strength MS and MS medium containing different concentrations of sucrose, and the time taken for the germination was also less (6–7 days) compared to the other types of media used (Table 47.1). Healthy seedlings with elongated shoot system were found on MS medium containing 15–30 g/L sucrose (Fig. 47.1). The percentage of seed germination was found to be lesser on B_5 and WPM media in all the concentrations of sucrose used (Samatha et al. 2016a, b).

Table 47.1 Effect of different types of media on in vitro seed culture of *Oroxylum indicum*

Type of medium	% Germination	No. of days for seed germination	Morphology of seedling	Seedling height (cm)[a]	
				Shoot length	Root length
1/2 MS + 30 g/L sucrose	100	6	Healthy elongated shoots and lengthy lateral roots	1.9 ± 0.10	1.02 ± 0.024
1/2 MS + 20 g/L sucrose	100	7	Feeble shoots and roots	1.7 ± 0.07	1.02 ± 0.024
MS + 30 g/L sucrose	100	6	Healthy elongated shoots and roots	2.0 ± 0.09	1.4 ± 0.023
MS + 20 g/L sucrose	100	7	Healthy elongated shoots and roots	2.1 ± 0.10	1.35 ± 0.011
MS + 15 g/L sucrose	100	7	Healthy shoots with many lateral roots	1.7 ± 0.07	1.35 ± 0.011
B_5 + 30 g/L sucrose	25	30	Feeble shoots and roots	0.9 ± 0.11	0.6 ± 0.026
B_5 + 20 g/L sucrose	25	30	Feeble shoots and roots	0.9 ± 0.11	0.5 ± 0.023
B_5 + 15 g/L sucrose	25	28	Feeble shoots and root	0.7 ± 0.12	0.6 ± 0.026
WPM + 30 g/L sucrose	15	25	Feeble shoots and root	0.9 ± 0.11	0.8 ± 0.02
WPM + 20 g/L sucrose	15	28	Feeble shoots and roots	0.7 ± 0.12	0.5 ± 0.023
WPM + 15 g/L sucrose	15	30	Feeble shoots and roots	0.4 ± 0.07	0.8 ± 0.02

[a]Mean ± standard error

47.3 In Vitro Zygotic Embryo Culture

In vitro zygotic embryo culture plays a vital role in rapid multiplication of this endangered tree species in a short time. In vitro zygotic embryo culture was successful in improving the percentage of germination in *O. indicum.* To reduce the time required to obtain a plantlet and also to avoid the problem of seed dormancy, an attempt has been made on in vitro zygotic embryo culture for the conservation of the species.

The zygotic embryos of *O. indicum* were cultured on MS medium fortified with different concentrations of BAP/Kn/TDZ (Fig. 47.2). Emergence of healthy embryonal axis from zygotic embryos with shoot and root within 2–18 days was observed on all the concentrations of BAP/Kn/TDZ used. But an early germination was noted at 5.0 mg/L BAP/Kn/TDZ (Table 47.2). The percentage of germination was enhanced gradually from lower concentration of BAP/Kn to 5.0 mg/L BAP/Kn/

Fig. 47.1 (**a–h**) In vitro seed germination in *O. indicum*. (**a**) Germination of the seeds on WPM + 30 g/L sucrose; (**b**) Germination of the seeds on B_5 + 30 g/L sucrose; (**c**) Germination of the seeds on MS medium + 30 g/L sucrose; (**d, e**) Development of seedling on MS medium + 30 g/L sucrose with fully expanded cotyledonary leaves after 30 and 45 days of germination respectively; (**f**) Acclimatization of plantlet on soilrite in a plastic pot; (**g**) In vitro raised plantlet showing lengthy root system; (**h**) Acclimatized plantlet in an earthenware pot containing garden soil

TDZ (Samatha et al. 2013a). Absolute percentage of germination was found at 5.0 mg/L BAP/Kn/TDZ. The zygotic embryos cultured on MS medium fortified with 1.0–7.0 mg/L BAP/Kn/TDZ induced multiple shoots. Maximum number (20 ± 0.82) of multiple shoots per explant was developed at 5.0 mg/L BAP. MS medium fortified with BAP alone showed superiority over Kn/TDZ for the induction of multiple shoots in *O. indicum*. Of all the concentrations of cytokinins tested, 4.0–5.0 mg/L BAP were found to be effective for induction of more number of multiple shoots/explant (Samatha et al. 2013a).

47.4 In Vitro Regeneration

Plant regeneration/organogenesis in *O. indicum* is carried out through callus-mediated and direct regeneration (Fig. 47.3). For callus-mediated regeneration, leaf and cotyledonary leaf explants were cultured on MS medium containing 30 g/L

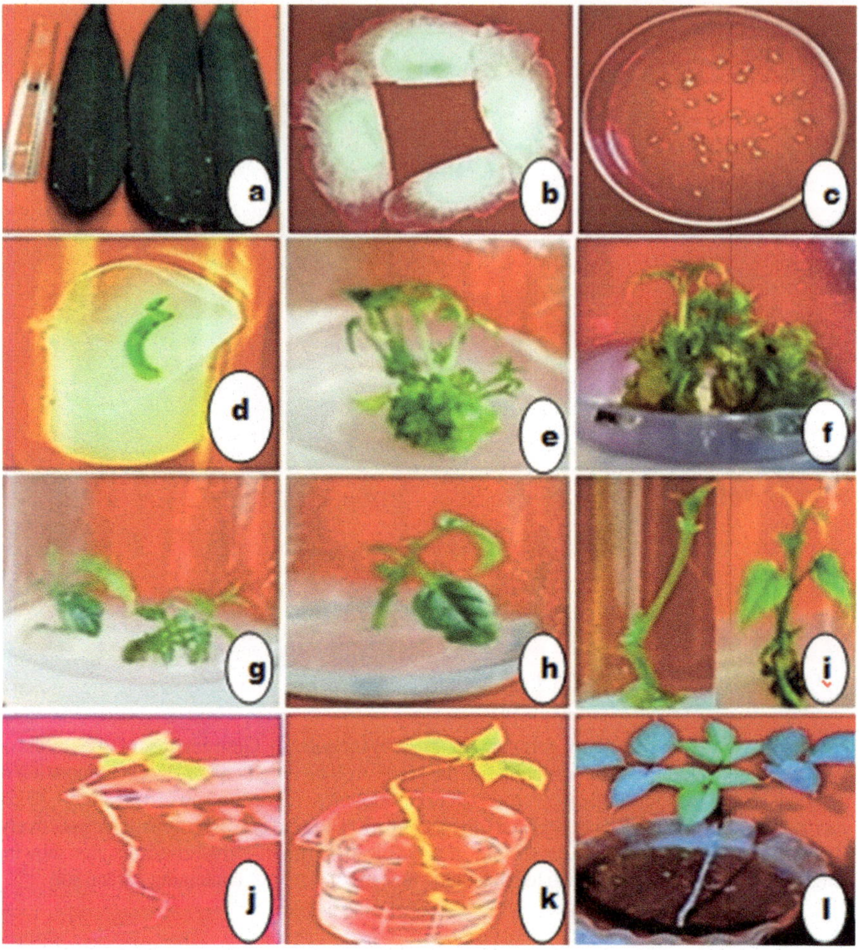

Fig. 47.2 In vitro zygotic embryo culture in *O. indicum*. (**a**) Fruits; (**b**) Seeds; (**c**) Isolated zygotic embryos; (**d**) Germination of zygotic embryo; (**e**, **f**) Multiple shoots formation on MS + 3.0 mg/L TDZ after 6 weeks respectively; (**g**, **h**) Individual micro-shoots; (**i**) Elongation of individual shoots on MS + 1.0 mg/L GA_3; (**j**, **k**) In vitro raised plant with well-developed root system; (**l**) Acclimatized plantlet in an earthenware pot containing garden soil

sucrose fortified with different concentrations (1.0–5.0 mg/L) of 2,4-D/IAA/IBA/NAA. Callusing response was found to be varied from auxin to auxin and also with explant to explant in *O. indicum*. Callusing response was higher in cotyledonary leaf explants cultured in all concentrations of auxins and used except at high and low concentrations of PGRs tested. Maximum frequency of responding cultures and high amount of callusing were recorded at 2.0 and 3.0 mg/L auxins used. The texture of callus was also found to be varied based on the type of auxin and the explant used. The calli consisting of green nodular/friable were cultured up to 3 cycles on the same concentrations of PGRs used (Samatha et al. 2016a, b).

Table 47.2 Effect of BAP, Kin and TDZ on zygotic embryo culture in *O. indicum*

Concentration of PGR (mg/L)	% of germination	Time taken for germination (days)	Average no. of shoots/ explant (±SE)[a]
BAP/KIN/TDZ			
1.0	40	7	4 ± 0.67
	30	10	3 ± 0.45
	26	8	3 ± 0.47
2.0	64	9	7 ± 0.27
	72	8	7 ± 0.27
	39	10	4 ± 0.58
3.0	82	5	9 ± 0.75
	80	6	8 ± 0.72
	45	11	6 ± 1.02
4.0	90	3	16 ± 1.02
	84	4	12 ± 0.42
	51	14	7 ± 0.52
5.0	100	2	20 ± 0.82
	100	2	10 ± 1.07
	65	18	8 ± 0.42
6.0	80	9	6 ± 0.45
	74	9	9 ± 0.75
	43	13	6 ± 0.25
7.0	76	10	3 ± 0.54
	65	11	4 ± 0.67
	24	9	4 ± 0.34

[a]Mean ± standard error

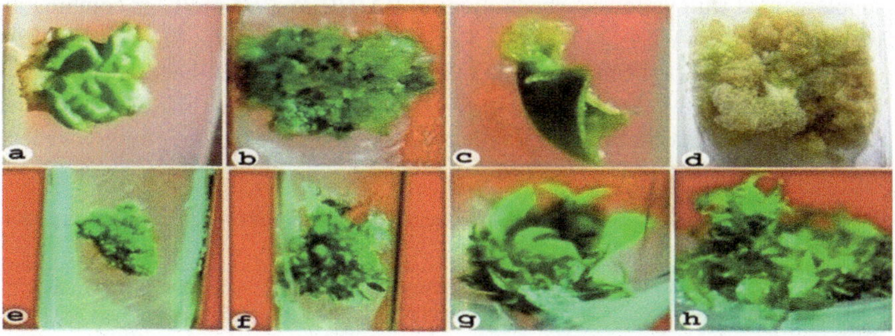

Fig. 47.3 In vitro regeneration in *O. indicum*. (**a–h**) Callus induction from leaf,cotyledonary explants and direct regeneration from leaf explants of *O. indicum.* (**a**, **b**) Callus induction and Proliferation of callus from Leaf expalnts at 3.0 mg/L 2,4-D respectively; (**c**, **d**) Callus induction and high amount of friable callus from cotyledonary leaf explants at 2.0 mg/L 2,4-D; (**e**) Formation of adventitious shoot buds on MS + 1.0 mg/L BAP; (**f**) Multiple shoot induction on MS + 5.0 mg/L BAP; (**g**, **h**) Development of multiple healthy shoots on MS + 0.5 mg/L 2,4-D + 5.0 mg/L BAP after 4 and 6 weeks respectively

For callus-mediated regeneration the freshly isolated, green nodular/friable calli were used after third subculture, obtained from leaf and cotyledonary leaf explants. These calli pieces were cultured on MS medium fortified with different concentrations of BAP/Kn (1.0–4.0 mg/L) alone and also in combination with 0.5 mg/L IAA. But we could not achieve the shoot buds induction even after 3 cycles of cultures except the formation of green meristemoids (Samatha et al. 2016a, b).

For direct regeneration and to know the efficacy of PGRs for induction of shoot organogenesis, the leaf explants were cultured on MS medium supplemented with various concentrations (1.0–5.0 mg/L) of BAP/Kn alone and also in combination with 0.5 mg/L 2,4-D (Samatha et al. 2013a).

Adventitious shoots were induced after 2 weeks of incubation in all the concentrations of BAP and Kn as a sole PGR. Maximum percentage (100%) of response with more number of shoots formation (45 ± 0.04) per explant was found at 5.0 mg/L BAP compared to Kn (20 ± 0.13). As the concentration of the PGRs increased, there was gradual enhancement in the shoot bud induction from 2.0–5.0 mg/L BAP/Kn. Repeated subculture of explants with shoot buds on fresh shoot proliferation medium helped to achieve continuous production of healthy shoot buds. Achievement of direct organogenesis is a breakthrough in tree species. Thus the direct organogenesis was successfully established in *O. indicum* (Samatha et al. 2016a, b).

47.5 Somatic Embryogenesis

Somatic embryogenesis and plantlet formation was established in *O. indicum.* Cotyledonary leaf explants were cultured on MS medium containing 30 g/L sucrose supplemented with various concentrations of BAP (0.25–0.75 mg/L) in combination with 0.5 mg/L IAA/2,4-D (Table 47.3). Somatic embryos were induced in all the combinations and concentrations of PGRs used. Callus was initiated at cut ends of the explant after 2 weeks of culture in all concentrations of BAP + IAA/2,4-D tested (Samatha and Rama Swamy 2015). The gradual initiation of somatic embryogenesis

Table 47.3 Effect of IAA/2,4-D + BAP on induction of somatic embryogenesis from cotyledonary leaf explants in *O. indicum*

Concentration of PGRs (mg/L)	% Cultures with somatic embryogenesis	Average no. of embryoids/ explant ± (SE)[b]	% Somatic embryos conversion into bipolar ± (SE)[a]
IAA + BAP			
0.5+ 0.25	58	13 ± 0.01	10 ± 0.02
0.5+ 0.50	68	35 ± 0.16	25 ± 0.01
0.5+ 0.75	89	58 ± 0.70	48 ± 0.16
2,4-D + BAP			
0.5+ 0.25	63	23 ± 0.16	08 ± 0.01
0.5+ 0.50	78	45 ± 0.27	43 ± 0.03
0.5+ 0.75	95	63 ± 0.18	78 ± 0.21

[a]Mean ± standard error; data scored after 6 weeks

Fig. 47.4 Somatic embryogenesis and plantlet development from cotyledonary leaf explants in *O. indicum*. (**a**) Induction of somatic embryos on MS + 0.5 mg/L 2,4-D + 0.75 mg/L BAP; (**b**) Development of different types of embryoids; (**c**) A group of bipolar embryos; (**d**, **e**) Germination of somatic embryo and elongation of shoot after germination, respectively

was noticed after the third week of incubation. Highest percentage (89%) of response in inducing somatic embryogenesis was found at 0.25 mg/L BAP + 0.5 mg/L IAA. Maximum frequency of somatic embryo conversion was also noted at the same combination of PGRs. The explants containing somatic embryoids were subcultured on the fresh medium containing the same combination of PGRs for the maturation of somatic embryoids from globular to different forms depending upon the PGRs present in the medium (Fig. 47.4). Whereas, the morphogenic event was absent when the globular stage embryos were cultured on MS medium without PGRs. Maximum frequency number of bipolar embryos were germinated and converted into plantlets at 0.2 mg/L IAA + 0.25 mg/BAP, followed by 0.3 mg/L IAA + 0.25 mg/L BAP. Thus somatic embryogenesis and plantlet formation have been successfully achieved in *O. indicum* (Samatha and Rama Swamy 2015).

In vitro multiplication through micropropagation has been attempted in *O. indicum* for conservation of the species using shoot tip, nodal and cotyledonary node culture.

47.6 Shoot Tip Culture

For meristem culture, shoot tip explants were cultured on MS medium containing 30 g/L sucrose fortified with different concentrations (1.0–5.0 mg/L) of BAP/Kn and TDZ (0.2–1.0 mg/L) individually. The shoot tips cultured on MS medium without growth regulators (MSO) showed the elongation of single shoot. High percentage (79%) of response and increased number of multiple shoots formation per explant was recorded at 3.0 mg/L BAP compared to all other concentrations of BAP/Kn and TDZ used (Fig. 47.5). Maximum number of multiple shoots/explant (12.0 ± 1.08) was found at the same concentration of BAP. In the present investigation, BAP had shown superiority over Kn and TDZ for induction of more number of multiple shoots from shoot tip cultures in *O. indicum* (Samatha et al. 2016a, b).

47.7 Nodal Culture

For clonal propagation, the nodal explants of *O. indicum* were cultured on MS medium supplemented with different concentrations of BAP/Kn (1.0–5.0 mg/L) and TDZ (0.2–1.0 mg/L) as a sole PGR. As the concentration of BAP increased up to 5.0 mg/L, the percentage of response was also increased and a maximum of 89% response was observed at 5.0 mg/L BAP. More number of multiple shoots (15.0 ± 0.10) formation per explant was found at 5.0 mg/L BAP, followed by 4.0 mg/L BAP compared to all other concentrations of Kn and TDZ tested. In the present investigation, BAP showed superiority in inducing more number of multiple shoots compared to Kn and TDZ in *O. indicum* (Samatha et al. 2016a, b).

47.8 Cotyledonary Node Culture

The cotyledonary nodal explants of *O. indicum* were cultured on MS medium supplemented with different concentrations (1.0–5.0 mg/L) of PGRs viz., BAP/Kn/TDZ. The cotyledonary nodal explants cultured in all concentrations of PGRs supported the induction and proliferation of axillary bud break in *O. indicum.* As the concentration of BAP increased up to 5.0 mg/L, the percentage of response was also increased and a maximum response of 88% was observed at 4.0–5.0 mg/L BAP. Maximum percentage of response and more number of multiple shoots (10.0 ± 0.42) formation per

Fig. 47.5 In vitro micropropagation in *O. indicum*. (**a–c**) Initiation of multiple shoots formation from shoot tip, node and cotyledonary nodal segments on MS + 3.0 mg/L BAP after 3 and 4 weeks of culture respectively and further multiplication; (**d–e**) elongation of healthy shoots on MS + 3.0 mg/L BAP + 1.0 mg/L GA_3; (**f**) In vitro rooted plantlet on ½ MS + 1.0 mg/L IBA; (**g**) Hardening in the culture room; (**h**) Healthy and established in vitro raised plantlet

explant was found at 3.0 mg/L BAP, followed by 4.0 and 5.0 mg/L BAP. In the present investigation, 5.0 mg/L BAP showed superiority for induction of more number of multiple shoots compared to all other concentrations of BAP/Kn and TDZ used. The explants with multiple shoots were shifted on to MS medium supplemented with 1.0 mg/LGA_3 + 3.0 mg/L BAP for further shoot proliferation and elongation. After 4 weeks, the individual shoots from these cultures were excised and cultured on MS medium augmented with 1.0 mg/L GA_3 for elongation. These elongated shoots were used for in vitro rooting (Samatha et al. 2016a, b).

Thus, in vitro micropropagation has been successfully achieved in *O. indicum* by using nodal, shoot tip and cotyledonary node explants on MS medium supplemented with different concentrations of BAP/Kn/TDZ. BAP had shown superiority in inducing more multiple shoots from all the explants of *O. indicum*, followed by Kn/TDZ.

In vitro rooting was standardized in the present investigations from the micro-shoots developed through nodal, shoot tip and cotyledonary node cultures (Fig. 47.5). It was quite difficult to standardize the protocol for rooting from in vitro developed micro-shoots in *O. indicum*. Poor rooting was observed on ½ strength MS medium supplemented with NAA. Moderate rooting was observed on ½ strength MS medium supplemented with IAA. Whereas profuse rhizogenesis was observed on ½ strength MS medium supplemented with IBA. MS medium supplemented with 1.0 mg/L IBA showed the best hormonal concentration for induction of profuse rhizogenesis. Even the best rooting efficiency was also recorded in the same concentration of IBA with lengthy and healthy roots. Thus, the protocol developed for in vitro rooting in the present investigations is a breakthrough, because, it is quite difficult for inducing in vitro rooting in tree species like *O. indicum* (Samatha et al. 2013a).

Ex vitro rooting was also observed from regenerated shoots of *O. indicum*. The in vitro derived micro-shoots when shifted directly into the pots containing garden soil developed roots within 20–25 days. The achievement of ex vitro rooting in *O. indicum* is important because it reduces the cost of micropropagation and also within short duration transferred to field conditions. Ex vitro rooting is a promising method as there is a reduction in cost by avoiding the in vitro rooting using by auxins, reduction in labour and the time of establishment from laboratory to soil (Samatha et al. 2013a).

Hardening/acclimatization plays an important role in tissue culture technology for establishment of plantlets developed in vitro, for Lab-to-Land program. In conservation and multiplication of especially endangered, medicinal and economically important forest tree species, this technique plays a vital role for an establishment of plantlets within a short period. Acclimatization of in vitro and ex vitro rooted plantlets has been established in *O. indicum* for the first time. The potted mixtures containing red soil + sieved sand + vermicompost (1:1:1) showed 80% survival of in vitro regenerated plants and 85% of survival for ex vitro rooted plants. Thus, the in vitro regenerated plants were successfully established and also identified the *elite* trees among the regenerants.

47.9 Conservation of Medicinal Plant *Oroxylum indicum* L. Kurz

Medicinal plants are the valuable sources of herbal products which are disappearing at an alarming rate. Medicinal plants can provide biologically active molecules and lead structures for the development of modified derivatives with enhanced activity and/or reduced toxicity. The medicinal plants and their derivatives have long been recognized as source of therapeutically effective medicines as they contain secondary metabolites which are potential sources of drugs. Plant-based products are healthier, safer and more reliable than synthetic products (Benli et al. 2008), and hence the WHO estimated that 80% of the population of developing countries rely on traditional medicines, mostly plant drugs, for their primary health care needs, and modern pharmacopoeia still contains at least 25% drugs derived from plants and many others which are synthetic analogues built on prototype compounds isolated from plants.

Habitat loss and deforestation combined with overharvesting due to the surge in popularity of herbal remedies and supplements in recent years has resulted in dwindling populations of important medicinal plants around the world.

Realizing the threat of extinction, there is a need to develop conservation strategies and quick propagation protocols. As such, in vitro culture technology appears to be a promising alternative to ensure the regeneration and conservation of all the plant species (Ahuja et al. 2002).

Forest tree improvement programs by selection and breeding techniques began in India recently. Traditional breeding methods in trees are limited by their large size and long life cycles. Hence, clonal forestry is now gaining increasing recognition as a quicker alternative for tree improvement. In vitro micropropagation techniques are being increasingly applied to supplement conventional methods of vegetative propagation. The benefits of this technique include high multiplication rates, generation of disease-free stocks and stress-tolerant varieties and also long-term storage of valuable germplasm. It has application in horticulture, agriculture and forestry and currently it is expanding worldwide. This method plays a great role in conservation and large-scale production of endangered, economically/commercially and industrially important forest tree species.

We have attempted to multiply and conserve an endangered and medicinally important forest tree species of *O. indicum,* a plant used in Indian system of medicine for variety of purposes, through in vitro culture technology . In view of its medicinal and economic importance, *O. indicum* is being overexploited and is becoming an endangered. This tree possesses a wide range of medicinal properties such as antinflammatory, anticancerous, anti-helminthic, anti-leucodermatic, anti-rheumatic and anti-anorexic, antimicrobial, antibacterial, analgesic, antioxidant, anti-angiogenic and anti-tussive. Hence it is being used in various traditional medicines such as *ayurveda*, *tribal*, *folk* and modern medicine too for the treatment of various ailments (Raghbir et al. 2008; Anonymous 1998; Samatha 2013; Ravikumar and Ved 2000; Samatha et al. 2013a, b; Samatha and Rama Swamy 2013, 2015; Samatha et al. 2017).

In conclusion, it is mentioned that the effectiveness of plant regeneration depends on the type of growth regulators and source of explants used in the experiment. Thus, the characteristics of the responded explants depend on the type of explant, culture medium and the micropropagation protocols developed or standardized including the present work for the plant regeneration and thus opens new prospective that could facilitate conservation and mass multiplication of the vulnerable and valuable medicinal forest tree, *O. indicum*. The various problems related to conventional propagation and the high demand of planting material of medicinal plants can be addressed by efficient and economical in vitro propagation in a short span of time.

Acknowledgments We thank the University Grants Commission, New Delhi, India for providing financial assistance under Rajiv Gandhi National Fellowship as JRF/SRF (Ref. No. F. 14-2(SC) 2009(SA-III)).

References

Ahuja A, Sambyal M, Koul S (2002) *In vitro* propagation and conservation of *Atropa acuminate* Royle ex Lindi—an indigenous threatened medicinal plant. J Plant Biochem Biotechnol 11(2):121–124

Anonymous (1998) Uttaranchal Mein Jari Butiyan: Paridrashya wa Sambhavanayain, Workshop Papers, vol 2(40). Centre for Development Studies, U.P. Academy of Administration, Nainital (in Hindi), p 132

Benli M, Bingol U, Geven F, Guney K, Yigit N (2008) An investigation on the antimicrobial activity of some endemic plant species from Turkey. Afr J Biotechnol 7(1):1–5

Fujita MS, Tuttle MD (1991) Flying foxes (Chiroptera: Pteropodidae): threatened animals of key ecological and economic importance. Conserv Biol 5(4):455–463

Jain SP, Singh J, Singh SC (2003) Rare and endangered medicinal and aromatic plants of Madhya Pradesh. J Econ Taxon Bot 27(4):925–932

Raghbir C, Gupta VS, Nisha S, Neeraj K, Bikram S (2008) *In vitro* antioxidant activity from leaves of *Oroxylum indicum* (L.) vent—a North Indian highly threatened and vulnerable medicinal plant. J Pharm Res 1:1

Ravikumar K, Ved DK (2000) 100 Red listed medicinal plants of conservation concern in southern India, vol 5(1). Foundation for Revitalization of Local Health Traditions, Bangalore, pp 4–67

Samatha T (2013) Micro propagation and phytochemical studies of an endangered and medicinally important forest tree species *Oroxylum indicum* (L) Kurz. PhD thesis, Kakatiya University, Warangal, TS

Samatha T, Rama Swamy N (2013) Direct *in vitro* organogenesis and plantlet formation from leaf explants of *Oroxylum indicum* (L) Kurz an endangered medicinal forest tree. Int J Adv Res 1(8):431–438

Samatha T, Rama Swamy N (2015) Conservation of an endangered ethno medicinally important forest tree species *Oroxylum indicum* through somatic embryogenesis. Int J Biol Pharm Allied Sci 4(3):1064–1073

Samatha T, Srinivas P, Rajinikanth M, ShyamSundarachary R, Rama Swamy N (2013a) Embryo culture an efficient tool for conservation of endangered medicinally important forest tree *Oroxylum indicum* L. Kurz. Int J Bio-Technol Res 3(2):45–52

Samatha T, Sampath A, Sujatha K, Rama Swamy N (2013b) Antibacterial activity of stem bark extracts of *Oroxylum indicum* an endangered ethnomedicinal forest tree. IOSR J Pharm Biol Sci 7(1):24–28

Samatha T, Shyamsundarachary R, Rama Swamy N (2016a) Conservation of an endangered medicinal forest tree *Oroxylum indicum* (L) Kurz through *in-vitro* micropropagation—a review. Eur J Med Plant 17(2):1–13
Samatha T, Shyamsundarachary R, Rama Swamy N (2016b) *In vitro* micropropagation of *Oroxylum indicum* (L) Kurz an endangered and valuable medicinal forest tree. Indo Am J Pharm Res 4(6):5210–5218
Samatha T, Shiva Krishna P, Rama Swamy N (2017) Anti angiogenic activity of *Oroxylum indicum* L. Kurz a medicinal tree. Int J ChemTech Res 10(5):276–280

DNA Barcode: The Genetic Blueprint for Identity and Diversity of *Phyllanthus amarus* Schum. et. Thonn

48

M. Ushakiranmayi, M. Vijayalakshmi, and P. Sudhakar

Abstract

Since the identification of the plant species is an important event, the present work was carried out to identify the medicinal plant *Phyllanthus amarus* taking the genetic variability among the species into consideration. The chloroplast genomic DNA was isolated and tested for quantity and quality. 2 μl of the template DNA, master mix, molecular-grade water and the forward and reverse primer were added and the final volume was made to 50 μl, and the amplification was performed in a gradient thermo-cycler and tested for amplification. The product was purified and sequenced and the sequence was submitted to GenBank. The sequence was subjected to multiple sequence alignment, and the phylogenic tree was constructed using MEGA 4 bioinformatics tools. The genomic DNA of the chloroplast was isolated, the highly conserved non-coding intron region of *t-RNA L* of the chloroplast was amplified, the amplified fragment was sequenced and the sequence in FASTA format was subjected to Multiple Sequence Alignment. The sequence showed no homology with any other sequence reported in the GenBank and this was the first highly conserved non-coding intron sequence of the plant *Phyllanthus amarus* that was deposited and the phylogenic tree was constructed. The identification of plant species based on the morphology is acceptable to an extent but the genetic blueprint gives you the exact information for the identification of the plant species. DNA barcode is a standardized and cost-effective molecular identification system used for plant identification. Hence the medicinally valuable plants needs the accurate identification among the species belonging to the same genus.

M. Ushakiranmayi · M. Vijayalakshmi (✉)
Department of Botany and Microbiology, Acharya Nagarjuna University, Guntur, Andhra Pradesh, India

P. Sudhakar
Department of Biotechnology, Acharya Nagarjuna University, Guntur, Andhra Pradesh, India

S. M. Khasim et al. (eds.), *Medicinal Plants: Biodiversity, Sustainable Utilization and Conservation*, https://doi.org/10.1007/978-981-15-1636-8_48

Keywords
Phyllanthus amarus · DNA barcode · Molecular identification · Medicinal plants

48.1 Introduction

A DNA barcode, in its simplest definition, is one or more short gene sequences taken from a standardized portion of the genome used to identify species (Kress and David 2008). Barcoding is a new concept (Floyd et al. 2002; Hebert et al. 2003) that aims to provide rapid, accurate and automatable species identification by using the standardized DNA region as the tag (Hebert and Gregory 2005). Identification systems based on DNA have the potential to facilitate both the identification of known species and the discovery of new ones (Blaxter 2003). DNA barcoding is based on the premise that sequence diversity within a short, standardized segment of the genome can provide a "biological barcode" that enables identification at the species level. A DNA barcode is an aid to taxonomic identification which uses a standard short genomic region that is universally present in target lineages and has sufficient sequence variation to discriminate among species (Hebert et al. 2003, 2004; Hajibabaei et al. 2007). In practice, a DNA sequence from such a standardized gene region is taken from an unidentified plant. This sequence is then compared to a library of reference sequences from known species. A match of the sequence from the unknown species to one of the reference sequences can provide a rapid and reproducible identification.

The idea of a standardized molecular identification system emerged progressively during the 1990s with the development of PCR-based approaches for species identification. Molecular identification has largely been applied to bacterial studies, microbial biodiversity surveys (Woese 1996; Zhou et al. 1997) and routine pathogenic strain diagnoses (Sugita et al. 1998; Wirth et al. 2006). According to the study of Pierre et al. (2006) the ideal DNA barcoding system should meet the following criteria. The DNA barcoding of the sequences should be variable sufficiently to discriminate different species. It should be highly conserved exhibiting less variability within the species. The DNA region selected for the study should have as much information as possible for its phylogenetic study, related to its taxa. The most important of all is the availability of the highly conserved priming site for the easy amplification of the DNA target sequence from the mixture of the sample DNA. Unfortunately, such an ideal DNA marker does not exist.

Plant researchers have proposed several different barcode regions. These focus on coding and non-coding regions located primarily in the plastid genome. Kress and Erickson (2007) suggested that two non-coding regions (the nuclear ITS region and the plastid trnH-psbA intergenic spacer) may have the potential to be universal plant barcodes. Other combinations involving three plastid regions have also been proposed by a working group that includes the Royal Botanical Gardens, Kew, UK. These comprise the trnH-psbA region and portions of two coding regions (matK and rpoC1) or three coding regions combined (matK, rpoB and rpoC1) (Chase et al. 2007).

Three highly conserved transfer RNA genes [tRNA genes for threonine (UGU), leucine (UAA) and phenylalanine (GAA)] are found in tandem, separated by spacers of several hundred base pairs (bp). The high variability of the two spacers and the intron in trnL have led to the wide use of trnT-trnF sequences in addressing relationships at the species and genus levels (Taberlet et al. 1991; Van Ham et al. 1994; Sang et al. 1997; Small et al. 1998; Bakker et al. 2000). Moreover, the region was quite informative in phylogenetic studies of families like Asteraceae (Bayer and Starr 1998), Arecaceae (Asmussen and Chase 2001) and Rhamnaceae (Richardson et al. 2007) and orders like Laurales (Renner 1999) and Magnoliales (Sauquet et al. 2003).

The term "DNA barcode" is used here to refer to a DNA sequence-based identification system that may be constructed of one locus or several loci used together as a complementary unit. DNA barcoding is already emerging as a tool on the modern taxonomist's work bench despite the debate and controversy among some scientists over the feasibility and utility of genetic identifiers in taxonomic and other applied studies (Hebert et al. 2004; Will and Rubinoff 2004; Rubinoff et al. 2006).

Phyllanthus amarus Schum. et. Thonn leaves were collected from the scrub forest of Acharya Nagarjuna University, Guntur district, Andhra Pradesh. The Voucher specimens (No. 2404) were identified with the help of regional floras (Pullaiah et al. 2002) and deposited in ANU, Botany and Microbiology Department, Guntur.

48.1.1 DNA Isolation

Total genomic DNA was isolated from the tender leaves of *Phyllanthus amarus* by using Rapid Genomic DNA extraction kit (HELINI Biomolecules, Chennai, India). The procedure adopted was rapid method. The genomic DNA was isolated from 100 mg of the tender leaves by adding 400 μl of Lysis buffer [1.5 M NaCl, 10 mM Tris-HCl (pH 8.0), 5% SDS], then the leaves were ground finely using the micro pestle and incubated at 65 °C for 15 min, the solution was cooled at room temperature (28 ± 2 °C), 130 μl of potassium acetate (pH 5.2) was added and mixed, and the solution was incubated for 5 min at room temperature. The sample was centrifuged at 10,000 rpm for 10 min, the clear supernatant was transferred into fresh 1.5 ml tube, an equal volume of isopropanol was added and gently inverted for 5 min and incubated at room temperature. After incubation, the sample was centrifuged at 10,000 rpm for 10 min and the supernatant was discarded. The pellet washed with 75% ethanol was dried at room temperature for 15 min. The pellet was then dissolved in 100 μl of TE buffer. The DNA sample was further purified from RNA contamination by adding 10 μl of RNase (10 mg/ml) and the sample was incubated at 37 °C for 1 h.

48.1.2 Quantification of DNA

Quantification of the DNA was carried out by analyzing the UV absorption of the nucleotides, which provides a simple and accurate estimation of the concentration of nucleic acids in a sample.

$$\text{DNA concentration }\left(\mu g/\text{ml}\right)=\frac{\text{OD }260\times100\ \left(\text{dilution factor}\right)\times50\ \mu g/\text{ml}}{1000}$$

48.1.3 The Universal Primers

Two primers (Forward and Reverse) were designed by comparing their sequences with the sequences available from the NCBI data bank and the primers were synthesized by HELINI Biomolecules, Chennai. The sequences of the forward and reverse primer were 5′CGAAATCGGTAGACGCTACG3′ and 5′GGGGATAGAGGGACTTGAAC3′.

48.1.4 Amplification of the Target Sequence

PCR was performed using Helini DNA polymerase master mix (1 U Taq DNA polymerase, 10× *Taq* buffer, 10 mM dNTPs and PCR-grade water). Primer solution contains 10 μM forward primer/μl and 10 μM reverse primer/μl. The reaction mixture contains 1.5 μl of forward primer, 1.5 μl of reverse primer, 2 μl of template DNA (10 ng/μl), and 20 μl of nuclease-free water in a 25 μl PCR reaction. Gradient Thermo-cycler (Corbett Research, Australia) conditions were optimized with initial denaturation of 3 min at 94 °C followed by 30 cycles of denaturation at 94 °C for 30 s, annealing temperature of 50 °C for 30 s and DNA polymerization at 72 °C for 1 min and the final extension at 72 °C for 3 min. To test success of the amplification, 10 μl of the PCR product mixed with 6× gel loading dye along with 10 μl of the 1 kb DNA ladder was electrophoresed in 1.8% agarose gel for 45 min at 100 V and the gel was stained with ethidium bromide. The gel was visualized under UV transilluminator, photographed and documented with gel documentation (UV Tec., London).

48.1.5 Purification of the PCR Product

The PCR amplified target sequence was purified using Helini Mini-Elute kit according to the manufacturer's directions. 40 μl of PCR product was mixed with five volumes of binding buffer and passed through spin column. The column was washed with wash buffer and again centrifuged for 1 min to remove residual wash buffer. DNA was eluted with 25 μl of TE buffer and 5 μl of eluted DNA was loaded in 1.8% gel along with 1 kb DNA ladder and checked for the presence of band as well as its approximate concentration.

48.1.6 DNA Sequencing

The purified PCR product was sequenced by MWJ, Bangalore, India according to the manufacturer's protocol.

48.1.7 Phylogenetic Analysis

The sequenced information of the non-coding intron of trnL (UAA) was subjected to multiple sequence alignment using Clustal W2 and then the phylogenetic tree was constructed with the Molecular Evolutionary Genetic Analysis (MEGA Version 4.1).

48.2 Results

48.2.1 DNA Isolation, Amplification and Sequencing

The leaf sample of plant *Phyllanthus amarus* was collected, total genomic DNA was isolated and the DNA was subjected to quantitative and qualitative study. The isolated DNA subjected to agarose gel electrophoresis (0.8%) showed no traces of RNA contamination (Fig. 48.1) and the quantity of the isolated DNA was determined as 10 ηg/μl by UV absorbance at 260/280 nm using UV visible Spectrophotometer.

The target sequence was amplified by using 2 μl of the DNA as template. After amplification, the PCR product was analyzed on the agarose gel along with the 100 bp marker DNA ladder (Helini). The gel showed the amplified fragment of 556 bp and the reaction showed no traces of contamination and the amplified fragment was purified by rapid purification kit (Helini) (Fig. 48.2).

48.2.2 Sequencing

The purified product was sequenced commercially by MWGAG Biotech, Bangalore and the sequence of the amplified intron is given in Hologram (Fig. 48.3) and also as FASTA sequence (Fig. 48.4). The sequence-amplified intron was subjected to the NCBI database and the accession number EU 861193.1 was assigned.

48.2.3 Phylogenetic Analysis

The phylogenetic tree was constructed using the online tree builder of the Molecular Evolutionary Genetic Analysis (MEGA Version 4.1). The FASTA sequence of the tRNA L intron of *Phyllanthus amarus* was subjected to BLAST algorithm and the intron sequence showed 97% identity with *Phyllanthus embilica.* The evolutionary history was inferred using the Neighbor-Joining method (Saitou and Nei 1987).

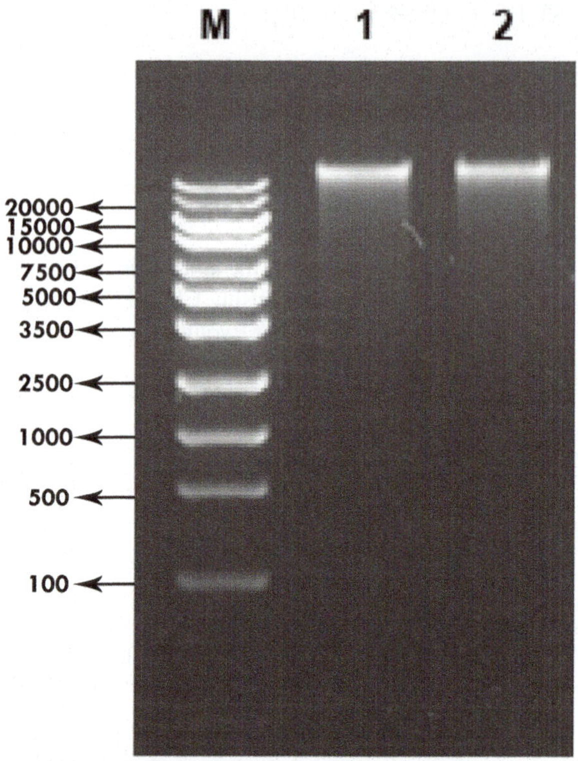

Fig. 48.1 Gel showing the isolated genomic DNA of *Phyllanthus amarus* (M—Marker, Lane 1 and 2 Genomic DNA)

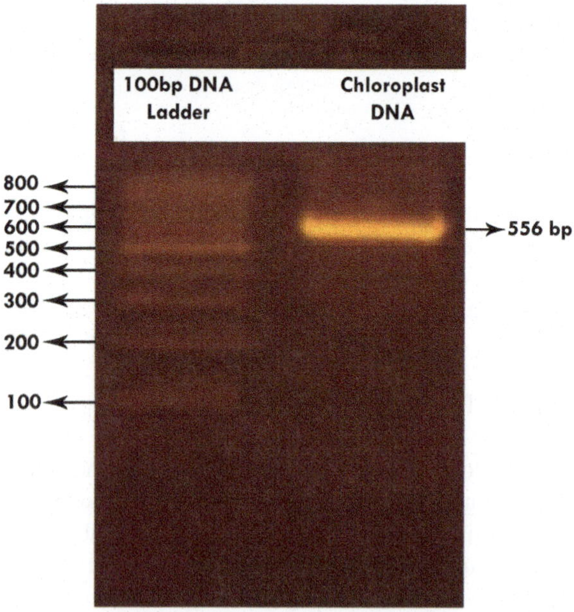

Fig. 48.2 Gel showing the 556 bp amplified DNA fragment along with the 100 bp marker DNA ladder

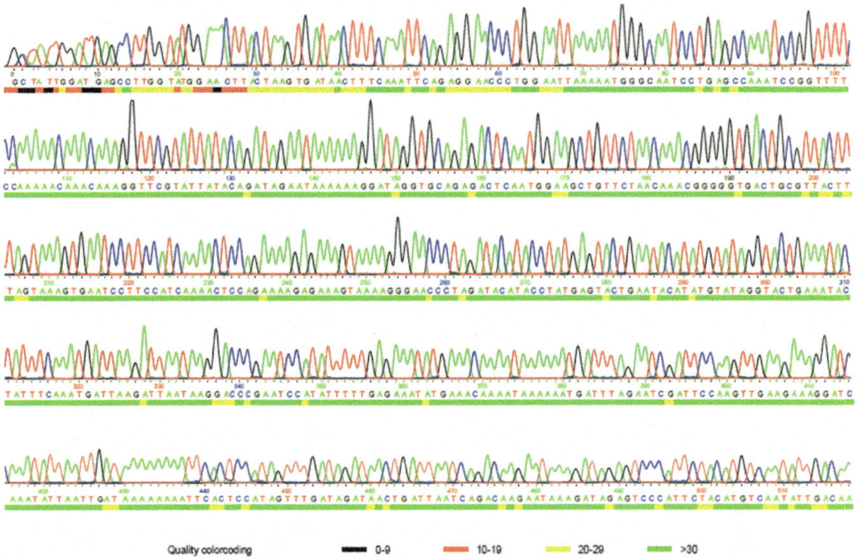

Fig. 48.3 Hologram of the sequence of the amplified non-coding t-RNA L intron

Fasta sequence

GCCTTGGTATGGAACTTACTAAGTGATAACTTTCAAATTCAGAGGAACCCTGGAATT
AAAAATGGGCAATCCTGAGCCAAATCCGGTTTTCCAAAAACAAACAAAGGTTCGTA
TTATACAGATAGAATAAAAAAGGATAGGTGCAGAGACTCAATGGAAGCTGTTCTAA
CAAACGGGGGTGACTGCGTTACTTTAGTAAAGTGAATCCTTCCATCAAAACTCCAGA
AAAGAGAAAGTAAAAGGGAACCCTAGATACATACCTATGAGTACTGAATACATATG
TATAGGTACTGAAATACTATTTCAAATGATTAAGATTAATAAGGACCCGAATCCATA
TTTTTGAGAAATATGAAACAAAATAAAAAATGATTTAGAATCGATTCCAAGTTGAA
GAAAGGATCAAATATTAATTGATAAAAAAAATTCACTCCATAGTTTGATAGATAACT
GATTAATCAGACAAGAATAAAGATAGAGTCCCATTCTACATGTCAATATTGACAAC
AAGGAAATTTATAGTAAGAGGAAAATCCGTCGACTTTAGAAATCGTGA

Fig. 48.4 FASTA sequence of the amplified intron

The optimal tree with the sum of branch length = 12.03807230 was shown. The tree was drawn to scale, with branch lengths in the same units as those of the evolutionary distances used to infer the phylogenetic tree. The evolutionary distances were computed using the Maximum Composite Likelihood method (Tamura et al. 2004) and are in the units of the number of base substitutions per site. All positions containing gaps and missing data were eliminated from the dataset (Complete deletion option). There were a total of 544 positions in the final dataset. Phylogenetic analyses were conducted in MEGA 4 (Tamura et al. 2007) (Fig. 48.5).

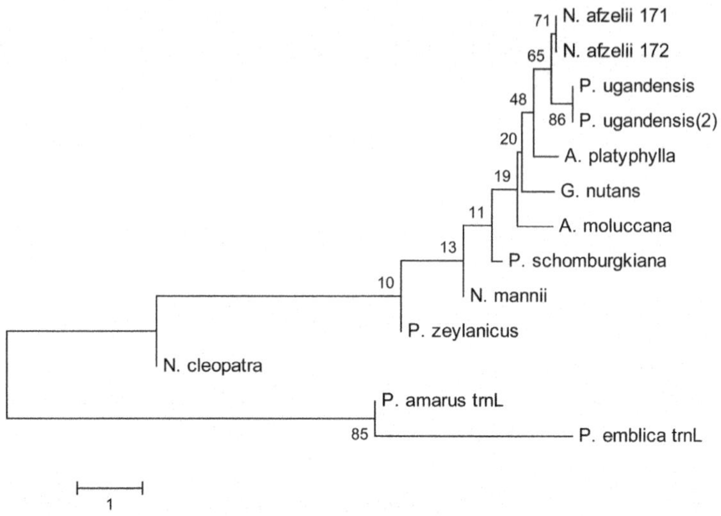

Fig. 48.5 Phylogenic tree constructed by Neighbor-Joining method using Molecular Evolutionary Genetic Analysis (MEGA Version 4.1)

48.3 Discussion

DNA barcoding practice has been designed as a concept of system to facilitate the recognition and identification of the species. But the challenge in barcoding is the identification of a suitable region in the genome of the plant mitochondria and plastid which is highly conserved within the species but highly variable among different species (Newmaster et al. 2008).

A great deal of confusion exists among scientists regarding plant identification, which makes evaluation of published information difficult. *P. amarus* is considered by some authors as a variety of *P. niruri* (Piyada et al. 2008). According to Clarice et al. (2006), it is difficult to identify the species correctly, such as the case of *P. niruri* and *P. amarus* (Unander et al. 1992), and to establish which of the species has useful medicinal properties with the available ethno-botanical and ethno-pharmacological data. In herbaceous *Phyllanthus*, floral characters have limited utility for the identification of species and supra-specific taxa, due to the small size of the flowers and the relative narrow range of variation in number and shape of parts. The studies conducted for the molecular identification of *Phyllanthus amarus* were Random Amplified Polymorphic DNA (RAPD) (Newmaster et al. 2008) and sequence characterized amplified regions (SCARS) (Jain et al. 2008). DNA barcoding is the most powerful, robust rapid and inexpensive diagnostic technique in which a short conserved DNA sequence among species is used for the identification of the unknown species or to confirm the identity of the known species.

According to the studies of Pierre et al. (2007) the chloroplast trnL (UAA) intron was selected as the target for the phylogenetic study (Chen et al. 2005). In the earlier plant molecular systematics era, chloroplast DNA (cpDNA) was surveyed through restriction site polymorphism studies. Simultaneously, non-coding regions of the chloroplast were being explored for lower level taxonomic studies because of the assumption that non-coding regions are under less functional constraint than coding regions and they should provide greater levels of variation for phylogenetic analyses (Gielly and Taberlet 1996). The regions to be exploited for barcoding studies include *trnT trnL-trnL-trnF* regions (Taberlet et al. 1991) and the *atpB-rbcL* intergenic spacer (Manen et al. 1994). Following these pioneering studies, the use of non-coding cpDNA regions has continually increased and are now routinely employed for studies of phylogeny at intergeneric and interspecific levels.

The present study was focused on the evaluation of the phylogenic relationship and identification of *Phyllanthus amarus* by amplifying the non-coding conserved *trna L* intron whose sequence is conserved within the species and variable between different species. The molecular identification of *P. amarus* was carried out only by Random Amplified Polymorphic DNA (RAPD), and sequence characterized amplified regions (SCARS) (Piyada et al. 2008). The study has been taken up to identify *P. amarus* based on the sequence diversity of the non-coding *trna L* intron. Based on the studies of Pierre et al. (2006) the non-coding *trna L* intron of *P. amarus was* amplified using the universal primer designed against the target *trna L* intron using the primer 3 online primer-designing bioinformatics tool (Steve and Helen 2000). The amplified region of the non-coding intron was found to be 556 bases. The amplified product was sequenced and the FASTA format of the sequence of the *trna L* non-coding intron was deposited in the NCBI with the accession number EU 861193.1. The non-coding sequence deposited can now be used for identification of *P. amarus.* Study of the phylogenetic relationship revealed that the sequence did not show any identity with the non-coding t-RNA L intron regions of reported plants.

References

Asmussen CB, Chase MW (2001) Coding and noncoding plastid DNA in palm systematics. Am J Bot 88:1103–1117

Bakker FT, Culham A, Gomez-Martinez R, Carvalho J, Compton J, Dawtrey R, Gibby M (2000) Patterns of nucleotide substitution in angiosperm cpDNA trnL (UAA)-trnF (GAA) regions. Mol Biol Evol 17:1146–1155

Bayer RJ, Starr JR (1998) Tribal phylogeny of the Asteraceae based on two non-coding chloroplast sequences, the trn L intron and the trn L/trn F intergenic spacer. Ann Missouri Bot Gard 85:242–256

Blaxter M (2003) Molecular systematics: counting angels with DNA. Nature 421:122–124

Chase MW, Cowan RS, Hollingsworth PM, van den Berg C, Madrinan S, Petersen G, Seberg O, Jorgsensen T, Cameron KM, Carine M, Pedersen N, Hedderson TAJ, Conrad F, Salazar GA, Richardson JE, Hollingsworth ML, Barraclough TG, Kelly L, Wilkinson MA (2007) Proposal for a standardized protocol to barcode all land plants. Taxon 56:295–299

Chen SY, Xia T, Wang YJ, Liu JQ, Chen SL (2005) Molecular systematics and biogeography of Crawfurdia, Metagentiana and Tripterospermum (Gentianaceae) based on nuclear ribosomal and plastid DNA sequences. Ann Bot 96:413–424

Clarice AM, de Oliveira PL, Lilian AM (2006) SEM observations on seeds of some herbaceous *Phyllanthus* L. species (Phyllanthaceae). Rev Bras Farmacogn 16:31–41

Floyd R, Abebe E, Papert A, Blaxter M (2002) Molecular barcodes for soil nematode identification. Mol Ecol 11:839–850

Gielly L, Taberlet P (1996) A phylogeny of the European gentians inferred from chloroplast trnL(U/AA) intron sequences. Bot J Linn Soc 120:57–75

Hajibabaei M, Singer GAC, Hebert PDN, Hickey DA (2007) DNA barcoding: how it complements taxonomy, molecular phylogenetics and population genetics. Trends Genet 24:167–172

Hebert PDN, Gregory TR (2005) The promise of DNA barcoding for taxonomy. Syst Biol 54:852–859

Hebert PDN, Ratnasingham S, deWaard JR (2003) Barcoding animal life: cytochrome c oxidase subunit 1 diverges among closely related species. Proc R Soc B 270(Suppl 1):S96–S99

Hebert PDN, Penton EH, Burns JM, Janzen DH, Hallwachs W (2004) Ten species in one: DNA barcoding reveals cryptic species in the neotropical skipper butterfly *Astraptes fulgerator*. Proc Natl Acad Sci U S A 101:14812–14817

Jain N, Kumar AS, Singh S, Singh SPK, Kumar S (2008) SCAR markers for correct identification of *Phyllanthus amarus, P. fraternus, P. debilis* and *P. urinaria* used in scientific investigation and dry leaf Bulk Herb Trade. Biochem Mol Biol 74:296–301

Kress WJ, David LE (2008) DNA barcoding—a windfall for tropical biology? Biotropica 40:405–408

Kress WJ, Erickson DL (2007) A two-locus global DNA barcode for land plants: the coding rbcL gene complements the non-coding trnH-psbA spacer region. PLoS One 2(6):e508

Manen JF, Natali A, Ehrendorfer F (1994) Phylogeny of Rubiaceae-Rubieae inferred from the sequence of a cpDNA intergene region. Plant Syst Evol 190:195–211

Newmaster SG, Fazekas AJ, Steeves RAD, Janovec J (2008) Testing candidate plant barcode regions with species of recent origin in the Myristicaceae. Mol Ecol Notes 8:480–490

Pierre T, Coissac E, François P, Ludovic G, Christian M, Alice V, Thierry V, Gerard C, Christian B, Eske W (2006) Power and limitations of the chloroplast trnL (UAA) intron for plant DNA barcoding. Nucleic Acids Res 36:1–8

Pierre T, Eric C, Francois P, Ludovic G, Christian M, Alice V, Thierry V, Gerard C, Christian B, Eske W (2007) Power and limitations of the chloroplast trnL (UAA) intron for plant DNA barcoding. Nucleic Acids Res 35:1–8

Piyada T, Nantawan K, Duangkamol M, Sumontip B, Pranom C (2008) Development of species specific SCAR markers for identification of three medicinal species of *Phyllanthus*. J Syst Evol 46:614–621

Pullaiah T, Ramakrishnaiah V, Sandhya Rani S, Rao PN (2002) Flora of Guntur District. Regency Publications, New Delhi

Renner SS (1999) Circumscription and phylogeny of the Laurales: evidence from molecular and morphological data. Am J Bot 86:1301–1315

Richardson JE, Hollingsworth ML, Barraclough TG, Kelly L, Wilkinson MA (2007) Proposal for a standardized protocol to barcode all land plants. Taxon 56:295–299

Rubinoff D, Cameron S, Will K (2006) Are plant DNA barcodes a search for the Holy Grail? Trends Ecol Evol 21:1–2

Saitou N, Nei M (1987) The neighbor-joining method: a new method for reconstructing phylogenetic trees. Mol Biol Evol 4:406–425

Sang T, Crawford DJ, Stuessy TF (1997) Chloroplast DNA phylogeny, reticulate evolution and biogeography of *Paeonia* (Paeoniaceae). Am J Bot 84:1120–1136

Sauquet H, Doyle JA, Scharaschkin T, Borsch T, Hilu KW, Chatrou LW, Le Thomas A (2003) Phylogenetic analysis of Magnoliales and Myristicaceae based on multiple data sets: implications for character evolution. Bot J Linn Soc 142:125–186

Small RL, Ryburn JA, Cronn RC, Seelanan T, Wendel JF (1998) The tortoise and the hare: choosing between non-coding plastome and Adh sequences for phylogeny reconstruction in a recently diverged plant group. Am J Bot 85:1301–1305

Steve R, Helen JS (2000) Primer3 on the WWW for general users and for biologist programmers. In: Krawetz S, Misener S (eds) Bioinformatics methods and protocols: methods in molecular biology. Humana Press, Totowa, NJ, pp 365–386

Sugita T, Nishikawa A, Shinoda T (1998) Identification of trichosporon asahii by PCR based on sequences of the internal transcribed spacer regions. J Clin Microbiol 36:2742–2744

Taberlet P, Gielly L, Pautou G, Bouvet J (1991) Universal primers for amplification of three noncoding chloroplast regions. Plant Mol Biol 17:1105–1109

Tamura K, Nei M, Kumar S (2004) Prospects for inferring very large phylogenies by using the neighbor-joining method. Proc Natl Acad Sci USA 101:11030–11035

Tamura K, Dudley J, Nei M, Kumar S (2007) MEGA4: molecular evolutionary genetics analysis (MEGA) software version 4.0. Mol Biol Evol 24:1596–1599

Unander DW, Webster GL, Blumberg BS (1992) Usage and bioassays in *Phyllanthus* (Euphorbiacae):a compilation III. The subgenera *Eriococcus, Conami, Gomphisium, Botryanthus, Xylophylla* and *Phyllanthodendron*, and a complete list of the species cited in the three-part series. J Ethnopharmacol 36:103–112

Van Ham RCHJ, T'hart H, Mes THM, Sandbrink JM (1994) Molecular evolution of non-coding regions of the chloroplast genome in the Crassulaceae and related species. Curr Genet 25:558–566

Will K, Rubinoff D (2004) Myth of the molecule: DNA barcodes for species cannot replace morphology for identification and classification. Cladistics 20:47–55

Wirth T, Falush D, Lan R, Colles F, Mensa P, Wieler LH, Karch H, Reeves PR, Maiden MCJ, Ochman H, Achtman M (2006) Sex and virulence in Escherichia coli: an evolutionary perspective. Mol Microbiol 60:1136–1151

Woese CR (1996) Phylogenetic trees: whither microbiology? Curr Biol 6:1060–1063

Zhou J, Davey ME, Figueras JB, Rivkina E, Gilichinsky D, Tiedje JM (1997) Phylogenetic diversity of a bacterial community determined from Siberian tundra soil DNA. Microbiology 143:3913–3919

Microbiological and Physicochemical Quality of Potable Water in Valasi, Agency Area, Andhra Pradesh

49

K. Laxmi Sowmya and D. Sandhya Deepika

Abstract

The present study was undertaken to evaluate the water quality of the Valasi panchayat of Ananthagiri mandal in Visakhapatnam district with affable means. The physicochemical and microbial studies are the most important means by which we are able to test the potability of water. The isolation and characterization of the pathogenic microorganism from the water sample collected were the main areas emphasized in the study. In this study, drinking water samples were collected from a stream, a hand bore and a well for a period of 2 years, i.e., from April 2011 to March 2013. The various constituents monitored include the physicochemical characters, the bacterial parameters like total plate count (TPC), most probable number (MPN) and isolation and identification of pathogenic bacteria. The physicochemical characters of all the three drinking water samples were within the recommended permissible level of WHO. The total plate count was above the WHO guidelines values (<10 CFUs/ml) in the three water samples studied and the highest count was during August and September. The bacteria isolated were *E. coli, Salmonella, Shigella, Staphylococcus, Group D Streptococcus, Vibrio cholerae* and *Klebsiella pneumoniae.* The samples were inoculated and were incubated at 37 °C for 24 or 48 h for appropriate bacterial growths. Poor quality of drinking water in the area gave rise to major risk factor for the large-scale of waterborne diseases. Thus we can use this study for the assessment of water and to resolve the hygienic problems of water.

Keywords

Drinking water · Quality assessment · Pathogenic bacteria · Valasi panchayat

K. L. Sowmya · D. S. Deepika (✉)
Department of Botany, Andhra University, Visakhapatnam, Andhra Pradesh, India

S. M. Khasim et al. (eds.), *Medicinal Plants: Biodiversity, Sustainable Utilization and Conservation*, https://doi.org/10.1007/978-981-15-1636-8_49

49.1 Introduction

India is already complaining about water shortage, not to mention many villages which lack safe drinking water. In the list of 122 countries rated on quality of potable water, India ranks a lowly 120. Although India has 4% of the world's water, studies show that the average availability is shrinking steadily. It is estimated that by 2020, India will become a water-stressed nation. Nearly 50% of villages still don't have any source of protected drinking water.

According to the 2001 census 68.2% households have access to safe drinking water. The department of drinking water supply estimates that 94% of rural habitations and 91% urban households have access to drinking water. But according to experts these figures are misleading simply because coverage refers to installed capacity and not actual supply.

The ground reality is that of the 1.42 million villages in India, 1,95,813 are affected by chemical contamination of water. The quality of ground water which accounts for more than 85% of domestic supply is a major problem in many areas as none of the rivers have water fit to drink. 37.7 million people over 75% are children are afflicted by waterborne diseases every year. Overdependence on groundwater has brought in contaminants; fluoride being one of them. Nearly 66 million people in 20 states are at risk because of the excessive fluoride in water.

However it is bacteriological contamination that leads to diarrhoea, cholera and hepatitis, which is widespread in India. A bacteriological analysis of the water in Bangalore revealed that 75% of the bore wells were contaminated with iron; hardness and salinity are also a concern. Nearly 12,500 habitats have been affected by salinity. In Gujarat it is a major problem in coastal districts. Often babies die of dehydration, and there are major fights in villages for fresh water. Some villages have seen 80% migration due to high salinity. Health is not the only issue; impure water is a major burden on the state.

49.2 Study Area

Ananthagiri (18°17′14″N, 83°6′43″E) is about 60 km away from Visakhapatnam and lies on the top of the Eastern Ghats. The area of the Ananthagiri mandal is roughly 50 km^2 and the entire area is inhabited by aboriginal tribes. Of the 25 panchayats in Anantagiri mandal, Valasi panchayat with 20 km^2 area was selected for the present study. It comprises 14 villages with a population of 4000, which includes 49% women and 51% men. The total population present in this panchayat includes 800 literates. The different tribal types present in this panchayat are "Konda Dora, Parena Karja, Petege, Bagatha, Valmiki and Gadaba" and most of them depend on agriculture. The mean temperature is 36 °C and receives 1171.0 mm normal annual rainfall. Based on their economic status they live in different types of houses such as sheet houses, tiled houses and slab houses. Drinking water sources include 9 hand bores, 4 wells and a small stream running from the hills. The stream is the main source of drinking water. Primary health centre is in the range of 5 km. The panchayath consists of 3 main 2 mini anganwadis and three primary schools. Crops grown in this panchayath are rice, millets, and flax seeds.

In the present study, water samples were collected from three sources, i.e., a well, a hand pump and stream once in a month for a period of 24 months from April 2011 to March 2013, in white plastic bottles, which were previously rinsed with distilled water and sterilized with 70% alcohol. At the collection point, the containers were rinsed thrice with the sample water before being used to collect the samples. The collected samples were placed in a thermocol box. The temperature in the box was maintained at 4 °C by using ice packs. The microbial isolation was done by streak plate method on nutrient agar and on selective media for their identification (Sherman Cappuccino 2009). The final identification of resulted isolates was done by the biochemical tests in accordance with the Bergey's Manual.

49.2.1 Arithmetic Mean

The arithmetic mean of a group is simply the mean or average of the observations. This is calculated by dividing the sum of all the observations by the number of observations.

$$X = \left(\Sigma Xi\right) / n$$

Xi = Value of observation i and n = total number of observations.

49.2.2 Variance (v) and Standard Deviation (σ)

The average of squared differences from mean is called variance.

$$v = \frac{\Sigma\left(Xi - X\right)^2}{N}$$

Standard deviation is the square root of the variance. It is the most important measure of variability

$$\sigma = \sqrt{v}$$

Xi = value of the observation i, X = arithmetic mean, N = Total number of observations, v = variance.

The data were subjected to a software program named PAST (Hammer et al. 2001), which generates diversity indices. The formula designed for various indices are described below. These indices were used to obtain estimation of species diversity, species richness and species evenness.

1. Shannon and Weiner (1949) and Simpson (1949) diversity index value were obtained by using the following equation:

$$(\text{Shannon's index})\ -\sum_i\left(\frac{n_i}{N}\cdot\log_2\left(\frac{n_i}{N}\right)\right)$$

$$(\text{Simpson index})\ \frac{\sum_i n_i(n_i-1)}{N(N-1)}$$

2. Species richness (R_1 and R_2) obtained using following equation:

$$R_1=\frac{S-1}{\ln N}\ \text{(Margalef 1958)}$$

$$R_2=\frac{S}{\sqrt{\sum_i n_i}}\ \text{(Menhinick 1964)}$$

where
R = Index of species richness
S = Total number of species
N = Total number of individuals

3. Species evenness was determined by using the following expression.
Shannon's equitability (E_H) can be calculated by following equation.

$$-\frac{\sum_i\left(\frac{n_i}{N}\cdot\ln\left(\frac{n_i}{N}\right)\right)}{\ln N}$$

Equitability assumes a value between 0 and 1 being complete evenness.

4. Dominance index is used to characterize the most conspicuous and abundant species with its relative importance related to the degree of influence it has on ecosystem components.

$$\text{Dominance index}=1-\left(\frac{\sum_i n_i(n_i-1)}{N(N-1)}\right)$$

5. The Berger–Parker Dominance Index is a simple measure of the numerical importance of the most abundant species.

$$\text{Berger}-\text{Parker Dominance Index}=\frac{n_{\max}}{N}$$

49.3 Microbiological and Physicochemical Quality of Potable Water

Water samples collected from Valasi panchayath for a period of 2 years, i.e., during April 2011 to March 2013, were analysed for physical, chemical and bacteriological characteristics. The physical characteristic measured is p^H. Among the chemical characteristics, total dissolved solids (TDS) and fluoride contents were measured. For total number of viable bacteria total plate count (CFU/ml), for faecal and total coliforms, most probable number (MPN/100 ml) and for isolation and identification of bacterial staining, biochemical and growth on selective media were performed.

The mean p^H value of stream water was 7.25 with a range of 7.13–7.31. In bore water it was in the range of 7.04–7.39 with the mean p^H value 7.266. In well water it was in the range of 7.12–7.69 with mean p^H value 7.37. The p^H value in the three water samples is in the safe limit as recommended by WHO.

The amount of total dissolved solids of the stream water was on the average 293 mg/l and fluoride content on the average was 1 mg/l. The amount of total dissolved solids of the bore water on the average was 592 mg/l and fluoride content on the average was 0.49 mg/l. The amount of total dissolved solids of the well water on the average was 902 mg/l and fluoride content on the average was 0.53 mg/l. Both the values in the three samples were in the permissible limits as recommended by WHO.

During the study period all the three water samples (i.e. stream, bore and well) showed the presence of the seven pathogenic bacteria: *Escherichia coli, Klebsiella pneumoniae, Salmonella typhi, Shigella dysenteriae, Staphylococcus aureus, Group D Streptococcus,* and *Vibrio cholerae* (Tables 49.1 and 49.2).

49.3.1 Stream Water

In stream water the total plate count fell in the range of 29.33 ± 6.5–67.33 ± 2.08 CFUs/ml. The water sample showed the maximum number of CFUs (67.33 ± 2.08 CFUs/ml) in rainy season and minimum number was noted in summer (29.33 ± 6.5 CFUs/ml) (Fig. 49.1).

The MPN values for coliforms present in all the water samples are presented in Fig. 49.2. In stream water the MPN index ranged from 7–15/100 ml. The maximum MPN index was recorded in (15/100 ml) rainy season. The minimum MPN index was recorded in (7/100 ml) summer season.

49.3.1.1 Percentage of Pathogenic Bacteria

Analysis of stream water revealed that the predominant bacterium was *E. coli* (20.0195%) followed by *Staphylococcus aureus* with 17.305%, *Group D Streptococcus* with 11.408%, *Pseudomonas aeruginosa* 10.88%, *Shigella dysenteriae* with 8.0479%, *Salmonella typhi* with 8.0229%, *Vibrio cholerae* with 7.9389% and *Klebsiella pneumoniae* with 7.82525% (Table 49.3).

Table 49.1 Morphological and cultural characteristics of organisms

Morphological and cultural characters	Organism	Disease caused by the organism
Gram negative rod, forms circular, low convex mucoid, opaque colonies with entire marginal growth on nutrient agar. Green metallic sheen colonies were observed on EMB agar.	*Escherichia coli*	Causal agent of gastroenteritis, urinary tract infections, and neonatal meningitis.
Gram positive coccus, non-spore forming and non-motile bacteria. It forms circular, low convex with margin, smooth, medium opaque colony on nutrient agar. It forms yellow-coloured colonies on mannitol salt agar.	*Staphylococcus aureus*	*S. aureus* incidence ranges from skin, soft tissue, respiratory, bone, joint, endovascular to wound infections. It causes a range of illnesses, from minor skin infections, such as pimples, impetigo, boils (furuncles), cellulitis folliculitis, carbuncles, scalded skin syndrome, and abscesses, to life-threatening diseases such as pneumonia, meningitis, osteomyelitis, endocarditis, toxic shock syndrome (TSS), bacteremia, and sepsis. It is still one of the five most common causes of nosocomial infections and is often the cause of postsurgical wound infections.
Gram positive coccus. It forms thin, even growth on nutrient agar. Black- (or) brown-coloured colonies were observed on bile esilin agar.	*Group D Streptococcus*	*Group D Streptococcus* causes urinary tract infections, meningitis, neonatal sepsis, spontaneous bacterial peritonitis, septic arthritis, and vertebral osteomyelitis diseases.
Gram negative curved rod. It forms abundant, thick, mucous white-coloured colonies on nutrient agar and yellow-coloured colonies on TCBS agar.	*Vibrio cholerae*	*Vibrio cholerae* is responsible for the occurrence of cholera.
Gram negative rod. It forms slimy, white somewhat translucent, raised growth on nutrient agar and dark pink-coloured colonies on MacConkey agar.	*Klebsiella pneumoniae*	*Klebsiella pneumoniae* is responsible for pneumonia, thrombophlebitis, urinary tract infection (UTI), cholecystitis, diarrohea, upper respiratory tract infection, wound infection, osteomyelitis, meningitis, and bacteremia and septicemia.
Gram negative rod. It forms thin even grayish growth on nutrient agar and dark green colonies on SS agar.	*Salmonella typhi*	*Salmonella typhi* causes typhoid.
Gram negative rod. It forms grayish growth on nutrient agar and colourless colonies on SS agar.	*Shigella dysenteriae*	*Shigella dysenteriae* is the bacillary dysentery causing bacterium.

Table 49.2 Biochemical characteristics of isolates

Test								
Catalase	+	–	+	+	+	+	+	+
Oxidase	–	–	–	+	–	–	+	–
Motility	–	–	+	+	–		+	–
Indole	–	–	+	+	–	+	–	–
Methyl-red	–	+	+	–	+	+	–	–
Voge-Proskauer	+	–	–	+	–	–	–	+
Citrate utilization	–	–	–	+	+	–	+	+
Urease	+	–	–	–	–	–	–	+
Hydrogen sulphide	–	–	–		+	–	–	–
Starch hydrolysis	–	–	–	–	–	–	–	–
Nitrate utilization	–	–	+	+	+	+	–	+
Gelatin liquefication	–	–	–	+	–	–	–	–
Lactose fermentation	–	A	AG	AG	–	–	–	AG
Glucose fermentation	A	A	AG	AG	AG	A	–	AG
Sucrose fermentation	A	A	A(+)	AG	AG	A+/–	–	AG
Organism	*Staphylococcus*	*Streptococcus*	*E. coli*	*Vibrio*	*Salmonella*	*Shigella*	*Pseudomonas aeruginosa*	*Klebsiella*

A acid production only, *AG* acid and gas production
+/– = Variable reaction, + = Positive, – = Negative, (+) = Late Positive

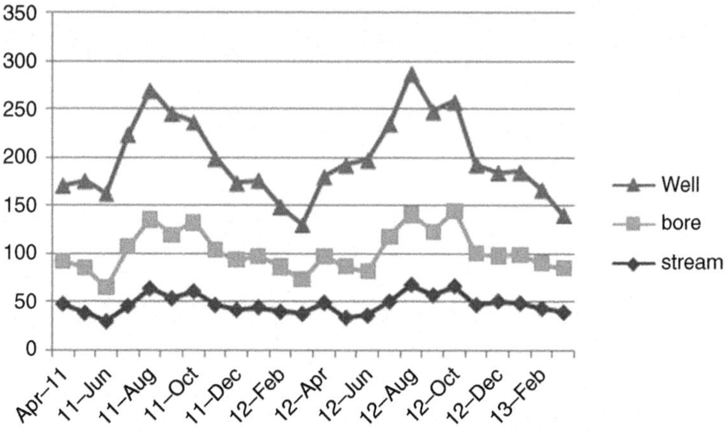

Fig. 49.1 Total plate count (CFU/ml) of bacteria in three water samples

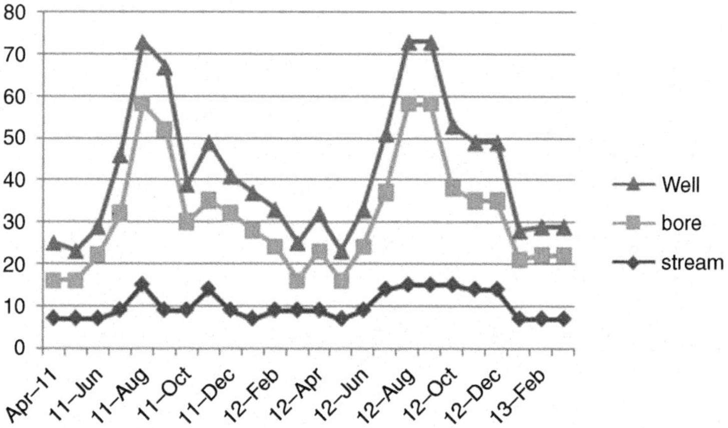

Fig. 49.2 Most probable number (/100 ml) of coliforms in three water samples

1. ***Escherichia coli***: The average percentage of *Escherichia coli* for both years was 20.019%; in the first year it was 18.93% and in the second year it increased to 20.68% (Table 49.4). In the summer it was 16.82% then increased to 17.35% by the next year. In the rainy season it was 21.26% then increased to 24.19% by the next year and in the winter it was 17.35% then increased to 20.5% by the next year (Table 49.5).
2. ***Staphylococcus aureus***: The average percentage of *Staphylococcus aureus* for both years was 17.305%; in the first year it was 15.4% and in the second year it increased to 18.57% (Table 49.4). In the summer it was 13.67% then increased to 16.3% by the next year. In the rainy season it was 16.82% then increased to 19.98% by the next year and in the winter it was 15.77% then increased to 19.45% by the next year (Table 49.5).

Table 49.3 Percentage of pathogenic bacteria in stream water

Month	% Organisms								Total plate count of stream
	E. coli	*Staphylococcus*	*Group D Streptococcus*	*Klebsiella pneumoniae*	*Pseudomonas aeruginosa*	*Shigella*	*Salmonella*	*Vibrio cholerae*	
Apr-11	20.8	14.5	12.5	4.166	10.41	6.25	6.25	8.33	47.33 ± 2.08
May-11	23.6	15.7	10.52	5.26	13.15	7.89	7.89	1052	38 ± 4.58
Jun-11	23.33	16.66	13.33	6.66	10	6.66	6.66	10	29.33 ± 6.5
Jul-11	19.5	15.21	13.04	10.86	10.86	8.69	8.69	6.52	45.33 ± 3.5
Aug-11	18.75	15.625	15.625	4.68	9.375	10.93	10.93	9.375	63.33 ± 2.08
Sep-11	18.86	13.207	15.09	5.66	11.32	9.43	11.32	11.32	53 ± 2
Oct-11	16.39	13.11	16.39	4.61	13.11	9.83	8.1	9.83	60.33 ± 2.5
Nov-11	19.148	14.89	12.76	4.25	8.51	6.38	6.38	8.51	46.33 ± 3.05
Dec-11	21.42	16.66	14.28	4.76	11.9	7.14	7.14	7.14	41.33 ± 3.5
Jan-12	17.77	15.55	13.33	11.11	11.11	8.88	8.88	6.66	44 ± 4.58
Feb-12	17.5	22.5	7.5	7.5	10	7.5	7.5	7.5	39.33 ± 2.08
Mar-12	21.05	21.05	7.89	10.52	7.89	7.89	7.89	7.89	37.33 ± 4.04
Apr-12	20.4	18.36	10.2	8.16	10.2	8.16	8.16	10.2	48.66 ± 0.5
May-12	20.58	17.64	8.8	11.7	14.7	5.88	8.8	5.88	33.33 ± 2.08
Jun-12	25	22.22	8.33	5.55	8.33	8.33	5.55	5.55	35.66 ± 2.08
Jul-12	20	20	10	10	16	6	6	8	50 ± 2
Aug-12	20.58	16.17	13.2	4.4	8.823	10.29	10.29	11.76	67.33 ± 2.08
Sep-12	17.54	14.03	14.03	5.26	14.03	8.77	10.52	10.52	56.66 ± 1.52
Oct-12	18.18	13.6	13.6	4.54	13.6	10.6	10.6	12.12	65.845 ± 1.19
Nov-12	21.27	14.89	10.6	8.51	10.6	8.51	8.51	6.38	46.66 ± 2.08
Dec-12	19.6	19.6	7.8	15.6	9.8	5.8	5.8	7.8	50.33 ± 2.51
Jan-13	20.4	20.4	8.1	12.2	10.2	6.12	6.12	4.08	48.33 ± 3.05
Feb-13	20.9	23.25	9.3	11.6	6.97	6.97	6.97	4.65	43 ± 2
Mar-13	17.9	20.5	7.6	10.25	10.25	10.25	7.6	7.6	39 ± 1

Table 49.4 Percentage of pathogenic bacteria in Valasi panchayath during the year April 2011–March 2013

	Stream		Bore		Well	
Percentage of organism	April 2011–March 2012	April 2012–March 2013	April 2011–March 2012	April 2012–March 2013	April 2011–March 2012	April 2012–March 2013
E. coli	18.93	20.68	17.68	17.98	16.59	18.03
Staphylococcus	15.4	18.57	16.16	16.47	13.1	13.78
Group D Streptococcus	12.6	10.85	13.44	11.79	10.8	12.59
Klebsiella pneumoniae	6.3	8.93	8.01	8.3	8.93	8.25
Pseudomonas aeruginosa	10.34	11.56	12.08	13.29	12.16	13.53
Shigella	8.05	8.41	8.91	8.91	10.29	10.98
Salmonella	8.05	8.41	9.52	9.52	12.05	11.06
Vibrio cholerae	8.41	8.62	8.3	10.88	10.46	11.06

3. ***Group D Streptococcus***: The average percentage of *Group D Streptococcus* for both years was 11.408%; in the first year it was 12.6% and in the second year it decreased to 10.85% (Table 49.4). In the summer it was 8.93% then decreased to 7.36% by the next year. In the rainy season it was 17.87% then decreased to 16.3% by the next year and in the winter it was 11.04% then decreased to 8.93% by the next year (Table 49.5).
4. ***Pseudomonas aeruginosa***: The average percentage of *Pseudomonas aeruginosa* for both years was 10.88%; in the first year it was 10.34% and in the second year it increased to 11.56% (Table 49.4). In the summer it was 8.41% then increased to 8.93% by the next year. In the rainy season it was 13.14% then increased to 16.3% by the next year and in the winter it was 9.46% for the both the years (Table 49.5).
5. ***Shigella dysenteriae***: The average percentage of *Shigella dysenteriae* for both years was 8.0479%; in the first year it was 8.05% and in the second year it increased to 8.41 (Table 49.4). In the summer it was 5.78% then increased to 6.83% by the next year. In the rainy season it was 11.56% for the both the years and in the winter it was 6.83% for the both the years (Table 49.5).
6. ***Salmonella typhi***: The average percentage of *Salmonella typhi* for both years was 8.0229%; in the first year it was 8.05% and in the second year it increased to 8.41% (Table 49.4). In the summer it was 5.78% then it was increased to 6.31% by the next year. In the rainy season it was 11.56% then increased to 12.09% by the next year and in the winter it was 6.83% for the both the years (Table 49.5).
7. ***Vibrio cholerae***: The average percentage of *Vibrio cholerae* for both years was 7.9389%; in the first year it was 8.41% and in the second year it increased to 8.62% (Table 49.4). In the summer it was 7.36% then decreased to 6.31% by the next year. In the rainy season it was 11.04% then increased to 13.67% by the next year and in the winter it was 6.83% then decreased to 5.78% by the next year (Table 49.5).

Table 49.5 Seasonal variation percentage of pathogenic bacteria in Valasi Panchayath during the year April 2011–March 2013

Percentage of organism	Stream						Bore						Well					
	Summer		Rainy		Winter		Summer		Rainy		Winter		Summer		Rainy		Winter	
	1	2	1	2	1	2	1	2	1	2	1	2	1	2	1	2	1	2
E. coli	16.82	17.35	21.26	24.19	17.35	20.5	14.96	16.78	19.95	19.95	18.14	17.23	14.3	16.34	22.21	21.96	13.27	13.27
Staphylococcus	13.67	16.3	16.82	19.98	15.77	19.45	13.6	14.96	18.14	18.14	16.78	16.3	11.74	12.25	15.83	16.34	11.74	12.76
Group D Streptococcus	8.93	7.36	17.87	16.3	11.04	8.93	8.16	9.07	18.14	17.23	16.3	9.07	9.7	11.23	13.78	16.34	8.93	10.21
Klebsiella pneumoniae	5.2	7.36	7.36	7.36	6.31	12.09	8.16	8.16	7.7	7.25	8.16	9.52	7.1	7.1	9.95	10.21	9.7	7.4
Pseudomonas aeruginosa	8.41	8.93	13.14	16.3	9.46	9.46	7.7	11.33	16.3	16.3	12.24	12.24	10.7	12.76	16.34	15.8	9.95	11.49
Shigella	5.78	6.83	11.56	11.56	6.83	6.83	6.34	6.8	13.15	13.6	7.25	6.34	8.42	9.44	14.3	14.81	8.17	8.68
Salmonella	5.78	6.31	11.56	12.09	6.83	6.83	6.34	6.8	13.6	15.41	8.61	6.34	8.93	9.7	13.78	13.78	8.42	9.7
Vibrio cholerae	7.36	6.31	11.04	13.67	6.83	5.78	4.98	7.7	11.79	16.3	8.16	8.61	8.68	10.21	14.81	14.81	7.91	8.17

8. ***Klebsiella pneumonia***: The average percentage of *Klebsiella pneumoniae* for both years was 7.82525%; in the first year it was 6.3% and in the second year it increased to 8.93% (Table 49.4). In the summer it was 5.2% then increased to 7.36% by the next year. In the rainy season it was 7.36% for both the years and in the winter it was 6.31% increased to 12.09% by the next year (Table 49.5).

49.3.1.2 Bioindices

1. *Simpson Index*: The Simpson index average for both the years was 0.1218 (Table 49.6). In the summer it was 0.2881 then decreased to 0.2216 by the next year. In the rainy season it was 0.2971 then decreased to 0.2517 by the next year and in the winter it was 0.2134 then decreased to 0.1738 by the next year (Table 49.7).
2. *Shannon Index*: The Shannon index average for both the years was 2.9609 (Table 49.6). In the summer it was 2.01 then increased to 2.352 by the next year. In the rainy season it was 2.043 then increased to 2.266 by the next year and in the winter it was 2.437 then increased to 2.699 by the next year (Table 49.7).
3. *Menhinick Index*: The Menhinick index average for both the years was 0.845 (Table 49.6). In the summer it was 0.2434 then decreased to 0.1526 by the next year. In the rainy season it was 0.1629 then decreased to 0.1603 by the next year and in the winter it was 0.1617 then decreased to 0.1286 by the next year (Table 49.7).
4. *Margalef Richness Index*: The Margalef richness index average for both the years was 1.556 (Table 49.6). In the summer it was 0.9139 then increased to 0.9872 by the next year. In the rainy season it was 0.8506 then increased to 1.04 by the next year and in the winter it was 0.8252 then increased to 0.9103 by the next year (Table 49.7).
5. *Equitability Index*: The Equitability index average for both the years was 0.986 (Table 49.6). In the summer it was 0.6537 then decreased to 0.4105 by the next year. In the rainy season it was 0.5805 then increased to 0.8499 by the next year and in the winter it was 0.7659 then decreased to 0.5818 by the next year (Table 49.7).
6. *Berger-Parker Dominance Index*: The Berger-Parker Dominance index average for both the years was 0.1799 (Table 49.6). In the summer it was 0.4125 then increased to 0.7702 by the next year. In the rainy season it was 0.58 then decreased to 0.2569 by the next year and in the winter it was 0.2649 then increased to 0.5146 by the next year (Table 49.7).
7. *Dominance Index*: The Dominance index an average for both the years was 0.8781 (Table 49.6). In the summer it was 0.6698 then decreased to 0.3896 by the next year. In the rainy season it was 0.5903 then increased to 0.8138 by the next year and in the winter it was 0.7773 then decreased to 0.6187 by the next year (Table 49.7).

Table 49.6 Biodiversity indices of pathogenic bacteria in steam water

Month	Index						
	Simpson index	Dominance index	Shannon index	Menhinick index	Berger-Parker dominance index	Equitability index	Margalef richness index
Apr-11	0.1333	0.8667	2.834	1.265	0.25	0.9448	1.898
May-11	0.127	0.873	2.86	1.333	0.25	0.9533	1.953
Jun-11	0.1217	0.8783	2.851	1.512	0.25	0.9504	2.101
Jul-11	0.1185	0.8815	2.922	1.22	0.2093	0.9741	1.861
Aug-11	0.1262	0.8738	2.906	1.024	0.1967	0.9685	1.703
Sep-11	0.1192	0.8808	2.932	1.12	0.1961	0.9773	1.78
Oct-11	0.1227	0.8773	2.918	1.069	0.1786	0.9726	1.739
Nov-11	0.1294	0.8706	2.848	1.298	0.2368	0.9493	1.924
Dec-11	0.1309	0.8691	2.838	1.298	0.2368	0.9461	1.924
Jan-12	0.115	0.885	2.937	1.234	0.1905	0.9789	1.873
Feb-12	0.1311	0.8689	2.845	1.352	0.2571	0.9483	1.969
Mar-12	0.1294	0.8706	2.85	1.352	0.2286	0.95	1.969
Apr-12	0.1246	0.8754	2.902	1.18	0.2174	0.9674	1.828
May-12	0.121	0.879	2.866	1.414	0.2188	0.9554	2.02
Jun-12	0.1532	0.8468	2.725	1.414	0.2812	0.9084	2.02
Jul-12	0.133	0.867	2.852	1.155	0.2083	0.9508	1.808
Aug-12	0.1298	0.8702	2.892	0.9923	0.2154	0.9641	1.677
Sep-12	0.1202	0.8798	2.929	1.089	0.1852	0.9763	1.755
Oct-12	0.1225	0.8775	2.927	1	0.1875	0.9757	1.683
Nov-12	0.1243	0.8757	2.896	1.234	0.2381	0.9653	1.873
Dec-12	0.1351	0.8649	2.841	1.167	0.2128	0.9469	1.818
Jan-13	0.1417	0.8583	2.797	1.22	0.2326	0.9323	1.861
Feb-13	0.1444	0.8556	2.782	1.281	0.2564	0.9274	1.911
Mar-13	0.1206	0.8794	2.894	1.333	0.2222	0.9648	1.953

49.3.2 Bore Water

In bore water the total plate count fell in the range of 40.41 ± 1.15 to 71.5 ± 1.25 CFUs/ml. The water sample showed the maximum number of CFUs (71.5 ± 1.25 CFUs/ml) in rainy season (2012) and minimum number was noted in summer season (2011) (40.41 ± 1.15 CFUs/ml) (Fig. 49.1).

The MPN values for coliforms present in all the water samples are presented in Fig. 49.2. In bore water the MPN index ranged from 7–43/100 ml. The maximum MPN index was recorded in (43/100 ml) rainy season. The minimum MPN index was recorded in (7/100 ml) summer season.

49.3.2.1 Percentage of Pathogenic Bacteria

Analysis of bore water revealed that the predominant bacterium was *E. coli* (18.22%) followed by *Staphylococcus aureus* with 16.71%, *Group D Streptococcus* with 11.91%, *Pseudomonas aeruginosa* 11.88%, *Vibrio cholerae* with 9.31%, *Salmonella*

Table 49.7 Seasonal variation biodiversity indices of pathogenic bacteria in Valasi panchayath during the year April 2011–March 2013

	Stream						Bore						Well					
	Summer		Rainy		Winter		Summer		Rainy		Winter		Summer		Rainy		Winter	
Bioindex	1	2	1	2	1	2	1	2	1	2	1	2	1	2	1	2	1	2
Simpson Index	0.2881	0.2216	0.2971	0.2517	0.2134	0.1738	0.306	0.1842	0.4046	0.2483	0.3216	0.2814	0.3302	0.6104	0.4097	0.1862	0.2227	0.3813
Dominance Index	0.7119	0.7784	0.7029	0.7483	0.7866	0.8262	0.694	0.8158	0.5954	0.7517	0.6784	0.7186	0.6698	0.3896	0.5903	0.8138	0.7773	0.6187
Shannon Index	2.01	2.352	2.043	2.266	2.437	2.699	2.115	2.565	1.662	2.271	1.842	2.046	1.961	1.232	1.742	2.55	2.298	1.745
Menhinick Index	0.2434	0.1526	0.1629	0.1603	0.1617	0.1286	0.1907	0.1333	0.2194	0.4634	0.171	0.1858	0.1737	0.2308	0.1306	0.2765	0.1319	0.1711
Berger-Parker Dominance Index	0.3935	0.3002	0.4248	0.3112	0.337	0.2519	0.4688	0.2569	0.5451	0.3188	0.4228	0.3641	0.4125	0.7702	0.58	0.2569	0.2649	0.5146
Equitability Index	0.6701	0.7841	0.681	0.7555	0.8124	0.8998	0.7051	0.8551	0.554	0.7571	0.6141	0.6819	0.6537	0.4105	0.5805	0.8499	0.7659	0.5818
Margalef Richness Index	1.002	0.884	0.8987	0.8951	0.8971	0.8474	0.9367	0.8548	0.9732	1.229	0.9102	0.9302	0.9139	0.9872	0.8506	1.04	0.8525	0.9103

typhi with 9.205%, *Klebsiella pneumoniae* with 8.53% and *Shigella dysenteriae* with 8.07% (Table 49.8).

1. ***Escherichia coli***: The average percentage of *Escherichia coli* for both years was 18.22%; in the first year it was 17.68% and in the second year it decreased to 17.98% (Table 49.4). In the summer it was 14.96% then increased to 16.78% by the next year. In the rainy season it was 19.95% remained constant in the next year and in the winter it was 18.14% then decreased to 17.23% by the next year (Table 49.5).
2. ***Staphylococcus aureus***: The average percentage of *Staphylococcus aureus* for both years was 16.71%; in the first year it was 16.16% and in the second year it increased to 16.47% (Table 49.4). In the summer it was 13.6% then increased to 14.96% by the next year. In the rainy season it was 18.14% and remained constant in the next year and in the winter it was 16.78% then decreased to 16.3% by the next year (Table 49.5).
3. ***Group D Streptococcus***: The average percentage of *Group D Streptococcus* for both years was 11.91%; in the first year it was 13.44% and in the second year it was decreased to 11.79% (Table 49.4). In the summer it was 8.16% then increased to 9.07% by the next year. In the rainy season it was 18.14% then decreased to 17.23% by the next year and in the winter it was 16.3% then decreased to 9.07% by the next year (Table 49.5).
4. ***Pseudomonas aeruginosa***: The average percentage of *Pseudomonas aeruginosa* for both years was 11.88%; in the first year it was 12.08% and in the second year it increased to 13.29% (Table 49.4). In the summer it was 7.7% then increased to 11.33% by the next year. In the rainy season it was 16.3% and remained unchanged in the next year and in the winter it was 12.24% for both the years (Table 49.5).
5. ***Vibrio cholerae***: The average percentage of *Vibrio cholerae* for both years was 9.31%; in the first year it was 8.32% and in the second year it increased to 10.88% (Table 49.4). In the summer it was 4.98% then increased to 7.7% by the next year. In the rainy season it was 11.79% then increased to 16.3% by the next year and in the winter it was 8.16% then increased to 8.61% by the next year (Table 49.5).
6. ***Salmonella typhi***: The average percentage of *Salmonella typhi* for both years was 9.205%; in the first year it was 9.52% and it remained constant in the next year (Table 49.4). In the summer it was 6.34% then it was increased to 6.8% by the next year. In the rainy season it was 13.6% then increased to 15.41% by the next year and in the winter it was 8.31% then it was decreased to 6.34% by the next year (Table 49.5).
7. ***Shigella dysenteriae***: The average percentage of *Shigella dysenteriae* for both years was 8.53%; in the first year it was 8.91% and it remained constant in the second year (Table 49.4). In the summer it was 6.34% then increased to 6.8% by the next year. In the rainy season it was 13.15% then increased to 13.6% by the next year and in the winter it was 7.25% then decreased to 6.34% by the next year (Table 49.5).

Table 49.8 Percentage of pathogenic bacteria in bore water

Month	% Organisms								Total plate count of bore
	E. coli	*Staphylococcus*	*Group D Streptococcus*	*Klebsiella pneumoniae*	*Pseudomonas aeruginosa*	*Shigella*	*Salmonella*	*Vibrio cholerae*	
Apr-11	17.77	15.55	13.33	11.11	11.11	8.88	8.88	6.66	44.66 ± 3.05
May-11	19.14	14.89	12.7	10.6	12.7	8.51	8.51	6.38	46.33 ± 2.08
Jun-11	22.8	22.8	8.57	11.4	8.57	8.57	8.57	5.7	35 ± 4
Jul-11	16.12	16.12	16.12	4.83	14.5	9.67	8.06	9.67	61.33 ± 1.5
Aug-11	16.66	13.8	13.8	6.94	12.5	11.11	12.5	8.33	72 ± 3.6
Sep-11	15.15	15.15	15.15	4.54	13.6	10.6	10.6	12.12	65.66 ± 3.05
Oct-11	16.6	13.8	13.8	8.33	12.5	11.11	12.5	8.33	71.66 ± 3.05
Nov-11	17.54	14.03	14.03	5.26	14.03	8.77	10.5	10.5	56.66 ± 4.5
Dec-11	19.2	19.2	9.61	9.61	15.3	5.76	5.76	7.69	51.66 ± 3.05
Jan-12	18.86	18.86	15.09	11.3	11.3	7.5	11.3	5.6	53 ± 1
Feb-12	21.27	19.1	10.6	8.5	10.6	8.5	8.5	10.6	46.33 ± 1.5
Mar-12	22.22	22.22	8.33	11.11	8.33	8.33	8.33	8.33	35.66 ± 3.5
Apr-12	18.3	18.3	10.2	8.16	10.2	8.16	8.16	10.2	48.66 ± 1.5
May-12	18.86	18.86	9.43	9.43	16.98	5.66	5.66	7.5	52.66 ± 1.5
Jun-12	20	11.11	13.33	11.11	13.33	8.88	8.88	6.66	45 ± 4
Jul-12	14.7	14.7	14.7	4.4	13.23	10.29	11.7	11.7	67.66 ± 2.3
Aug-12	16.43	13.69	13.69	6.8	12.3	10.9	12.3	10.9	73 ± 2
Sep-12	15.15	15.15	12.12	4.5	13.6	10.6	10.6	15.15	65.845 ± 1.19
Oct-12	15.18	12.65	12.65	6.3	11.3	10.12	12.65	12.65	78.66 ± 2.51
Nov-12	18.5	18.5	9.25	9.25	1.66	5.55	5.55	9.25	53.33 ± 2.08
Dec-12	18.75	18.75	10.4	12.5	10.4	8.33	8.33	10.4	47.33 ± 2.08
Jan-13	19.60	19.60	9.8	11.76	15.68	5.88	5.88	7.8	50.33 ± 2.5
Feb-13	19.14	14.89	10.63	8.51	10.63	8.51	8.51	10.63	46.66 ± 2.08
Mar-13	19.5	19.5	8.69	8.69	10.86	8.69	8.69	10.86	45.33 ± 4.04

8. ***Klebsiella pneumonia***: The average percentage of *Klebsiella pneumoniae* for both years was 8.07%; in the first year it was 8.01% and in the second year it increased to 8.3% (Table 49.4). In the summer it was 8.16% and remained unchanged in the next year. In the rainy season it was 7.7 decreased to 7.25% by the next year and in the winter it was 8.16% increased to 9.52% by the next year (Table 49.5).

49.3.2.2 Bioindices

1. *Simpson Index*: The Simpson average for both the years was 0.1218 (Table 49.9). In the summer it was 0.306 then decreased to 0.1842 by the next year. In the rainy season it was 0.4046 then decreased to 0.2483 by the next year and in the winter it was 0.3216 then decreased to 0.2814 by the next year (Table 49.7).
2. *Shannon Index*: The Shannon index average for both the years was 2.960 (Table 49.9). In the summer it was 2.115 then increased to 2.565 by the next year. In the rainy season it was 1.662 then increased to 2.271 by the next year and in the winter it was 1.842 then increased to 2.046 by the next year (Table 49.7).
3. *Menhinick Index*: The Menhinick index average for both the years was 0.8456 (Table 49.9). In the summer it was 0.1907 then decreased to 0.1333 by the next year. In the rainy season it was 0.2194 then increased to 0.4634 by the next year and in the winter it was 0.171 then increased to 0.1858 by the next year (Table 49.7).
4. *Margalef Richness Index*: The Margalef richness index average for both the years was 1.55 (Table 49.9). In the summer it was 0.9367 then decreased to 0.854 by the next year. In the rainy season it was 0.9732 then decreased to 1.229 by the next year and in the winter it was 0.9102 then decreased to 0.9302 by the next year (Table 49.7).
5. *Equitability Index*: The Equitability index both the years was 0.986 (Table 49.9). In the summer it was 0.7051 then increased to 0.8551 by the next year. In the rainy season it was 0.554 then increased to 0.7571 by the next year and in the winter it was 0.6141 then increased to 0.6819 by the next year (Table 49.7).
6. *Berger-Parker Dominance Index*: The Berger-Parker Dominance index average for both the years was 0.17995 (Table 49.9). In the summer it was 0.4688 then decreased to 0.2563 by the next year. In the rainy season it was 0.5451 then decreased to 0.3188 by the next year and in the winter it was 0.4228 then decreased to 0.3641 by the next year (Table 49.7).
7. *Dominance Index*: The Dominance index average for both the years was 0.87812 (Table 49.9). In the summer it was 0.694 then increased to 0.8158 by the next year. In the rainy season it was 0.5954 then increased to 0.7517 by the next year and in the winter it was 0.6784 then increased to 0.7186 by the next year (Table 49.7).

Table 49.9 Biodiversity indices of pathogenic bacteria in bore water

Month	Index						
	Simpson index	Dominance index	Shannon index	Menhinick index	Berger-Parker dominance index	Equitability index	Margalef richness index
Apr-11	0.115	0.885	2.937	1.234	0.1905	0.9789	1.873
May-11	0.1184	0.8816	2.924	1.206	0.2045	0.9746	1.85
Jun-11	0.1337	0.8663	2.822	1.372	0.2353	0.9407	1.985
Jul-11	0.1251	0.8749	2.907	1.042	0.1695	0.969	1.717
Aug-11	0.1198	0.8802	2.954	0.9631	0.1739	0.9848	1.653
Sep-11	0.121	0.879	2.934	1	0.1562	0.9779	1.683
Oct-11	0.1184	0.8816	2.964	0.9562	0.1714	09882	1.648
Nov-11	0.1202	0.8798	2.929	1.089	0.1852	0.9763	1.755
Dec-11	0.133	0.867	2.852	1.155	0.2083	0.9508	1.808
Jan-12	0.1248	0.8752	2.903	1.099	0.1887	0.9677	1.763
Feb-12	0.1246	0.8754	2.902	1.18	0.2174	0.9674	1.828
Mar-12	0.1294	0.8706	2.85	1.352	0.2286	0.95	1.969
Apr-12	0.1212	0.8788	2.917	1.193	0.2	0.9722	1.839
May-12	0.1344	0.8656	2.845	1.143	0.2041	0.9485	1.799
Jun-12	0.1173	0.8827	2.927	1.234	0.2143	0.9758	1.873
Jul-12	0.1207	0.8793	2.936	0.9923	0.1538	0.9788	1.677
Aug-12	0.1183	0.8817	2.965	0.9494	0.169	0.9883	1.642
Sep-12	0.121	0.879	2.934	1	0.1562	0.9779	1.683
Oct-12	0.1185	0.8815	2.966	0.93	0.1622	0.9886	1.626
Nov-12	0.1322	0.8678	2.858	1.131	0.2	0.9526	1.789
Dec-12	0.1193	0.8807	2.929	1.167	0.1915	0.9764	1.818
Jan-13	0.1318	0.8682	2.858	1.143	0.2041	0.9527	1.799
Feb-13	0.1163	0.8837	2.938	1.22	0.2093	0.9792	1.861
Mar-13	0.1226	0.8774	2.908	1.206	0.2045	0.9692	1.85

49.3.3 Well Water

In well water the total plate count fell in the range of 56.33 ± 2.51 to 127 ± 2 CFUs/ml. The water sample showed the maximum number of CFUs (127 ± 2 CFUs/ml) in rainy season (2011) and minimum number was noted in summer season (2012) (56.33 ± 2.51 CFUs/ml) (Fig. 49.1).

The MPN values for coliforms present in all the water samples are presented in Fig. 49.2. In well water the MPN index ranged from 7–15/100 ml. The maximum MPN index was recorded in (43/100 ml) rainy season. The minimum MPN index was recorded in (7/100 ml) summer season.

49.3.3.1 Percentage of Pathogenic Bacteria

Analysis of well water revealed that the predominant bacterium was *E. coli* (17.24%) followed by *Staphylococcus aureus* with 13.56%, *Pseudomonas aeruginosa* 12.88%, *Group D Streptococcus* with 11.74%, *Salmonella typhi* with 10.69%, *Vibrio cholerae* with 10.57%, *Shigella dysenteriae* with 10.43%, and *Klebsiella pneumoniae* with 8.65% (Table 49.10).

Table 49.10 Percentage of pathogenic bacteria in well water

Month	% Organisms								Total plate count of well
	E. coli	*Staphylococcus*	*Group D Streptococcus*	*Klebsiella pneumoniae*	*Pseudomonas aeruginosa*	*Shigella*	*Salmonella*	*Vibrio cholerae*	
Apr-11	15	12.5	12.5	10	12.5	10	11.25	10	79.33 ± 7.5
May-11	17.39	15.21	10.86	8.69	13.04	10.86	10.86	8.69	91.66 ± 4.04
Jun-11	18.18	14.14	10.10	9.09	12.12	10.10	10.10	12.12	98.66 ± 6.5
Jul-11	16.9	11.86	13.55	8.47	13.55	10.16	10.16	11.86	117.33 ± 1.5
Aug-11	18.5	13.33	10.37	7.4	13.33	11.85	11.85	11.85	134.33 ± 8.08
Sep-11	17.3	12.59	11.02	7.87	12.59	11.02	11.02	12.59	127 ± 2
Oct-11	19.04	13.33	9.5	8.57	11.42	13.33	11.42	11.42	104.66 ± 4.5
Nov-11	18.55	16.49	10.3	10.3	12.37	10.3	10.3	8.24	96.33 ± 1.52
Dec-11	14.8	12.34	8.64	12.34	11.11	9.87	9.87	12.34	80.66 ± 1.5
Jan-12	15	12.5	12.5	12.5	11.25	10	11.25	10	79.66 ± 2.08
Feb-12	15.62	15.62	12.5	12.5	14.06	9.37	9.375	7.81	63.66 ± 2.3
Mar-12	17.24	13.79	13.79	5.17	13.79	8.62	10.34	10.34	57.33 ± 2.08
Apr-12	19.27	14.45	12.04	6.02	14.45	9.63	9.63	9.63	83 ± 2
May-12	18.69	13.08	9.345	9.345	13.08	11.21	11.21	11.21	106.33 ± 2.5
Jun-12	15.38	11.96	13.67	8.54	13.67	10.25	10.25	11.96	116.66 ± 2.08
Jul-12	16.94	11.86	11.86	8.47	13.55	11.86	11.86	10.16	117.33 ± 3.05
Aug-12	17	12.24	13.6	6.80	12.24	12.24	9.52	10.88	147 ± 2
Sep-12	18.25	14.28	11.11	7.93	12.69	11.11	11.11	12.69	125.33 ± 4.04
Oct-12	15.78	12.28	14.03	8.77	12.28	10.52	10.52	12.28	113.33 ± 4.04
Nov-12	19.35	15.05	10.75	8.60	12.90	10.75	10.75	8.60	93 ± 2
Dec-12	18.39	16.09	11.49	5.74	13.79	9.19	11.49	9.19	87 ± 2
Jan-13	18.39	13.79	11.49	9.19	13.79	9.19	9.19	9.19	86.66 ± 1.52
Feb-13	15.38	12.82	12.82	10.25	11.53	10.25	12.82	1025	77.33 ± 2.08
Mar-13	17.54	14.03	14.03	5.26	14.03	8.77	10.52	10.52	56.33 ± 2.51

1. ***Escherichia coli***: The average percentage of *Escherichia coli* for both years was 17.24%; in the first year it was 16.59% and in the second year it increased to 18.03% (Table 49.4). In the summer it was 14.3% then increased to 16.34% by the next year. In the rainy season it was 22.21% then decreased to 21.96% by the next year and in the winter it was 13.27% remained constant in the next year (Table 49.5).
2. ***Staphylococcus aureus***: The average percentage of *Staphylococcus aureus* for both years was 13.56%; in the first year it was 13.1% and in the second year it increased to 13.78% (Table 49.4). In the summer it was 11.74% then increased to 12.25% by the next year. In the rainy season it was 15.83% then increased to 16.34% by the next year and in the winter it was 11.74% then increased to 12.76% by the next year (Table 49.5).
3. ***Pseudomonas aeruginosa***: The average percentage of *Pseudomonas aeruginosa* for both years was 12.88%; in the first year it was 12.16% and in the second year it increased to 13.53% (Table 49.4). In the summer it was 10.7% then increased to 12.76% by the next year. In the rainy season it was 16.34% then decreased to 15.8% by the next year and in the winter it was 9.95% then increased to 11.49% by the next year (Table 49.5).
4. ***Group D Streptococcus***: The average percentage of *Group D Streptococcus* for both years was 11.74%; in the first year it was 10.8% and in the second year it increased to 12.59% (Table 49.4). In the summer it was 9.7% then increased to 911.23% by the next year. In the rainy season it was 13.78% then decreased to 16.34% by the next year and in the winter it was 8.93% then decreased to 10.21% by the next year (Table 49.5).
5. ***Salmonella typhi***: The average percentage of *Salmonella typhi* for both years was 10.69%; in the first year it was 12.05% then decreased to 11.06% in the next year (Table 49.4). In the summer it was 8.93% then it was increased to 9.7% by the next year. In the rainy season it was 13.78% remained constant in the next year and in the winter it was 8.42% then it was increased to 9.7% by the next year (Table 49.5).
6. ***Vibrio cholerae***: The average percentage of *Vibrio cholerae* for both years was 9.31%; in the first year it was 10.46% and in the second year it increased to 11.06% (Table 49.4). In the summer it was 8.68% then increased to 10.21% by the next year. In the rainy season it was 14.81% remained constant in the next year and in the winter it was 7.91% then increased to 8.17% by the next year (Table 49.5).
7. ***Shigella dysenteriae***: The average percentage of *Shigella dysenteriae* for both years was 8.53%; in the first year it was 10.29% then it increased to 10.98% by the second year (Table 49.4). In the summer it was 8.42% then increased to 9.44% by the next year. In the rainy season it was 14.3% then increased to 14.81% by the next year and in the winter it was 8.17% then increased to 8.68% by the next year (Table 49.5).
8. ***Klebsiella pneumonia***: The average percentage of *Klebsiella pneumoniae* for both years was 8.65%; in the first year it was 8.93% and in the second year it decreased to 8.25% (Table 49.4). In the summer it was 7.1% and remained

unchanged in the next year. In the rainy season it was 9.95 increased to 10.21% by the next year and in the winter it was 9.7% decreased to 7.4% by the next year (Table 49.5).

49.3.3.2 Bioindices

1. *Simpson Index*: The Simpson average for both the years was 0.1218 (Table 49.11). In the summer it was 0.3302 then increased to 0.6104 by the next year. In the rainy season it was 0.4097 then decreased to 0.1862 by the next year and in the winter it was 0.2227 then increased to 0.3813 by the next year (Table 49.7).
2. *Shannon Index*: The Shannon index average for both the years was 2.960 (Table 49.11). In the summer it was 1.961 then decreased to 1.232 by the next year. In the rainy season it was 1.742 then increased to 2.55 by the next year and in the winter it was 2.298 then decreased to 1.745 by the next year (Table 49.7).
3. *Menhinick Index*: The Menhinick index average for both the years was 0.8456 (Table 49.11). In the summer it was 0.1737 then increased to 0.2308 by the next year. In the rainy season it was 0.1306 then increased to 0.2765 by the next year and in the winter it was 0.1319 then increased to 0.1711 by the next year (Table 49.7).
4. *Margalef Richness Index*: The Margalef richness index average for both the years was 1.55 (Table 49.11). In the summer it was 0.9139 then increased to 0.9872 by the next year. In the rainy season it was 0.8506 then increased to 1.04 by the next year and in the winter it was 0.8525 then increased to 0.9103 by the next year (Table 49.7).
5. *Equitability Index*: The Equitability index both the years was 0.986 (Table 49.11). In the summer it was 0.6537 then decreased to 0.4105 by the next year. In the rainy season it was 0.5805 then increased to 0.8499 by the next year and in the winter it was 0.7659 then decreased to 0.5818 by the next year (Table 49.7).
6. *Berger-Parker Dominance Index*: The Berger-Parker Dominance index average for both the years was 0.17995 (Table 49.11). In the summer it was 0.4125 then increased to 0.7702 by the next year. In the rainy season it was 0.58 then decreased to 0.2569 by the next year and in the winter it was 0.2569 then increased to 0.5146 by the next year (Table 49.7).
7. *Dominance Index*: The Dominance index average for both the years was 0.87812 (Table 49.11). In the summer it was 0.6698 then decreased to 0.3896 by the next year. In the rainy season it was 0.5903 then increased to 0.8138 by the next year and in the winter it was 0.7773 then decreased to 0.6187 by the next year (Table 49.7).

Total plate count for bacteria performed for all water samples showed that the bacteria in all the samples were above the WHO guideline values (<10 CFUs/ml). The total plate count in all the three water samples was highest during the rainy season, i.e., August, and was due to the contribution of all the pathogenic bacteria. However the water samples from tap showed relatively higher plate count throughout the year. This may be due to the presence of sewage surrounding the well, which continuously seeps into the well water. This study is in conformation with the result

Table 49.11 Biodiversity indices of pathogenic bacteria in well

Month	Index						
	Simpson index	Dominance index	Shannon index	Menhinick index	Berger-Parker dominance index	Equitability index	Margalef richness index
Apr-11	0.1157	0.8843	2.986	0.9238	0.16	0.9954	1.621
May-11	0.1223	0.8777	2.96	0.8528	0.1818	0.9866	1.563
Jun-11	0.1225	0.8775	2.964	0.8208	0.1895	0.9879	1.537
Jul-11	0.1225	0.8775	2.971	0.7493	0.1754	0.9902	1.478
Aug-11	0.1255	0.8745	2.96	0.6937	0.188	0.9865	1.431
Sep-11	0.1232	0.8768	2.97	0.7243	0.1803	0.9899	1.457
Oct-11	0.1239	0.8761	2.96	0.7883	0.1942	0.9866	1.51
Nov-11	0.1251	0.8749	2.949	0.8251	0.1915	0.9829	1.541
Dec-11	0.1163	0.8837	2.982	0.93	0.1622	0.9939	1.626
Jan-12	0.1154	0.8846	2.988	0.9177	0.1579	0.9961	1.616
Feb-12	0.1174	0.8826	2.961	1.016	0.1613	0.9869	1.696
Mar-12	0.1202	0.8798	2.929	1.089	0.1852	0.9763	1.755
Apr-12	0.1269	0.8731	2.926	0.9001	0.2025	0.9752	1.602
May-12	0.1232	0.8768	2.964	0.7845	0.1923	0.9882	1.507
Jun-12	0.121	0.879	2.978	0.7559	0.1607	0.9926	1.484
Jul-12	0.1219	0.8781	2.974	0.7493	0.1754	0.9914	1.478
Aug-12	0.1256	0.8744	2.959	0.6786	0.1799	0.9863	1.419
Sep-12	0.1244	0.8756	2.964	0.7155	0.184	0.988	1.45
Oct-12	0.1206	0.8794	2.979	0.7628	0.1636	0.9931	1.489
Nov-12	0.1251	0.8749	2.947	0.8433	0.2	0.9823	1.556
Dec-12	0.1272	0.8728	2.925	0.8781	0.1928	0.9749	1.584
Jan-13	0.1232	0.8768	2.952	0.8835	0.1951	0.984	1.588
Feb-13	0.1157	0.8843	2.986	0.9238	0.16	0.9954	1.621
Mar-13	0.1202	0.8798	2.929	1.089	0.1852	09763	1.755

of Zaky et al. (2006) who reported increased bacterial content in the water of Manzala Lake, Egypt, which is polluted by drainage and sewage.

Sources of contamination may be due to open defecation, animal washing near the sampling points, washing of other household utensils near the sampling points, washing of clothes near the sampling points, graveyard is in the range of 1 km.

Diseases frequently observed were typhoid, dysentery, malaria.

49.4 Seasonal Variations

All samples were analyzed monthly to get the seasonal changes of the above-mentioned microbial groups for their bacteriogical quality. These investigations were carried out for a period of 24 months (2 years, i.e., April 2011 to March 2013). The result obtained from the three sources in the study area revealed that highest number of counts was recorded during monsoon and post monsoon and least counts were recorded in summer.

The highest microbial numbers during these seasons may be due to flooded rainy water, runoff from rural animal husbandry and agricultural grounds and open-air defecation along the margins of canal by rural people. Bahador et al. (2005) also investigated that population of faecal coliforms in the monsoon was maximum followed by winter and summer (Krishna Kumar et al. 2013). In the same way the microbial populations in this study was found very high during the August and September, i.e., end of monsoon and start of post monsoon. Present investigations revealed that open water is not safe for drinking purpose in any season as it was found to contain a variety of bacteria including the coliforms in high loads. Microbial populations found were variable from location to location and peak values were during monsoon and post-monsoon seasons. It might be due to percolation of water from sanitary land filled areas and leakages from septic tanks. The present findings are at the par with the observations of Khannan et al. (2001), who have reported high load of coliform counts for open well water. The open wells are found to be the most sensitive for contamination compared to the tube wells in the study area. Another reason for contamination of bore well water may be due to seepage of domestic waste water from nearby areas and from the soak pits adjacent to the wells. The fact that bore wells, which were exposed to environment and are on the upper strata, are expected to have higher microbial contamination than bore wells, High levels are found primarily where human and animal waste is allowed to runoff. It may be due to percolation of water from sanitary land filled areas and leakages from septic tanks. Thus environmental settings of the storage tanks contribute to its microbiological quality. Beside these general findings the chief reasons for deprived quality of treated well waters are intermittent water distribution. This causes temporarily negative pressure in the pipelines, which increases the chance of contaminant water to enter into siphon, which may carry microorganisms, bacteria, and viruses as well as contaminants present in the sewage, soil and groundwater, into the distribution system. Moreover, domestic sewage and municipal water distribution pipelines run in parallel which is increasing the extent of pollution. Based on these results it may be proposed that extreme measures need to be taken in sprawling metropolitan areas in order to improve and guarantee the quality of water being supplied to the population. The present cases may be considered as typical examples of what is happening in other cities in the developing world and may act as a catalyst for analyses. Thus it may hopefully bring improvements in the quality of water in the cities of developing world by switching on to continuous water supply. Since the objective of using bacterial indicators is to achieve good water quality, the data retrieved in the present study can be used in improving the drinking water system monitoring. A WHO expert committee strongly recommended that intermittent and low pressure service should be avoided. Present investigations revealed that open water is not safe for drinking purpose in any season as it was found to contain a variety of bacteria including the coliforms in high loads. Microbial populations found were variable from location to location. Present study also reports high bacterial loads in well water as compared to the other sources and peak values were during monsoon and post-monsoon season. In the same way the well water investigated for bacteriological quality clearly revealed that 100% of well water samples are not fit for human consumption in all seasons. It is clear that all microbial parameters showed a clear monsoon and post-monsoon maxima, whereas a little was observed between summer and winter season. The high contamination was due to the improper sanitation, maintenance and monitoring of the storage tanks. The deprived quality of drinking water may be considered as typical example for the quality of drinking water at other public places.

The isolates were initially differentiated on the basics of the cultural and morphological studies after which they were subjected to IMViC biochemical test. All the water samples were contaminated with *Escherichia coli, Klebsiella pneumoniae, Salmonella typhi, Shigella dysenteriae, Staphylococcus aureus, Group D Streptococcus, Pseudomonas aeruginosa* and *Vibrio cholerae*. These bacteria cause waterborne diseases like intestinal infections, dysentery, typhoid fever, cholera and other illnesses.

Treatment of water by Municipal Corporation should be done in such a way that impurities and pathogenic organisms should be completely removed or minimized to such an extent that it should not be hazardous to humans and animals (Bharti and Katyal 2011).

Human and animal wastes are the primary sources of different bacteria in water. The sources of bacterial contamination include run off from feedlots, pastures, dog runs and other land areas where animal wastes are deposited. Bacteria from these sources can enter in taps that are either open at the land surface, or don't have water tight casing or caps, or don't have seal in the annular space (the space between the wall of the drilled tap and the outside of the tap casing). Insects, rodents and animals entering the tap are other sources of contamination. Another way through which bacteria can enter the water supply is through inundation or infiltration by flood waters or by surface runoff. Flood water commonly contains high level of bacteria. Small depressions filled with flood water provide excellent breeding ground for bacteria (Ley and Samant 2003). In the present study area the places surrounding the drinking water sources are not hygienic. The open taps are surrounded by drainage, throughout the year. The daily household activities like washing clothes and cleaning utensils are being carried out at the hand bores. The stream water gets polluted in multiple ways. Cleaning the domestic animals and washing clothes in the stream and throwing domestic wastes into the stream contaminate the water throughout the year.

Chan et al. (2007) isolated pathogenic bacteria such as *E. coli, Streptococcus faecalis* and *Pseudomonas aeruginosa* from the water samples. Ajibade et al. (2008) confirmed the presence of the coliforms. They isolated different pathogenic bacteria viz. *Pseudomonas* sp. *Escherichia coli, Acetobacter* sp., *Maroxalla* sp., *Bacillus* sp., and *Klebsiella* sp. from the River water samples. As a result they concluded that the water of the four rivers in the park is not potable during the wet seasons. Omezuruike et al. (2008) Isolated different bacterial pathogens and these pathogens were identified to be *Staphylococcus aureus, Salmonella* species, *Escherichia coli, Pseudomonas aeruginosa, Enterobacter aerogenes, Bacillus* species, *Proteus* species, *Klebsiella* species, *Flavobacterium* species and *Acinetobacter* species. Usharani et al. (2010) in their bacteriological study showed that the total heterotrophic bacteria, total coliforms, faecal coliforms, faecal Streptococci and FC/FS ratio in the river water samples were found to be greater than the standard WHO limits. The generic distribution in the samples revealed that the presence of *Escherichia coli, Staphylococcus, Enterobacter, Streptococci, Bacillus* and *Micrococcus* were predominant in river water samples.

Open air defecation, may lead to contamination of the water supply system and results in outbreaks of diarrheal disease (Bora et al. 1997; Sarkar et al. 2007). The practices of tethering animals close to human dwellings and the consequent proximity to animal faecal matter further enhances the risk of contamination of drinking water (Howe et al. 2002; Licence et al. 2001).

49.4.1 Bioindices

Diversity indices can serve as a good indicator of the overall pollution of water. Bioindices of species diversity can be derived from species counts and are of three main categories, which are species richness (Margalef index), species evenness/dominance (Simpson index) and a combination of richness and dominance (Shannon–Wiener index) (Sigee 2004). These diversity indices have been developed by taking into account the number of species and their relative abundances, which means the higher the values of these diversity indices, the more the oligotrophic state of water bodies. Wilhm (1975) implied that a high value of diversity index (H') suggests a more healthy ecosystem, while a low value suggests a less healthy or degraded ecosystem. Margalef index (Margalef 1958) relates the number of species to the total number of individuals. The fall in the value of Margalef index shows the rise in the level of pollution. Mukherjee (1997) reported that higher species richness (R_1 and R_2) is characterized by larger food chain. The higher value of Shannon's index (H') indicated greater species diversity. The greater species diversity means larger food chain and more cases of inter-specific interactions and greater possibilities for negative feedback control which reduced oscillations and hence increases the stability of the community (Ludwick and Reynolds 1998). Pielou's evenness index (E) (1967) states that species evenness is a measure of diversity that quantifies how equal the community is numerically. Species evenness decreased with increasing size of the plankton population (Shinde et al. 2011; Adesalu and Nwankwo 2008). Berger-Parker dominance index (1970) is the number of individuals in the dominant taxon divided by number of individuals (n). It is the largest species proportion of all species in a community. This index is most strongly influenced by evenness of the indices (Shannon and Weiner 1949).

Various diversity measures have potential applications in aquatic ecosystems, mainly in conservation. It is often understood that species rich communities are better than species poor communities. Secondly, in environmental monitoring, it is assumed that the adverse effects of pollution will be reflected in the reduction of diversity or change in the composition of species abundance. Rosenberg (1976) and Patrick (1973) are of the opinion that enriched or polluted ecosystems display a reduction in diversity. Platt et al. (1988) used the Simpson's index in bio monitoring. Magurran (1983) comments that ecologists should be able to ask the question and formulate the hypothesis to help them understand and sensitively manage the natural ecosystem.

It is generally believed that ecosystems high in diversity are more stable. Since the Earth Summit held in Rio de Janeiro, Brazil, in 1992, interest in species diversity has been increasing, especially in terms of conserving natural ecosystems. However, the factors producing and maintaining species diversity have not yet been well documented.

49.5 Conclusion

Results of the present study indicated that water in Valasi panchayath is contaminated with various pathogenic bacteria and unfit for drinking. This study reveals that the increase in the microbial loads at the consumer points (i.e., stream, bore, and well) was due to the observed activities such as the direct washing of human clothing and other household utensils around the sampling point. The presence of animals and the intense agriculture-related activities going on around the consumer point could lead to contamination.

Thus, potable and domestic water should be harmless for the health of humanity and should have organoleptic properties and should be suitable for domestic use. Water quality should be controlled in order to minimize acute health problems of water-related disease in humans.

The following three-point approach is suggested for improving the quality of water supplied to the tribal communities of the panchayath studied.

- Investigate the source of contamination of pipe-borne water supplies to delineate the roles of the water delivery system and of household water storage system.
- Institute a system to monitor the quality of untreated water sources so that water collection can be restricted to uncontaminated sources and water treatment advisories can be issued appropriately.
- Educate the public on appropriate water handling, storage and treatment methods. It is evident that until these recommendations are implemented, the water supplied to the tribal communities in Valasi panchayath of ananthagiri mandal, Visakhapatnam district will continue to pose a health hazard to the population.

References

Adesalu TA, Nwankwo DI (2008) Effect of water quality indices on phytoplankton of a sluggish tidal creek in Lagos, Nigeria. Pak J Biol Sci 11:836–844

Ajibade WA, Ayodele IA, Agbede SA (2008) Microbiological characteristics of waters in the major rivers in Kainji Lake National Park. Afr J Environ Sci Technol 2(8):208–216

Bahador N, Baseri Salehi M, Patil DN, Kapadnis BP (2005) Seasonal variations of microbial pollution in surface waters of Pune city. Nat Environ Pollut Technol 4(1):53–56

Berger WH, Parker FL (1970) Diversity of planktonic foraminifera in deep-sea sediments. Science 168:1345–1347

Bharti N, Katyal D (2011) Water quality indices used for surface water vulnerability assessment. Int J Environ Sci 2(1):154

Bora D, Dhariwal AC, Jain DC, Sachdeva V, Vohra JG, Prakash RM (1997) *V. cholera* OI outbreak in remote villages of Shimla district, Himachal Pradesh. J Commun Dis 29:121–125

Chan CL, Zalifah MK, Norrakiah AS (2007) Microbiological and physicochemical quality of drinking water. Mal J Anal Sci 11(2):414–420

Hammer Ø, Harper DAT, Ryan PD (2001) Past: paleontological statistics software package for education and data analysis. Palaeontol Electron 4(1):9. http://palaeo-electronica.org/2001_1/past/issue1_01.htm

Howe AD, Forster S, Morton S, Marshall R, Osbrn KS, Wright P (2002) Cryptosporidum oocysts in a water supply associated with a cryptosporidiosis outbreak. Emerg Infect Dis 8:619–624

Khannan P, Sridhar SR, Sarvanan R, Rajasekhara Pandian M (2001) Isolation of Vibrio sp. and a few coliforms from portable water of Rasipuram and Tiruchengodu taluks, Tamilnadu. Ecotoxicol Environ Monit 11(1):43–46

Krishna Kumar K, Israel Y, Sowjanya P (2013) Geochemical assessment of ground water along Thandava river basin, Andhra Pradesh

Ley V, Samant J (2003) Bacteriological analysis of drinking water of Kolhapur City, Maharashtra, India, vol 24. Department of Environmental Sciences, SubCentre, Mumbai University, Ratnagiri, pp 689–694

Licence K, Oates KR, Synge BA, Reid TM (2001) An outbreak of *E. coli* O157 infection with evidence of spread from animal to man through contamination of a private water supply. Epidemiol Infect 126:135–138

Ludwick JA, Reynolds JF (1998) Statistical ecology a primer on methods and computing. Wiley-Interscience Publication, New York, pp 1–337

Magurran AE (1983) Measuring biological diversity, 2nd edn. Blackwell Science Ltd., Oxford

Margalef R (1958) Information theory in ecology. Gen Syst 3:36–71

Menhinick EF (1964) A comparison of some species-individuals diversity indices applied to samples of field insects. Ecology 45:859–861

Mukherjee B (1997) Environmental biology. Tata McGraw Hill Publishing Company Limited, New Delhi

Omezuruike OI, Damilola AO, Adeola OT, Fajobi EA, Shittu OB (2008) Microbiological and physicochemical analysis of different water samples used for domestic purposes in Abeokuta and Ojota, Lagos State, Nigeria. Afr J Biotechnol 7(5):617–621

Patrick R (1973) Use of algae especially diatoms in the assessment of water quality. Special technical publication 528. American Society for Testing and Methods, pp 76–95

Pielou EC (1967) The use of information theory in the study of the diversity of biological pollutions. In: Proceedings of the fifth Berkeley symposium mathematical statistics and probability, vol 4, pp 163–177

Platt T, Sathyendranath S, Caverhill CM, Lewins M (1988) Ocean primary production and available further algorithms for remote sensing. Deep Sea Res 35:855–879

Rosenberg R (1976) Benthic faunal dynamics during succession following pollution abatement in a Swedish Estuary. Okios 27:414–427

Sarkar R, Prabhakar AT, Manickam S, Selvapandian D, Raghava MV, Kang G (2007) Epidemiological investigation of an outbreak of acute diarrheal disease using geographic information systems. Trans R Soc Trop Med Hyg 101:587–593

Shannon CE, Weiner V (1949) A mathematical theory of communication. University of Illinois Press, Urbana, IL, pp 101–107

Shinde SE, Pathan TS, Raut KS, Sonawane DL (2011) Studies on the physicochemical parameters and correlation coefficient of Harsool-savangi Dam, District Aurangabad, India. Middle East J Sci Res 8(3):544–554

Sigee DC (2004) Freshwater microbiology: diversity and dynamic interaction of microorganism in the aquatic environment. Wiley, Chichester

Simpson EH (1949) Measurement of diversity. Nature 163:688

Usharani K, Umarani K, Ayyasamy PM, Shanthi K, Lakshmanaperumalsamy P (2010) Physicochemical and bacteriological characteristics of Noyyal river and ground water quality of Perur, India. J Appl Sci Environ Manag 14(2):29–35

Wilhm JL (1975) Biological indicators of pollution. In: Whitton BA (ed) River ecology, Studies in ecology, vol 2. Blackwell Sci. Publ., London, pp 375–402

Zaky MM (2006) Environmental factors influencing multidrug-resistant and harbouring plasmid DNA Aeromonas Hyrdophila isolated from polluted waters of Lake Manzala, Egypt. In: Proceeding of the 19th international conference on environment, health and sustainable development, September 2001. Alexandria, Egypt, pp 159–160

Index

S. M. Khasim et al. (eds.), *Medicinal Plants: Biodiversity, Sustainable Utilization and Conservation*, https://doi.org/10.1007/978-981-15-1636-8

Printed by Printforce, the Netherlands